KU-797-925

LADY MARGARET HALL LIBRARY
1967
OXFORD

302209091Q

THE COASTLINE OF ENGLAND AND WALES

THE COASTLINE OF ENGLAND AND WALES

BY

J. A. STEERS

Fellow of St Catharine's College and Professor of Geography in the University of Cambridge

CAMBRIDGE

AT THE UNIVERSITY PRESS

1964

PUBLISHED BY
THE SYNDICS OF THE CAMBRIDGE UNIVERSITY PRESS

Bentley House, 200 Euston Road, London, N.W. 1
American Branch: 32 East 57th Street, New York 22, N.Y.
West African Office: P.O. Box 33, Ibadan, Nigeria

First Edition 1946
Reprinted 1948
Second Edition 1964

Printed in Great Britain at the University Printing House, Cambridge
(Brooke Crutchley, University Printer)

To
MY WIFE

CONTENTS

LIST OF TEXT-FIGURES

PREFACE TO SECOND EDITION

IN the eighteen years since the first edition of this book, a great deal of research has been done on the coasts of these islands. Many new techniques have been used, and we are now beginning to connect features of the land with those of the sea bottom more effectively as a result of aqualung diving. The coast is the boundary between land and water, but it is a mistake to suppose that it represents more than an arbitrary separation at a given instant of time. We need to correlate features on either side of it if we are to understand coastal evolution fully.

In 1953 a very significant event happened on our east coast—the great storm surge of 31 January and 1 February. Surges are not uncommon, but this one reached unusual heights and did great damage. It was fortunate that it did not coincide with a particularly high spring tide. On account of the damage and loss of life it caused, millions of people in this country by means of the wireless and the press realized, perhaps for the first time, what havoc the sea can do in a few hours. The two reports of the Interdepartmental (Waverley) Committee, which was appointed to examine the cause of the floods and to suggest means whereby in the future the damage caused by surges could be minimized, also brought home to the nation the seriousness of sea erosion and flooding, as well as the cost of maintaining walls and sea defences in a proper condition.

There is no doubt that in one way or another the storm was responsible for a renewed interest in the coast. The many investigations made by official bodies, including the admirable work of the engineers of the Ministry of Agriculture Fisheries and Food and of the Ministry of Housing and Local Government in rebuilding and redesigning coastal defence schemes, the many experiments carried out in the Hydraulics Research Laboratory at Wallingford, and the investigations of various river boards who are responsible for sea embankments, especially on our east and south-east coasts, represent one aspect. On the other hand, numerous private workers, many from university departments interested in coastal research, have studied and analysed various parts of the coast. With these may be associated The Nature Conservancy, which has done much not only to advance our knowledge of the morphology of the coast, but also to emphasize the significance and importance of ecological studies.

In the additional pages I have tried, however inadequately, to give an account of this new work. To do so, and at the same time to keep this book within reasonable bounds both of size and expense, it has only been possible to add new chapters, and not to correct, amend, and reset the whole volume. For the same reason it has not been feasible to include photographs. Those in the earlier edition are no longer satisfactory, and to provide a new series would have made the price of the book prohibitive. A compromise was necessary. Throughout the book references are given to the relevant plates in *The Coast of England and Wales in Pictures*, which was published by the Cambridge University Press in 1960. I am only too conscious that such an arrangement is far from ideal. Photographs are undoubtedly a great help in appreciating the coast or any other land form, but they are not essential; if the reader of this book has access to the smaller picture book, so well and good, but I think and hope that he who has not will suffer no serious handicap because the diagrams and maps in this volume should be sufficient.

Any further edition of this book is unlikely; it has already attained a size which is somewhat unhandy. But I have thought it worthwhile to try to collect in one place sufficient material to give a reasonable and I hope accurate account, from the point of view of a physiographer, of coastal study in this country as it now is. Future writers will study certain parts of our coasts in greater detail and with new techniques, and will develop new methods of investigation. All new and worthwhile work provokes thought and shows only too often the inadequacy of what has already been written. The coast offers endless opportunities for scientific study and not least for team work of experts in different fields of study.

My thanks are due to Dr C. A. M. King for help in certain chapters, and to Miss G. Seymour who prepared the new text-figures.

J. A. S.

CAMBRIDGE
January 1964

PREFACE TO FIRST EDITION

IN the years between the two world wars there appeared several studies of parts of the British coast. Each of them threw light on local problems, but none took the form of a systematic approach to the whole subject. There seemed, therefore, to be room for a book on rather different lines from, and wider in treatment than, the local monographs, and also more comprehensive than the few existing works dealing with the coastline as a whole.

I had often contemplated writing such a book, but hesitated, partly on account of other preoccupations and partly because I did not feel familiar enough with the whole coastline. The prompting to begin came in the early part of the recent war when I knew that I was likely to remain in Cambridge. This was not due to any shortage of other work, but because it seemed right to use any spare time as constructively as possible. The development from this beginning was quite unexpected. Early in 1943 I was asked by the Ministry of Town and Country Planning to compile a report upon the whole coast. To put together the survey, which in many ways supplements the original manuscript of the book, I had to visit all parts of the coast; it was thus possible to see almost every mile of the English and Welsh coastline except for a very few prohibited areas, some of which I knew before the war. By the time I began this survey, most of the book was in type, but fortunately I was able to correct and emend the typescript before having a fair copy made. After this systematic inspection, two things at least became clear to me—first that I had obtained a great deal of additional first-hand information, and secondly that the shortcomings of this volume were increasingly apparent.

A book of this nature must of necessity be partly a compilation. However, I had seen a good deal of the coast, and had carried out detailed field-work in certain areas. Thus the approach to the descriptions and problems of the less familiar parts was considerably eased, and later, of course, the official survey became a valuable check. But when all has been said I remain very conscious of inequality of treatment. I have also found it far more difficult to write about cliffed coasts than flat coasts. Whereas in the latter there is often an interesting physiographical evolution to trace, and also a relationship between physical features, history, and archaeology, in describing the former it is hard to avoid a dull statement of rather obvious facts. Paradoxically, the more spectacular coasts from the point of view of the tourist, are often less interesting physiographically.

This book is a physiographical treatise, but I hope it will also serve as a geographical background to the assessment of coastal scenery, and so to the proper use and enjoyment of our coasts. This is hardly the place in which to discuss problems of coastal planning and preservation, but it is worth while drawing attention to the fact that a proper physiographical appreciation of our coasts is the necessary foundation of comprehensive planning. Because of this relationship I have asked Dr Dudley Stamp and Sir Patrick Abercrombie to write introductory notes. I am very grateful to both of them for doing so.

The arrangement of the plates follows that of the regional chapters (IV to XI). The inclusion of a number of photographs is essential, and I have been particularly glad to make use so often of the photographs of the excellent Dixon-Scott collection of the British Council, of the Geological Survey, and of Aerofilms Ltd. I am also indebted to Professor H. H. Swinnerton, Messrs Batsford (through whose courtesy the photographs by Mr W. F. Taylor are included), *The Times*, Mr J. Gibson, and Aero Pictorial Ltd. I have filled some gaps from my own collection. For the two colour photographs I am grateful to Dr Julian Huxley and Mr F. H. Kendon. I hope the reader will find that the insertion of all the photographs at the end of the book gives him a handy pictorial atlas of the coast. Mr S. H. Beaver and his assistants at the Ministry of Town and Country Planning have helped me very greatly indeed in finding many of these photographs.

Finally, it is my pleasant task to thank many friends for their kind and unstinted help. Each chapter has been read by someone intimately acquainted with the area or topic. I must thank in particular the following experts who handled the chapters noted in brackets after their names: Mr W. V. Lewis (III); Mr D. M. Wray (IV); Professor T. N. George (V, VI); Mr F. H. Edmunds (VII and VIII); Professor H. H. Swinnerton and Dr H. Godwin (X); Dr R. G. Carruthers (XI); Dr H. Godwin (XII); and Dr V. J. Chapman (XIII and XIV).

To my friend, Dr Dighton Thomas, I owe much in this piece of work as in others: he read the whole book in typescript and made many helpful comments and suggestions. Miss Enid Gibbons and Miss Shirley Carpenter gave much time and trouble to many of the diagrams and maps. For the others, Mr Britt-Compton and Mr D. Baldwin are largely responsible. I am also greatly indebted to the Royal Society, the Royal Geographical Society, the Geological Society, the Geologists' Association, the Norfolk and Norwich Naturalists' Society, the *New Phytologist*, and the Cambridge University Press for allowing me to make use of maps and figures which have appeared in their publications. The Controller of H.M. Stationery Office has allowed me to base figures on diagrams and maps which have appeared in official publications. I hope that I have fully acknowledged all the sources either under the figures or in the text.

To the Cambridge University Press I am very grateful for the care and courtesy which they have given to all the processes and details accompanying the production of this book in war time.

Finally, to my wife I owe a great deal, not only for making my original version much more readable, but also for encouragement and interest in all the stages in the making of this book.

J. A. S.

CAMBRIDGE
1946

In this reprint I have taken the opportunity of correcting a few slips, and also of inserting a new and up-to-date map of Blakeney Point. I have also added on page 547 some new figures of the rate of accretion on salt marshes.

I should like to thank friends and reviewers for noticing the slips.

CAMBRIDGE J. A. S.
July 1947

INTRODUCTORY NOTE I

By L. DUDLEY STAMP

Chief Adviser on Rural Land Utilization to the Ministry of Agriculture and Vice-Chairman of the Scott Committee

WITH the certain knowledge that the rehousing of the people of Britain according to modern standards must inevitably make further demands on the limited resources of land still open and rural, and with the ever-present object lesson afforded by the uncontrolled sprawl which took place largely in the inter-war years, the Government in 1941 appointed the Committee on Land Utilization in Rural Areas under the chairmanship of Lord Justice Scott. In the Committee's terms of reference there is an indication that the basic problem, the best apportionment of the limited land areas of our crowded country to the many and varied needs of our people, had been clearly appreciated. The Committee was charged 'To consider the conditions which should govern building and other constructional development in country areas consistently with the maintenance of agriculture...the well-being of rural communities and the preservation of rural amenities.' Proceeding on the simple principle that three of the primary needs of mankind are work, homes and food and that each of these makes specific demands on the land, the Report of the Committee, now popularly known as the Scott Report, deals with the location of industry and the provision of work in the countryside, with the fundamental problem of rural housing, and with the conditions under which agriculture can best fulfil its function of providing for all a balanced, nutritious and interesting diet.

But there remains a fourth primary need of mankind to be satisfied—the need for rest and recreation. To make adequate provision for this need is a necessity, yet when land is demanded for gardens, allotments, parks, playing fields and open spaces generally there is far too often the suggestion that such provision is a luxury we may not be able to afford, forgetting that it is on this that the physical and mental health of the nation so largely depends. It is, of course, true that we are mainly an urban nation, that nine-tenths of the people of England and Wales live on one-tenth of the land of the country. More and more do those town-dwellers turn to the countryside for their rest and recreation and the country-dwellers have had thrown upon them the duty and the responsibility—so often made difficult for them by the thoughtless rather than heartless attitude of the townsmen—of maintaining in good order the nation's estate, of preserving a live, vigorous, progressive yet beautiful countryside.

Britain has a heritage, unmatched in any part of the world, in the beauty and variety of its natural scenery and everywhere that heritage is in danger of being marred, if not ruined, by ill-considered building development. But, as the Scott Committee recognized, if that is true of the countryside as a whole it is doubly true of the meeting-place of country and sea—the coast of Britain. 'The movement', says the Report, 'which was leading more and more people to spend their week-ends and holidays in the countryside and on the coast as well as to use the more accessible country as a dormitory is likely to continue, more particularly if holidays with pay are extended to further sections of the community and if there is a spread of the five-day working week. This would lead to a need for more accommodation in existing holiday resorts; the creation of further holiday resorts; the building of holiday camps; the provision of additional Youth Hostels; the catering for the needs of week-enders—all of which would mean a very considerable building programme. Whilst these needs should —and indeed must—be provided for, the building which would result, if such development were unregulated, would ruin many more districts of the beautiful land of Britain, particularly along stretches of the coast line.' The problem of the preservation of the coast is a national rather than a local concern, for many of the most beautiful stretches lie in the poorest Rural Districts where the local authorities have neither the money nor the power to resist 'development' however undesirable and where, indeed, the temptation of a possible increase in rateable value resulting from any new non-agricultural building is too often irresistible. So we find the Scott Committee making a definite and specific recommendation 'that the coast of England and Wales should be considered as a whole with a view to the prevention of further spoliation'.

In this book, which is a physiographical and not in any sense a planning monograph, Mr Steers has, in fact, carried out the recommendation of the Scott Committee. Actually, he has done much more, for he has presented the story in a most attractive form to the general reader and has also made his official reports to the Minister of Town and Country Planning. For many reasons it is fortunate that these two tasks should have fallen to him. In the first place Nature is dynamic and never static; the coast is not a line fixed on a map but a zone where there is a constant daily interplay between natural forces, where problems of coastal erosion in one place change within a few miles to equally serious problems caused by silting and deposition. It is essential, therefore, that any comprehensive survey of the coasts should be carried out by one thoroughly familiar with coastal phenomena and it is in this field that a large proportion of Mr Steers's scientific work has been carried out. No one else, living on Our Unstable Earth, can combine the experience gained through actually watching the growth of sand-dunes and mud-flats of the Norfolk coast with a period of residence on the Great Barrier Reefs of Australia and another on the little Cays of the West Indies. In the second place such a survey had to be carried out by one entirely free from partisan or sectional interests, by one who is a member of neither a central government department nor a local authority and by one who by personal interests or residence has no axe of his own to grind. The book that follows is the work and includes the carefully weighed judgement of one man; it is an example of unbiased academic assess ment, shown at its best, of the problems raised in a matter of the highest national importance.

INTRODUCTORY NOTE II

By SIR PATRICK ABERCROMBIE

AND so at last we have a Survey of the Coastline of England and Wales: a scientific survey made by a geographer who is alive not only to the changing face of Nature but to human works, good and bad, of the past and human needs of the present and future. It is always fortunate for the Planner, if the survey which is to provide him with the data for his plan is not completely detached from or even disinterested in the use to which he intends to put it. This study of the coast leads inevitably and without a break to its practical application in use. The comprehensive picture it gives is the strongest argument for treating the coast as a national possession: it cannot be split up into sections, sectionally treated, according to the County or District Boundaries. The coastline, indeed, is the most precious gift from nature for the recreation of man in these islands.

> Two voices are there: one is of the sea,
> One of the mountains; each a mighty Voice:

but the sea is more ubiquitous than the mountains and its appeal is more popular. The mountains, as Mr Dower has pointed out in his Report on National Parks, should be reserved for 'Such as wish to enjoy and cherish the beauty and quietude of unspoilt country...': there are many strips of coast which should also be kept free from general development, and certain wild stretches which should be treated as National Parks, such as parts of Pembroke. But a seaside town is a legitimate use of the coast: Blackpool and Brighton, which would destroy the particular character of the Lake District, are easily absorbed by the immensity of the open sea.

This greater intensity of use, which is not only to be expected as a result of holidays with pay and better means of transport, but also to be encouraged, entails grave responsibilities. There has been much destruction of coastal beauty and interest through ignorant selfishness and sheer absence of any sort of systematic development: ignoble architecture and a shortsighted desire to obtain a private frontage have ruined the appearance of and denied access to the shore in many places: and ribbon building threatens its coastal counterpart.

With this study before him, it should be possible for the Minister of Town and Country Planning to direct his research Department to working out with the local planning authorities a scheme which will allow hundreds of thousands more to use the coast for their holidays without destroying its characteristic features. Country planning, of which coastal development is one aspect, has lagged behind town planning in positive direction. It has largely been a question of town growth and overspill being warned off certain land: Dr Orwin's recent study of rural problems for a sample area shows what positive planning for the agricultural country might be. It would be equally valuable to take a sample of coastal country, and, using Mr Steers's survey, supplemented with necessary details, show how it could be both developed and protected. The objectives of coastal use are, of course, considerably different from those of a farming countryside: the dominant factor of the former is perhaps urban recreation, of the latter rural industry. But the coastline and the farm land lie side by side and frequently interpenetrate: a careful reading of the descriptive chapters in this book shows how variable is the depth of the strip for coastal zoning; and there are many places where the farm fields come up to the edge of a vertical cliff whose foot is washed by the high tide.

I have no fear whatever of increased building on the coast, provided it makes use of the knowledge accumulated in this book and displays a proper sense of landscape design and compact siting. There is an opening for several new seaside towns and many coastal villages and holiday camps. Mr Clough Williams Ellis has shown how phantasy can seize a site of high picturesqueness and produce a coastal town of imaginative beauty, in which it is proved that cheerful architecture and bright colour can be controlled by a master hand. I am not afraid of a pier for a larger town, provided that pier can escape from the Victorian convention of a domed palace or moorish hall supported upon spidery legs: could not a pier be designed like a ship thrusting its bows into the sea; does anyone object to a liner anchored at the quayside?

This is perhaps wandering far afield or at sea: but here in this book is the basis of all planning, the background of knowledge which clearly presented (as here), and rightly absorbed (as it is hoped it will be) should be the stimulus of imaginative and practical design.

NOTE

References in bold type are to plates published in J. A. Steers, *The Coast of England and Wales in Pictures* (1960).

Chapter I

SOME GENERAL REMARKS ON THE STUDY OF THE COASTLINE OF ENGLAND AND WALES

It is probably true to say that in few other branches of physical geography does the investigator come across such rapid changes as when he is studying parts of a coastline. Whilst alterations in landscape as a whole are continuous and evolutionary, they are seldom noticeable and measurable in the space of a few years. The results of catastrophic floods in rivers are, of course, obvious; the effects of a great storm in the mountains, or on certain types of country such as the Bad Lands of the Dakotas, are striking; there is no need to do more than hint at the swift and perhaps cataclysmal events of a volcanic eruption or of a severe earthquake. The bursting of natural dams, the sudden emptying of glacial lakes, for example, the Märjelen See, are other instances of sudden and rapid changes. The reader will know of further examples. But in many ways these violent episodes differ greatly from the physiographic processes on a coastline. The great storm may there indeed have effects of far-reaching significance, but what might be called the 'normal' storm often leaves a severe mark, and it is in the everyday events of coastal evolution that we can trace some of the great changes in parts of our coasts.

The waves lapping on the beach stimulate the more or less incessant movement of sand and fine shingle. Offshore the tidal currents, aided by wave action, are always moving vast quantities of material on a shallow sandy shore. The flow and ebb of the tides carry great quantities of silt up and down the estuaries and other sheltered waters. Much of this is deposited in quiet corners and also at the turn of the tide, and mud banks grow continuously. Once they have attained a certain height, and if they are in suitable positions, vegetation is likely to develop on them and in turn promote their further growth. In Chapter XIV the rate at which certain salt marshes grow upwards is demonstrated. On cliffed coasts the continuous wear and tear of the waves brings about erosion. This is clearly accentuated in rough weather, and naturally always mostly concentrated at high water. In a few years an observer can easily see what has happened, although it is perhaps more difficult to see why it has happened, and hardest of all to measure accurately the change. It is easy to estimate that so many yards of cliff have disappeared in so many years, but to follow the process of erosion is less simple. On soft cliffs, for example, land water is often the prime cause of cliff falls: great masses of cliff may slide down on to the beach, and it will take some considerable time for the waves to remove this debris from the cliff foot and prepare a smooth profile once again. In harder rocks, especially where the rock type and nature of the bedding is suitable, the waves cause undercutting, and, sooner or later, falls occur because the upper parts of the cliff are unsupported. This material has usually to be cleared away before further serious erosion can take place. These points have been mentioned to show that, if the distance to the edge of the cliff top from some fixed mark farther inland be measured, the precise amount of erosion is not necessarily recorded. By such a method nothing is thus allowed for the cliff profile and the debris below. In the second case the collapse of part of a cliff, and not the actual erosion, is registered. Such comments may give the impression of straining the point, but it is important to make clear that precise measurements of erosion would not be easy to obtain even if there were a series of constant observers at many places along our coastline. *A fortiori*, there must be caution in assessing the amount of erosion that has gone on over some centuries, since exaggeration is extremely easy.

Furthermore, precision about the feature under measurement is essential. For example, if the sea be cutting into a chalk cliff which is part of a ridge running parallel with the sea, and rising inland, erosion will, as a rule, decrease in pace in so far as measurements from the cliff edge are concerned. As the cliff is cut back, it increases in height and more material has to be removed to allow the sea to cut back another yard at the base. The total volume of material eroded may possibly be equal to the amount removed when the cliff was lower, but the investigator should ask himself: is he measuring the loss by the recession of the cliff top, or foot, or by quantity of material removed over a certain period? The converse of this reasoning will apply should the sea have cut back beyond the crest line of a ridge, so that the continued recession of the cliff implies also a progressive lowering of its upper surface. It is worth while stressing this point, because the argument applies to any sort of ground or formation, and in making estimates of cliff recession over several centuries, for instance, since the Roman occupation of Britain, writers have failed at times to allow for such circumstances. It should never be forgotten, therefore, that figures and estimates are at the best approximate, and often quite unreliable.

The matter may indeed be pursued further. All students of shore features must some time or other make thorough use of maps, modern and ancient. In this country they are more than fortunate in having the admirable and reliable productions of the Ordnance Survey, and from the

1-inch maps, and still better from the 6-inch sheets, they can obtain very sound evidence. Before the time of the Ordnance Survey publications, however, cartography was far less accurate, and this comment also applies sometimes to the very early editions of some of the 1-inch maps. In studying cliff recession the various editions of the 6-inch map are reliable, but probably no other cartographical evidence is trustworthy, except certain local surveys. The measurement of the growth and extension of sand and shingle spits needs even greater caution. Striking changes may take place in these formations in single storms, although from a map it would be easy to conclude that over a period of four or five decades a spit has either advanced or retreated continuously. For example, in 1897 a mile or so of the southern end of Orford Ness was cut off, but since then it has apparently been lengthening once again. This is but one instance, but it emphasizes once more the need for continuous observations in order to collect reliable data: investigation should also take the form of maps made at short intervals. Older maps are often of great value, but they must be interpreted carefully. Saxton's and Speed's county maps may afford many clues to coastal evolution; but in the writer's experience they are often extremely misleading. After all, surveying in the sixteenth century was in an elementary state, and it is probably fair to conclude that the surveyors would have paid but little attention to waste coastal areas. But from time to time detailed and relatively accurate surveys of coastal districts crop up, made for special purposes, that of Norden (1601) for Sudbourne and Orford being a case in point. In general a map of this kind is accurate for showing the outline of a spit for a specific date, but it does not follow that all subsequent changes have been in one direction only. Cartographic evidence, where available, should always be considered with great attention; as far as possible, however, it should be checked by other findings, and in old maps at least regarded with a healthy suspicion.

At the same time old maps are often of very great value in elucidating certain historical matters of local interest. There is no need to elaborate the point here since it is treated fully elsewhere. Suffice it to say that the local historian should be prepared to consider geographical evidence and nature of physiographical change just as much as the geographer should avail himself fully of the historian's help. To take two obvious examples: no one writing the history of the Humber ports could possibly neglect the great geographical changes in Spurn Head, nor could the historian of Roman Britain afford to ignore the evidence of physical change in the Lower Thames region and especially around the Isle of Thanet. The full discussion of the controversial question of Caesar's landing place by Rice

Holmes[1] is an excellent example of the way in which geographical evidence is of value to the historian.

This interesting interdependence of history and geography leads to a far bigger subject—the inter-relations of the work of many sciences in the interpretation of coastal features. There is no point in discussing the limits of any one subject or science. It seems more important to realize that any particular writer is almost inevitably interested in one particular aspect of large-scale research; he cannot be omniscient. But that does not mean that he can afford to neglect the writings of others slightly, or even substantially, off his particular line of research. Salt marshes, for example, provide an admirable field of study for the physical geographer; he can there obtain quick and reasonably satisfactory measurements of accretion, of changes in creek development, of the effects of sand blown on to the marshes, or of encroaching shingle beaches and their results. These are related problems and are all of great interest, but they cannot possibly be segregated from the work of the ecological botanist. The physical nature and constitution of a marsh have direct and important effects on the vegetation, and, what is more, the vegetation features may provide the best pointers in suggesting the physical characteristics most worth study. The geographical and botanical methods of approach are different, but if field collaboration between workers is practicable the combined results are likely to be more satisfactory than single-handed research.

The overlap between physical geography and geology is obvious, although once again the method of approach is often extremely different. The stratigrapher is concerned mainly with the sequence and nature of beds in a cliff section and with their fossils, and in compiling this volume numerous stratigraphical papers have been consulted. It is rare, however, to find that their writers have given detailed, or indeed any, treatment of cliff erosion or of interesting physical features in the cliff face. Still less common is it to find any reference to the nature of the adjacent beach. Such a comment need not in any way imply adverse criticism. In the first place, geographers are deeply indebted to stratigraphical research for help; in the second, as already suggested, there is no intrinsic reason why the stratigrapher should burden himself with what may be extraneous details any more than the physical geographer should, or could, consider the refinements of stratigraphy. Experience suggests, however, that if the geographical and stratigraphical aspects of scenery could have been examined jointly the results would have been of far greater value than the existing water-tight studies. The complete survey of a problem, and

[1] See pp. 327, 328.

especially of those of the kind presented in this book, is seldom obtainable by the work of one investigator.

This point may be elaborated with profit by a brief allusion to the submerged forests and related features which are common phenomena around the coasts of England and Wales. These submerged forest beds are often closely associated with former land surfaces which sometimes contain the remains of ancient man, and more often those of his artefacts. Thus the archaeologist can often give substantial help in the dating of these finds. Of still greater value is the recent development of the science of pollen analysis. By means of an examination of the pollen grains contained in a deposit, the investigator can obtain excellent evidence not only of the vegetation then prevailing but also of the climate. Further details again, especially in areas such as the Fenland, are obtained from a study of the foraminifera. The inter-relations of the various methods of approach to certain coastal problems are well illustrated in the combined studies of the submerged land surface of the Essex coast.[1] It is greatly to be hoped that circumstances will allow of a like investigation of many physical episodes, since only by this kind of co-operation can the proper significance and proportions of the phenomena be obtained.

The mention of vertical movements brings to the forefront at once one of the most important and at the same time most difficult problems associated with our coasts. The investigator has not only to consider land surfaces now submerged but also raised beaches which clearly imply a time when the land stood lower than it does relative to the sea. The bulk of the discussion belongs rightly to Chapter XII; all that needs emphasis here is the essential value of the combined work of the archaeologist and the palaeontologist. The evidence obtained by joint studies of the remains in Minchin Hole and Paviland Cave in Gower, as well as in many other places, serves as a good example. The association of Azilian and Campignian man with different parts of the 25-foot raised beach is in itself a clear indication of the different ages of that beach in somewhat widely separated areas. The probable exclusion of the 100-foot beach from the upper part of certain Scottish lochs connects coastal work with that of the glaciologist.

It is time now to turn to another aspect of inter-relationships, to one indeed which has been deliberately suppressed in this book because of its rather specialized nature and also because of its inevitable detail, namely the connection between problems of silting and the work of the harbour engineer. Comments on this topic might properly extend to the community of interest between physiographers and those engineers and local surveyors who build piers and groynes, thus affecting the accumulation of

[1] See p. 403.

beach material. References to this type of problem are not infrequent in the following pages. A full account of it, including the nature and types of sea walls and similar structures, would require a separate volume, which would have to be written by an engineer conversant with the technicalities involved. What needs attention is the obvious and close interdependence between all these branches of science, and probably others as well. This volume is concerned pre-eminently with natural conditions and, except where locally relevant, reference to engineering problems is omitted. Nevertheless, the reader should bear the connection in mind. It is often of great importance from the purely natural point of view in that engineering projects may have no small effects on neighbouring parts of the coast. The work of the engineer in reclamation problems is important, and so also is that associated with the regulation of coastal sand dunes. It is in Holland that the results of this collaboration are plainest, but on a smaller scale the problems of the Wash are a good illustration of the point. The excellent working model prepared by the Great Ouse Catchment Board in Cambridge is an admirable example of the way in which important practical problems may be investigated in the laboratory. A further instance lies in laboratory research on the profiles of shingle beaches. A beginning has been made in the Cambridge Department of Geography, and, had it not been for the war, Brigadier Bagnold might have developed further his researches on desert phenomena and on coastal formations. It is probably only in this way that a true idea of the work of waves on beaches will ultimately be achieved. Whilst much excellent work has been done in the field, it must, by its very nature, be mainly of a qualitative kind. Laboratory experiments, even if the full conditions prevailing in the field cannot be represented, undeniably enable an analysis of various factors which is impossible on an open coastline. Whilst discussing this point it is worth mentioning the value of certain parts of the coast, not only for long-period experiments but also as training grounds for young observers. In the salt-marsh areas, especially those preserved by the National Trust and so probably under constant surveillance, the establishment of such equipment as tide-gauges, measuring stakes, gear used in investigating the movement of water under marshes or in dunes is rewarding, and also that required for ecological and other experiments. These can be set up and left for long periods without much fear of disturbance by the merely inquisitive or actually burglarious. Moreover, because some types of change tend to be rapid, and because the interactions of the processes of physical geography and ecology are so clear, salt marshes are amongst the best possible localities for demonstrating natural events and conditions to students.

It will be apparent to anyone who reads this book that the amount of information about different parts of the coast is very unequal. Unless, however, some papers have been inadvertently overlooked it seems that there is less literature on the stretches of coast from Berwick to the Tees mouth, from the Solway to Blackpool, and on parts of Devon and Cornwall than there is on other areas. Certain conspicuous features such as Dungeness with Romney Marsh, the Chesil Beach, the Gower peninsula, Pembrokeshire, and the Isle of Purbeck, to take a few rather random examples, have often been described and discussed. Only too often, also, the available information of the coastline occurs incidentally in books or papers which are concerned primarily with other matters: on the whole, cliffed coasts are less fully described than alluvial areas, but to this trend the work of E. A. N. Arber on north Devon stands out as a marked exception. Throughout the present volume there is constant allusion to the Reports and Minutes of Evidence of the Royal Commission on Coast Erosion (1907–11). These volumes are a mine of information, and once their somewhat complicated make-up is grasped, the factual detail sought after is readily found.

The *Memoirs of the Geological Survey* form another great source of sound and reliable knowledge. They have been consulted very freely by the writer, but since they have been written by many different people over a considerable period of time, they are by no means of equal value. Once again, the *Memoirs* dealing with cliffed and rocky coasts are the less illuminating, although there are some striking exceptions, as, for example, those dealing with Anglesey, Pembrokeshire, and the Isle of Wight. There seems therefore to be scope for a good deal of very interesting work on the intricate details of many miles of rock-bound coast. Much of the south coasts of Cornwall and Devon, for instance, deserve far more attention from the physiographer than they have yet received. The new series of *Regional Memoirs* is often of great value, but here, too, much seems to depend on the particular interests of the writer. Descriptions of the coastline itself are to be found in the Admiralty Pilots. They, also, give a vast amount of information about local tides, winds, and other phenomena directly affecting navigation. The Admiralty Charts are worth careful study and form a complement to the Ordnance Survey maps. The Reports of the Tidal Harbours Commission (1845–47) are likewise of value, but in a more limited way, since the Commission was not concerned with the coast as a whole.

Search through a large number of local journals has proved profitable, for in these much of the most interesting and useful matter has come to light. Important papers in them are often mentioned in journals belonging to a national society, but local details are often ignored. Once again, their usefulness varies. *Archaeologia Cantiana*, for example, contains many papers dealing with the coast, and this is to be expected if the nature and historical significance of the shores of Kent be borne in mind. But it is far more difficult to find information of comparable value for many other counties. Doubtless there are many reasons for this inequality, but one may well be that mentioned earlier, namely, that hard cliffed coasts appear to have attracted less attention; further, unless a coast has some historical traditions of significance it has but seldom appealed to the local antiquary or historian. This may be a somewhat sweeping statement, but it is the impression gleaned from a fairly thorough survey of the literature.

There are but few general works dealing with the coasts of England and Wales. W. H. Wheeler's *The Sea Coast* is mainly descriptive and differs from the approach in this book. E. M. Ward's *English Coastal Evolution* represents the first real attempt to tackle the subject as a whole. The treatment is rather unequal; for example, little if anything is written of the coasts north of the Mersey and Humber. *Tidal Lands*, by A. E. Carey and F. W. Oliver, is an admirable survey and contains some particular aspects of great interest, but its scope is necessarily limited. G. Ashton's *Evolution of a Coastline: from Barrow to Aberystwyth* is unfortunately largely based on theories now known to be untenable, and in addition is marred by the failure of the author to give proper references. Considerable if indirect use has been made of D. W. Johnson's *Shore Processes and Shoreline Development*. This was the first comprehensive book on the subject and, although the illustrative examples are only occasionally drawn from this country, its value is great.

The chapters which follow contain first an outline survey of the geographical and geological setting of our coasts. These are followed by a general summary of some of the more important factors and processes at work on our shores. The detailed and regional chapters, which form the bulk of the work, come next. The style of treatment varies inevitably, both with the nature of the coasts and the amount of information available. The regional limits to some of the chapters may at first sight seem a little artificial, but places such as the Point of Air in Flint, the Parrett mouth in Somerset, the Otter mouth in Devon, Reculver in Kent, Hunstanton, and Flamborough Head seem on several grounds to be satisfactory dividing points. Such divisions are, of course, largely a matter of opinion, and should not be taken as final, but they serve as a framework for a regional approach to coastal problems. The final chapters deal with special subjects —Recent Vertical Movements, Sand Dunes, and Salt Marshes.

Many writers have tried to classify shorelines. Some classifications depend merely on the form of the coast, some are descriptive and refer to the structure of the adjacent land, others are genetic, and the best are genetic and evolutionary—that is to say, they include descriptions of the way in which the coast as a whole came into being and also its subsequent changes. D. W. Johnson's classification is of this type: it has the further merit of being simple:

> i. *Shorelines of Submergence*, or those shorelines produced when the water surface comes to rest against a partially submerged land area; ii. *Shorelines of Emergence*, or those resulting when the water surface comes to rest against a partially emerged lake- or sea-floor; iii. *Neutral Shorelines*, or those whose essential features do not depend on either the submergence of a former land surface or the emergence of a former subaqueous surface; iv. *Compound Shorelines*, or those whose essential features combine elements of at least two of the preceding classes....Shorelines of Submergence and Shorelines of Emergence are explanatory terms; they are genetic rather than empirical; they do not carry any implications as to whether it is the land or the sea which moves, and do not even imply any vertical change of level in either; they are easily understood, and are not in danger of being confused with other terms applied to shoreline phenomena.

By a shoreline of submergence Johnson means (1) any shoreline formed by the partial submergence of a land mass dissected by river valleys, and (2) shorelines characterized by fiords. A typical shoreline of emergence is produced by the raising up of a submarine (or sub-lacustrine) plain. It will, at any rate in its early stages, be simple in outline. Neutral shorelines include such forms as deltas, alluvial plains, outwash plains, volcano shorelines, coral reefs, and fault shorelines.

In Great Britain as a whole the shores clearly belong to the first type—those of submergence. In north-western Scotland the fiord type occurs, and elsewhere the form resulting from the drowning of a dissected land mass. But classifications of shorelines are perhaps more usefully applied to bigger areas than England and Wales or even all of Great Britain. During the not far distant time when the moor-log was forming on the bottom of what is now the North Sea, the level of the sea relative to the land must have been of the order of 200 ft. lower than it is now. Such conditions might well have made a great difference to the appearance of these islands, but it does not disturb the general truth of the statement that all the shores of England and Wales are typical of submergence. This is most evident on the west and south-west. On the Channel and North Sea coasts the form of the coasts has been altered in historic time by the filling-up of estuaries and the formation of spits and bars, the whole effect being that of smoothing. If all these superficial features were to be removed, the typical submerged valleys would certainly be found, although perhaps not so pronouncedly as in the west. In any case the British Isles are, in geological time, but recently separated from the continent, and in this sense only form part of the submerged shores of north-western Europe. A map of the geography of the Pliocene period shows that the land area of the British Isles then stretched outwards roughly to the boundary of the continental shelf. Without examining in detail the facts and theories on the formation of this shelf, it is safe to say that when the land extended so far outwards all the river courses must have been much longer. But with the gradual rise of sea level, whether continuous or punctuated with occasional downward movements in the Ice Age, the fundamentally drowned nature of our coasts would persist. The comparatively local and small isostatic movements of northern Britain in no way invalidate the general truth of this statement. The way in which England and Wales in particular, apart from the British Isles as a whole, acquired their present form belongs to the study of stratigraphical geology. The treatment of it in detail is outside the scope of this book, but the short synopsis given in Chapter II will serve as an outline.

Nor is it needful to follow out in any great detail the evolution of submerged or emerged shores. To do so would be merely to recapitulate what has already been fully and carefully described by Johnson. It is enough to remember that on any submerged shore the waves at once begin to cut back salient points and to sweep material into the re-entrants. At first the only obvious changes would be in the waves eating more rapidly into the softer than into the harder rocks. But in course of time headlands will be cut back and cliffs, fronted by planes of marine denudation, formed. The drift of beach material, which generally tends to increase in amount as further cutting-back of the coast proceeds, will lead to the formation of beaches and of spits and bars of sand and shingle, partly across re-entrants and especially in the upper parts of bays and gulfs. In later stages cliffs and beaches will either be cut back or pushed back, until they form long sweeping curves typical of maturity or old age. It does not follow that the cliffs will be by then worn back to gentle slopes, and it is essential to distinguish between young, mature, and old stages of coastal outline as distinct from cliff profile. Many of our chalk-cliff coasts are mature in outline, but in an early stage of profile development. Naturally, the rate of evolution of outline or profile varies greatly with the nature of the rocks. The Pembrokeshire coast and most of that of Cornwall and Devon are still in an early stage: and the coastal pattern in these areas may perhaps best be summed up as crenulate. Holderness and East Anglia, areas of very soft rock, are worn back to smooth outlines, although the cliffs are often steep. The chalk coasts are about intermediate between these two.

On an emerged shore the sea usually deepens very gradually, and one feature common to this type of coast is the offshore bar built by wave action. It is hard to think of a true example of this in Great Britain. The bar off the Culbin Sands in the Moray Firth and the bars on the north coast of Norfolk, of which Scolt Head Island is the most conspicuous, are indeed genuine offshore bars, but they are not due to the initial decrease in depth resulting directly from the emergence of a shallow sea floor. True, the waters of the Moray Firth and those of the North Sea off the part of Norfolk in question are shallow, but this seems to be due largely to the amount of sand and fine material held up by the east-west coasts of Morayshire and Norfolk. These have, as it were, acted as great natural groynes, and the shallowing, and later formation of offshore bars, are secondary effects. The result, however, is to all intents and purposes identical with what may be regarded as the normal evolution of a bar on an emerged coastline. Indeed, the future evolution of Scolt Head Island—its gradual retreat to the mainland shore and with this retreat the obliteration of the marshes within, and the likely deepening of the water on the seaward side —is probably in every way in harmony with the theoretical evolution outlined by Johnson for ordinary offshore bars. The examples cited, therefore, afford admirable instances of the danger of applying generalized explanations too readily to coastal features.

In considering, therefore, any one part of our coasts its particular state of evolution must be borne in mind. It is subject to constant change which may be slow or fast according to the nature of the rocks of which it is formed. Along the rock-bound western coasts little if any change will be traceable over a century or more, while on the softer Channel and east coasts alterations are very often apparent in a few months. It is this constant state of evolution, particularly in its long-term effects, that gives especial interest to the study of the shores of England and Wales.

Chapter II

THE STRUCTURE AND PHYSIOGRAPHY OF ENGLAND AND WALES

(1) INTRODUCTORY

A brief summary of the physiographic evolution and present physiography of England and Wales seems necessary as a background to the particular study of the coast. Such an outline should not only make sufficient the short accounts of the geology of certain parts of the coastline which follow, but should also knit all of them together. The need is still greater because we still await a comprehensive book on the physiography of England and Wales: when that is written subsequent writers on special topics will find their work greatly helped. In this study, which might reasonably be thought of as a final volume of the geological and geographical history of our country, it is clear that examples of nearly all of the main periods or systems of rocks reach our coasts at some place or another. It is therefore appropriate to try to give a brief general historical account of them. A further argument for this lies in the number of people who are interested in shoreline problems, but who are not necessarily geologists. Whilst the geographer willingly acknowledges the great debt he owes to geological writings, it is a fact that the geologist is apt to make things rather difficult for the general reader, especially in the matter of terminology. The latest edition of a well-known geological textbook shows an awareness of this shortcoming: 'It is devoutly to be wished that some day British geology may extricate itself from the bog of pedantic nomenclature in which it is at present involved, but there is as yet little sign of so desirable a consummation.' In this chapter the geologist may find important matters dismissed far too cursorily, but readers with less technical knowledge will find it of some general help and guidance.

(2) BRIEF RÉSUMÉ OF THE GEOLOGICAL HISTORY OF ENGLAND AND WALES

Geologists divide the history of the earth into five eras: the Pre-Cambrian or Archaean in the rocks of which the remains of living organisms are seldom found, the Primary or Palaeozoic, the Secondary or Mesozoic, the Tertiary or Kainozoic, and lastly the Quaternary or Holocene. All of these are subdivided into periods or systems, and these again into smaller divisions of geological time. Since it is often necessary to refer to these smaller divisions, more detailed tables are given on pp. 39–43.

We do not know very much of the geography of Britain in Pre-Cambrian times, but the visible rocks of this longest of all the eras fall into three main groups. (1) The majority are crystalline or metamorphic rocks, a term which means that their original form has been changed, usually as a result of heat or pressure or by both. These rocks are hard and resist weathering very effectively, but often they are now much worn down because of the enormous length of time during which erosive agents have been able to work on them. It is important to note that since many of these rocks are altered sediments, they are no longer thought to represent the original crust of the earth. (2) The second group consists of many volcanic and other igneous rocks. (3) There are also many sedimentary rocks that have not been metamorphosed, and consequently still preserve much of their normal appearance. In this Pre-Cambrian era there were probably at least three great periods of orogenesis or mountain building which were accompanied by manifestations of volcanic activity. The main area of these Pre-Cambrian rocks is in the north-western Highlands of Scotland, and is therefore beyond our concern: but they also occur in Wales, mainly in Anglesey and the Lleyn peninsula of Caernarvonshire where they form considerable coastal tracts.

In the Cambrian, Ordovician, and Silurian periods of the Palaeozoic it is probable that a continental area lay north-west of what is now Scotland, and that a broad trough (or geosyncline) of sea stretched across the British Isles. In the Ordovician, in particular, there was much volcanic activity: some of these volcanoes were on the sea floor and emitted great amounts of lava: others, on land, sent forth masses of dust and ashes. It is thus clear that the main sediments of these three periods consist of clays, silts, sands, and conglomerates, in which are interspersed many lava flows and beds of ashes. At the same time dykes and sills of igneous rock were often intruded into these sediments from below. Most of these sediments have been hardened and recrystallized by pressure caused by mountain building, so that instead of muds and sands occur slates and quartzites. The igneous masses are usually still harder and go to form mountains such as Snowdon and Cader Idris. On the Caernarvonshire coast smaller masses form the headlands of Criccieth, Pen-ychan, and Llanbedrog.

Near the end of the Silurian period earth movements took place, and there were formed great mountain chains which have had a profound effect on the subsequent history of these islands. Although the main results are seen in Scotland (whence their name—Caledonian) they also affected the scenery of North and Central Wales and the Lake District.

The movements were for the most part due to pressure directed towards the north-west or north-north-west, so that the mountain chains ran north-north-east to south-south-west. This is still the dominant trend of the rocks in North and Central Wales and the Lake District. These mountain-building movements began near the end of the Silurian, but they continued well into the Devonian period, and finally completely obliterated the great geosynclinal trough of earlier Palaeozoic times. As a result, nearly the whole of what is now Great Britain became a land area traversed by great mountain ranges. In the intervening lowlands and valleys mountain streams swept down vast amounts of coarse material which gradually formed the rocks known as the Old Red Sandstone. But in the south of England different conditions existed. South of, roughly, the line of the Bristol Channel and the Thames open-sea conditions obtained, so that in this area instead of the continental deposits of the Old Red Sandstone lie the muds, silts, and sands of the Devonian seas. In north Devon and on the northern margin of the Bristol Channel there is some mingling of the two different types of deposit. The nature of the Old Red deposits suggests that conditions were not only continental, but also arid. Further, their great thickness implies a vast and probably rapid wearing down of the Caledonian mountains. In any case the geological record makes it plain that at the end of the Devonian period these mountains were nothing but stumps, and that at the beginning of the next period, the Carboniferous, the sea once again invaded nearly the whole area of England and Wales, although parts of Scotland remained land. In the waters of this Carboniferous sea relatively clear conditions prevailed and were favourable to the growth of corals and other clear-water organisms. Hence, in this sea was accumulated the great mass of material which now forms the Carboniferous Limestone. Whilst the best examples of this occur in inland areas such as north-western Yorkshire and parts of Derbyshire, there are some admirable coastal exposures in the Tenby and Gower peninsulas of South Wales. Moreover, although Scotland is beyond the range of this study, it is relevant to note that the great continental area between Scotland and Scandinavia, which apparently still existed in the Carboniferous, provided much sediment for the more northern seas of this period. Thus, instead of the comparatively clear-water limestones which are characteristic of England, there are alternating beds of thin limestone, sandstone, and shale forming the Calciferous Sandstone series. These materials were probably brought down by rivers from the northern continent, and in the middle of the Carboniferous period there is no doubt that a huge river, flowing from the north, swept over all northern and much of central England. It brought with it vast quantities of sands and gravels which invaded and filled up large parts of the sea of the Carboniferous Limestone. These new deposits, which resemble those of a huge delta, now form the rocks known as the Millstone Grit. The delta was the logical and direct precursor of the swamp conditions of the Upper Carboniferous period during which our Coal Measures were formed. It filled up large tracts of the former sea and there resulted great flats on which the Coal Measures vegetation grew, perhaps not unlike vast mangrove swamps in the tropical parts of the world to-day. Some writers have also compared them to the vast fresh-water regions known as the Dismal Swamps of Virginia (U.S.A.). Although conditions for these swamps first developed in Scotland while the open sea of the Carboniferous Limestone lay over much of England, they gradually extended southwards as far as the ridge of higher land which from at least Devonian times onwards lay approximately along the line of the present Bristol Channel and Thames. Every now and again the coal forests were overwhelmed by further supplies of sand and mud brought down by the Millstone Grit river. At other times slight changes in the relative levels of land and sea led to an invasion of the forests by sea water. Such changes explain the relatively thin marine bands in our Coal Measure rocks.

Throughout this time the higher land areas were being constantly worn down, and by the end of the Carboniferous period they were not far above sea level. Moreover, the evidence of the rocks formed at this time implies a renewal of arid conditions.

Once again, however, normal evolution was upset by the great earth-movements, which are rightly termed the Carbo-Permian, because they affected the Carboniferous and older rocks, and were succeeded by the Permian period. These movements are also known as Armorican or Hercynian, after the Armorican peninsula of Brittany and the Harz Mountains of Germany; both of these areas are structurally part of the mountain ranges produced by the Carbo-Permian orogeny. In these islands four main sets of folds derive from these movements. (1) In the north and north-west they led to an accentuation of the older Caledonian folds. (2) In Central Wales new folds, for example, the Vale of Towy, were formed generally parallel to the older Caledonian folds. (3) Most characteristic of all are the east-west folds so clearly seen in South Wales and the Mendips. The last group of folds is continued underground to reappear in the Kent coalfield and across the Channel in those of Belgium and the Ruhr. Farther south the same Carbo-Permian movements gave rise to complex and intricate folds and fold systems in Cornwall and Devon. In this area the folding was accompanied by the intrusion from below of the vast masses of granites that now form Dartmoor, Bodmin Moor, Carn

Menellis, Land's End, and other moorland areas of Cornwall. (4) In other parts of the country certain north-south folds are of this age, for instance the general uplift of the Pennines (certainly *not* a simple fold chain), and also the Malvern Hills.

Thus, in the early part of Permian times, England and Wales formed a very mountainous country, and in the intermont basins the coarse deposits of the early part of the Permian period accumulated. These deposits were often of the nature of coarse screes: others were conglomerates carried down by mountain streams. There are good examples of the latter type of deposition in the Permian rocks of eastern Devonshire. At the same time over much of what is now the continent of Europe a Caspian-like sea existed. It covered the area of the present North Sea and reached as far west as the eastern coastlands of northern England to-day. In it were laid down the rocks which we now call the Magnesian Limestone and which distinguish so finely the Durham coast. The waters of this sea also seem to have penetrated through the Pennine area, because the Magnesian Limestone also occurs to the west of those hills. All the general evidence derived from the Permian deposits suggests that the climate of the islands was then arid, and it is significant that most of the continental rocks of this period are red in colour and many contain wind-rounded sand grains. Because of their prevailing red tinge, these deposits were at one time called the New Red Sandstones, and indeed in one sense they represent conditions similar to those obtaining in Old Red Sandstone times. The Permian is accepted as the latest division of the Palaeozoic period, but in point of fact there is no very marked break between it and the Trias, the earliest division of the Mesozoic. In northern Europe the Trias is threefold, hence the name, but in England the middle division (the Muschelkalk) is absent. The lowest deposits, the Bunter, in this country were laid down in much the same way, and in much the same places as the Permian. The succeeding Red Marls (the Keuper) also accumulated in shallow basins in an arid area and are mainly responsible for the prevailing red colour of the soils in the Midlands of England. Every now and again such a basin would dry up, so that deposits of rock salt and gypsum formed and ripple marks and rain pittings on semi-dry mud were made. In the central areas of England there stood up as islands higher parts of the older rocks, fragments of the Carbo-Permian mountains, while around them lapped the Triassic seas. To-day, however, erosion is, in places, removing the Triassic deposits and once again exposing the pre-Triassic surface, for example, in parts of Charnwood Forest.

Once again, however, the sea broke into the Triassic basins, and the fresh-water fishes living in them were killed. Their remains are now preserved for us in the earliest of the succeeding deposits, the Rhaetic, thus explaining the frequent occurrence of Bone beds in the lowest part of the formation. Actually, the Rhaetic period seems to have lasted but a short time in the geological sense. Indeed, the older upstanding masses of hard rocks were by this time almost worn down, and in consequence no longer yielded coarse deposits, but rather fine sands and muds. After the Rhaetic period, conditions of more open water favoured the formation of certain types of limestone and muds, and these initiated the important Jurassic period which now needs attention. That period falls roughly into three divisions. In the Lower Jurassic the deposits known as the Lias were laid down; they consist mainly of clays and muds, muddy limestones, and sometimes of sands. Large-scale mountain building did not occur in the Jurassic period, but the nature of the deposits clearly shows that there were small folding movements. This is very plain in the Middle Jurassic deposits which consist of limestones, sandstones, and clays apparently accumulated in relatively tranquil but separate basins. It is partly for this reason and partly on account of later earth movements that the scarps formed of the harder Jurassic rocks are discontinuous, although very well developed in some places. Thus they are not strictly comparable with the much more continuous scarplands formed by the Cretaceous beds, especially the Chalk. The Middle Jurassic limestones are often very oolitic. In the Upper Jurassic period conditions were more favourable to the formation of clays and sands, but occasionally limestones were deposited: the well-known Portland Stone belongs to this time. A great deal of the existing scenery of England is directly derived from the differences between the harder limestones and the soft clays of the Jurassic period.

As it drew to a close the sea retreated north-eastwards, and in the Wealden district a large lake or big deltaic plain formed which stretched into France. Sand and clay deposits gathered therein and now form such rocks as the Hastings Sands and the Weald Clay. In the north-east of England normal marine conditions still prevailed, and the Speeton Clay of Yorkshire was laid down. Between these two areas lay a ridge, which was finally covered by the sea in Lower Greensand times. The Wealden lake and the open sea of the Speeton Clay were before long invaded by the marine waters of the early part of the Cretaceous period. The lands surrounding the Cretaceous seas were almost certainly low, and, as the purity of their waters proves, little or no terrestrial sediment was carried to the true Chalk sea. But before the Chalk sea gained its full extent, the surrounding lands certainly yielded local sands and muds to form the Lower Greensand, during which time the waters of the southern lake and northern sea united. In the succeeding period of deeper water the Gault

Clay was deposited. After this, however, the great masses of Chalk indicate very clear water in which ooze accumulated, probably not unlike the *Globigerina* ooze of the modern oceans. But the Chalk seas were not deep as is proved by the presence of many relatively shallow-water fossils. It has been effectively argued that since land-derived detritus is absent the surrounding lands were not only low, but also subject to an arid climate so that there were no rivers able to carry any detritus to the seas. The Chalk sea over Britain was an extension of that over north-western Europe, and it is difficult to give any definite western and northern limit to it in these islands. It certainly reached Antrim, in Ireland, because there we find Chalk under the basalts of Tertiary age. The great Welsh peneplain (p. 182) also is thought by some to have originated through the erosion of the waves of the Chalk sea. As a result of later earth movements and erosion, the Chalk is now worn back to a well-defined escarpment running from Salisbury Plain to Yorkshire. These Miocene earth movements are treated below.

The rocks higher than the Chalk follow apparently without much discordance. Actually a long period of time intervened between the laying down of the Chalk and subsequent deposits, and there is also a very great change in character between the Chalk and the Tertiary beds which begin with the Eocene.

It was in the Eocene period that these islands began to take on the appearance with which we are now familiar. It seems that most of the British region was by then raised up into a land mass, and that only in the south and south-east of the country did marine conditions prevail. Into this Eocene sea of the south-east a great river drained from the west, from an area now beneath the Atlantic Ocean. This is clear from the nature of the ensuing sediments. The river naturally laid down fluviatile deposits in the western part of the basin, which now is in two parts as a result of the Miocene folding, the London and the Hampshire Basins. Farther eastwards normal marine clays and muds were formed. Stamp has shown that as the marine waters of the basin alternately encroached westwards on the fluviatile area and withdrew from it, there was laid down an alternating series of deposits of marine and terrestrial origin. The marine beds are known as the Thanet Sands, London Clay, and Bracklesham Beds; the terrestrial ones are the Reading Beds and Bagshot Sands.

The Oligocene deposits cover but a small part of the Hampshire Basin and Isle of Wight. If they were laid down elsewhere, they have since been eroded away. The Tertiary earth movements, which reached a climax in the Miocene, and which formed what are now the greatest mountain systems of the globe, began to make themselves felt in this country in the

Eocene during which period the dome of the Weald began to rise. Although the Alpine folds never seriously involved the British Isles, they have nevertheless left a great impress on the country. They appear as definite folds in the south of England running through the Isle of Purbeck and the Isle of Wight: the folding of the Weald and its western extensions (in, for example, the Vales of Kingsclere and Pewsey), together with the folding exhibited by the base of the Lias (see p. 191), belong to this time. Further, the Hampshire and London Basins were thus formed, and the Chilterns and Gog Magog Hills, with their dip-slopes towards the Thames, are in part the result of the same movements. In the north-western part of the kingdom renewed folding gave place to great outbursts of volcanic activity now so evident in the great basalt flows of Antrim and some of the western isles of Scotland, and also in the granite intrusions of the Mourne Mountains and in parts of Scotland. These, however, hardly concern the theme of this book.

In the Miocene, therefore, the British Isles were mainly if not wholly a land area. But in the succeeding Pliocene period the sea still remained in the London Basin and in the present coastal parts of Essex and East Anglia. In this sea the Crag deposits were laid down. These now form much of the heathy country of east Suffolk and Norfolk, but their physiographical, as apart from their stratigraphical, distinction from the glacial sands and gravels is not always obvious.

(3) THE ICE AGE

Before closing this summary some account of the Glacial period is necessary. In actual time it was far shorter than any of the periods so far considered, but its effects were very far-reaching and have exercised a profound influence not only on our inland scenery but also on our coasts. The mountain areas of the country, especially the Highlands of Scotland, the Southern Uplands, the Lake District, the Snowdonian area, and the Pennines each became the centre of great ice sheets which spread out far from their origins and covered nearly the whole of England and Wales approximately as far south as the line of the Bristol Channel and the Thames. It would be impossible even to outline the effects of this glaciation in summary, and it seems best to say only that on the higher ground it produced such typical forms as cirques, U-shaped valleys, smooth rock surfaces, and many other features. On the lower ground it left great spreads of boulder clay, sands, and gravels, which swathed hill and dale alike. Much of this, especially in the valleys, has been removed by subsequent erosion, but vast coverings remain. Around our coasts many cliffs are still plastered with boulder clay, and in some places such as Holderness and north Norfolk the cliffs themselves are wholly formed of it.

There is also another point of great significance concerned with the Ice Age. In Chapter XII there is a discussion of the relative levels of land and sea which fluctuated a great deal during this time. We need not enter into a discussion of the various theories on this phenomenon and it will suffice indeed to describe the actions but shortly. When the ice caps, not only of the British Isles but also of the Northern Hemisphere generally, grew to a maximum, it naturally followed that the supply of moisture required for their maintenance came from the oceans and seas. In consequence the level of the water bodies fell, but rose again when the ice melted. Since we have good reason for believing that there were possibly four advances and retreats of the ice sheets in eastern England at least, it is clear that from this cause alone there must have been several oscillations of the sea relative to the land. In consequence several marine platforms, raised beaches, and related features were produced. As the ice sheets did not necessarily grow to the same degree during each advance, it follows that the oscillations of sea level were not identical each time. Thus, if a complete record could have been preserved, it would be clear that from this cause alone the structures then produced would have evolved at different levels. But an oscillation of sea level also implies changes in the lower part of river valleys, and we therefore find that many valleys still show terraces produced by these alterations of sea level. Erosion during the various stages of the glaciation and also subsequently has materially worn away most of these beaches and terraces: others are buried beneath cliff talus and other recent deposits. Hence, it is clear that at the present time only a fragmentary record is available, and one of the greatest obstacles in the path of studying such remains lies in obtaining any sound correlation. The tackling of this problem through the combined approach of geographers, geologists, archaeologists, and pollen analysts is discussed in Chapter XII, but other difficulties still remain. The great ice caps naturally added much to the weight of the land masses on which they rested. Without going into a detailed study of the structure of the globe, it is correct to state that as the weight of the ice increased so also did the continents press downwards; conversely when the ice sheets waned, the relief from pressure caused the continents to rise again. Hence, as well as there being comparatively simple oscillations due to rise and fall of sea level derived from the waxing and waning of the ice sheets, the quite independent movements of the land masses also need consideration. The combined effect is often extremely confusing, the more so since the ups and downs of the lands took place slowly, by no means synchronizing with the rise and fall of sea level. In addition, there were in certain parts of the world entirely separate earth movements of a purely tectonic nature. These, however, hardly concern our own islands.

It has sometimes been assumed that the local British ice caps were not of sufficient magnitude to cause separate land movements in these islands, but the evidence of the 100-foot beach of Scotland seems to disprove this (p. 485). The more important point to bear in mind, however, is that these islands lie on the fringe of the European land mass and are really part of it: the intervening seas of the Channel, Straits of Dover, and the North Sea are but flooded parts of the continent. It is easy to appreciate that the down- or up-warping due to the great mass of ice on the Scandinavian-Baltic area would cause local warping many miles away from that centre. Such a process may be compared very roughly with the pressure exerted by a finger on a rubber ball: the main indent is obviously under the finger, but a considerable region round about is slightly depressed. This outer and depressed area corresponds approximately to the position of the British Isles relative to the main centre of the European glaciation. Further, since the crust of the earth did not respond quickly to these stresses, and since the Ice Age, in the geological sense, is but an event of yesterday, there need be no difficulty in connecting the known upward movement of part of the Baltic coast with the final recovery from the weight of the ice cap. It also follows for the same reason that lesser movements of a like nature may still be in progress elsewhere; thus it is justifiable to attribute the slight downward movements in the Low Countries and eastern and southern England, which are known to have taken place since Roman times and which are probably still in progress, to a like process (p. 496).

It will thus readily be seen that the Ice Age has had an effect on our scenery, and especially on our coasts and low-lying river valleys out of all proportion to its length and significance in the geological time-scale as a whole.

(4) THE CONTINENTAL AND MARINE PERIODS OF GEOLOGICAL HISTORY

Before turning to a discussion of the physical geography of England and Wales, it is worth summing up very shortly, in a somewhat different way, what has already been written in this chapter.

Of Pre-Cambrian times there is relatively little knowledge. The best examples of rocks of this age are in north-western Scotland where one great series, the Torridon Sandstone, is unmetamorphosed and rests with a great discordance on intensely altered schists and gneisses. In Anglesey, also, at least two pre-Palaeozoic periods of diastrophism are recognized.

The Torridon Sandstone definitely accumulated under true continental conditions, and thus, obviously, much of what is now Scotland was then part of a land area which probably extended eastwards to include Scandinavia. The southern limits of this continent are unknown, and it is not clear whether the Pre-Cambrian volcanics of the English Midlands were laid down on its surface or in a sea washing its shores. Certain red rocks of the Longmynd (Shropshire) suggest, however, that continental conditions prevailed there. The foregoing sequence of geological events is known as the *First Continental Period.*

Marine conditions set in with the Cambrian period, forming the great geosyncline referred to above, and they also succeeded the Torridon Sandstone in Scotland. In the geosyncline, the region which really comes within the scope of this book, sedimentation, with occasional breaks, went on for an enormous length of time. In the Ordovician period volcanic activity was marked, and near the end of the Silurian various red deposits heralded the beginning of the Old Red Sandstone. Thus ended the *First Marine Period*, although south of the Severn-Thames line marine conditions continued.

In the *Second Continental Period* the Caledonian earth movements took place, and produced the characteristic grain of so much of Wales, the Lake District, and Scotland. The Old Red Sandstone rocks accumulated, and in northern England and Scotland there was again much volcanic activity. Only in the Devon-Cornwall region did marine conditions prevail.

At the end of Devonian times the sea overflowed much of England as well as of Scotland and Ireland, and thus the Lower Carboniferous must be regarded as the *Second Marine Period.* This did not last long because, as we have seen, the deltaic conditions of the Millstone Grit, and the swamps of the Coal Measures succeeded relatively soon. But at the end of the Carboniferous red rocks characteristic of arid continental areas predominated once more. The *Third Continental Period* coincided largely with the great Hercynian or Armorican mountain-building movements, which produced the east-west trend lines of South Wales and the Mendips, as well as in the underlying rocks of southern England: further, there appeared then the north-south trends of the Pennines and Malvern Hills. There was, finally, much igneous activity in Devon and Cornwall (see p. 210) connected with the Hercynian mountain building. The *Third Marine Period* came in rather suddenly with the Rhaetic which, as has been shown, invaded the Keuper and other red deposits resting with such an abrupt unconformity on the Carboniferous rocks. Nearly all the Jurassic and Cretaceous deposits are marine, and the fact that they often formed in basins was due to the comparatively gentle earth movements prevailing at the time of their deposition. This marine period had no very clear ending, but gradually gave place to the *Fourth Continental Period* which, as far as Britain was concerned, was related, even if somewhat distantly, to the great Alpine movements. Shallow seas existed in the southern part of England in the Eocene and Oligocene periods, but in the Miocene these islands had largely assumed the form with which we are familiar to-day. The existing drainage system dates back to the Miocene, and despite many modifications, especially during the Ice Age, remains more or less intact. The Ice Age is included in this fourth continental period in which we are still living.

(5) THE PHYSIOGRAPHICAL REGIONS OF ENGLAND AND WALES

(*a*) *Introductory*. The physiography of England and Wales now claims attention. The treatment may, however, appear rather unequal, since inland areas receive short comment only unless their physical features have some bearing upon those of coastal areas. Discussion of the physiography of Scotland and Ireland is limited to incidental reference.

England and Wales may be divided into two main regions. A line drawn between the mouths of the rivers Exe and Tees separates a lowland district to the east from a highland district to the west. On a geological map this line very roughly separates the old Palaeozoic rocks from the softer Mesozoics and Tertiaries. To the north and west of it lie the remains of the Pre-Cambrian, Siluro-Devonian (Caledonian), and Carbo-Permian (Hercynian) highlands. True, these are but the worn-down stumps of their former selves, but they nevertheless make up what we can rightly call mountain Britain. In the central plains of England, for example in Charnwood Forest, there are further fragments of these ancient chains. But farther east lie the true scarplands which result from the outcrops of the harder Jurassic and Cretaceous strata, and the clay bands which form the low-lying vales. These rocks, as well as the Triassic rocks of the Midlands, all rest upon a now hidden platform of the older Palaeozoic rocks which are occasionally found in borings, and also in the mines of the Kent coalfield. All the Jurassic and Cretaceous rocks have been gently tilted to the south and east by the Alpine movements, and it is this fact, together with the effects of erosion, that has given them their present scarp form. In the south of England the Alpine orogeny (p. 19) produced definite folds, like those in the Weald, the Isle of Wight, and Isle of Purbeck.

It will probably be most convenient to describe concisely the structure and physiography of the main regions of England and Wales. Those contiguous with the coast will be treated first and rather more fully, but to make the account complete the section includes a brief summary of the inland regions.

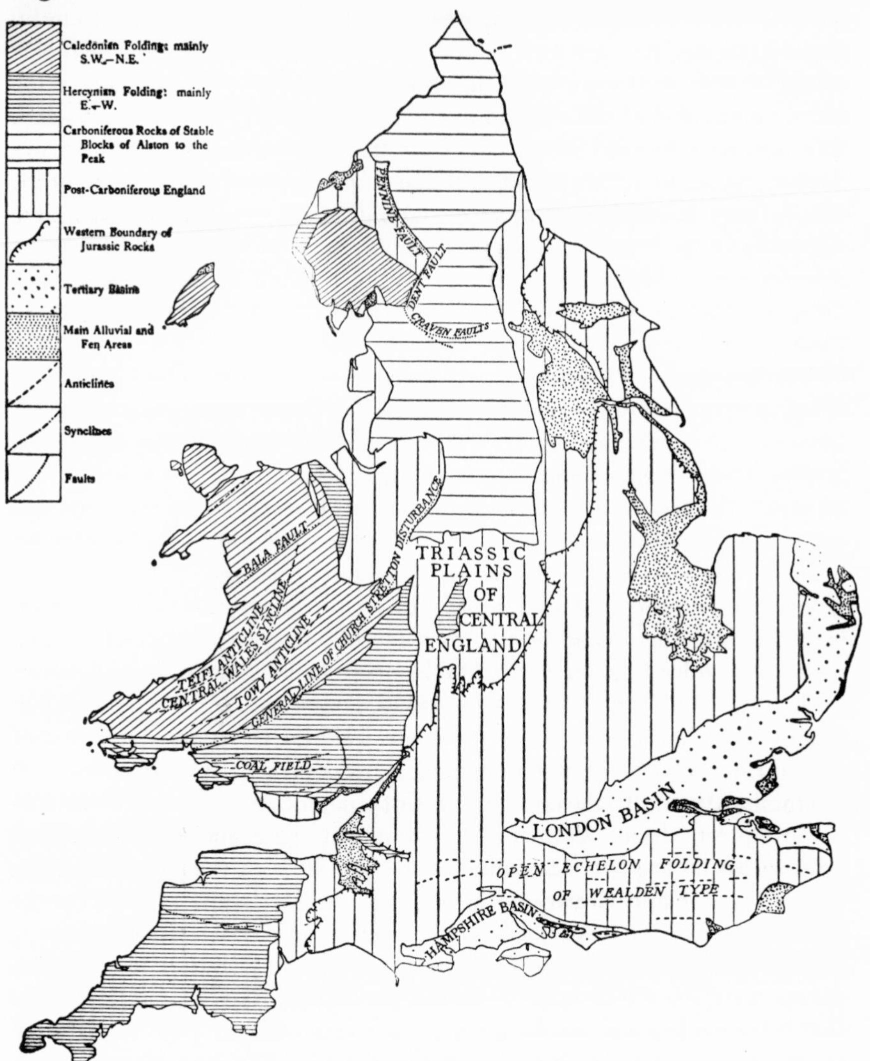

Fig. 1. Morphological sketch-map of England and Wales

(*b*) *The Lake District.* This is a folded area belonging to the Caledonian period of mountain building. The folding probably began in Ordovician times, but reached its climax in the Siluro-Devonian period. The country has thus acquired a characteristic south-west to north-east grain and at one time the Isle of Man was undoubtedly joined to the Lake District in a single mountain chain. These Caledonian mountains suffered great erosion, and in Lower Carboniferous times they were certainly surrounded, even if not completely submerged, by the sea. The Carboniferous rocks, as a result of later erosion, now form an almost complete girdle round the older rocks; and in the depressions of the Eden-Solway and Morecambe Bay, and also on what is now the Cumberland coastal area, Permian and Triassic rocks are preserved. The Alpine movements of Tertiary age seem also to have affected the region, because at some late stage a local uplift occurred there. Whether or no this was connected with a great mass of igneous rock which domed up the surface without actually reaching it, does not matter. More important is the fact that this doming led to the formation which we see to-day—a central mass of older rocks, Skiddaw Slates, Borrowdale Volcanics, and Upper Slates, showing the Caledonian trend, surrounded by Carboniferous and younger rocks which all dip outwards from the dome. This doming also brought about the radial system of drainage which is such a marked feature of the district. On account of recent movements, mainly connected with the Ice Age, the actual coasts of the area, especially in and near Solway Firth, consist of upraised carse lands of recent deposits.

(*c*) *The Lancastrian Plain*. This area is really a continuation of the Midland Plain of England as far as its topographical form is concerned, and it includes, indeed, much of Cheshire. It is invaded by two broad tongues of old high land, the Forests of Bowland and Rossendale which coincide with anticlines of Millstone Grit and Lower Coal Measures. But the true Lancashire plain is an undulating lowland, most of which consists of glacial and recent deposits. Occasional outcrops of Bunter Sandstones and Keuper Marls occur, showing the substratum on which the superficial deposits rest. The plain is bordered on the east by the Pennines, and on the south by the Welsh Massif. Northwards much is submerged under Morecambe Bay, but the Lake District is its true boundary in this direction.

(*d*) *Wales*. Chapter v shows that Wales may be divided into a Caledonian part (to which may be added the relatively small areas of Pre-Cambrian rocks) and an Armorican or Hercynian part, but it is better treated here as a whole. It thus includes all that region west of the English Midlands, although the southern part of the county of Glamorgan is formed of much newer rocks. It should also include a considerable area in Shropshire which, structurally, is part of the Welsh Massif. The north-western district of Wales forms the oldest part, and there the Pre-Cambrian crystalline rocks predominate which make up most of Anglesey and a substantial part of the Lleyn peninsula. Both of these regions were subjected to two periods of Pre-Cambrian folding, which was, in a sense, rejuvenated in Caledonian times, and also after the Carboniferous deposits were laid down. This accounts for the general south-west to north-east strike, and also for the narrow bands of Carboniferous Limestone pinched in between the belts of older rocks. Anglesey has now been worn down to a lowlying peneplain, and there is little in its topography to suggest the long and extremely complicated history through which the island has passed.

The rocks associated with the folding of North and Central Wales are mainly the Cambrian, Ordovician, and Silurian. There was, however, also great volcanic activity here, especially in Ordovician times, and the rugged nature of much of the area is directly due to these ancient igneous masses. They form, for example, the summits of Snowdon and Cader Idris, and a host of smaller mountains and hills such as Yr Eifl in the Lleyn peninsula, and the smaller masses such as Llanbedrog and Graig Ddu of the coastal areas. Chapter v contains a reference to the great Harlech dome of Cambrian rocks in Merionethshire which forms the core of the main anticline. Nearer Central Wales the igneous masses become less conspicuous, which in part explains the less rugged topography, and in the same direction the age of the folding becomes less. In Carboniferous times North and Central Wales formed part of a land area called St George's Land, but South Wales was then covered first by the sea of the Carboniferous Limestone and then by the swampy areas of the Coal Measures period. These Carboniferous rocks were later buckled during the Hercynian mountain-building period, and the resulting folds run from the coast of St Bride's Bay in Pembrokeshire right through the whole of South Wales and beyond. Doubtless such important movements must have affected the old and hard rocks of St George's Land, but they were certainly unable to bring about any considerable folding of that area. Instead they may have caused local movements along older folds and faults, and indeed the folds in South Wales owe their form and trend largely to this resistant northern mass. Farther east they wrapped round it, so to speak, and so produced the north-south folds of the Malvern Hills. Further complications where north-south and east-west folds met appear plainly in the nodes and basins in the Forest of Dean and in the Bristol district. Between the east-west folds of South Wales and the north-south folds of the Malverns is a triangular area covered with Old Red Sandstone rocks. Where these are hard and resistant they result in the bleak and barren moorlands of the Brecon Beacons, but where they are soft and lowlying they produce the fertile basins of Hereford. The Vale of Glamorgan consists mainly of Liassic rocks, and is more closely related to the scarplands of England than to the general structure of Wales.

The complicated structure of Wales is naturally reflected in its coastline, but not perhaps as much as might be expected. Over almost all the land the ice sheets held sway, and large areas were swathed in boulder clay, much of which still remains in the coastal area. The Lleyn peninsula, with its Pre-Cambrian and Early Palaeozoic rocks, shows the Caledonian trend, which is also apparent in the great sweep of the southern and eastern part of Cardigan Bay. It is really only in South Wales, and especially in Pembrokeshire, that the intimate connection between structure and coastal scenery is obvious.

(*e*) *The Plain of Somerset*. This lowland is separated from the Midland Plain by the Bristol and Mendip region. The only comment necessary is that the plain is very close to sea level and is in many respects similar to the Fenland (see p. 425), but it includes islands of older rocks.

(*f*) *Devon and Cornwall*. These counties constitute another area of old rocks, very different from the Lake District and Wales. The extreme south-west was folded in a complicated manner in Hercynian times, with accompanying igneous intrusions (see p. 210), and the main direction of strike is east-west. During the Permian period it was a hilly or mountainous country, and the deposition of Permian and Triassic rocks occurred in the intermontane basins. Since then, however, erosion has removed nearly all trace of these ancient mountains, and has exposed the granite masses which are now salient features. Exmoor, composed of tough Devonian rocks, stands relatively high on account of their resistant hardness. There are traces of several high-level erosion surfaces in both counties, and the coast is very diversified and crenulate in pattern. Incidentally, the rocks often have a direct influence on the coastal detail, a characteristic which is clear enough in the numerous headlands of igneous rock and in the granite cliffs of the Land's End district. Recent drowning of the coast has helped to produce many fine and branching inlets like those at Falmouth, Salcombe, and Plymouth.

(*g*) *The Jurassic Scarplands*. Although this zone lies mainly in the interior of the country, the scarplands affect the coastline of north-eastern Yorkshire at one extremity and parts of the Dorset coast at the other. They stretch in a great belt between these two littoral areas. Usually the beds dip fairly gently to the east or south-east and the harder beds stand up as scarps facing west or north-west. The scarp slopes are fairly steep, but because the tracts of hill and scarp were formed from various rock groups in different areas the scarps themselves are not continuous (see also p. 17). Where the dip is very slight, they are scarcely important elements in the scenery. The following regional subdivisions of this belt have been outlined by Stamp and Beaver, and it is clear that only the first and the last concern the coastline. (1) The Northern Basin of deposition in which lie the Cleveland and Hambleton Hills. A well-marked scarp is formed by thick sandstones of the Inferior and Great Oolites and accentuated by the fact that the underlying beds of Middle and Upper Lias are resistant. Deep valleys intersect the area. The coast of this part of Yorkshire runs more or less at right angles to the trend of the main scarps. (2) The Market Weighton anticline of Yorkshire does not produce a scarp, and thin Jurassic beds are overlapped by those of Cretaceous age. (3) In Lincolnshire and Northamptonshire the marked ridge of Lincoln Edge stands up between the Liassic clays of the west and the low ground to the east, while farther south there are three separate scarps. That of the Rhaetic west of Grantham is short, and is succeeded eastwards by clays of the Lower Lias. Then comes the Ironstone Ridge of Melton Mowbray formed of the marlstones of the Middle Lias and followed eastwards by the plain of the Upper Lias. There is a third scarp which overlooks the Vale of Catmoss and the Welland valley. In Warwickshire and Northamptonshire well-marked scarps are absent because, since the rocks are relatively resistant, the whole is uniformly an upland area. In Oxfordshire the limestones of the Lower Lias form an upland and the marlstones account for the well-marked feature of Edgehill. (4) The most spectacular scarps of all appear in Gloucestershire where the Cotswolds stand some 600 ft. high. The scarp depends mainly on the Midford Sands and limestones and grits of the Inferior Oolite, while the dip-slope is capped by the limestones of the Great Oolite and the Forest Marble. The scarp may be stepped in places on account of the marlstones of the Middle Lias, and farther west there is a minor scarp produced by the Rhaetic. To the east lie the extensive lowlands of the Jurassic clays, occasionally interrupted by a minor scarp in the Corallian. (5) In Dorsetshire the low ground of the Jurassic rocks consists of the Lower Lias and Oxford Clays, while the Inferior and Great Oolites form the higher undulating country. It is the scarps of the Upper Greensand and Chalk, however, which really dominate the scenery. Throughout the whole length of the Jurassic outcrop the lowlands of the Great Clay Vale are formed of the Kimmeridge Clay as in the Vale of Pickering, or of the Oxford and Kimmeridge Clays as in Lincoln Vale and the Fens. The valley of the Great Ouse lies mostly in the Oxford Clay, while the Vale of Aylesbury, in the Upper Jurassic and Lower Cretaceous clays, is continued westwards into the Vale of Oxford and the Vale of the White Horse. In the two last, however, there are interruptions due to the small Corallian scarp and the outcrops of Portlandian rocks which account for features such as Shotover Hill.

(*h*) *The Chalklands* on the whole form conspicuous ridges, and stand up well above the surrounding clays. In many places, however, and especially in Norfolk and Suffolk, they are covered by a thick mantle of glacial deposits. The Chalk generally is thick, 600 to more than 1000 ft., and shows little variation, so that the scarps are much more continuous than those formed by the Jurassic rocks. Salisbury Plain is the nucleus of the Chalk: the Dorset Chalk ridge, the South Downs, the North Downs, and the main escarpment running via the Chilterns to the Norfolk Heights, the

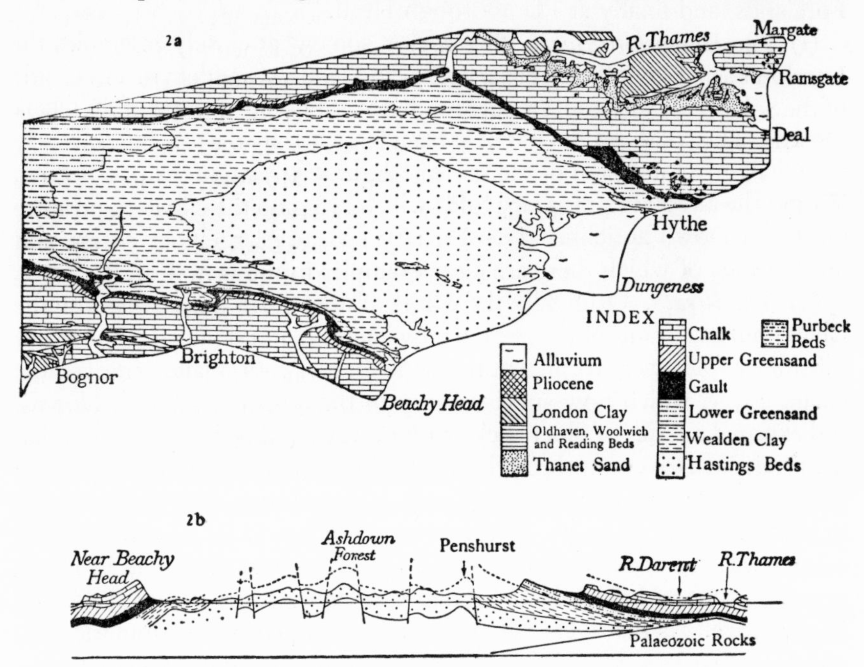

Figs. 2*a*, 2*b*. Geological map and section of the Weald (after F. H. Edmunds, *British Regional Ecology*, 'The Wealden District', and S. W. Wooldridge and D. L. Linton, *Inst. British Geographers*, 1939)

Lincolnshire, and Yorkshire Wolds all radiate from that place. The various hard bands in the Chalk often give a stepped appearance to the main Chiltern escarpment, and indeed sometimes produce what are really three separate steps. The Lower Greensand sometimes forms minor scarps like those in Bedfordshire and farther west (see also the Weald). Only in the Breckland of Suffolk and Norfolk does the scarp feature fail entirely, although it is not at all pronounced between the Breckland and the Wash. The main escarpment is breached by many water gaps, amongst which that of the Thames at Goring is distinctive, and farther north, those marked by the Wash and the Humber. There are many other lesser water gaps as well as dry or wind gaps. The folding to which the Wealden area has been subjected has given rise to the pronounced inward facing scarps of the Downs, and between Guildford and Farnham the process has been intensified by thrusting and faulting to produce the Hog's Back ridge. When the Chalk reaches the sea it forms bold cliffs and headlands like those which distinguish the coast just north of Swanage, at the Needles and at Culver Cliff in the Isle of Wight, at Beachy Head, and the South and North Forelands, and finally at Flamborough Head.

(*i*) *The Hampshire Basin.* This region somewhat closely resembles the London Basin (*q.v.*). The Tertiary clays and sands, so evident on either side of the Solent and Spithead, are surrounded by the higher ground of the Chalk Downs. The southern rim has now been breached by the sea, but formerly the Chalk of Dorset and of the Isle of Wight was continuous (see p. 300). Within the basin are subsidiary folds which bring up the Chalk to produce the Ports Down anticline. The Tertiary beds are largely of coarse and mixed sands of which there are clear examples in the New Forest.

(*j*) *The Weald.* Originally the whole of the Weald was covered by the Chalk, but as a result of the Alpine movements it was domed up and many subsidiary folds were formed at the same time. Erosion later removed the Chalk cover, which now survives only in the North and South Downs, and exposed a series of older rocks within. It should be borne in mind that the Straits of Dover cut through the Weald, the eastern end of which is in the Boulonnais area of northern France. There is no need in this summary to enter into the question of the river development of the Weald: the subject was first made classic by the work of W. M. Davis, and of recent years its scope and detail have been greatly extended by others, especially Wooldridge. As the rivers cut down and developed strike tributaries, the underlying rocks were exposed. These now run in general east-west bands, and their disposition is clearly seen in Figs. 2*a* and 2*b*. In the centre lies the relatively high ground of the Forest Ridges made up chiefly of the Ashdown, Tunbridge Wells, and Hastings Sands; the last run out to sea and form the well-known cliff sections near Hastings. Surrounding them is a low vale formed of the Weald Clay, on the outer border of which, especially in the north, follows the escarpment of the Lower Greensand; this in the west passes into an undulating, dry, and heathy district, sometimes called the Western Heights. Between the Greensand ridge and the North Downs is the Vale of Holmesdale formed chiefly in the Gault Clay, while the Lower Greensand and Gault reach the coast near Folkestone. The Downs have already been mentioned. The other two features to which attention may be called are the Pevensey Levels and Romney Marsh which are fully described elsewhere.

(*k*) *The London Basin.* This region is a unit both from the geographical and geological points of view. Geologically it is a great syncline, bordered to the north and south by the dip-slopes of the Chalk of the Chilterns and of the North Downs. In the basin are the sands and clays of Tertiary age which rest on the underlying Chalk, while borings have shown that the Palaeozoic platform underlies all at no great depth. The main axis of the syncline is followed fairly closely by the Thames, but the basin is interrupted by minor tectonic features which, together with the varied nature of the Tertiary deposits, give rise to several different types of scenery. The London Clay of Eocene age is the most important of these Tertiary beds, and its presence is evident on the shoreline of the Thames estuary. In the north of the basin, glacial clays, sands, and gravels are conspicuous.

(*l*) *East Anglia (with the Fenland, Eastern Lincolnshire, and Holderness).* This region strictly consists of the counties of Cambridge, Norfolk, and Suffolk, but since the Chalk areas have already been described it is only the actual coastlands which are mainly formed of Pliocene crags and of glacial beds which need brief consideration here. These beds also extend into northern Essex, and together with the crags form heathy country. Where they reach the coast, however, they are liable to considerable erosion. Inland the main rock is the Chalk, which is usually covered by a thick mantle of glacial deposits. A short reference to the Fenland, eastern Lincolnshire, and Holderness is perhaps in place in this section. The Fenland is a former great gulf eroded out of the Jurassic clays. Subsequently it was filled by ice which left thick deposits of boulder clay, while comparatively recent incursions of the sea have given rise to the Fen clays and silts. These processes are, however, discussed at length in Chapter x, so that further immediate comment is unnecessary. The coastlands of Lincolnshire and Holderness are likewise fully treated in the same chapter. The 'substratum' in both is almost entirely boulder clay, and, since both were non-existent as land areas in pre-Glacial times, it is best to discuss their evolution when considering their coastal features.

(*m*) *The North-East.* East of the Pennines there is relatively low ground. Indeed, along the Yorkshire dales tongues of low ground extend well into these highlands. The prevailing eastward dip brings the Carboniferous Limestone and Millstone Grit beneath the Coal Measures, and these, in their turn, lie below the Magnesian Limestone, Bunter Sandstones, and Keuper Marls. In general, therefore, there is a series of north-south belts, each with fairly distinct physiographic characteristics. In Northumberland the Carboniferous rocks form most of the surface south of the igneous Old Red Sandstone area of the Cheviots. The various limestone and other bands meet the coast in places, but much of the coast is formed of dunes or outcrops of the Whin Sill. The Lower Coal Measures usually produce barren land and moorlands, whereas more fertile conditions accompany the gentler relief features of the Middle and Upper Coal Measures. The Magnesian Limestone produces interesting scenery, especially where it is cut by river valleys, and the escarpment associated with this formation faces westwards. Farther south, for example, in Nottinghamshire, the sandstones of the Bunter give elevated, sandy, and rather barren or well-wooded tracts. The low ground of the Keuper Marls is distinct from the Tees mouth to the Trent: it coincides with the Vale of York and the Axholme district, in both of which it is masked by glacial and superficial deposits.

(*n*) *The Pennines*. These hills and the remaining regions needing comment are wholly inland; this position justifies brief reference only in a book about the coast, and not necessarily in proportion to their area. The Pennines are not a chain of mountains: it is more accurate to call them an upland region, consisting mainly of Carboniferous Limestone, Millstone Grit, and Coal Measures. Older rocks appear occasionally below, like those, for example, around Ingleborough. The Pennine area was uplifted in Hercynian times, and now the Carboniferous Limestone and Millstone Grit make up most of the high land. To the west they are in large part margined by a fracture belt formed by the Pennine, Dent, and Craven Faults. Eastwards the dip is gentler and the older beds of the Carboniferous pass under the Coal Measures and later rocks. The higher part of the Pennine area forms an excellent example of mountains produced partly by faulting and partly by circumdenudation.

(*o*) *The Midlands*. The greater part of the Midlands is lowland: to the north-west they are connected with the Lancastrian Plain through the Midland Gate, and to the north-east, by way of the Trent valley, to the Vale of York. To the south and east the Midlands are bordered by the scarplands of the Jurassic rocks, to the west by the Welsh Massif, and to the south-west by the Bristol region and the Mendips. Over much of the area the Upper Trias or Keuper Marls predominate and are responsible for the prevailing red soils. The Bunter is usually identifiable with rather higher ground as in Cannock Chase. The Triassic deposits enclose several islands of older rocks including the Pre-Cambrian outcrops of Charnwood Forest, the Leicestershire coalfield, the ancient rocks of the Nuneaton ridge, the Warwickshire coalfield, the old rocks of the Lickey Hills, the coalfields of Staffordshire, Ironbridge, and Forest of Wyre, and the old rocks of the Wrekin.

(*p*) *The Bristol-Mendip Region*. The Midland Plain joins on to this region through the Vales of Gloucester and Berkeley. Around Bristol and the Mendips the topography is very varied and, like that of the Midlands, consists largely of islands enveloped in newer rocks of Triassic and Liassic age. The islands in this case, however, occupy most of the area, and consist of extensive stretches in the Mendips of the Carboniferous and Old Red Sandstone rocks, which were folded in Hercynian times, and of smaller patches of the same rocks together with outcrops of Silurian.

(6) THE COASTLINE

The division of England and Wales into two main parts separated by the Exe-Tees line not only defines two major areas of different structure, but also two very different types of coast. Apart from Lancashire, the Wirral peninsula, and the upper waters of the Bristol Channel above Cardiff and Bridgwater, the western coasts are mainly hard and rocky. The south and east coasts, in contrast, are formed of softer rocks. But although this distinction is sound, it by no means gives the whole of the picture. Structure alone would suggest, for instance, that the Cumberland and Furness coast should be rock-bound. In point of fact, however, much of that part is bordered by flats of recent origin, and the cliffs, where they exist between Maryport and St Bees, are either in the Coal Measures or Trias. The Welsh littoral as a whole may be regarded as a cliff coast, but in many places, especially in Cardigan Bay, extensive sand spits and flats have grown out in front of the former cliffs. Moreover, it will not do to assume that the cliffs themselves are due simply to marine erosion (see p. 137). Again, in Pembrokeshire and South Wales there is a clear and definite relation of cliff form to both marine erosion and structure, but here, as elsewhere, it should be borne in mind that the cliffs are not merely the product of recent wave action. On the contrary, the changes of level associated with the Ice Age have played a great part, and likewise subaerial erosion both past and present. In Devon and Cornwall the cliff scenery is magnificent, and the crenulate pattern of much of the coast may be ascribed largely to marine erosion. The detailed discussion in Chapter VII, however, will show that the cliff form is often much more the result of sub-aerial erosion than of wave action. That the sea is cutting back into the land is true enough, but a great deal of detailed work on cliff profiles has yet to be completed. Study often shows that the sea has really, in modern times, only slightly modified the form of the cliffs. Alterations in level of the order of 200–300 ft. which almost certainly took place in the Ice Age have played a most important role, and they are distinct from still greater changes implied by the remains of high-level platforms so common, for example, in the Cornubian peninsula.

The true Channel coast beginning east of Sidmouth is a very different area. Changes of level have taken place there also, but the relative, and often absolute, softness of the rocks has allowed modern wave erosion to cut back quickly inland and has produced the long sweeping curves so characteristic of this coast. Harder masses project as headlands, and in Dorset and the Isle of Wight the relation of structure to erosion is clear enough. The intricacies of the coastline are not, however, by any means all due to modern conditions: the flooding, and often the filling up with alluvium, of former valleys account really for a large amount of the actual detail. This is plain in Poole Harbour, Southampton Water, and in the several irregular inlets between Gosport and Chichester. The smoothing of the coast by the growth of sand spits and flats characteristic of Cardigan Bay, has, on the Channel coast, proceeded much farther. Features such as Chesil Beach, Langley Point, and Dungeness now give the coast a far more regular form than it possessed even a few centuries ago.

Exactly the same comment applies even more emphatically to the east coast between Flamborough Head and the South Foreland. Only very occasionally do solid rocks reach the sea, and when this occurs they form promontories such as Flamborough Head, Hunstanton cliffs, and the chalk of the North and South Forelands. Nearly all the intervening stretches are formed of boulder clay, glacial gravels, or soft Pliocene deposits. All yield readily to both atmospheric and marine erosion, but, naturally, high boulder-clay cliffs, like those in parts of Holderness and near Cromer and Sheringham, retreat more slowly because so much more material has to be removed. It is evident, too, not only in cliffs of these types, but also in many other rocks, like the Lias near Lyme Regis, that land drainage, in making gullies, is often a more effective means of loss than direct marine erosion. Throughout the length of the east coast the former intricate details have been largely smoothed out by the growth of spits and bars. This is most evident in Norfolk and Suffolk, where only a few hundred years ago long arms of the sea penetrated far inland in the Broads and other places. The Essex coast, however, still retains much of this irregularity.

North of Flamborough Head as far as Berwick, the nature of the coast varies a good deal relative to the type of rocks in its hinterland. The Jurassic rocks of the Cleveland district in many ways resemble those of Dorset, but the connection between structure and marine erosion is not always as evident in Yorkshire as in the south because the strike is not so nearly parallel with the coast, and also because of the presence of much boulder clay on the Yorkshire cliffs. The Tees mouth is on the lower ground of the Keuper, and to the north of it lie the older rocks, first the Magnesian Limestone in Durham, and then the various members of the Carboniferous in Northumberland. Although these Carboniferous rocks form cliffs in places, it is the outcrops of the Whin Sill and the many dune-fringed bays which in some ways are more characteristic of the bleak Northumbrian coast.

The next chapter gives a treatment of the various shore processes. It is, however, relevant to make an immediate, if brief, reference to the question of the exposure of our coasts to winds and consequently to waves. The prevailing winds are the westerlies, whence it follows that most of the western shores of these islands, facing the Atlantic Ocean, are severely battered. North of Pembrokeshire, however, the land mass of Ireland begins to afford some shelter and this increases considerably to the north of Anglesey. Storms in the Irish Sea are, however, often severe, and on the flat and soft coast of Lancashire, which in many ways resembles that of East Anglia, considerable damage may follow them.

From Land's End to Morecambe Bay beach material travels as a rule to the north and east: this occurs along the Bristol Channel, Cardigan Bay, the south coast of Caernarvonshire, and Liverpool Bay, although south of Formby local travel of beach material is toward Liverpool. St Bees Head is another divide: to the south of it material travels to Walney Island, whereas northwards it moves to the Solway. Along the whole of the south coast waves from the Atlantic have considerable effect, but clearly it lessens a great deal up Channel, and east of Portland Bill local Channel waves are often of greater significance than the oceanic. The *general* tendency is for beach material to travel eastwards, but there are several exceptions to this. On the east coast there is complete shelter from the westerlies. The winds and therefore the waves which are significant in coastal changes here are usually from the quarter between north and east, although south-east waves are by no means negligible, especially on the coasts of Suffolk and Essex. Except for that part of Norfolk west of Sheringham, the main direction of travel of beach material is southwards, and its effects are obvious at Spurn Head, Yarmouth, Orford Ness, and at many other places. On the coast between Hunstanton and Sheringham, however, it usually, but not always, moves westwards.

It is right to anticipate here two important points discussed in Chapter III. In the first place it is unsafe to generalize too much about the details of movement of beach material. On the Channel coasts it is usually to the east, and on the east coast it is usually to the south, but there are important and significant exceptions. In the second the reader may think that tidal action has been forgotten. Modern research shows, as a rule, however, that although tidal currents are often important, they do not appear to have much effect on the travel of *beach* material (see p. 52). Because the flood current runs in a certain direction on a stretch of coast, and because spits and bars run in the same direction, there are no good grounds for assuming that the one is the cause of the other. It is a bad example of arguing *Post hoc, ergo propter hoc.*

APPENDIX

STRATIGRAPHICAL GEOLOGY

The object of Stratigraphical or Historical Geology is to give an account of the history of the earth and of the evolution through which it has passed. The stratigrapher thus reconstructs the geographies of former periods, and his material is found in the rocks which form the crust of the earth. These are either sedimentary or igneous. The former were laid down as layers or strata, usually under the sea or other water surface, but sometimes on land. The latter have either been poured out on to the surface of the earth by some form of volcanic action, or intruded into strata or into other igneous rocks already formed.

Two fundamental principles govern the stratigrapher's work: (1) in any series of strata, the uppermost is the newest, unless, as a result of folding or thrusting, overturning has taken place; (2) rocks can be dated and correlated by means of their included organisms (fossils). In estimating the age and order of rocks, the second principle is the more important. Since rocks were obviously laid down in separate basins of deposition, it is clear that no stratum can be traced for an unlimited distance. Furthermore, the water basin varied in depth, like those of the present seas and oceans, and the type of sediment that was accumulating naturally changed according to different influences, for example, the supply of material, distance from land, and depth of water. Hence, it follows that in the distant past a sand deposit (a beach) might have graded outwards into a mud, and the mud again into a limestone just as they may succeed one another to-day. Then as now all were being formed contemporaneously, and such processes have taken place times without number.

Later, the rocks were raised up to form dry land, but erosion wore them away, and only parts remain. If, then, there were only one principle on which to work, that of superposition, it might well happen that in one particular area the order would prove to be so and so, whereas in a not far distant locality for very good reasons there would be found a quite different series. Yet if there were substantial proof that a certain clay bed in area *A* was really part of a clay bed in area *B*, or was equivalent in age to some other stratum, the finding would be extremely valuable, because there would then be some means of correlating the sequence in the two localities. Fossils enable us to do this. It is obvious to-day that certain organisms live on sandy shores, others in muddy estuaries, and others again in the clearer waters of the open sea. Some mixture, however, is more than likely to take place, and in any case floating or free-swimming creatures probably occur in all or at least large parts of a water area, and when they die their hard parts (usually shells) will come to rest in each type of deposit. If then there be proof that a certain organism, *Z*, is found in a given bed in both areas *A* and *B*, the correlation of strata is in part accomplished. But greater precision than this is really necessary. Some organisms have lived for very long periods of time, are found in many rocks, and consequently do not afford any very useful chronological data. Others, on the contrary, are now known to have spread from their place of origin rapidly in the reckoning of geological time, to have covered wide areas, and to have died out comparatively quickly. Hence, their remains are now distributed over a big area, but through only a relatively thin vertical thickness of strata. These fossils are of very great value in establishing correlations. We may assume that the time taken for their dispersal was short compared with the whole period of deposition of the rocks wherein they accumulated. This period is represented by the vertical thickness of strata in which they are found. We may also conclude that any rocks containing such quickly dispersed fossils were contemporaneous. Modern geological research has established many such species which are known as zone fossils. A zone may be defined as 'a layer of deposit, of limited but variable thickness, characterized by a very definite assemblage of species (or only one species), which distinguishes it from all other deposits'. A zone, however, only represents a very short time in the geological scale, and a system such as the Jurassic contains many of them. But even so, the same two principles apply. While one part of the Jurassic system in one locality may be correlated with another in a different place by means of zone fossils, the whole series of fossils belonging to the Jurassic period is also reasonably distinct, although it will be realized that in some parts of the world there is only a gradual transition from species in older rocks to those in newer rocks. We may be certain, however, that when we speak of Jurassic rocks in Britain and America they are broadly contemporaneous.

Stratigraphical Geology originated mainly in this country, and the names of the main systems given on p. 39 are nearly all of north-west European origin. Assume for a moment the impossible case of a marine area persisting on the earth's surface since the beginning of geological time, and receiving a continuous supply of material; a bore through all the deposits contained in it would give an unbroken sequence from the earliest times to the present day. But the seas and parts of the oceans have not evolved in this way. Land and sea have frequently changed places, and seas, and even oceans, have never had more than a relatively limited areal extent. Strata laid down in them have been raised up into land: erosion has worked on these new lands, and has partially or wholly destroyed them. These again have sunk beneath the waves and new deposits have been laid down on them, only once more to be followed by up-raising. Hence, in this way we find large-scale breaks—unconformities—between great series of rocks. Major unconformities of this type obviously imply a great length of time, and are often used as the dividing lines between systems or groups. Minor unconformities result from like processes on a smaller scale: for example, the raising of a limited area of sea floor into the zone of wave agitation would account for them, assuming that erosion has taken place, and possibly a cessation of sedimentation, before renewed subsidence gave conditions favourable to further deposition. Still lesser breaks may result from the cutting-off of the supply of sediment to a given area as the result of changes in the main currents of a sea, or because a large river ceases to bring down much detritus. Breaks of this kind all form useful dividing lines in the geological time-scale, but it must be remembered that none of them can be world-wide. Supposing that at the present time there were continuous deposition going on in the seas around our islands, and supposing also that a land bridge were raised up between Great Britain and the continent in the Channel area, there would be a marked break in sedimentation in that place, but deposition would go on continuously elsewhere. In this country, for example, there is a great unconformity between the Carboniferous and the Permian, but this is not found, for example, in India.

Further detail is unnecessary here, and it remains to state only that systems are divided from one another by breaks of major magnitude, while those within systems produce the smaller divisions. Shorter periods of time are represented by fossil zones, the temporary cessation of sedimentation, or other minor factors.

The varying types of sedimentary rocks depend mainly on the conditions under which they were laid down. Conglomerates or pebble beds and sands are characteristic of shallow water; muds belong to deeper water, and thick limestones suggest clear rather than deep water. Here too, however, caution is much needed. Limestones, for example, may be laid down in actual contiguity with a shoreline if other sediment be missing, but at the same time shelly limestones and sands may represent conditions like those found in modern oyster, mussel, or cockle beds. Again, certain sands and coarse angular rocks imply terrestrial conditions such as existed in Torridonian, Old Red Sandstone, and Permo-Triassic times.

It should be borne in mind that fossil remains also give great help in deciding the physical conditions under which strata were formed. In existing coral reefs there is evidence of particular circumstances in our own time and we may not unreasonably assume that ancient reefs imply like conditions. Occasional scattered fossil corals, however, must not lead to the assumption of tropical habitats in the past which we now associate with reef builders. In considering the evidence of fossils, the whole assemblage of a fauna should be envisaged, not merely the distribution of one or two species.

Igneous rocks may be of volcanic origin and appear as lava flows or great masses of dust and ashes. Others, such as the basalts, were poured out from fissures on to a land surface without much, if any, explosive activity. Igneous activity is often submarine: many Ordovician igneous rocks are of this nature. Igneous rocks have often been intruded from below, and penetrate the upper strata more or less vertically in the form of dykes or follow along the bedding planes in the form of sills. If, therefore, we find a well-marked sill between two beds, it does not mean that the underlying bed is older than the sill and the overlying bed newer. The sill is newer than either, since it penetrated after both were laid down. A surface lava flow is, however, newer than the surface rock. Often great masses of igneous rock were pushed upwards from below, but did not reach the surface, although they may have domed it up, and subsequent erosion may have exposed them. Such a process explains in a general way such features as Dartmoor and the Land's End granite. When igneous rocks come into contact with sedimentaries, changes due to heat take place in the latter which are often baked, recrystallized, and hardened. Still greater transformations affect both sedimentary and igneous rocks when, owing to earth movements, violent folding and crushing of the rocks take place. Under such conditions vast masses of rock are altered or metamorphosed. Muds are turned into slates, sands and sandstones recrystallize into quartzites, limestones often recrystallize to form true marble, and granites may become gneisses. Naturally our oldest rocks have suffered most in this way.

Earth movements of this violence are usually mountain-building or orogenic processes. Mountain building, for reasons that do not come within the scope of this chapter, has been limited to certain fairly definite periods. Two or even three such periods occurred in Pre-Cambrian times, and there was another great orogenesis near the end of the Silurian and lasting into the Devonian, known as the Caledonian orogeny. The next upheaval took place at the end of the Carboniferous and before the Permian and is termed the Armorican or Hercynian. The last period of mountain building reached its climax in the Miocene, and formed the great Tertiary mountains now called the Alpine System. It must not be supposed that in any period the production of folds was sudden or rapid. Premonitory symptoms and minor folding preceded the climaxes, and Swiss geologists have reason to believe that the Miocene folding in the Alps began in a gentle way as early as Carboniferous times. Moreover, apart from these major periods of mountain building, there have been episodes of local and minor folding, for example, during the Jurassic (see p. 17).

For a full discussion of principles and details the reader must consult one or other of the many excellent textbooks on geology. The foregoing notes and the following tables may, however, give the non-geological reader a better appreciation of the terms used in this book. One further word is perhaps in place: eras and periods or systems such as Cambrian, Jurassic, etc., do *not* connote equal lengths of time. It is more than likely, for example, that the Archaean or Pre-Cambrian period was far longer than the whole Palaeozoic, and that the Palaeozoic in its turn lasted longer than the Mesozoic and Kainozoic together.

Era	Period or System
Quaternary	Recent or Holocene Pleistocene
Kainozoic or Tertiary	Pliocene Miocene Oligocene Eocene
Mesozoic or Secondary	Cretaceous Jurassic Rhaetic Trias
Palaeozoic or Primary	Permian Carboniferous Devonian and Old Red Sandstone Silurian Ordovician Cambrian
Archaean or Pre-Cambrian or Eozoic	

Further subdivisions:

1. Pre-Cambrian: may be separated into (1) a series of highly metamorphosed rocks such as crystalline schists and gneisses, and (2) sedimentary rocks, usually unconformable to (1). The Pre-Cambrian rocks are unfossiliferous except for certain obscure remains.

2. Cambrian: the general classification is:

Upper Cambrian	Transition Series	=Tremadoc Slates[1]
	Olenus Series	=Lingula Flags
Middle Cambrian	*Paradoxides* Series	=Menevian Slates Upper Harlech Beds Solva Series of S. Wales
Lower Cambrian	*Olenellus* Series	=Lower Harlech Beds Caerfai Series of S. Wales

(Words in italics in this and other tables are generic names of fossils: in this case the Trilobites.)

[1] Often now regarded as Ordovician.

3. Ordovician: Bala { Ashgillian / Caradocian }
Llandeilian
Llanvirnian
Arenigian
} Each of these groups is further subdivided into zones

4. Silurian: Ludlovian or Ludlow Series / Wenlockian or Wenlock Series } = Salopian

Valentian or Llandovery Series { Upper / Lower

Most of the Silurian has been zoned by graptolites, but this classification has led to complications since authorities are not agreed about the total number of zones.

(The Downtonian, consisting of grey and red sandstones, is by some authorities put at the top of the Silurian instead of the base of the Old Red Sandstone.)

5. Devonian and Old Red Sandstone:

Upper Devonian { Fammenian / Frasnian

Middle Devonian { Givetian / Eifelian

Lower Devonian { Coblentzian / Gedinnian

Old Red Sandstone { Farlovian / Orcadian / Dittonian / Downtonian

In Cornwall and south Devon, and in north Devon and west Somerset, the epochs given above correspond to names given in Chapter VII as follows:

	Cornwall and S. Devon	N. Devon and W. Somerset
Fammenian	Green and purple slates, etc.	Lower Pilton Beds Baggy and Marwood Beds Pickwell Down Sandstone
Frasnian	Calcareous Slates Limestones of Torquay, Plymouth, etc.	Morte Slates Ilfracombe Beds Coombe Martin Beds
Givetian	Hope's Nose Limestone	Hangman Grits
Couvinian or Eifelian	Limestones Staddon Grits	
Emsian } = Coblentzian Sigenian }	Meadfoot Beds: Looe Beds	Lynton Beds Foreland Sandstone
Gedinnian	Dartmouth Slates ?	— —

6. Carboniferous: Upper { Coal Measures / Millstone Grit

Lower or Carboniferous Limestone = Avonian or Dinantian.

The Lower Carboniferous is zoned in south and central England as follows:

Zone		
Dibunophyllum Zone	D	=Visean
Seminula Zone	S	
Syringothyris Zone	C	=Tournaisian
Zaphrentis Zone	Z	
Cleistopora Zone	K	

In Northumberland, where the Lower Carboniferous is more sandy, the usual division is:

Bernician	Calcareous Division
	Carbonaceous Division (Scremerston Series)
Tuedian	Fell Sandstone Series
	Cementstone Series

The Millstone Grit is classified with the Coal Measures. The classification of the Upper Carboniferous is difficult, and in attempting it the observer should take account of all floras and faunas. The deposits as a whole consist of sandstones and shales with seams of coal, but details differ materially between one coalfield and another. It may be summarized very briefly as follows:

Upper Westphalian	Radstockian or Upper Coal Measures
	Staffordian or Transition Coal Measures
Middle Westphalian	Westphalian (*sensu stricto*) or Middle Coal Measures
Lower Westphalian	Lanarkian or Lower Coal Measures and Millstone Grit

7. Permian:

Permian	Magnesian Limestone Series	In Durham and north Yorkshire
	Marl Slate Series	
	Lower Sand Series	

8. Trias:

Trias	Keuper Series
	Bunter Series

9. Rhaetic:

Upper Rhaetic	Watchet Beds
	Langport Beds (='White Lias')
	Cotham Beds
Lower Rhaetic	Westbury Beds
	Sully Beds

10. Jurassic:

Upper Jurassic (Upper Oolites)	Purbeck Beds
	Portland Beds
	Kimmeridge Clay
	Corallian Rocks and Ampthill Clay
	Kellaways Rocks and Oxford Clay
Middle Jurassic (Lower Oolites)	Bathonian or Great Oolite
	Bajocian or Inferior Oolite
Lower Jurassic (Lias)	Upper Lias
	Middle Lias
	Lower Lias

(The Lias is zoned throughout, mainly on the basis of species of Ammonites.)

In the south-western district the Middle Jurassic is further sub-divided as follows:

Bathonian	Cornbrash Forest Marble and Bradford Clay Great Oolite and Stonesfield Slate Fuller's Earth
Bajocian	Inferior Oolite Midford Sands (in part)

In Yorkshire:

Cornbrash
Upper Estuarine Series
Scarborough Limestone
Middle Estuarine Series
Millepore Limestone
Lower Estuarine Series
Dogger

11. Cretaceous.

The Lower Cretaceous in the Wealden area:

Lower Greensand		Folkestone Beds Sandgate Beds Hythe Beds Atherfield Clay
Wealden	Weald Clay Hastings Beds	Tunbridge Wells Sand Wadhurst Clay Ashdown Beds Fairlight Clays

(Spath includes the Gault also in the Lower Cretaceous.)

No exact correlation is possible with Yorkshire. In that county the Speeton Clay rests directly on the Kimmeridgian and the only sign of erosion is a bed of phosphatized nodules at the base. The Portland and Purbeck Beds are absent. The Speeton Clays are marine and the corresponding beds in southern England are of fresh-water origin. In Yorkshire the Speeton Clay is overlapped (west and south) by the Upper Cretaceous. The Lower Cretaceous reappears in Lincolnshire where the sequence is as follows:

Carstone, i.e. Lower Greensand (in part)
Tealby Limestone and Roach Ironstone
Tealby Clay
Claxby Ironstone and Spilsby Sandstone

Upper Cretaceous	Upper Chalk = Senonian Middle Chalk = Turonian Lower Chalk = Cenomanian Upper Greensand and Gault = Albian

12. The Tertiary as a whole is usually subdivided as follows:

Upper Tertiary or Neogene	Pleistocene Pliocene Miocene

- Lower Tertiary or Palaeogene
  - Oligocene
  - Eocene

Further subdivisions are:

- Eocene
  - Barton Beds
  - Bracklesham Beds
  - Bagshot Beds
  - London Clay
  - Lower London Tertiaries
    - Blackheath and Oldhaven Beds
    - Woolwich and Reading Beds
    - Thanet Sands
- Oligocene of the Hampshire Basin
  - Hamstead Beds
  - Bembridge Marls
  - Bembridge Limestone
  - Osborne Beds
  - Headon Beds

The Miocene is unrepresented in Britain.

- The Pliocene
  - Cromer Forest-bed Series
    - Upper Fresh-water Bed
    - Forest Bed
    - Lower Fresh-water Bed
  - Weybourn Crag
  - Chillesford Beds
  - Norwich Crag
  - Red Crag
  - Coralline Crag
  - Lenham Beds

REFERENCES

L.D. Stamp and S.H. Beaver. *The British Isles*, 1933.
P. Lake and R.H. Rastall. *Text Book of Geology*. 5th ed. 1941.
J.E. Marr. *The Principles of Stratigraphical Geology*, 1905.
L.D. Stamp. *Introduction to Stratigraphy*, 2nd ed. 1933.
L.J. Wills. *Physiographic Evolution of Britain*, 1929.

Chapter III

GENERAL CONSIDERATIONS

(1) INTRODUCTORY

The subject matter of this chapter could easily be expanded to form a large book. D. W. Johnson showed its scope in his *Shore Processes and Shoreline Development*, published in 1919, a comprehensive attempt, and the first of its kind, to bring together information hitherto scattered through a vast number of papers. But, in point of fact, he did far more than this: his book[1] contains a great amount of original work, and has had much influence on all later investigations. Some of his views in the course of a quarter of a century have been superseded, and naturally a great deal of new information, based on recent field work, is now to hand. Nevertheless, any writer on shorelines must readily and fully acknowledge the debt that he owes to Johnson.

In this chapter there has been no attempt to review fully all the processes at work, but at the same time the reader may find here a sufficient account of the more essential factors which should be constantly in mind either when working in the field or when reading about coastal physiography. Since this book deals with England and Wales, references are obviously more pertinent if directed towards those parts of the British shores. Further, some of the material that might rightly be considered relevant to this chapter is to be found in other parts of this volume, in order to avoid repetition wherever possible. Thus, as the regional chapters contain detailed discussions of particular coastal features, it is more appropriate in an introduction to leave these details aside. Both published and unpublished work of Mr W. V. Lewis, also of the Cambridge Geographical School, has been freely used. His research has helped physiographers considerably in understanding certain processes concerned with wave action.

(2) WAVES AND WAVE ACTION

When wind blows over water, waves are formed. In a typical wind wave in open water the individual particles travel through a circular orbit, the movement is forward at the wave crest and backward in the trough (Fig. 5). This motion is plain enough if a light floating body be observed. Further,

[1] One of a projected trilogy, the third of which was not published before the writer's death. The second volume is entitled *The New England-Acadian Shoreline*.

it will be seen that the body makes one complete revolution whilst the wave travels forward one wave length. From this it follows that the velocity of propagation of the wave is greater than the orbital velocity in

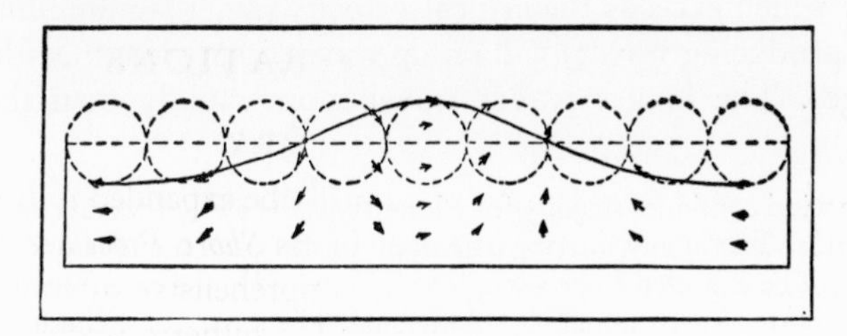

Fig. 3. Wave motion

the proportion that the wave length is greater than π times the wave height. As will be shown later, the wave form is transmitted by pressure, and the wave energy is partly potential, and partly kinetic. The potential energy is due to the height of the wave crest above the general sea level, and the kinetic energy to the velocity of the water particles in their orbits. The effect of waves decreases rapidly with depth. White has suggested the general rule that for each additional $\frac{1}{9}$ of a wave length below the mid-height of the surface wave, the diameter of the orbit is decreased by a $\frac{1}{2}$. This clearly implies that the ability of waves to agitate the sea bottom depends primarily on the length of the wave.

Much has been written about waves, both from a general and a mathematical point of view.[1] It is not necessary here to delve deeply into this difficult subject, since the primary concern of this book is with wave effects on a coast rather than with their nature and origin. Brief references to Jeffreys's work are, however, in place since his views correlate theory and observation satisfactorily:

If waves are once formed on water, the main air current, instead of flowing steadily down into the troughs and over the crests, merely slides over each crest and impinges on the next wave at some point intermediate between the trough and the crest. The region sheltered from the main air current contains an eddy with a horizontal axis, while smaller eddies exist along the boundaries between this eddy and the main current. If such a theory is correct, the pressure of the air will be greater on the slopes facing the wind than on those away from it; for the deflexion of the air upwards when it strikes the exposed slopes implies a reaction between the air and the water.[2]

[1] Sir G.B. Airy, *Encyclopædia Metropolitana*, 5, 241; Sir G.G. Stokes, *Math. and Phys. Papers*, 1, 197; D.W. Johnson, *Shore Processes and Shoreline Development*, 1919 —this book contains a very full bibliography; V. Cornish, *Ocean Waves*, 1934—this volume includes a chapter by H. Jeffreys.

[2] *Proc. Roy. Soc.* A, 107, 1925, 189.

This means that once the wave has been formed differential pressure helps to increase its amplitude. Jeffreys's hypothesis also helps very well to explain the variation in the length of wind-formed waves. Any wind the velocity of which exceeds the critical velocity (i.e., the minimum velocity capable of producing waves) will set up waves of any length within certain finite ranges. The longer waves travel more rapidly than the shorter ones.

When the wind continues steadily for a long time, the swifter waves continually gain on the slower, so that at any intermediate point of the wave train many different types of waves might be expected to be superposed, as indeed is the case. But a new feature appears: the shorter waves are wholly obliterated. The theory of sheltering affords two explanations of this fact which probably co-operate in producing the change. In the development of the theory all terms depending on the squares and products of the deviations from steady motion have been ignored. It has, however, been found that when a wave is actually formed its amplitude will increase exponentially with the time, so that a stage must be reached where the neglect is no longer justifiable. The nature of the change is well known. The form of the waves ceases to be purely sinusoidal, the crests become sharper and the troughs flatter than in a simple sine curve.

There is a limit to the height that such waves can attain; in a gravity wave it is fixed by the fact that the crests become definitely angular, the angle being 120°. A short wave superposed on a long one will be of smaller height than the latter when both are fully developed. Hence the short one will be sheltered from the wind for the whole time except when it is near the crest of the long one and on the exposed side. Its opportunities for growth are therefore very much less than if it were alone, while the damping effect of viscosity will be at least as great. Further, when it is on the sheltered side of the long wave, the splashing of the water from the crest of the long wave will tend to fill up the trough of the short one. On both grounds, therefore, it may be expected that the short wave will die out.[1]

This theory explains the way in which waves of different lengths may be formed and so also the variation in length and regularity of waves reaching a particular piece of coastline. In general, waves set up by local winds are irregular in size and periodicity and are also short waves: waves due to a distant storm are far more uniform in length and periodicity.[2]

This question also involves the length of 'fetch' of open water over which the waves are travelling. This point has been taken up by many writers, and Vernon Harcourt concluded that the height of waves is simply proportional to the square root of the length of fetch.[3] The available fetch varies greatly round the coasts of England and Wales, and it is undoubtedly this factor which primarily determines the direction from which the largest waves approach a given shore. It is enough to point out here the exposed

[1] *Proc. Roy. Soc.* A, 107, 1925, 201.

[2] See, e.g., J.S. Owens and G.O. Case, *Coast Erosion and Foreshore Protection*, p. 16.

[3] *Harbours and Docks*, p. 17.

position of western Cornwall to Atlantic breakers, the importance of waves coming in from the quarter between north and east on the Norfolk coast, and the relation of the trends of the coasts enclosing Cardigan Bay in respect to open water.

The action of waves breaking on the shore is all important in the study of coastal details. It is a subject about which much has been written, but also one on which precise information is often lacking. Certain general points have often received notice, for example, beaches are apt to be built up by certain waves, but combed down and destroyed by others. Too much stress has also probably been laid on waves of translation in which the whole of the water movement is forward in the direction of the waves. Waves of translation exist, but only in extremely shallow water, and it is doubtful whether any ordinary wave breaking on a shingle beach is ever of this type. They are most likely to occur when normal oscillatory waves break on a shoal and then continue their progress over a very shallow-water area. Before invoking them in explanation of beach features the experiment of throwing light floating objects into the waves just before they break will show whether the motion of the water be that peculiar to oscillatory or translation waves. Probably few physiographers could claim with assurance to have seen translation waves in action on beaches.

When a wave reaches shallow water its orbit becomes elliptical and its speed decreases, but the period remains the same. Further, the ellipticity of the orbit increases with depth, and on the bottom becomes in fact a straight line. A wave rolling inshore will be seen to increase in height and diminish in length, its front slope will become steeper, attain verticality, and finally curl over to fall with a thud on the beach where it breaks and sends a swirling mass of water, called the 'send' or 'swash', up the beach. This swash sweeps material with it, but its power soon ceases since it is working against gravity and friction, and at the same time much water is sinking into the beach itself. When the upward movement has ceased, the remainder of the water retires directly down the beach as the backwash, the speed and power of which increase downwards with the slope. Usually the slope is steeper on the higher parts of the beach, and becomes quite gentle lower down, especially on a wide foreshore. On a gently sloping beach the swash is often sufficiently more powerful than the backwash and sweeps material upwards against gravity and so steepens the beach until equilibrium is attained. On a steep beach the backwash, together with gravity, is likely to be the more powerful. This tendency of waves to form a sand or shingle slope adjusted to their action has been noticed and discussed by several writers. Little, however, was said about the effect of different types of waves in connection with changes of beach profile until W. V. Lewis[1] investigated the matter. His work, even if subject to emendation, is really valuable because he has tried to solve the important problem of beaches being built up under certain conditions and of their erosion under others.

Lewis is concerned primarily with what he calls destructive and constructive waves. The idea that some waves build up whilst others erode is certainly not new, but Lewis goes into the matter much more deeply than previous writers. In his research on destructive waves the main feature is the power attributed to the backwash compared with that of the swash which spreads up the beach in a lifeless fashion, although the actual amount of water is large. Waves giving this effect appear to be characterized by an intense orbital motion associated with the actual breaking. This means that the waves try, as it were, to complete this motion and so plunge vertically downwards, or even curl under seawards a little—a point noted by Cornish. Waves up to 6 ft. high on breaking have indeed been seen to do this. Hence their energy is expended almost wholly in this manner and the swash travels, somewhat ineffectively, only a few yards up the beach. Lewis has counted ten to twenty of these waves in a minute, but as a mixed series of large and small waves may break at varying intervals, he does not regard the timing of the frequency[2] as satisfactory.

Constructive waves break with much greater regularity and, according to Lewis, have a frequency of five to eight in a minute. The breaker plunges far less vertically than in the case of a destructive wave. Consequently much more of the wave energy is transmitted to the swash which, while it may be less in volume than in destructive waves, is swift-running and effective. The backwash is weaker because the comparatively small amount of water in the swash is further diminished by percolation and, once the backwash has begun, by friction.

The length of a wave is roughly proportional to the square root of the wave period. From what has been said already it follows that the length of constructive waves is greater than that of destructive waves of similar height. The nearly vertical plunge of a destructive wave is induced by the great mass of water brought in by its predecessor, which allows the new wave to travel farther inshore, and plunge vertically on to a stretch of water without any extensive swash. For much the same reason the backwash is intensified, that is to say because the greater amount of water on the beach implies a much less effective loss through percolation and friction. In a constructive wave the orbital motion is more elliptical, and from ob-

[1] See, e.g., *Geogr. Journ.* 78, 1931, 131. I am also much indebted to Mr Lewis for access to his unpublished work.

[2] Frequency also varies with the size of the waves.

servation it appears to break at an earlier stage of its orbit than a destructive wave. Possibly this is due to the greater amount of water brought to the beach by these waves which have a fairly flat plunge and so transmit a considerable horizontal motion to the swash. Constructive waves also appear to have a far better defined moment of breaking so that all the orbital motion seems to end with one plunge on to the beach.

The action of storm waves now needs attention. Their general effect on shingle beaches has been noted by many writers, but it will be necessary to refer only to Coode who observed that on the Chesil Beach after a great storm in 1852 the slope of the beach was reduced from 1 in $3\frac{1}{2}$ or 4 (during offshore winds) to 1 in 9 or $9\frac{1}{2}$ after an onshore gale.[1] It is plainly noticeable on several parts of the East Anglian coast that strong onshore winds comb down the beach, whereas the normal action in steady calm weather is a gradual building up. In other words, a flatter slope between high and low water marks usually prevails after an onshore storm. Cornish, in explanation of these points, emphasized that percolation has less effect in weakening the power of the backwash in large than in small waves, and that, in consequence, the backwash during a storm can sweep material down a flatter slope. Storm waves of this type usually result from gales fairly near to the coast, and are *not* the product (i.e., ground swell) of a distant gale. Hence, reverting to Jeffreys's views, it is probably right to regard these waves as high in proportion to their length, and therefore in that respect also they would resemble normal destructive waves. At the same time it must be borne in mind that *big* ground-swell waves may have a destructive action because friction and percolation will then have less effect in weakening the great volumes of swash and backwash, and gravity will thus work to make the latter dominant. Finally, it should be remembered that big storm waves also throw up some shingle to the highest part of a beach.

Many doubts and difficulties still remain. As Jeffreys has pointed out, the shortest waves will be the product of local winds. But it is not at all improbable that several sets of waves are superimposed, and so measures of frequency are difficult to interpret. Lewis, again, writes: 'Thus the repeated observations of destructive action with onshore winds might well be due to the fact that the responsible waves were the product of...local winds and therefore of shorter wave length and very irregular in size and period,...whereas the constructive action noted with offshore and light winds probably meant that the waves considered were the product of distant winds and therefore of long wave length....In the case of the writer's [W. V. Lewis] observations destructive waves were undoubtedly

[1] *Mins. Proc. Inst. Civ. Engs.* 12, 1852, 520.

the result of local winds, whereas the best examples of constructive waves were noted when the wind had dropped some twelve hours after a storm had passed.'[1] This is not, however, the view of all physiographers, and Coode, in his paper on the Chesil Beach, asserts that constructive and destructive actions of waves have no connection with wave frequency. As a result of his own observations he stated that with nine breakers a minute there was very slight constructive action, but that with seven or less there was destructive action. This comment is somewhat opposed to the main mass of evidence, and Lewis suggests that the waves breaking at seven or less per minute were probably very large ones—the destructive ground swell mentioned above. If this be so, the nine or more breakers a minute would be smaller waves with a higher frequency, which, by comparison, would be constructive ones. This idea is also supported by the exposed position of the Chesil Bank, on which far larger waves prevail than on most shingle beaches around these islands.

It follows from these comments on constructive, destructive, and storm waves, that the profile of a shingle beach is continually changing as a result of the interplay of these different waves. A further point is of great importance, namely, the fluctuating water level as determined by the cycle of spring and neap tides. The general effects of it which have been propounded by Lewis, whose deductions are extremely useful, can be summarized clearly with the help of Figs. 4, 5 and 6.

In Fig. 4 *CD* represents the beach profile after a period of constructive waves. Suppose destructive waves to begin working on this profile from low water mark upwards. The swash sweeps but little material upwards, and the backwash carries a good deal downwards. Thus the new profile *EHK* is formed. As the water level rises, *EHK* travels up the beach and is represented successively by stages such as *FLM* and *GNP*. At and near high water the ridge *C* is rapidly cut back: if the waves be small, they may only reach, say, the position *R*. In the figure, *AX* represents the limit to which cutting proceeded at high water, and this slope is left by the falling tide. On the other hand, the minor slope *NP* retreats downwards with the breakers, so that at the next low water the profile will resemble the simple curve *AB*. The steep upper slope *AX* is often noticeable: it is usually unstable, and pressure with the foot often causes a small slide of pebbles. It must be borne in mind, of course, that the whole process undergoes endless variations according to the size of the waves.

In Fig. 5 *AB* represents, as above, the slope left after a period of destructive waves. Assume again the conditions of low water. With constructive waves the swash sweeps up more than the backwash carries down the

[1] W.V. Lewis, unpublished work.

beach. Hence a small ridge begins to form like that at *C*. As the tide rises the ridge travels up the foreshore and is likely to increase in size—say, at *E*. At high water, the water offshore is deeper and waves are bigger, and since the water remains at or near this level for some little while, there is probably time for a ridge to be built at *G*. When the tide falls, such a ridge is left undisturbed, and the profile *GC* is produced as in Fig. 5. At extreme low water the new profile will probably approximate to the curve *GCD*, but in point of fact this curve may be slightly ridged. Bigger waves

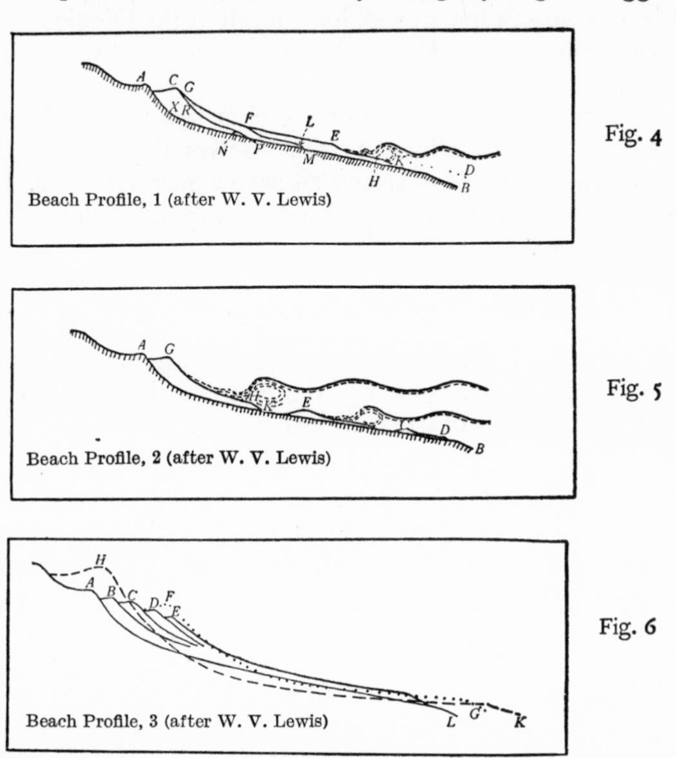

Fig. 4

Fig. 5

Fig. 6

will lead to no significant change of this profile, except that the ridge *G* may be larger. As Lewis remarks, it is not easy to distinguish between a ridge formed in this way and one left by *large* destructive waves rearranging the shingle of an earlier ridge.

In Fig. 6 *AL* represents the beach profile resulting from destructive waves at spring tides. If storm waves are at work, they will cut into any earlier ridges at or near *A*, and carry shingle down below low-water mark. If, however, during the succeeding ordinary and neap tides constructive waves prevail, ridges such as *B*, *C*, *D*, and *E* will be built, falling in altitude with the falling tides. In the following spring tides, constructive waves might readily reform ridges *E* and *D* into a major ridge, *F*, although this, in time, would be likely to suffer erosion from destructive waves. But if storm waves were to succeed, most of the material in the ridges *B*, *C*, *D*, and *E* would be combed down, even if some shingle, during the onset of these powerful waves, were thrown up to build a higher ridge such as *H*. At low water a profile resembling *HK* would emerge, representing a gentle slope formed by the powerful backwash. In the succeeding tides, assuming, of course, calmer conditions, much of the shingle swept down to the neighbourhood of *K* would be rebuilt to form smaller ridges in front of *H*.

This brief account is enough to make clear that the profile of shingle beaches is very variable. Different types of waves, the ever-changing level of the tides, and the effect of storms all have a great influence. It is time now to turn to the supply of material and the way in which beach material travels alongshore.

(3) MOVEMENT OF MATERIAL ALONGSHORE

The movement of material alongshore takes place in various ways. Much has been written on this subject in the past and there have been many supporters of the theory of transport by currents.[1] Others have supported the idea of beach-drifting as a result of wave action, and of late years this process has gained recognition as the more important of the two. Some allusion is necessary to the considerable effects of currents, but it may be said at once that there has been a good deal of inaccurate writing about them. A current can move much material in proportion to its speed and to the nature of the surface over which it is flowing, but it cannot by itself throw up beach material above high water mark. Nor can a tidal current have any influence once the waves have broken on a beach. Thus the explanation of such coastal features as Dungeness by the meeting of currents,

[1] In this context 'currents' usually mean the horizontal translation of water due to the tides. For a discussion on this matter see G.K. Gilbert, *5th Ann. Rept., U.S. Geol. Surv.* 1883–4, p. 85; F.P. Gulliver, *Proc. Amer. Acad. Arts and Sci.* 34, 1899, 180; I.C. Russell, *U.S. Geol. Surv. Monograph XI*, 1885, p. 90; W.M. Davis, *Physical Geography*, 1898, p. 364; J.E. Marr, *Scientific Study of Scenery*, 1926, p. 335; V. Cornish, *Geogr. Journ.* 2, 1898, 538; W.H. Wheeler, *The Sea Coast*, 1902, p. 199; E. de Martonne, *Traité de Géographie physique*, 2, 1926, 980; A. de Lapparent, *Leçons de Géographie physique*, 1907, p. 278; E. Haug, *Traité de Géologie*, 1, 1927, 477; Royal Commission on Coast Erosion and Afforestation, 3, pt I, 1911, 12; J.A. Steers, *Geogr. Journ.* 83, 1934, 485; J.S. Owens and G.O. Case, *Coast Erosion and Foreshore Protection*, 1908, p. 51; D.W. Johnson, *Shore Processes and Shoreline Development*, 1919; R.S. Tarr, *Amer. Geologist*, 22, 1898, 3; J.E. Woodman, *ibid.* 24, 1899, 330.

eddy, tidal, or any other kind, is quite inadequate. Even supposing the shingle of Dungeness to be carried to the neighbourhood by currents, it could not have been built up into the numerous shingle ridges by current agency. Yet such statements are still implicit in official publications.

When waves break on a beach they may do so directly onshore. If that is the case, an up-and-down movement of the shingle and sand with the swash and backwash alone will result, and there will be no clear lateral movement of any consequence. This is plain enough on bay-head beaches, on to which material may be carried; when once it is on the beach, however, it can only be moved up or down, or temporarily away in a direction roughly perpendicular to the trend of the beach. On the contrary, on an open shore, the waves frequently approach the beaches obliquely. Imagine

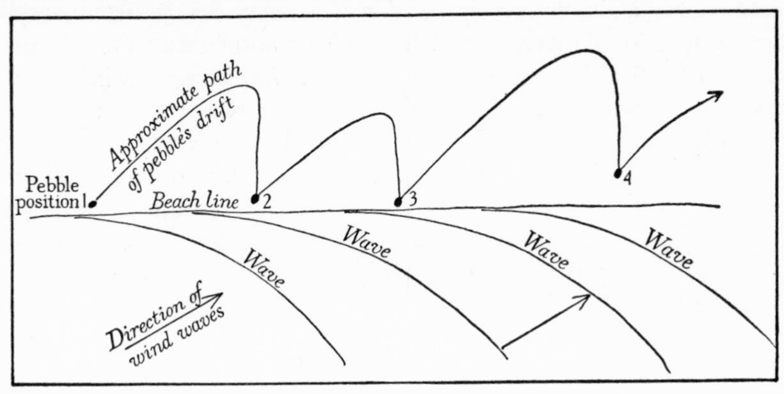

Fig. 7. Beach-drifting

for the moment the progress of one wave. As it rolls in towards the beach, the landward end of it first comes into shallow water and the wave speed is retarded. The rest of the wave 'catches' up to some extent, and the result is the familiar sight of a wave breaking as its successive portions reach the beach. Consider next any one part of this breaker. It has an oblique impact, and the swash runs obliquely up the beach. Any pebbles, sand, or other material capable of being moved by that swash are also swept up the beach at the same angle. How far they will be carried depends on the nature of the wave, the power of the swash, the slope of the beach, the size of the material, and other more local factors. When the swash dies out, the backwash begins. This, however, is directed down the steepest slope of the beach and is likely, therefore, to be direct rather than oblique. By it much, although not all, of the material moved up by the swash is carried down the beach, and the tracing of any particular pebble shifted by wave action will thus show its lateral movement over a certain distance. Successive waves all have similar effects, and the result is a very considerable lateral translation of beach material. It is in this process, called beach-drifting, that we must seek the major explanation of the movement of beach material (Fig. 7).

It is also important to bear in mind that the beach as a whole is not permanent. It consists of sand, shingle, and other particles resting on a platform cut in the underlying rocks which may be of any type. Thus, after a severe storm, it is not uncommon to see that the whole of a beach in a particular locality has been removed, although it is later rebuilt in normal weather. Again, on an open beach it does not follow that the lateral movement is always in one direction. The waves breaking on a shore are the product of certain winds, and after a period of marked drifting in one direction, winds, and consequently waves, may set in from another quarter and give a reverse drift. This observation raises the important question of dominant and prevalent winds. A prevalent wind may be defined as the normal wind expected in a given place. Thus around the shores of England and Wales south-westerly and westerly winds are prevalent, and their effects upon particular areas are discussed in other chapters. What matters immediately is that, along the shores of the English Channel, for example, the south-westerlies naturally cause a general eastward movement of material, although there are exceptions. The same reasoning applies to the shores of the Bristol Channel and Cardigan Bay. On the east coast, however, the same winds are offshore and so have little or no effect on coastal physiography, except locally, in such places as north Norfolk where there is a considerable stretch of shore trending west and east. The dominant winds are those which, although they may be less frequent, have, nevertheless, a far greater effect than the prevalent ones on the shores and beaches. On the western coasts of Great Britain prevalent and dominant winds usually coincide in direction, a further reason for the movements of material alluded to above. But on the east coast, however, the dominant winds are, as a rule, those coming from off the greatest stretch of open water relative to any part of that coast, that is to say, from the quarter between north and east, although locally, south-easterly winds are of significance. It is, therefore, safe to say that the commonly observed southerly movement of beach material along the eastern coasts of England is due almost entirely to these, locally, dominant winds. The complexities of movement which may emerge from detailed coastal observations as, for example, on the north Norfolk coast, are described in Chapter IX.

That beach-drifting is not only an effective but also often a rapid process is easily shown. If on a beach a number of marked pebbles, or a distinctive type of pebble, or broken brick, be put down and watched, especially during a tidal period between neaps and springs, it is quite easy to measure the distances travelled by many of the pebbles from day to day. Experience has shown that the best results are usually obtained if the marked stones be first put down near the lower edge of the shingle. Those placed on sand are apt to be buried or lost, and those set high up on the shingle are only occasionally touched. It is for this reason, also, that the experiment is most rewarding when the tides are making up to springs. The following table, taken from observations made on the foreshore of Scolt Head Island, illustrates the working of the process in normal weather.[1]

Date	Direction of movement of pebbles	Average daily movement of pebbles	Maximum movement	Wind		General direction from which waves approached shore	Remarks
				Direction	Average velocity		
June 23	W.	5–10 yd.	64 yd. W.	E.N.E.	c. 5 m.p.h.	N.E.	Waves small until 24th; 25th–29th increasing in size. Brick fragments used throughout
24	W.	5 "	80 "	N.E.	c. 5 "	E.N.E.	
25	W.	10 "	200 "	N.E.	c. 10 "	E.N.E.	
26	E.	5–10 "	120 "	N.N.W.	c. 20–25 "	N. by W.	
27	E.	4 "	120 "	N.	c. 15 "	Almost directly onshore	
28	—	0 "	120 "	N.	c. 18 "	"	
29	E.	5–10 yd.	120 "	N.N.W.	c. 20–25 "	N. by W.	

The figures given in this table are not exceptional, although movement will obviously be less in calmer weather. Several other similar experiments made at Scolt and Blakeney Point further illustrate the process. It is also worth noting that when waves come in obliquely and with a strong wind the movement is very rapid. Some experiments made in 1939 on the long beach known as the Palisadoes enclosing Kingston Harbour (Jamaica) showed this very well. Thus, when the trade wind was blowing freshly the movement was considerable with each wave. Unfortunately, the nature of the beach and the very small tidal range (less than 1 ft. even at springs) soon caused the loss of the marked stones.

Just outside the breakers true current action exists. It is clear already that the effect of waves, as far as orbital motion is concerned, decreases downwards rapidly. But in the relatively shallow water just offshore, especially on a gently shelving coast, waves certainly stir up a good deal of fine material. Anyone bathing in water about breast high when small waves are rolling in will observe not only the disturbance but also the stirring up of much material, assuming that the offshore zone is sandy. In addition it is often noticeable that the whole body of water has a lateral movement in one or other direction along the beach. This movement is

[1] *Scolt Head Island*, edited by J.A. Steers, 1934.

a true current, usually, if not always, tidal in origin. The current itself will vary much in velocity according to locality and other circumstances, but even with a very gentle one the material stirred up by the waves will be carried along in saltatory fashion, and in this way large amounts of sand are moved alongshore. If the current be swift, it can by itself move material, but in this connection it is as well to remember that experiments made under laboratory conditions are much modified in nature. Grains of sand or pebbles placed singly or in small groups in an artificial runway can be moved much more easily than like particles from out of a fairly compact mass of similar material on the sea floor. There is no need to enter here into much detail about current movement, but it is relevant to point out that currents cause the shifting of bottom material in two ways: first, by direct thrust against any particle projecting up above the general level of the bottom, and secondly, by downward currents due mainly to eddies. For this reason large stones lying on a soft bottom act as distinct aids to erosion. A further point of immediate interest, also, is the need to know the speed of the currents on the bottom in order to obtain accurate deductions. This may differ a good deal from the surface speed which is measured more easily. Thus, in water of moderate depth the bottom speed is, on an average, about 85 per cent of that of the surface speed.[1]

Experiments by Owens have also shown other curious anomalies. He found 'that where sand exists in quantity all currents up to 2·5 feet per second, or 1·7 miles per hour, are ineffectual in moving shingle, whereas at about 2·5 feet per second the current suddenly acquires the power of moving stones up to nearly three inches in diameter over a sandy bottom.' Further complications are introduced if the bottom be irregular or if the stones have large flat sides. His conclusion is of great interest: '...since the sea bottom is nearly always irregular, and stones are seldom perfect spheres, the effect of currents alone, unless of exceptional velocity, is chiefly limited to the transport of fine matter such as sand and mud.'[1]

Briefly, then, a current may move material by rolling it along the bottom, or in suspension in the mass of the water. Often, as noted earlier, suspension is intermittent since the material is stirred up by the waves. Currents flowing over a bed of heterogeneous material also cause sand waves or ripples which usually move in the same direction, although under certain conditions anomalous movements have been detected. Without entering upon the intricacies of ripple formation and other effects of current action, enough has been said to show that the major influence in shifting material along a shore is that of waves approaching obliquely, whereas currents

[1] In J.S. Owens and G.O. Case, *op. cit.* Chapter III, which contains a very interesting discussion on the Transporting Power of Running Water.

usually provide a minor agent of transport. Moreover, it is necessary to realize that, especially on coasts where there is a fairly large tidal range, since the currents may change in direction with the state of the tide, it is quite likely that material moves in one direction at or near high water, and in the opposite direction towards low water. This phenomenon is quite irrespective of any movement due to oblique wave action on the beach. At high water waves break on the upper parts of the beach and move material in a certain direction, while outside the breakers that of the current may or may not be the same. At low water the waves leave the higher parts of the beach untouched and only affect the lower part. It is a common feature of many beaches that, in normal weather, shingle only occurs on the upper parts, and consequently is only worked on by waves at high water. The lower parts of the beach are affected by tidal or other currents at high water, but mainly by wave action at low water, and hence may be subject to contrary movements. On a steep-to shingle beach the same processes are at work, but only the shingle is affected. From these observations, the danger of generalization is clear, and also the need to investigate carefully the local factors operating on any stretch of coast. It should never be assumed that conditions remain constant over long distances. Broad statements to the effect that, since the tide moves southwards down the east coast of England, spits of sand and shingle point in the same direction have no meaning. In the first place, for reasons already given, the 'tides' do not throw up material as beach ridges, and secondly, the twists and turns in a coast imply changes in the action not only of the currents, but also, and more significantly, of the directions from which dominant waves approach. The latter statement is very well illustrated between the Wash and Yarmouth.

Before leaving this matter it is essential to direct attention to currents close inshore. Admiralty and other publications give a great deal of information about currents in general, but it may be that a statement referring to a current only a mile out from the shore is inapplicable to the inshore one. Further, it is always as well to notice how and when the inshore current changes in relation to the height of the tide. Information of this type is known to local people, but is seldom available in printed form. The nature of the problem can be simply illustrated from an experiment made at Scolt Head Island.[1] Continuous observations were made from a boat anchored just outside the breakers, that is to say, observations were made every hour (on occasions less) during two calm periods of 24 hours, one period at neaps the other at springs. A vertical float was used of such a kind as to give the true water movement unin-

[1] *Scolt Head Island*, 1934.

fluenced by slight winds and nothing was registered until the boat had settled itself and was pulling steadily at its anchor. The directions taken by the float were always along the coast, either to east or to west, and the speed of this lateral current was obtained by attaching the float to a fine cord. It was easy to check carefully the length of cord which ran out during any observation. The two curves are shown in Fig. 8. Their irregularities are due to the short time over which the experiment was made, a disadvantage which could have been eliminated had circumstances allowed its continuance over a longer period. These observations at Scolt make it quite clear that the beach at that place owes nothing directly to the influence of tidal currents. The shingle on the upper part of the beach is only touched near the time of high water, during (see Fig. 8)

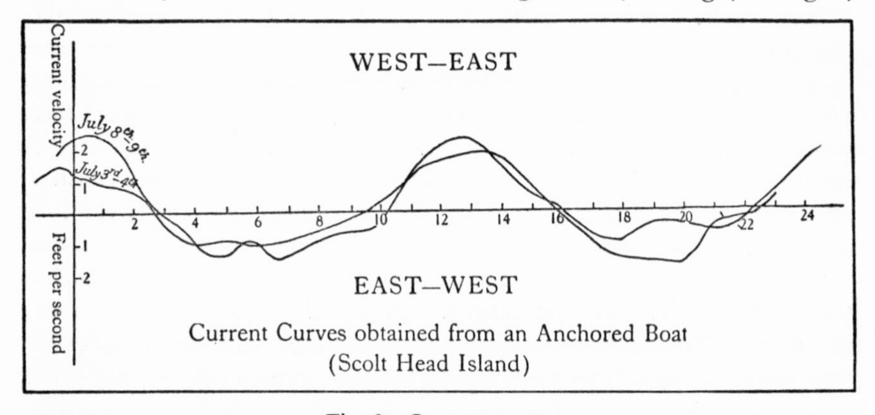

Fig. 8. Current curves

two or three hours on either side of which the currents are running east, or against the growth of the beach. It is true that while the tide is rising the current is flowing west for some hours, but to all intents and purposes, by the time that that part of the beach on which shingle normally rests is covered, the current has turned and is running to the east. Again, the velocities as shown on the graph make it abundantly clear that they are not of sufficient strength to move shingle even if other circumstances were favourable.

It must not, however, be concluded that tidal currents never have any effect. On the contrary, they are often of great importance. Tidal races such as those in the Pentland Firth, or the lesser ones between the Pembrokeshire coast and off-lying islands, are very powerful and have great scouring effect. Similarly the ebb currents from certain rivers are often rapid. A current of about seven knots at springs runs out of the narrow entrance between the mainland and the southern end of Orford Ness. It is often said that anchor chains of small craft lying within the harbour are brightly polished by being scoured by shingle while these currents are running. The ex-current from the Mersey is all important in keeping clear the estuary where it is narrowest, and many other examples could be cited.

(4) SAND AND SHINGLE SPITS: ORIENTATION OF BEACHES

It will be clear now that in the writer's opinion beach-drifting is a process infinitely more important than any other in the transportation of beach material. It is therefore time to turn to two important matters and study the application of this agent, which involves wave and not current action. It is a commonplace that many sand or shingle spits run often for considerable distances across the mouths of re-entrants or deflect rivers, and although opinions have differed, a number of writers in a somewhat uncritical way have invoked the explanation of longshore drift under current action. It has already been shown that currents cannot possibly throw up shingle ridges above high water mark, and it is a feature of many of these sand and shingle spits that they consist largely of shingle ridges on their upper parts. Lewis[1] has given a good deal of attention to the problem of the elongation of spits, and his findings corroborate substantially those of the writer. He has considered spits growing forward in shallow water, like Blakeney Point, and others, like Dungeness, which push out into deep water. Whilst the speed of progradation differs—those in shallow water growing more rapidly—the same process, possibly with slight local modifications, applies to both.

The normal beach-drifting of ordinary oblique waves carries material to the end of the main line of a spit where large quantities are deposited. Sooner or later this material is thrown up into ridges by wave action, especially by moderate-sized constructional waves at periods of spring tides. Thus, fresh and more or less permanent ridges are built overlapping the end of the main line of the spit, and the whole is lengthened, the new extension now becoming an integral part of it; the beach-drift continues along the freshly formed piece of spit just as along the older part. The same process takes place in deep water, except that far more shingle has to be carried along and thrown up to make any appreciable addition. 'Thus whatever variety might be presented by tidal currents, size of waves, coastal configuration, offshore slope and depth of water, longitudinal extension seems to consist essentially of the addition of ridges overlapping the end of the spit in a direction largely dictated by the dominant (storm) waves. The fact that this action takes place equally on the shores of a lake

[1] *Proc. Geol. Assoc.* 49, 1938, 107.

and on the tidal shores round the British Isles demonstrates its complete independence of either tidal or large planetary currents. The spasmodic nature of the additions when examined at frequent intervals is just what one would expect from the vagaries of storm waves...' (Lewis, unpublished). It seems quite possible also that the true tidal currents carry a good deal of fine material to the end of the spit and beyond it, especially if the dominant one runs in the direction of the growth of the spit. If this process takes place, then the effects of wave action already described are enhanced. But at the same time it can happen that spits will lengthen in the manner outlined even if the dominant tidal current, and consequently the transported finer material, be contrary to the direction of growth of the spit. The process of lengthening may merely take longer.

Associated with the whole question of wave action and beach-drifting is another interesting matter first pointed out clearly by Lewis. If an inspection be made of the main spits of shingle around our coasts on large-scale maps it is clear that many of them have a tendency to run somewhat outwards and away from the main trend of the coast. This is especially plain in Cardigan Bay: both Morfa Dyffryn and Morfa Harlech, the two main coastal forelands, are good illustrations. On the east coast, Blakeney Point and Scolt Head Island also exemplify the same feature. Again, on the south coast, Dungeness and the smaller Hurst Castle spit trend outwards from the coast, and it would be easy to cite a number of other instances. Many beaches between headlands are arranged, too, so as to show a clear tendency to run at right angles to the main direction of wave approach, a characteristic discussed more fully below. Others are set back a little—for example, the spits across the mouths of the Dyfi and Mawddach in Cardigan Bay. There is no need to enter into details of the examples listed here: those cited above, and many others, are specifically described in subsequent chapters. But in spite of the local individuality of each, two important points need emphasis. In the first place, all the *main* parts of these spits or beaches face very closely the direction of approach of the dominant waves acting upon them—a fact in itself highly suggestive of their wave-formed origin. Secondly, this direction may or may not be that of the prevalent waves. In the latter case there would have been some retardation in the speed of beach-drifting, and at the same time an addition of material to the foreshore of the spit. This extra material would sooner or later have been thrown up on to higher ridges by dominant and storm waves until such a time as the spit became of sufficient size for permanency. It is in this way that the general orientation of ridges may be explained: their subsequent longitudinal extension follows as a result of normal beach-drifting along the lengthened spit.

Up to now there has been emphasis upon the direction of the *main* parts of beaches or spits. It is, however, most noticeable that running back from the main spit there is often a series of lateral ridges each of which was formerly the end of the spit. In the past they have often been attributed to 'longshore currents', but observations over a long period encourage the opinion that they are primarily due to wind, and therefore wave action, from a direction generally contrary to that of the growth of the spit. For a moment it pays to jump a stage. At Scolt Head Island and Blakeney Point the lateral ridges are very well developed. One or two of them have been definitely curved back at their extremities in recent years by waves

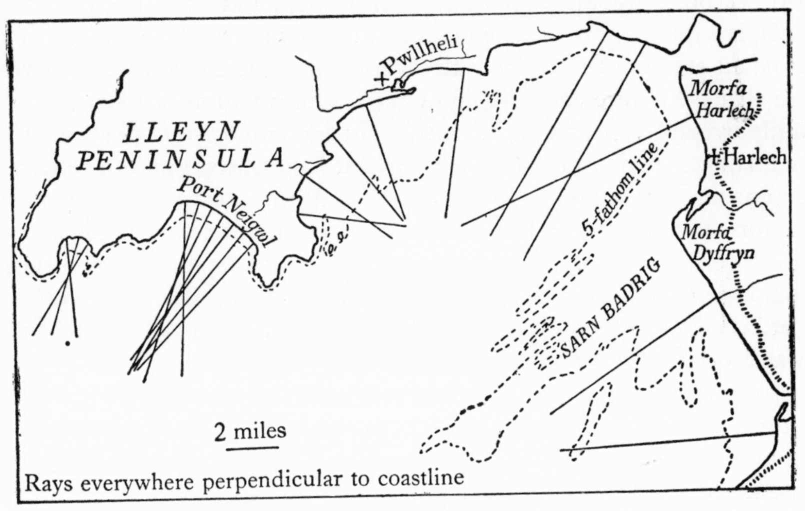

Fig. 9. Orientation of beaches in Cardigan Bay (after Lewis, *op. cit.*)

in the harbour. At Blakeney, one lateral ridge running directly back from the main bar is recurved westward at its end, a development which is also due to waves and storms in the harbour at high water. This example is interesting because the minor recurving at the ends of laterals is nearly always backwards rather than forwards. But to return to the main beach, it is evident that material can accumulate at and near its end, and that this may be built into a normal longitudinal extension of the beach. It may also, however, just as easily become an agent in forming a lateral ridge as a result of contrary winds and waves at the critical time. These contrary waves will try to form a ridge, roughly at right angles to their direction of approach, and their potency will obviously depend upon the available fetch relative to the particular locality. The depth of water and amount of material, the relative amounts of sand and shingle, and other factors will, in addition, all have their effects. At Scolt Head Island, for example, where there is a small amount of shingle relative to the sand, and where the amount of open water to the west and north-west of the island is limited, the new ridges are small and widely spaced. At Orford Ness, in contrast, where there is abundant shingle and no sand, where also there is a good stretch of open water to the south and south-east, the result is quite different. Dungeness is, perhaps, a better example than either of the above of a shingle foreland growing out into deep water.

Such varying factors and others not mentioned here should certainly be borne in mind. It is, moreover, well to remember that the incidence of storm or other waves, from a direction contrary to that of the growth of a spit, is a quite unpredictable matter; it may or may not coincide with an abundant supply of material at the end of the spit. Thus it will easily be seen that the nature and spacing of the lateral ridges must themselves be highly irregular. In addition, it is relevant and important to point out that the unattached end of a spit, especially if it be growing out into relatively deep water, is more than likely to be turned inwards, to some extent, by the normal waves acting upon it. If these conditions were to prevail over a long period of time, the continued growth of the spit, or its recurving, would seem to depend largely on the amount of material supplied to the unattached end by beach-drifting. An abundant supply would probably mean a general lengthening: a shortage might mean a recurve. The whole problem needs systematic attention and, although the general principles outlined above are probably correct and relevant, there is no doubt that far more attention must be given to local examples. A number of the mistakes that have been made in the past about coastal features are probably due to the neglect of observations in specific small areas. No one would claim as yet enough knowledge of coastal physiography to be dogmatic, but the factual information which gives confidence for some generalization is more likely to accumulate if local features be mapped and examined in detail, and, what is of great importance, observed over a long period of time. Regularity of growth is certainly not a law. The maps shown in Fig. 80, for instance, show how the distal end of Scolt Head Island has varied over a period of years. With this lack in mind it is the greater pity that observation and mapping had perforce to cease with the war.

The orientation of sand or shingle beaches in wide bays and open coasts also appears to conform to a pattern largely determined by the direction of approach of the dominant waves. Details of particular examples are better left for later chapters, but to illustrate the general point the bay beaches on the northern shore of Cardigan Bay, and those on the somewhat exposed shores of Somerset, serve very well. The relation of these and others to the direction of approach of dominant waves is clear on

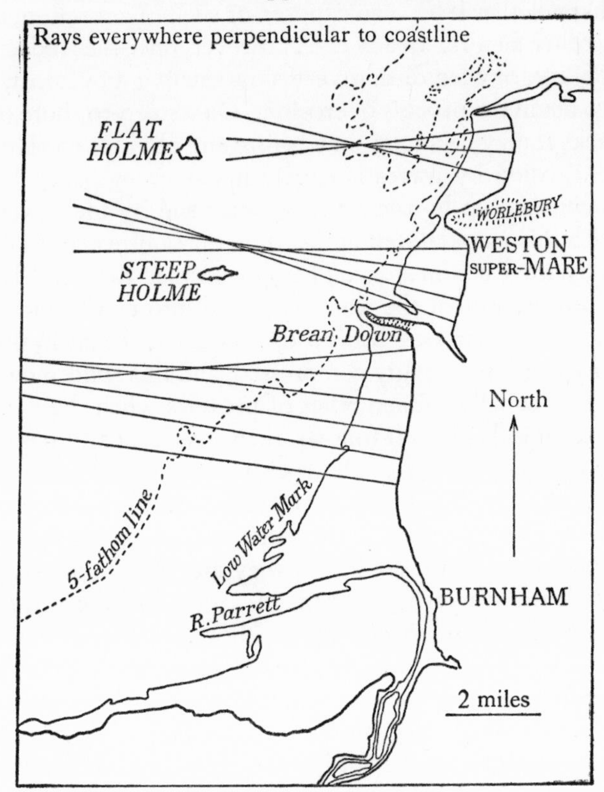

Fig. 10. Orientation of beaches, near Weston-super-Mare (after Lewis, *op. cit.*)

Figs. 9 and 10, taken from one of Lewis's papers. Although much detailed study is lacking, there will be substantial agreement with his main conclusion that these beaches are orientated at right angles to the direction of approach of dominant waves.

(5) EROSION: FORMS OF CLIFFS

So far, comment in this chapter has been confined mainly to the effects of wave action on shingle and sand beaches. It is time to turn now, however, to some aspects of marine erosion in general. It might perhaps have been more logical to have discussed these at an earlier stage, but it is neither easy nor wholly desirable to maintain a rigid compartmentalism in dealing with shorelines. It will be clear from the outline in Chapter 1 that the general nature of a piece of coastline depends greatly upon whether it be primarily an emerged or submerged shore; whether it be formed of hard or soft rocks; whether it be in an exposed or sheltered position, and indeed upon many other factors. In any case, however, once the land and sea have attained stability, or even during very slow relative movement, the waves have time to begin the process of erosion. On a steep-to shore of hard and resistant rocks it may be a long time before any obvious erosive effects are noticeable. Erosion by waves is clearly much strengthened by the sand and stones churned up by the moving water and by the waves battering the cliffs with this material, but at the same time quite appreciable erosion takes place by wave action alone. The blows which they give have a great disrupting power, so that the water is forced into cracks and crevices of the rocks and weak parts are broken off. Further, the air in these cracks, crannies, or fissures is strongly compressed, and this sudden compression, followed by an equally sudden relief of pressure when the wave recedes, has great mechanical effect. It follows, then, even on hard rocky and deep-water coasts, that that part of the rock surface which comes under the influence of wave action is gradually eaten away. In softer rocks the process is more rapid. If the water be shallow offshore, the same effects are present as on a deep-water coast, but they are much more obvious. The zone in which the waves break is much wider, and the waves, off very shallow coasts, break some distance offshore. They erode the bottom, loose material is produced, and this, under wave action, has great cutting power. Hence, it is not long before a fairly level submarine platform is chiselled out. On hard steep-to shores it requires a much greater time before even a narrow notch is cut by the waves, but sooner or later, provided that the levels of land and sea do not alter, this will happen. Between these two extremes there is every possible gradation. The point, therefore, to be emphasized is that, given time, the waves will form a shore platform. In early stages this will be formed almost wholly of bare rock, but the material eroded from it must, to a large extent, be scattered over the platform. It will increase in amount with time, and is the very stuff[1] with which the waves can the more effectually cut and widen and deepen the platform.

This loose material, and especially that near and at the landward side of the platform or terrace, forms the embryo beach. For reasons already discussed, we know that beach material travels along the shore, and thus, in early stages, it is likely to be found in greater quantities in re-entrants. The beach and underlying platform will usually be more pronounced

[1] Temporarily it may be present in such abundance as to protect the platform from erosion. Such a phase is not likely to be long-lasting.

where there is a big tidal range and also in areas of more easily eroded rocks. In the early stages of the platform's history, material under wave action is usually swept up toward the land, but often much finer stuff is carried seaward into deeper water. The waves are constantly working on this platform and in time wear it downwards as well as enabling it to be cut farther inland. Off coasts like those of Cornwall and Devon the platform is often inconspicuous, but off softer coasts such as those of the Mesozoic rocks of south-eastern England it is wider; here again, however, the question of exposure is important. It is interesting to speculate on what would have been the nature of the coasts of England and Wales, had the Atlantic had full play on the east coast.

There is no point in discussing here the details of the terminology of beach features. It is enough to bear in mind that the platform gradually widens and deepens, and that on its upper and landward side material accumulates to form our beaches, whereas seawards, finer material may be swept off the cut platform to form a kind of submarine embankment which, as far as height and contour are concerned, has the effect of extending the platform seawards. It should also be clear that the platform will vary much in width according to locality, and that its nature will depend upon the underlying rocks. In soft clays it will probably be completely covered with sand brought by beach-drifting and tidal currents from elsewhere: if the supply of sand be scanty, the clay will show. On harder shores occasional knobs or ridges of the more resistant rocks will show up above the sand. Dykes of igneous rocks are apt to appear as walls, though in some places they weather more readily than the material through which they cut. Should a change of sea level have taken place (see Chapter XII), the platforms, assuming that they have been uplifted, will appear as raised beaches which may or may not retain their covering of loose material.

A platform also implies a cliff, and the forms of cliffs are infinite. The chief formative factors are the nature of rocks, which may be hard or soft, finely or massively bedded; the inclination of the strata composing the cliff, which may be horizontal or directed seaward or landward; and the topography of the land into which the sea is cutting, which may be a gently sloping upland, a country diversified by many small valleys, or an area of boulder clay. The degree of exposure and the size and force of the waves acting upon it are also important.

The cliff coasts of Great Britain should be reckoned one of the country's most treasured possessions. They are of every variety and form, including the great chalk promontories of Flamborough Head and Beachy Head, the granite masses of Land's End, the sandstones of St Bees Head, the fine series of limestones of Tenby and Gower, the Old Red Sandstone and Carboniferous cliffs of western Pembrokeshire, the newer limestones, sandstones, and clays of Dorset, and the glacial cliffs of Holderness and Norfolk. These magnificent cliff regions are all treated in some detail in other chapters, and there is no need here to do more than call attention to some general features. Sections of the coast formed of almost horizontal beds usually form steep and nearly vertical cliffs, like those near Bosherston in Pembrokeshire and Hunstanton in Norfolk, especially if the bedding be massive. Where the bedding is thinner and where alternations of hard and soft beds occur, vertical cliffs are still frequent, but their retreat is usually faster owing to the undercutting of the softer beds. The Lias cliffs near Lyme Regis illustrate this point. Where the beds dip fairly steeply seawards the profile is generally coincident or nearly so with the plane of the dip. Many instances of this condition could be quoted, but the cliffs at Giltar Point, Tenby, will serve as an example. The chalk cliffs of England are, however, often almost vertical rather as a result of the homogeneity of the rock than of the type of structure. The great overthrust zone and nearly vertical bedding to be seen just north of Swanage in Dorset does not greatly affect the profile. Cliffs with beds dipping inland are generally less precipitous since the latter are in stable equilibrium: undercutting cannot easily occur in such a way as to lead to cliff falls. Boulder-clay cliffs are again often nearly vertical, their retreat in fact is more often due to land water than to direct marine erosion, although the latter factor is very important. Large falls are frequent and come about very often because considerable masses slide forward on soft clay or sand beds which may be inclined at all angles as a result of minor contortions in the cliffs brought about by ice pressure (see p. 374). This fallen material is gradually carried away by the waves, but for the time being protects the cliff foot, and clean exposures in boulder-clay cliffs are somewhat uncommon for these reasons. In strongly folded rocks, including beds of varying thickness and hardness, occurs perhaps the most interesting cliff scenery in England and Wales. The sea, in its constant wear and tear, soon finds out the weaker parts and so produces coves and many other intricate forms. Such features are obvious in many places—outstanding examples being the Lizard coast in Cornwall and St Bride's Bay in Pembrokeshire. Strong bedding and jointing naturally aids the sea in its attack, and these planes of weakness may be at any angle. In relatively horizontal beds the joints are vertical and the sea eating into them cuts long gashes. If, as is often the case, there are two sets of joints at right angles to one another, the sea first cutting into the one series finds little difficulty in later excavating the second series. In this way stacks are formed. Since the waves act only up to a certain vertical limit, dependent mainly on their size, the exposure of the coast, and the tidal range, it follows that in high cliffs they will only affect the lower parts and thus the 'stacks' are in early stages often connected by arches with the upper cliffs. It must be remembered that subaerial erosion is active all the time, so that arches are not particularly common features because they are being attacked from above as well as below. The cliff profile, indeed, depends very largely upon the relative rates of marine and subaerial denudation, and, although the factors outlined above are of great importance, all are subordinate to these two influences just mentioned. It is presumably the nice balance between the two conditions which explains the verticality of the chalk cliffs.

Studies of cliffs from a physiographical point of view are comparatively rare in this country. There are many general statements to be found in papers and books, and these are often sufficient to explain the main features. But with the important exception of some of the *Memoirs of the Geological Survey*, details of great interest are too frequently overlooked. It may, therefore, be of interest, even in a chapter of general considerations, to turn to certain features in the cliffs of north Devon which have been admirably described by Arber[1] (see also p. 214). The five main headlands are the Foreland, Bull Point, Morte Point, Baggy Point, and Hartland Point. The first appears to be the remnant of a former northward-projecting ridge. At Bull Point erosion only follows the change in direction of the watershed, and Arber feels that the hardness or softness of the rocks plays no decisive part[2] in cliff structure. Morte Point consists of slates, which have not been proved more resistant than those in Rockham Bay to the north or in Morte Bay to the south. Bull Point is also formed of the same slate series. Baggy Point is a westward-projecting ridge, part of a former secondary watershed: the Pickwell Down, Baggy, and Pilton beds are all represented in it, and they include sandstones, shales, and slates. There is, however, no clear indication of any kind that the lithology of these beds affects the form of the point. Saunton Downland, a mile or two to the south, is similar. At Hartland, as at Bull Point, erosion merely conforms with the change in direction of the watershed. Therefore, of these five major headlands, two are really right-angled turns in the shoreline, and three are relics of former watersheds. 'In none of these cases is it apparent that the character or hardness of the rocks has any special significance in explaining the existence of these promontories. On the other hand, it is obvious that the sea is cutting back the land in conformity

[1] *The Coast Scenery of North Devon*, 1911.

[2] This is clearly open to question: in many places rock hardness is all important.

with the pre-existing sculpture of the country, due to long-continued atmospheric denudation.'[1]

The minor headlands along the same stretch of coast are more complex in origin, and the relative hardness of rocks, the physiographic features of the land, and other factors such as have been discussed above, all play a part. Gullies are cut along the weak parts or along joint or other planes, and the points between the gullies, at first blunt, become sharper with time and may eventually be worn away. The headlands due mainly to the existence of hard rocks are Cockington Head, Windbury Point, Blackstone Point, and Higher and Lower Long Beaks. Erosion in the mouths of the coastal streams leads to the formation of bays, for example, Crackington Haven and Combe Martin, and, conversely, one wall of a river valley (Heddon's Mouth) projects seaward. Hence these minor points also depend a good deal on the nature of the pre-existing land surface. Although in the earlier part of this discussion of erosion the general principle has been propounded that softer rocks are more easily worn away, there are exceptions. In the reefs at Titchberry Water the sea has certainly destroyed the softer shale beds and left those formed of sandstones as salients. But in the buttress reefs between Speke's Mill and Welcombe the shales often stand higher. This peculiarity may be due to the absence or scarcity of joints in them, or possibly to the position of the beds in the folded structures, but conjecture alone is possible. Near Bude and also near Bull Point there are great buttress reefs of sandstones and shales tilted at a high angle and projecting outwards from the cliffs. Their origin still remains somewhat obscure, but they may be the sole remains of the most stable part of a seaward-running anticline. In general, anticlinal folds are stable, a feature which is clear enough at Tut's Hole and in the great anticline near Bude. Later, however, the beds in the core of an anticline may be cut out, and a cave may be formed like those near Clovelly and Mouthmill, or that near Broad Haven in St Bride's Bay. Contorted rocks on the north Devon coast seem to wear back evenly. On the many immature wave-cut platforms of this coast the surfaces are usually jagged and there is often an abundance of large debris. The blocks were quarried out by the sea working along major joint and bedding planes, leaving only minor planes in the block which may thus have been enabled to withstand weathering. The rocky reefs projecting from the surface of the platforms depend for their characteristics mainly on the type of rock, the nature and dip of the beds, the cleavage, and other fairly obvious factors. Cleavage certainly is responsible for the nature of many of the Morte Slate reefs, and reefs which are much denuded often become choked with sand.

[1] Arber, *op. cit.*

Other details of the effects of erosion will be found in those sections of this book which deal specifically with Pembrokeshire, Cornwall, and Dorset. The important point that emerges from the brief reference to Arber's work is the reiteration of the dangers of generalizing, and the risk of concluding that what appears to be obvious from ordinary reasoning, is necessarily true. Every part of a coast demands specialized and detailed study.

It is obvious that the rate of erosion will vary much from place to place, and that, other things being equal, it is likely to be most rapid in soft incoherent rocks such as boulder clay. In other chapters there is special emphasis on this point: for the moment it is enough to recall the great losses on such parts of the coast as Holderness, Norfolk, and Suffolk, and those sections of the Devon-Dorset coast where landslips have been severe. If buildings are destroyed, loss of land by coastal erosion is especially apt to cause comment in the press, but the continual small losses in agricultural land are collectively of far greater importance, although they seldom receive much public notice. Precise measurements of losses of erosion are wanted in many parts of our coasts. They would be of considerable scientific interest, and would indeed call attention to a matter that is often serious from points of view other than the physiographical. The Royal Commission on Coast Erosion (1911) published some interesting material on this matter, largely derived from comparisons between the 6-inch Ordnance Survey Maps of different dates, and also from details supplied by persons whose knowledge of a particular stretch of shore was reliable.

Erosion, however, should not be treated without a corresponding reference to accretion. In Chapter XIV the rapid growth of salt marshes is discussed. Accretion is far less spectacular than is erosion, and often passes unnoticed except by those immediately affected. In most of our estuaries and big inlets the process is in action; witness the great areas of reclaimed land in the Humber, the Wash, the Thames, Southampton Water, the Somerset Levels, Morecambe Bay, and other places. The general conclusions of the Royal Commission are, indeed, worth quoting:[1]

> The evidence as regards the total superficial area gained and lost in recent years on the coasts and in the tidal rivers of the United Kingdom shows that far larger areas have been gained by accretion and artificial reclamation than have been lost by erosion. The most reliable figures that could be put before us with regard to gain and loss in recent years were those furnished by the Ordnance Survey Department, which showed that, within a period on the average of about thirty-five years, looking at area alone without regard to value, about 6,640 acres have been lost to the United Kingdom, while

[1] Third (and Final) Report, 1911, p. 158.

48,000 acres have been gained. It should, however, be remembered that the gain has been almost entirely in tidal estuaries, while the loss has been chiefly on the open coast. Moreover, the gain...has been due, not so much to the agency of the sea as to the deposition of sediment brought down by rivers from their drainage areas....It is not, however, possible from the evidence to state what has been the comparative value of the area lost and gained, and we must point out that while the gain of land has, in superficial area, greatly exceeded the loss, erosion has been serious in many places, particularly on parts of the east coasts of England and of Ireland. The erosion, moreover, would have been far more serious if extensive works of defence had not been constructed by local authorities, railway companies, and others, at great cost, though, on the other hand, *such works in many places have been responsible for erosion of the neighbouring coasts by interfering with the normal travel of the beach material.*[1] On the whole we think, however, that, while some localities have suffered seriously from the encroachment of the sea, from a national point of view the extent of erosion need not be considered alarming.

It is not proposed in this book to discuss the building of sea walls, groynes, and other defensive measures. They are naturally of great interest from an engineering point of view, but not in themselves strictly relevant to the physiographer's approach, although reference may be very useful to a physiographical discussion on particular examples. At the same time it can hardly be over-emphasized that the problem of defence against sea attack is a national one, even if a national policy has never yet been evolved. Admirable defences have been built by certain municipal authorities, railway companies, and private owners, but the point made by the Commission remains of the utmost significance. The small-scale limited protection of one place nearly always means a transference of difficulties to a neighbouring area, a problem which is most evident on our south and east coasts along which material usually travels in a definite direction. Lowestoft, for instance, is bound to suffer on account of the measures taken at Yarmouth, and similar effects are plain enough in some localities on the Sussex coast.

[1] The italics are mine.

Chapter IV

THE SOLWAY TO THE DEE

(1) INTRODUCTORY

Between the upper part of Solway Firth and the estuary of the Dee the solid rocks ranging from the older Palaeozoics of the Lake District to the top of the Trias have comparatively little direct influence on the coastal scenery, except between Maryport and St Bees and in parts of Furness. Nevertheless, the older rocks largely hidden by superficial formations form both the foundation and the background of the coastal tract[1] (Fig. 11).

The Palaeozoic rocks of the Lake District include the Ordovician Skiddaw Slates, Borrowdale lavas and ashes, and the Coniston Limestone. They are followed by a monotonous series of grits and shales of Silurian age. All these rocks have been strongly folded, and the folds as well as the outcrops trend generally east-north-east to west-south-west. Later masses of igneous rock of Devonian age ranging from granites to gabbros are intruded into the older series. Although neither the sedimentary nor the igneous rocks of the Lake District have any direct connection with the coast, erratics from them are common on the Cumbrian and Lancastrian shores. These older rocks are succeeded unconformably by the Carboniferous Limestone, Millstone Grit, and Coal Measures, which, in their turn, are followed by the New Red Sandstone, composed mainly of sandstone in its lower part, but giving place upwards to thick shales and marls. Because of the doming of the Lake District in Tertiary times, the Carboniferous and newer rocks now form partial rings around the older Palaeozoics and for the same reason there is now a radial drainage system.

Along the coastal tract from the Solway to the Dee Triassic rocks predominate although mainly hidden by superficial deposits. Near Whitehaven, however, they emerge and form the fine headland and cliffs of St Bees. They also reappear at intervals on the Lancashire coast, but fail to form any extensive section north of the Wirral. The Coal Measures, which form part of an inner ring to the Trias, reach the coast between Maryport and St Bees Head. It is possible that they may also occur under Morecambe Bay, but in Furness the Carboniferous Limestone forms the inner ring and occurs near the mouth of the Duddon and other estuaries.

[1] For an account of the general geology of the area, see T. Eastwood, 'Northern England', and D.A. Wray, 'The Pennines and Adjacent Areas,' both published in *British Regional Geology*, 1935 and 1936.

It is this formation which accounts for the scenery of the northern part of Morecambe Bay. The Millstone Grit just emerges near Heysham, and

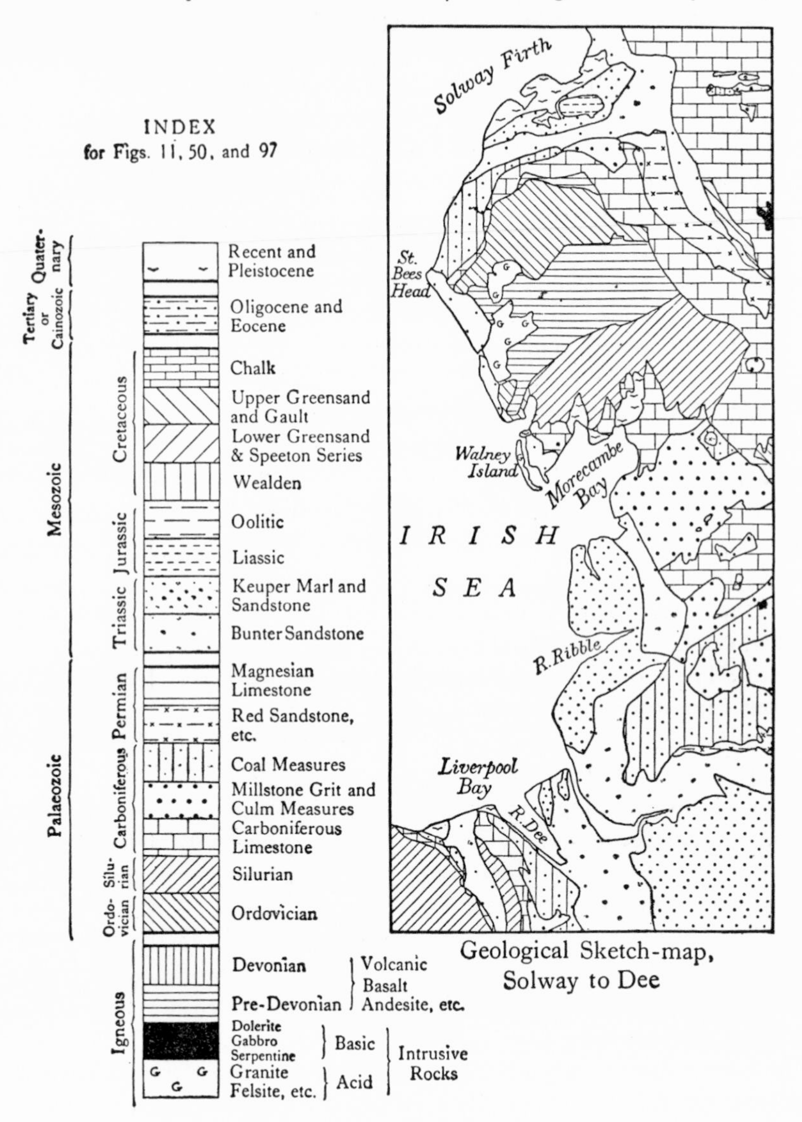

Fig. 11. Geological sketch-map

landwards from Lancaster gives rise to high ground, while the Silurian rocks come into the coastal region only near Ulverston. Near Millom is the conspicuous mountain, Black Combe, composed of Skiddaw Slates which are only separated from the sea by a few hundred yards of Triassic and superficial deposits.

The effect of the glacial[1] and alluvial beds on the coastal scenery is the subject matter of this chapter, but the well-known boulder-clay cliffs at Fleetwood and Blackpool need perhaps immediate specific mention. Finally, along much of the coast described in the following pages dunes and sand are plentiful: they rest indiscriminately on any type of rock, and are often backed by peat and alluvium which sometimes run up to old cliffs cut in boulder clay.

(2) THE SOLWAY SHORE (BOWNESS TO MARYPORT)

The southern shore of the Solway Firth is low and flat, but is of considerable physiographical interest since it reveals much evidence of recent changes of level.[2] The basal rocks are Triassic, but seldom occur at the surface on account of the covering of boulder clay, warp, and peat. The actual coast lands consist mainly of terrace alluvium on which are situated peat bogs like those of Wedholme Flow and Bowness. The channels of the estuary and of the rivers flowing into it wind about extensive flats of

[1] There is still uncertainty in correlating the schemes of the boulder clay in this area. In the Lake District and Solway Firth five main glacial episodes are recognized (*Geol. Mag.* 1942, p. 374):

5. Retreat phenomena, lakes, channels, sands and gravels, and laminated clays. Scottish Re-advance Boulder Clay.

4. Retreat phenomena, lakes, channels, sands and gravels, and laminated clays (=Middle Sands of Carlisle). Boulder clay of ice of Lake District-Edenside maximum and North Pennines.

3. Gravels and laminated clays. Boulder clay of 'Early Scottish glaciation' (including Lake District ice).

2.}
1.} ? Weathered boulder clay of upper Caldew valley.

In the southern coastal districts of the Solway the Upper Boulder Clay corresponds to (5) in the table: it alone affects the coast, but sands and gravels occur above and below it. Farther south, in the Gosforth district, two boulder clays are distinguishable on the low ground. The upper (4) is sometimes covered with a mere skin of the clay (5) associated with the Scottish re-advance.

In Furness there are two drift deposits, the Irish Sea drift which covers the south end of the peninsula and contains granites and other rocks from Eskdale and the coastal region, and the local drift with rocks of local origin only. The clay of the Irish Sea glaciation appears also along almost the whole coast of Lancashire. It corresponds with (4) of the table. The relation of the glacial to the later superficial beds is discussed in appropriate sections.

[2] See especially *Mems. Geol. Surv.* 'The Geology of the Carlisle, Longtown, and Silloth District', E.E.L. Dixon and others, 1926.

pale yellow, mainly quartz, sand. A good deal of carbonaceous matter is usually present since the sand banks are often littered with coal dust which has travelled up from the Workington and Whitehaven coalfield area. A feature, not only of this shore, but of all that as far south as Morecambe Bay is the occurrence of numerous scars of shingle and boulders. These are all associated with underlying boulder clay and merely represent the coarser material of hummocks or spurs eroded by the waves. They are, on a small scale, like Llys Helig and similar features off the Welsh coast. In relatively sheltered areas salt marsh occurs at the inner edge of the flats, but unfortunately no systematic and up-to-date ecological account of the Solway marshes appears to have been published. In 1885 and 1888, however, Hodgson[1] produced what amounted to a list of plants on the Solway shore and from this it is clear that most of the common marsh types are present, including *Aster tripolium*, *Salicornia* spp., *Triglochin* spp., *Glaux maritima*, and some *Statice limonium* near the former viaduct. The marshes consist of fine brown laminated sandy loam, mainly grass covered, a feature which has led to systematic grazing and to turf cutting for export. There is noticeable accretion, and locally some erosion.

The marshes give place inwards to terraced warps of pale yellow and brown sandy loams, two of which are particularly clear. The first and lower terrace is well developed between the mouth of the Eden and Port Carlisle, and reappears farther south in Moricambe Bay, where on Newton and Skinburness marshes it is separated from the present marsh by a bank 3 to 4 ft. high, sometimes broken by smaller steps. The second and higher terrace behind is far more extensive, and often runs for some distance up the river valleys. To the south-south-east of Cardurnock the second warp terrace appears in close relationship with the raised beach. Inland it is usually resting on this beach, but at Longwood occurs also below the shingle of the beach, thus showing that beach and terrace are contemporaneous.

The raised beach itself is well developed in many places. In the Cardurnock peninsula it rests against the truncated mass of boulder clay at Bowness, and is more or less continuous from the site of the former viaduct to Anthorn. It contains a number of ridges of gravel, all generally corresponding to the trend of the modern beach, and rims or ridges sometimes run out into the warp. West of Moricambe Bay the beach is finely exposed at and near Grune Point, and is continuous from there as far south as Beckfoot, in which area it forms the seaward border of warp flats. North of Beckfoot it rests on the second terrace and sometimes against

[1] *Trans. Cumberland and Westmorland Assoc.* No. XI, 1885–6, p. 114 and No. XIV, 1888–9, p. 49. I noticed some *Armeria maritima* in 1943.

hidden masses of boulder clay, while near Silloth it is composed of ridges with recurved ends (**124**).

The *Geological Survey Memoir*[1] ascribes the north-easterly growth of the beach to 'the dominant north-easterly tidal drift'. This phrase, however, should be interpreted as suggesting the effect of wave action rather than that of tidal currents *sensu stricto*. At Grune Point the spit is now being extended by the present waves, and it is possible to make a fairly

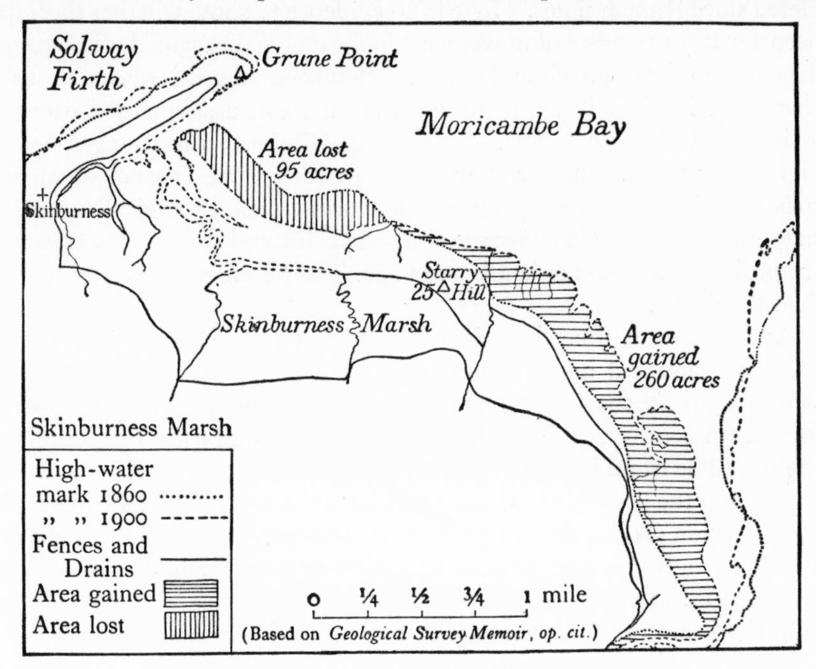

Fig. 12. Skinburness Marsh

exact estimate of the difference of level involved. The maximum elevation of the beach is $15\frac{1}{2}$ ft., which corresponds very well with the height of the second terrace. At Silloth and Skinburness also its level corresponds to the seaward border of the second terrace. Near Grune House, about three-quarters of a mile south-west from Grune Point, its elevation is however the same as that of the first terrace, while the present alluvium (marsh) at Skinburness and Newton corresponds with the new and lower extension of the beach at Grune Point. In the peninsula of Cardurnock the beach and second terrace interdigitate. This western piece of the beach is part of the so-called Neolithic 25-foot beach. It is not possible, though, to say with any precision how much accretion is due to storms and how much to

[1] *Op. cit.*

elevation. 'The manner however in which the surface levels of the raised beaches and warps fall off gradually from the older areas seaward, the fact that the spit at Grune Point is still growing forward at a fairly rapid rate (about 90 yards in 40 years) and the presence of small terrace steps on the modern warp flat, all seem to indicate that the uplift has been continuous to very recent times. This conclusion is in agreement with that of Sir A. Geikie that the elevation of the Carse-lands of Scotland has been completed since Roman times. There is no evidence to show whether there has been a halt (or reversal of movement) in the emergence which followed the deposition of the gravels and warps.'[1] Before leaving this question of the change of level, it is worth noting that at times a submerged forest is visible on Cardurnock flats. Near Starry Hill its top is 10–12 ft. below high-water mark. The forest probably extends some distance along the coast, but its exposure depends on tidal scour. The geological surveyors suggest that it extends under the warp deposits, and that forest and warps are probably all pre-Tardenoisian in age. The raised beach is later in date than the forest.

At the present time there is a fair amount of erosion on part of the Solway shore, and north of Grune Point it is apparently related to the swirl set up by the tides. The destruction of Skinburness began in the fourteenth century, and the decay of Port Carlisle in the middle of last century. The cause is the same in both areas, that is to say, changes in the position of the Solway channels. But here as in many marsh areas erosion and accretion go on side by side, and it is estimated that between 1860 and 1900 although 95 acres were lost to Skinburness Marsh, 260 acres were gained (Fig. 12).

There is not much blown sand east of Grune Point and, although some occurs at Herd Hill and the Cardurnock peninsula, dunes are absent: there is, however, an abundance between Beckfoot and Silloth. Near Silloth many dunes have been removed in recent years on account of building, but nevertheless two periods of dune formation are traceable. On the seaward side there are new dune ridges which are partly marram covered, while east of the main road lie old dunes fixed by grass and heather. The two sets of dunes are separated not only by the road but also by a strip of raised beach which is visible near Beckfoot, but sand-covered nearer Silloth golf links. Older and brown dunes are, indeed, distinguishable as far north as Silloth docks. The Beckfoot dunes run as far south as Dubmill Point, and there are also sandhills along the shore of Allonby Bay. Those in the southern part of the bay stand somewhat back from the shore and are older than those in the north.

[1] *Ibid.*

Northwards and eastwards from Silloth is the narrower part of Solway Firth, a tidal estuary rather than true coast. The banks of the Firth are low, but the country is open and affords extensive views. The whole setting is attractive and the marsh development good. The change of level of water with the tides, and the alternate baring and covering of the wide sand and marsh flats give, together with the remoteness of the locality, a peculiar charm.

From Beckfoot to the neighbourhood of Workington the raised beach is almost continuous and forms gently rolling land. Its ridges are generally parallel with the present coast. In this part its lower limit is taken at the top of the highest modern storm beach, and it rises inland to about 30 ft. O.D. Since high-water mark is 11·7 ft. O.D. and since about another 5 ft. must be allowed for the storm beach, the actual elevation of the raised beach must average some 14 ft. Near Allonby it seems possible to divide it into two, the lower reaching 25 ft. O.D. No such division, however, can be made farther south, and at Siddick there is a gentle slope up from the modern beach to 30 ft. O.D. The old cliff cut in boulder clay can often be traced inland, and is sometimes plainly fringed with isolated patches of raised beach. The cliff is also well seen in occasional morainic hills near the coast. Swarthy Hill, about two miles south of Allonby, is a good example. The raised warp of the Solway shore also occurs in this tract, a step only a few feet high separating it into two distinct divisions. The upper part reaches 28 ft. O.D., and the relations between the warps and the beach are similar to those farther north. Here, too, are traces of the submerged forest, for example, off Stinking Crag and Bank End. The geological surveyors, deducing from the position of the raised beach, estimated that the maximum depression was to 30 ft. below O.D., or 18 ft. below high-water mark. In the Solway and about as far south as Allonby, elevation took place in stages and so helped to produce the terraced warps. Farther south still, the elevation was gradual and continuous, and no good dividing lines are available between the raised and present beaches.

The whole of this stretch of coast is backed by lowland covered with boulder clay. There are few indentations and the trend of beach material is northwards, a feature as obvious here as it is farther north on account of the drift of coal, coal-sand, and dust. The river Ellen enters the sea at Maryport, and about a mile inland and up the valley there is one end of a trough filled with alluvium trending north-north-east and reaching the sea at Bank End (Fig. 13). A similar trough runs south-south-west from very nearly the same point in the Ellen valley to reach the coast at Flimby. These may both have been former mouths of the river at different stages of its history. It is suggested that the Bank End channel, which is cut into St Bees Sandstone, originated as a glacial overflow, while the southern channel is certainly later, and of interdrumlin nature. It may have become disused on account of the northward drift of beach material, working in with a new mouth of the river, cut through the sand hills at Maryport. The present mouth is obviously the most direct exit for the river. At Siddick there is a somewhat comparable former channel running to Northside on the Derwent which now reaches the sea at Workington. This channel is cut through the raised beach and may again have been a former mouth of the river.

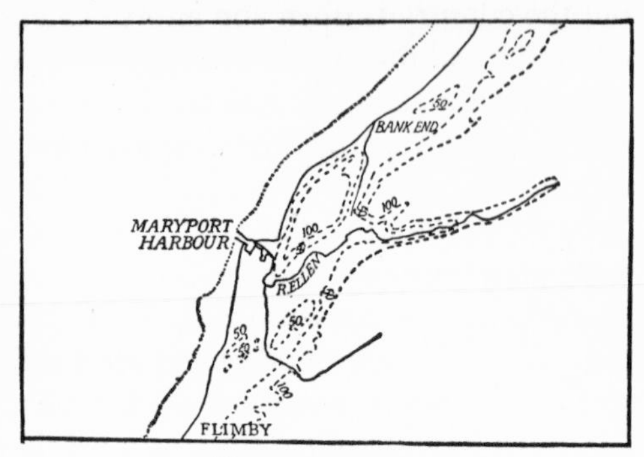

Fig. 13. Maryport
(based on Ordnance Survey)

(3) THE COALFIELD COAST AND ST BEES HEAD[1]

The character of the present beach changes at Maryport.[2] To the north of that town it is sandy and fringed by shingle on its landward side. South of Maryport there is less sand, and scars, although present, are less characteristic and give place to boulders, often of considerable size and weight. Throughout, the beach almost certainly rests on boulder clay, but occasionally outcrops of solid rock occur, for example, the reefs of St Bees Sandstone between Bank End and Maryport, and the productive Coal Measures of Fothergill. In addition to the movement of coal already mentioned, the northward drift is also reflected in the pebbles of blast-furnace slag and mine refuse. On the whole, accretion is in excess of erosion, the former resulting largely from the dumping of slag and refuse. At Bank End, however, there is local, and often severe, erosion of the raised-beach deposits. South of Maryport there is much less blown sand, but a mass of old dunes lies in the loop of the river Ellen at Maryport, while others occur near Fothergill and north-east of Siddick. The coast south of Maryport is locally much spoiled by old tips which, at first glance, make it appear higher than it is. The tip heaps, in fact, form local cliffs.

[1] Strictly, still part of the Solway Firth, which is usually taken to begin at St Bees Head.

[2] *Mems. Geol. Surv.* 'The Geology of the Maryport District', T. Eastwood, 1930.

From Workington to Whitehaven the coast is bolder in feature and mainly cliffed, and the surface of the land is usually plastered with boulder clay. South of Workington there is a low-lying tract, between Moss Bay and the harbour, which is filled with warp, while the higher ground of Chapel Bank forms an 'island' between this hollow and the sea. In this area the rocks dip generally seawards, but faults cause the repetition of certain outcrops. As far south as Harrington, where there is a small harbour, the direction of beach drift, still to the north, is traceable by the amount of slag gravel which often predominates over, and even obscures, the brown sand worked out from the glacial deposits. South of Harrington ridges of natural gravel and sand beaches rest on the Coal Measures. The cliffs behind, all in the Coal Measures, are much intersected by faults, and several small ravines have been cut along these lines of weakness. The Picnic Rock near Cunning Point outcrops on account of faulting, and there is a deep ravine south-east of Parton station.

From Maryport to Whitehaven the coast is completely ruined by industrialism. Coal mines occur close to or on the coast and tip heaps and refuse spoil the whole landscape. There is some small amount of fairly open agricultural country just to the south of Harrington. This coast bears to the rest of the Cumberland coast almost the same relation that the industrial parts of Durham and Northumberland do to the north-east coast.

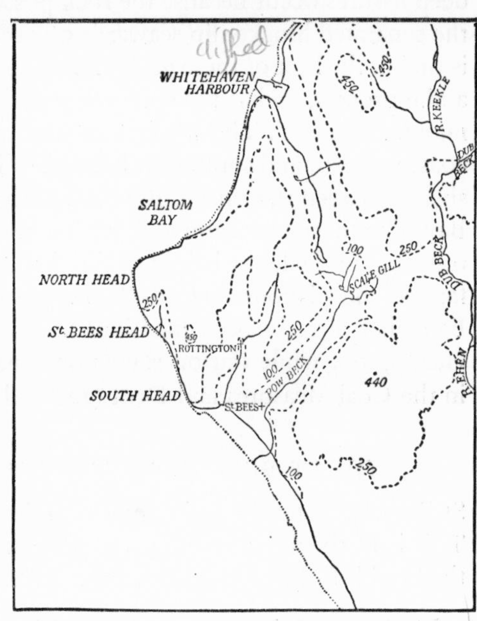

Fig. 14. St Bees Head (based on Ordnance Survey)

Before touching on the magnificent cliffs of St Bees Head there should be some mention of the valley running behind Preston 'Isle' and connecting Whitehaven with St Bees. The existence of this valley is in part explained by the one-time course of a river rising north of Arlecdon and following the line of the present Dub Beck which reached the sea at St Bees. It is thought, however, that glacial erosion and also the plugging of the old valley by glacial deposits between Low Wreah and Scale Gill led to the diversion of the post-Glacial Dub Beck and its tributary, the Keekle, to the Ehen valley (Fig. 14).

To return to the coast: around Whitehaven Harbour the Whitehaven Sandstone (Upper Coal Measures), let down by a fault, is well exposed, and south of the harbour forms the high ground of the northern part of Preston 'Isle' except between the faults at Raven Hill. It is very evident in the cliffs as far as Barrowmouth where it finally disappears under the Brockram. In Saltom Bay the Brockram is succeeded by the Magnesian Limestone, and that in its turn in perfect conformity by the St Bees Shales and Sandstone. At the headland the average dip is to the south-west and south-south-west, and is usually less than 12°. The cliffs are fine, often falling sheer for 200 or 300 ft. They are the only prominent cliffs of solid rock along the whole coast described in this chapter. They are momentarily broken in Fleswick Bay where a small stream has cut down to sea level.[1] In the bay is a well-known beach of shingle which includes a number of gem stones of some beauty but little value. South Head consists for the most part of flaggy and false-bedded sandstones. From Fleswick Bay as far as St Bees there is a wave-cut platform showing interesting carious weathering and carrying numbers of fallen blocks. At the south-western corner of the headland there are good examples of differential erosion: deep fissures occur because the rock possesses two sets of major joints and the separated blocks slip seawards on account of the dip. Pattering Holes is one of the best of these gashes. Isolated masses of rock are produced by a like process, and as the sea further undercuts the cliff, more subsidences may be expected. The whole headland projects farther seawards than the coast to north and south of it because of its greater hardness: the drift and shales between Parton and Barrowmouth are more easily eroded. Fleswick Bay, like those at Parton and Whitehaven, is due mainly to stream action resulting in the lesser height of the cliffs. The raised beaches and warps are far less conspicuous south of Workington; the warp at the back of Chapel Bank has already been mentioned. Traces of raised beach occur round Harrington, while Parton stands on a narrow shelf, hemmed in by cliffs in the Coal Measures and Drift (**122** and **123**).

(4) ST BEES HEAD TO THE DUDDON ESTUARY

St Bees Head marks an important divide in the direction of coastal drift. This is to the north in the parts of the coast so far discussed, but beyond the head it is directed southwards. At St Bees itself, the sandy beach rests

[1] This part of the coast was described in general terms by J. D. Kendall, *Trans. Cumberland and Westmorland Assoc.* 5, 1879–80, 97.

on boulder clay, and is backed by storm-beaches which effectively obstruct Rottington Beck and Pow Beck (**121**). For some miles to the south-east the coast is almost straight and without pronounced features while the cliffs are low and cut in glacial material. Typical sections are:

(1) The seaward side of the railway cutting due west of Nethertown village which shows[1]

Gravel, 15 ft., on brown sand 5 ft.	20 ft.
Red boulder clay	8 ft.
Sands and gravels, bouldery in part	31 ft.

(2) Herding Nab, Seascale, showing

Coarse sand and fine gravel, partly obscured by blown sand	3 ft.
Red sandy boulder clay with few stones	8–10 ft.
Gravel, bouldery to fine	5 ft.
Red boulder clay, stiffer than that above	8–10 ft.

An interesting feature is the direct course of the Calder to the sea in marked contrast to that of the Ehen. The latter, by no means a weak stream, is deflected southwards for about two miles by a narrow shingle and dune spit. The line of the old cliffs is very plain to landward of the river. Whilst it is easy to explain a southward diversion of the Ehen, the reason for the course of the Calder is not so evident and the matter is worth investigation (Fig. 15).

South of Seascale lies the joint mouth of the Irt, Mite, and Esk. The Irt is turned about two miles to the south-east by the broad spit of sand and shingle ending in Drigg Point. The single mouth for the three rivers is interesting on account of the southerly deflection just mentioned, and, at present, a northerly deflection of the Esk also. There is unfortunately little trustworthy information about the early courses of these rivers and of the Ehen. On Speed's map of 1676 the Ehen is shown with a straight course to the sea, the lower valley appearing as an estuary, 'and this is possibly true, for not only has a small amount of uplift taken place during historic times, but also the growth of the spit would tend to cause the deposition of alluvium behind it'.[2] The Irt at one time presumably had its mouth opposite Kokoarrah Island. Speed shows only a short spit, and if his map is correct the Irt had a separate and almost direct outlet to the sea. The Mite and Esk shared a mouth, and the present spit stretching north from Eskmeals station then seems to have run towards the west. In 1836 Monk's map shows a deflected Ehen, an extended Drigg Point, and a rather longer

[1] *Mems. Geol. Surv.* 'The Gosforth District', F.M. Trotter and others, 1936, p. 102.
[2] *Ibid.*

Eskmeals spit northerly in direction. Although old maps often help greatly in elucidating the past history of coastal features, they are not by any means always reliable. In this case, however, the geological surveyors

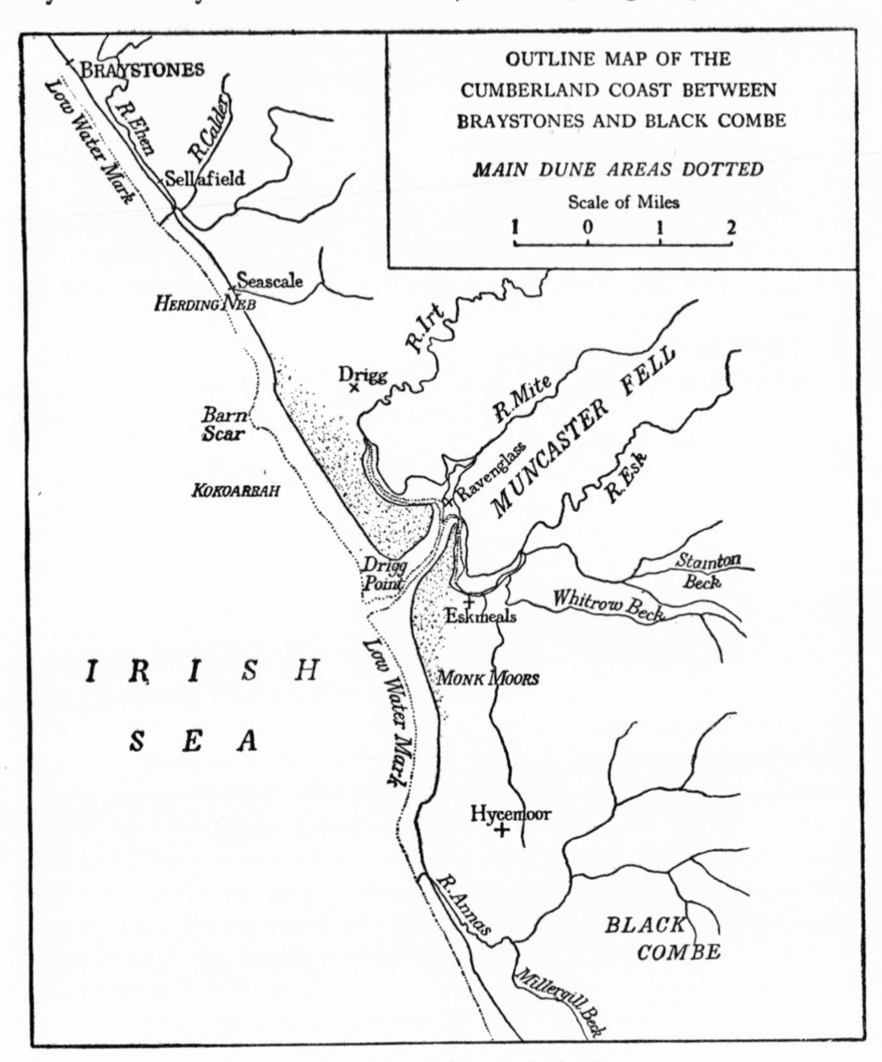

Fig. 15. Outline map of the Cumberland coast (based on Ordnance Survey)

consider the maps of Speed and Monk reasonably dependable since others of intermediate and later dates corroborate them. It appears, therefore, that the deflections of the Ehen and Irt have taken place within the last

300 years. At one time the Esk also flowed southwards within the Eskmeals spit and through Newbiggin and Monk Moors. There is no certainty about the date of this southward course, but it was evidently abandoned some centuries ago.

It is a pity that there are no detailed maps of this area made by a physiographer, since careful cartographical work on the spits, ridges, dunes and other features might produce useful information about past land forms. For example, had the Irt, Mite, and Esk (**120**) at one time a common mouth near the southern end of what is now the Eskmeals spit? Moreover, a fuller investigation of the effect of waves is also desirable. In the *Memoir* we read 'Currents are variable off this coast, but the mean results are a northerly drift offshore and a southerly drift inshore. The latter is responsible for piling up some of the gravel spits'.[1] This is a loose statement, and makes no distinction between the all-important effects of wave action and those due to tidal and other currents.[2]

There are traces of a submerged forest off Drigg and Monk Moors, and the section at Drigg is given as follows:[3]

Peat, etc.	$4\frac{1}{2}$ ft.
Sandy clay, etc.	2–3 ft.
Sandy peat	$\frac{1}{4}$–$\frac{1}{3}$ ft.
Sandy clay, etc.	*c.* 1 ft.
Sandy clay	

Godwin examined samples from the top 6 in. of the highest peat, and was able to establish its age as post-Boreal. Remnants of raised beach also occur, the best exposure being inside Drigg Point near Saltcoats (28 ft. O.D.), while another mass appears near Eskmeals. The beach at Drigg probably belongs to that so clearly traceable north of Maryport, but there is less certainty about the other smaller patches which may be newer. There are warps associated with the past and present beaches like those farther north—the lowest stage forming the existing salt marshes which carry a light growth of grasses. Drigg and Eskmeals spits are mostly covered by dunes, which, on Drigg, reach a width of about half a mile. The greater part of the coastal district south of Seascale[4] is a barren waste of sand showing dunes in the usual stages of development except in so far

[1] *Mems. Geol. Surv.* 'The Gosforth District', F.M. Trotter and others, 1936, p. 102.

[2] A survey of the marshes inside the spits would also be of interest.

When near Eskmeals station in 1943, I noticed *Aster tripolium*, *Suaeda maritima*, *Glaux*, *Salicornia* spp., *Spergularia* spp., *Plantago* spp., *Armeria maritima*, *Glyceria maritima*, *Triglochin maritimum* and *Juncus* spp.

[3] *Mems. Geol. Surv.* 'The Gosforth District', F.M. Trotter and others, 1936, p. 102.

[4] *Trans. Cumberland and Westmorland Antiq. Assoc.* 22, 1922, 101.

as they have been disturbed by buildings and experiments connected with the recent war. Landwards, the newer dunes give place to an older series, fixed by grass and heather. The whole coast from Nethertown southwards deserves careful attention from physiographers and ecologists.

From the southern end of the Eskmeals spit as far as Duddon Sands there is a flat area, formed principally of glacial deposits capped with a good deal of blown sand between the sea and the high mass of Black Combe (**119**).[1] The small Annas River is deflected northwards, but the reason for this is not clear. South of the river there are cliffs in boulder clay, seldom of any height, but at one place reaching 82 ft. The beach below them is strewn with boulders. Towards Haverigg Point there is an accumulation of gravel ridges and dunes which have deflected Whitcham Beck and Haverigg Pool south and east. In a sense Hodbarrow Point, an outcrop of Carboniferous Limestone at the mouth of the Duddon estuary, marks the limit of the coastal section so far described in this chapter. Incidentally, wave erosion at Hodbarrow has produced a smoothly sculptured rock platform on the foreshore by rolling about the pebbles derived from the boulder clay.

Although there is nothing that is of outstanding physiographical significance on the coast from Eskmeals to Silecroft it is by no means without attraction. The boulder-clay cliffs are often locally ravined. Usually the cliffs are grass covered nearly to sea level and erosion by the waves is apparently confined to particular places such as the cliffs due west of Bootle station. The actual coastal and cliff strip is higher than the ground traversed by the railway, but all the land seaward of Black Combe is undulating and hummocky. The marked contrast between the huge, almost threatening, mass of the mountain and the green of the fields of the boulder-clay strip is striking (**118**).

(5) MORECAMBE BAY[2]

(*a*) *Formation.*[3] The type of coast changes greatly at the Duddon estuary. In the wide sense of the term the whole of Morecambe Bay is a depressed area corresponding to the Eden valley and the Solway, north of the Lake District. That much of Morecambe Bay was once occupied by Triassic rocks is clear from the patches which remain around Barrow-

[1] *Quart. Journ. Geol. Soc.* 68, 1912, 402.

[2] The coast from Barrow eastwards is usually taken as Morecambe Bay. I have somewhat arbitrarily included certain features of the Duddon estuary and Walney Island.

[3] J.E. Marr, *Quart. Journ. Geol. Soc.* (Pres. Address), 62, 1906, lxvi, and *The Geology of the Lake District*, 1916, pp. 142–3. (See also **117**.)

in-Furness, east of Cartmel Sands, and at Heysham. Marr has also pointed out that from Millom to Tebay the junction between the Lower Palaeozoic and later rocks is fairly regular. The Carboniferous strata are here let down in great wedges, with generally northward-pointing apices, against the older Palaeozoics. To some extent these wedges are due to early faulting, but they have also been greatly influenced by subsequent movements. The combined effect has been to produce conditions leading to the formation of estuaries along tracts of country underlain by newer rocks—the estuaries of the Duddon, Leven, Winster, Gilpin, and Kent. It is also clear, however, that when these and other main valleys were deepened the country stood higher than it now does, and the Isle of Man may have been joined to the mainland with St George's Channel as dry land. The exact height of the land relative to present sea level is not known, but important evidence exists in Park House mine near Dalton-in-Furness. Here the drift is very thick along a narrow strip and reaches 537 ft.[1] at one spot. The mines are not far above sea level and the clear suggestion is of a drift-filled valley at least 450 ft.[1] below present sea level. This valley, moreover, is held to be of subaerial origin, and consequently shows that the last *major* movement was one of depression. This downward trend drowned any valleys which were unfilled with drift to a level higher than the existing sea level and so produced the estuaries of to-day, including not only those mentioned above, but also those of the Irt, Mite, and Esk. It is not certain when this movement occurred: Marr remarks that probably it was not immediately before the Ice Age since there is other evidence to show that this part of the country was then not more than a few feet different in level from what it is now. 'All we can say is that before the filling-in of these valleys by drift, the land was at considerably higher level than its present one, and at some subsequent date it was depressed to its present position.'[2] (This reasoning also applies to the Eden part of the Solway depression.) Subsequently the estuaries silted up to a great extent and large peat mosses formed. There followed some minor movements which need no immediate attention. The work of ice has been largely responsible for many features in the estuaries, and for some time after their formation they probably presented a more diversified appearance than is the case to-day since they must have been island-studded. The recent changes of a relatively minor and artificial character which took place when the Furness section of the London Midland and Scottish Railway was built are outlined on p. 91.

[1] Mr Eastwood tells me that the greatest thickness of drift near the Duddon known to the Survey is about 300 ft. The figures given above presumably include collapsed Triassic material, etc. which is not true drift.

[2] *The Geology of the Lake District*, 1916, p. 143.

The shores of the bay vary a good deal. The upper parts of the estuaries are silted up and the higher land rises abruptly. All the estuaries are beautiful, but the greater amount of forest land gives added charm to that of the Leven. The Furness peninsula has for the most part low shores of boulder clay. South of Cark there is a good deal of reclaimed marsh which stands in marked contrast to the steep western side of the mass of Carboniferous Limestone forming Humphrey Head. The finest coastal scenery is undoubtedly near Silverdale and Arnside. The woods and small valleys, and the marshes running right up to the high ground give an unusual picture. As in all places where there are extensive sands exposed at low water, the tides in their changing cycles add enormously to the beauty of landscape, and this is especially true in these mountain-enclosed estuaries.

South-eastwards from Silverdale the shores between Carnforth and Morecambe are flat and broken by drumlins often partly eroded by the sea. The Morecambe-Heysham peninsula is built-up along much of its western side. Facing the Lower Lune the country is rural and sinks down to fine marshlands. South of Overton is a comparatively remote district which finds its counterpart on the south side of the river between Glasson Dock and Cockerham. The small peninsula on which stands Cockersand Abbey, of which but little remains, is flat, but ends in a low cliff now protected by a wall. There is a good deal of shingle on the beach. Round the head of the Cocker channel are some salt marshes which lead to the far more extensive Cockerham and Pilling marshes referred to below.

This small part of Lancashire from Overton to Cockerham, shut in as it is between Blackpool-Fleetwood and Heysham-Morecambe, is not only a particularly pleasant district, but also one that, as it were, stands aside from the usually crowded coasts of Lancashire.

(*b*) *Walney Island.* At this point it is suitable to discuss the origin of Walney Island[1] (see Figs. 16 *a*, 16 *b*). Its total length is about eight miles, and it is narrow and extended both northwards and southwards by a series of recurved shingle ridges. Most of it consists of agricultural land, but Barrow has extended on to the middle part of the island. Walney Island acts as a breakwater to Barrow-in-Furness. The intervening channel between it and the mainland varies in breadth from a quarter of a mile to two miles, but at low tide the water-covered part may be only a few hundred yards wide throughout much of its length. The tidal range is considerable, that is to say, 26–33 ft. The island is mainly composed of boulder clay and alluvium, and much of the latter has been dyked. The seaward side of Walney Island is comparatively straight and simple in outline, but the

[1] In this account I have drawn freely upon an unpublished paper by M.M. Spencer, who kindly allows me to use the information.

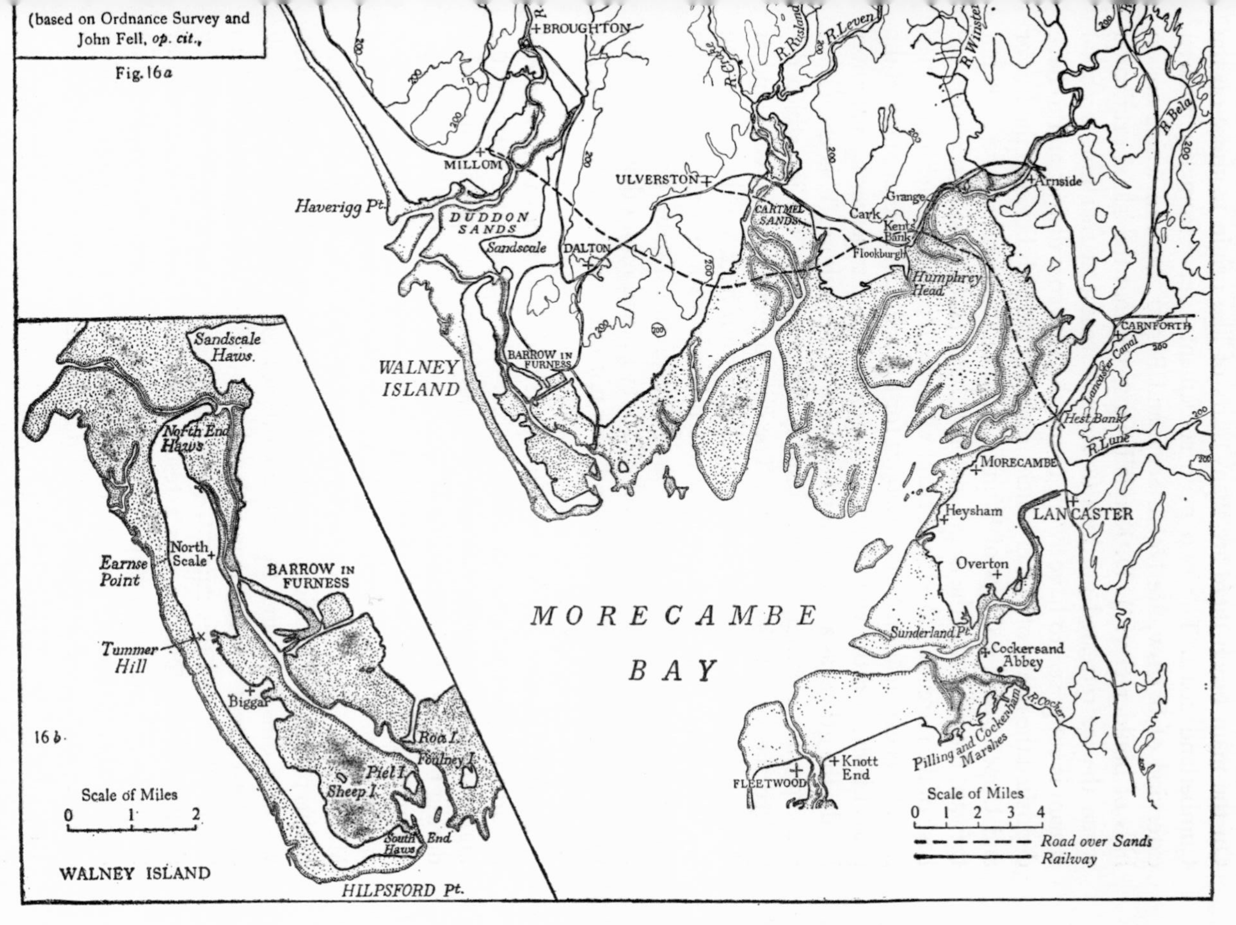

Figs. 16a, 16b. a, Morecambe Bay; b, Walney Island

landward side is diversified by numerous bays and marshes. The beaches are shingly, and marsh grows rapidly in the enclosed eastern recesses.[1] On the main beach there are many scars of like origin to those on the Cumberland coast. The two main settlements, apart from the modern extension of Barrow, are North Scale and Biggar, both of which are on hills of boulder clay. J. D. Kendall suggested that the island originated when the ice retreated northwards from Furness; at that time a glacial stream is supposed to have swept round the Hawcoat district and to have deposited the banks of gravel which now form the beds between the Lower and Upper Boulder Clays of Walney Island. Spencer, however, has put forward the view that the island may be a large esker which grew from the base of the retreating Duddon glacier when sea level was rather higher than now. This explanation would account for the deltaic nature of many of the gravel beds, and the wide spreads of gravel joined together by narrower bands, the gravel as a mass being continuous throughout. This gravel was subsequently covered, and presumably disturbed, by the ice which deposited the Upper Boulder Clay. The cliffs are cut in these clays and at Trough Head the intermediate gravel crops out between the Lower and Upper Clays. Spencer states that raised-beach platforms are apparent in various places, including Biggar Bank and Earnse Point. These indicate a post-Glacial change of sea level of 10–15 ft., and imply that during the time of higher sea level the island was not one compact formation, but rather a string of islets.

Despite the glacial origin of Walney Island, it assumes to-day several features characteristic of an offshore bar. The lower parts between the higher masses of boulder clay were probably formed as spits and tombolos during the time of higher sea level, for example, the part of the island near Tummer Hill. Erosion is serious along the western side; between 1879 and 1935 some 40 ft. disappeared near Sandy Hook Lane, and 20 ft. from Middle Hill. The rate of erosion is very variable. The main tidal current of this part of the Irish Sea sets inshore from an approximately west-north-westerly direction, and the stream divides off Earnse Point. A weak counter-current sets to the northward around North End Haws and into Scarth Bight, and this current, which keeps well inshore, is increased by the flow of the Duddon. As a result there is a rotary current north of Shope Tree Scar, the mean position of which varies according to the state of the tide. South of Earnse Point a relatively strong south-running current sweeps round Hilpsford Point and Scar. The eddy here produced

[1] *Salicornia* spp., *Glyceria maritima*, *Armeria maritima*, *Triglochin maritimum*, *Spergularia* spp., *Limonium* spp., *Juncus* spp., *Artemisia* spp., *Obione portulacoides*, and *Phragmites* were noted in 1943.

seems to have been largely responsible for the formation of Hilpsford Bank, and the southward drift of beach material is thus in part explained. Indeed, in view of the general position of the island, beach-drifting in this direction is to be expected. How far the northern recurved ends are due to local beach-drifting in an opposite direction is uncertain. A lesser and opposite movement from the main one on islands such as Walney is, however, not uncommon, especially as in this case there is a large area at the north end of the island in the mouth of the Duddon estuary and covered only at high water. Waves travelling into the estuary naturally cause local northward and north-eastward drifting. Although on a smaller scale, the local drift of material at the eastern end of Scolt Head Island in Norfolk (see p. 358) is directly comparable to that of Walney. The sand hills on the island call for no particular comments: they are lightly grass covered, and the sand blows up from the flats exposed at low water.

Barrow Island, like Walney, consists of boulder clay covered by sands and gravels. On Old Barrow Island only the sand and gravel are visible. Piel Island is mainly made up of shingle, but there is some sand and gravel on its south-eastern side. Roa, Foulney, and Sheep Islands are of shingle only.

(*c*) *Shoreline Vegetation at Sandscale and Walney.* The shoreline vegetation of Morecambe Bay has not yet been investigated fully. There is only one published paper dealing with the Walney district.[1] Pearsall's work shows that the shingle plants include *Salsola kali*, *Atriplex prostrata*, *Crambe maritima*, *Polygonum Raii*, *Rumex crispus*, *Cakile maritima*, *Glaucium luteum*, *Eryngium maritimum*, *Mertensia maritima*, *Arenaria peploides*, and *Agropyrum junceum*. In the dunes of Walney and Sandscale, a large triangular dune area at the mouth of the Duddon, five main vegetation stages are recognized: foredune, open-dune association, fixed-dune association, dune grassland, and heathy grassland. In the first *Elymus arenarius* is often predominant, and *Psamma* (and others) in the second. On the more stable dunes *Festuca rubra*, *Phleum arenarium*, *Taraxacum erythrospermum*, *Viola pesnaui*, etc. are common. At a relatively late stage *Lotus corniculatus* and *Galium verum* abound, and on the mainland *Salix repens* comes in with these plants. Damp slacks with *Salix* seem to be absent on the northern part of Walney, and Pearsall suggests that the drier conditions of the island dunes may be due to the fact that they rest on shingle ridges. Certainly the water-table is higher at Sandscale than at Walney. 'The general resemblance of the Sandscale dunes to those at Southport is distinctly striking, the principal difference perhaps being the relatively greater extent of *Salix repens* at Sandscale. The contrast between

[1] W.H. Pearsall, *The Naturalist*, 1934, p. 201.

the Walney dunes and those of Sandscale is equally marked. At Walney *Salix repens* is scarce, and instead *Festuca rubra* var. *arenaria* passes from a constant to a dominant species. That the difference is mainly due to the drainage conditions can hardly be doubted, for there are definite patches of dune grassland developed at Sandscale, mainly on sand overlying the shingle bank at the south-eastern end of the dunes.'

(*d*) *The Drift Deposits*. Much if not most of the lower land surrounding Morecambe Bay is drift covered, often to a considerable depth.[1] Hence, around the bay, cliffs in boulder clay or related deposits are common, for example north and south of Aldingham, Leonard Hill, and Point of Comfort, all of which are on the eastern side of the Furness peninsula. At Wadhead, on the western side of Cartmel Sands, the sea has cut away about half of a gravel and sand knoll. There are several drift sections also visible on the coast between Ulverston and Carnforth, and the hard Lower Boulder Clay shows up well to the west of Grange-over-Sands. At Hest Bank the sea has cut fine sections across the drumlins, and between that place and Morecambe large stones are usually in groups forming scars which suggest either old drumlin-like mounds or merely places where the boulder clay rises rather higher than usual. Nearly all the peninsula between the Lower Lune and the sea is drift covered, and the seaward side has been eroded into boulder-clay cliffs. Here, as in Furness, two distinct clays are usually visible: the lower is stiff and compact and packed with boulders and stones, and the upper is decidedly reddish with fewer stones. The upper clay is conspicuous near Bolton-le-Sands, and the lower at Lower Heysham. Sunderland Point, subject to considerable erosion, is a knoll of drift joined to the mainland by sand dunes and salt marsh.[2] At Heysham itself there is an outcrop of the Permian rocks which are cut back into low cliffs. Offshore, at high-water level, a thin peat bed rests on clay, which in its turn lies on boulder clay.

(*e*) *Records of Lost Villages*. There do not appear to be many records of former settlement conditions in Morecambe Bay, apart from some interesting information about lost villages collected by Fishwick.[3] Herte (Hertye) and Fordebottle-in-Furness, as well as Argarmeles and Arnoldsdale (Aynesdale) in West Derby are known to have been lost. No refer-

[1] The following papers deal with the drift: D. Mackintosh, *Quart. Journ. Geol. Soc.* 25, 1869, 407; W.T. Aveline, *Mems. Geol. Surv.* 'The Geology of the Southern part of the Furness District', 1873; A. Crofton, *Trans. Manchester Geol. Soc.* 14, 1875–6, 152; G. Grace and F.H. Smith, *Proc. Yorks Geol. Soc.* 19, 1914–22, 401.

[2] T.M. Reade, *Proc. Liverpool Geol. Soc.* 9, 1900–4, 163, and Rev. A. Crofton, *Trans. Manchester Geol. Soc.* 14, 1875–6, 152.

[3] *Trans. Hist. Soc., Lancs and Cheshire*, 49, 1898, 87.

ence to Fordebottle occurs after 1537. It seems to have stood on low ground between Aldingham and Barrow. Herte Island was somewhere on the shore of Dalton parish and probably formed one of a group of islands near Piel. Both villages were definitely in existence at the end of the fourteenth century, but for centuries all record of them has disappeared. There was a severe storm on this coast on 12 December 1553, which is known to have damaged Walney Island considerably, and it is possible that the *final* disappearance of Herte coincided with it. Crivelton, probably the Cliverton of Domesday, is also supposed to have been washed away, but this assumption is incorrect. It seems to have been renamed Newton, which still exists. Argarmeles was quite flourishing in 1361, but from evidence based on a pleading in the Duchy Court in 19 Henry VII 1503–4 seems to have been destroyed at the end of the fourteenth century. It is likely that Arnoldsdale or Aynesdale was lost much about the same time.

(*f*) *Reclamation; the Roads over the Sands; Milnthorpe.* The area of maximum tidal range in this part of the Irish Sea lies between Fleetwood and Morecambe Bay. The tidal currents carry with them the vast amounts of sand which have gathered in the bay. This material is mainly derived from Triassic rocks and boulder clay: it is chiefly quartzose and much rounded by attrition. This sand is still accumulating and is free from silt, and for this reason is of negligible value if reclaimed. Thus, in his evidence before the Royal Commission, Stileman[1] stated that, of the 1000 or so acres already reclaimed, about 500 had almost been abandoned as unprofitable. This is unfortunate, and especially so for the various private owners, who, after the building of the Furness railway across the estuaries in 1857, were able to reclaim large areas very cheaply owing to the great accretion behind the embankments.

The broken nature of the coast has naturally always made communications difficult and tedious. Hence it is not surprising that before the railway was built, and before modern roads were made, there were recognized crossing places on the sands.[2] These are generally firm, although they are intercepted by channels, and quicksands occur in many places. Nevertheless wheeled traffic was regular across the sands from Hest Bank to Kents Bank, about a mile and a half south of Grange-over-Sands. In the middle of the Cartmel peninsula the route divided. One branch went via Cark and a little south of the present railway to Ulverston; the other

[1] Third (and Final) Report, Roy. Comm. Coast Erosion, pp. 67, 130. See also vol. 1, Appendix 23, under Lancashire, and also supplement to Appendix 23.

[2] J. Fell, *Trans. Cumberland and Westmorland Antiq. and Arch. Soc.* 7, 1884, 1; E.C. Woods, *Trans. Hist. Soc. Lancs and Cheshire*, 87, 1935, 1.

maintained a more direct line via Flookburgh to Conishead Priory via Chapel Island, and thence across the Furness peninsula and the Duddon Sands to a point near Aldringham. The chapel on Chapel Island, formerly known as Harlesyde, is the oldest building connected with the traffic across the sands of which any record remains. The time available for the crossing of the sands, particularly on the route from Hest Bank to Kents Bank, was from two to three hours before low water until two to three hours before the next high water, in all some five hours. Much of course depended on the state of the tide and the weather. Coaches ran regularly up to 1857, when the railway was opened. In earlier times it is interesting to note that Robert Bruce, in his invasion of England, crossed the sands in 1322. John Wesley records his crossing in May, 1759. In 1837 George Stephenson proposed the building of a straight wall from Poulton (now Morecambe) to Humphrey Head so as to enclose about 40,000 acres for reclamation.[1] On account of the changes in the channels and the danger of quicksands regular guides were available to help passengers. But even so there were many serious accidents and the writer of an undated pamphlet was prompted to suggest the building of a series of refuges. The sands are still crossed by people with local knowledge, and in the war of 1914–18 a troop of 50 cavalry once made the crossing.

Largely because of the shallow water and the vagaries of the channels, there are now no ports between Barrow and Heysham, but in the past there seems to have been a fair amount of navigation. The Vikings certainly used the Kent estuary, although the first documentary mention of the port is in the St Bees Register, 1282. However, up to about the end of the eighteenth century or later, Milnthorpe, the only port in Westmorland, was still active.[2] The building of the canal from Lancaster to Kendal dealt Milnthorpe a severe blow, but already by that time most ships were becoming too large for its harbour, and the diversion of traffic following the opening of the railway in 1857 confirmed its decay. It is worth noting that before the Arnside viaduct was built the bore, which now seldom exceeds 18 in. in height, could reach $3\frac{1}{2}$ ft. if strengthened by a favouring wind.

(*g*) *The South-eastern part of Morecambe Bay*. The finest development of true salt marshes in Morecambe Bay is in its south-eastern corner at Cockerham and Pilling. They extend east and west for about four miles, and vary in width from a quarter of a mile to a mile. They are not quite natural, since in the extreme western part *Salicornia* spp. (Samphire) is

[1] He also proposed a similar scheme for enclosing much of the Duddon estuary.

[2] W.T. McIntyre, *Trans. Cumberland and Westmorland Antiq. and Arch. Soc.* 22, 1922, 101.

cultivated, and much of Cockerham marsh is exploited for turf. This is favoured by allowing sheep to crop the grass, a practice which also works to the detriment of other plants such as aster, lavender, and arrowgrass. In parts the marshes have an intricate system of creeks and pans, while near the river Cocker is a transitional area between salt marsh and fresh-water fen. All the Morecambe Bay marshes are very flat and the impression of flatness is intensified by the close cropping of sheep.

The Royal Commission stated in their report of 1911 that a rise in level of 4 ft. in the forty years preceding their inquiry held good for certain parts of the Pilling and Cockerham marshes.

(6) THE DEPOSITS OF THE COASTAL AREAS OF LANCASHIRE

Apart from the outcrop of Red Sandstone of Permian age at Cockersand Abbey there are no solid rocks on the coast between Heysham and the Wirral peninsula. All this part of Lancashire is covered with drift, peat, and thin superficial deposits which appear to rest on a rock plane gently inclined seawards. Where this underlying platform is undulating the drift deposits tend to be thicker. de Rance[1] was of the opinion that the superficial beds as a whole were deposited during subsidence. The rivers, especially the Ribble, have excavated their lower valleys almost entirely in these beds, only occasionally does the Ribble cut down to the underlying material. In general, in coastal areas between the Mersey and the Wyre the underlying rock surface is 50–60 ft. below low-water mark, while such cliffs as occur are for the most part cut in boulder clay and intercalated sands. Before discussing the coast as it is to-day it is appropriate to give a brief account of the boulder-clay deposits and those succeeding them. The boulder clay is divided into Lower and Upper which are at times the more easily distinguished because Middle Sands and Gravels occur between them. The sands seem to have been laid down in water held in front of the earlier ice sheet as it retreated. Since they are undisturbed it seems that they were frozen before the succeeding ice advanced over them. As a rule in the Fylde the sands and gravels at least underlie the Upper Boulder Clay, but in places this has been eroded to expose the sands.[2] The Upper Boulder Clay is fairly constant in character and contains Lake District rocks. After the deposition of the clay, streams

[1] *Mems. Geol. Surv.* 'The Superficial Deposits of the Country adjoining the coast of South-West Lancashire', 1877.

[2] Mr Eastwood has kindly pointed out that the absence of the Middle Sands does not necessarily imply erosion. He thinks non-deposition is the explanation in many cases.

washed sand and shingle out of it and these accumulated in certain places, thus forming the Preesall Hill Shingle and the Shirdley Hill Sand. When the shingle and sand were laid down, the land rose a little relative to sea level, so that fluvio-marine or estuarine conditions prevailed, and gave rise to the *Scrobicularia* and *Cyclas* clays. During this period slight elevation of the surface every now and again led to the formation of soil and the growth of trees, but since the upper surface of the clay was irregular a number of swamps formed on it in which thick deposits of peat accumulated. Because of the conditions prevailing there are now several intercalations of peat with the *Scrobicularia* and *Cyclas* clays. After the formation of the clays and peats much alluvium gathered along the coastal areas and the rivers. This deposit was tidal or fresh water according to locality, but, naturally, in the somewhat brackish waters of the estuaries there are transition regions, for example, Freckleton Marsh, now embanked and drained.

The effect of these various formations on the present coast and the changes in conditions in historic time form the subject matter of the following section.

(7) THE WYRE ESTUARY, MARTON MERE, AND THE FYLDE COAST

In the region of the Wyre estuary the glacial drift ends in a line of cliffs ranging from 40 to 85 ft. in height. To north and south the boulder clay is bounded by a small escarpment facing lowland covered with peat. Three main peat areas, separated by tongues of drift, may be distinguished, Pilling Moss in Over-Wyre, Marton Moss farther south, and Rossall Plain south of Fleetwood. Pilling Moss is now bounded seawards by a wide belt of fluvio-marine deposits. These largely conceal the peat which at one time extended out into Morecambe Bay, though small patches are still visible resting on boulder clay when the overlying marine sand is swept away. The Preesall Hill Shingle, it will be remembered, underlies the peat. During its formation there was much denudation in the boulder clay, and it was then that the cliffs around Preesall Hill, and the scarps between Preesall and Cockerham, were shaped.

To return, however, to the estuary of the Wyre: until about two centuries ago the mouth of the river was being pushed steadily eastwards by sand and other matter gathering at, and east of, Rossall Point. This movement has now been arrested by the outcrop of hard boulder clay at Knott End, and also, quite recently, by the building of sea walls. Larbreck may be reckoned the head of the estuary, although the highest spring tides reach as far as the mouth of the Brock. The estuary is fringed by salt marshes and by occasional bluffs of Upper Boulder Clay like those at Liscoe which are 30 ft. high, and Little Thornton and Staynall which are 15 ft. high. Greenwood[1] distinguishes three zones between the actual banks of the river: the true channel, the mud flats and banks, and the salt marshes which are covered only by high spring tides. These marshes are comparatively narrow, are fairly mature, formed of firmly consolidated material, and characterized by a mature creek system. The lowest zone is on the river side of the true marsh. Mud banks are forming here and *Salicornia* spp. is the main colonizer, although a small patch of *Suaeda maritima* occurred near Wardleys in 1935. The second zone, of low marsh, is composed of fairly soft sticky mud, and is frequently covered by the tides. It may be an entirely new formation or a re-growth, following erosion, at a level below that of the high marsh. It is not completely covered with vegetation, but carries marsh grasses together with *Aster tripolium*, *Suaeda maritima*, and *Obione portulacoides*. The third zone forms the inner and high (true) marsh: it is now well drained and made up of relatively dry and firm brown mud which is more or less completely plant covered. *Glyceria maritima* dominates amongst the grasses, but common also are *Armeria maritima*, *Statice* spp., *Triglochin maritimum*, and patches of rush and sedge. This zone stands from 1 to 3 ft. above the low marsh, and the two are either separated by a small cliff less than a foot high, or more often by a broken slope. A fourth zone occurs at the inner margin of the marsh and is characterized by rushes and reeds. It is actually a little lower than zone three and so is damper.

Before discussing the present coastline of the Fylde, a short comment is in place about Marton Mere,[2] the small lake about two miles east of Blackpool. The whole area is low, and before the tidal silt was deposited in the Wyre estuary, tidal water was able to flow up the Skippool valley to Marton Mere. If the level of the land relative to the sea *before* the growth of the peat was much the same as it is now, there would have been about 12 ft. of water in the Marton depression. But from the evidence of the Preesall Hill Shingle and the Shirdley Hill Sand de Rance thought that the land was then some 15 ft. lower than at present, and thus nearly 30 ft. of water stood in the Skippool valley at high spring tides. In this way the terraced appearance of the district is explained, as well as the re-sorting of the boulder clay so often noticed. A like argument indeed applies to the depression between Preesall Carrs and Lytham Moss.

[1] R.H. Greenwood, *The Wyre Estuary* (unpublished). I am much indebted to Mr Greenwood, who allows me to use his work freely.

[2] de Rance, *Mems. Geol. Surv.* 'The Geology of the Country around Blackpool', 1875.

Thus during the deposition of the Preesall Hill Shingle and the Shirdley Hill Sand there was a continuous channel between the Wyre and the Ribble through the low ground between Mythop and Great Plumpton. In much the same way it can be shown that a similar channel existed between the Wyre at Rawcliffe and the Lune estuary at Pilling. The drainage of this district is still difficult and is carried to the sea by a sluice near Pilling.

The Fylde coastline[1] north of Blackpool consists mainly of cliffs cut in the Lower Boulder Clay, but nearly all is now obscured by walls and buildings or grass. The finest section, before the extension of Blackpool to its present size, was at Eagberg Brow, Norbreck. Outcrops of the 'mid-glacial' sands were frequent, but after the wash of heavy rains were usually obscured by clay. Until the building of the sea walls erosion was often serious, and the former extent of the cliffs can be gauged by the isolated and consolidated masses of sand and stones from the glacial gravels which now form off Norbreck, the Pennystone, Higher and Lower Gingle, the Silkstone, and other rocks. de Rance also suggests that the peat plain of Lytham and South Shore swept round this part of the coast to Rossall and Fleetwood. The groynes along the coast demonstrate clearly that northwards of Blackpool at least there is a pronounced northerly drift of beach material. The existence of North Wharf, the bank beyond Fleetwood, may also support this, and at the same time suggests that the drift cannot turn into Morecambe Bay possibly because of the tides in the Lune estuary. Between Fleetwood and Cleveleys there is an abundance of shingle on the foreshore, and at Rossall Point there are recurved ends. South of Cleveleys, however, the amount of shingle is less and it practically ceases at North Shore. There is none between this point and South Shore, the beach consisting wholly of quartz sand. To-day there are sea walls for about ten miles between Fleetwood and South Shore, and the only part of the coast still unprotected lies between South Shore and St Anne's, where there are some dunes, the Starr Hills, mainly covered by *Psamma arenaria*, which afford a short but welcome break in a built-up area. Sea walls again protect the coast from St Anne's to Lytham, at which point salt marsh begins. As a general rule shingle and other beach material to the south of Blackpool travels towards the Ribble. This is to be expected, say, from St Anne's towards the Ribble, but it is not easy to explain the apparent change in the direction of drift on the straight coast at, and near,

[1] de Rance, *op. cit.* 1875 and 1877; R.K. Gresswell, Ch. 3, 'A Scientific Survey of Blackpool and District', *Rept. Brit. Assn. Adv. Sci.* 1936. Blackpool itself is called after a small stream, now the Spen Dyke, which flows from Marton Mere over the peat from which it derives its black colour.

Blackpool. Shirley[1] notes the widespread opinion that the shingle occurring on this coast comes from the northern cliffs, and there may, of course, be variable movements at Blackpool itself and on either side of it. It appears, however, to be one of several instances along our coasts where the precise reason for the change of drift direction is not known. Throughout the area the dominant winds are just south of west, and the main tidal current runs to the north. There is no difficulty in explaining the drift towards Fleetwood, nor is there any mystery in a drift up the Ribble estuary, but it would be interesting to have the reasons for the precise location of the divide carefully analysed.

One further point needs attention. A good deal has been written about the Roman settlement, Portus Sentantiorum, which is '*known to have existed in this neighbourhood*'.[2] It is usually assumed that the site of the port was near the entry of the Wyre into the Lune estuary. The problem of its identification seems to be closely related to that of the trackway known as the Danes Pad. The latter is mentioned by Burrows as an 'alleged route...and (I) nowhere found the slightest resemblance over or under the ground of a paved road'. There is much uncertainty about the whole matter. Mawson later says: 'I have dealt at some length on the point of the position of the Roman port because, *if it existed*, it solves so many problems as to the nature and meaning of other features,'[2] especially the Danes Pad. The two phrases in italics, both of which appear in the same article, speak for themselves. Until more precise data are available there is little to be done beyond bearing the problem in mind despite the 'facts' discussed by many writers who have given what perhaps may be described as undue attention to Ptolemy's map of Great Britain.

(8) THE RIBBLE ESTUARY: MARTIN MERE

All the area surrounding the lower part of the Ribble is covered with peat, and in de Rance's view its estuarine bed rests on this deposit. On the north side, in Lytham Moss, the peat is not more than 8 ft. thick and is bounded landwards by slopes of Upper Boulder Clay which rise from below it. There are also isolated knolls of boulder clay, and where the peat is lowest fresh-water *Cyclas* clay is seen to intervene between it and the drift. On the peat there is much modern alluvium which near Lytham runs inland for some distance: much has been reclaimed, but much remains natural salt marsh. The dunes already mentioned, between South Shore

[1] In an unpublished essay (by J.A. Shirley), from which I am allowed to quote.

[2] E.P. Mawson, Ch. 2, 'A Scientific Survey of Blackpool and District', *Rept. Brit. Assoc. Adv. Sci.* 1936.

and St Anne's, rest on the peat or on shingle ridges now some way inland. They form a natural embankment west of Lytham called the Double Stenner, and it is interesting to note that the St Anne's dunes often carry plants not native to this country. Bailey[1] has shown that they grew from seeds derived from grain siftings and sweepings of American rather than continental ships. They have been brought to this district with fodder for the poultry which, before the war of 1914–18, were often reared in the dunes. The extension of the towns has since brought great changes and consequently these plants are not likely to become true colonists.

The discharge of the Ribble is variable, and even now the channels are subject to change.[2] The river brings down much detritus in floods, and considerable reclamations were made in 1806, before which year the estuary proper extended to Preston. The building of walls has lessened the tidal scour in general, but within the walls themselves has increased it. The construction of the walls also caused the rapid accumulation of material outside them, thus producing marshes which are sooner or later inned. These reclaimed marshes are best seen on the south side of the estuary. The bed of the estuary consists of fine sand resting on boulder clay, although, as noted above, peat may intervene. The flood current runs for four hours; the ebb for eight. The flood carries with it much detritus from off the sand flats. At the same time, a good deal of sand is washed into the channels by water flowing off the enclosing banks, and this leads to much accumulation which the ebb is unable to wash away. It is for these reasons that banks and channels alter with almost every tide. One feature, however, remains constant—the Bog Hole off Southport. The reason for this is uncertain, although Dickson suggests it may be because peat underlies the sand. A study of old and fairly recent charts suggests that the Bog Hole is deeper now that it is not a through channel, a change which may possibly result from the eddy set up by the flood: this would clearly not occur if the Hole were an open channel. There are several interesting charts of the estuary, the first dating back to 1689. It is clear from them that, before man interfered, there were only two channels, both being navigable for small ships. In general, the northern channel was more liable to change than the one which formerly ran nearer to Southport.

On the south side of the estuary there is, or rather was, a feature comparable to Marton Mere behind Blackpool. Inland from Crossens is the

[1] *Mem. and Proc. Manchester Lit. and Phil. Soc.* p. 47, 1902–3, No. 2; *ibid.* p. 51, 1906–7, No. 11; *ibid.* p. 54, 1909, No. 15.

[2] E. Dickson, in 'Southport Handbook', issued separately for British Association Meeting, Southport, 1903.

basin of the old Martin Mere.[1] The stream which once flowed out at Crossens was blocked by alluvium and so the mere was formed, and flooded a certain amount of forest land (Fig. 17). The mere was shallow, nowhere more than 20 ft. deep, and its outline varied with the rainfall. Judging from remains found in it, the mere was probably formed between the time of the Roman occupation and the Norman Conquest. It is shown

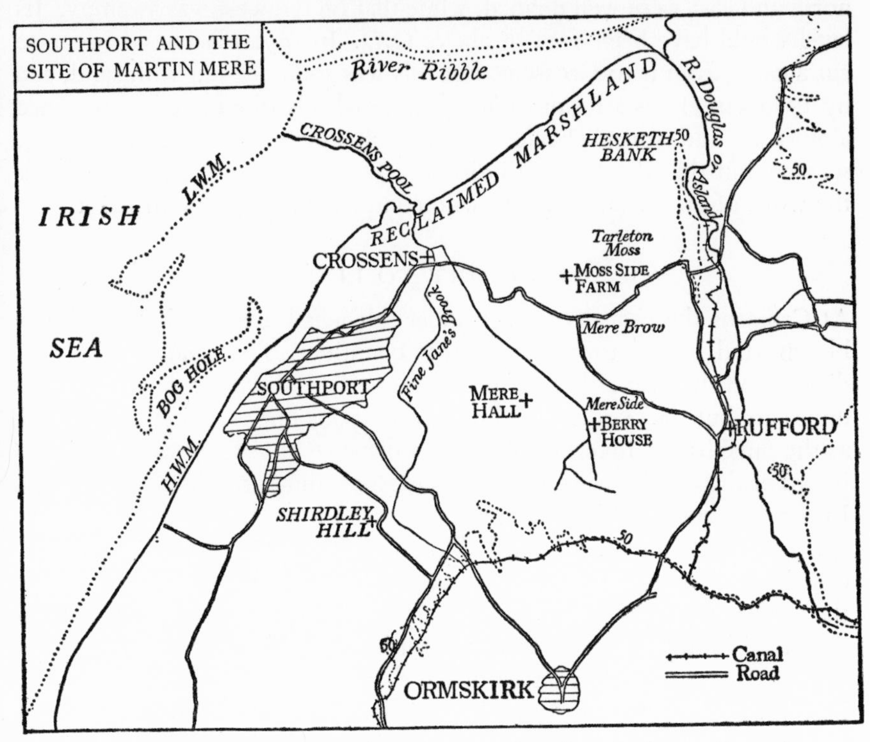

Fig. 17. Southport and site of Martin Mere (based on Ordnance Survey)

on a map of 1598, and from then until 1745 its outline on the maps remains fairly consistent. Leland says that it measured four miles by two, which agrees reasonably with the cartographic evidence, and all the early maps show an outlet channel to the Douglas (or Asland) River near Rufford. Camden, however, speaks of it as having a direct outlet to the Ribble, while Blome, in 1673, gave it two outlets, to the river and to the sea. Brodrick, who has carefully examined the area, thinks that under normal

[1] *Ibid.* 'Martin Mere and its Antiquities', H. Brodrick; also E. Dickson, *Proc. Liverpool Geol. Soc.* 8, 1897–1901, 454. See also *Mems. Geol. Surv.* 'The Geology of the Country around Southport, Lytham, and South Shore', C. E. de Rance, 1872.

conditions the outlet would have been to the Douglas, but that in flood times the Douglas would have overflowed into the mere which, in consequence, had to discharge at Crossens. The maps show three islands in it, of which various interpretations have been given: the only certainty is that one of the islands coincided with the existing high ground at Berry House, two miles due west of Rufford. The boundaries of the mere to north and east were well defined, while that on the west was swampy. Its level would have been 9–11 ft. above O.D. In 1692 Fleetwood began to cut a canal from it to Crossens, but this was soon choked with sand, and by 1760, largely as a result of an exceptional tide, the mere had regained almost its old form. Eccleston in 1778 attempted another drainage scheme, but this also met with disaster. The present outlet at Crossens is the work of Sir T. D. Hesketh, who completed his scheme in 1849.

(9) SOUTHPORT TO LIVERPOOL

At Crossens the peat disappears under salt-marsh clay, and at Southport blown sand begins to occur in great quantities. The sand usually rests directly on the peat although locally it overlies alluvium. The Southport dunes are but the northern end of a continuous line reaching as far south as the outskirts of Liverpool, that is, a distance of sixteen miles. The belt is two or three miles wide, and in places, some dunes reach 80 ft. in height. In the Southport area at least, the older and inner dunes show stratified lower layers, and were obviously blown on to marshy ground. de Rance[1] called these basal layers the *Bithynia* sand. The marsh below is wet with the surface water of the moss on which the dunes rest, and this water is described locally as the Ream. At Southport the sand blows up rapidly, and the Royal Commission estimated that by 1911 about 5,000 acres had been grassed over since 1839. There are local differences of opinion about the effects of marine and fluviatile agencies, a not unusual occurrence on our coasts when two important towns are not far apart. Southport citizens on the one hand are apt to blame the Ribble Navigation Company and the Preston Corporation for the gathering of the sand, asserting that it is due to training walls. The Preston authorities, on the other hand, incline to the view that the accretion at Southport is the result only of local coastal works. The dune belt begins abruptly just north of Southport, and north of the promenade the coast, as shown on the maps of 1845, is more than 2,500 ft. inside the most seaward embankment existing to-day.

From Southport to Liverpool docks the shore where there are no sea walls consists almost exclusively of sand backed by dunes. The average

[1] *Op. cit.* 1872; 'Southport Handbook', *op. cit.* p. 61.

height of the dunes, except for small and local outcrops of clay and peat, is 20–30 ft., but between Birkdale and Ainsdale and at Formby they are nearer 50 ft. As the springs' tidal range is 30 ft., there is a very wide foreshore which, as Gresswell[1] has demonstrated, is characterized by broad sand rolls 3–5 ft. high and about 500 ft. from crest to crest. They seem to be permanent features, and are about two miles in length. They are elongated at right angles to the direction of approach of the dominant waves and are probably formed by ordinary rather than storm waves. Similar features are by no means rare on other sandy foreshores, but those on the Lancashire coast seem to be peculiar in their durability. In the northern section between Birkdale and Woodvale the dunes show no clear arrangement, but to the south of Woodvale they are in distinct ridges, at any rate as far as Freshfield, the artificial growth of which has been encouraged during the century. At Southport and in its neighbourhood there is little or no erosion, but considerable loss has occurred at Freshfield and Formby Point. South of the mouth of the Alt the coast is liable to erosion, but it becomes very artificial, because from Blundellsands onwards it is quite built-up and only a narrow line of dunes separates houses and gardens from the sea.

There is no detailed documentary information about this piece of coast before the sixteenth century, at which time it seems to have been a barren waste of dunes and peat mosses. Gresswell has analysed the available data thoroughly and has come to the following conclusions. (1) In the sixteenth and seventeenth centuries there was a bay south of Churchtown (the north-eastern part of Southport). (2) The offshore sandbanks were as fully developed at the end of the seventeenth century as they are to-day, although their form has varied since then. There is every reason indeed to suppose that they are much older, but there is apparently no conclusive evidence about the date of the spread of the dunes inland. (It is worth noting that de Rance and others, quoting earlier writers, who were, however, not necessarily authorities, suggest that there were no dunes at Formby before approximately 1690. They are also of the opinion that the dunes blew up from off a bank which was driven shorewards and which led to the formation of Formby harbour. It certainly appears that the church at Formby was moved inland to its present position on account of sand in 1746, and that by 1750 streets, gardens, and orchards were becoming buried.) (3) In 1598 there was a bay at Bankhall and Litherland, but this had become filled in by 1689. (4) Since 1698 at least, the river Alt has flowed in a southerly direction across the beach: it was dammed in

[1] *Geogr. Journ.* 90, 1937, 335; see also (R. Alt) C.B. Travis, *Proc. Liverpool Geol. Soc.* 13, 1919–23, 52.

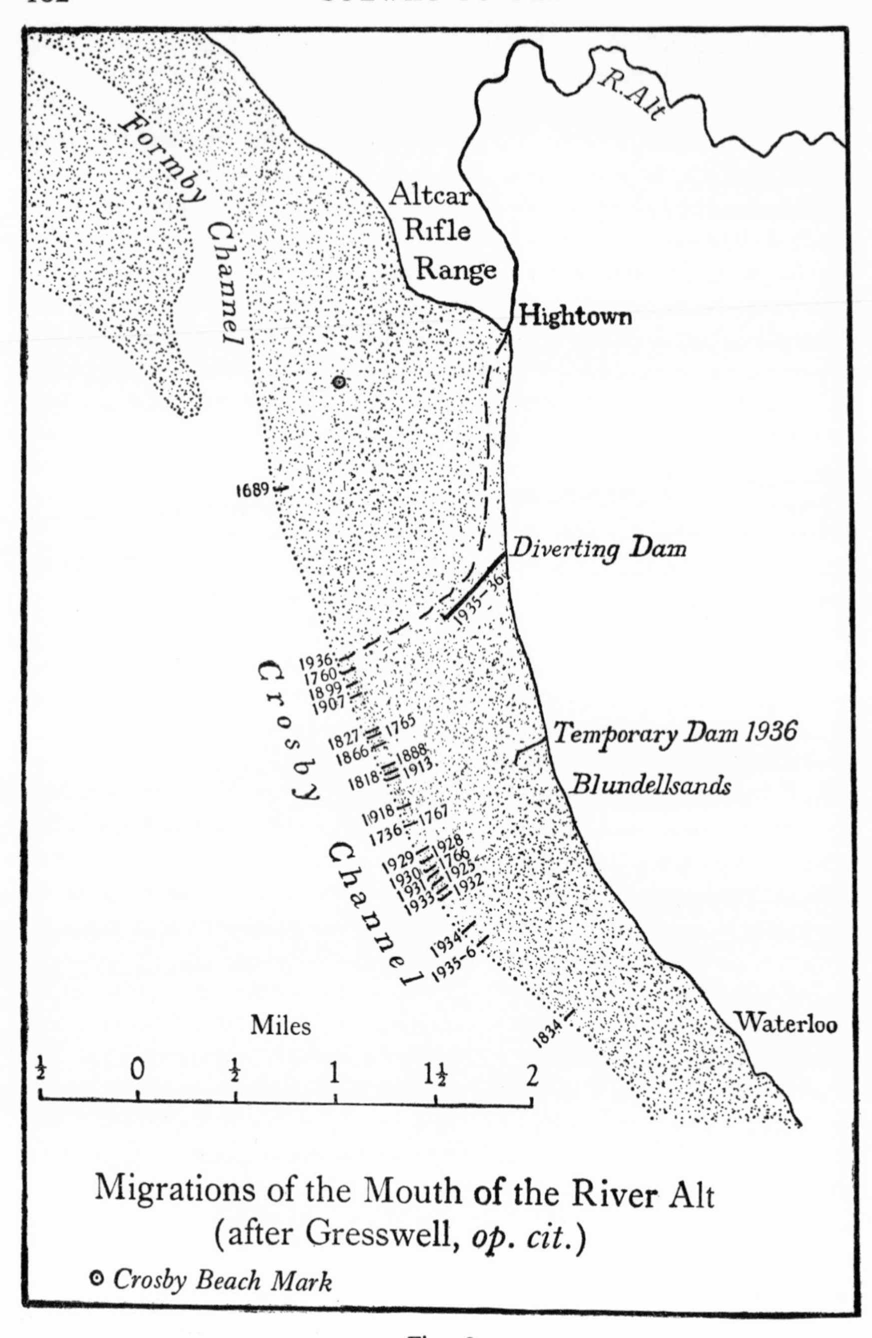

Migrations of the Mouth of the River Alt (after Gresswell, *op. cit.*)

⊙ *Crosby Beach Mark*

Fig. 18

1936 (see Fig. 18). Since 1845 there has been continuous erosion of the dunes by the sea only at the southern part of Hightown: slight erosion at Seaforth was also recorded between 1845 and 1892, and at North Ainsdale from 1892 to 1906. Up to 1906 the greatest accretion was at Freshfield, but this was due in part to artificial circumstances. Curiously enough at the place where accretion was once greatest there is now most erosion: thus at Formby Point gains of up to 12 ft. a year have been replaced by losses of 23 ft. or more, and the erosion at Hightown has spread as far as Blundellsands. Erosion is always due to westerly storms coinciding with a high tide. The direction of approach of the dominant waves is also from the west, and in consequence the direction of beach drift nowadays is north and south from Formby Point. The river Alt in its course across the sand stops much of it from blowing inland, and accordingly the material lost by wave action is not replaced. This process, together with the deepening action of the river which causes a lowering of the beach, accounts for the erosion at Hightown and Blundellsands. The erosion at Formby is ascribed to a general lowering of the shore, so that waves reach the dunes more easily (**116**).

Before leaving this section of the coast one or two minor points of interest are worth noting.[1] The modern town of Southport arose on the site of the old fishing village of South Hawes, and its steadily increasing importance led to the larger and older town of Meols being renamed North Meols. The original Meols, now usually called Churchtown, is a part of Southport. There are records of human activity here in the Middle Ages: for example, Roger de Poictou, in the reign of William II, gave to the monks of Lancaster the tithes of 'Melis'. The encroachment of the dunes south of Southport led later to the ruin of much agricultural land, but an interesting tradition remains that the earliest potatoes grown in England were produced here as the result of a ship being wrecked on the coast and its cargo distributed by the waves. The dunes also contain numerous shells which the carbonic acid in rain water slowly dissolves. This gives the bicarbonate of lime which, when it is precipitated, renders the sand largely impermeable, thus leading to the formation of slacks and ponds in the dunes. It is also worth while calling attention to the fact that this long stretch of dunes fringing one of the most populous parts of the British coasts is still left almost unspoiled by huts and bungalows.

[1] 'Southport Handbook' (*op. cit.*) under H. Brodrick and E. Dickson; see also de Rance, *op. cit.* 1877 (this reference is very full and should be consulted on many matters concerning the whole Lancashire coast).

(10) THE MERSEY AND THE DEE ESTUARIES

(*a*) *The Buried Peneplain and Recent Deposits.* Before considering local details, it may be better to treat both banks of the Mersey and the Dee and the whole of the Wirral peninsula as a unit. In pre-Glacial times there extended a plain, probably of marine denudation,[1] from north-east Wales to south-west Lancashire, cut in the softish rocks of the Trias. At a later date it was overspread by boulder clay which, on the whole, evened out the surface by filling up hollows and by grading the slopes of hills. Borings show that some of the hollows reach well below sea level. The pre- and post-Glacial drainage systems differed considerably. The streams of Wirral now radiate from a low divide covered with boulder clay, and at Liverpool and elsewhere this boulder clay is divided by gravels and sand into upper and lower divisions. Natural sections, as distinct from those made artificially during dock excavations, are rare, but that between Dingle and Otters Pool is informative. The cliffs, 30 ft. high, consist of red boulder clay capped in places by Shirdley Hill Sand, the whole resting on Triassic sandstone. Other sections need mention later. The general sequence of post-Glacial beds visible during the work at Old Liverpool Dock was:

6.	Silt with Sand at base	3 ft.
5.	Blue Clay	6 „
4.	Forest Bed and tree trunks	1 „
3.	Blue Clay	10 „
2.	Peat or Forest Bed	1 „
1.	Upper Mottled (Bunter) Sandstone at 40 ft. below H.W.M.	

At other places, for example Wallasey Pool, only one forest bed at 30 ft. below H.W.M. was found, and the well-known forest beds at Leasowe are not necessarily absolute evidence of recent subsidence. They are not more than 10 ft. below the level of the alluvium which lies in the rear of the dunes, and the geological surveyors suggest that the great sand flats offshore may have helped in their formation. At Leasowe the sequence visible in 1913 was:[2]

4.	Upper Forest Bed	3 ft.
3.	Blue-grey Silt	1–2½ „
2.	Blue-grey Silt including patches containing *Scrobicularia piperata* and *Cardium edule*	1½ „
1.	Lower Forest Bed	½–1 „
	Boulder Clay	

[1] de Rance, *op. cit.* 1871, p. 158.

[2] *Mems. Geol. Surv.* 'The Geology of Liverpool with Wirral', C.B. Wedd and others, 1923.

Considering the whole range of the forest beds and the levels at which these appear to have formed, it seems reasonable to conclude that at the close of the Glacial period the land in this district was anything between 50 and 100 ft. higher than now. It must also have extended considerably farther seawards, and the Dee and Mersey estuaries were probably wooded slopes.

(*b*) *The Cliff and Beaches at the Point of Air, and the Alluvium of the Dee.*[1] There is a well-developed, ancient, but post-Glacial, cliff cut in the Gwespyr Sandstone (the top of the Lower Carboniferous in this area) between Ffynnon-Groew and Talacre, and lying behind the low ground of the Point of Air. The cliff base is at most a very few feet above the level of high water of ordinary spring tides, but the clays, which form most of the Point of Air, are up to 10 ft. above that level, and the shingle ridges are a little higher still. In the Dee estuary there are great masses of recent alluvium. Near the mouth their thickness is not merely due to depression, nor, for that matter, do the estuarine clays and shingle ridges at the Point of Air mean a definite, if but small, re-elevation. On the contrary, it is pointed out that a flooded river, held back by strong winds, might rise sufficiently high to deposit mud at the present level of these flats. The old cliff probably dates from a late stage in the downward movement, and is but little earlier than the deposits on the flat ground in front of it: no further comment on it can be made with any certainty. The shingle ridges are low and narrow and run roughly east-north-eastwards at a considerable angle to the old cliff. 'Thus they appear to mark successive stages in the growth of this projection of low ground, from a narrow coastal fringe gradually increasing in width outwards towards north and east. They evidently represent a sequence of storm-beaches formed at the margin of the protecting spit as it rose in elevation and grew outward, the rise of surface probably being helped by the temporary drift of blown sand behind the beaches.'[2] They all rest on the estuarine clay, but while those to the west are on older parts of the clay, it follows that they themselves are more ancient than the higher parts of the clay farther east. Whatever may be the precise post-Glacial date of the old cliff, there is no doubt that the Point of Air took many centuries to form. Since 1684, the first year in which reliable data occur, the effects of erosion on the point have been traceable and the process still continues.[3]

[1] *Ibid.* and *Mems. Geol. Surv.* 'The Geology of the Country around Flint, Hawarden, and Caergwrle', C.B. Wedd and W.B.R. King, 1934.

[2] *Ibid.* 1934.

[3] The Point of Air dunes are high and well formed. There is also a good deal of salt marsh. It is regrettable that the whole area is now ruined by indiscriminate building of huts and bungalows—to say nothing of a colliery at its extreme south-eastern end.

The alluvium in the Dee estuary needs further examination. It has been forming since the post-Glacial depression at least, or even earlier, and is still accumulating. At Hawarden bridge it is more than 50 ft. thick below the river bed, and east of Queensferry it extends to 60 ft. below O.D. The alluvium consists of loosely stratified silty, quartzose, and fine-grained sand with here and there seams of gravel. Nearer the mouth the upper beds are of slippery black mud. In the past every high tide flowed up to Chester, and charts dated as late as 1720 show that the estuary was open, and that a winding and shifting channel drained off the water. Small ships were able to reach Parkgate,[1] although silting had ruined Chester haven even before the middle of the fifteenth century. A new channel was projected by the River Dee Company in 1677, begun in 1733 and finished in 1754. It was reputed to have a minimum and constant depth of 13–15 ft. About 1870 a bank was built from Burton Point to Connah's Quay, but this was broken before it was finished, and much of the marsh above it is still unreclaimed. To-day all is inned from the iron works below Hawarden bridge to a point above the embankment made by the Hawarden-Neston railway. Since all this alluvium took a long while to form, it is impossible to maintain an accurate chronology, and it is thus difficult to date the shingle ridge between Denhall House and Burton Point. This is almost certainly later than the deep-lying alluvium, but younger than the superficial beds. As its elevation and general nature resemble closely the ridges at the Point of Air, it may be that the Burton-Denhall ridge is contemporaneous with the rather newer ridges of Flintshire.

(*c*) *The Coastal Features of the two Estuaries*. The water-front in the Liverpool district is now almost entirely artificial[2] but until the seventeenth century the Pool was the main feature of interest. It was a tidal creek which narrowed inland from its mouth, and the Mersey water penetrated upwards for about a mile at high water. It covered an area of about 50 acres, but reclamation, beginning more than two centuries ago, caused its final disappearance in the middle of the eighteenth century. The stages of reclamation are uncertain, although a map of 1725 by Chadwick and Eyes does not show the Pool. Lying between the Haymarket and Old Dock areas, it seems to have been a shallow stretch of water, and there were also several smaller streams. The little river Stirpool, for example,

[1] In the sixteenth century vessels used Shotwick instead of Chester: by 1778 this place was derelict and gave place to Parkgate. This was also a harbour in Elizabethan times from which ships sailed to Ireland even as late as the early part of the nineteenth century. Now the promenade is faced by salt marsh.

[2] R. Stewart-Brown, *Trans. Hist. Soc. Lancs and Cheshire*, 82, 1930, 88; Sir J.A. Picton, *Proc. Liverpool Geol. Soc.* 6, 1888–92, 31.

fed the Pool. All these streams are now obliterated as features of the landscape, but it is worth noting that the Otters Pool was of some importance and formed a small harbour. In general, the banks of the Mersey show low bluffs of boulder clay with a few exposures of solid rock. On the south bank, however, between Ellesmere Port and the river Weaver there are extensive marshlands, now largely reclaimed. The promontory of Ince is in the Lower Pebble Beds of the Bunter, and Stanlow Point is

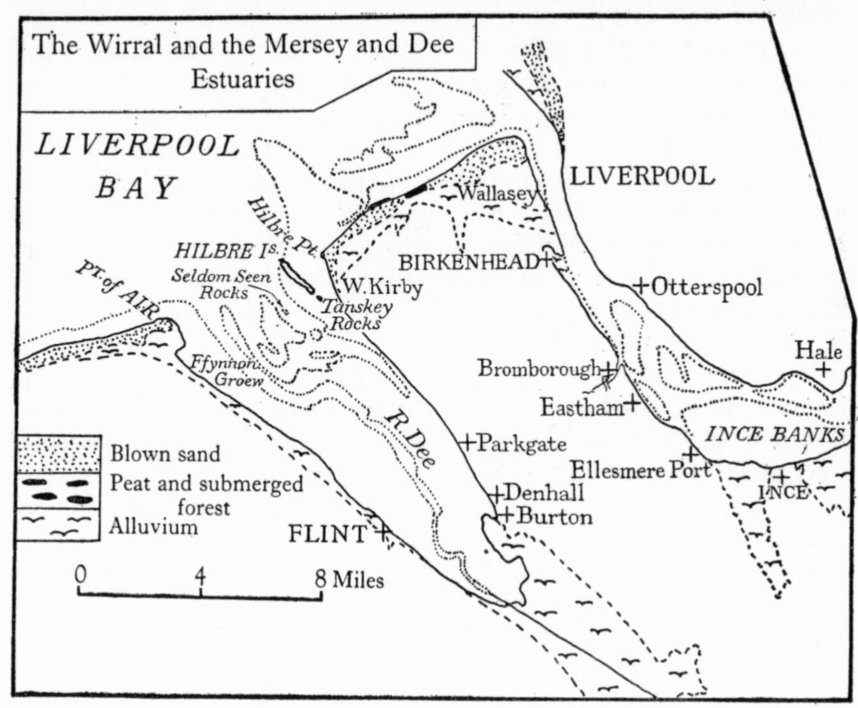

Fig. 19. The Wirral and the Mersey and Dee Estuaries (based on Geological Survey)

similar but lower, while at Frodsham Score there is natural marsh outside the embankment. Morton is of the opinion that the old land surface near Ince is in fact that of land cultivated before 1294. In that year the monks of Stanlow left because of the great floods described in the Chronicle of St Werburgh. The present marsh surface is higher than formerly on account of continuous deposition, but whether the change in level is in part due to subsidence is a debatable point. Further losses of land at Ince, Whitby, Eastham, and Bromborough are recorded between 1344 and 1362. The building of the Manchester Ship Canal led to the final reclamation of large areas of marsh including that in the Gowy valley. From Eastham

Ferry to Bromborough Pool the cliffs show an almost continuous section of the Pebble Beds.[1]

The Mersey upstream from Eastham Ferry as far as Runcorn, is, apart from Ellesmere Port and neighbourhood, mainly open and unspoiled by industrial development. The escarpment, particularly prominent in Helsby and Frodsham Hills, is a fine feature and affords admirable views over the tidal and sandy estuary to the still rural part of Lancashire near Speke and Hale **(115)**.

Fortunately the Mersey has a narrow neck and faces north. Thus it is not exposed to west and north-west gales and, since it is much wider above the mouth, there is a strong tidal scour. Near the mouth on the Wirral side are some interesting features now much obscured and altered by man (Fig. 19). Bromborough Pool was still open in 1922 to the tides which flowed up about one and a half miles. Birkenhead Pool (also named Birket or Tranmere Pool) is now mainly occupied by the Cammell-Laird works. This pool was originally about 300 yd. wide at its mouth and ran inland for a mile; it was dammed early in the nineteenth century. The larger Wallasey Pool was an open arm of the sea up to the middle of last century. It was nearly three-quarters of a mile wide at its mouth and ran inland for two miles, but was shallow enough for easy wading at low water. It is now enclosed and converted mainly into docks. Cliffs of boulder clay, partly masked by the promenade, occur to the north of the Pool at Seacombe and Egremont, and this locality, together with New Brighton, was presumably at one time an island. To-day low-lying ground, indeed recent marshland, extends from the head of Wallasey Pool to the neighbourhood of the Leasowes, and up to the days of artificial protection it was not uncommon for it to be overflowed at big tides. Before the enclosing of Wallasey Pool and the building of Leasowe embankment there was always a fear that the sea would break through and make a permanent channel to the Mersey. The embankment was made in 1829, was extended westwards in 1850, and still later at both ends. It was breached in a storm in the winter of 1919–20. Erosion is still prevalent to the west of Leasowe, and eighteenth-century charts show two lighthouses, one of which stood well seaward of the present embankment. At New Brighton the Bunter pebble beds form hard reefs, which appear again at the mouth of the Dee at Hilbre Point and Hilbre Island: these indeed are features of some interest. The reef of Hilbre Island follows the strike, and extends to Little Eye, Tanskey Rocks, and Caldy Blacks. Red Stones, however, is cut off

[1] For the Wirral coast in general see W. Hewitt, *The Wirral Peninsula*, 1922. See also J. Lomas, *Proc. Liverpool Geol. Soc.* 9, 1900–9, 332; Rev. A. Hume, *Trans. Hist. Soc. Lancs and Cheshire*, 11, 1859, 219; W.G. Travis, *Proc. Liverpool Geol. Soc.* 13, 1919–23, 207; de Rance, *op. cit.* 1871.

from Hilbre Point by a fault downthrown to the east, and a similar fault accounts for the Seldom Seen rocks. In general, between New Brighton and Hilbre Point the coastal area is made up of post-Glacial beds showing the following sequence:

Recent	Blown Sand
Post-Glacial	Recent Tidal Silt Upper Peat and Forest Bed[1] Laminated Clay and Sandy Tidal Silt Lower Peat and Forest Bed[1]
Glacial	Boulder Clay

The dunes are extensive and now form well-known golf links including those of Hoylake, West Kirby, and Wallasey.[2] Owing both to erosion and inland travel the dunes now rest on land formerly marshy. South of West Kirby the Wirral coast is for the most part fringed by a boulder-clay cliff, although sandstones crop out at Burton Point. The western coast of the Wirral, partly because of the cliff, but still more because of the fairly open and heathy high ground just inland and the fine views across the estuary to the Welsh mountains, stands in marked contrast to the Flintshire coast. From the Fynnon-Groew to Connah's Quay is an almost continuous built-up area, rendered locally worse than it might be by derelict industrial property. Flint Castle standing on the marshes is a local but welcome interruption. The marshes fringing the river are, in themselves, good, but their setting spoils them. The combination of main road and main railway running close to the sea has from many points of view certainly had unfortunate effects not only in Flintshire, but nearly everywhere in North Wales.

To-day the channels of the Mersey and Dee are strongly contrasted. The Dee is almost silted up owing to the building of sea walls. The earliest, possibly, was a causeway built by Earl Hugh of Chester. The walls prevent scour, and the steady widening of the estuary seawards also prevents the ebb tide from developing a strong scouring effect. The Mersey, on the contrary, with its narrow mouth and inner basin favours a strong scour. It is for these reasons that, as a port, Chester has decayed whilst Liverpool and Birkenhead, aided by the dredging offshore, have advanced rapidly.

(*d*) *The Controversy concerning Ptolemy's Map*. Ptolemy's map of Great Britain shows only two inlets in this part of the coast instead of the three of the Dee, Mersey, and Ribble[3] (Figs. 20 and 21). Many papers have

[1] Attention has been called to the forest beds on p. 104.

[2] Hereabouts the dunes are now, as it were, 'preserved' inside the sea wall. They are still natural between Hoylake and West Kirby.

[3] *Mems. Geol. Surv.* 'The Geology of Liverpool with Wirral', 1923; T.M. Reade, *Geol. Mag.* 9, 1872, 111.

been written on this matter, and it is often maintained, on Ptolemy's evidence, that in Roman times the Mersey estuary did not exist. There is, moreover, no known reference by the Romans to the Mersey, a silence which gives a negative support to Ptolemy's cartography. This omission is interesting because the Romans definitely had stations on the Dee and Ribble. Mellard Reade called attention to a deep channel, now drift-filled, under the town of Widnes, whereas the present river cuts through the

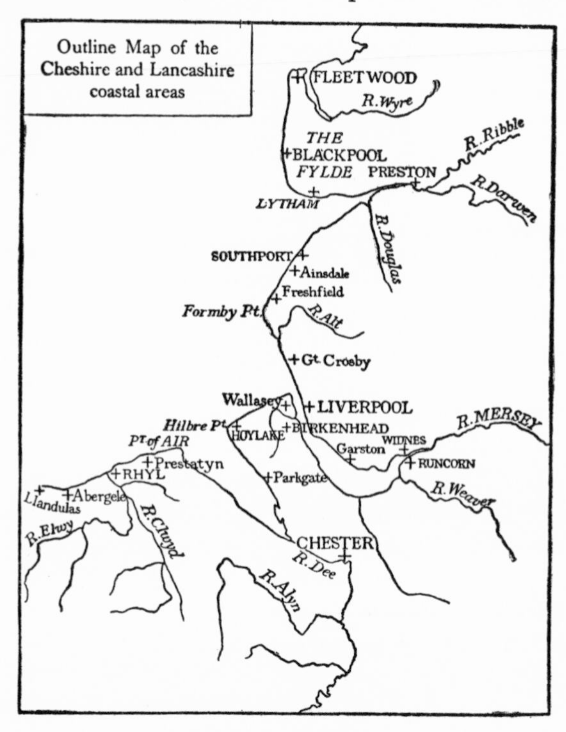

Fig. 20. Outline map

rock channel at Runcorn. Further, Lomas has argued that the part of the Mersey between Runcorn and the sea is a post-Glacial diversion. He thinks that the old course ran from Ince across the southern end of the Wirral peninsula along the general line now followed by the Shropshire Union Canal, the artificial channel appearing to coincide fairly well with a pre-Glacial channel. There are, however, no substantial proofs of these changes. If it were not for docks, embankments, and buildings, there would be many more square miles under tidal influence in the Mersey than is now the case, since the land slopes gently down to the estuary on all sides. As the geological surveyors pointed out, it is highly probable that a river valley, bordered by marshes, and now represented on account of subsidence by submerged forests, existed from very early times. With gradual subsidence and encroachment from the sea the river valley would in time have become a big tidal estuary scoured out and widened, in its narrower

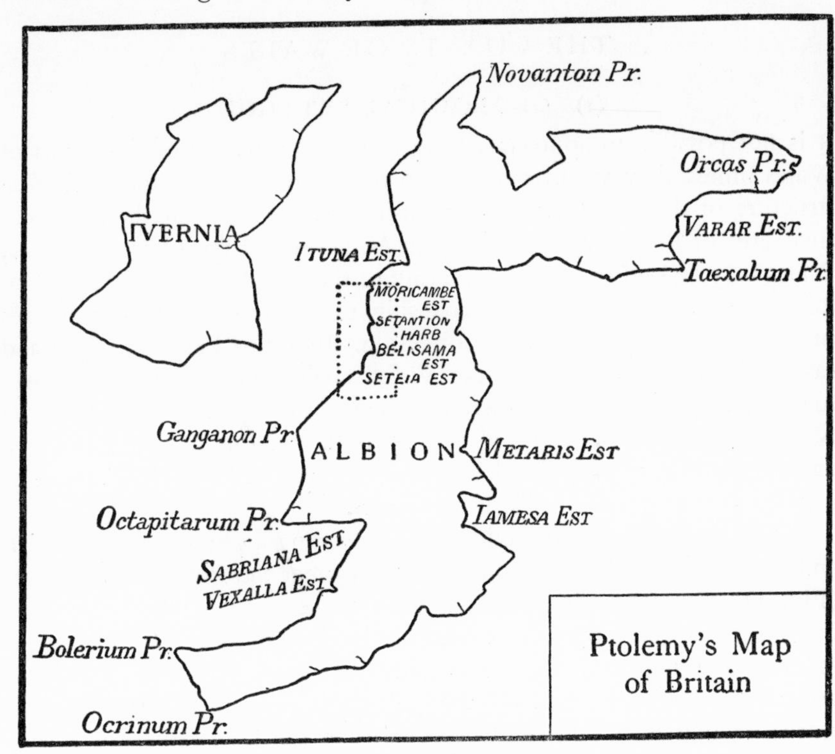

Fig. 21. Ptolemy's map

parts, by the tide. If this has been the history of the river, it is quite probable that in Roman times a deep estuary at Liverpool did not exist, and accordingly the stream which ran out there did not call for any particular note on Ptolemy's map.[1]

[1] On Ptolemy's map the northernmost of the two estuaries is named Belisama. Superimposed on a modern map Belisama is situated about midway between the Mersey and the Ribble. There is little doubt that the southern estuary is the Dee. Does Belisama, therefore, correspond to Mersey or Ribble? If the arguments in this section are reasonably correct, it follows that Belisama and the Ribble were one and the same, and on the grounds suggested the Mersey was not shown on Ptolemy's map. At the same time, however, I cannot help feeling that the attempt to equate actual indentations on Ptolemy's map and those of to-day is a work of supererogation, especially in view of the fact that Ptolemy's work, however good it may be for its time, is clearly a generalization. It is, for example, well known that on Ptolemy's map Scotland is turned about 90° out of its correct orientation.

Chapter V

THE COASTS OF WALES

(1) GEOLOGICAL SETTING

It is not possible to appreciate fully the general nature of the coasts of Wales unless they are first related to the main trends of the geological structure of the country.[1] Leaving aside for the moment the relatively small area of Pre-Cambrian folding in Anglesey and the Lleyn peninsula, Wales may be said to belong primarily to the Caledonian and Armorican fold systems. From north Pembrokeshire the trends of the rock structures run in a general north-easterly direction. In south Pembrokeshire and eastwards therefrom through the Welsh coalfield the trends are nearly east-west. Pembrokeshire, therefore, marks the meeting place of those two great orogenies. Here, also, they both run out to sea, forming a particularly fine, if local, development of a ria coastline; the best feature in it is the drowned valley of Milford Haven.

It will be convenient to divide this study roughly between Caledonian and Armorican Wales, and to add to the former the Pre-Cambrian areas (Fig. 1). Corresponding roughly with the county of Merionethshire, but of greater extent than that county, is the great mass of Cambrian rocks forming the barren and sparsely inhabited Harlech dome. Sweeping around this in a great arc from the Lleyn peninsula up to Conway, eastwards beyond Bala, and south-westwards to the Dyfi estuary is a wide spread of Ordovician rocks, largely sedimentary in origin, but containing numerous and extensive outcrops of contemporaneous igneous rocks. Beyond this again the Silurian forms another and outer arc running as a broad band from the north coast near Conway to Corwen and Llangollen, then as a narrower mass to Lake Vyrnwy, after which it widens out again to form most of Central Wales, and also to form the coastal cliffs from Borth to south of New Quay. Ordovician rocks reappear on the south stretching from the neighbourhood of St David's, Newport, and Cardigan to Carmarthen, Llandilo, and north-eastwards as far as Rhayader. There are also Pre-Cambrian and Cambrian rocks near St David's.

This very generalized description will emphasize that, apart from more

[1] There are many works dealing with the general geology of Wales: for the present purpose the two volumes of *British Regional Geology* (Geological Survey and Museum) —'North Wales', B. Smith and T.N. George, 1935, 'South Wales', J. Pringle and T.N. George, 1937—are admirable and contain bibliographies of important papers.

local structures, we have a great anticlinal area behind Harlech; further in Central Wales there is first the main anticlinal axis running from a point near the north-east corner of St Bride's Bay to Plynlimon, followed on its south-eastern side by the Central Wales syncline. The Ordovician rocks reappear along another anticline corresponding with the Towy valley and running north-eastwards to a few miles east of Rhayader. A relatively narrow outcrop of Silurian rocks trending north-east to south-west through Llandovery disappears east of Carmarthen[1] under the Upper Palaeozoics which form the eastern shore of St Bride's Bay. The northern limit of these Palaeozoic rocks runs through Haverfordwest almost due east to a point about two miles south of Carmarthen, and thence north-eastwards to near Llangammarch Wells.

South of this line is the Armorican region of folds aligned nearly east-west. In southern Pembrokeshire and the Gower peninsula the relation between structure and coast scenery is usually very clear. Carmarthen Bay and Swansea Bay lie mainly in the Coal Measures. Then, in south Glamorgan we enter a newer series of rocks—the Trias, Rhaetic, and Lias—which form the coastline between Porthcawl and Cardiff. The coal basin of South Wales, together with its continuation west of Carmarthen Bay in Pembrokeshire, is primarily a great syncline, and the Millstone Grit, Carboniferous Limestone, and Old Red Sandstone form girdles around it. These are much interrupted, however, by indentations of the coast on the southern side of the basin. It is to the comparatively small-scale folding of these rocks that the coasts of southern Pembrokeshire and Gower owe their variety and beauty.

If we turn once again to North Wales, we see the extent of the Pre-Cambrian rocks. Between Nevin and Bardsey Island there is a continuous band of them, but apart from anticlinal inliers in inland Caernarvonshire, the classic area for the study of the Pre-Cambrian is in Anglesey where the schists and gneisses of the Mona complex form most of the island, and the adjacent Holy Island as well.

To describe in any detail the structure of Anglesey and of the neighbouring parts of Caernarvonshire is unnecessary for our purpose: the map shows the main features, including the outcrops of Cambrian and later rocks parallel to the Menai Straits. The only other points to which we will call attention here are the outcrops of Carboniferous Limestone forming the Great Orme and the coast near Llandulas, the Triassic rocks of the Vale of Clwyd, and the Upper Carboniferous rocks between the Clwyd and the Dee.

In the old Palaeozoic rocks, that is to say mainly in 'Caledonian' Wales,

[1] The Silurian reappears locally in Pembrokeshire.

there are large outcrops of contemporaneous igneous rocks with many intrusives. The inland exposures of these do not come within the scope of this study, but in the Lleyn peninsula particularly, and again between St David's and Fishguard, igneous rocks, mainly in the form of intrusions, have a great effect on the detail of the coastline (see below).

Boulder clay is spread over wide areas and in places covers the cliffs and forms the actual coastline. Here again, however, it is best to leave an account of its effects on the coastline for later discussion. But we may note one topic which is treated in detail farther on; there is reason to think that extensive boulder-clay areas have sunk under Cardigan Bay, and that some of the great spits of shingle, such as that across the Dyfi estuary, derive their material largely from this source.

From this brief description of the geological setting, it will be clear that in North Wales there is no close relationship between tectonic structure and the coast. Moreover, a comparatively recent uplift has exposed a plane of marine or subaerial denudation cut indiscriminately across rocks of many types. This again serves to obscure the connection, especially in Anglesey and the region around the Menai Straits. In Tremadoc Bay, and as far south as Barmouth, extensive accumulations of sand and marsh have grown out in front of the Cambrian rocks of the Harlech dome, so that these rocks form the *present* coastline only very occasionally. In view of these complications it will be best to describe the coast of North Wales in sections: it is chiefly in South Wales that the relationship of coastal scenery and structure is clearly evident.

(2) THE COAST OF NORTH WALES

The north coast of Wales does not present any particularly interesting features.[1] This study begins just west of the Dee estuary which has been discussed in the previous chapter. Between Rhyl and Colwyn Bay the shore is largely fringed by glacial deposits, alluvium, and blown sand; they cover the lower ground, and glacial beds also plaster much of the higher ground. Around the lower part of the Vale of Clwyd, between Abergele and Llandulas, and around Colwyn Bay a drift terrace ranging from 150 to 200 ft. above sea level separates the hills from the sea. This drift is banked against the pre-Glacial cliff. Hence, the coast itself with its shingle beaches is formed from this glacial material, but the broad structural outline of the coast is determined mainly by faulting. The faults have usually thrown down rocks that yield easily to erosion, and as far west as

[1] *Mems. Geol. Surv.* England and Wales, 'The Geology of the Country around Flint, Hawarden, and Caergwrle', C.B. Wedd and W.B.R. King, 1924, and *ibid.* 'The Geology of the Coasts adjoining Rhyl, Abergele, and Colwyn', A. Strahan, 1885.

Llandulas the old, pre-Glacial, coast follows closely the limits of the harder Silurian and Lower Carboniferous rocks.

At Llandulas, Abergele, and Dyserth pre-Glacial valleys, now partly filled with boulder clay, breach this old cliff. Bordering the Clwyd and Gele rivers, and also between Rhyl and Prestatyn, there are extensive alluvial deposits, the upper layers of which consist of tidal silt overlying sand. Near Prestatyn they are formed of fresh-water marsh-clay and peat which often formed in shallow meres in dunes and sandbanks. Below these, and reaching to minus 40 ft. O.D., are peat beds. Another small strip of alluvium at the north-western end of Colwyn Bay is continuous with the marsh from Llangwystenin; this strip is in part below high water springs, and so has to be protected by a bank and sluice.

There is a line of dunes, narrow and artificially encouraged, between Abergele and the Clwyd, but east of the river the sand spreads and covers part of the boulder clay. Here it forms dunes 40–50 ft. high, thus protecting Llangwystenin marsh. The winds carry sand eastwards, and also cause easterly beach-drifting, a fact illustrated in the eastward deflection of the rivers. The sand and shingle, also, generally increase eastward, and they are helped in this off Prestatyn by stakes and wattles; they terminate in the storm beaches of the Point of Air which, on its inner side, encloses marshland.

Farther west the Little Orme and Great Orme are headlands of Carboniferous Limestone, the latter with its magnificent cliffs being virtually an island (**113**). It is joined to the mainland by an isthmus of blown sand. West of the glacial valley of the Conway are the Ordovician rocks with intrusives which near Dwygyfylchi and Llanfairfechan form headlands. But in a sense Conway Bay may be considered with the Menai Straits. Before, however, touching on the origin of the Straits the problem of Llys Helig and the Anglesey coastline need attention.

The whole length of coast from Llandudno to the Point of Air is unfortunately spoiled in various ways. Natural conditions at Llandudno, Penrhyn, Rhos, and Colwyn Bay are disturbed by extensive sea walls and promenades. Headlands, including the Little Orme and Tan Penmaen, and also Penmaenmawr itself, are badly disfigured by extensive quarrying. Farther east, along the dune and marshland coast, widespread and unplanned masses of huts and bungalows detract greatly from the natural landscape. Forethought and consideration for the future could very easily have given pleasure to all the people who visit this area, and at the same time could have kept it in a far more natural and beautiful condition (**114**).

(3) LLYS HELIG

On various parts of the coasts of Wales there are traditions or legends of lost lands: on pp. 148–52 the question of Cantref-y-Gwaelod is discussed in some detail. On the north coast there is a similar story associated with the patch of seaweed-covered stones visible at low water about one mile out to sea off Penmaenmawr, and called Llys Helig. These have often been described as the ruins of a palace belonging to Helig ap Glannog whose lands, according to legend, were inundated by the sea at some time between the fourth and ninth centuries A.D. North[1] has recently investigated the evidence of this story, and his paper ought to suppress the imaginative and ill-founded tales that are still current. The actual story of the inundation is only about a century old; nevertheless, it appears to be regarded by many as history based on the authority of *An Ancient Survey of Pen Maen Mawr* by Sir John Gwynn who identified a patch of stones in Conway Bay with Helig's palace. He refers to a certain 'manuscript', attributed to Sir John Wynn of Gwydir (died 1626): the authenticity of this writing was taken for granted, but North, after examining the evidence, concludes that it derives only from the copy of a copy, and that it was not even intended for a survey of Penmaenmawr. The parts relating to Llys Helig are pure imagination. According to the *Ancient Survey* Helig and his people, when the supposed sudden inundation occurred, fled to Trwyn 'r Wylfa, in Dwygyfylchi parish, and this hill is still often referred to as the 'hill of mourning'. Imagination has also 'reconstructed' a river Ell between Caernarvon and Anglesey as well as a great sandy stretch called Traeth Ell. So-called 'reconstructions' by Hall and Ashton showing the supposed ancient coast (see Fig. 22) are entirely fictitious. Certainly there was once land here, but this was before the close of the post-Glacial submergence. If Llys Helig had been a human habitation, it would imply a subsidence of at least 40 ft. so that the Menai Straits would not have existed as a continuous channel. But the island of Anglesey was definitely separated from the mainland long before the sixth century. The Menai Straits were dry land when some at least of the submerged forests existed, but they were undoubtedly open in Roman times —and indeed long before. The earliest genuine references to Helig belong to the thirteenth century (**112**).

Unfortunately the story of the submergence may not even be genuine folklore, although it may possibly have a similar background to that of the tale of Cantref-y-Gwaelod.

[1] Supplement to the *Llandudno, Colwyn Bay and District Field Club Proceedings*, Llandudno, 1940.

The theme of the argument there outlined may apply also to Llys Helig. It is, however, as well to point out that the legends of the submergence need not refer specifically to either place: they may have been introduced. However, since we know that the last downward movement of the coast was part of the post-Glacial submergence, it is reasonable to suppose that folk tales have been handed down from, possibly, Neolithic people and in course of time have been made to apply to specific objects popularly supposed to be of human construction. Moreover any 'evidence' based

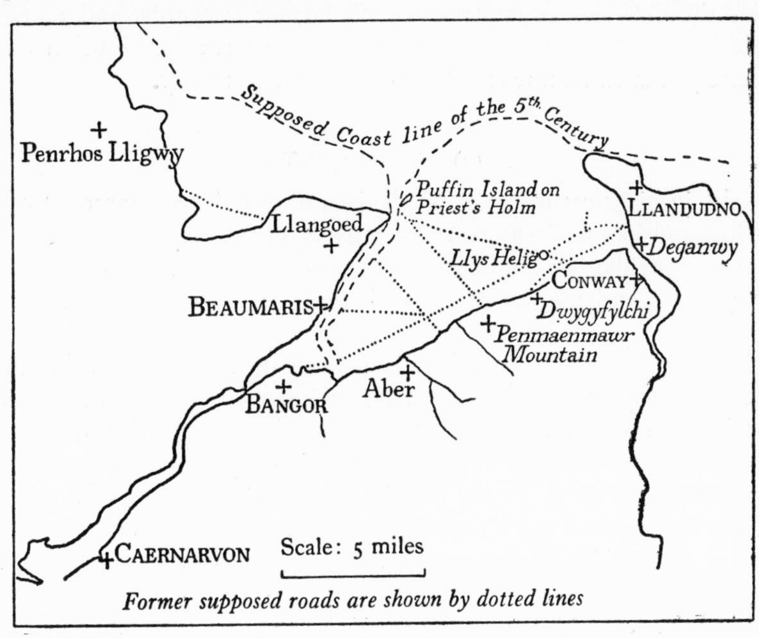

Fig. 22. Supposed reconstruction of part of the coast of North Wales (cf. Ashton, *op. cit.*)

on Ptolemy's maps is emphatically worthless, for they were far too inaccurate to show anything of value in this connection. The various sketches of the 'walls' of Llys Helig drawn by Ashton and others are equally valueless. North cogently points out that since approach to the site must be made in a boat, the appearance of straight narrow lines of stones, often at right angles to one another, is largely an illusion due to the low angle of vision of the observer. Unfortunately there does not seem to be an air photograph to illustrate this point convincingly. Examination of the area shows that the seaweed does not rise from straight lines of stones, which incidentally are of all sizes and quite irregular distribution.

There is little doubt that these stones are the remains of a moraine eroded away by the waves: the finer material has been washed out and the stones left. They are all of local material, and, as Grimes points out in an appendix to North's paper, of a type quite unsuited to building for 'they are smooth rounded masses, irregular in shape and varying greatly in size. They show no sign of shaping or dressing and without mortar could only be used for very rough low walls.' In short, Llys Helig is in some ways similar to the Sarns in Cardigan Bay and to the scars of Cumberland. In earlier times Wales certainly extended farther seawards, and moraines would be scattered on this now submerged surface: their erosion would produce just such features as those of Llys Helig to-day.

(4) ANGLESEY

Greenly, in his great memoir[1] on the island, has given an account of the coastline and its main features. The northern coast may be regarded as a strike section: the east and west coasts are dip sections. The origin of the Menai Straits and the narrow strait between Holy Island and Anglesey present special problems.

The main drainage of the island is towards the north-east or the south-west, and many of the bays are really the seaward ends of these valleys. Most of them were formerly graded to a sea level lower than the present and so the true rock floor of the valleys is deep down and commonly covered with boulder clay and alluvial material. The largest of the valleys is now almost filled with Malldraeth Marsh. Originally the tide flowed as far as Hirdrefaig, leaving at ebb a flat expanse of sand and mud, like that on Malldraeth Sands to-day. In 1788–90 a dam was built across the bay at the 'Yard' and the Cefni was straightened and embanked for about six miles. At the present time the floor is marshy and mostly pasture land with an occasional lagoon, and the remnants of the old meanders are still visible. There is also a great amount of blown sand in this valley; it has partly overwhelmed cultivated land, and its depredations were recorded in an Act of Parliament of 10 Elizabeth. Incidentally, sand has still further polished the glaciated rocks bordering the valley. Although on a smaller scale, and with local differences in detail, other valleys such as that at Aberffraw and Dulas Bay are similar.

On the north-west the bays have their longer axes trending with the strike of the rocks, because it is in this direction that the waves can most

[1] *Mems. Geol. Surv.* 'The Geology of Anglesey', 2 vols. 1919; and by the same author, 'Some Recent Work on the Submerged Forest in Anglesey', *Proc. Liverpool Geol. Soc.* 1928.

easily attack the land. But here, too, the bays are only the mouths of normal land valleys, valleys which are transverse rather than longitudinal. They seldom contain rivers, as they are often pre-Glacial channels blocked with boulder clay, for example, Bull Bay. In Cemlyn Bay there is a good example of a bay bar, about half a mile long, formed by storm-beach shingle impounding a tidal lagoon. In Cemmaes Bay the sea is working laterally along recesses weakened by pre-Glacial decay and also in Ordovician shales. Two pre-Glacial drift-filled valleys enter the bay, and as many as eleven different formations outcrop in it.

In Glacial times much boulder clay was added to the coasts of Anglesey: this is now being eroded fairly rapidly, a process helped by the downward movement of the land in post-Glacial times. The effects of this erosion are plain at Llangwyfan church which now stands on an islet of boulder clay. Speed (1610), on his map (if it be correct), shows it as part of the mainland: now it is more than 180 yd. away (**109**).

The most interesting coastal features are perhaps the straits between Holy Island and Anglesey and the Menai Straits. The first was etched out in pre-Glacial times along the line of the Namarch faults by two streams. During the Glacial period, north-east to south-west hollows were scooped out here and elsewhere on Anglesey, and where these hollows intersected the faults pools were formed. The isolation of the island followed as a result of later submergence. Glacial action was also largely responsible for the many tributary creeks, and for the depression which divides Holy Island from Rhoscolyn 'Island'.

The Menai Straits (Fig. 23) are more complicated in their origin, but nevertheless belong to the longitudinal valley system of the island. It would only require the removal of some 60 ft. from the

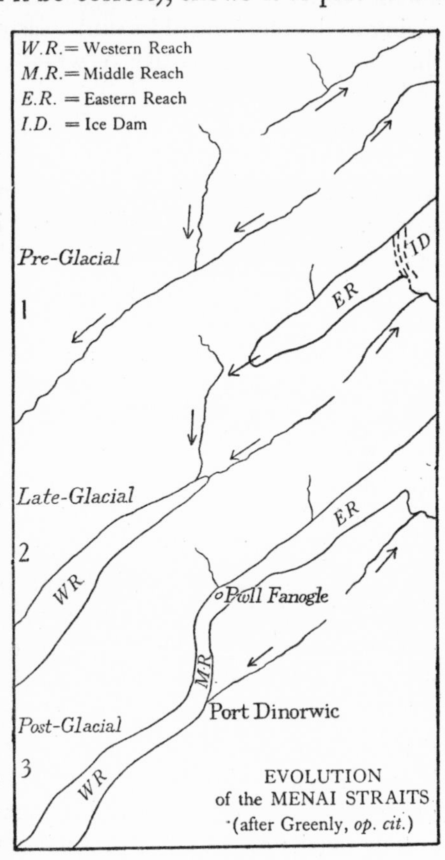

Fig. 23. Menai Straits

watershed to make Malldraeth Marsh and Red Wharf Bay another through valley. The Menai Straits result from the unification of three valleys, and the problem of the formation of the Straits is really that of the middle reach, that is, roughly from the railway bridge to Port Dinorwic. The north-east and south-west parts are parallel, but quite independent, valleys. Between the road and rail bridges soundings show a definite, but now submerged, watershed. On the Anglesey side, near Pwll Fanogle is a hollow, some 77 ft. deep, which is about 50 ft. more than the average depth of the straits (**108**). As appears from a map, the middle reach runs nearly north and south; its direction was determined pre-Glacially. When the ice had retreated from the north-east and south-west reaches they must have appeared much as they are to-day. The water in the south-west reach ran out to sea, but that in the north-east reach, still ponded by ice, found a land barrier almost a 100 ft. high at its southern end. At a later stage, owing to the blocking of its north-eastern end by ice of the Ogwen glacier, this valley became a glacial lake, which eventually overflowed at its southern end where it cut a channel in the relatively soft Carboniferous rocks. This overflow water soon found itself in the old channel of the Gaerwen River and was consequently diverted to the south; thus the middle reach of the straits was cut. These events took place when the land stood higher than it does to-day. The deep hole at Pwll Fanogle is cut in soft marls, and in Greenly's opinion is of the nature of a great pothole made by the Gaerwen when it became ice free and flowed out into the newly formed middle reach. Later the land was lowered relative to sea level, and the straits took on their present appearance. There is a strong tidal race through them, and extensive sand flats have gathered on both sides of the southern entrance forming Morfa Dinlle and Newborough Warren. It is, perhaps, worth noticing that apart from the actual tidal channel of the straits, these two sand areas lie in a position very similar to Morfa Harlech and Morfa Bychan in Tremadoc Bay. It is clear that each has grown outwards towards the strait and the form of the sand spits is due to the general effects of south-westerly winds and beach-drifting. There is no physiographic description of these two spits, but a study of all available recent maps suggests that they have grown into the straits in a manner comparable to similar formations elsewhere. Morfa Dinlle is bordered by shingle which fans out in normal fashion at the distal end. It is a very flat area and its southern end is dominated by Dinas Dinlle. Newborough Warren is a fine mass of dunes, still mainly unspoiled, and thus unlike Morfa Dinlle. It, too, is fringed by a certain amount of shingle.

The details of the coast of Anglesey are interesting and intricate. The cliffs rise to great heights only near Holyhead Mountain, especially at

North Stack and South Stack. Near Penrhyn Mawr and towards Carmel Head, and along parts of the north coast, their height is considerable, but it is not so much the altitude of a few sections that gives beauty to the island's coast: it is rather the detail in plan. Tre-Arddur and Rhosneigr are both characterized by quite low cliffs, but the amazing combination of sandy cove, rocky headland, and islets or stacks offshore, gives to these places a fascination, probably quite unequalled in England and Wales. Only to a rather less extent is this true of the north coast: Cemlyn Bay, Cemmaes Bay, Hell's Mouth, Porth Wen, Bull Bay, Amlwch Harbour, and Porth Eilian are all fine features, each one possessing certain assets peculiar to itself. The more sheltered side of the island contains Dulas Bay, Lligwy Bay, Traeth Bychan, Benllech, and Red Wharf Bay with its fine sands, but somewhat dangerous tidal conditions. The north-east corner of the island, with extensive outcrops of Carboniferous Limestone, stands rather apart from the rest, but has also its own great charm. The flatter coast near Beaumaris and Lavan Sands lies in strong but pleasing contrast to the higher ground of the interior. Almost the whole length of the Menai Straits is defined by gently shelving banks and pleasant country or private estates reaching the water's edge. There has already been a brief mention of the extensive dunes at Newborough, Aberffraw, Trwyn Trewan and other places. They too give much variety to the coast. But part of the beauty of the coast at least derives from the magnificent view of the Snowdon Mountains which rise directly from sea level and which give a magnificent background to the local views. Indeed, taken as a whole, the coastline of Anglesey and Holy Island is one of great beauty and of much interest to the physiographer. The intricate effects of erosion on hard rock or boulder clay, on high cliffs or low; the forms of blown sand; the problems of marsh development; and the close relationship of glacial phenomena, all contribute to make these coasts exceptional, even if not unique, in England and Wales (**110, 111, 107**).

(5) THE LLEYN PENINSULA

The northern coast of the Lleyn peninsula south-westwards from the Menai Straits is composed of Ordovician, Cambrian, and Pre-Cambrian rocks. There is much boulder clay which forms low cliffs between Dinas Dinlle and Yr Eifl. Occasionally, as at Aber Desach, the cliffs disappear and are replaced by a boulder beach and some blown sand. The coastal fringe is low and without great interest. The high ground near Yr Eifl and the headland of Trwyn y Gorlech and Careg y Llam are in the Ordovician area, and are themselves formed of igneous rock. Their form is spoiled by local quarrying. The main interest is however in the

Pre-Cambrian coast between Porth Dinllaen and Aberdaron Bay. Porth Dinllaen was once considered as a possible alternative to Holyhead as a packet station for Ireland. Matley[1] has described the geology of the district, and points out that whilst much of the area is drift covered, the coast is nevertheless usually faced by fine but low lines of cliff. The drift in Porth Nevin and Porth Dinllaen has filled old bays to a depth well below present sea level, and has obstructed the drainage system. In other parts cliffs in the drift often reach 100 ft. in height. The headland of Dinllaen itself is a particularly good exposure of spilites, and is composed mainly of lavas many of which show pillow structure. The coast of Pre-Cambrian rocks is generally rather low, but at the same time steep and rocky, and the rocks are partly sedimentary partly igneous, broken here and there, as at Porth Oer, by the drift falling to sea level. All the rocks are much metamorphosed, folded, and sheared, and frequently intersected by dykes of Palaeozoic age. In general, along the Pre-Cambrian coast, there is a belt of lower ground adjacent to the shore. The details in plan of the coast are of interest, but in profile it is rather dull. Many of the headlands are quite low, and fields run to the cliff edge. South of Porth Oer, however, the cliffs rise in height, and at and near Braich Anelog, and Braich y Pwll are of considerable beauty.

The highest cliffs are at Parwyd in the Arenig shales and flags, and Aberdaron Bay lies in Ordovician rocks[2], but the high cliffs at the head of the bay are cut in boulder clay which shows characteristic erosion features. On the eastern side of Penrhyn Headland, itself composed of Arenig beds exposed in inaccessible cliffs, the sediments dip under the Gallt y Mor sill, detached parts of which also form the islets of Ynys Gwylan-fawr and Ynys Gwylan-bach. The wide bay of Porth Neigwl is largely filled with glacial deposits, perhaps as much as 200 ft. thick. The little Soch River presumably at one time had its mouth in the east side of the bay. Even to-day it comes to within half a mile of the bay, and is only separated from it by boulder clay and blown sand, the lowest part of which is but 10 ft. above the level of the river. The Soch, however, turns back on itself and has cut a deep gorge through Ordovician rocks. This is a typical glacial overflow channel. The whole phenomenon must be considered in relation to a terrace-like feature,[3] at about 50–56 ft. above sea level, which occurs in the eastern part of Porth Neigwl and at intervals along the south coast of Lleyn probably as far as Criccieth (Fig. 24). This terrace sinks seawards and rises to its maximum height around small 'islands' which stand out

[1] *Quart. Journ. Geol. Soc.* 84, 1928, 440.

[2] C.A. Matley, *ibid.* 88, 1932, 238.

[3] C.A. Matley, *Proc. Geol. Assoc.* 47, 1936, 221.

from the platform. Near Pwllheli the feature may reach two miles in width, and the small streams of the district have entrenched themselves in it (106). The beds which compose it all suggest a connection with glacial conditions. At the north-eastern corner of Porth Neigwl it consists of boulder clay capped by laminated clays with sands and gravel, the whole resting on Cambrian rocks. Between Pwllheli and Criccieth it appears to be built mainly of fluvio-glacial sands and gravels. Its surface in this district is often very irregular, a fact which Matley suggests may be due

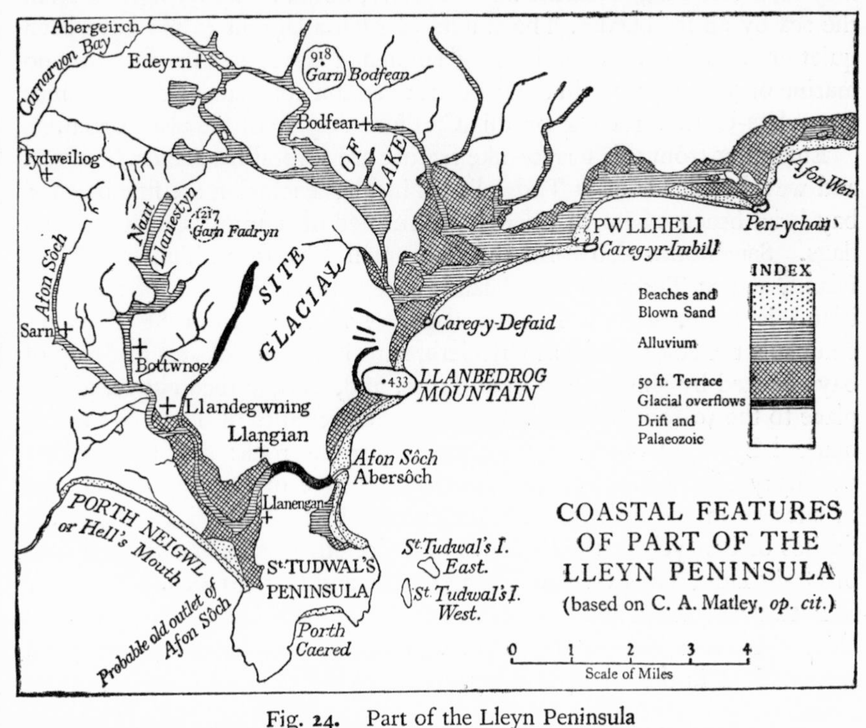

Fig. 24. Part of the Lleyn Peninsula

partly to its deltaic nature and partly to later erosion. It may also form the low ground between Criccieth Castle and station. It does not occur generally in Cardigan Bay, and although features of a somewhat similar height are found at Llandudno and possibly near Caernarvon and Bangor, the evidence suggests that it is not a true marine terrace but is in some way connected with glacial conditions. The fact that no marine organisms have been found in it strengthens this view. Matley suggests that the Welsh ice, at this stage advancing westwards across the northern part of Cardigan Bay, 'reached the coasts between Pwllheli and St Tudwal's Peninsula before the Irish Sea ice had withdrawn from the west coast of Lleyn and so permitted the formation of a glacial lake....' At first the waters of this lake reached some 255 ft. above present sea level, as is suggested by certain inland features: later, by retreat of the ice, the surface was lowered to the 50-foot contour during which stage the deposits of the terrace feature were laid down. This explanation, as far as it goes, only applies to southern Lleyn: the features of similar height in the Menai district require a separate explanation. It is also possible, as suggested by Dixon, that the terrace may have formed in a marine lagoon partly, but not wholly, shut off from the sea by an ice barrier. The waters of such a lagoon would have been quiet enough for the deposition of laminated clays, and the fact that no marine organisms occur in the deposits need not contradict this view since many late-Glacial gravels[1] on this Cardigan Bay coast are also barren.

Emerging from this terrace-like feature are several headlands. The east and west headlands of St Tudwal's peninsula,[2] enclosing the fine bay and bay-head beach of Porth Caered, are formed of Tudwal sandstones and flags. Sandstones also form the two small islands, St Tudwal's West Island, and St Tudwal's East Island; except that about half of the latter island is composed of Lingula flags. Eastwards from St Tudwal's peninsula the coast, as far as Criccieth, runs in a series of arcs made up of bays backed by shingle and sand dunes, and, as described above, giving place to the 50-foot 'terrace'. The eastern extremities of these bays are all marked by small outcrops of resistant igneous rock: first the granite-porphyry of Llanbedrog,[3] next the keratophyre of Careg y defaid, and in succession the dolerite of Careg yr Imbill at Pwllheli Harbour, the felsites of Pen-ychan, the rhyolites of Criccieth Castle, and the dolerites of Graig Ddu, Ynys Cyngar, Careg Cnwc, and Gareg Goch.

Between Pwllheli and Llwyngwril there are many excellent examples of shingle beaches, sand-dune areas, and marshland, including the great reclamations in the Glaslyn valley.

(6) PWLLHELI TO PORTMADOC[4]

From Pwllheli to the headland of Pen-ychan there is a remarkably smooth stretch of shore. Near Pwllheli[5] the sandy beach is backed by a narrow

[1] Not certainly marine gravels.

[2] T.C. Nicholas, *Quart. Journ. Geol. Soc.* 71, 1915, 83.

[3] C.A. Matley, *ibid.* 94, 1938, 555.

[4] The account of the coast between Pwllheli and Fairbourne is based on a paper by the author in *Geogr. Journ.* 94, 1939, 209.

[5] Artificial alterations made in Pwllheli Harbour are not discussed. Reference may be made to Lewis Morris's *Places and Harbours in St George's Channel*, 1748, and Act. Parl. 48, Geo. 3.

line of dunes, but to the east these dunes are more irregular and wider. About half a mile east of Abererch station the dune belt is interrupted by an outcrop of boulder clay. Up to this point there is little or no shingle, but eastwards shingle increases in amount, and at Pen-ychan there are three or four high storm ridges. In this eastern part of Morfa Abererch there is also a good development of old, grey dunes now some distance from the sea. The old cliff line can be traced inside them along the high ground of Pen-ychan and towards Pwllheli. The whole bay shows clearly the eastward movement of both beach material and blown sand, and the headland of Pen-ychan acts as a great groyne. The bay also illustrates the effect of exposure. At its western end there is no direct incidence of waves from the south-west, since this part is protected by Careg yr Imbill, the tied island south of Pwllheli Harbour. But near Pen-ychan the coast receives the full force of the winds and waves, and the amounts of shingle and sand are arranged in direct proportion to this attack.

North of Pen-ychan Point there is an extensive outcrop of the Lower Boulder Clay[1] which reaches to the Afon Wen, a small stream which trickles over the beach. The clay is subject to erosion, and the material travels eastwards, except for that held up by extensive sea walls and groynes near the station. The three small lakes, marked on the Ordnance Map just east of the station, lie on a marshy area which before the railway was built was open salt marsh. A culvert allows the sea to flood part of this area during storms. *Phragmites* and *Scirpus lacustris* are common.

The cliffs of boulder clay and glacial gravels with occasional peat beds east of the station provide abundant shingle, which in its eastward movement has deflected the Afon Dwyfawr for about a mile. On the east side of the river is another outcrop of boulder clay which has held the mouth close to its present position for a long time. The deflecting shingle bar is covered with low dunes, and north of the river are several older ridges in the marsh between the river and Llanystumdwy. The boulder-clay outcrop east of the river's mouth is almost continuous as far as Criccieth Castle. Traces of a submerged forest, possibly of the same age as that at Llanaber, are occasionally visible about a quarter of a mile east of the Dwyfawr mouth. Criccieth cliffs are subject to considerable erosion,[2] and the Borough Surveyor (Mr W. Williams) has found that the best protection is obtained by building groynes outwards from them making an angle of about 85° with the coast to the west.

[1] T.J. Jehu, *Trans. R. Soc. Edinburgh*, vol. 47, p. 17. Erosion is serious in the boulder-clay cliffs east and west of Afon Wen.

[2] See W. Ashton, *The Evolution of a Coastline: Barrow to Aberystwyth*, p. 236.

The boundary between this section of the coast and that to the west is the outcrop of rhyolite on which stands Criccieth Castle (105).[1]

East and west of the castle, Fearnsides notes the occurrence of the oldest known Pleistocene deposit of the neighbourhood as a shore talus or 'head' banked against the castle rock. It is made of angular unworn rock and rests undisturbed on a wave-eroded shelf only a foot or two above the level of present high tides. The boulder clay seen in the cliff to the east of the town also underlies the esplanade, makes the small headland of Merllyn, and provides abundant shingle for the beaches. Approximately three-quarters of a mile east of the station begins a typical shingle embankment which runs as far as Graig Ddu. At Rhiw-for fawr the railway cuts through solid rock, and an old but small cave can be seen on the landward side. Eastwards of Rhiw-for fawr the shingle dams Llyn Ystumllyn, a former arm of the sea. There is no record of the building of the embankment; the fact that Saxton and Speed show open water is certainly not conclusive evidence of the change having taken place since the sixteenth century. The stream now draining the lake is diverted eastwards artificially and empties into the sea at Graig Ddu:[2] part of its lower course is enclosed. Mr Owen of Croesor thinks the lake was drained to its present state about 1840, but there is no definite record. The former mouth, probably a percolation through the shingle, was near Rhiw-for fawr, and the change to the present outlet was made apparently between *c.* 1763 and *c.* 1785.

Near Graig Ddu the shingle increases both in amount and in the size of the pebbles: on a smaller scale it resembles the eastern end of the shingle alongside Morfa Abererch. Some older ridges are inside the present storm beach near the railway halt, while recent storms have breached the outer shingle in several places, and spread out fans.[3]

Llyn Ystumllyn is now a fresh-water marsh, grading in places to a shallow lake. In storms some sea water may find its way over the bar, but appears to have only a small effect. The dominant plants are *Scirpus lacustris* and *Phragmites communis*, with a good deal of *Salix cinerea* along the eastern side. In September 1938 the *Salix* was often under about 2 ft.

[1] The first certain reference to this castle occurs in 1239. 'Provisional guide to Criccieth Castle', H.M. Office of Works, 1934.

[2] There is much artificial drainage in the basin of the lake.

[3] On 23 November 1938 the shingle was severely breached, and the railway bridge, under construction, much damaged (see *The Times*, 24 November 1938). W. Ashton, *loc. cit.* p. 237, says the embankment was artificially built about two hundred years ago. I do not know what evidence there is for this statement. The shingle appears to be natural, but may be piled on an old wall. There is every reason to expect that a natural bank would form here.

of water. There remain large stretches of open water, almost without vegetation. The bog bean, *Menyanthes trifoliata*, dominates several isolated areas.

The gulf of Llyn Ystumllyn formerly ran up to the main road near Pentre'r-felin and at that time only a narrow isthmus joined the Moel y Gest peninsula to the mainland. It is also interesting to note that the lake, which lies at the centre of an anticline, is in the Ffestiniog beds (Lingula Flags) which are not particularly resistant.

The Black Rock, Graig Ddu, is a narrow dolerite mass which runs a little farther seawards than the Ffestiniog and Maentwrog beds west and east of it. In the latter, caves are eroded along igneous sills. The beach to the east is hard and sandy,[1] and between it and the high ground is a dune system now called Morfa Bychan. The general shallowing of Tremadoc Bay, discussed below, caused extensive sand flats to form, and the prevailing winds blowing over them have built dunes in the former irregular bay. Ynys Cyngar is a rocky headland which has been tied to the mainland by sand bars. The Glaslyn channel south of Portmadoc has been, and is still, subject to considerable changes, and is difficult to navigate. At present it cuts into that part of Morfa Bychan between Ynys Cyngar and Gareg Goch. The remainder of the coast up to Portmadoc Harbour[2] is rocky, with small sandy bays. There is little need for comment except that the vagaries of the Glaslyn channel are apt to cause local changes. The Ordnance Survey maps do not show the present position of the channel.

(7) PORTMADOC AND THE GLASLYN VALLEY

Before the enclosure and the building of the embankment by W. A. Madocks in 1811, the tide reached nearly to Aber Glaslyn bridge, and the whole of the Glaslyn valley resembled the Mawddach and Dovey estuaries. The idea of enclosing the Glaslyn valley arose as early as 1625 when Sir John Wynn of Gwydir wrote to Sir Hugh Myddleton on the matter. It is not necessary to discuss the various schemes put forward from time to time, but it should be borne in mind that many minor embankments were built before 1800. These banks enclosed areas of 50–1000 acres and were made of soil, sand, and turf. In places trees were planted. The details of the smaller embankments, as well as those of Madocks's bigger scheme, have been described in an unpublished thesis by W. M. Richards, now of

[1] The Borough Surveyor of Criccieth told me that just east of Graig Ddu shingle can be found at 6 or 7 ft. below the sand. So far as is known this is very local.

[2] At H.W.O.S.T. the rise of the tide at Portmadoc Quay is $12\frac{1}{2}$–$13\frac{1}{2}$ ft.; at neaps, 7–8 ft. The tides flow for about four and a half hours at the quay, and ebb about eight hours at springs and neaps in normal weather.

Portmadoc, entitled 'Dissertation on the history of Traeth Mawr and the industrial results of the formation of the embankment' (1925).[1] The enclosure map appertaining to Madocks's scheme is at the County Offices in Caernarvon. It is dated 1825, and shows clearly that considerable areas between Portreuddyn Castle and Tremadoc[2] had been enclosed before 1800. For example, W. E. Oakeley, of Tan-y-bwlch, reclaimed the Penmorfa area (1900–2000 acres) between 1798 and 1800, and on the eastern side of the Glaslyn valley a large area extending from Gareg to Gareg-hylldrem, and including Ynysfor and Ynys Fach, was enclosed soon after 1770. It is also probable that farmers and landowners had completed much minor embanking at a still earlier date, but records of this cannot now be found. Certainly no serious attempts were made before 1770.

The Act of the Enclosure of Traeth Mawr was passed in 1807 (47 Geo. III, sess. 2, cap. 36, August, 1807). The embankment ran from Penrhyn isaf to Ynys Towyn, and was engineered and designed by Thomas Payne, of Fenny Compton. It was foreseen that the pent-up waters of the river would complicate the building of the bank, and so the river was diverted westwards to its present course, and a bridge and flood-gates constructed. The bank was finished in September 1811, but in February 1812 it was breached by a gale. This breach, about 100 yd. wide, was closed by the end of 1814. Fig. 25 shows how the wall altered the appearance of the coast. The former island-studded estuary of sand banks and shallow water was converted into an area of flat land available for various agricultural purposes. It added a large area of not very valuable land to Caernarvonshire, but it hardly increased its natural beauty.

The building of the towns of Tremadoc and Portmadoc followed the enclosures at Penmorfa and the Portmadoc embankment. The harbour at Portmadoc was also built, chiefly as a convenient place from which to

[1] Mr Richards kindly lent me a copy of his thesis, to which I am much indebted for various details; a copy is kept in the Library of the University College of Wales, Cardiff.

[2] Fearnsides, *Quart. Journ. Geol. Soc.* 66, 1910, 142, notes that at the beginning of the nineteenth century there was a bathing place at Tremadoc. Speaking of the area between Graig Ddu and Portmadoc he says that no definite raised beach has been identified 'but the promontory of Graig Ddu seems to show a wave-cut platform about ten feet above most tides, and at Ynys Cyngar, Careg Cnwc, Gareg Goch...I think I can recognize a similar but less deeply cut wave-notch. The various Ynys about Tremadoc also show wide rock-platforms a little above the back of the Morfa, but it would require careful measurement to determine their relationships to the tide level outside the embankments.' The caves in Tan y graig are still fresh: Her Fyncourt was occupied in Roman times, and at Wern kitchen-middens show that the silting-up of the whole marsh was not more ancient. Fearnsides also thinks that Morfa Bychan began to grow at the same time.

export slates which had previously been brought down by pack-horse and waggons from Blaenau Ffestiniog to the Traeth Bach. Quays for this purpose existed at Tyddyn Isaf, Cemlyn, Gelli-Grin, Pentrwyn Garnedd, and Cei Newydd, all on the river Dwyryd. Thence the slates were taken in barges to Ynys Cyngar to be loaded on to larger ships. In August 1807 Madocks obtained authority to build a pier and quay at Ynys Cyngar, but as a result of the Traeth Mawr bank the rivers Glaslyn and Dwyryd cut a deep channel from a point near the present Ballast Island to Ynys Cyngar. This led to Madocks's construction of a new harbour, that is, Portmadoc, in 1821 (Act 1 and 2 Geo. IV, cap. 115). Hence the Traeth Bach quays were doomed, but they did not fall into complete disuse until a tramway had been built along the embankment from the quarries. The Act for the so-called Slate Railway was obtained in 1832.

Portmadoc was, and still is, a difficult harbour, and can only be used by small ships. The water is shallow, and the channels liable to big changes. After 1906 the Dwyryd took a separate course to the sea, and to-day the combined estuary of the Glaslyn and Dwyryd is drained by two channels.

(8) THE DWYRYD ESTUARY AND TALSARNAU

The Penrhyndeudraeth peninsula separates Traeth Mawr from Traeth Bach. Its coast is picturesque and rocky, with small sandy bays. The Dwyryd reaches the sea through a narrow glacial gorge, the natural beauty of which is spoiled by the road and railway bridge. Above the gorge the valley widens out, and was formerly tidal as far as Maentwrog. The river has been embanked, and much of the estuary reclaimed, but a certain amount of open marsh still exists in its lower parts. Small ships were built in this estuary in quite recent times, for example, *The Ocean Monarch* at Pentrwyn Garnedd in 1851. Lower down, at Abergafran, several ships were built between 1834 and 1858.

South of the bridge (Pont Briwfad) there is an extensive series of marshes. The village of Talsarnau[1] on the main Harlech-Maentwrog road almost certainly 'obtained its name from the junction of several trackways at a point where stands the small hostel called "The Ship Aground", whence a lane leads to the place of crossing of the Traeth Bach'.[2] This trackway was only passable at low water, and led north of Ynys Gifftan to the by-road, still marked on the current popular edition of the 1-inch Ordnance Survey, immediately south of the 'n' in Minfordd (**104**).

The foreshore from Pont Briwfad to the sea wall running from Llan-

[1] *Sarn* means a causeway.

[2] *Inventory of Ancient Monuments*, *Merionethshire*, No. 519, p. 163.

fihangel y traethau to the main road at Glan y wern is a unit. Before reclamation the tides flowed almost to the main road, but the sea wall now runs west of the railway. The reclaimed marshes are either cultivated or grazing land.

At Talsarnau the whole marsh is really a sward with a narrow *Juncus* belt near the sea wall. The sward is of the higher type, and consists mainly of *Festuca rubra*, *Agrostis alba*, and *Glaux maritima*. *Plantago coronopus* is found at the highest levels to some extent replacing *Glaux*. There is not much sward belonging to the lower levels,[1] but there has been recent rapid growth of the marsh which now reaches Ynys Gifftan. In this part the level is slightly lower, the sward is largely of the lower type, and *Glyceria maritima* dominant. The marsh is grazed by sheep, and, apart from creeks and pans, presents a close-cut lawn-like appearance. Turf is often removed. There is virtually no mud, and approximately the upper six inches form a soil intertwined with grass roots and of finer texture than the sand below. This is very evident along any creek, but more particularly in the northern part, where a meander of the Dwyryd is cutting into the marsh. The underlying sand is washed out and the more coherent soil layer falls down in blocks, often very rectangular in form. Along the main stream the fallen blocks almost have the appearance, at a distance, of an artificial revetting of the bank. The true marsh creeks are unlike those in Norfolk. This is partly due to the different vegetation, especially the absence of *Obione portulacoides*, and also to the absence of mud and fine silt. At the bottom of the creek there is nearly always a narrow and shallow drainage trench, which in the larger ones is usually clear of obstructions. The ease with which the upper coherent layer is undercut leads to natural dams forming in the creeks far more easily than in those of the Norfolk marshes. The result is that the smaller creeks are seldom continuous, but are interrupted by blocks of sward: only the higher tides rise over these dams. As time goes on the dams increase, and the original creeks are converted into a line of separate holes or pans lying in a slight depression of the marsh surface. The effect is rather like that described by Yapp in the Dovey marshes, but the lesser amount of silt at Talsarnau is an important differentiating factor. Round or oval primary pans are usual, and these are shallow and often plant covered, especially where the drainage is good. The sward is only covered by spring tides (**104**).

South of Penrhyndeudraeth station there is a similar area, but *Juncus* is more abundant. Changes in the relative rate of growth of these two areas are likely to be influenced by the swings of the Dwyryd west of

[1] See *Journ. Ecology*, vol. 5, 1917, in which Prof. Yapp describes the Dovey marshes.

Briwfad bridge. The narrow marsh fringe north of Llanfihangel y traethau is at present rapidly extending northwards. Here other plants are more common, especially *Armeria maritima*, *Aster tripolium*, and *Limonium humile*.

(9) MORFA HARLECH

Morfa Harlech and Morfa Dyffryn form the two most prominent coastal features of the northern part of Cardigan Bay. Morfa Harlech is a broad, triangular, sandy foreland largely devoid of any clear structural features (Fig. 26). Before discussing its probable origin, it will be well to give a general description of it. As it has grown northwards from Harlech the description follows as nearly as possible that direction and order (**103**).

The spit begins at the railway embankment, and gradually extends away from the old coastline, which is marked by a line of high cliffs formed of Cambrian rocks. In the first half a mile or so the spit is simply a dune area bordered by a belt of coarse cobble-stones. This shingle has travelled northwards about 450 yd. since the 6-inch survey of 1901,[1] and is the only visible shingle in the whole area. It may be pointed out here that the name Cefn-mine, probably a corruption of Cefn maene (=at the back of the stones), indicates a pebbly area under the marsh near that farm. Mr Morris, of Gwrachynys, remembers seeing stones near the marsh road bridge over the stream which flows eastwards from near Cefn-mine, but they could not be found in 1938. Their significance is uncertain: they may represent the remains of an isolated outcrop of boulder clay comparable with the small boulder-clay outcrop in Pwllheli Bay.[2]

Since no other surface shingle exists on Morfa Harlech, its growth cannot be traced by methods similar to those used for studies of, for example, the Culbin Sands or Scolt Head Island. The foreshore is bordered by normal, but wind-eroded dunes. Numerous blow-outs are found north of Cefn-mine, and between it and the shore are two lines of dunes separated by a slack which is wet in places.[3] The inner dunes are encroaching on the pasture land within, and the rate of movement since 1901 at the

[1] Figures such as these (also for Morfa Dyffryn) are very rough. It is not clear where the Ordnance Survey would take the outer edge of sand or shingle, nor is it certain whether in a revision of the map such points would be considered. Hence these figures must be treated with caution.

[2] On the other hand, it is more than probable that, before the railway embankment was built to protect the boulder-clay cliffs south of Harlech, a great deal of shingle was carried northwards. If so, this shingle now underlies Morfa Harlech; it is certainly not visible on the surface.

[3] All low-lying areas are, of course, much wetter in winter and, although on a smaller scale, somewhat resemble the winter lochs in the Culbin Sands (*Geogr. Journ.* 90, 1937, 498–528).

point where it is fastest can be roughly estimated at 12 ft. a year. On the ground they resemble a parabolic dune, but this term, if it is relevant, applies only to the north-eastern corner. There, near bench-mark 16·0 on the map, the blown sand spreads outwards as a flat area, and overlies former swampy ground. To the west are two well-marked lows with characteristic vegetation: in the one the dominant plant is *Equisetum variegatum* and in the other *Hydrocotyle vulgaris*.

There is also a fresh-water swamp in a deeper hollow that contains water at all times of the year,[1] although its level varies much with the season. Its vegetation is interesting, the most abundant plants being *Sparganium ramosum*, *Scirpus lacustris*, *Potentilla comarum*, and *Menyanthes trifoliata*.

North of this swamp the dunes decrease in height and fan out, but less clearly than if they rested on shingle ridges. The innermost and longest line has partly accumulated along an old wall. The southern part of this line consists of old grey dunes with an abundant growth of *Rosa spinosissima*. Low and wet areas, with hummocks of blown sand, separate this and the other dune lines from one another. The next one is formed of low grey dunes, and the outermost lines are of low shifting foredunes. This is an obvious case of a normal prograding shore, in which the various stages of evolution are not clearly marked because sand, often blown with violence, has not kept to definite lines. The outermost dune line can be traced for about a mile to the south, and forms a clear and new belt outside the older and higher foredunes.

The outermost zone of embryo dunes is interesting. Two factors at least play a part in its formation, but precisely how they work is by no means clear. In the first place along the whole foreshore of Morfa Harlech, especially at its north-western end, large amounts of seaweed are swept up on to the beach and left as hummocks. These become sand covered, and appear to start dune formation. But examination of many slightly older dunes lying close to, or actually in, this belt did not reveal any traces of seaweed. This may be in part accounted for by the fact that the seaweed soon dries and is blown away, while the sand which gathered in its rear may remain. It is not an entirely satisfactory explanation, but the connection between seaweed heaps and sand accumulation is so immediately clear that one must cite it as a cause of the development of embryo dunes. The second factor is that *Glyceria* takes root on the wide, sandy foreshore, outside any defined dune belt, gathers sand round it, and so begins to form a dune. The more exposed of these hummocks are often washed away, but this process is an important one.

[1] Ordinary lows may be enlarged blow-outs; the swamp seems to rest in an original hollow.

The vegetation of the older Harlech dunes need not be described in detail. At the northern end of the area *Agropyrum* is usually the pioneer, and there is abundant *Euphorbia paralias* on the young foredunes. Marram grass (*Psamma arenaria*) here, as in other coastal areas of the British Isles, is the dominant plant of all the dune systems, but on the older dunes *Rosa spinosissima* often covers large areas. *Erodium cicutarium*, also common, usually comes in rather earlier.

The curving round of the dune lines to the north-east resembles similar formations elsewhere. It is partly controlled by the fact that the northward growth of Morfa Harlech is now largely restricted by the swings of the river channel. A comparison of the few reliable maps[1] covering the last hundred years shows that the north-westward growth of Harlech Point has been some 220 yd. It seems unlikely that the spit will grow much farther in that direction.

Eastwards of the recurved dune ends, and between them and Llanfihangel y traethau, is a good example of marsh development. The pioneer plants on the bare sand are *Glyceria* and *Salicornia* spp. *Aster tripolium* follows, and is succeeded by a belt consisting chiefly of *Spergularia marginata*, two varieties of *Suaeda maritima*, and *Plantago maritima*. *Glaux* is abundant near the upper limits of the *Salicornia*. These three zones form the lower level. The upper part of the marsh consists mainly of a sward formed of the grasses *Festuca rubra*, *Agrostis alba*, and *Glyceria maritima*. *Juncus maritimus* is abundant at the highest levels. *Glaux* is ubiquitous except at the lowest level of both zones. Changes in the channel of the Dwyryd have caused local erosion of the marsh near Glan y Morfa farm.

Within the whole dune belt the rest of Morfa Harlech is a flat plain formed of sand.[2] West of the marsh road from Harlech to Llanfihangel there is a great deal of recently blown sand, and the area comes roughly under the category of sandy pasture. East of the road the soil is better, and, although it is used mainly for grazing, some is cultivable. Some of the lower areas, however, are still wet and marshy. For reasons that are discussed below we may assume that Morfa Harlech has grown from the south. The available evidence and comparison with present-day processes suggest that Tremadoc Bay was continually shallowed by material carried into it by the marine processes sweeping sand and other fine sediment into

[1] The maps used were: original 2-inch to the mile O.S. drawing of 1819, the 1-inch O.S., 1838, and the current edition of the 1-inch maps. Earlier maps and charts are not sufficiently reliable on which to base measurements.

[2] The numerous drainage channels (artificial and natural) cut in the Morfa all show its sandy nature. No boreholes have been put down to establish the depth of this sand, but it is probably considerable, especially in the northern part.

this part of the coast.[1] Outside the high-water mark of Morfa Harlech sand now gathers on the flats. This dries and is carried inwards by the wind to form dunes or hillocks around any suitable obstacle. It may be a long time before such hillocks unite to make a dune belt, but sooner or later this occurs. An examination of maps covering the last century shows that, whereas there has generally been seaward growth, there have also been times of retrograde movement. In the past, when the Morfa was less well developed than now, such vicissitudes were probably more frequent. Eventually the Morfa extended so as to include the former island of Llanfihangel y traethau.

The Act for the embanking of Morfa Harlech was passed in 1806, and soon afterwards a sea wall was built from the north-eastern corner of the high ground at Llanfihangel to Glyn Cywarch on the main road. A similar wall enclosed Talsarnau marshes. Before that time the tides flowed some way up inside the dune belt of Morfa Harlech towards Harlech Castle. Unfortunately there is no known record which describes the condition of the Morfa at that time. Ordinary spring tides reach to within about 4 ft. of the top of the wall; occasional storm tides may just overlap it. Apart from patches of blown sand and numerous field enclosures formed by sand and turf walls, the whole area is almost flat, and even now, if the wall were taken away, abnormal tides would reach far in towards Harlech.

Much may be learned from a study of air photographs[2] of the area which have been carefully examined for past changes. The area west of the marsh road is sand covered, and the photographs do not reveal any detail. But between the marsh road and the main road photography shows numerous traces of former marsh creeks which are invisible on the ground. These creeks[3] were traced from the mosaic, which was on a scale of 1/8160, and then transferred by pantograph to the map. It will be seen that they conform to the usual intricate pattern of marsh creeks. The present channel, taking the spring water from Harlech Castle, is inserted: in parts it is obviously canalized, but in others its course appears to be natural. Allowing for the fact that the whole area is now pasture or arable land, and that

[1] Prof. W.G. Fearnsides tells me that analyses of the deposits do not suggest that they come from the rivers. On the other hand, it is possible that some of the deep material underlying the surface originated in this way. The surface material is wind- and sea-borne from the open coast and offshore zone. The rivers do not bring any appreciable amount of detritus to the foreshore to-day.

[2] I am indebted to the Air Ministry for obtaining for me a complete set covering Morfa Harlech.

[3] It would have been possible to insert more creeks on this map, as others are faintly visible on the mosaic.

in places a good deal of blown sand rests on it, it is reasonable to infer that the creek system originally drained towards the one main creek into which the water from Harlech Castle now runs. The map shows that although there are exceptions there is nevertheless a general 'grain' in this direction. It would be unwise to press this point too far, but together with other arguments and evidence outlined below, it supports the hypothesis of a northward growth of the foreland, and bears directly on the view that the former port or harbour of Harlech was situated within a northward-growing sand spit. There seems therefore to be a strong case for supposing that, as the spit grew from south to north-north-west, it enclosed a slightly lower area which, up till the time of the final embanking in 1808, was largely open marsh and intersected by creeks. It is almost certain, in spite of the lack of records, that small enclosures had been made at an earlier date, but these would not invalidate the argument here outlined.

We must now turn to the problem of Harlech Castle, which was built by Edward I in 1286.[1] It is generally assumed that, as it possesses a water gate and stands in a position similar to Criccieth Castle, it had formerly easy connection with the sea. But in spite of a careful search, it has proved impossible to find any early document which clearly states the condition of the coastline when the castle was built. Many records exist about the castle, and some of the more important are printed in various volumes[2] of *Archaeologia Cambrensis*. In vol. 1 of the first series there is mention of a manuscript dated 23 September 1564, which twice refers to the castle as being near the sea and having havens, creeks, and landing-places on either side of it. Historians generally conclude that the castle was built with access to the sea in order to make victualling and garrisoning easier.

Mr C. L. Wayper, at my request, kindly made a careful search in the Record Office; although he was unable to find any complete account of former conditions, the following references are of considerable interest.

Public Record Office, C. 54. 141, Membrane 5, 15 April 1325. Beaulieu, 'To the bailiffs and community of the port of Hardelowe (Harlech)—to cause proclamation to be made forbidding anyone inflicting damage, wrong or annoyance by land or sea upon the men of Flanders, merchants and others, and that the bailiffs shall answer that all men treat favourably the said men coming into the realm with their goods and merchandise by that port.'

On 12 December 1325 (C. 54. 141, Membrane 16) there is another

[1] This is the date of construction of the masonry stronghold. A ditch was cut in 1285 and from the records it seems that a pre-Edwardian castle existed. See the official (H.M. Office of Works) guide to the castle.

[2] Apart from vol. 1, see especially vol. 3, 1848; vol. 4, 1858; vol. 13, 1867; vol. 6, 1875; vol. 13, 1913. See also *Trans. Hon. Soc. Cymmrodorion*, sess. 1921–2, p. 63.

reference to Harlech as a port, and on 10 May 1324 (C. 54. 141, Membrane 11) there is an order given at Westminster to the effect that all ships carrying 40 tuns of wine are to be prepared and found without delay so that they be ready to set out on the King's service on three days' summons. This is addressed to Sandwich and other ports including Hardelowe.

In C. 54. 145, Membrane 22, 28 June 1328, an order, given at Evesham, to Southampton and other ports, states that owners and masters prepare all ships of upwards of 40 tuns to be ready to defend the ports against malefactors from France; Hardelowe is again included. In (Record Office) Accounts E. 101. 16/40 Hardelowe is referred to as a port, and in Exchequer Accounts E. 101. 19/13, 9 to 12 Edward 3 are particulars of the report of Nicholas de Acton, who was appointed to guard the coast of North Wales and to arrest ships. It is a document concerning the arrest of vessels 'de portubus Carnarv. et Hard.'

The following quotation from the official guide to the castle by Sir Charles Peers and W. J. Hemp will serve to sum up current opinion:

> The purpose of the way from the marsh was undoubtedly to secure direct and safe access to the harbour or landing place, which, as is clear from medieval references, must have been at the foot of the rock. Throughout the active history of the castle the harbour was of vital importance, as the sea provided the most easy and natural—sometimes the only possible—means of communication with bases of supply. The present appearance of the marsh makes it difficult to realize that a harbour once existed at this spot. There is reason to believe however that the river Dwyryd, which flows directly into the head of the bay, formerly ran across the marsh at the foot of the high ground, passed close under the castle rock, and fell into the sea some little distance to the south-west of it.

These general conclusions seem correct, except for the lack of evidence that the Dwyryd ever took the course suggested above. Such an opinion is derived mainly from W. Ashton's book, *The Evolution of a Coast-line*, which, although of interest, is often misleading in its interpretations. Ashton bases his view first on local hearsay evidence of a former stream bed about a mile north-east of Harlech, and secondly on the fact that Talsarnau village grew up at a crossing of the stream. Thirdly, he attaches great importance to the inset map of Harlech Castle on Speed's map of Merionethshire. It is doubtful, however, whether Speed's map can be relied on for detail such as this, and the broad river Ashton cites on this inset plan is probably the open water of Traeth Bach and not the southward-flowing Dwyryd.[1]

[1] It can be argued with like ease and conviction that the arm of the sea running inland towards Harlech and east of Llanfihangel y traethau on Speed's county map of Merionethshire, represents a former creek. The inset map hardly agrees with the smaller-scale county map, and neither, in my opinion, supports Ashton's interpretation.

Other evidence comes from studies of the air mosaic and of the shore processes and features in this part of Cardigan Bay, which show that the drift of beach material is northwards. In the thirteenth century, probably shortly before Edward I built the castle, we may reasonably suppose that what is now Morfa Harlech began as a small spit south of Harlech. As time went on the Morfa gradually grew until it reached the condition it was in before the embanking in 1808. It is clear from what can be seen now and from the air mosaic that marsh creeks, probably of some size, existed within the dunes. Thus the havens and creeks referred to may well be those of the marsh within the spit, which towards the end of the thirteenth century had grown only a short distance from its starting-point. With the conquest of the Welsh assured and also with the continued growth of Morfa Harlech the castle declined in importance. But probably for some centuries it was still possible for small boats to come round the north of the growing spit, which could only have reached the high ground of Llanfihangel very gradually. They could even have sailed round the high ground itself, and have made their way up towards Harlech. After the final embanking all such traffic would have been shut off and the marsh would soon have attained something of its present form. Possibly the Dwyryd formerly flowed south of Llanfihangel as it did to the south of Ynys Gifftan as late as 1816; but there is no evidence of this, and a study of the local physiography suggests that a course north of Llanfihangel is more probable.

The hypothesis outlined here is consistent with the use of Harlech as a port in the reign of Edward I, and with the processes of coastal evolution. It also agrees with the fact that in the fifteenth century and later there is no further mention of Harlech as a port, although this may be due to purely historical reasons. The existence of a tidal creek or creeks joining the Dwyryd between Talsarnau and Llanfihangel and its gradual silting-up seems best to explain the few known facts.

Before leaving the Harlech area a word should be said about the 'cliff' which runs from Harlech to and beyond Talsarnau. At one time the sea washed its foot, but it is not a simple cliff cut at present sea level. It is a vast structure: the amount of erosion required to produce it would be enormous, and it is in one of the most sheltered parts of Cardigan Bay. The cliffs immediately south of Harlech, now protected by the railway embankment, are cut in boulder clay, and are probably due to marine erosion. It is also pertinent to point out that the high ground behind Morfa Dyffryn, which at one time must have been washed by the sea, shows little, if any, evidence of cliffing, although it was originally in a much more exposed position. There is minor cliffing on Lasynys and a normal amount on what was the exposed western side of Llanfihangel.

(10) MORFA DYFFRYN

Morfa Dyffryn is in many ways similar to Morfa Harlech (Fig. 27). It is an extensive sand flat attached to the old coast, and fringed on its outer side by dunes although in the southern part of the area there is little dune growth. From the railway embankment at Llanaber a tongue of shingle extends northwards as far as the main exposure of the submerged forest, but the supply of shingle is now largely cut off by the railway works. Furthermore, between Llanaber and Tyddynmawr the shingle has in recent years been pushed in some places about 80 yd. landwards over the marsh, and it has also been breached, so that a gap of some 300 yd. now virtually separates it into two main masses. Within this southern area, and bordered inland by the railway, is a fresh-water swamp containing abundant *Scirpus lacustris* and *Phragmites*. Occasional incursions of salt water flow into it in storms and with exceptional tides. North of this swamp the shingle extends along the coast as far as the mouth of the Afon Scethin, which it has slightly deflected to the north. The sand on the foreshore, in its northward movement, also helps in this deflection. For about half a mile south of the river there is a narrow and wind-eroded dune belt, which the shingle protects from direct sea attack. There is some shingle north of the Scethin mouth, and this has advanced rather more than a mile since the survey for the 6-inch Ordnance Survey map of 1901.

For about a mile north of the Scethin the dunes form a fairly simple foredune system, much cut up by both marine and aeolian erosion. They are advancing inland almost throughout their length, and the flat ground within them is best described as sandy pasture. But north of Dyffryn golf links the dune system widens and forms the best example of wind-blown sand along Cardigan Bay. There is first an outer and much eroded line of foredunes, and in one short stretch they are of peculiar interest. The prevailing wind is south-westerly, and thus the western side of the dunes presumably should suffer most. But along the stretch in question, although small blow-outs have intersected the dunes, it is their leeward sides which are eroded away leaving a narrow ridge 20–30 ft. high, which falls seawards as a marram-covered slope. The lee side is precipitous and is being destroyed. It appears as if the wind, after making the blow-outs, had widened them sideways, and so caused the removal of the sheltered side of the ridge. The new sand blown from the foreshore replenished the windward side and, with the help of marram grass, is maintained *in situ*. Interesting and bizarre forms are seen in the dunes all along this seaward side of the main dune area. In one place a high pyramid of unvegetated sand was a prominent landmark in 1938, but as it is unprotected, it cannot last long. The inner edge of these dunes is advancing over a large low, containing much *Juncus acutus* and *Salix repens*, which separates them from a prominent inner line of moving dunes. These, like the corresponding feature on Morfa Harlech, are formed largely of bare sand and are advancing over the pastures within.[1] They represent an old foredune system outside of which the present foredunes have originated on a prograding shoreline. The inner dunes are often pitted in their older and better vegetated parts by deep and narrow blow-outs, most of which seem to have been started by rabbits.

At the northern end of this system there is a broken mass of old dunes enclosing small lakes. These dunes formed alongside the former course of the Afon Artro before it was deflected in about 1819 to its new mouth north of Mochras Island. The lakes or ponds lie in low hollows to which small streams drain, and are comparable to the fresh-water marsh on Morfa Harlech. In their deeper parts (i.e., more than 1·5 ft. in September 1938)[2] *Scirpus lacustris* and *Sparganium ramosum* are dominant. *Chara* is found nearly all over the floor of the ponds. The shallower marginal parts of the deeper pools have the same flora as the shallow pools nearby—namely *Heleocharis palustris* (dominant) and *Litorella lacustris*, with *Chara* covering the bottom. The old mouth of the Afon Artro is blocked at its seaward end by the foredunes which run in an unbroken line until they merge with Mochras, and within them the level falls to the marshy ground described below. Apart from the lakes already alluded to, there are also several well-marked *Salix* lows,[3] and alongside the lakes there are one or two small lows which would be wet in winter.

Mochras Island is formed of boulder clay and is an old moraine. It bears to Morfa Dyffryn a relationship similar to that existing between Llanfihangel and Morfa Harlech. The clay supplies abundant material for boulders and coarse shingle which spread over all the beach alongside it. This shingle travels northwards, and the north-eastern part of Mochras is a long shingle spit (recently protected) with dunes which curves round in typical fashion at the mouth of the river. At low water only a shallow channel exists, and the shingle really continues north of the river, where it forms another shingle-and-dune spit running northwards by Llandanwg

[1] The approximate rate of travel where progress is fastest, as deduced from our survey and the current edition of the 6-inch map, is about 20 ft. a year. But since the Ordnance Survey does not indicate clearly the inner edges of dune belts, estimates such as this are open to correction.

[2] As on Morfa Harlech, there is much more water in these ponds in winter.

[3] Lows with *Juncus acutus* occur; they may be of rather less height than the *Salix* lows.

church to join the boulder-clay cliffs south of Harlech. It is therefore correct to say that the shingle which begins at the southern end of Mochras is continuous with that outside Morfa Harlech, the places where it is less in amount being accounted for by artificial changes.

Within Mochras is a broad sand flat which fills at high spring tides. Along its southern edge the vegetation is of much interest, and is largely of fresh-water origin. It is occasionally flooded by the tides and so represents a transitional stage. The following species are co-dominant and grow in juxtaposition: *Scirpus lacustris*, *Phragmites communis*, *Scirpus maritimus*, and *Juncus maritimus*. These grade northwards into normal salt marsh which is growing over the sand flats. *Salicornia* spp. and *Statice* (*Limonium*) *humile* are common, but are scattered and do not form a close cover. The best development of salt marsh in this area occurs north of the Afon Artro and near Llanbedr and Pensarn station, where the transition is clear from the lowest or algal zone, through scattered *Salicornia*, to a *Glyceria* sward, and finally to the *Juncus* belt. A small amount of inning has been done north of the river, but the sea wall follows the eastern side of the flat within Mochras and runs back to the main road. Outside this wall, and facing Mochras, the salt marsh, though narrow, is well developed and shows steps or 'cliffs' resulting from wave erosion at high water. The straight stretch of road known as Sarn Hir is in itself an embankment and was built before 1806.[1]

The whole area within the Dyffryn dunes closely resembles the corresponding area on Morfa Harlech and need not be described in detail. North of Dyffryn village the ditches drain to the north. South of the Afon Scethin the drainage is more irregular, and a small and unnamed stream breaks through the shingle about a mile south of the Scethin. Llandanwg church was certainly not built when the sea was so close to it. It is now disused, and, but for clearing, would be nearly covered by dunes. There is no known record which describes its position relative to the sea when it was built.[2]

The submerged forest series near Llanaber was excellently exposed in 1938. Part of it has been described by T. N. George,[3] who says that the underlying 'silt and clay rest at a height of about six to eight feet below high-water mark at its landward edge, but sloping noticeably seawards'. Later he writes: 'For although the peat is not present above the silts at Barmouth and Llanaber,...the presence of rootlets in the silts...is conclusive evidence of the one-time extension of an ancient soil over that

[1] I do not know the precise date.

[2] See *Arch. Camb.* vol. 90, 1935.

[3] *Proc. Swansea Sci. and Field Nat. Soc.* I, 1933, 187, 188.

ground, since removed by denudation.' In 1938 there was a considerable area of peat exposed. The top surface of it was levelled and connected with a bench-mark on the main road. This established the fact that the top surface of the inner edge (i.e. the highest part) of the peat was between 13 and 14 ft. above Ordnance Datum. An excavation was made through the peat at this point, and showed that the whole layer was about $5\frac{1}{2}$–6 ft. thick. Throughout this particular section the peat was typical of a shallow-water and possibly brackish lagoon, and consisted mainly of the remains of *Phragmites* and similar plants. The shingle had been removed from this upper surface. Seawards there was a small cliff, about 2 to 3 ft. high, and then came the main exposure. On the upper surface of this lower platform there are numerous alder stools, many of them in the position of growth. Careful observation and excavation showed that they were all in a layer between 18 in. and 2 ft. above the underlying blue clay. Ordinary neap tides would scarcely cover this lower layer; spring tides would reach up to, and probably slightly above, the topmost layer. Behind the peat is a lagoon, and it is probably correct to say that much of the peat accumulated under conditions somewhat similar to those now prevailing in the lagoon. It seems unlikely that this peat bed is identical with that exposed at lower levels on the coast of Cardigan Bay (**102**).

South of Llanaber Halt, which stands on a minor salient of the coastline, there is at Barmouth a slight residual travel of beach material to the south. This appears to be local, and is often mildly evidenced by the groynes along Barmouth promenade. The reason for this travel of material remains obscure especially as the alteration in the angle of the coast at Llanaber seems too small to be the effective cause: it is a feature more noticeable on the ground than on a map. Barmouth itself stands on a narrow flat, probably similar in origin to that of Morfa Dyffryn. Although no authentic records are available, it is known that a certain amount of erosion has taken place at the southern end of the town since it has been a place of any importance. In the years 1550–1603, when it was called Abermowe, it consisted of four houses, and possessed but two little boats.[1] From Ynys y Brawd, at the mouth of the Mawddach, a pebble ridge runs westward, and consists of glacially transported material with many marine shells. It has been suggested that it is a relic of a raised beach.[2]

[1] E. A. Lewis, 'Welsh Port Books, 1550–1603', Cymmrodorion Record Series, No. 12.

[2] T.N. George, *loc. cit.*

(11) FAIRBOURNE AND RO WEN

Ro Wen is a typical storm beach of coarse pebbles, and near its distal end shows several minor recurved ridges (Fig. 28). It is unlike Morfa Harlech and Morfa Dyffryn, but is the counterpart of the spit across the mouth of the Dovey. It has been built by northward-directed beach drift, and as a result of storm attack is set well back from the general trend of the coast from Llwyngwril to Fairbourne. Throughout most of its length it is covered by simple, but much eroded and patchy dunes. This is often due as much to human interference as to natural causes.

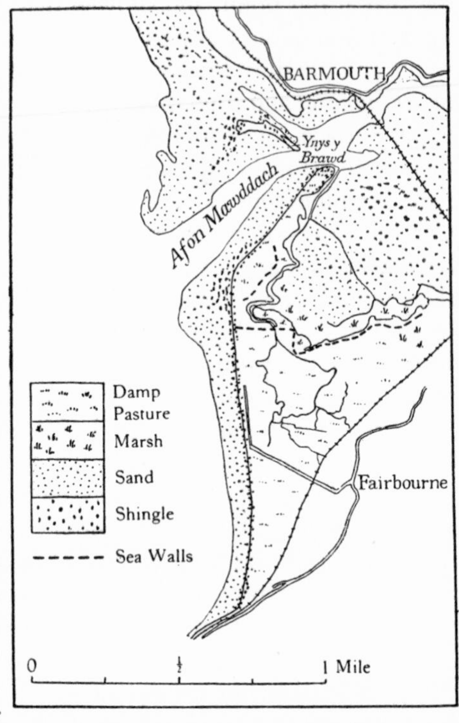

Fig. 28. Fairbourne Marshes and Ro Wen

Within the spit is an interesting development of salt marsh which is more muddy than most others described in this chapter: this characteristic is probably due to the river. Sheep graze on the unreclaimed marsh and give it a lawn-like appearance. The marsh rests on bare sand flats colonized over large areas by *Salicornia* spp., near the inner edge of which a fair amount of silt has accumulated. The next higher zone includes *Glyceria* as the dominant plant, with *Suaeda maritima*, *Salicornia* spp., *Armeria*, and *Glaux*. There is no clear separation between these two lower zones. A slight break of some 4 or 5 in. introduces a third zone characterized by *Armeria* (dominant), *Festuca*, *Spergularia*, *Glaux*, and *Suaeda maritima*. The highest zone, about 1 ft. above that just described, is shown by the dominance of *Festuca*. The other important plants are *Armeria*, *Plantago coronopus*, *Glaux*, *Spergularia*, and *Agrostis alba*. On the bare mud flats nearer Barmouth Junction station are three patches of *Spartina Townsendii*. Whether these were introduced naturally or artificially is not known. On a small scale this marsh is similar to those of the Dovey. Pans and creeks occur, but are not very well developed. Reclamation has been carried out, and most of the area is now agricultural land. These marshes were once continuous with those at Arthog, but are now separated by the road and railway. Near Arthog station is a fine development of an *Aster* marsh, and near Fegla Fawr is an excellent example of a raised bog with characteristic vegetation, *Molinia* being the dominant plant (**101** and **100**).

(12) FAIRBOURNE, TOWYN, AND THE DOVEY (DYFI) ESTUARY

The estuary of the Mawddach is in Cambrian beds, and on the north side is the Harlech dome. The estuary, like those of the Dyfi, Glaslyn, and Maentwrog valleys, is one of the drowned embayments of Cardigan Bay. The coast south of Fairbourne[1] is steep; a direct continuation of the cliffs on the south side of the river. The Gamlan Beds outcrop from Fairbourne to Llwyngwril, and south of Llwyngwril the Ffestiniog and Dolgelly Beds; they form a cut platform at beach level. At Llwyngwril there is a large gravel fan which supplies much shingle to the beach, and presumably has helped to form Ro Wen. The Dysynni River finds its outlet through extensive shingle masses piled up into a storm-beach. Here, again, the outlet is deflected somewhat to the north by the prevailing drift in this part of the bay. Within the shingle is a wide area of alluvium, and Broadwater is a natural lake-like expanse of the river lying partly in the alluvium and blocked back by the shingle. The Dysynni and Fathew valleys are also drowned, and now filled up with alluvial deposits.

The Dyfi estuary lies in the slates, grits, and sandstones of the higher Ordovician and Lower Silurian rocks, all of which are deficient in lime. All are strongly folded, and the general strike is north-north-east to south-south-west. The rather straight north shore of the estuary probably coincides for some four miles eastward from Aberdyfi with the Llyfnant fault.[2] Furthermore, in this area the long southward slopes are in accord with the pitch of the folds. Before discussing the nature of the estuary a problem treated more fully later on needs preliminary comment. Between Towyn and Aberystwyth we see well the two plateau features that characterize so much of Wales. The coastal plateau forms a strip up to eight miles wide south of the Ystwyth: near the coast it is between 400 and

[1] B. Jones, 'The Geology of the Fairbourne-Llwyngwril District', *Quart. Journn. Geol. Soc.* 89, 1933, 145. Locally, especially just south of Fairbourne, the cliffs are very steep and often undercut. The beach consists mainly of coarse boulders and is very difficult to traverse.

[2] O.T. Jones and W.J. Pugh, 'The Geology of the Districts around Machynlleth and Aberystwyth', *Proc. Geol. Assoc.* 46, 1935, 247; see also (same authors) *Quart. Journ. Geol. Soc.* 71, 1915, 343.

600 ft. high and rises inland. Its inner margin is not well defined, and it gives place inland to the high plateau which rises to more than 2000 ft. in the north and sinks to about 1800 ft. southwards. It is dominated by Cader Idris and other higher hills. In general the rivers flowing to Cardigan Bay are associated with the lower surface, and those flowing to the Severn with the higher one. A later uplift caused the rivers to incise themselves in these plateaux so that often all traces of the early mature stage of the valleys has been lost.

The history of the Dyfi estuary is probably similar to that of the others farther north. O. T. Jones[1] holds that the coastal area was given its present peneplain form during the Tertiary period, and that later uplift and valley down-cutting ensued. Since the area was glaciated, ice has somewhat modified the detail of the valleys, and it is clear that large amounts of boulder clay were left in them. It remains to be proved, however, whether the Dyfi valley was excavated to its present depth by glacial erosion or by the river when the land stood higher than it now does. The submerged forest at Borth clearly proves some submergence, and when the railway bridge at Barmouth was built a peat bed was found at 55 ft. below H.W.O.S.T. Jones considers three possibilities:

(1) If the subsidence were mainly post-Glacial, then at the end of the Glacial period the Dyfi valley would have been filled with boulder clay with, presumably, an uneven surface. The sea would have cut cliffs in this and the boulders in it would soon have led to the formation of a shingle barrier enclosing shallow flats. The main river here (as in the other estuaries) would have flowed in its lower course over the boulder clay which at that time certainly extended some distance out into Cardigan Bay.

(2) If no subsidence took place, there would have been steady erosion of the boulder-clay cliffs and, as before, storm-beaches would have been formed and would have moved gradually inland.

(3) If subsidence and deposition more or less balanced, there would have been no extensive submerged area at any stage. Both the storm beach and the head of the estuary, and also the boulder-clay cliffs on the coast to north and south, would have moved slowly landwards, but all the while the general appearance of the area would not have varied much. In other words, a subsidence taking place under these conditions would not have attracted much attention. But if subsidence were greater than alluviation, the head parts of the estuary would sooner or later have been drowned, and in time quite an extensive area would have been flooded; once the downward movement ceased, however, sedimentation would gradually have led to dry land being formed. If this were the case, the

[1] In R.H. Yapp, D. Johns, and O.T. Jones, *Journ. Ecology*, 5, 1917, 65.

nature of the sediments would clearly yield a reasonable account of the estuary's history. But a factor applicable not only to the Borth forest, but also to others, is that relatively rapid subsidence seems to have brought them to their present levels. Moreover, once the estuary had filled with sediments to a level nearing that of the water, salt-marsh plants would have spread as they do to-day. In general, what is known of the estuary and the Borth forest suggests that the subsidence was spasmodic.

The boulder clay in the cliffs clearly implies that that material formerly covered large areas of Cardigan Bay: the evidence deduced from the Sarns points to the same conclusion. There is every reason for thinking that the boulder-clay surface was extensive, and that a submergence such as brought the Borth forest to its present position would cause much to disappear. Then, as the submergence probably did not exceed 200 ft. as a maximum, it follows that the drowned area would only have been covered by relatively shallow water which would have enabled wave erosion to use the included boulders and to build them up into the present storm beach. Once a protective beach had formed and stable conditions had been attained, silting would have continued and eventually the present marsh surface would have developed, diversified by the ynys (islands) of solid rock which here and there stand out from the tidal flats.

The submerged forest has been investigated:[1] it is often completely sand covered and suffers much from erosion. At Borth and Ynyslas the general nature of the peat deposits is similar, brown peat overlying clay and gradually merging into it. Pollen analysis applied to the peat gives the following (generalized) sequence. The lower half has a dominance of alder pollen, followed first by a phase of birch dominance, then a pine maximum, and finally a return to high values for alder and birch. The basal clay is penetrated by *Phragmites* and shows a very low ratio of tree to non-tree pollen of monocotyledonous type. It is suggested that the development of the 'forest' took place as follows: (1) the formation of fen over brackish-water basal clay with the growth of alkaline peat with *Phragmites*; (2) fen alder woods grew on this peat; (3) growth of fen woods *in situ* especially alder, then birch and finally pine; and (4) *Sphagnum* peat possibly forming a raised bog (cf. Cors Fochno) above the forest peat. Unfortunately, dating cannot be precise, except that the forest is post-Boreal, and pollen analysis gives no evidence of the way in which the forest came to its present position.

Separating the forest from the salt marshes and raised bog (Cors Fochno) within, is the great storm beach of large pebbles, capped and extended in its northern parts by sand dunes (Fig. 29). This beach is

[1] H. Godwin and L. Newton, *New Phytologist*, 37, 1938, 333.

throughout most of its length a single ridge, but it gives off short laterals towards its distal end. It is continued by extensive sand flats on which dunes have accumulated; these are largely but by no means completely grass grown. Like Ro Wen and the shingle spreads on Morfa Dyffryn and Morfa Harlech, it has grown northwards: indeed, the same agencies which are held to be responsible for those formations, have been active in this storm beach.

Within the ridge there are two distinct features, Cors Fochno and the Dyfi saltings. A boring put down in 1933 through Borth Bog showed

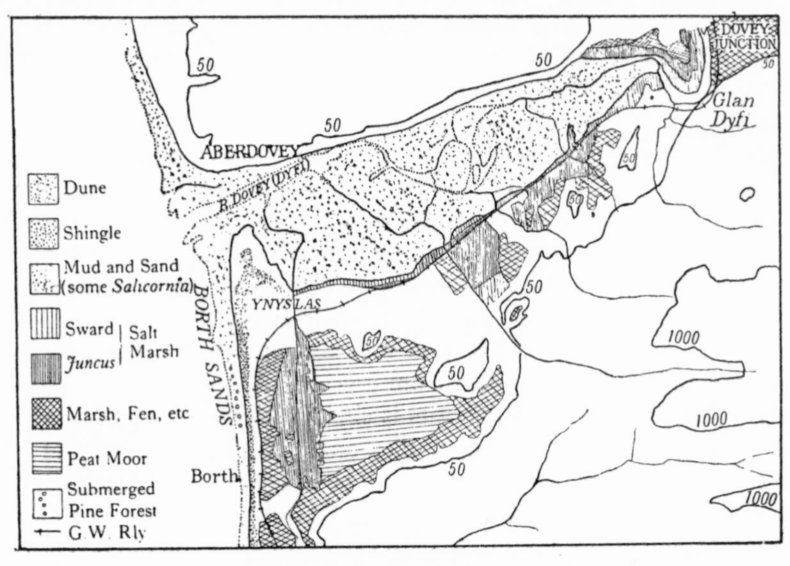

Fig. 29. The Dyfi Estuary (based on R. H. Yapp, *op. cit.*)

results generally similar to those in the submerged forest. There are (1) a clay base containing foraminifera, (2) the lowest peat samples show a low tree/non-tree pollen ratio, suggesting (as the non-tree pollen is monocotyledonous) a fenland area into which distant tree pollen was carried, (3) the samples next in height show a high tree/non-tree ratio with a well-marked preponderance of alder and birch (as at Ynyslas the alder maximum precedes that of birch), and (4) the three upper samples show a fall to low tree/non-tree ratios, thus corresponding generally with the *Sphagnetum* of the submerged forest. It appears, therefore, that the forest and the bog followed similar lines of evolution. The higher pine values at the base of the bog suggest a date of the same age as that of the Ynyslas peat. The curves for *Ulmus* and *Tilia* also imply that the upper part of the bog is later in date than the corresponding part of the forest sequence, and may indicate the beginning of the sub-Atlantic period.

Allowing therefore for local details, the forest and bog are contemporaneous; presumably, also, the forest is merely the continuation of the bog seawards, and is now exposed on account of the cutting back of the former boulder-clay coast and the over-rolling of the shingle bar. Incidentally, the belt of *Pinus sylvestris* across the middle of the moor may be the direct descendant of the pines in the submerged forest. Whilst there is naturally no certainty about this, it is noteworthy that opinion used to limit *P. sylvestris* to Scotland in these islands. The trees, however, may easily have colonized the moor from some plantation.

We must now turn to the salt marshes of the estuary.[1] These are of considerable interest since they were amongst the first in this country to be studied carefully. Again, allowing for local differences they are characteristic of Welsh coastal marshes and serve as an example for useful comparison with those of East Anglia. The setting of these marshes in the estuary surrounded by hills and mountains and enclosed by the storm beach[2] is particularly beautiful.

Two main types of vegetation on the marshes are distinguishable: a close dwarf sward with numerous channels and salt pans, and a taller variety mainly characterized by *Juncus*. Both rest on estuarine silt which here is more sandy than in many estuaries, the proportion of sand increasing seawards. Near the surface the silt is brownish, but it becomes greyer below. Five main plant associations are recognizable:

5. *Juncetum maritimi* } highest
4. *Festucetum rubrae* }
3. *Armerietum maritimae*
2. *Glycerietum maritimae*
1. *Salicornietum europaeae* lowest

Of these 2, 3, and 4 form the sward. No. 1 is an open association on bare silt: it is often absent, especially near the head of the estuary. No. 2, the lowest zone of the sward, is mainly characterized by *Glyceria* with many dwarf *Salicornias* and some other plants in the higher parts. Usually it occupies but a small area except where the vegetation is extending over bare silt. It is a pioneer association, open at first, becoming closed later. *Armeria* is the dominant plant of the middle zone, but in the lower and badly drained parts *Glyceria* is dominant. This grades easily into the fourth zone in which *Festuca rubra* is dominant, but it is often possible to sub-

[1] R.H. Yapp, D. Johns, O.T. Jones, *Journ. Ecology*, 5, 1917, 65.

[2] The storm beach corresponds with the offshore bars of Norfolk and the Moray Firth.

divide this belt into the *Lower Festucetum* which contains much *Armeria*, and the *Upper Festucetum* in which *Festuca* is dominant, but is mixed with abundant *Agrostis*, *Juncus Gerardii*, and *Plantago coronopus*. *Juncus* dominates the highest zone, but downwards it grades into the *Armerietum* and the *Lower Festucetum* in such a way as to bring in plants like *Spergularia media*, *Triglochin maritimum*, and *Aster tripolium*, which are rare in the *Upper Festucetum*. The rushes in fact allow these plants to grow much bigger than is possible on the sward. Needless to say, although the zones are all clearly shown, there are no sharp transitions between them. The vertical range from the *Glycerietum* to the *Juncetum* (both inclusive) is rather less than $4\frac{1}{2}$ ft. The *Juncetum* is only occasionally submerged, the average spring tide just covers the *Lower Festucetum*, and ordinary neap tides do not cover the sward at all. In striking contrast to Norfolk it will be noticed that neither *Obione* nor the various sea-lavenders grow in the Dyfi.

As in other areas of this kind *Salicornia* plays a relatively small part in helping directly to build up the marsh. *Glyceria* is the first real colonizer and becomes tufted as the level rises. Later *Armeria* and other plants arrive. The primary hummocks grow upwards and outwards and so sooner or later coalesce. The water ebbing and flowing over the surface is gradually limited to more and more definite channels, and frequently the vegetation spreads in such a way as to enclose shallow hollows, or pans. In the channels coarser material is often deposited and so a truly uniform level is never attained. It will, however, be better to discuss the formation of the detailed features in this and other marshes in a separate chapter, with the exception of one point which needs stressing here. On account of the wanderings of the main channel of the Dyfi in the estuary, and also of the erosion produced by small waves during times of high water in the estuary, small cliffs are commonly formed in the marsh deposits. These are by no means peculiar to the Dyfi marshes, but they are perhaps more than usually conspicuous there. The first measurements of the rate of marsh accretion in these islands were also made in these marshes: their nature and significance is discussed in Chapter XIV.

(13) THE SARNS AND CANTREF-Y-GWAELOD

Before continuing with the survey of the coastal features south of the Dyfi it is appropriate to give an account of the Sarns[1] in Cardigan Bay. They have given rise to a great deal of controversy in which imagination has

[1] 'The Origin of the Welsh Legends', O.T. Jones, *The Welsh Outlook*, Jan. 1941; 'Archaeology and Folk Tradition', H.J. Fleure, *Proc. Brit. Acad.* 17, 1931; 'Cantref-y-Gwaelod', D.J. Davies, *Cardiganshire Antiq. Soc.* To obtain a correct impression, the Sarns should be studied on Admiralty Charts Nos. 368 and 1411, Cardigan Bay.

played a large part. There are five of these 'causeways': Sarn Badrig, or St Patrick's Causeway, runs seawards from Mochras Island for an extreme length of twenty-one miles, nine miles of which may be exposed at a very low ebb. It is formed of loose rounded stones and boulders and is narrow. Sarn y Bwch, or the Buck's Causeway, is midway between Aberdyfi and Barmouth. It is visible at ebbs for about one mile from the shore. Sarn Cynfelyn is about two miles north of Aberystwyth and is seven or eight miles long, while Sarn Dewi near Llandewi-Aber Arth is only about a quarter of a mile long. Sarn Cadwgan is a mile south of Sarn Dewi, and is about one and a quarter miles long. Tradition has it that some of these sarns represent remains of masonry.

The whole question of the sarns is related to the legend of the lost land of Cantref-y-Gwaelod. The first record of the inundation of this land seems to be that in the Black Book of Carmarthen which belongs to the early twelfth century. The Black Book puts the blame for the flooding on one Margaret or Meredig 'who, at times of feasting, allowed the waters of a magic well, under her charge, to overflow the country', while popular tradition usually charges Seithennin, a renowned drunkard, with neglect of the sluices. By the sixteenth century, however, the legends varied a good deal, although more definite charges are made against Seithennin than anybody else: he was supposed to have allowed the flooding to take place when he was drunk. It was not until 1662 that Robert Vaughan of Hengwrt, in *British Antiquities Reviewed*, connected the story of the inundation with Sarn Badrig. The same story is taken up, with slight variations, by Lewis Morris (1702–65), in his *Celtic Remains*, and by W. D. Pughe in his *Cambrian Bibliography* (1803). T. L. Peacock, in his novel *The Misfortunes of Elphin*, also describes the 'scene', and again blames Seithennin. Thus in the eighteenth and nineteenth centuries there was a wealth of myth and legend all tending to assert that the lands of Cantref-y-Gwaelod had disappeared in historic times as a result of carelessness. It is not therefore surprising to find that a specific date is assigned to the event, and in Sir S. R. Meyrick's *History of Cardiganshire* A.D. 520 is given as the fatal year. Historic fact, however, does not support such a statement. The Romans do not appear to have occupied any part of the present coastal areas between 55 B.C. and A.D. 409: no reference is made to the lost land in Antonine's *Itinerary* (second century) nor by Ptolemy, neither does Giraldus Cambrensis (1188) make any mention of the catastrophe.

We are consequently faced with an interesting problem. All modern scholars are emphatic in their view that the Sarns are natural formations and not in any way whatever to be associated with ruined walls or buildings. On the other hand, we know quite definitely, from studying the submerged forests of Borth and other places in Cardigan Bay, that a considerable subsidence has taken place. Furthermore there is every reason to believe that much of Cardigan Bay was occupied by a low-lying boulder-clay plain before the subsidence took place. What then is the significance of the legend of Cantref-y-Gwaelod? It is true this was much embellished from the sixteenth century onwards, and that a tremendous amount of nonsense was written by people who had vivid imaginations. But are there any means, after discarding the more recent additions to the tales, of correlating folklore and physiography?

It seems clear that somehow a tale has come down to us from antiquity which refers to a former extension of land in Cardigan Bay. Legends of the overflowing of wells or lakes are very common in the folklore of Ireland, Wales, Cornwall, and Brittany. Some of them, like that of Cantref-y-Gwaelod, also refer to flooding of lowland areas by the sea. Now these are really fairy tales, or rather folk-tales, and occasionally a definite basis of fact can be substantiated. By way of example I cannot do better than quote O. T. Jones:[1]

> The numerous interesting features which these tales have in common seem to suggest the existence in past times of people who were supposed to live in lakes, were small of stature, disliked iron, possessed few articles of furniture, had not learnt the art of making bread, disliked the greensward being broken up by the plough, were successful in tending animals, had a limited ability to count, and probably used a language of their own which no one else understood. They appear also to have reckoned descent in the female line. If this suggestion is accepted, these characteristics are consistent with a pastoral people in a primitive stage of culture, who were not acquainted with the use of iron, but who attributed great importance to stocks of cattle, sheep and goats, and knew little or nothing about tilling the ground and the growing of corn.

Folk-tales, touched up and modified, may be handed down over long periods of time, and despite such alterations they can still contain the germ of truth. The Bronze Age inhabitants of these islands were often associated with pile dwellings, and this probably explains why so many of the Celtic tales have the theme of people living in or near lakes. The later Iron Age peoples, with their superior weapons, must have terrified these earlier groups, and in this reaction we may possibly see the reason for the frequent references to iron in the legends.

We may now turn to the anthropological side of the problem. Fleure and James[2] made a careful study of the peoples of certain specified areas in western Wales. They found a distinct type associated with the more remote hill lands: short and dark and rather long headed. They are still

[1] *Op. cit.*

[2] *The People of England and Wales.*

most frequent in those districts in which traces of Neolithic man in particular are common. A second type is connected with the Mawddach valley, Towyn, and some other places. These are in the area associated with Bronze Age pottery, and may possibly be identified with a Bronze Age race. The distribution of these people suggests that they were later comers than the dark hill folk, and, in fact, pushed in between them like a wedge. Later, the use of iron provided implements for the clearing and tilling of the valleys, and also sharply distinguished the lowlanders from the very poorly equipped hill peoples. It may be that this inequality led the hill people to withdraw to the remoter parts and to adopt or elaborate magical practices in self-defence.

So it is possible that many of these tales go back at least as far as the first contacts of the iron-users with the earlier inhabitants of Wales: others may date back to the still earlier contact of the Bronze Age peoples with their Neolithic predecessors. We have no record of occupancy for some considerable time after the Iron Age, but, as suggested above, in or before the Neolithic period the boulder clay reached some distance seawards, and the present rivers would have wound sluggishly over this area and their actual mouths would have been well out in what is now Cardigan Bay. We may therefore reasonably suppose that the watersheds between these streams were low and formed of boulder clay. The post-Glacial submergence included the Neolithic period, but there is no reason for assuming thereafter any serious continuation of that movement. Hence we may reasonably assume that the major part of the submergence was completed before the advent of the Bronze Age. Fleure and James have shown that anthropologically many modern Welshmen are the direct descendants of Neolithic man, and it is quite possible that their ancestors should have witnessed the submergence of these boulder-clay lands. In brief, it would be quite unjustifiable to dismiss these legends as mere nonsense in their entirety.

Submerged forests seem to imply a fairly rapid submergence (p. 488), and this would have led also to the flooding of the valleys, which were later silted up. In addition, this downward movement would also have brought the low boulder-clay watersheds beneath the sea. Wave erosion would have attacked them, washing the finer material away and leaving the coarser behind. In this way there were possibly developed the long ridges we now call the Sarns. But it is difficult to see why they should be so narrow in relation to their length. Possibly the original watersheds were low and broad, and with the washing away of material stones would be scattered over a wide area. Later storm waves might have piled them up in ridges just like our storm beaches to-day. Whatever the precise mechanism, the most satisfactory correlation of geology, anthropology, and folklore is probably that outlined above. This opinion is further strengthened by the fact that Sarn Badrig and Sarn Cynfelin are almost midway between the Mawddach and the Dyfi, and the Dyfi and the Rheidol-Ystwyth respectively. It suggests the possibility of their being continuations of the watersheds between these rivers on the now submerged boulder-clay plain.

(14) THE CARDIGANSHIRE COAST

From Borth as far as Llangranog the Aberystwyth Grits (Silurian) form steep and fairly high cliffs except where boulder clay swathes them or reaches the sea in extensive flats. South of Llangranog as far as Cardigan, sandstones of a greenish-blue or blue-grey colour are seen in the cliffs.

South of Borth the coast is rocky and the cliffs again are steep, their slopes rising upwards to the platform. For about a mile the coast cuts into the plateau margin obliquely, but farther on the cliffs and coastal slopes are cut in it directly. Careg y Delyn is an outstanding rock: immediately south of it the big, sheer cliffs reach the plateau top. The cliffs are actually formed in rocks dipping steeply westwards. This type of structure dominates the cliff, and great grit-slabs can be seen 'plunging seawards to form the cliff-faces which are broken up by joints and plastered with scree'.[1] To the south again the dip is to the east-north-east, and the cliffs no longer show the dominating effect of bedding planes. They are steeper, but not as high as those to the north, and there is a less well-marked coastal slope. The Wallog mouth shows much boulder clay, but only as a veneer, since the cliffs show solid rock below capped by glacial material. Between the river Wallog and Clarach Bay the cliff structures are complicated. Clarach Bay is wide, and the present stream is a misfit owing to modifications which took place in the Ice Age. The river shows a slight northward diversion because shingle and dune ridges have grown across its mouth. Thence to Aberystwyth the cliffs are large and sheer-faced again with the strike running approximately north and south. There are a variety of structures to be seen in these cliffs, but as Challinor points out, they do not control the cliff form as a whole. Every now and again there are small examples of the way in which structure has guided marine erosion (**99, 98, 97**).

The Ystwyth and Rheidol are large valleys. Aberystwyth might more

[1] J. Challinor, 'Some Coastal Features of North Cardiganshire', *Geol. Mag.* 68, 1931, 111.

appropriately be called Aber Rheidol, but possibly the Ystwyth[1] once flowed nearer the site of the town than it now does. Erosion is often serious at Aberystwyth, and was particularly severe during the great storm of 23 November 1938, when part of the northern promenade was destroyed. The course of the Ystwyth has changed from time to time between Llanychaiarn church and the sea. St Mary's church, which appears to have been standing in Elizabeth's time, was swept away some three centuries ago. The Allt Wen cliffs form a rocky coast: the westerly dip of some 30° mainly governs their slope, although there is local steepening caused by marine erosion at their base. In this part landslips are not uncommon, and take place mainly along joints and bedding planes. This is due to the fact that hereabouts the Aberystwyth Grits are markedly bedded and jointed. This type of cliff then gives way to glacial material for about two miles: the solid rock can be seen below in places. The cliffs, about 100 ft. high, near Cwm-Ceirw farm are entirely of boulder clay. They are nearly vertical and often undercut and are always much gullied and cut into mounds and pillars. Upwards the cliffs merge into the plateau surface, and that in its turn into the true coastal slope. No pre-Glacial cliffs are now visible (**96** and **95**).

There is a small hanging stream just north of Monk's Cave, and from that place to the neighbourhood of Llanrhystyd the fairly steep cliffs are again cut in rock in which the bedding is approximately horizontal. It is a wild stretch of coast and hanging valleys are common. Between Llanrhystyd and Morfa Mawr there is a well-marked boulder-clay platform up to half a mile wide, and sloping gradually downward from the 100-foot contour to low cliffs faced by a certain amount of shingle beach, the inner boundary probably coincides with the pre-Glacial cliff: this boulder-clay spread is almost continuous with that between Aber-Arth and Aberayron.[2] Cliffs 50 ft. high occur here, and north of the Afon Arth the glacial deposits are sometimes 50 ft. thick. The low cliffs are often obscured and overridden by shingle. There are two boulder clays involved, Lower and Upper.[3] They are occasionally separated by intermediate sands and gravels, but usually the upper clay rests directly on the lower. Boulder clay also occurs in Mwnt Bay, on both sides of Aberporth inlet, Traeth Dyffryn,

[1] The Ystwyth is deflected northwards by a great shingle beach. The beaches north of the harbour (i.e. the joint mouth of the Ystwyth and Rheidol) are mainly of sand and small shingle.

[2] At Aberayron shingle accumulates to the south of the harbour just as at Aberystwyth. The town itself is an excellent example of early planning, and contains many very attractive houses and vistas, especially the green around the upper harbour.

[3] K.E. Williams, 'The Glacial Drifts of Western Cardiganshire', *Geol. Mag.* 64, 1925, 205, and cf. Jehu, *infra*.

Traeth Saith, Penbryn, Traeth Bach, Trwyn Croi, Cwm Tydi, Castell Bach, and the Afon Soden mouth; and New Quay Bay and Little Quay Bay are both subject to much erosion in their boulder-clay cliffs. The Afon Drowy has carved its channel through about 100 ft. of glacial deposits. In general, where boulder clay occurs the beaches are formed largely of cobbles, and big storm ridges fringe their upper limits (**94**).

Near Llangranog the Ordovician-Silurian junction occurs, and on Lochtyn peninsula (Pen y Billies) gnarled mudstones form cliffs 100 ft. high, and underlie the massive grits of the peninsula of Ynys Lochtyn. At the east end of Traeth yr Ynys the cliffs of thin bedded grits reach 300 ft. Hendriks, writing of this district, says: 'In following the secondary folding along the coast, a conspicuous feature is the relative stability of the syncline. A transient effect of this is the production of a series of coves of the Lulworth type, such as may be seen in Traeth yr Yscland, and again north of Cwm Tydi. This structure further accentuates the approximate parallelism of strike, pitch and coast-line though...it is to the slight divergence between strike and coast-line that the main interest of the cliffs is due.'[1]

The Cardiganshire coast really falls into two distinct parts. North of New Quay are boulder-clay flats and cliffs and beaches of coarse cobbles. South of that town are dark cliffs of shales and sandstones, broken in several places, including Cwm Tydi, Llangranog, and Aber-Arth by deep and narrow valleys, which give character and great beauty to the coast. Some of the cwms have sandy beaches or bays at their seaward ends: others have shingle bars, of which Cwm Tydi is a good example with its tiny impounded lagoon, and also Afon Soden. Details of cliff form vary naturally from place to place, but coves and minor headlands are extremely well developed at Aberporth. This type of coast really extends into Pembrokeshire. The estuaries of the Teifi and Nevern are but larger inlets, somewhat resembling those in the northern half of Cardigan Bay and possessing mainly sand beaches which partly block the mouths. Not until the neighbourhood of Fishguard is reached does the prevailing dark colour of the cliffs give way to lighter tints due to the presence of igneous rocks.

Charlesworth touches on certain features belonging to the coastline between Aberayron and Fishguard. He contends that the Irish Sea ice stood over much of Cardigan Bay during the time of maximum advance of the Newer Drift period. In so doing it ponded back the waters of the

[1] E.M.L. Hendriks, 'The Bala-Silurian Succession in the Llangranog District', *Geol. Mag.* 63, 1926, 121. The erosion features near Cwm Tydi are certainly very reminiscent of those at Stair Hole.

rivers and impounded large lakes: Lake Manorowen in Fishguard Bay and its immediate hinterland, Lake Nevern in the Nevern valley at Newport, Lake Teifi, the largest, in the Teifi valley, and Lake Aeron behind Aberayron. There were also many other minor lakes. It is not relevant here to discuss the history of these lakes, but it is worth noting that, as in other examples of marginal drainage, overflow channels were cut, and the deep channels behind both Dinas Head and Cemmaes Head are ascribed to this cause. As the discussion of Charlesworth's paper showed, there is no complete agreement on some of his views, but in these two cases the nature of the channels appears to be adequately explained. Cowper Reed[1] spoke of Dinas Head as having been an island: certainly the channel is only a few feet above sea level and a slight downward movement would easily make it an island again. He also speaks of the Slade at Fishguard as a former sea inlet, and suggests that the watercourses, tunnelling, and rock platforms at Pwll y Wrach require a comparatively recent change of sea level to explain their existence. Boulder clay often caps the cliffs or descends to beach level between Strumble Head and Cardigan. It is the source of the flints which occur in all the gravels and beaches. They must have been derived in the first place from the Chalk of the north-east of Ireland and presumably also from rocks now hidden under the sea (**93**, **92**, **91**).

(15) THE PEMBROKESHIRE COAST

Although the Cardiganshire coast shows much of interest, it cannot compare in scenery with that of the neighbouring country of Pembrokeshire. Probably nowhere else in the British Isles is there so much variety in such a comparatively small area. The country inland is pleasantly undulating, but gives little hint of the beauty of the coast. Between Fishguard and St David's we find a remarkably interesting stretch where intrusive igneous rocks in the Ordovician explain the character of much of the coastal scenery.[2] Strumble Head (Pen Caer) is a diabase mass, and from that point to St David's nearly all the prominent features on the coast are of igneous origin. Reference to Fig. 30 will make this point clear, but the following summary account will accentuate it. Trwyn Castell is formed of rhyolites, Abereiddy Bay and Traeth Llyfn are cut in dark shales, the headland north of the Traeth is diabase; Porth Eger is in shales, Penclegyr is diabase. Between Trwyn Ellen and Trwyn Llwyd and also at Pwll Whiting, the coast is formed mainly of shales, but Castell Coch

[1] *Quart. Journ. Geol. Soc.* 51, 1895, 149.

[2] 'The Geology of the District between Abereiddy and Abercastle', A.H. Cox, *ibid.* 71, 1915, 237.

(rhyolite) and Ynys Daullyn and Ynys Castell (diabase) are again all of igneous rock. Rarely in such a short distance can the distinctions in hardness and resistance of the rocks show up so well.[1] The post-Glacial change of sea level has been responsible here as elsewhere for certain detailed features as well as for the drowned valleys; there is a good example of the latter in Abercastle Harbour. A curious channel also runs behind Pen Clegyr from Porth Gain to Abereiddy. Cox suggests that it is part of a valley both beheaded and truncated by the sea. If it were not for the storm beach in Abereiddy Harbour, it would be partly inundated even to-day. Cox does not agree with Jehu in ascribing it to sea erosion, but notes that there is no evidence to show which way the drainage ran in the valley. It would not be out of place, perhaps, to suggest a glacial overflow channel as an explanation. Throughout this stretch of coast the cliffs are precipitous, and the valleys are broadly open and choked with glacial drift or alluvium. The same type of coast continues as far as St David's Head:[2] the dolerites of Pen Bery and Carn Llidi stand up as great monadnocks from the plateau, and with other outcrops of intrusive rocks give variety to the landscape. In Porth Melgan and Whitesand Bay drift comes down to and below sea level, filling hollows in the peneplain. The northern headland enclosing St Bride's Bay is rhyolitic (**90**).

Ramsey Island[3] is merely a detached part of the Pembrokeshire plateau. Its coastal outline and steep cliffs reflect clearly the relations between erosion and geological structure. The coast facing Ramsey Sound, through which a strong tide race runs, is fairly straight so that only small irregular bays have been eroded in the softer (Ordovician) shales: these are separated by harder masses of quartz-porphyry. On the west coast the full force of the Atlantic is felt and has produced features like the deep and gloomy inlet of Aber Mawr between the igneous masses of Trwyn Llundain and Carn Ysgubor. 'At the southern end the results of erosion are even more manifest. Here the sea has literally swept out the narrow belts of shale, leaving the isolated porphyritic masses of Cantwr, Gwelltog, Eilun and Bery.'[3]

[1] Prof. T. N. George tells me that even in minute detail there is an astonishing parallelism between the coastal crenulations and the variations of rock type. The contrast between the dark sand and shingle ridge in Abereiddy Bay with the fine light-coloured sands in Traeth Llyfn is very striking. The two bays may be compared to Aber Bach and Aber Mawr a little farther east. Both of these are very beautiful and have large storm beaches at their heads.

[2] 'The Geology of the St David's District', A.H. Cox, J.F.N. Green, and O.T Jones, *Proc. Geol. Assoc.* 41, 1930, 241; 'The Glacial Deposits of Northern Pembrokeshire', T.J. Jehu, *Trans. Roy. Soc. Edinburgh*, 41, 1906, 53.

[3] 'The Geology of Ramsey Island', J. Pringle, *Proc. Geol. Assoc.* 41, 1930, 1.

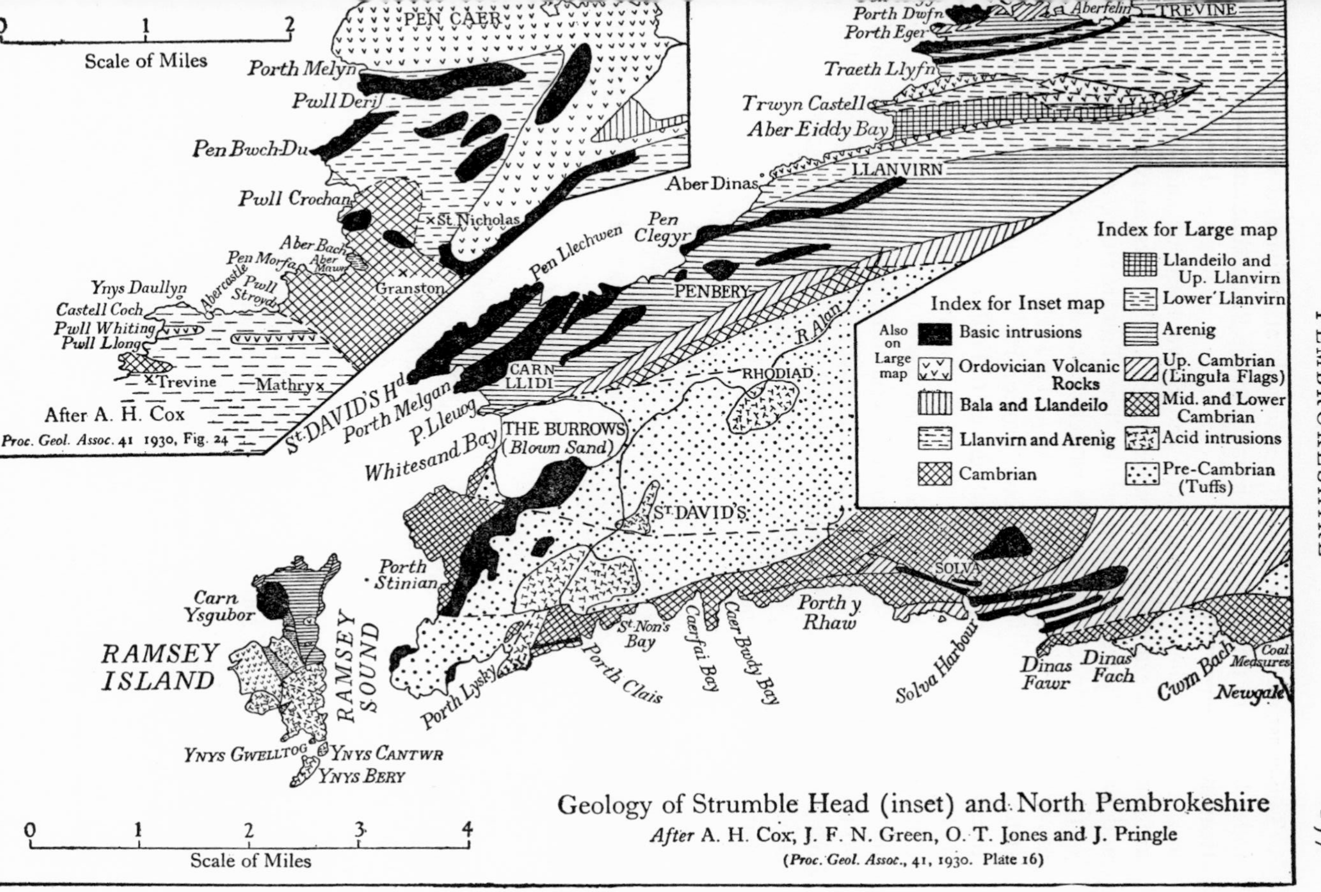

Geology of Strumble Head (inset) and North Pembrokeshire

After A. H. Cox, J. F. N. Green, O. T. Jones and J. Pringle

(*Proc. Geol. Assoc.*, 41, 1930. Plate 16)

Fig. 30

The coast of St Bride's Bay is remarkably attractive and varied.[1] It is, indeed, hard to think of coastal scenery in the British Isles more beautiful than the lichen- and plant-covered cliffs near St David's and Solva. These are mainly in the Cambrian beds and Solva Harbour and Porth Clais are both good examples of drowned valleys. The bays, St Nons, Caerfai, and Caer Bwdy are very picturesque. The cliffs are steep, and the colours of the rocks range from purples to yellows and greys. Stacks and islets, caves and minor indentations abound and are related to rock hardness and structure. In the softer shaley beds erosion is often rapid and leads to the tumbled cliffs of Porth y Rhaw: here an old cliff camp is being destroyed and gives a rough measure of the encroachment. The river system of the county is probably early Pliocene in age, or even before this. Drift now fills the valleys. The streams were rejuvenated by a pre-Glacial uplift, and so their courses are now incised in the lower platform. The meandering nature of the streams in the drowned valleys is due to silting. The rock floors of the streams, however, are really graded to a sea level lower than that of the present, and hence their present lower courses are choked with alluvium. Along the northern shore of St Bride's Bay beach material travels eastward, and the storm beach at Newgale thus accumulates, at least in part. It dams back Newgale Marsh and Bathesland Water and is about one and a half miles long. It is formed of well-rounded stones of local origin, mainly Cambrian and Carboniferous sandstones and porphyrites. The beach, which is faced by a wide expanse of sand, is still working inwards (**89***a* and *b*).

St Bride's Bay is formed in the relatively soft Carboniferous rocks which, except for the narrow outcrop of Ordovician at Druidston Haven, fringe the eastern shore. These rocks show a beautiful variety of structures and also show extremely well on a small scale the nature of differential erosion in rocks of varying hardness. Most of the minor headlands are in sandstone beds, and the re-entrants are in shales. This feature is seen to advantage between Broad Haven and Little Haven: the sandstone outcrop on the north side of Little Haven is an anticline of which the nose has been hollowed out into a cave. A little farther north the Sleek Stone is a fine monoclinal fold in the sandstones, and shows a reversed fault on its northern limb (**88**). In the extreme south-east corner of the bay the Falling Cliff is in the shaley beds, and is separated near Talbenny by a fault from the Pre-Cambrian intrusive mass which forms imposing cliffs especially

[1] The geology of the southern part of this bay and of Milford Haven is fully described in (*a*) 'The Geology of the South Wales Coalfield, Pt. XI, The Country around Haverfordwest', *Mems. Geol. Surv.* 1914 (A. Strahan and others), and (*b*) Pt. XII, 'The Country around Milford', *ibid.* 1916 (T.C. Cantrill and others).

at Borough Head and Ticklas Point. The Stack Rocks are an outlying part of the intrusion. From Mill Haven to Musselwick the Old Red Sandstone forms fine cliffs of deep red colour: most of the smaller inlets in this stretch are associated with minor faults. In Musselwick Bay the contrast between the bright red-brown of the Old Red and the black of the Ordovician shales is striking. The Ordovician has but a narrow outcrop, and on the northern side of the little peninsula running out to Skomer Island the relations between rock hardness and minor faulting are shown to perfection. Those rocks near sea level are mainly keratophyres with sediments of Ordovician age above. There are many small nearly north-south faults, and each of the numerous and often deep inlets coincides with one of them: in one or two cases they are really minor rifts. The best marked of these bounds that part of the peninsula known as the Deer Park which is formed largely of igneous rocks. The continuation seaward of the southern headland of St Bride's Bay, like that to the north, is marked by off-lying islands. Volcanic rocks form Skomer almost entirely, except for some quartzites which are best seen in the narrow neck separating North and South Havens. As the sediments are faulted on both sides, there is no doubt that the faults, together with the bedded nature of the quartzites, have played a large part in guiding the erosion which made the two havens. Midland Island is also mainly composed of volcanic rocks, and its separation from Skomer on the one side and the mainland on the other may have been due to the same processes that formed the two Skomer havens and the deep gut between the Deer Park and the main part of the peninsula. Farther seawards Grassholm Island is formed of basalts and keratophyres, and Skokholm, a little to the south, of Old Red Sandstone (**87** and **86**).

South of the Deer Park peninsula there are again many interesting features in detail, but there is no space for a lengthy discussion here. Gateholm Island is really a great stack of Old Red Sandstone hardly yet severed from the mainland. The fine sweep of Marloes Bay shows numerous stacks, and the rocks are very varied,[1] because the structure brings several formations in contact with the coast within one mile. The Three Chimneys are three vertically projecting sandstone beds separated by soft mudstones.

A little to the south of Marloes Bay is the Dale peninsula. It is almost an island, and is separated from the mainland by a deep valley in which Dale itself lies. The Ritec fault is probably continued along the line of Milford Haven near the Stack Rock and then along this valley. The Dale valley receives a small tributary valley on its southern side, and this con-

[1] The dip is often steeply seaward, hence the cliffs often correspond with the bedding planes.

tains a stream flowing to Milford Haven. Undoubtedly land at the west end of the Dale valley has been lost by erosion, and formerly the valley was a strait. Now it is drift filled, and Head rests on the drift. The Head is well seen in the western cliffs: it dips to the centre line of the channel below sea level to an unknown depth. Post-Glacial subsidence may explain this, and if, as is likely, the channel itself is pre-Glacial, tidal erosion, acting together with the downward movement, has excavated it to its present depth. At the same time faulting, as suggested above, is the prime cause of the channel in the sense that faulting provided a line of weakness along which erosive agents could work more easily. Milford Haven with its numerous branches, or Pills, is an excellent example of a ria. Study of a topographical or geological map will show that between St Bride's Haven and Dale, between Walwyn's Castle and Herbrandston, and in other places, the whole area is much more peninsulated than appears at first sight. Removal of alluvium and artificial banks would make the 'sea-line' much more intricate than it is to-day. Traces of raised beach, which become numerous in South Wales, are found in parts of the haven, and usually the cut platform is best developed in west-facing headlands. Head rests on the platform which on the one side of the haven terminates at about high-water mark: it is several feet higher on the other side.

St Bride's Bay marks the beginning of the area of Armorican folding, and before touching on detailed features it is necessary to give a brief account of the structure of the coastal area of South Wales from Milford Haven to east of Swansea. There can be few, if any, more suitable districts in which to study the relationships between structure, coastal form, and erosion. In both the Tenby and Gower peninsulas the inland surface is undulating, forming parts of the plateaux of Wales already mentioned and of which the origin will be discussed later. The sections (Figs. 31 *b*, 32 *b*) show better than a written account that the folding is fairly regular and that the various rocks outcrop in long belts running nearly east and west. Assuming the absence of drift, it is also clear that the same type of structure occurs in both peninsulas, and that the rocks must continue under Carmarthen Bay. This and Swansea Bay have been cut mainly in the softer Carboniferous rocks, but it would be a mistake merely to suppose that they are due to modern marine erosion. Considerable traces of submerged forest appear in each, and we may suppose that both are low-lying areas partly worn down by subaerial, and partly destroyed by marine, denudation before the post-Glacial submergence brought them below sea level and allowed erosion to etch out their present detailed forms. Moreover, blown sand has accumulated in vast stretches and so has modified the scene further. If the sections (Figs. 31 *b*, 32 *b*) are compared carefully

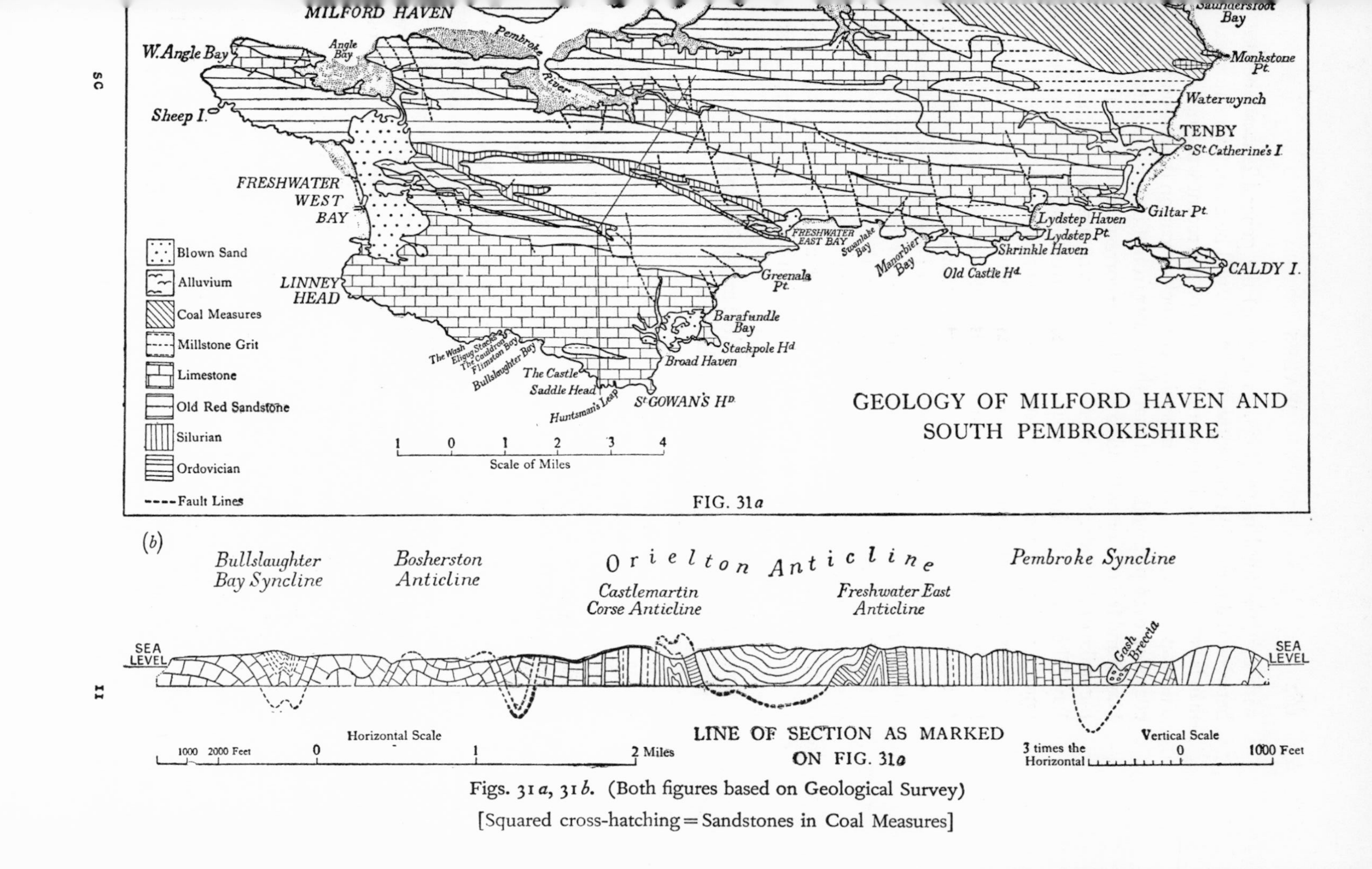

FIG. 31*a*

Figs. 31*a*, 31*b*. (Both figures based on Geological Survey)

[Squared cross-hatching = Sandstones in Coal Measures]

with the structural maps (Figs. 31 *a*, 32 *a*) the general correspondence between the formation of the two peninsulas will be clear. The structural map is, so far as Gower is concerned, based on the recent work of George. The maps show that similar successions of rocks occur on the east and west coasts of each peninsula, and that the coasts from Linney Head to Broad Haven and from Worms Head to Port-Eynon correspond more nearly with the strike of the beds. Freshwater West Bay may be compared with Rhossili Bay. The general form of the coasts is clearly structural, but details depend upon a variety of factors. Exposure to wave attack is an important one and is brought out in some of the differences between a west- and an east-facing bay in the same formation, for example, West and East Freshwater. Other factors include bedding, whether thin or massive; dips, whether to or away from the beach; jointing; cleavage; the relative hardness or softness of rocks; and their solubility. It is worth remembering, too, that in Gower at least, and probably in other parts, there are traces of several raised beaches, and the forest beds occur at intervals all along the coast. Therefore we must obviously consider also the effects of these relatively small vertical movements in addition to the other factors just mentioned.

In the Tenby peninsula[1] the control of the major fold and fault structures is plain in the general position of the larger bays and inlets, for example, West Angle Bay, Pembroke Pill, the two Freshwater Bays, Lydstep Haven, and Manorbier Bay. This control appears also in the river courses and in many less conspicuous indentations of the southern coast, particularly where the cliffs follow the strike, as between Freshwater East and Old Castle Head.

Perhaps the best way of appreciating the more detailed features is to consider in turn the three parts of the coast formed of the Old Red Sandstone, and of the Lower and the Upper Carboniferous rocks. There is first of all the Carboniferous Limestone coast between Linney Head and Broad Haven. The upper surface of the cliffs is a nearly level plateau varying between 100 and 160 ft. above sea level near the sea. The cliffs are usually vertical, or almost so, and have been cut into a fantastic variety of forms. Leach emphasizes that many of these features result not merely from direct erosion of the sea on limestone cliffs, but also from erosion acting on a rock already honeycombed by water courses, now dry, and formed by subaerial denudation. The formation of stacks is beautifully shown at Eligug: their bedding dips at low angles, but strong vertical

[1] *Mems. Geol. Surv.* 'The Geology of the South Wales Coalfield, Pt XIII, The Country around Pembroke and Tenby', E.E.L. Dixon, 1921, and 'The Geology and Scenery of Tenby, etc.', A.L. Leach, *Proc. Geol. Assoc.* 44. 1933, 187.

joints are mainly responsible for their formation. Near Flimston Castle the sea has cut caves and has cleared out and partly undermined a great hole with vertical sides—The Cauldron. The Green Bridge, a natural arch, is a half-made stack. The Wash (near Flimston) is a bedding plane well exposed and separated from the cliff behind by a rift cut out along an overthrust. Where the cliffs are more argillaceous, and where mudstones occur interbedded with the limestones, the cliffs show a smooth, rounded profile, as near Arnold's Slade. Smaller gaps, such as that at St Gowan's Chapel, have been worn out along small faults, or other planes of weakness. Some of these are certainly old, because they are filled in places with Triassic debris and stalagmite. Many of the fissures, including that known as the Huntsman's Leap, are due to erosion along minor faults (**84** and **85**).

The Old Red Sandstone cliffs are well developed at either end of the Orielton compound anticline, especially in the two Freshwater Bays. Since there is but little calcareous matter in these rocks, solution plays a negligible part. Towards Old Castle Head and Sheep Island, recession of the coast is to some extent restricted by the partly submerged basal conglomerates; where the sea has advanced, it has done so along small faults and other tectonic structures. Swanlake and Manorbier Bays and Precipe inlet are largely, if not wholly, due to erosion along groups of small faults, and the fissures on each side of Manorbier Bay are due to the erosion of marl beds between harder sandstone beds. The caves, like the bays on this stretch of coast, are formed along joints or small faults. Between Freshwater East and Old Castle Head the strata are nearly vertical and strike along the coast giving a striped effect where the beds of different colours (mainly red and greyish green) outcrop. In Freshwater West the Old Red Sandstone, with some still older rocks, forms a beach platform. It is backed in places by the raised beach, and towards the northern end of the bay there is an exposure of submerged forest. The structure is much obscured by blown sand forming Kilpaison, Gupton, and Brownslade Burrows. Minor details in the Old Red Sandstone cliffs enclosing the bay to the north are connected with small faults (**83**).

Before commenting on the features of the Upper Carboniferous coast north of Tenby, there should be some reference to the interesting historical changes in the Ritec valley.[1] The caves of Hoyle's Mouth and Longberry Bank, at about the 50-foot contour and overlooking the Ritec, are now between one and two miles inland. Leach found *Scrobicularia* clay just below the former cave, and a Neolithic or Bronze Age canoe was also discovered, implying that in those times the cave was near open-water level. An

[1] H.C. Darby, 'The Tenby Coast', *Int. Geogl. Cong.* Cambridge 1928, *Rept. of the Proceedings*, p. 466.

ecclesiastical document of the eleventh century makes it clear that the Ritec was then an open estuary as far up as St Florence. Later maps, inaccurate though they may be in detail, bear this out, and even in the mid-seventeenth century there must have been quite a wide estuary, since in 1643 it was sufficient to obstruct the Parliamentary forces. Reclamation has, however, been carried out, first by the construction of banks, and secondly by the enclosure of the Ritec in a culvert. This permitted the gathering of sand along the course of the old estuary. The earliest bank was at Gumfreston; a later one, also of unknown date, reached to a point near Holloway Bridge. In 1811 Sir J. Owen built a bank along the line of the present railway, and the actual railway embankment was made in 1865. Between Tenby and Giltar Point is a wide dune belt, faced, over part of its length, by a storm beach, and throughout by a wide sandy foreshore. The shingle is mainly of local origin, but it also includes some material derived from submerged glacial deposits, the raised beach, and ballast. The south-west winds carry much sand towards Tenby, but Darby thinks the shingle is more or less stationary and does not travel in that direction. Inshore, the tidal current at flood runs to Giltar. Ground-swells from the Bristol Channel approach this foreshore at right angles, largely owing to the local effects of Caldy and St Margaret's Islands. The general result of normal south-west wind waves is to drive the sand towards Tenby, and these are much more effective than the inshore flood currents. Moreover, at low water, when the beach is exposed, these same winds blow sand at Tenby, while strong backwash is apt to shift the sand away from the area immediately north of Giltar Point. The fact that the storm beach approaches more closely to Tenby than the foreshore shingle 'is due to the fact that during storms the area of eastward drifting encroaches upon the area of broadside feeding under the lee of Giltar owing to the strength of the wind. This is so, because the stretch of coast protected by Giltar Point is reduced; some of the shingle actually begins to travel east with the sand. This is piled up into a storm-beach which cannot be worked back during normal conditions.' So much for the present; in the past, following the post-Glacial submergence, a bay bar was built from Giltar to Tenby and forms the nucleus of the present dunes. The area behind silted up, and the Ritec kept its own outlet. Then enclosure followed: the marsh and dunes widened, but there is no evidence that the dunes are still growing. It is worth noting that on a smaller scale there are good bay-head beaches and often storm beaches in other smaller bays—for example, Lydstep, Manorbier, Swanlake, and Freshwater East. Broad Haven was also once an arm of the sea, now sand-choked and artificially blocked.

Caldy Island[1] is merely a separated fragment of the mainland. The rocks striking out to sea between Old Castle Head and Lydstep reappear on the island. The shaley beds behind Lydstep Haven would, if prolonged eastwards, occupy Caldy Sound. During the post-Glacial submergence Caldy became isolated, and we may assume that the shale area was relatively softer than the beds to north and south, so forming lower ground. With the subsidence, erosion would remove this fairly easily, and the strong tidal currents in Caldy Sound to-day strengthen the opinion that the process just described accounts in part at any rate for the present formation. The raised beach is very apparent in the island, especially near the western end (**82** and **81**).

The Upper Carboniferous coast is north of Tenby and runs eastwards as far as Ragwen Point, an outcrop of the basal quartzites of the Millstone Grit. Telpen Point is formed by the massive Farewell Rock (i.e. the top of the Millstone Grit series). Between these two headlands a wide bay has been eroded in the Middle Millstone Grit shales. From Amroth to Saundersfoot and Monkstone Point there are many faults and folds, so that shales and sandstones are often repeated in the cliffs. Since the shales are frequently crushed there are cave-like openings cut into them in many places. The sandstones are harder, and evolve from promontories or points through stacks to reefs, as can be seen at Monkstone. Near Waterwynch is a thrust plane, the course of which is marked by grooves and alcoves. Monkstone, together with the promontories forming Bowman's Point, First Point, and Second Point, is carved in highly contorted sandstones. Darby[2] comments on an interesting factor in beach accumulation at Saundersfoot. The shingle on the beach now is but a remnant of what was there in the latter half of the last century. At that time Saundersfoot exported much coal by sea, and ships visiting the place in ballast dumped it overboard, and so helped to form a beach. Now that it is wasting away, and the sea is eating in behind it, old mooring-posts have been revealed, making it clear that there was an earlier period of erosion before the accumulation of the 'ballast' beach. Between Saundersfoot and Ragwen Point are several beaches of coarse boulders. Because of the eastward travel of beach material the boulders are more numerous and the beaches are higher at their eastern than at their western ends. This is very clear at Amroth.

Before leaving the Tenby peninsula it is worth noting that the *Patella* raised beach is visible at Giltar Point, at several places on Caldy, at Freshwater West, Whitesheet Rock, and Lydstep Head. The effect of recent

[1] See A.L. Leach, *Arch. Camb.* 6th ser., 16, 1916, 155, and *Proc. Geol. Assoc.* 45, 1934, 189.

[2] *Op. cit.*

erosion in producing the detail is, on the whole, small, but there seems little doubt that during raised-beach times it was decidedly more potent. Incidentally, all the valleys and other main features of both inland and coast were in being during the raised-beach period; but the later submergence has drowned the lower ends of the valleys, so adding greatly to the variety of coastal detail. The fact that many of the rivers, often in their drowned parts, cut across hard ridges can only be explained by invoking superimposed drainage. In fact the river system is independent of the structure.

(16) CARMARTHEN BAY, GOWER, AND SWANSEA BAY

Eastwards from Ragwen Point lies the joint mouth of the Taf and Towy and one of the great dune and marsh areas of South Wales. The former cliffs in the Old Red Sandstone run inland behind Laugharne Burrows[1] and Marshes. On the east side of the river are Towyn Burrows and Pembrey Burrows, which pass into the largely silted-up bay and marsh on the north coast of Gower. The main dune areas have many features in common and these will be discussed later on (**80** and **79**).

Apart from the dunes, there is not much coast between Pendine and Gower. The Llanstephan peninsula is rather shut in; nevertheless, it has a short but attractive coastline of cliffs passing on either side to the pleasant, open and unspoiled estuaries of the Taf and Towy. To the east of the Towy there is a low flat followed by the railway, but this gives place rapidly to high ground between approximately Ferryside and Kidwelly. Before the formation of Towyn Burrows, this section of the coast was far more indented, and arms of the sea penetrated the two Gwendraeth valleys, and the gulf between Gower and Llanelly was wider and more open. In addition to natural marsh, there is a fair amount of reclaimed land near Loughor (see p. 171).

The coast of Gower,[2] which shows so many similarities to that south of Milford Haven, has been fully studied by George whose recent mapping supersedes that of the Geological Survey. Gower illustrates the effects of Armorican folding remarkably well. Compression directed in general to the north-north-east caused folding and over-thrusting of the Upper

[1] *Mems. Geol. Surv.* 'The Geology of the South Wales Coalfield, Pt X, The Country around Carmarthen', A. Strahan and others, 1909.

[2] T.N. George, 'The Structure of Gower', *Quart. Journ. Geol. Soc.* 96, 1940, 131; *Mems. Geol. Surv.* 'The Geology of the South Wales Coalfield, Pt IX, West Gower and Pembrey', A. Strahan, 1907; T.N. George, 'The Coast of Gower', *Proc. Swansea Sci. and Field Nat. Soc.* 1, 1933, 192.

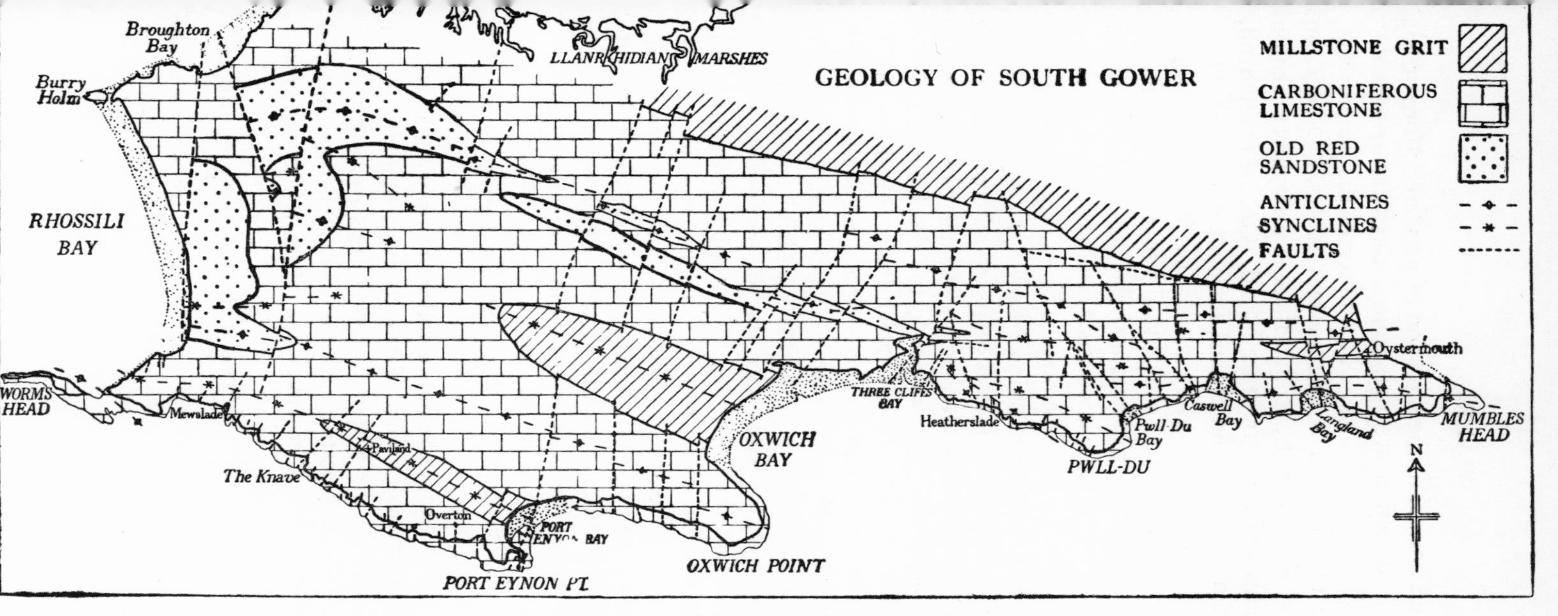

Fig. 32*a*. (After T. N. George, *op. cit.*)

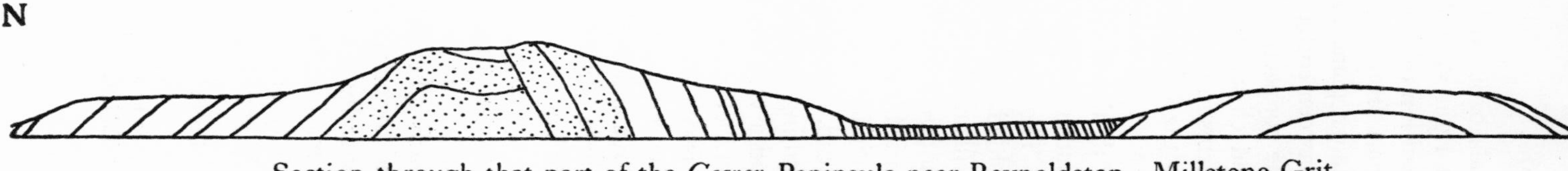

Section through that part of the *Gower* Peninsula near Reynoldston. Millstone Grit, shaded; Carboniferous Limestone, uncoloured; Old Red Sandstone, dotted.

Fig. 32*b*. (After T. N. George, *op. cit.*)

Palaeozoic rocks to run normal to the direction of pressure. More or less at right angles to this direction are numerous faults, the general disposition of which is shown on Fig. 32. The interior of the peninsula is, apart from the ridge of Old Red Sandstone and Rhossili Downs, an almost plane surface at about 200 ft., so that there is little direct relation between surface features and rock structure inland. On the other hand, the connection between structure and coastal forms is remarkably clear, and closely resembles that in the Tenby peninsula. In this account it seems most suitable first to show how the general structure has given rise to the important and interesting features, and then to review George's work on the raised-beach sequence—a sequence not in complete harmony with that found elsewhere in this country. The question of the high plateaux will be treated in a separate section (see **78**, **77**, **76**, **75**).

The greater part of Gower is a plateau of Carboniferous Limestone folded in a somewhat complex manner. The synclines of shales in the Millstone Grit series are primarily responsible for the bays at Oystermouth, Oxwich, and Port-Eynon, while Oxwich Point coincides with an anticline in the limestone. South of the Port-Eynon syncline there are also anticlinal remnants preserved in the low neck of land running to Worm's Head. Mumbles Head lies at the nose of the Langland anticline. The faulting was largely contemporaneous with the folding, and it, too, has had a very marked effect on the coastline. The thrust faults are roughly parallel with the folds, the tear faults mainly at right angles to them. In Caswell Bay a thrust is exposed in the cliffs, and both the bay and the Caswell valley have been cut along the line of a transverse fault. Brandy Cove and Pwll-du Bay are cut along similar faults. Fracture influence in Three Cliffs Bay is obvious: the fault is a tear-fault. On the east side of the bay pressure mainly produced thrusting; on the west folding occurred. Hence there is not much correspondence between the two sides: it is 'an example of the growth *pari passu* of folds, thrusts, and cross-faults'. From Port-Eynon Point to Tears Point the minor indentations of the coast are nearly all closely related to small faults.[1] The estuary of the Burry River and Swansea Bay are both in the Upper Carboniferous rocks which are generally softer and more easily destroyed by erosion, whether subaerial or marine. These two major inlets, however, are far from being the result merely of present conditions.

The raised beaches[2] of Gower are of great interest and present difficult

[1] From Port-Eynon to Worm's Head the minor indentations of the coast are nearly all eroded along transverse faults, some of which are large, or master, joints. These include Paviland, Mewslade, and Red Chamber.

[2] T.N. George, *Proc. Geol. Assoc.* 43, 1932, 291; see also Chapter XII.

problems. They are all later than the 200-foot platform, which was probably formed in Pliocene times. An erosion period followed during which the steep cliffs of the present shore were cut. No contemporary deposits of this time are known to exist, but we must stress the point that the cliffs are not only formed by modern erosion. The earliest of the beaches is the *Patella* beach named on account of the number of limpets in its deposits. It is banked against the cliffs and is in very much the same place as the modern beach. Indeed, apart from changes due to submergence and erosion in the synclinal bays, the form of the shore in *Patella* times was similar to that of to-day. The beach can be traced along the limestone borders of Port-Eynon, Oxwich Bay, and elsewhere. It is difficult to identify it outside the actual peninsula, but the nature of the northern cliffs suggests that the Loughor estuary and Swansea Bay were eroded in *Patella* times.

The next episode was the formation of certain bone-bearing breccias, such as that in Minchin (Mitchin) Hole. The fossils suggest a correlation with the Taplow terrace of the Thames and indicate a temperate climate, probably of late Acheulian or early Mousterian age. This may also possibly be equated with the inter-Glacial period between the Chalky Jurassic and Upper Chalky Drift deposits of East Anglia.

The *Neritoides* beach is just a little younger than the breccia. This is shown by the uncemented state of the breccia when the beach sands were laid down on them, and also by the fact that *Rhinoceros hemitoechus* recurs in the cave earths just above the beach. The time during which this beach was built was apparently short, and it seems to have had no real effect on the form of the coastline.

The gravels of the Older Drift of South Wales follow; they are regarded by George as probably Middle or Upper Mousterian in age, or even a little younger. Since Paviland cave was occupied in Aurignacian as well as in the early part of Solutrean times,[1] there is at least a suggestion that the area was ice free from the Mousterian until about the end of the Palaeolithic. The Newer Drift of Wales is approximately Magdalenian in age.

The Heatherslade beach and platform is to all intents and purposes coincident with the modern beach. Its height and the absence of diagnostic fossils make correlation difficult, and unfortunately its relation to the Newer Drift is not clear. So far no other beach of this height is known in south-western England, but George equates it with the 25-foot beach, which is known to decline in altitude southwards. The Heatherslade beach clearly implies a long period of stability during which also the greater part

[1] In a personal note Prof. George tells me that Solutrean has been queried. The period may be Aurignacian III.

of the modern beach platform was gradually cut. It is usually 10–15 ft. lower than the *Patella* beach. All that can be said with certainty is 'that it is younger than the Older Drift of South Wales. But to refer it to a period so far back in the Pleistocene as to ante-date the Newer Drift raises difficulties possibly more acute than those arising from correlation with the Neolithic beach.'[1] The deposits of the beach were elevated, cementation occurred, and later subsidence brought it back to its present level. There are also many traces of a submerged-forest series around and near Gower.

The Gower cliffs therefore show five main features: the level plateau surface, the steep fall to the *Patella* beach, the *Patella* beach platform, the fall to the modern beach, and finally the modern beach platform. Since the last is planed across inclined limestone beds and is often very perfect in form, it clearly demands a long period of erosion. There are good reasons for thinking, therefore, that it is not wholly modern, and that it was largely cut in Heatherslade times to be brought later once again to that level. The actual configuration of the Gower coasts is thus the result of several oscillations, and although structural control is clear the more detailed forms are evidently the outcome of many shifts of sea level relative to the land.

A walk along the coasts of Gower is, for the reasons already given, of great physiographical interest. It is also one of great beauty and variety. The magnificent range of Carboniferous Limestone cliffs between Worm's Head and Port-Eynon is probably the finest in the peninsula. They are high and cut into coves and headlands. Usually the sea is now cutting into the lower part of the cliffs only, and these extend a little seaward as a low platform (the *Patella* beach). The views from the headland overlooking Worm's Head are excellent, both eastwards along these cliffs, and northwards across Rhossili Bay with its very marked platform at about 100 ft. This feature is of uncertain origin, and is banked against the high Old Red Sandstone mass of Rhossili Downs. The marsh on the northern coast lies under the old cliffs, and its setting is impressive. The great sweeps of Port-Eynon and Oxwich Bays with their sand dunes and woods, break the general trend of the south coast east of Port-Eynon Point and add much to its interest. Farther east, the alternation of smaller bays and lines of good cliffs, with in nearly all places the lower platform still prominent, make the coast perhaps less imposing, but nevertheless most attractive. The eastern side, facing Swansea Bay, is spoiled by the railway which lies seaward of the road, and gives a completely artificial aspect.

North of the Gower peninsula there is an interesting development of

[1] George, *op. cit.* For a fuller discussion of the Gower beaches see Chapter XII.

salt marsh[1] which has grown up behind Whiteford Spit, a pebble ridge surmounted by dunes. The general direction of the spit is determined by the prevailing beach-drift, and it appears to conform well with Lewis's views (p. 60). On the other hand, the pebbles, mainly[2] of limestone, forming it may be glacial in origin and come from an old and now destroyed terminal moraine. The plant associations on the marshes are comparable with those in the Dyfi. There are four main zones: (1) the outermost, submerged some part of each day, and characterized by *Salicornia herbacea*, *Glyceria maritima*, and *Suaeda maritima*; (2) a more closed association between the limits of high waters at springs and neaps, the main plants being *Glyceria maritima*, *Glaux maritima*, *Spergularia salina*, *Plantago maritima*, together with those of zone 1; (3) a closed association at the limit of high water springs with *Glyceria maritima* dominant, but grazed short by sheep, and with *Plantago maritima* and *P. coronopus*, *Statice limonium*, *Salicornia vulgare* and *herbacea* co-dominants (this part of the marsh contains many salt pans); (4) the innermost zone with *Juncus maritimus* and *Glyceria maritima* dominant. The subdominants are similar to those in the other three zones, but are less luxuriant, and less maritime. There are also brackish-water areas.

The plants on the shingle beach and on the two sand-dune areas (i.e. the actual sea shore and on Llangenydd Cliff) are typical of comparatively young dunes and need not be specified here. Outside the marshes the estuary is mainly filled with the Llanrhidian Sands. The former cliff line is clearly visible.

The lower part of Swansea,[3] as well as the flat ground bordering the sea westwards of the town, was formerly a sand-dune district, and the surface was levelled, partly by the tipping of rubbish. The Tawe[4] reaches the sea at Swansea. Its valley has been eroded in solid rock, and the rock floor is at least 40 ft. and probably more than 100 ft. below O.D. In the lower part of the valley the alluvial deposits have been investigated by George, who notes that the peat bed (no. 6) is not well developed, but corresponds to the peats just below O.D. found in many dock excavations between Llanelly and Barry. It implies a subsidence of at least 20 ft., and,

[1] E.J. Bowen, 'A Survey of the Flora of the North Gower Coast...', *Proc. Swansea Sci. and Field Nat. Soc.* 1, 1930, 109. *Obione* occurs on the north of the estuary.

[2] Some pebbles probably came from the south and so were not carried by ice.

[3] See (1) *Mems. Geol. Surv.* 'The Geology of the South Wales Coalfield, Pt VIII, The Country around Swansea', A. Strahan, 1907; (2) 'Shoreline Evolution in the Swansea District', T.N. George, *Proc. Swansea Sci. and Field Nat. Soc.* 2, 1938, 23, and 'Superficial Deposits at the Mouth of the River Tawe', T.N. George and J.C. Griffiths, *ibid.* 2, 1938, 63.

[4] See Chapter XII (O.T. Jones, *Quart. Journ. Geol. Soc.* 98, 1942, 61).

as George remarks, 'its location indicates that the river was not flowing along or near its present course when it was being formed'. The Tawe is only one of several similar drowned valleys reaching the sea in Swansea Bay. Just east of it is Crymlyn Bog, a wide and marshy breach in the Pennant scarp for which there is as yet no satisfactory explanation: no stream runs into it. The Neath is tidal as far as Aberdulais, and its present mouth is at Briton Ferry, but an equally good outlet would be along the western side of Coed-yr-iarll. The Avon now runs out to sea at Port Talbot: its older mouth was just over a mile to the south. The present course was taken when the docks were made. Dock excavations made it quite clear that the river mouth formerly fluctuated over an alluvial flat. South of Porthcawl is the Ogmore which drains the Vale of Glamorgan. Although the deposits in the lower courses of these rivers have not been studied, it is probable that we may take those of the Tawe as generally indicative of what may be found in the others. When the submerged forests were growing in what is now Swansea Bay it would have been in the nature of an alluvial flat, and it may be that several of these rivers united in a common mouth. Local tradition has it that a forest called Silverwood once occupied the western part of the bay, but this is doubtful, and changes of that extent in historical time are unlikely. Thus, both Carmarthen and Swansea Bays originated as drowned parts of several rivers; they are now largely filled up, partly with glacial material, but the submarine contours still indicate the general outlines of the former extension of the streams fairly well. Near Porthcawl erosion in historic times has certainly caused considerable change. Tusker Rock 'in living memory' is said to have been a sheep pasture. This is almost certainly an exaggeration, but only in point of actual time. The Tusker Rock is probably a former limestone hillock, possibly capped by Trias. The shallow sea between it and the mainland, according to North, was here filled with easily eroded material which has now been swept away. It may perhaps be compared with Sully Island farther east, except that there there is still an island, and not merely a rock awash. Woodward, in 1887, pointed out that the severance of this island was due to the quick erosion of the Triassic beds which still connect it at low-tide level with the mainland; the Carboniferous Limestone of the island has resisted attack. Barry Island was first separated from the mainland by denudation of the Rhaetic and Red Marls, while the promontories enclosing Whitemore Bay are made of Carboniferous Limestone. The island is now artificially connected with the mainland.[1]

[1] It has been suggested to me by Prof. George that the effect of present-day erosion in stripping the Mesozoic sediments off the Palaeozoic (Carboniferous Limestone)

A point of general interest that may be made here is that the industrial coast of South Wales is far less spoiled than those of Cumberland, Durham, and Northumberland. A journey along the main road or railway may easily leave a feeling of depression, but landwards, the industrial area gives place almost at once to hills and mountains, and seawards, the extensive sand dunes, apart from one or two war-time factories, are remarkably clear. It is somewhat striking to leave the main road near the middle of Port Talbot and go down to the harbour. Northwards from the harbour mouth lie some two miles of open sand hills. Even the congested district between Briton Ferry and Swansea has some open coastline.

(17) SOUTH GLAMORGANSHIRE

In South Glamorganshire the coastline[1] is no longer formed of Carboniferous and other older rocks. The cliffs are mainly of Trias, Rhaetic, and Lias formations, and are generally much softer than the Palaeozoic rocks farther west. At the same time in Southerndown cliffs the Carboniferous Limestone often forms the actual foreshore, and is overlain by the Liassic Sutton Stone.[2] At Cwm Nash and Nash Point, limestones of the lower part of the Bucklandi zone of the Lower Lias account for the small headland (although there is no obvious reason for Nash Point in terms of variation in rock type) and this zone also forms the cliffs and beach as far as Cwm Col-hugh. The sequence is interrupted by many faults, and several small bays such as St Donat's and Stradling Well are the result of quicker erosion of the higher and more argillaceous beds brought down by faulting (73). Stout Point is formed of harder limestones, but the fall in cliff level near it can be correlated with the coming in of more argillaceous beds. Around the mouth of the Ddaw (Thaw) there is a spread of alluvium, with blown sand and much coarse shingle which has travelled from the west; the area is all part of the old estuary. The cliffs from Cold Knap to St Donat's are mainly cut in the Lower Lias limestones. Beyond Laver-

cores of, e.g., Sully Island and Barry Island is to re-expose to some extent the Mesozoic landscape. These islands were islands in the Triassic sea just as much as they are islands to-day.

[1] 'Notes on the Coastline from Penarth to Porth Cawl', H.B. Woodward, *Rept. Brit. Assoc. Adv. Sci.* 1888 (Bath), p. 900; 'The Liassic Rocks of Glamorgan', A.E. Trueman, *Proc. Geol. Assoc.* 33, 1922, 245; 'The Liassic Rocks of the Cardiff District', A.E. Trueman, *ibid.* 31, 1920, 93; 'The Geology of the Cardiff District', A.H. Cox, *ibid.* 31, 1920, 45.

[2] In Southerndown cliffs the basal Lias is nearly as hard as the underlying Carboniferous Limestone, and the nearly vertical cliff shows scarcely a step at the plane of unconformity. This line of cliffs is one of great beauty and is seen particularly well from the headland near Dunraven Castle.

nock Point westerly gales and waves have little effect. At Lavernock itself the lithological variations of the Liassic beds give rise to detailed coastal features. 'For instance, small headlands are formed by the relatively hard limestones of the *Ostrea* beds, and between these headlands the Lavernock Shales, brought down to the beach by the syncline, have been worn back more rapidly and give rise to a shallow bay. Nevertheless, the upper part of the cliff in the centre of the syncline, where limestones again predominate, is once more almost vertical. At the extreme ends of the section where the Rhaetic beds form the cliffs and foreshore, the cliffs are again worn back, producing the cove at Lavernock, in the east, and St Mary's Well Bay in the west.'[1] At Penarth Head (Lias, Rhaetic, and Keuper) there is a constant but slow fall of material. From this point eastwards the coast really forms part of the Severn Estuary.

(18) THE DUNE AREAS OF SOUTH WALES

There has been frequent reference in this chapter to individual dune areas on the coast of South Wales, and it is, perhaps, useful in conclusion to comment generally on those formations as a whole. They are all marked on the map (Fig. 33), and it is noticeable at once that nearly all occur in

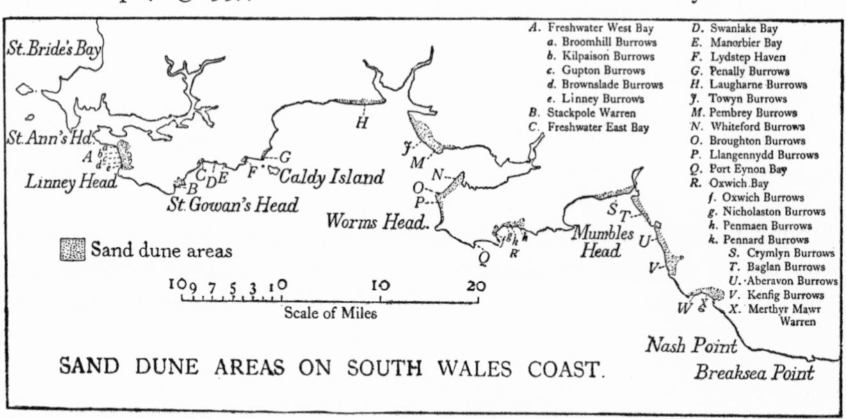

Fig. 33. Sand Dune areas (W = Newton Burrows)

bays, especially in those facing west or south-west. The size of any dune mass is controlled mainly by the size of the bay and by the nature of the land behind the dunes, and bearing this in mind it is rather curious that St Bride's Bay contains no dunes worth consideration. Those at Whitesand Bay are fed directly from the beaches in front of them. The dunes of South Wales begin in Freshwater West Bay, where from north to south

[1] Trueman, *op. cit.* 1922.

they there go under different names—Broomhill, Kilpaison, Gupton, Brownslade, and Linney Burrows. They are all fed from the sands in the bay, but the outcrop of rock in its central part suggests that Brownslade and Linney Burrows receive their material mainly from that part of the bay called Frainslake Sands. Incidentally the shingle beach in the bay is probably derived from submarine masses of boulder clay, since the shingle of the beach contains flint and igneous material not likely to have been derived from the coastal rocks.

There are smaller masses of sand in Stackpole Warren, Freshwater East Bay, Swanlake Bay, Manorbier Bay, and Lydstep Haven. In the first of these the sand drift is both westward from Barafundle Bay and east-north-eastwards from Broad Haven. In the second the sand blows up directly from the beach under the action of easterly and south-easterly winds. In Manorbier Bay, and probably in Swanlake Bay, the drift is directly due to the south-west winds, and in Lydstep Haven due mainly to easterly winds: in this haven it has spread on to the higher ground in the west. Penally Burrows, between Giltar Point and Tenby, have been mentioned already; the sand in general drifts north-west from the beaches and some has reached the higher ground.

In Carmarthen Bay there are much larger expanses of dunes. On the north side are Laugharne Burrows. Between Watchett Pill and Ginst Point Cantrill[1] has examined certain shell mounds. He also records the finding of flints and sherds and comments: 'All things considered...I am disposed to refer the first definite occupation of the burrows...to the Early Iron or Late Celtic period, i.e. to a time beginning say 400 B.C. and extending up to the Roman occupation of the district.' There are records of the burrows having been occupied in the fourteenth century. According to tradition, Laugharne Marsh behind the dunes was reclaimed by a Dutchman in the fifteenth century, and the causeway from Kingaddle is supposed to be of the same date. The western part of the foreshore of these burrows is known as Pendine Sands, which are hard and firm. The ridges at Pendine and Laugharne run east and west and reach some 50 ft. in height, while some of the sand at Pendine when in contact with limestone cliffs is compacted into a calcareous sand rock.[2] On the east side of the bay are Towyn and Pembrey Burrows, where some antiquarian investigations[3] in 1910 resulted in the finding of shell mounds and pottery, as well as two metal buckles probably of the fourteenth century. These sand ridges generally trend east-west or a little east of south and in places

[1] T.C. Cantrill, *Arch. Camb.* 6th ser. 9, 1909, 433.

[2] *Mems. Geol. Surv.* 'Carmarthen', A. Strahan and others, 1909.

[3] *Trans. Carmarthen Antiq. Soc.* 7, 1911–12, 3.

reach 50 ft. in height. Before 1902 there were several shallow pools in the dunes,[1] but Strahan records that by this date all except one had been more or less sand filled. The sand is banked up against an old sea cliff. On the south side of the Loughor estuary there are some dunes at Whiteford Burrows (see p. 171), where the dunes and underlying spit have advanced about one and a half miles into the estuary. There are also dunes at Broughton and Llangenydd, but they need no special comment, nor, indeed, do the bay-head dunes at Port-Eynon, Oxwich, Nicholaston, and Penmaen. At Penard, however, there is some interesting historical material. The first castle was built there about 1100, and the present building and St Mary's church[2] probably date from 1270. The dunes seem to have been in existence in 1316, by 1478 they were advancing dangerously, and in 1528 church, glebe, and other land were inundated by sand. The church was found again in 1861, and it is worth noting that the old church at Nicholaston was also overrun by the Penmaen dunes. Davies[3] states that in the early part of the fourteenth century it stood in a village called Stedwerlango.

Another series of dune areas is found on the eastern side of Swansea Bay. Those at Crymlyn are rather narrow, and shut off Crymlyn Bog from the sea, while at Baglan, Aberavon, and Margam they stand on the seaward side of, and protect, flats over which they have made some advance. Higgins[4] states that danger from moving sand could not have existed in about A.D. 470 when the Hermitage of Theodoric was possibly first built. The dunes were certainly there in 1205, and by 1344 had extended inland as far as the main road between the Kenfig and Avon rivers. In 1336 the inhabitants of Margam Abbey complained of losses due to inundations by the sea, and Gray,[5] who excavated the Hermitage in 1898, concluded that the existing building dates from 1227 and that it was overwhelmed by sand about 1300. 'Some years ago, when removing sand from the dunes...at Port Talbot docks, a stone in an upright position was exposed at a depth of 20 ft., and when examined about 1928, the date 1626 was found plainly cut near the top. If this date is authentic, it provides valuable information as to the movement of the sand, for it shows that these deposits had accumulated at this spot to the height of over 20 feet in the period from 1626 to a few years ago. It has also been shown that

[1] *Mems. Geol. Surv.* 'West Gower and Pembrey', A. Strahan, 1907.

[2] W.L. Morgan, *Antiq. Soc. East Gower*, 1899, p. 210.

[3] J.D. Davies, *A History of West Gower*, 1894, pp. 441–4; also *Mems. Geol. Surv.* 'Swansea', A. Strahan, 1907.

[4] L.S. Higgins, 'An investigation...of the Sand Dunes...on the S. Wales coast', *Arch. Camb.* June 1933 (an important paper).

[5] T. Gray, *Arch. Camb.* 1903, p. 121.

considerable movement has taken place in the dunes near the sea within the last hundred years.'[1]

Kenfig Burrows are situated between the Kenfig River and Porthcawl, but the river really makes no true break in the dunes. In fact, physiographically and ecologically, the whole mass of dunes from Crymlyn to Porthcawl is one. Orr[2] has studied them from an ecological point of view, and his botanical remarks, which apply to Kenfig Burrows (*sensu lato*), are generally applicable to the vegetation of most of the South Wales dunes. The sand at Kenfig rests on Keuper Marls which occasionally are seen,[3] but the land around Kenfig Pool consists partly of boulder clay resting on a Triassic conglomerate. Three main plant associations are present: (1) *Ammophila arenaria* predominates on the outer, moving dunes, and it is interesting to note that *Agropyrum junceum* is rare, and *Elymus arenaria* apparently quite absent. (2) There is a *Salix repens* belt; this plant forms a fairly continuous carpet in the more sheltered dune valleys, but in more exposed places it collects sand and forms hummocky dunes. (3) *Pteris aquilina* (bracken) covers many acres of fixed dunes, and is a useful sand binder. The marram grass reaches inwards to Kenfig Pool, while the *Salix* covers a broad alluvial area (seaward of the pool) and also extends laterally. At Kenfig the *Salix repens* association is regarded as a transitional type: on its seaward side the subordinate species associated with it are those belonging to the Ammophiletum, whereas inwards, the sand is fixed, and the included species are those peculiar to stationary dunes.[4] On the drift line the usual plants occur—*Cakile maritima*, *Arenaria peploides*, and *Salsola kali* (74).

Kenfig Pool, in 1912, was a part of the burrows not invaded by sand. The water is fresh, and the pool was of the same outline in 1876, but, according to Orr, was of rather different form in 1814. Originally it was a marsh, and at that stage was drained by a stream flowing northwards to the Kenfig River. There is now no visible outlet, but Orr suggests

[1] Higgins, *op. cit.* p. 33, and A. Stuart, 'Petrology of the Dune Sands of South Wales', *Proc. Geol. Assoc.* 35, 1924, 318.

[2] M.Y. Orr, 'Kenfig Burrows', *Scot. Bot. Review*, 1, 1912, 209.

[3] Probably the solid foundation is very varied—Trias, Millstone Grit, and Carboniferous Limestone, and conceivably Coal Measures.

[4] Plants in the Marram association include, *Erodium cicutarium*, *Senecio Jacobaea*, *Euphorbia paralias*, and *Carex arenaria* (all common). With *Salix* are found *Rubus caesius* and *Ammophila arenaria* (both common). Low-lying hollows, submerged in winter, develop their own plant societies as also does Kenfig Pool. In the *Pteris* zone, where this plant is sparse or absent, *Ammophila* is dominant. *Salix* and *Rubus* spp. also help in fixation, as also does *Festuca rubra*. Mosses and lichens are common. There is no sea buckthorn on the burrows.

a northward filtration of the waters through the sands towards the river.[1]

The so-called Via Julia, possibly of the first century A.D., is supposed to have passed through the north-eastern part of the burrows. The settlement itself was on the river at a road crossing and was probably occupied by the end of the ninth century.[2] The castle certainly existed in 1152, and in 1184 the port was in active use. The church also dates from before 1154. In 1262 the new (Mawdlam) church was built on higher ground, and the pool is known to have existed in 1365.

> In spite of these changes, the old church of St James was in use up to 1397, the so-called Via Julia was still open in 1344, and the Castle was still occupied in 1403. We later have evidence of the final obliteration of the site by sand.... The evidence...makes it clear that in this region there was no apparent danger from moving sand at the close of the twelfth century, but that within a hundred years the sand in the immediate vicinity of the town itself was a danger, and that by 1485 moving sand was affecting the area beyond the town and had besanded the King's highway where it crossed the north-eastern corner of the area. It also shows that while the inner margin has become 'fixed' by a plant cover, considerable movement has recently been taking place near high-water mark.[3]

The last dune areas on this part of the Welsh coast are those known as Newton Burrows and Merthyr Mawr Warren at the mouth of the Ogmore River. A good deal of recent archaeological work has been carried out in this area and, although there are gaps, knowledge of it is more complete than of other similar places. In the following brief account Higgins's findings form the main source of information. All prehistoric discoveries come from the natural surface, and so suggest that up to near the end of the 'Neolithic' period sand had not gathered here; this opinion, however, is inconclusive since the sand might have been there and even so the finds would easily have collected on the surface during the erosion of the dunes. There certainly was blown sand present on the coast during the Beaker period (*c.* 2000–1500 B.C.), but it probably did not cover the whole flat. In the Middle Bronze Age the sand had reached the limestone scarp, and in the Early Iron Age there was some 6 ft. of sand at the scarp foot. In the Romano-British period a water-way probably passed through the area, and by A.D. 800 a certain amount of fixation of the sand by plants had taken place, especially in the western part. In the Norman period there was further stabilization, and still later a good deal was cultivated.

[1] Prof. George tells me (1943) that there is a small visible outlet occupied by a stream after rains, but temporarily choked after drought. The stream flows direct to the sea.

[2] Higgins, *op. cit.*

[3] Higgins, *op. cit.* p. 35: full references are given in Higgins's paper.

But sand was again in movement in the late sixteenth century, or earlier, because the road was overwhelmed by 1578. Between 1514 and 1573 the sand seriously encroached upon the river, and by 1700 the top of the scarp was besanded. Smaller variations in stability of the sand took place in the nineteenth century, and heavy movement occurred during some gales in 1897 and other years.

The evidence from nearly all the dune areas of this coast is thus fairly consistent. In the historic period no danger from moving sand was suspected up to the early part of the thirteenth century. But in the fourteenth century a comparatively sudden change occurred, and the records all imply that storms blew up much sand which greatly damaged property and cultivated land. The thirteenth and fourteenth centuries were times of great storminess, notably on this coast, but also elsewhere, and especially on our south and east coasts and in the Low Countries. Whether or no destruction was helped by actual vertical movements of the coast remains to be proved, but it is interesting to connect this possibility with the disturbances of this date suspected in Romney Marsh and Dungeness.

The petrology of the dune sands of South Wales has been investigated by Stuart[1] who shows that, with some small but significant exceptions, their composition is similar throughout, and that the bulk of the material was derived primarily[2] from a complex of igneous and metamorphic rocks. Muscovite (white mica) is conspicuous by its absence, and since the potential source of supply of this mineral is near at hand there are grounds for thinking that it remains beneath water level. Naturally, the frequency of particular minerals varies with geographical conditions. In Swansea Bay zircon, dolomite, rutile, chert, and chlorite are mainly concentrated between Port Talbot and Singleton, while minerals derived from the Pennant Series (except dolomite) have been concentrated by south-westerly winds in the north-east part of the bay. Minerals carried down by the rivers are more widely distributed, and it is interesting to note that, although the tin industry in South Wales is modern, tin fragments are common in all dunes, and may be found as much as 50 ft. below the surface.

[1] *Proc. Geol. Assoc.* 35, 1924, 316.

[2] Probably not immediately, but at secondhand via local sediments and glacial deposits, according to George.

(19) CHIPPING FLOORS, COOKING PLACES, AND CLIFF CAMPS

The remains of flint-chipping floors[1] found in south Pembrokeshire are allied to the early occupation of the dunes. Cantrill examined many of them and believes that, while this industry was primarily Neolithic, some of the sites conceivably were occupied in later Palaeolithic times, and, at the other end of the time scale, by Bronze Age people. He also argued that when Palaeolithic man was in Wales the land stood about 100 ft. higher above sea level than it does now, and that the flints were obtained from beaches long since submerged. Consequently Azilian and Neolithic man had to be content with the diminishing supplies as the beaches crept inwards with the submergence. Leach[2] maintained that some of the flint-working sites were on the now submerged land deposits on the Pembrokeshire coast, and that the collections of flint flakes obtained from them indicate places where early man worked the raw material into his implements. He also deduced that these sites were occupied before the full growth of the forest trees and also that in the absence of definite information they are rather earlier than Late Neolithic or Bronze Age stations. Further information is obviously necessary before anything very definite can be said about these chipping floors.

Of rather similar age are certain sites designated 'cooking places'. One near Marros, described by Leach, is clearly older than the storm beach which nearly overwhelms it. Cantrill and Jones[3] sum up their views as follows: '...we believe that the heaps of burnt stones we have found in South Wales are prehistoric hearths or cooking-places, where "stone-boiling" was performed; that as the only relics hitherto found associated with them are flint-flakes, their age is presumably Neolithic; but that such methods survived into the Bronze Age and Late Keltic times is probable, and the age of any given hearth can be determined only by the discovery of datable relics clearly attributable to the users of the hearth.'

In numerous places in South Wales, including Pembrokeshire, there are many 'camps' on headlands,[4] but precise dating is difficult since very little excavation has been carried out, but some are almost certainly Iron Age. That on Black Point is alleged to be Neolithic.[5] The scanty available evidence suggests that camps and flint sites are coeval, but the range of dates for the camps is wide. North remarks that they afford evidence of

[1] T.C. Cantrill, *Arch. Camb.* 6th ser. 15, 1915, 157.

[2] A.L. Leach, *Proc. Geol. Assoc.* 29, 1918, 146, and *Arch. Camb.* 11, 1911, 433.

[3] *Arch. Camb.* 6th ser. 11, 1911, 253. J.P. Gordon Williams, *Arch. Camb.* 81, 1926, 86, and see also *Journ. R. Anthrop. Inst.* 28, 1922, 73.

[4] F.J. North, *The Evolution of the Bristol Channel*, National Museum of Wales, 1929, p. 66.

[5] T.C. Cantrill, *op. cit.* 1915.

erosion during a period of relatively stable sea level. When they were built they were presumably near the coast, and usually were four-sided. 'For the edge of the cliff to have been anywhere near its present situation, the development of cliff and shore-platform must have proceeded for a considerable time, with little or no change of relative sea-level, and...there are reasons for supposing that stable conditions were reached by the close of Neolithic times...the...evidence suggests that the Camps must have been erected some time after the land attained its present relative level, that is to say, not very long before or after the beginning of the Christian era.'[1] In the light of more recent work on the Neolithic period and also of the whole post-Glacial period this view must be treated with some reserve. In any case it must be maintained that there is very little direct evidence of the age and nature of these camps. Furthermore, those referred to by North as indicating erosion are all on receding cliffs of Lias.

(20) THE PLATEAUX

This chapter on the Welsh coasts closes with a summary account of the several plateaux which have been mentioned so often in the foregoing sections.

In North Wales, especially in Anglesey and the neighbouring parts of Caernarvonshire, three platforms are recognized by Greenly.[2] Between 200 and 300 ft. above sea level, and covering some 400 square miles, is the Menaian platform; it formerly extended considerably beyond its present limits, but since its outer parts were in softer rocks these have been destroyed by erosion. From out of the remaining parts of the Menaian platform nine monadnocks rise, no two of them have the same structure and composition, but all have in common a resistant summit rock. They have been remodelled by glaciation, are generally rugged and nearly all conform fairly well with the 500-foot level; they have been named by Greenly the Monadnock platform. At 430 ft. is another shelf—the Tregarth shelf—seldom more than a mile wide, which Dewey maintains is due to marine erosion.

The Menaian platform is probably late Tertiary in age, the Monadnock is a little older, but both are almost certainly Pliocene. The Menaian is thought to be due to subaerial denudation, and the three may be regarded as successive denudations around the North Wales highland. In Cardiganshire O. T. Jones[3] has recognized two major platforms, the High

[1] North, *op. cit.*

[2] *Op. cit.* see also H. Dewey, 'On the Origin of some Land Forms in Caernarvonshire', *Geol. Mag.* 5, 1918, 143.

[3] *The Physical Features of Central Wales.* Nat. Union of Teachers, Souvenir of Aberystwyth Conference, 1911, p. 25.

and the Coastal. The former has a very even surface, rising to over 2000 ft. in the north, and sinking to about 1800 ft. in the south. It is dominated by such summits as Cader Idris. The Coastal platform south of the Dyfi forms a strip five to eight miles wide south of the Ystwyth; near the coast it is 400–600 ft. high and rises gently inland, but its inner margin is not well defined. Challinor[1], who has examined these land forms by the method of projected profiles, is of the opinion 'that if the hill-top surface approximates to any regular surface at all, it is to a curved surface rather than to a series of plane surfaces'. In general he is opposed to the view that these are uplifted peneplains.

It is well known that Sir Andrew Ramsay long ago first drew attention to the high plateau and ascribed its origin to marine denudation; he regarded the Brecon Beacons as an ancient line of cliffs. The high plateau extends over most of Wales, from Denbighshire to Cardigan, Carmarthen, Radnor, and perhaps Brecon. It truncates highly folded rocks of all types and is now deeply dissected by rivers. But whatever the importance and interest of this feature, detailed discussion of it is not relevant here since it scarcely concerns the present coast.

In South Wales, 'coastal plateau' is a somewhat vague term because the area is really made up of several different plateaux. The best defined is in Pembrokeshire and Gower, where it is cut right across highly folded Palaeozoic rocks, and bears no relation whatever to geological structure. In Gower it is well developed at about 200 ft.; in Pembrokeshire it is rather lower, and the present sea cliffs are cut into it (but see the *Patella* and other beaches in Gower). In the Vale of Glamorgan and in Pembrokeshire there is a higher platform at about 400 ft. which is much more dissected than the lower one. At about 600 ft. there is still another platform to be seen in Pembrokeshire and Cardiganshire, and Carn Llidi and Pen Bery may be outlying monadnocks of it. The Old Red Sandstone hills of Gower and of southern Carmarthenshire also approximate to this height.

The coastal plateaux of South Wales are later than the folding of the Mesozoic rocks, are probably Pliocene in age, and may compare with platforms of similar height found in other parts of the country. There are, however, varying views on the origins of these features. Goskar and Trueman[2] maintain that the surfaces found in South Wales were produced by erosion as the land was rising, 'the lower platforms having been cut so far as to remove all traces of a higher level at certain places, so that the lower platforms may adjoin the higher hills of the inland areas'.

Miller,[3] in discussing the platforms in Carmarthenshire and Pembroke-

[1] *Geography*, 15, 1929–30, 651. [2] *Geol. Mag.* 71, 1934, 468.
[3] *Geogr. Journ.* 90, 1937, 148, and see also *Proc. Yorks. Geol. Soc.* 24, 1938, 31.

shire, concluded that the 600-foot feature was due to subaerial denudation, and that it was peneplained by rivers to a sea level of about 400 ft.; he adds, however, that such a view does not necessarily preclude the idea of a drainage system developed on a surface of marine denudation at an earlier date. Sea level remained for a long time at the 400-foot level, leading to widespread erosion and cliffing. The sea later fell to 200 ft., and so an emerged coastal plain formed over which the rivers lengthened their courses and gradually graded themselves to the new sea level. The plain was therefore cut into isolated plateaux. Subsequently the sea fell lower than the present level, the land extended beyond its present limits, and the sea cut cliffs in the emerged coastal plain. Still later the sea rose, the estuaries were drowned and cliff erosion was rejuvenated.

In Miller's opinion the 600-foot platform is really the subaerial contemporary of the 400-foot plain, but this view is not always tenable in the Swansea district. In the Vale of Glamorgan the two features (i.e., the 200- and 400-foot levels) are separated by a step, and there is no suggestion of one grading into the other. In other words they must here, at least, be regarded as distinct entities and the 600-foot level is absent.[1] These two older features, however, are largely destroyed by subsequent erosion, and so are not always easy to follow in detail. The lower (200-foot) plateau is naturally the better preserved, but since it is by no means everywhere of the same height, it cannot be regarded as a normal wave-cut bench eroded to a long still-stand period of sea level. George pictures it as having had an infinity of shorelines. At this stage Swansea and Carmarthen Bays were not in existence, although they were in part lowland areas due to river erosion. Later than all these platforms are the raised beaches of Gower and other parts of the coast.

The full succession in Gower, and the rather lesser detail of Pembrokeshire are not evident all round the Welsh coasts. After all, this is not surprising, since erosion, especially in boulder-clay areas, would have cut away the features even if they had been present at one time.[2] But even if platforms higher than the 200-foot feature do not greatly influence the present coastline, that coastline is, nevertheless, the outcome of a complicated series of processes. Classifications of shorelines are often attempted and the simpler ones are usually the better, but it is clearly difficult to regard much of the Welsh coast as merely either submerged or emerged. It is true that the post-Glacial submergence has left an obvious imprint upon Wales; to regard the shorelines treated in this chapter as typically submerged would, to say the least of it, be inadequate.

[1] *Proc. Swansea Sci. and Field Nat. Soc.* 2, 1938, 23. In Gower the 200- and 600-foot platforms are separated by a step.

[2] They may also be masked by boulder clay.

Chapter VI

THE BRISTOL CHANNEL AND THE THAMES ESTUARY, AND THE SHORE FEATURES OF THE UPPER PART OF THE BRISTOL CHANNEL

(1) THE EVOLUTION OF THE BRISTOL CHANNEL

Before discussing in detail the shore features of the upper parts of the Bristol Channel and of the Lower Severn it is useful to examine the general nature and origin of this great indentation, and also to make some comparison between it and the London Basin. It is important at the outset to note the two definite parts of the Severn estuary: there is first the valley of the Avon and Lower Severn, and of this depression, which trends north-east and south-west, the Avon is the true head water; secondly, there is the Bristol Channel proper, with its east and west trend. Any comment on its history means allusion to events a long way back in geological time.

In the earlier part of the Palaeozoic the Bristol Channel region did not exist as such, but was merely part of a much larger area, and there is nothing that we know of that time that throws any light on the formation of the future depression. But the Silurian sediments which still remain suggest that the part of the district between the river Usk and Cardiff was occupied by an open sea. The nature of the sediments also indicates that a sea lay to the south and east of a ridge of land which was later submerged. In the Devonian period continental conditions prevailed over most of what are now the British Isles, except that in the south, and especially the area to-day covered by Devon and Cornwall, open-sea conditions prevailed. It was at this time that the Old Red rocks of South Wales were formed, possibly in a lagoon separated by a narrow ridge from the open sea to the south. This ridge, however, is quite hypothetical. After the Devonian period the waters gained on the land, and the Carboniferous sea lapped round the higher parts of the earlier land which survived as islands. Jukes-Browne, in his palaeogeographical restorations, postulated a land mass near the end of Carboniferous Limestone times called St George's Land, which occupied most of Wales and parts of central England. Its southern shore lay not far north of the northern edge of the present South Wales coalfield.

At the end of Carboniferous Limestone times parts of the land were raised, and rivers flowing from off the emerging surface helped to produce the sediments now known as the Millstone Grit. Later, elevation ceased and once more slow and intermittent subsidence took place, providing conditions for the formation of the Coal Measures, although towards the end of this time the coal seams became thinner and less persistent. This trend implies a climatic change which foreshadowed a period of great earth movements, the Hercynian orogeny. In South Wales and the Mendips the Hercynian folds run generally east and west, even though they run north and south in other places. This folding involved all the Carboniferous and earlier rocks in South Wales, and subsequent erosion removed a great part of the Coal Measures. This denudation implies that there was then relatively high ground in the coalfield area, whereas to the south of it the land seems to have been low lying, or, at most, undulating. At the same time, farther south, in Cornubia, the land stood relatively high. Thus are reconstructed conditions which are compatible with a depression in the Bristol Channel region. There is interesting evidence of them to be found at places such as Barry where the waves of the Triassic period formed a definite shore platform or wave-cut bench and a cliff in the Carboniferous Limestone. This feature clearly implies change of level, which may possibly have begun earlier in Glamorgan than farther west, because in that direction similar beach deposits are known of Liassic age. It seems certain, then, that not long after the Hercynian folding took place the Bristol Channel region was a broad and low-lying depression, although such a statement emphatically does not apply to the north-east and south-west parts of the upper estuary which developed later. Furthermore, the river which appears to have flowed along this low ground rose in the *west*, and continued its course over the present Bridgwater flats.[1] The evidence for this eastward-flowing river is found partly in the nature and distribution of some of the pebbles in the Lower Triassic conglomerates of Devon and Somerset. Moreover, when the sea, probably an inland one, into which this river flowed, expanded in Rhaetic and Liassic times, it did so from the east and south, a process suggested by the lateral changes that can be traced in the deposits of these times. In any case, in early Mesozoic times there is no doubt that the sea encroached on the undulating ground to the south of the coalfield, and probably turned the lower part of the eastward-flowing river of early Triassic age into a wide estuary. Later, either on account of subsidence, or of the breaking of a barrier, the open sea of the Rhaetic broke into this lake or inland Triassic sea, and the bone bed was formed. One possibility of tracing both the subsidence and levelling by marine action at this time lies in the manner in which the

[1] It is clear that many of the essential physical features of South Wales had developed by this time.

Triassic and Liassic deposits of a coastal type bank round the old hills and later islands of Carboniferous Limestone.

> It is the way in which Triassic, Rhaetic, and Liassic rocks alike, take on a littoral or shallow water aspect when traced north and west in Glamorgan, that indicates the proximity of the shore of the sea in which they were deposited....[1]
>
> Quite early in the Liassic period, the whole of the area with which we are more immediately concerned, with the exception of the elevated tract of the coalfield, had been buried beneath a covering of sediment. This preserved the Triassic desert topography, for further denudation of the older rocks was, of course, impossible, and as the Mesozoic era wore on, the subsidence, with its consequent encroachment of the sea, continued, until by the close of the Cretaceous epoch, when the Chalk was being deposited, the whole of South Wales, in common with the greater part of the remainder of the British area, was beneath the sea. By this time, strata belonging to various other subdivisions of the Mesozoic rocks, up to and including the Chalk, had been laid down, and the old western surface was still more deeply buried beneath a great but unknown thickness of sediment.[2]

During parts of this time, however, the east-flowing river maintained itself and, when the subsidence was at a maximum, it would have had a wide estuary. When subsidence was interrupted by periods of uplift, as, for example, during the deposition of the Portland Stone, the sea retreated eastwards once again. But in Chalk times there was a general downward movement, and the Chalk sea probably enveloped nearly all the western land mass, and, as is known, reached at least as far as Antrim in north-eastern Ireland.

At the end of the Cretaceous period there was renewed uplift, a forerunner of the later Alpine orogeny. In South Wales this early movement exposed much of the Chalk to denudation, so that probably a great mass of it was worn away before the earliest Tertiary strata of England were formed. Denudation continued during the Tertiary era, with the result that, apart from the Lower Lias, all the Mesozoic strata which had been over the southern part of Wales were removed. It is difficult, if not impossible, to trace the actual changes that took place during such periods of erosion, and in order to get some idea of what was happening, geological events in other parts of the country need examination. It is unprofitable, however, to treat this matter at all fully in this chapter, since not only is it one of some difficulty, but also rather distracting from the main theme. Suffice it to say that at the beginning of the Pliocene period this country stood at a lower level, relative to the sea, than it now does, so that early in it the Bristol Channel region definitely subsided. The extent of subsidence is difficult to determine; the nature of the evidence depends much

[1] North, *The Evolution of the Bristol Channel*, p. 26.
[2] *Ibid.* p. 27.

on the study of surfaces of marine denudation, and also on that of Pliocene deposits like those at St Erth in Cornwall. Probably the total sinking was of the order of 400 ft. Later, as the evidence from the platforms implies, uplift took place in stages; taking the country as a whole, however, emergence was probably in the nature of a gentle warping, since in this process allowance must be made for the deposits of later Pliocene age so well developed in East Anglia. As North writes: 'Although there is no definite evidence forthcoming from the Bristol Channel region, it is reasonable to assume that, as a result of the uplift, the coast line retreated far to the west, and the Bristol Channel became the valley of a *westerly* flowing stream.'[1]

One general result, first of the extension of the Mesozoic seas and secondly of the Tertiary Alpine movements, was probably to give the surface of southern Britain a general slope to south and south-east, and it was on this surface that the present river system of the country originated. Many physiographical changes due to capture, glacial diversion, and other causes have, however, taken place since those times. Some are certainly due to original puckerings on the Chalk surface, and this comment applies possibly to the Bristol Channel area, so that east-west folds occur as well as others running north-east to south-west. It is conceivably to the latter group that the Avon-Severn[2] owes its course, and its later headward cutting led to the diversion of many of its present right-bank tributaries. The rivers on this Chalk surface presumably originated in the Miocene. The submergence at the beginning of the Pliocene has already been quoted. It was at this time that the Bristol Channel region was drowned, and it is possible that water communication was continuous thence across England to the North Sea. Indeed, the nature and distribution of the Pliocene deposits suggest it (see below). Thus, during a part of Pliocene times, the Severn estuary and the London Basin may conceivably have been a continuous open-water area. 'A general uniform uplift alone would probably have been sufficient to divide the Pliocene channel into an easterly flowing Thames and a westerly flowing Bristol Channel, by restoring in some measure the conditions before the subsidence, when there were two valleys draining in opposite directions, but other factors are also involved—factors concerned in the larger problems relating to the drowning of many of the westerly opening valleys, the isolation of Ireland, and the final disappearance of the old North Atlantic Continent.'[3] The larger problems alluded to can, however, be left without comment: it is the

[1] North, *op. cit.* p. 40.

[2] This would not be the case if Buckman's theory is correct: the Avon-Severn would then be a secondary strike stream.

[3] North, *op. cit.* p. 89.

reference to the Bristol Channel which matters, explaining its main features as we see them to-day. There remains still for consideration the process of erosion in this area and it will be clear that, especially after the Hercynian and Alpine periods of mountain building, vast masses of strata were thus removed. It is, however, a topic which is treated more conveniently later.

(2) THE LONDON BASIN

The suggested connection of the Pliocene sea across England has not yet been proved, and the views of Wooldridge and Linton are worth quoting: 'As has already been pointed out the western limits of the Pliocene sea beyond the London Basin have not been determined by direct evidence, but every available indication suggests that the waters passed beyond the limits of the basin into the Wessex region....Direct stratigraphical evidence of this further trespass of the Pliocene sea over southern England still awaits discovery.'[1] This point deserves emphasis in order to stress the idea that, although the basins of the Severn and Thames may be closely related, they were not necessarily actually joined at any time by salt water.

The evidence for the formation of the London Basin in pre-Eocene times is meagre, but what there is suggests that there was a syncline in many ways similar to its successor. It was also probably drained by a longitudinal consequent stream and appears to have been asymmetrical, with its south side steeper than the north. In the Eocene there were two periods of deepening, roughly coincident with the Thanetian and Ypresian transgressions. These alternated with periods during which continental conditions spread seaward. Apart from the major folding movements which produced the syncline, there were also minor warpings during the Eocene along lines roughly transverse to the main axis. These minor movements naturally affected the thickness and lithology of the sediments. The present disposition of the strata in the London Basin results mainly from post-Oligocene movements, but these in point of fact produced relatively few new structures. 'Their energy was for the most part expended in augmenting features of a far anterior date. It is true that not a few folds or faults whose formation is attributed to post-Oligocene times have no obvious Eocene analogues, but there is every reason to believe that this is due in part to the inevitable absence of detail in the maps....'[2] In any event, there is probably a very close relationship be-

[1] S.W. Wooldridge and D.L. Linton, 'Structure, Surface, and Drainage in South-East England', *Inst. Brit. Geographers*, 10, 1939.

[2] S.W. Wooldridge, 'The Structural Evolution of the London Basin', *Proc. Geol. Assoc.* 37, 1926, 162.

tween surface structures in the basin and those in the underlying Palaeozoic floor. Wooldridge compares these superficial movements to those of a thin skin of incompetent material resting on a thick parquet floor subjected to readjustments in both vertical and horizontal directions. Furthermore, 'the London platform bears the same relation to the Kent coalfield as the Welsh massif does to the South Wales coalfield, and hence it seems reasonable to expect some degree of analogy between the structure of the Welsh massif...and that of the rocks under London'.[1]

In late Miocene times there seems to have been a long period of quiescence in so far as tectonic movements were concerned. It was a time of great planation. The Diestian sea, in which the Lenham Beds were deposited, seems to have filled the basin up to the bounding Chalk escarpments, and in Wooldridge's view came into being as a result of some slight down-warping like that which caused transgressions in Eocene times. The more or less undisturbed nature of the Diestian Beds is reckoned proof that there was no further warping of the London Basin during the Pleistocene. That region is definitely both a tectonic and a stratigraphical unit, and the salient features of its history are thus summarised by Wooldridge: '(*a*) The syncline has grown spasmodically from small beginnings, maintaining throughout its growth the same major features of form and structure. (*b*) The successive phases of deepening have coincided with, and are to be regarded as the cause of, the several temporary incursions of the sea. (*c*) The syncline has been drained throughout by a longitudinal consequent river system.'[2]

Topographically the estuaries of the Severn and Thames are perhaps the most prominent features of our coastline.[3] The history of the former appears to go farther back in geological time than that of the latter, but it must be borne in mind that the underlying floor of Palaeozoic rocks is not exposed in the east of England, and that consequently the evidence of those rocks is indirect in tracing the history of the Thames Basin. Undoubtedly much detailed work on the Bristol Channel region is needed before its structural evolution can be traced as certainly as Wooldridge has done for the London Basin.

(3) O. T. JONES'S VIEWS ON THE BRISTOL CHANNEL AND RELATED PROBLEMS

Before discussing actual coastal details it is useful to turn to O. T. Jones's views on the evolution of the Bristol Channel. In his Presidential Address to Section C of the British Association at Bristol, 1930, his attention was

[1] Wooldridge, *op. cit.* p. 189. [2] *Op. cit.* p. 193.
[3] The coastal features of the Thames estuary are described in Chapter IX.

given admittedly to certain aspects only of this subject, but his paper includes also many valuable comments on the structure of south Britain.

In his opinion the planation in Triassic times was a very important episode. There was then certainly intense erosion which probably took place under arid continental conditions. If the relation of the New Red Sandstone rocks to the older Palaeozoics be considered, it becomes clear that in some places all the Carboniferous and Old Red Sandstone had been removed, since the New Red Sandstones lie directly on the Wenlock. Thus, probably 8000 ft. of strata were removed from the anticline in Palaeozoic rocks running between Cardiff, Llandaff, and Cowbridge: hereabouts the Keuper rests directly on the edges of the older rocks. It is arguable, further, that the Welsh plateau was eroded subaerially at this time; in any case Jones maintains that the then 'Channel' was eroded as a broad depression flanked by erosion scarps having approximately the same trend. The Trias was deposited against these scarps. The trend of the evidence is certainly in favour of a depression caused by erosion and not by tectonic movements.

It is clear in the first place that after the Triassic period there was subsidence, during which the Rhaetic deposits obviously indicate the flooding of an almost level plain, and in the second, that the Mesozoic rocks probably extended far over the Palaeozoic areas of Wales. This, in its turn, may imply, as Ramsay believed, that the Welsh plateau is due, not to subaerial, but to marine planation, and that features such as the Brecknock Beacons represent old cliff lines. It is also recognized that the Miocene movements were very important in southern Britain, and that they undoubtedly affected the underlying Palaeozoic rocks. The structures of the London Basin and Weald are not simple despite the accounts which often appear in textbooks and which tend to mislead. In the Weald there are numerous folds arranged in echelon, but no one fold can be traced for more than a few miles.[1] On examining once more the Bristol Channel region, however, even greater complications are apparent. The Trias, Rhaetic, and Lias are the Mesozoic formations which reach its shores: the Lias is conformable to the Rhaetic, and that formation rests discordantly on the older rocks where the Trias is absent. Jones argues that in spite of much loss on account of erosion, if the Keuper occurs, the Rhaetic and Lias may be assumed to have followed it, but that the converse of this deduction is *not* true.

If this assumption be allowed, it is then possible to estimate the level of the base of the Lias in places from which that formation has been re-

[1] W.B.R. King, *Quart. Journ. Geol. Soc.* 77, 1921, 135, shows that a similar arrangement of folds occurs in Picardy.

moved, but in which the Rhaetic remains. Moreover, another approximation, if less accurate, can be obtained, should only the Keuper survive. To illustrate this point Jones's scheme worked out in the Mendip region is interesting to follow. There the Lias-Rhaetic junction is about 850 ft. O.D., and falls to north and south. In the central Mendips the junction stands at some 500 ft. near Wells and Shepton Mallet, and reaches sea level between three and four miles farther south. At Glastonbury the Lias is found at −200 ft. O.D., and in the Polden Hills at +200 to 300 ft. To the north of the Mendips the plane at first falls, then rises, falls again to the river Avon, and then rises once more. 'On a line of section drawn, therefore, from the northern suburbs of Bristol through North Hill, to Glastonbury and beyond, we have two well-marked regions of elevation, viz. one in the North Hill region and the other in the central Mendips; and three synclinal depressions, viz. at Glastonbury, between North Hill and the Mendips, and in the Avon Valley.'[1] All this is taken to mean post-Liassic folding[2] along impersistent axes similar to those in the Weald and northern France. Hence arises the question: is this folding, as postulated by Jones in the west, of the same nature and age as that in the east and in Picardy? Unfortunately, the obvious Miocene folds of south-eastern England cease suddenly at the outcrop of the Jurassic rocks, apart from a few shallow folds in the Middle and Upper Jurassic strata. This occurrence may possibly be explained in two ways: (1) if a series of rocks with an easterly dip, like the Jurassic, be folded along east-west axes, the result will be less pronounced than in the case of horizontal beds; (2) in any folded area there appear, at right angles to the main direction of the folds, culminating points forming great domes and cols. In the domed regions the movements are usually concentrated on one or two chief axes, while in the cols there are as a rule many minor axes, no one of which is predominant. In the central Weald anticlines are very clear, but they are hardly recognizable in the Chalk plateau between Winchester and Kingsclere. Farther west, the folds reappear in the Vales of Pewsey, Warminster, and Wardour. 'Despite the fact that the Middle and Upper Jurassic outcrops appear to be almost devoid of folding transverse to their outcrops, there is a remarkable general correspondence between the main folds that traverse the base of the Cretaceous and those that affect the base of the Lias farther west.'[3] Jones therefore argues that the varying level of the Lias-Rhaetic junction implies a continuation in the Bristol Channel area of the Miocene

[1] Jones, *op. cit.* p. 19.

[2] The folding may very well be intra-Liassic: see Tutcher and Trueman, *Quart. Journ. Geol. Soc.* 81, 1925, 595.

[3] Jones, *op. cit.* p. 22.

folds of south-eastern England. The London Basin itself may be represented in the west by the depression between North Hill and the Mendips, but more likely in the Avon valley syncline to the west of Bath.[1]

Should this westerly extension of the Miocene folding be accepted, the different effects in the two areas would be explained largely by the nature of the Palaeozoic floor, and since that in the west was the more thinly covered the floor itself should reveal the folding more clearly than in the east. But if the London Basin and Bristol Channel were determined by Miocene movements, what happened to the north and south of the latter? Strahan[2] assumed that the rivers of South Wales rose on a cover of Mesozoic rocks, and that the directions of some of them resulted from a south-easterly tilt; the Neath, the Tawe, and others, however, run to the south-west, and he held that the latter direction was the result of renewal of movement along certain lines of disturbance. The streams cut back along these lines and beheaded some of those flowing to the south-east. But other streams, for example the Loughor, add complexities, since their directions are exceptional. Jones agrees with Strahan on most points, but thinks it possible that the effects of two movements are involved, that is, an earlier tilt to the south-east, and a later one to the south-west and south: his argument may indeed justify further development. The high plateau of Central Wales may have been down-warped in a southerly direction far enough to reach the level of the coastal platform, and the relations between these two features suggest the warping to be about 1900 ft. in eighty miles. If this happened, it would certainly have had a great effect on the streams, and would easily explain the trend of those flowing to the south-west.[3] What is more, such a warping of the high plateau implies that it was domed along a broad area running roughly east and west in north-central Wales. If that direction be continued eastwards, 'it meets the axis of elevation that has caused the great swing in the strike of the Cretaceous rocks in the Fen Country. Moreover, the transverse watershed in the Midlands that divides the southern streams of the Warwick Avon drainage from the northern streams of the Trent system lies approximately on the same line.'[4] Much additional work on this particular problem is required, but it is recognized that the south-westerly inclination of the plateau to the Bristol Channel is analogous to the south-easterly slope of the Chalk to the London Basin. It is useful to turn for a moment

[1] It is important to note that in south-western England intra-Jurassic and pre-Upper Cretaceous movements are sometimes important.

[2] A. Strahan, 'The Origin of the River System of South Wales', *Quart. Journ. Geol. Soc.* 58, 1902, 207.

[3] See p. 193 (T.N. George).

[4] Jones, *op. cit.* p. 25.

to the region south of the Bristol Channel. The syncline of the Hampshire Basin seems to be prolonged between the Blackdown Hills and the Dorset coast. In central Devon the streams converge towards the centre of the plateau basin before finding their exits in the Exe, Torridge, and Taw, an arrangement rather like that obtaining in the Hampshire Basin. Jones suggests that in central Devon the Palaeozoic surface has been warped into a synclinal form thus explaining the disposition of the river systems, and that this movement was the effect of the Miocene folding on the hard Palaeozoic areas. Should his view of the warping of the central plateau of Wales and the central plain of Devon be accepted, Exmoor may owe its height rather to a doming-up of the area than to greater resistance of its rocks. This hypothesis is supported by an examination of the main east-west watershed of southern England. From the Straits of Dover to the Vale of Pewsey it seems to be controlled mainly by the position of the most important folds in the Chalk, and trends generally north of west. Westwards from the Vale of Pewsey it runs south-west to the south of Yeovil, west along the north flank of the Blackdown Hills, and thence to the centre of Exmoor. Thus Jones regards 'the Bristol Channel as having come into existence as a definite basin by folding during the Miocene period, and that the present form of the surface in Devon and South Wales owes its origin to warping during the same period of an ancient surface of erosion'.[1]

In giving this outline of Jones's interesting and significant views, the main object has been to show the possible relationships between the two largest inlets of the English and Welsh coasts. It is quite clear, however, as Jones himself said at the end of his address, that he had 'propounded more problems than [he] had solved'.

George[2] is not in agreement with the views expressed by Jones, and is of the opinion that 'the illusory appearance of tectonic warping on the lower ground may then be attributed to the emergence of a submarine floor that, in face of the effects of repeated rejuvenation displayed by all the [South Wales] streams, is best interpreted as a series of platforms intermittently uplifted and separated by lines of degraded cliffs....' George doubts also if any relics of a warped surface are still to be found on the high plateau, and holds that the chief changes in the drainage pattern as well as the main rejuvenation phases of the streams took place during the formation of the coastal plateau, which is composite, and also long after the cessation of any warping. Each successive rejuvenation caused deeper incision of the streams, and Strahan's views are regarded as inapplicable

[1] *Op. cit.* p. 27.

[2] 'The Towy and Usk drainage pattern', *Quart. Journ. Geol. Soc.* 98, 1942, 89.

to the streams of the coastal plateau 'for there can be little doubt that the extensions southwards of the ancient Cothi-Towy into the Loughor and of the Brechfa streams into the proto-Gwendraeth were a consequence of the emergence of the unwarped "600-foot" platform and took place in Pliocene times; while the principal events of diversion and rejuvenation were even more recent. This conclusion receives strong support if it is accepted that, as O. T. Jones has contended, the Bristol Channel and its environs were considerably deformed by the Miocene earth movements'. The whole subject is one of great complexity, however, and the discussion on George's paper showed that there is still much to be achieved before full agreement becomes possible.

Those who wish to follow in far greater detail the structure of south-eastern England must consult other works, especially *Structure, Surface, and Drainage in South-East England*, by S. W. Wooldridge and D. L. Linton. It is, however, relevant to make one quotation from their account:

> There can, however, be little doubt that important differences of structure exist between the areas north and south of the Lower Thames. Comparison with the emergent areas in Belgium, and in Devon and Cornwall, suggests that the southern area is traversed by powerful folds of generally E–W trend. North of the Thames there is neither direct evidence nor indirect suggestion of E–W folding in the floor. The projection of the floor structures of the Midlands towards London suggests rather that N.W.–S.E. elements are dominant.... We should anticipate no close general relation between the form and structure of this floor, for it is essentially a warped peneplane bevelling the structures of its constituent rocks. Nevertheless both form and structure combine to emphasize the magnitude of the contrast in underground conditions which takes effect at the line of the Lower Thames valley. Structurally indeed this line is probably the continuation of one of the major tectonic boundaries of Europe, that which separates the *Palaeo-Europe* and *Meso-Europe* of Stille. In the latter, the E–W Armorican folds advance wave-like towards the earlier folded and compacted Palaeo-Europe and break in thrusts upon its edge.

This factor has had much effect on the folding in the cover of newer rocks. Clearly, south of the important structural line demarcated by Wooldridge and Linton, the surface rocks all show good folds, but this is not the case north of the Thames.[1]

This is not the place to comment further on the complicated history of south-eastern England, and in any case it would be impossible to relate in detail the geological evolution of this region to that of the west. Admittedly, in Pliocene and later times, there were in both estuaries many stages of planation which are represented by the coastal platforms of the Bristol Channel and the various terraces of the Thames. The history of these events is of great interest, but for the most part only indirectly

[1] For details, see Wooldridge and Linton, *op. cit.*

affects the true shoreline problems.[1] The possible correlation of events in the Bristol Channel, Severn, and Thames estuary[2] is better shown in Fig. 34, after Wills, than in further description. The interest of this theme

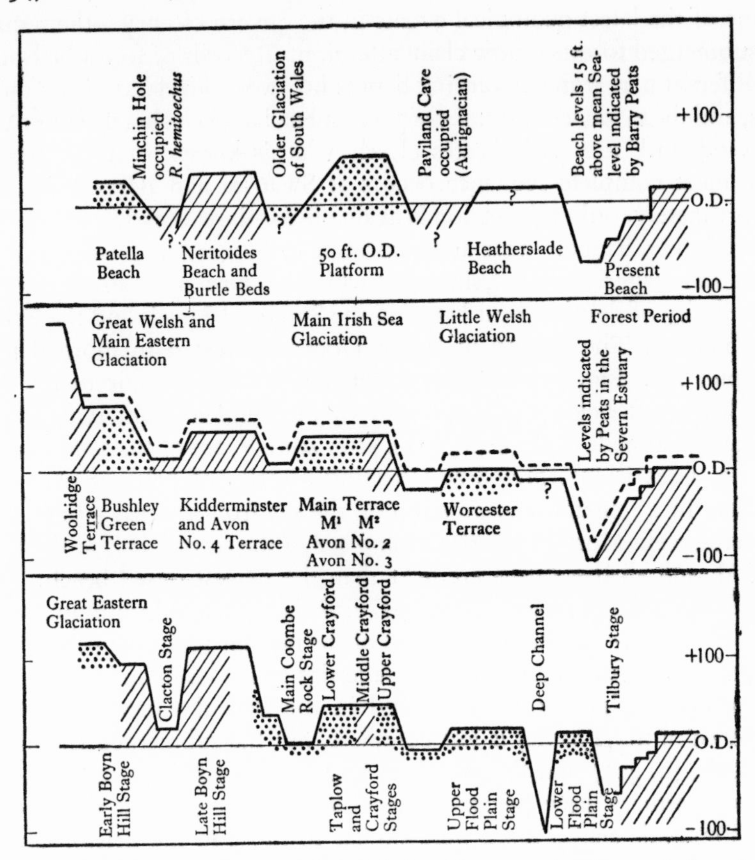

Fig. 34. Graphs showing suggested correlations between beaches and deposits of Gower, the Bristol Channel, and the Thames (after L.J. Wills, *op. cit.*)

is undoubted, but a full discussion of river terraces is outside the scope of this volume, and would introduce so much new matter that it has seemed better to leave it alone. For further treatment of the problems of the Severn and Thames the paper by Wills cited in the footnote is very important.

[1] See L.J. Wills, 'The Pleistocene development of the Severn from Bridgnorth to the sea', *Quart. Journ. Geol. Soc.* 94, 1938, 161.

[2] Details of the Thames estuary are given in Chapter IX, and the effect of submergence on it is demonstrated.

(4) THE PEAT AND FOREST BEDS OF THE SEVERN ESTUARY

Some of the latest geological events in the Severn estuary—the sequence of submerged forests—now claim attention. Records of forest beds found in different places and at varying depths below sea level have been numerous, and the evidence about these phenomena has been shortly summarized by Godwin.[1] Fig. 35 indicates clearly what is known of these deposits. The most complete sequence occurs at Swansea and Barry Docks but, unfortunately, only one section is available for the succession on the south side of the Channel—that at Combwich in Somerset. Further work in this area is eagerly awaited, since that which has been done already suggests a similarity of development between the Somerset Levels and the Fenland (see below). The section at Barry Docks is treated in more detail in Chapter XII. Fig. 35 also shows that the peat beds occur to the depth of at least 50 ft. below O.D., that they vary in number in different localities, and also that beds of the same level, after zone V, are of the same age. The whole series is interpreted as reflecting a series of halts or even slight retreats in a general transgression of the sea from at least −50 ft. to the present level. Most of this rise took place in the second half of the Boreal period, and all these beds are later than the newest raised beach. In the North Sea and Fenland there has probably been a continuous transgression from −170 ft. towards that of the present time (see p. 493). This is probably a case of a genuine eustatic rise of sea level, and consequently lower peat beds may yet be found in the Bristol Channel area.[2] The evidence in the Thames is less complete, but the submerged land surface in Essex (p. 403) and the peat beds and Romano-British surfaces in the Thames estuary now below sea level at any rate suggest that the later physiographic stages there were similar to those in the Bristol Channel. Godwin[3] states that 'the stratigraphic data reinforce the view that the structure and development of the region (i.e. the Somerset Levels) are very similar in principle to those recognized in the East Anglian Fenland'.

Therefore, both in the Bristol Channel and in the Thames there appears to have been during and since Pliocene times a long period of down-cutting punctuated with considerable halts or minor transgressions during

[1] H. Godwin, 'A Boreal Transgression in Swansea Bay', *New Phytologist*, 39, 1940, 308.

[2] See note, p. 171, which calls attention to the alluvial deposits in the Tawe valley. These deposits strongly suggest that the rise of sea level in the Bristol Channel has been more or less continuous from 200 ft. below O.D.

[3] *New Phytologist*, 40, 1941, 131.

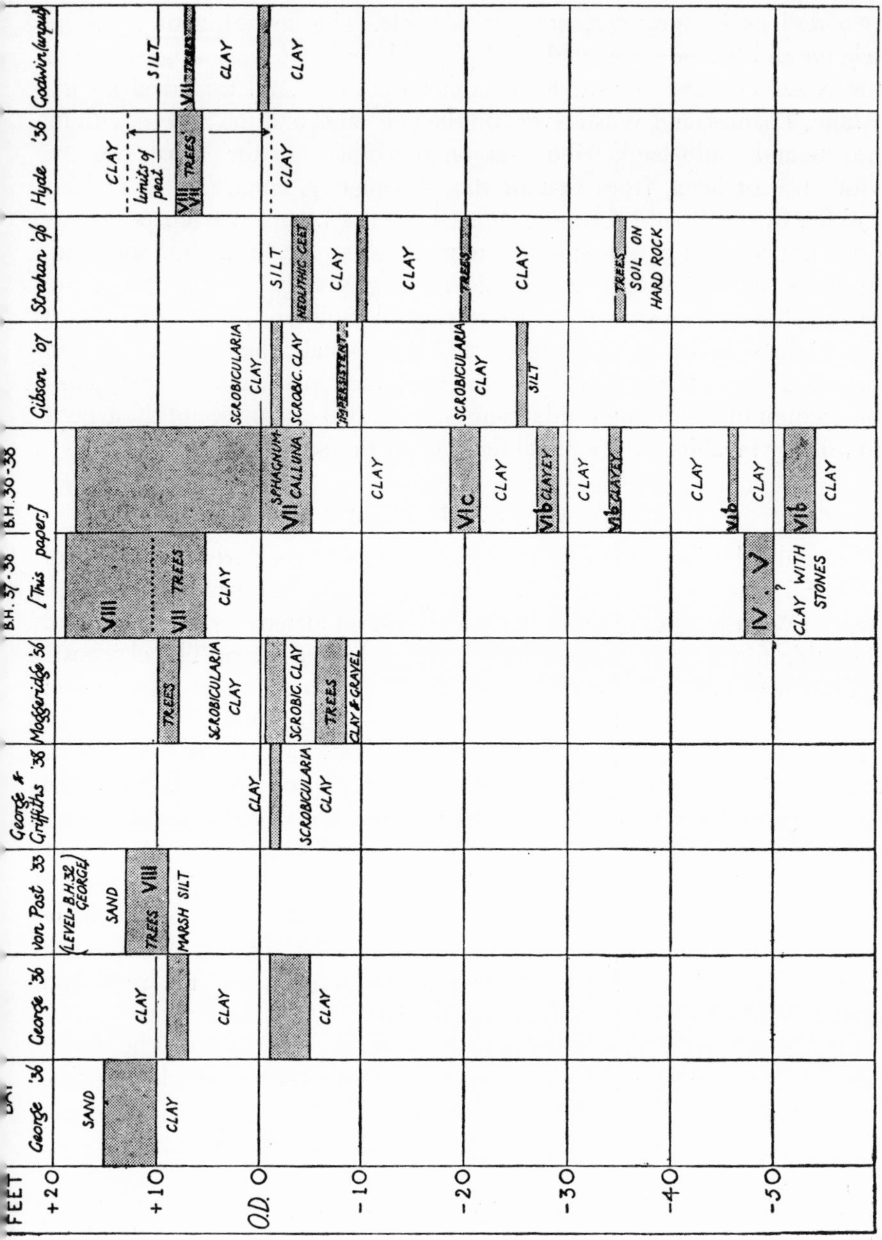

Fig. 35. Sequence of submerged forests in the Bristol Channel

(From H. Godwin, *New Phytologist*, **39**, 1940, 308.)

which platforms and terraces were built. *Exact* correlation between the two regions is, however, not yet possible. This lowering of sea level relative to the land continued until probably most of the southern part of the North Sea and the Bristol Channel were dry land traversed by the Rhine, Thames, and Wash rivers on the one hand, and the Severn with its north- and south-bank tributaries on the other. In the North Sea the difference of level, from that of the present day, amounted to at least 170 ft.: in the Bristol Channel direct evidence obtained from peat beds does not suggest a difference of more than 50 or 60 ft., but there are grounds for assuming it was considerably greater because of the nature of the alluvial deposits in the Tawe valley.[1] Still later, mainly in Boreal times, the sea rose again in stages, and the several submerged-forest and peat beds were then formed. Slight variations in their levels may exist on account of the different tidal ranges in the two estuaries, but these need in no way invalidate the general findings of this section.

(5) COASTAL DETAILS

The actual details of the coastal features of the upper parts of the Bristol Channel finally call for examination (Fig. 36). Upstream from Cardiff are the Wentlloog Levels which consist of bluish marsh clay of tidal origin. The tide formerly reached the old cliff line behind them. The clay rests on Keuper Marl, and the existing tract is a product of post-Glacial, and quite possibly of post-Boreal, times. There is no blown sand in this area, and the mud itself is not of the kind to yield to wind action at low water. The Caldicot Levels are similar to the Wentlloog. Both have often suffered from severe inundations, due usually to the breaking of banks during high tides and storms. The greatest recorded flood occurred in the winter of 1606–7. To-day the levels are reclaimed and cut up into fields by hedges and dykes. Seawards they are fringed by a sea wall, outside of which are either bare mud flats or natural salt marsh. The contrast between the nearby urban development of Newport and Cardiff and the rural and remote character of the levels is striking.[2]

Upstream, Sedbury and Aust cliffs serve as convenient limits.[3] They were, indeed, at one time continuous. Aust Cliff is now a truncated ridge face of Triassic and Lower Jurassic rocks which are bent into a gentle but faulted anticline, and the faults each coincide with slight promontories.

[1] See footnote 2, p. 196.

[2] Gold Cliff, a hill nearly 50 ft. high, dominates the south-western part of Caldicot Levels. It is an inlier of Keuper and Rhaetic rocks.

[3] For further details of the tidal reach above the Severn Tunnel see L.J. Wills, *op. cit.*

This arrangement is apparently explained by the fact that the faults let in the relatively hard Lower Lias beds which also help to protect the

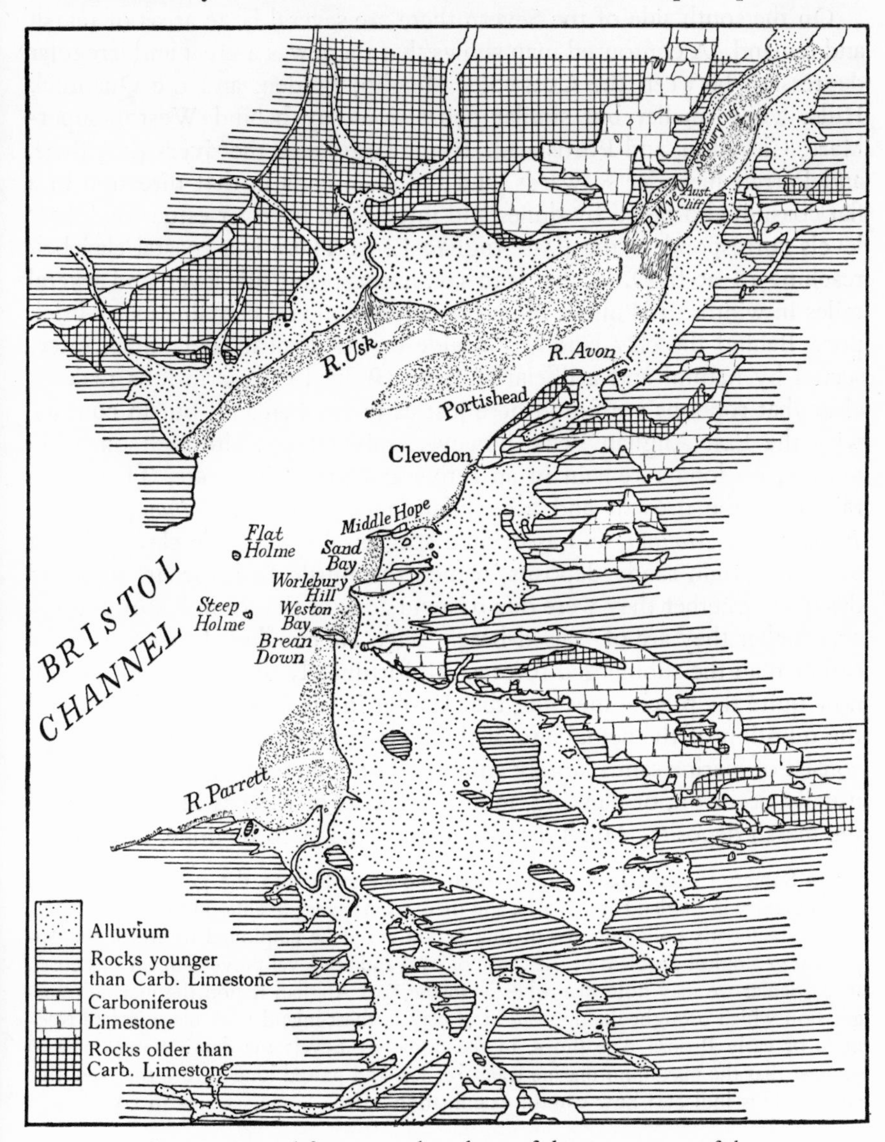

Fig. 36. Coastal features and geology of the upper parts of the Bristol Channel (based on Geological Survey)

underlying rocks from erosion. It may, however, depend in part on the hardening of the rocks by pressure due to faulting. The marked colours of the beds, the steepness of the cliffs, and their general setting make both Aust and Sedbury cliffs well worth a visit.

On the south side of the Severn there are several large areas of marsh and fenland. A contoured map shows that there was a great and irregular depression between the Cotswold, Mendip, Polden, and the Quantock Hills, while smaller but similar depressions lay behind Weston-super-Mare, Clevedon, and Portishead. In the larger area the rivers Exe, Brue, and Parrett meander seawards, and the plain ends in that direction in a flat coastline bordered by dunes and intersected by river estuaries. These levels correspond to the East Anglian Fens. There is a coastal clay belt reaching 18–20 ft. O.D., that is to say, mean spring-tide level, and several miles in width. It is under intensive cultivation and contains many brick pits. Behind the clay land lies a wide extent of peat land, much intersected by natural and artificial streams. Bulleid's[1] descriptions make it clear that roddons are found here just as in the Fenland, but in contrast with the East Anglian Fens, however, raised bogs, although once extensive, no longer remain in an active and increasing state. They were raised by the growth and accumulation of *Sphagna*, *Calluna*, *Scirpus*, *Eriophorum*, *Molinia*, and associated species. Between the clay and peat districts certain sandy deposits, named the Burtle Beds, occur. Ussher[2] discusses whether they were due to incursions of the sea in historic times, or whether they are of 'raised-beach' age. He concluded that they were earlier than the submerged forests. Unfortunately this statement has not now quite the precise meaning that Ussher in his day assumed for it, and the question of the actual age of the beds must remain open.[3] The beds consist of marine sands largely composed of comminuted shells, and they usually occur on, or flanking, mounds of Lias or Keuper Marl outliers in the marsh deposits. Much detailed work remains to be done, but

These Somerset results have already been shown to fit into the general pattern of forest history in England and Wales, and there seems every likelihood that the sensitive and characteristic pollen curves of this region could be employed to attack the important geological problem of the cause of land and sea-level movement on this coast. In particular we may note that whereas in the East Anglian region the marine transgression which brought the deposition of clay farthest inland took place in late Neolithic or early Bronze Age times, in the Somerset Levels the data (though sparse) suggest that the maximum marine transgression had passed long before the Neolithic period. This indicates how careful examination of the Somerset area in comparison

[1] A. Bulleid, 'Ancient Trackway in Meare Heath...', *Proc. Somerset Arch. Nat. Hist. Soc.* 79, 1933, 19.

[2] W.A.E. Ussher, *ibid.* 60, 1914, 17.

[3] Reference to Fig. 34 suggests (Wills) that they may be the equivalent in age of the *Neritoides* beach of Gower.

with the Fenland might yield clear evidence of an isostatic component in the post-glacial land and sea-level movement of the southern half of Britain.[1]

The Severn estuary waters are notably muddy, and in the main channel the waters ebb and flow at speeds varying between six and twelve miles an hour twice daily. In Sollas's[2] view this implies that they erode more than they deposit. He argues that there is in them a storage of suspended mud, resulting from the accumulations of days, weeks, or even months. Withdrawals seaward are always lessening this amount, but it is made good by material brought down by the numerous rivers. In the past, deposition on the levels was far greater than it is to-day, when the flats are built up and are seldom flooded. Modern canalization is thought to bring the mud up higher than formerly, and naturally, also, much gathers in sheltered places as off Weston-super-Mare and Berrow. The lower channels of the tributary rivers are also very muddy.

The coast has several other points of interest. The shores are nearly always low and flat, and there is an extensive belt of dunes between Burnham and Brean reaching a breadth of three-quarters of a mile at Berrow.[3] Elsewhere, although the wind is favourable, dunes are poorly developed. True salt marshes are usually situated at the river mouths, especially in the Parrett, Brue, and Avon, but smaller marshes occur in the Yeo, Axe, and Kenn. The mud is brought in by the tides, not by the rivers, and 10–20 ft. in vertical section are exposed at low water. In these respects the area somewhat resembles the Bay of Fundy. The sequence of vegetation can be briefly summarized as follows:[4]

1. Between the limits of neap tides is an association of *Salicornia herbacea*.

2. In the area washed only by springs and storms, *Glyceria maritima* and *Triglochin maritimum* predominate. The zone is not very well defined.

3. In the mud areas which are scarcely ever tide-washed occurs the general salt-marsh association.

There are also salt-marsh pastures which are artificially drained and much browsed.

Berrow Flats have only recently been colonized. Thompson[5] analysed

[1] Godwin, *op. cit.* 1941, p. 131.

[2] W.J. Sollas, 'The Estuaries of the Severn and its Tributaries', *Quart. Journ. Geol. Soc.* 39, 1883, 611.

[3] The natural dunes between Burnham and Brean are good, but unfortunately the area is much built over.

[4] C.E. Moss, 'Geographical Distribution of Vegetation in Somerset...', *R. Geogr. Soc. Special Publ.* 1907.

[5] H.S. Thompson, 'Changes in the Coast Vegetation near Berrow...', *Journ. Ecology*, 10, 1922, 53.

the evidence and concluded that, although an exact date cannot be given, 1910 is probably the earliest year for postulating the beginning of plant growth: most of the plants came later, approximately since 1918. It is interesting to note that *Zostera* is rare on this coast, and that when Thompson wrote (1922) there was very little *Suaeda maritima* and no *Spergularia*, *Statice*, or *Armeria*.

Changes other than those caused by the spread of vegetation are also taking place. The mouth of the Parrett is deflected northwards by a spit which has incorporated Fenning Island, and, at the present time, is growing towards Stert Island.[1] Near Stert Point there is a fair amount of shingle arranged in the usual recurved ridges characteristic of such formations. Stert Island seems to have been separated from Stert Point about 1790, and since 1802 the new point has advanced nearly one-third of a mile. The existence of Fenning Island, which began as a mud bank, was recorded in 1802, it was well defined in 1891, and some time between that date and 1904 became incorporated in Stert Point. Both point and island are now subject to a good deal of erosion, the shingle being derived from the Lias cliffs to the west. Between Stert and Stogursey, sea walls protect the coast, but it is a threatened area; in fact the whole of this low-lying section is subject to dangerous floods during storms. At Chilton Trinity there are several shingle banks indicating old shallows or a former position of the coast. Before 1739 the Parrett had two mouths, but the northern one was in that year blocked by ice and the current diverted entirely into the main channel. The formation of mud-and-stone balls on Stert Island, due to the rolling of these materials in water, has been noted (**71** and **70**).

The headlands[2] of this coast were, with the exception of the Clevedon-Portishead ridges, recently islands. Brean and Uphill form the western end of the Mendips, the former being composed of Carboniferous Limestone dipping gently to the north, and showing a steep scarp face to the south. Worle Hill is another limestone feature; the rocks dip to the south, but there is not a continuous succession since a thrust traverses the whole length of the ridge. Near Spring Cove there are igneous rocks, and at about 25 ft. above high-water mark is a patch of raised beach. The surrounding alluvium does not rest directly on the limestone, as there is a thin belt of Keuper in between.

Woodspring, also known as Swallow Cliff or Middle Hope, is another limestone ridge including some igneous rocks: these are mainly tuffs,

[1] O.D. Kendall, 'The Coast of Somerset (1)', *Proc. Bristol Nats. Soc.* 8, 1936, 186.

[2] S.H. Reynolds, *A Geological Excursion Handbook for the Bristol District*, 2nd ed. 1921, and see also his paper on 'Erosion of the Shores of the Severn Estuary', in *Proc. Bristol Nats. Soc.* 1, 1906, 204.

but there is one basalt exposure. A fragment of raised beach also occurs, and is associated with a raised-shore platform. At Clevedon two Palaeozoic ridges meet, each mainly formed of Old Red Sandstone overlain by Carboniferous Limestone. In the Clevedon-Portishead ridge the beds dip to the south-east, and are fringed by Triassic rocks. The Dolomitic Conglomerate forms a shore screen, and where it is breached the erosion in the underlying Old Red rocks is more rapid. The structures at Portishead are more complicated. The east-west ridge is named Eastwood; part of its seaward margin is composed of Pennant Sandstone which is fringed by the Dolomitic Conglomerate which dips north-west and is faulted against the limestone forming the main part of the ridge. On Portishead Down the Old Red Sandstone forms the coast and contains some fine examples of unconformable junctions between it and the Dolomitic Conglomerate. Near Black Nore Farm recent erosion is noteworthy: the dip is to the south-east.

Brean Down, Worle Hill, and Middle Hope are also good examples of tied islands (**72**). Sand spits have now united them with the mainland, but in each case the drainage has been diverted northwards. The river Axe reaches the sea between Brean Down and Weston-super-Mare, and the views across the marshlands and former islands from St Nicholas's church are very striking. The river Yeo is deflected to find its mouth in the marshland immediately east of Middle Hope. It is worth noting that shingle only occurs now (i.e., in 1944) at the back of the beach for about half a mile southwards from Brean Down while, nearer Burnham, new marsh is growing outside the dunes. Weston Bay is mainly sandy, and in Sand Bay there is only a narrow fringe of shingle, roughly in the southern half of the bay. The Clevedon-Portishead ridges were never islands, but they certainly separated extensive shallows and marshes to north and south. Further, although the coastal hills alone have been mentioned, it is worth bearing in mind that Brent Knoll is in every way similar to them—an island now enclosed in the marsh. In addition to technical likenesses, the whole area of the Somerset Levels is very reminiscent of the Fenland, and the position of Bridgwater recalls that of Boston, Wisbech, or King's Lynn.

The deflection of the rivers is illustrated most clearly perhaps in the mouth of the Parrett (see p. 202). The narrow neck of land joining Fenning Island to the marsh behind is virtually a shingle bar. There is no other feature quite like it in England and Wales, and the wide mud flats on the outer side stand in striking contrast to the shingle: the bar really begins about a mile west of Stolford Farm. The great sweeps of the Parrett between Bridgwater and the sea are well seen from Pawlett Hill. There has been a considerable amount of new drainage work in the last few years and a main drain now reaches the sea just south of Huntspill.

Before passing on to the coast of south-western England there should be a brief allusion to the high erosion levels of the Bristol area. That between 200 and 300 ft. is the most widespread and suggests a long still-stand of the sea. Higher levels at 550–600 ft. and 750–800 ft. are fairly common, the latter especially in the Mendips. The smoothness of the Bristol plateaux suggests a marine origin, but Trueman[1] does not exclude subaerial processes: in any case the Bristol Channel had come into existence before the cutting of these flats. The raised beaches are lower and newer. Palmer[2] regards the higher traces of raised beaches found between Portishead and Brean Down as belonging to the '50-foot' stage. The lower ones (referred to above) appertain to the 10-foot pre-Glacial (i.e. *Patella*) stage, although the base of the deposit or the level of the associated rock platform is often about that of the present high-water mark.

[1] A.E. Trueman, 'Erosion Levels in the Bristol District...', *Proc. Bristol Nats. Soc.* 8, 1938, 402.

[2] L.S. Palmer, 'On the Pleistocene Succession of the Bristol District', *Proc. Geol. Assoc.* 42, 1931, 345.

Chapter VII

SOUTH-WESTERN ENGLAND

(1) GEOLOGICAL SETTING[1]

The peninsula of Devon and Cornwall is mainly composed of rocks ranging in age from Pre-Cambrian to the Lias: the western part of Somerset should also be included. Cretaceous rocks cover parts of Devon, and outliers of Tertiary strata reach as far west as St Erth and the Lizard. The eastern boundary here taken is somewhat arbitrary; it is roughly from the high ground to the west of the Parrett to Budleigh Salterton. The southern boundary is nearly enough that of the Mesozoic rocks in so far as they touch the south coast and form from the Otter eastwards the features characteristic of the Channel coast, while on the north side the boundary omits an area that may fairly be regarded as belonging to the narrower part of the Bristol Channel and the Lower Severn.

The peninsula consists largely of high moorland areas, Exmoor and the granite masses of Dartmoor, Bodmin Moor, St Austell Downs, Carn Menellis, Land's End, and adjacent smaller masses are particularly characteristic. The coastal cliffs are magnificent, and in north Devon especially are perhaps the most imposing in England and Wales. The sedimentary rocks naturally vary a good deal in hardness, particularly the grits and sandstones of the Lower Devonian and Culm Measures. The shale bands usually form the lower ground. The coastline reflects the nature of these rocks, but an enormous number of dykes, sills, and bosses of igneous rock form many of the more prominent headlands. Although there is a close relation between geological structure and coastal scenery, it is not usually so evident as in South Wales.

The whole area is included in a great compound syncline striking more or less east and west, and with smaller folds to north and south of it (Figs. 37 *a* and 37 *b*). It includes in its more southerly part five elongated domes each corresponding on the surface with a great granite mass: Dartmoor, Bodmin Moor, St Austell Moor, Carn Menellis, and Land's End. The oldest rocks are those of Pre-Cambrian age in the Lizard, and in the Bolt and

[1] For a general description of the geology, see *British Regional Geology*: 'South-West England' (Geol. Surv. and Museum), H. Dewey, 1935; E.M.L. Hendriks, 'Rock Succession and Structure in South Cornwall', *Quart. Journ. Geol. Soc.* 93, 1937, 322; and *Handbook of Cornish Geology*, E.H. Davison, 2nd ed. 1920. The sheet memoirs are given, where relevant, in subsequent footnotes.

Start region of south Devon. In the Lizard the mica schists are the oldest, and are followed by hornblende schists. Into these two were intruded the great mass of serpentine, nearly circular in outline, but in places eroded

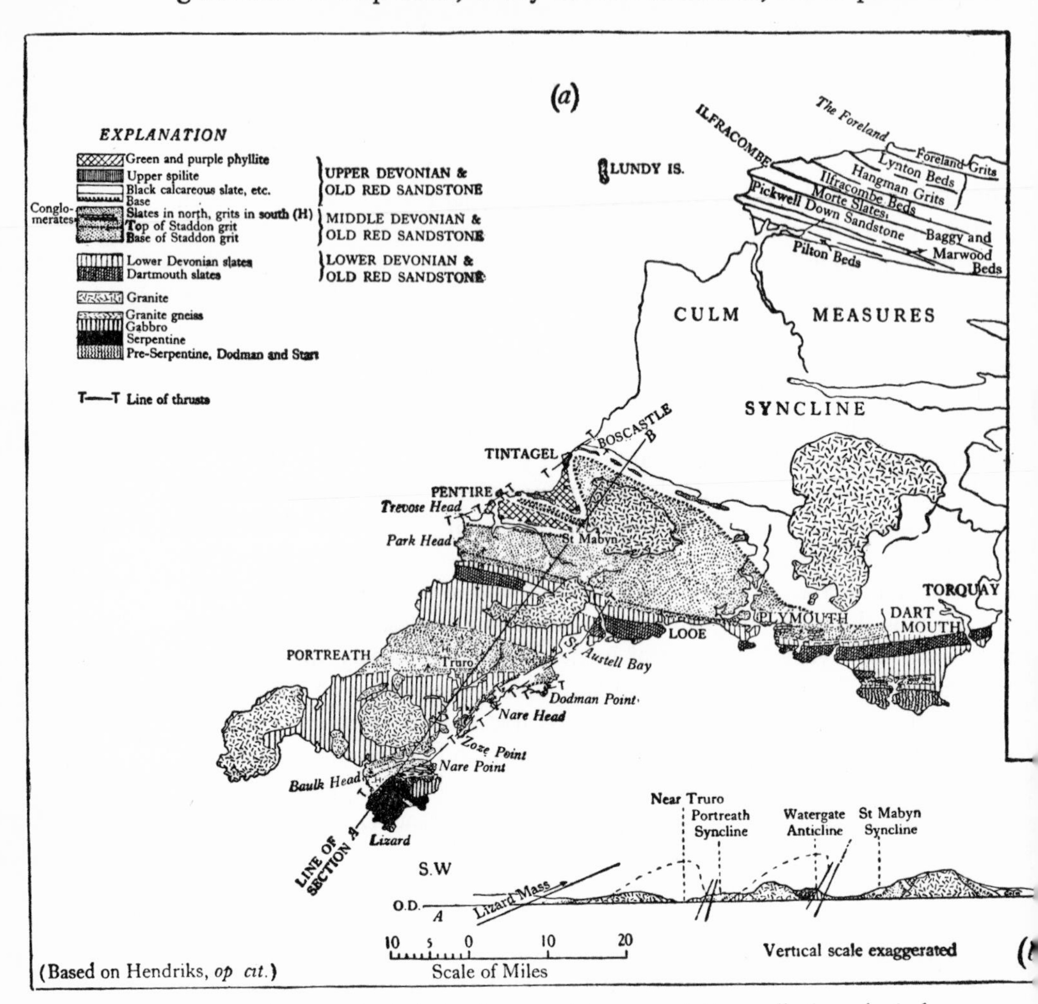

Figs. 37*a*, 37*b*. *a*, Geological sketch-map of Devon and Cornwall; *b*, geological section through Cornwall; vertical scale exaggerated

by the sea. Three types are distinguishable (see Fig. 38). Still later a mass of gabbro was intruded, and there are also numerous gabbro dykes in the serpentine. After the gabbro, dykes, first of dolerite and epidiorite, and secondly of granite and granite gneiss, penetrated the earlier-formed rocks. In south Devon the green schists, mica schists, and quartz schist of the

Start and Bolt may roughly correspond to the oldest of the Lizard rocks. Actually their origin is unknown; they may possibly be altered Lower Devonian rocks, but more probably are Palaeozoic or pre-Palaeozoic remnants.

Immediately north of the Lizard district is a narrow belt of country whose structure is very complicated. Reference to Fig. 39 will show that it consists of two main parts—the Gramscatho Beds and the Meneage

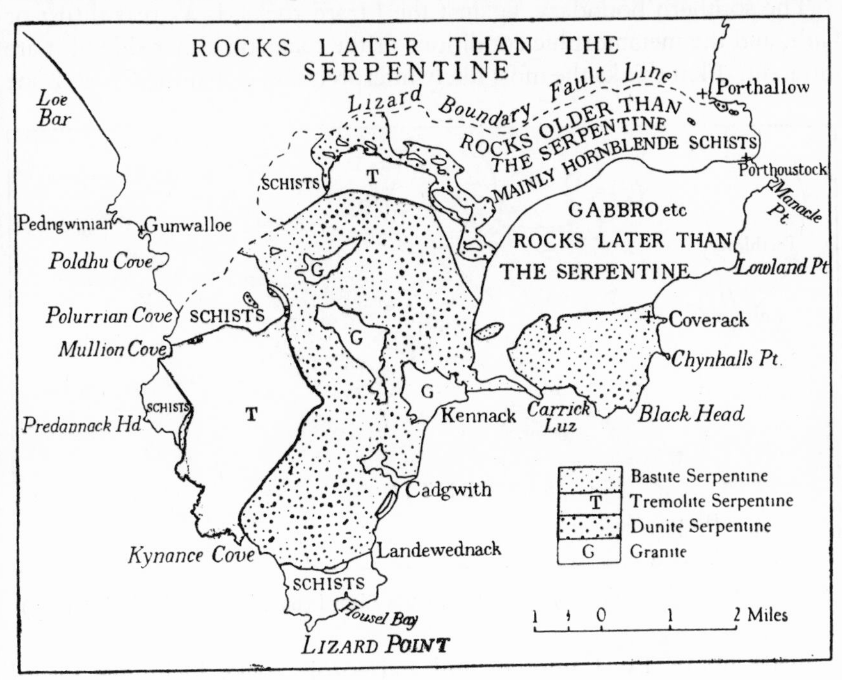

Fig. 38. Geology of the Lizard (based on Geological Survey)

Crush Zone. In 1912 Hill gave the following sequence in descending order: Devonian (including Manaccan Conglomerate), Unconformity, Veryan (Ordovician, probably Caradocian), Portscatho, Falmouth, and Mylor. Of major importance in this sequence were the rocks, mainly shales and grits, around Gunwalloe and Nare Point, which he regarded as Devonian. At Nare Point the association of these rocks with the Menaver Conglomerate suggested to him the unconformable base of the Devonian. In *c.* 1928 Hendriks reported her discovery of fossil plants in the Manaccan, Veryan, Portscatho, and Falmouth Beds. Flett[1] confirmed this, and the Falmouth, Portscatho, and Manaccan Beds of Hill's sequence

[1] Sir J.S. Flett, 'Geology of Meneage', *Summary Geol. Progress*, Pt II, 1932, p. 1.

are now put into one group, and further, although the Veryan has somewhat distinctive characters, the whole is now renamed the Gramscatho Beds. The strip of country marked on Fig. 39 as the Meneage Crush Zone is much cut up by faults and thrusts, and the rocks occur in lenticles generally elliptical in shape and often of considerable size. It is in fact a great breccia, materials of which have a roughly east-west strike, and dips, where visible, at high angles to the south.

The southern boundary, against the Lizard rocks, is a great thrust or fault, and the metamorphic conditions of the rocks on either side of it are different. Flett thinks the most likely interpretation is that the Crush Zone

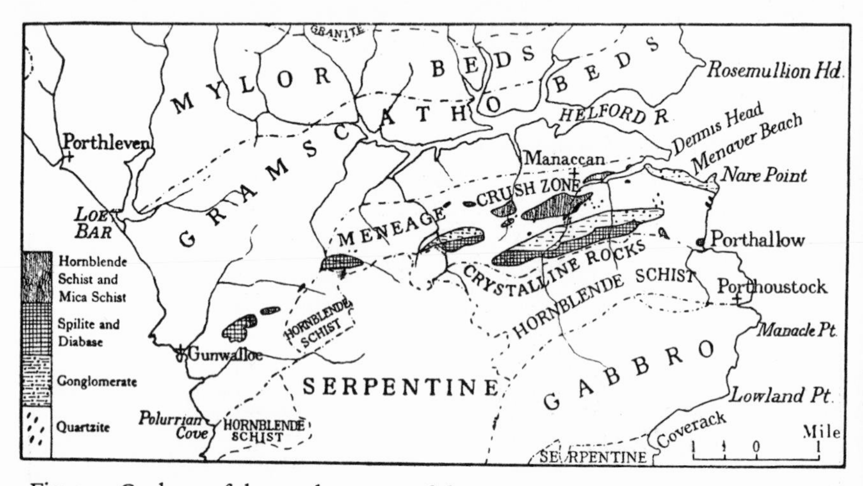

Fig. 39. Geology of the northern part of the Lizard (based on Geological Survey)

is made up of Pre-Cambrian and Palaeozoic rocks including Lizard Schists, Ordovician, Silurian, Devonian, and possibly even Carboniferous rocks, and that these were pushed to the north and broken by the Hercynian movements. The banded structure which is also characteristic of the Crush Zone may reflect an original sequence of rocks. 'At the base we have the crystalline rocks (of the Lizard): on these rested the Ordovician rocks: next came the fossiliferous Silurian: then the pillow lavas and cherts with the Conglomerate, and in close association, the Hendriksi Beds.' The other fragmental rocks in the breccia cannot yet be determined with certainty. It is interesting to note that the many small coves in this zone where it meets the coast are usually cut in shales, grits, or cherty breccias.

The Lower Devonian rocks consist mainly of grits, sandstones, and conglomerates. They lie to the north of the groups just mentioned and also outcrop in a wide band in west Somerset and north Devon. In south

Devon and Cornwall they have been divided into the Dartmouth Beds, the Meadfoot Beds, and the Staddon Grits. Their outcrops run generally east and west, but an anticline north of the Start-Bolt area repeats the sequence. In north Devon they come up from under the Culm Measures, form the cliffs between Morte and Minehead, and also appear again in the Quantock Hills. They include in this area (in ascending order) the Foreland Grits, Lynton Beds, and Hangman Grits.

Middle and Upper Devonian beds are made up mainly of slates and sandstones and, more locally, of limestones. Shaley areas are more characteristic of north Cornwall, while the limestones are most clearly seen in south Devon, especially in the Torquay area. They stretch in a broad band from a point between the northern part of Watergate Bay and Boscastle, around Bodmin Moor to the southern half of Dartmoor and south Devon to Plymouth. The Middle Devonian encloses Dartmoor on the south and both it and the Upper Devonian reappear to the east of the moor. In north Devon,[1] the Middle Devonian includes the Ilfracombe Beds. Although the Morte Slates appear to overlie the Ilfracombe Beds, their age remains doubtful; their fossils suggest the Lower Devonian. The Upper Devonian includes the Pickwell Down Sandstones, the Baggy Beds, and the Pilton Beds. In the Upper Devonian there are very extensive developments of spilitic lavas, especially in north Cornwall. These lavas are very noticeable in Pentire Point: in fact on the coast they nearly always form headlands, whereas inland they make hill ranges.

The Carboniferous rocks of both counties are known as the Culm Measures because soft sooty coal, known in Devon as culm, is sometimes found in them. They are limited to the north of the region, extending inland in a broad band from the area between Boscastle and Barnstaple. They are usually divided into the Lower and Upper Culm Measures, the former making a narrow zone only, on either side of the Upper Measures. The Lower Measures are mainly dark shales with local slates, lavas, limestones, cherts, and grits, whereas the Upper are largely slates and sandstones. On the coast the monotony of the Upper Measures is more than relieved by the magnificent examples of folding so well displayed in the cliffs.

Before describing the newer rocks, the structure of the area which has been affected by two periods of folding needs attention. The Caledonian folds, aligned along north-east to south-west axes, can be traced in the pre-Devonian rocks, but not in the greater part of the region. It is probable, however, that these movements influenced the granite masses and

[1] In general the Devonian rocks of south Devon and Cornwall are marine, whereas those of north Devon indicate interdigitations of marine and Old Red Sandstone facies.

elvans since they follow similar trend lines. The Devonian and Carboniferous rocks show very obviously the effects of the Armorican orogeny. They are all folded along east-west lines, and are often overfolded and overthrust. It is to these movements that the main structures of both counties are due. Roughly, the southern part of the region is an anticlinorium, and the northern part a synclinorium. The anticline in the Dartmouth Slates visible in Watergate Bay, although its base is not exposed, is regarded by Hendriks as the key. This axis trends east-south-east towards Fowey and Looe, and perhaps to the north of the boundary of the Start district of Devon (see Figs. 37 *a* and 37 *b*). Near the end of the period of the Armorican movements great granite masses were intruded into domes, and metamorphosed the surrounding sedimentary rocks. They were later followed by the intrusion first of aplite and then of quartz-porphyry (elvan) dykes. A later, but comparatively minor, outbreak of volcanic activity occurred in the Permian period.

The New Red rocks were formed under arid conditions. Screes of huge proportions gathered on hill sides to form breccias and breccio-conglomerates, while marls and sandstones were laid down in areas of fresh water. It is these rocks which have given the red-coloured earths of Devon. Then, at some time during the Jurassic period, an eastward tilt was given to the area, and in Cretaceous times cherty sandstones were laid down unconformably on the older strata. These were later buried under the Chalk sea, and eventually various Tertiary beds were deposited (see below).

The coast of the south-western peninsula is like that of Pembrokeshire, but differs from that of the rest of the country in that it is far more crenulate and indented. From the point of view of the cycle of shoreline evolution it is in a distinctly young stage. On the other hand, the rocks of which it is formed are, for the most part, hard and old, and it is partly for this reason, combined with the deep and often narrow valleys made by the rivers partially drowned in their lower courses, that it owes its indented nature. Many of these small coastal streams debouch into picturesque coves with sandy beaches. The majority of these have been seized upon for the building of huts and bungalows which, to say the least, spoil the scenery. There are only a few sandy coves that remain unspoiled. Coves and bays with boulder or pebble beaches have not proved so attractive, and nearly all of them remain in their natural state. In addition to these minor inlets, there are also larger ones, including the estuaries of the Taw and Torridge, the Camel, the Fal and other rivers flowing into Carrick Roads, Plymouth Sound, and Salcombe Harbour. These in their various ways are very beautiful and give great charm to this coast.

The lines of cliffs in Somerset, Devon, and Cornwall are very diverse in form, and certainly include some of the finest in the country. Opinions are bound to differ, but some students of our coast, including the present writer, regard the cliffs between Clovelly, Hartland, and Morwenstow as unrivalled in England and Wales. The way in which cliff forms vary from place to place will be discussed below, but brief attention must unfortunately be called in this place to the buildings which occasionally ruin the sky line. The beautiful coast near Tintagel is greatly spoiled by the larger structures in Trevena, and the bare, open country of the Lizard peninsula is not improved by Lizard Town.

(2) THE PARRETT TO PORLOCK

From the river Parrett as far as Minehead the Rhaetic and Lower Lias beds form the coastline except where recent accumulations have gathered. Near Watchet there is a fair amount of erosion, and a rocky platform, varying in width from 200 to 600 yd. and covered at high water, extends along part of the coast. In places it is covered by mud,[1] sand, or shingle. It is a clear example of a plain of marine denudation, and when the land stood somewhat higher was trenched by the streams which now traverse it. The cliffs are nearly all wasting to some extent, especially to the east of Watchet Harbour; it is probable that a small hamlet, Easenton, formerly stood about a quarter of a mile seaward of the present coast.[2] Where the Red Marls form the cliff, they show a steep seaward dip, and inland are faulted against the clays and limestones of the Lias. This fault is a line of weakness, and is only 20–80 yd. inwards from the cliff face. A little farther east the sea has cut through the marls and is now rapidly eating into the Lias, thus forming a bay. The beach material travels to the east and north-east, and in doing so allows, here and there, as, for example, just east of Doniford, a little gain. The Doniford stream has been diverted a little eastwards, but is working back towards its old outlet. In St Audrie's Bay the Lias and Rhaetic beds form cliffs about 100 ft. high on the western side: they are fronted by reefs. Farther east the bay is cut in soft Red Marls and many small landslips are common.

Between Blue Anchor and the landward end of the Parrett shingle bar, the cliffs and shore features are of considerable interest and beauty. The fine waterfall in St Audrie's Bay, the steep cliffs and shore platform often showing well-marked structural features both in section and plan, and the

[1] Dr V.J. Chapman, in a personal note, suggests that this place must be near the seaward limit of the Severn mud, in so far as it is deposited on the foreshore. Certainly the mud has an effect upon the marine vegetation growing on the rocks.

[2] W.A.E. Ussher, 'The Geology of the Quantock Hills and of Taunton and Bridgwater', *Mems. Geol. Surv.* 1908.

slight deflection of the small stream which flows out at the now disused but picturesque harbour of Lilstock, are all noteworthy. But even more remarkable are the fine setting of the coast with the Quantocks running down to the sea near St Audrie's Bay, the high ground crowned by Dunster Castle at the back of the shingle-fringed flat between Blue Anchor and Minehead, and finally the high country inland from Minehead (**69**).

In the Minehead district the major traits of the landscape owe their origin to faulting[1] which originated in pre-Triassic times. The lowlands are all formed of soft rocks which are preserved between the dip-slope of one fault block and the scarped edge of its neighbour, while the uplands consist of the fault-blocks themselves. All the blocks are tilted to the north or north-east, and the step-faulting is very regular. The ancient valleys form the present-day lowlands of Porlock, Cleeve, and Stogumber. Although the whole region was covered by the Rhaetic and Liassic seas, there is no doubt that the present features of the scenery were roughed out in Triassic times. In the lowland areas recent deposits have accumulated, and the coastal flats are now covered with fine marine and fluvial deposits, as well as by spreads of river sands. The Head occurs at the foot of slopes. The level and easily flooded land within Porlock Bay and the large flat between Minehead and Dunster consist of marine and other fine alluvium. These flats are now protected from the sea by recent storm beaches, of which the one in Porlock Bay is a fine example. The river gravels which reach over the flats also extend up the valleys, and the earlier ones form terraces at about 200 ft. between Porlock[2] and Holnicote, 150 ft. at Minehead, and 50 ft. on the coast at Watchet. Between Porlock and Minehead the high ground and cliffs are formed of the hard Devonian slates. Hurlstone Point consists of the Foreland sandstone, and at Greenaleigh is a small pitching anticline in the same beds. There is a great unconformity between the Devonian and the Trias in this district. The convex slope of the high ground to the sea is very noticeable just west of Minehead. True cliffs only exist near sea level where they are usually fronted by a beach of coarse boulders which, at Selworthy, give place seawards to a sandy beach. The shingle travels eastwards and there are large masses at Greenaleigh.

(3) PORLOCK TO BOSCASTLE

(*a*) *Porlock to Croyde*. In studying the coast westwards from Porlock as far as Boscastle, the student of coastal forms is particularly fortunate in

[1] A.N. Thomas, 'The Triassic Rocks of North-West Somerset', *Proc. Geol. Assoc.* 51, 1940, 1.

[2] An incipient salt marsh is forming in the area enclosed by the lock gates.

that he can make extensive use of Arber's work.[1] It is curious indeed that the magnificent and long stretches of cliffed coasts occurring in these islands should have stimulated so few physiographical studies concerning them. Arber's book, however, gives an admirable and lucid account of the north Devon coast, and also treats in a most interesting way the nature of marine erosion on various types of cliff and structure.

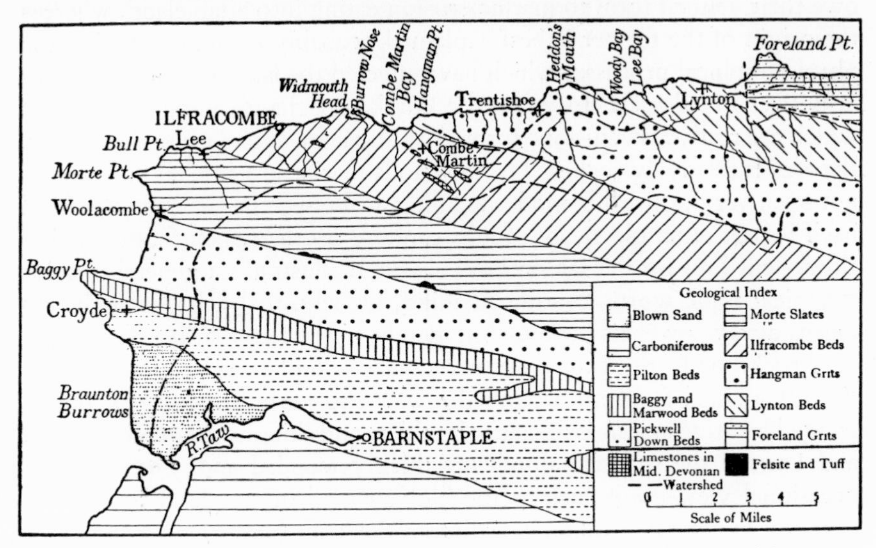

Fig. 40. Geology of Lynton and Barnstaple (based on Arber, *op. cit.*)

Reference to Fig. 40 will show the general sequence of rock types in the area. There is still some discussion as to whether it is the real or only the apparent sequence; from the point of view of this study, however, that is of minor importance since it is concerned primarily with the nature of the coastline as exemplified in the various rock types. The following table shows the order of the beds, and their general nature. In passing it is worth noting that the Devonian in this region forms, as a whole, a trough:

Upper Devonian	Pilton Beds	(mainly)	Slates
	Baggy Beds	,,	Sandstones
	Pickwell Down Sandstone	,,	,,
Middle Devonian	Morte Slates	,,	Slates
	Ilfracombe Beds	,,	,,
Lower Devonian	Hangman Grits	,,	Sandstones
	Lynton Beds	,,	Slates
	Foreland Grits	,,	Sandstones

[1] E.A.N. Arber, *The Coast Scenery of North Devon*, 1911.

Throughout all this stretch of coast there are only a few spaces without cliffs, the chief being in Barnstaple Bay, but there are many small gaps. The cliffs fall into two main types called by Arber the Flat-Topped and the Hog's-Back types. Allowing for local exceptions the cliffs in the Lynton, Ilfracombe, and Mortehoe areas are of the hog's-back type, and those around Clovelly, Hartland, and Boscastle are flat topped. The latter owe their general form to marine erosion eating into a tableland, whereas the origin of the former is best explained probably through studying the physiographical processes which have affected the land mass as a whole.

Between Porlock and Lynton the Foreland Grits give hog's-back cliffs which are really a steep slope formed by subaerial denudation. Moreover, the area is relatively sheltered from serious marine erosion, and so there is not even very noticeable modern cliffing at the sea margin. The Foreland itself is part of an old range of hills running north and south, now being slowly cut into by the sea on both sides. It illustrates a general physiographical topic worth bearing in mind, namely, that a headland by no means always owes its prominence merely to the fact that it is formed of hard rock. At Porlock the eastward travel of shingle has led to the formation of a small foreland around the weir and harbour. Porlock Bay is fringed by large amounts of coarse boulders and shingle. Between Porlock and Lynton the seaward slopes are thickly wooded and, except for an occasional view-point such as the high ground of Foreland Point, it is not easy to get a general appreciation of the coast (**68**).

After flowing through deeply incised and very beautiful valleys, the East and West Lyn rivers unite and reach base level at Lynmouth, where there is a large accumulation of stones, partly derived from the cliffs by marine action and possibly partly of deltaic origin. They certainly push the low-water mark considerably seawards.

The Valley of the Rocks to the west of Lynton is a dry valley; its seaward wall is much weathered and is formed of impure and soft sandstones. The weathering shows plainly along joint and bedding planes, and because of this and also on account of the disposition of the beds which are nearly horizontal, the sandstone tends to weather out in rectangular blocks. It is a feature of subaerial, not marine, weathering. The seaward slope is a typical hog's-back cliff, and Hanging Water is characteristic of streams draining such cliffs (**67**).

Between Lee Bay and Combe Martin there is a remarkably fine stretch of coast characterized by hog's-back cliffs. Another feature which adds great beauty to all this coast is the fine interior country running back to Exmoor. In Lee Bay and Woody Bay the slopes are heavily forested, but nearer Heddon's Mouth, a deep scree-scarred ravine cut to sea level, the cliffs and their tops are open. Very fine views of the coast can be obtained from Trentishoe and the high ground on either side of Heddon's Mouth. Farther west the Great and Little Hangman are only slightly less spectacular. Approach to the beach, usually of boulders, all along this stretch is difficult. Hills such as the two Hangmans, Holdstone Down, and Trentishoe Burrows result mainly from the unequal weathering of a tableland, since, in this neighbourhood, the Hangman Grits resist denudation more effectively than elsewhere. The incoming of the Ilfracombe Series, bluish-grey or silvery slates and shales with bands of impure limestone, alters the nature of the coast.

Westwards of Trentishoe the cliffs are lower; their slopes are more gradual and their summits are often cultivated. The coast is diverse and broken by promontories separated by narrow coves. There is also usually a beach. These contrasts are explained by the differences in the characters of the rocks and the ways in which they react to weathering. Watermouth is a fine example of a drowned river mouth. The lower part of the stream ran nearly parallel with the coast, and its north-eastern wall has been breached, thus accounting for the islets of Sexton's Burrow and Burrow Nose. The breaching process is being repeated at Small Mouth Cove. Newberry Water is another good example of a drowned valley mouth.

The Morte Slates give distinctive cliff scenery; the cliffs are much if not deeply indented, and between tide marks there are reefs consisting of large masses and plates of rock tilted almost vertically and broken off obliquely. These rocks have sharp edges and a glistening surface, while quartz veins add to their striking appearance. Off Damagehue Cliff they are very jagged. Bull Point is a right-angled turn in the coast: it is not a true promontory, since it also forms the western wall of Burnett's Water. At Morte Point the cliffs are fairly low and possess a gentle slope; they are of the hog's-back type, and only the lower 50 ft. are actually undercut by the sea. The reefs and stacks off the point add much to the scene. At Woolacombe the coast is low, and there is much blown sand in the rear of a fine sandy beach and reaching some way inland and on to the old cliffs. Baggy Point prevents any material travelling into Morte Bay from the south. On the point itself the beds are steeply inclined, and great slabs of sandstone rise sheer from the sea for some 200 ft. Sand accumulates in Croyde Bay, and both at Baggy Point and Saunton the raised-beach platform is plainly visible: at Saunton the beach deposits contain a large granitic erratic which has probably come from Scotland (**66** and **65**).[1]

(*b*) *Barnstaple Bay*. Between Croyde and Westward Ho! cliffs are absent, and instead this section of the coast is distinguished by the greatest

[1] Arber, *op. cit.* p. 82.

development of sand dunes in Devon and Cornwall. At Appledore there is also the well-known pebble ridge or popple. Although several papers dealing with these features have appeared, the whole complex has not yet been worked from a physiographical point of view, and there remains plenty of scope for a comprehensive investigation. The Braunton dunes and their vegetation have been described by Watson,[1] and reference to Fig. 41 will make the following summary clear. The first zone on the seaward side consists of flat sands which are more or less bare, but occasional plants of *Salsola* occur near high-water mark. Within this zone is a second distinguishable belt of scattered and comparatively small foredunes which occasionally reach about 15 ft. in height, and which are more or less linked to form an eroded foredune. *Ammophila* is the most abundant plant, while *Agropyrum* is absent. The third zone is formed of large sand hills up to 50 ft. high which make a fairly continuous ridge parallel to the sea. They also show a tendency to lie at right angles to the prevalent south-westerly winds. Again, *Ammophila* is dominant; indeed, the unstable nature of the dunes makes it difficult for slower-growing plants to establish themselves. Within these dunes is a line of slacks characterized particularly by *Riccia crystallina* (liverwort). Smaller dunes sometimes interrupt the line of the slacks, and in wet weather temporary lakes are formed. Those parts which are less liable to flooding carry the more prolific vegetation.[2] Behind the belt of slacks follows another line of dunes generally parallel with the first line, but the second is more uneven,

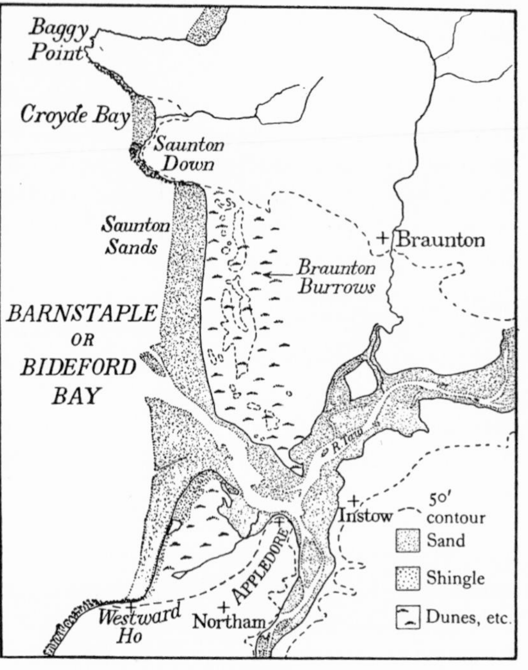

Fig. 41. Sketch-map of the Taw-Torridge Estuary (based on Ordnance Survey)

[1] W. Watson, 'Cryptogamic Vegetation of the Sand Dunes of the West Coast of England', *Journ. Ecology*, 6, 1918, 126.

[2] This zone has been altered to some extent artificially to make a golf course.

and often sends out 'flanges' at right angles to the main line. The height varies, but in places reaches 100 ft. The seaward side is nearly all loose sand, so that *Ammophila* is the only constant plant on this ridge, and actually the only plant found on the exposed side. A second belt of slacks succeeds this ridge, but is less well defined than the one mentioned above; it is often broken by branches of the dunes. There are a few permanent pools, and many others exist for a large part of the year. The variable nature of the water supply, and the artificial drainage of part of this belt, cause a curious mixture of plants. Landwards, there is another ridge of more or less fixed dunes, but these are not so high as the other two main lines. They carry a closer covering of characteristic plants, but *Ammophila* is again plentiful. Finally, there are flats broken by scattered sand hills: this zone gradually merges into sandy pasture land.

Details of the plants found are mentioned in Watson's paper; it is of greater interest here to stress the parallel nature of the several lines of dunes and of intervening stretches of low ground. This suggests a prograding shore, and, given time, the outer and somewhat discontinuous foredune will develop into a full-grown series like the others. At a later date, if conditions are favourable, they may be succeeded by yet other lines.

South of the joint mouth of the Taw and Torridge is another alluvial area, bounded on its seaward side by the shingle ridge which is mainly composed of large pebbles and boulders, which has grown to the north-east. It is clearly a product of wave action: the pebbles, often as large as a man's head, could not be moved by currents, the more so as they rest high on the beach and are only worked on by waves at times of high water. Spearing[1], in 1884, said the shingle ridge was about 50 yd. wide, and that its crest line was some 20 ft. above the sandy beach, and about 7 ft. higher than the general level of the burrows within it. He also remarks (referring to a paper by Pengelley) that formerly its dimensions were much greater. It is now almost one and a quarter miles long, and stands well back from the line of cliffs. There is no doubt that the ridge has encroached on the burrows, and occasionally it is breached by the sea. Rogers,[2] from a study of maps dated 1863, 1866, 1873, 1884, and 1908, remarks that since 1863 the south-western end of the ridge has been driven inwards about 200 yd., but that between 1903 and 1908 the scouring action of the waves was

[1] H.G. Spearing, 'On the Recent Encroachment of the Sea at Westward Ho!', *Quart. Journ. Geol. Soc.* 40, 1884, 474.

[2] I. Rogers, 'On the Submerged Forest at Westward Ho!' and an appendix on the 'History of Northam Burrows', *Rept. and Trans. Devon Assoc.* 40, 1908, 249 (see also C. Reid, 1908, reference at end of this chapter).

rather less, implying a smaller loss of land and a fairly stationary position of the ridge. It carries a few typical recurves at its distal end. Rogers also notes that the lake called Sandy Mere is of fairly recent origin. In 1884 it covered two acres, but when the encroaching ridge reached it the sea broke through and since then it has expanded to cover nearly fourteen acres. Within the ridge are the burrows on which the golf course is laid out. The first reference to them is that by Risdon in 1630 (**64**).

Within the mouth of the two rivers there are great mud flats and marshes in various stages of development. It seems that the area has never been investigated, and it would probably well repay study. Barnstaple (or Bideford) Bay represents the low ground of the combined valleys which formerly ran much farther westwards. The width and depth of the bay are possibly related to softer Mesozoic rocks, quickly affected by erosion, which at one time may have extended over the area now covered by the bay. Even now the bay is in places bordered by raised-beach and submerged-forest deposits, both of which are easily removed. It is common to find masses of peat thrown up from below sea level on to the pebble ridge.

(*c*) *Barnstaple Bay to Boscastle.* South of Barnstaple Bay the country consists of the Culm Measures. In the cliffs between Cornborough and Abbotsham there are many limestone bands, and a well-marked cut platform. This platform is more or less continuous to Peppercombe, and farther west occurs in patches. Where best developed it is remarkably level-cut and shows a variety of structures which, as seen from the cliff top resemble nothing so much as specially selected exercises in geological mapping. The inner edge of the platform carries a coarse boulder beach[1] throughout its length. The boulders tend to move north-eastwards. The anticline in thick sandstone beds at Cockington Head projecting clear from the cliffs is noteworthy. At Peppercombe there is a small outlier of Triassic rocks, and the cliffs are lower. Between Buck's Mill and Clovelly the cliffs are of the hog's-back type, often with long and heavily wooded seaward slopes the lower parts of which are much tumbled.[2] The street of Clovelly was originally a watercourse and, just to the west of the village, the features produced by erosion in folded rocks are well seen in the cliffs; Blackchurch Rock, for example, illustrates an advanced stage in the erosion of a syncline. In the Hartland district the rocks are mainly alternating sandstones and shales of Upper Carboniferous age. They form magnificent flat-topped cliffs, and near Hartland Quay produce perhaps

[1] V.J. Chapman suggests (personal note) that as only the very biggest pebbles bore any vegetation, movement must be considerable at times.

[2] Arber, *op. cit.*

the finest coastal scenery in the whole of England and Wales. All this area is quite unspoiled, and the combination of first-rate cliffs, interesting physiographical features (see p. 222), good inland views, and a large assortment of wild flowers on the cliff tops makes the Hartland district preeminent. Hartland Point itself is a right-angled turn which conforms with a change in the direction of the coastal watershed (see below), and sharp reefs extend outwards from it. Some of the more interesting features of this part of the coast are treated below (p. 220). In Warren and Marsland cliffs, which are very fine, there are several examples of anticlines and synclines, and magnificent contortions are visible along Broadbench beach: this stretch of coastline also exemplifies faulting in alternate and equally thick layers of shales and sandstones. Just to the north of the Tidna stream differential erosion in shales has led to the formation of great buttresses projecting from the cliffs, and Higher Sharpnose is a prominent hog's-back ridge. At Lower Sharpnose Point the shales stand almost vertically, but the cliffs between it and the Coombe valley, although reaching 300 ft., are not remarkable. The sandy beach extending for some considerable distance north of Bude Harbour is, however, somewhat unusual; it appears to be due to the lowness of the cliffs, and 'the amount of material worked out of the cliffs by the sea is probably less than elsewhere to the north and south, and such rock debris as occurs has been ground down by the sea to a greater degree than in the case where boulders are thickly strewn along the shore'.[1] At Bude[2] itself there is a great deal of blown sand, and the Strat is partly sand choked. The open down-like cliff tops to north and south of the town are very pleasing (**61**).

The Upper Carboniferous sandstones and shales continue as far south as Rusey, and between that place and Boscastle the Lower Carboniferous, consisting mainly of shales with subordinate sandstones and cherts, comes in and forms cliffs of the hog's-back type, whereas the Upper Carboniferous rocks are associated with the flat-topped type. The beach near Widemouth, where the cliffs disappear for a short space, is often boulder strewn, and bands of nearly vertical sandstone and shales produce great buttress reefs. A particularly fine anticline is cut into by the sea near Upton, and in Millook Mouth and Haven are excellent examples of contorted rocks. It is probably right to associate the hog's-back cliffs near Pencannow with the end of a high ridge separating two deep valleys only

[1] Arber, *op. cit.* Bude sand has been used for three centuries or more for agricultural purposes: the beach is still, however, replenished by natural means. South of Bude, rock platforms rise from sandy beaches, e.g. Widemouth. The coast is similar north of Newquay. The power of Atlantic breakers all along this coast must be borne in mind.

[2] Arber, *op. cit.*

a short distance apart. Landslips are common between Cambeak and Rusey, and stacks are conspicuous at Samphire Rock and Northern Door where the sea has cut out a small promontory by lateral working: where the promontory is narrowest the sea has cut through it. Near Boscastle there are many deep and narrow clefts in the coast. Gull Rock is a stack, due to the sea having cut off a part of a small promontory; the same kind of thing has also happened at Boscastle on the north side of the Valency, a larger stream at base level, and Penally Point is once more in the process of providing a second stack. The whole range of cliffs from Foxhole Point (near Millook), by Dizzard Point, Cambeak and to beyond Boscastle is very fine. There is much variety of form and detail, and the tumbled and often grass-grown seaward slopes, together with many stacks and islets offshore, give a landscape of great beauty and interest .

(*d*) *The Coastal Waterfalls*. The stretch from Lynton to Boscastle, apart from Barnstaple Bay, is characterized more than any other part of the British coast by a fine series of coastal waterfalls. These falls are all in hanging valleys the lower parts of which have been cut into by the sea. In short, they represent the balance between marine and subaerial forces. Only the larger streams have reached base level near their mouths. The watershed runs close to and roughly parallel with the coast: hence most of these streams are short, and the nature of the falls depends largely upon whether they are draining cliffs of the hog's-back or flat-topped type. In general, streams cascading down hog's-back cliffs give but short and uninteresting falls like those at Glenthorne and Woody Bay. Flat-topped cliffs, often standing nearly vertical, give fine falls: Litter Water falls 75 ft., and shows a clear case of the sea cutting inland faster than the stream can cut downwards. On the other hand, flat-topped cliffs are naturally more prone to landslips, and Hobby Water and Cleave Fall are examples of falls obliterated in this way. A further influence on the nature of the fall is the direction taken by the stream above the fall in relation to such factors as strike, dip, hardness, jointing, and degree of slope of the rocks. Generally, the streams tend to keep to the softer rocks.

A few examples will serve to illustrate these points. Litter Water is the best example of a vertical fall in a simple stage. Here the relative hardness of the beds has not much effect since the cliff itself is also quite vertical. If the section over which a stream falls is inclined, and if the beds vary in hardness, modifications may occur. The dip-slope often tends to capture the stream, and a soft bed is worn out so that the water often flows along and down, instead of over, the ledge of hard rock. In this way a further fall is produced. Milford Water (Fig. 42) is a fairly big stream and shows a very interesting series of falls, five in number. The fall farthest from the shore is nearly 54 ft. long and is a dip fall. The stream then turns at right angles at the pool at the base of the fall, to form the second drop. This is a gutter fall (*B–C*) 132 ft. long, and cut along a bed of shale: it is really older than the first fall because the stream has cut along the strike in such a way that the length of the first, or dip, fall has increased. At *C* there is another right-angle bend, and the stretch *C–F* is an immature canyon. Near *C* is the third fall, 20 ft. long, and running contrary to the dip; it also ends in a pool cut along the strike. The stream then soon flows in the direction of the dip, and a small inclined fall of about 9 ft. is initiated. The last fall, 11 ft. high, carries the water directly to the shore. Arber notes that the point *C* was probably determined by some plane of weakness in the westward wall of the gutter, so that the stream found it easier to make a turn; when it first did so, however, it was 50–60 ft. higher than now.

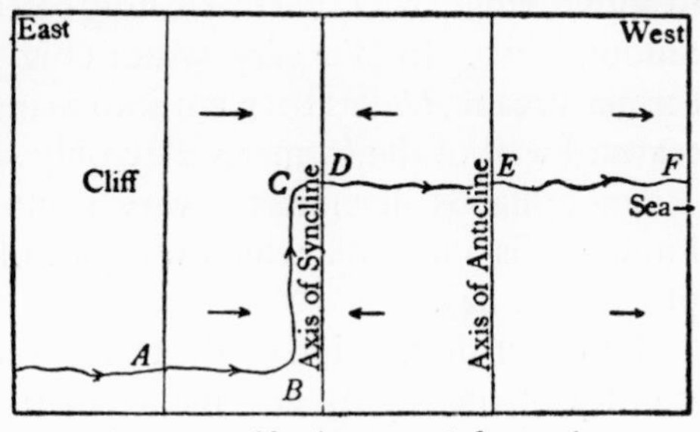

Fig. 42. Milford Water (after Arber, *op. cit.*) (63)

Blegberry Water is a nearly sheer dip-stream fall: the Abbey River makes a good and mature canyon about 64 yd. long and 15 ft. deep: it is formed along the strike of sandstone beds in an anticline. Marsland Water illustrates a senile canyon with a degraded inland fall. In all there are 77 streams, 33 of them end in some sort of fall; 33 are at base level at their mouths; the remainder are altered artificially.

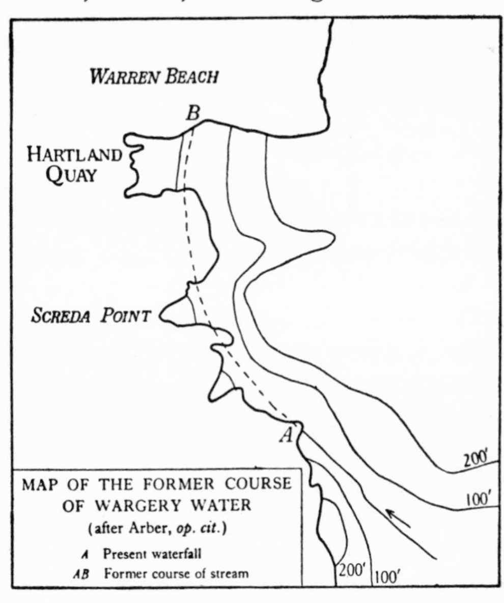

Fig. 43. Wargery Water

Two or three of the streams also afford other interesting features apart from falls. These particular streams, instead of heading direct to the coast, turned parallel to it for some little distance from their mouths. The sea has continued to erode their seaward flanks, and has cut right through the wall in some places so that now parts of the former valley are seen on either side of a small bay. One example, Watermouth, was referred to above, while two other interesting cases are those of Wargery Water and Smoothlands. In Wargery Water (Fig. 43) the present fall is at *A*; the former stream *AB* has been cut into as indicated in the sketch; in short, the seaward wall of the former valley only remains in a few places to-day.

Smoothlands illustrates a very similar course of events: as its name implies, it is a flat area which was part of the valley before dissection took place (Fig. 44).

The brief description of the features shown in Figs. 43 and 44 can give no idea of the beauty and nature of their setting in the field. The high

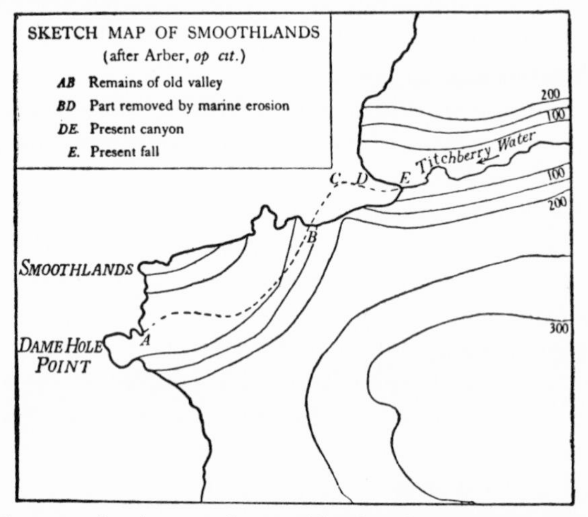

Fig. 44. Sketch-map of Smoothlands (after Arber, *op. cit.*)

cliffs, topped with grass and overspread with wild flowers, reach back to a fine stretch of hilly ground largely farmed, but broken by heath and woods. The whole is cut by numerous deep ravines, usually densely forested in part. The varying hardness of the rocks and their great variety of structures have allowed the sea to fret the coast into great intricacy of detail. The streams have not all cut down to sea level, hence the numerous falls. In and near Wargery Water the half-cut-through ridge of St Catherine's Tor and the higher outer parts of Screda Point, alternating with those places where the sea has already cut through the westward wall of the old valley and has reached its floor, all unite to make a scene, not only of great physiographical interest but also of great beauty. A little farther south the view north-eastwards from the extremity of Marsland Cliff is made far more interesting because of the deep valleys debouching but a short distance to the north of the cliff. The same kind of effect is seen from Higher Sharpnose Point (**62**).

The coast eastwards from Hartland Point is naturally somewhat more sheltered, and there is less intricacy of detail. On the other hand, the high bold cliffs often inclined to crumble, occasional beaches, for example Shipload Bay, and, nearer Clovelly, the thick woodlands stretching far inland, give a type of coastal scenery second only to that south of Hartland Point. These northern cliffs are fringed throughout by a boulder beach, and the beach especially and even the cliff top are not at all easy of access.

At this point it may be convenient to anticipate somewhat a description of the coastal features and to refer to Dewey's views on certain of the river gorges of north Cornwall. Arber has contended that the coastal falls of north Devonshire represent the balance between marine and fluvial erosion. But Dewey[1] introduces another and important factor, namely, the effect of uplift after the Pliocene. How far this may have affected the short north Devon streams it is difficult to say, especially as planation surfaces are not so evident in that area. The gorges to which Dewey refers show sinuous courses in plan, indicating not a slow continuous uplift but a raising of the land relative to the sea in two stages. Naturally, where the uplifted plain was undercut by the sea it was bounded by cliffs over which the streams formed falls; these streams cut downwards and backwards so rapidly as to produce gorges. The gorges show conspicuous pot-hole action, and it is argued that it is this type of erosion which actually produced the gorges. The Pentargon, just north of Boscastle, represents an early stage; the stream forms a cliff waterfall now cutting a deep ravine in Carboniferous grits and shales, but it does not extend far inland. The Valency at Boscastle has striking scenery; the valley has precipitous and craggy sides and the stream mouth is drowned, forming a delightful and picturesque harbour. Rocky Valley near Bossiney began as a coastal waterfall which has retreated several miles inland. Dewey asserts that it began its work when the land was raised in post-Pliocene times 'by cutting a deep gorge in the plateau, and has since continued ripping its way backwards and downwards through the bare rock until it has breached the old cliff line bounding the plateau'.[2] The sides of the gorge are nearly vertical and formed of a highly metamorphosed sericite-phyllite. The Trevena or Tintagel valley is similar, but the stream reaches the sea as a fall 40 ft. high. At Trebarwith the brook has cut down to sea level (**60**)

[1] H. Dewey, 'On the Origin of some River Gorges in Cornwall and Devon', *Quart. Journ. Geol. Soc.* 72, 1916, 63.
[2] *Op. cit.*

(4) BOSCASTLE TO NEWLYN

A little to the south of Boscastle the Culm Measures give place to Upper and Middle Devonian with included igneous rocks. At Tintagel the structure is complicated. '...It may be said that the sediments and contemporaneous igneous rocks of Upper Devonian age were folded into an anticline pitching north-west; that the beds were crumpled on the south side of the nose of this fold; that the crumpling increased in amplitude until overthrusts replaced the minor folds towards the north-west of the crumpled area; and that denudation and erosion have since revealed these structures....'[1] Reference to Figs. 45 *a* and 45 *b* will illustrate the general structure, and also serve as a basis for the following brief notes on the nature of the coastline. The lavas form precipitous cliffs; the overlying blue-black shales form the islands known as Saddle Rocks; the Tredorn phyllites weather along joints and produce stacks such as Short Island and Grower and also Lye Rock and the Bossiney Sisters. The nose of the anticline is completely worn away in Bossiney Bay and the Woolgarden phyllites give very fine cliffs. The structure of Tintagel Haven is complex and may be, in part at least, associated with thrust faulting of late Carboniferous age. Tintagel island is flat topped, a cut-off piece of the plateau within. The position of the island allows magnificent views to north and south along the coast which in this district is remarkably fine. It is much to be regretted that the buildings in Trevena seriously spoil the sky line (**59**).

Southwards from Tintagel the dip is generally northward: this holds good as far as the Watergate Bay anticline. The more prominent features of the coast are nearly all formed of hard igneous rocks but, from north of Port Isaac where lavas disappear, almost to Tintagel, a fairly even sweep of cliff is cut in slates: indentations are small, and occasionally valleys debouch on to short stretches of foreshore. Tough greenstone forms Stepper Point, the Gulland Rock, Newland, and the Mouls. It is also present in Rump's Point and Cliff Castle. Porthquin Bay lies between lava crags, and Varley Head is exceptional in that it is formed of slates. Pentire Point,[2] Cataclews, and Porthmissen Points, Trevose Head, the Quies, and Park Head are all formed of resistant igneous rock, and the pillow lavas of Pentire Point are well known. In contrast the intervening bays are cut in the softer shales, for example, Constantine Bay and Mother Ivey's Bay. South of Park Head the coast shows minor indentations; just

[1] H. Dewey, 'On Overthrusts at Tintagel', *Quart. Journ. Geol. Soc.* 65, 1909, 265. See also D.E. Owen, 'The Carboniferous Rocks of the North Cornish Coast and their Structures', *Proc. Geol. Assoc.* 45, 1934, 451.

[2] H. Dewey, 'The Geology of North Cornwall', *Proc. Geol. Assoc.* 25, 1914, 154.

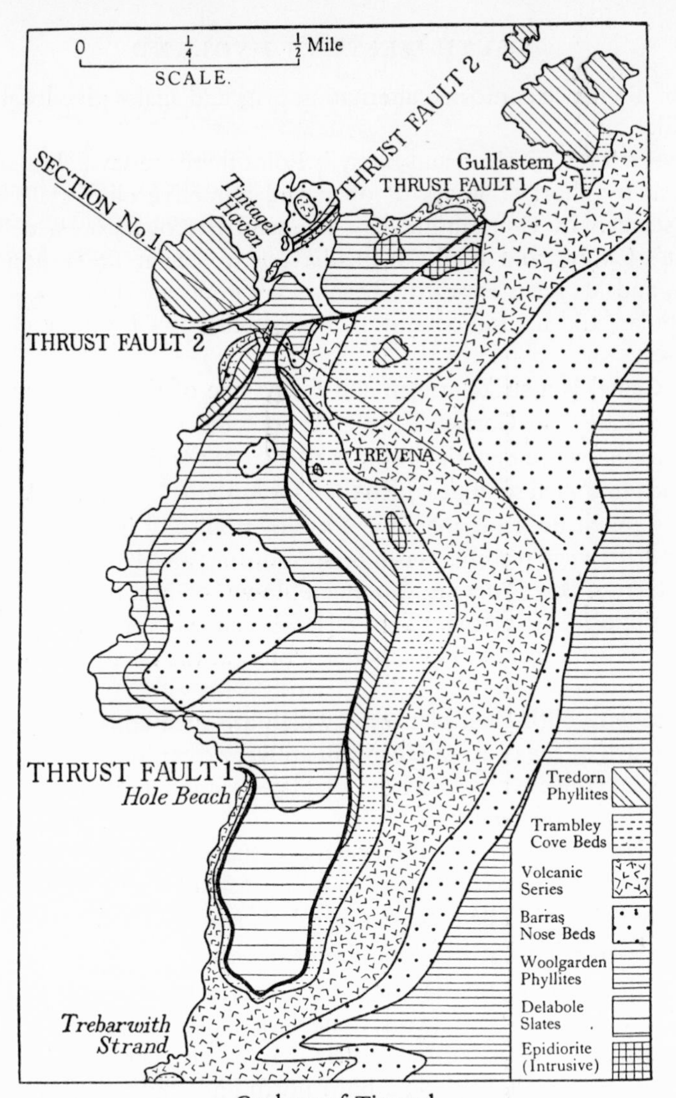

a, Geology of Tintagel

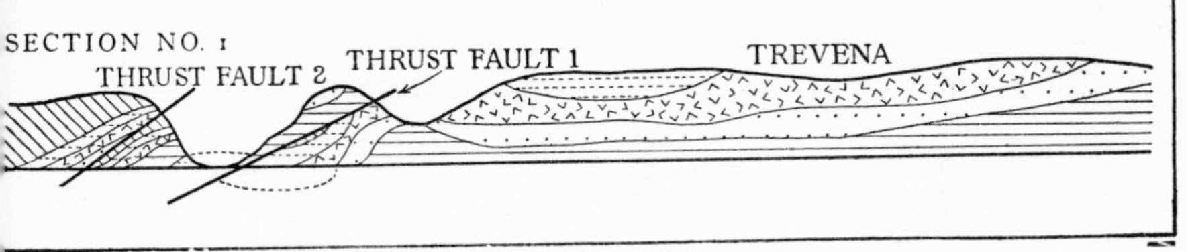

b, Section through Tintagel Island (58)

Figs. 45 *a*, 45 *b*. (Based on Dewey, *op. cit.* 1909)

north of Trenance contorted alternating grits and shales give local variety of detail.

Between Trevose Head and Berryl's Point there are several sandy coves separated by short stretches of open and attractive cliffs off which are many stacks. In Constantine Bay the sand has overwhelmed St Constantine's Church and Well. In this bay, as well as at Treyarnon, Porthcothan, and Mawgan Porth the beaches are good and in consequence all are popular, and natural conditions have been upset by the rapid spread of houses.

The Camel is a strike stream and forms one of the many examples of drowned valleys in Cornwall (**57**).[1] It contains a great deal of sand due largely to the waste material in times past from the tin workings. There is also much fine shell sand of which the Doom Bar is formed. This sand was formerly much used for manure. Physiographically it is interesting because it drifts very readily, as can be seen during a walk from Constantine Bay to Harlyn Bay. There is also much blown sand on the east side of the Camel opposite Padstow. The raised-beach platform and overlying deposits are well seen at Trebetherick Point, at the north end of Daymer Bay (see p. 479).

The several high-level platforms which are such conspicuous features around much of the coast of Cornwall are very clear in this area: they will be discussed collectively, however, in a later section and not as they occur in various parts of the coast.

In the Newquay[2] district the essential structural feature is the anticline in the Dartmouth Beds; it trends roughly east and west, and is visible in Watergate Bay (**56**). The crest of this fold is near Horse Rock which itself is an isolated hog's-back stack pointing seawards. The cliffs, formed largely of black shales along the bay, are often nearly vertical and over 200 ft. high. The existence of headlands, none very prominent, is often explained by vein quartz or silicified rock rather than by any particular stratum. It should be realized that the dips both north and south of the Watergate anticline are extremely confusing, and near Newquay the indented coast may run with, or across, the strike. The 'strait' separating Porth Island has been widened and smoothed artificially, but it probably originated along a joint. A study of the Newquay coast shows numerous faults and

[1] See *Mems. Geol. Surv.* 'The Geology of the Country around Padstow and Camelford', C. Reid, G. Barrow, and H. Dewey, 1910. There is some development of salt marsh especially near Trewornan, a few miles up the Camel.

[2] See *ibid.* 'The Geology of the Country near Newquay', C. Reid, and J.B. Scrivenor, 1906. The sand beach between Newquay and Watergate and beyond is admirably seen from Towan Head.

the bigger caves usually correspond with them. The Criggars are rocky ledges composed of silicified shales and limestones. Towan Head projects partly because it is formed of the same material as the Criggars, and partly because a few volcanic dykes in it have hardened the shales. On the other hand, the shale cliffs in Fistral Bay weather rapidly. East and West Pentire Points and Kelsey Head appear to project on account of local hardness of the rocks: they are not due to the presence of igneous masses. The Gannel, a drowned feature, recalls Watermouth near Ilfracombe (**55**).

In Holywell and Perran Bays there is a great deal of blown sand: that in Perran Bay has steadily worked inland and covered St Piran's church. The dunes reach to 225 ft., but probably the actual depth of the sand does not exceed 150 ft. As at Padstow the sand is mainly formed from shells and so is very light and mobile. In most of the dune areas *Ammophila* is dominant. Many of the smaller and narrower inlets of north Cornwall show a sand barrier near their heads and situated at the junction of salt and fresh water. The barrier is due to the action of wind-blown sand, but seldom attains any height. The plants are somewhat different from those of the true dune areas.[1] At Porthcothan (south of Trevose Head), *Salsola kali* is first and also dominant in its zone: with it occur *Cakile maritima*, *Arenaria peploides*, *Atriplex laciniata*, etc. This passes into an *Agropyrum-Eryngium* association, which is still further developed at the summit of the barrier where *Ononis repens* is usually common. At Porthcothan the barrier is planted with tamarisk, and the main features are similar at Treyarnon (about one and a half miles to the north). If the Constantine dunes may be regarded as typical of Cornish dunes from the point of view of vegetation, we note that there is very little *Agropyrum*, and that *Ammophila* is dominant on the newer or steeper dunes. Landwards such plants as *Ranunculus*, *Sedum*, *Galium*, *Leontodon*, *Carex*, etc. come in with *Erodium* and *Potentilla*. *Ononis* with *Carex*, *Potentilla*, *Solanum*, and *Senecio*, form a transition to the sandy plains behind the dunes. The only other dune area of any size in north Cornwall is on the east side of St Ives Bay, where the dunes have buried St Gothian's chapel, and by 1907 had banked themselves around the walls of Millook churchyard. There is a small sand area farther north at Porth Towan, hence the name.

The most prominent features on the coast between Newquay and St Ives are St Agnes Beacon and St Agnes Head formed in much altered and disturbed silty and sandy slates. The Pliocene sands rest in a semi-circle around the beacon: these are discussed later in the section dealing with plateau levels and Tertiary deposits. North of St Agnes Head, the

[1] L.A.M. Riley, 'Some Notes on the Coast Flora of North Cornwall', *Journ. Royal Inst. Cornwall*, 17, 1907–9, 248.

granite, against which the Staddon Grits are disturbed, forms Cligga Head, and farther south Navax Point is composed of the slates of the Falmouth Series. Between St Agnes Head and Navax Point the Falmouth and Portscatho Series[1] form the cliffs. Differential erosion is plain and characteristic features are several stacks and small islands, including Gull Rock, Samphire Island, and Crane Island. Godrevy Island and adjacent rocks are in sandstones, while Dead Man's Cove is cut in the relatively soft core of dark slates and on each side it is faulted against sandstone, thus illustrating the combined effect of tectonic movement and rock hardness in the formation of coastal detail. Between Perranporth and Porth Towan, although the cliffs are good, the coastal district and views are much spoiled by the numerous remains of former mining activities. The country near the coast is flat and rather open, thus making the derelict buildings associated with the mines stand out gauntly against the sky line. Beyond Porth Towan, and especially between Portscatho and Godrevy Island, compensation is found in the beautiful and rugged cliffs of Reskajeage and Hudder Downs. Fortunately the interior is also unspoiled. On the west side of Godrevy Point there is a good deal of blown sand originating from Gwithian beach and burrows—a mass of high and well-developed dunes which reach as far south as the Hayle River **(54)**.

West of St Ives Bay[2] there comes first a stretch of greenstone, patches of which also occur farther along the coast, but from St Ives as far round as Newlyn the great granite mass which forms the Land's End promontory is of primary concern.[3] In this stretch granite cliffs are developed, but St Ives Head (The Island) and Gurnard's Head are in the greenstone. The 'island' was isolated in raised-beach times, but is now joined to the mainland by the raised beach and Head which remain in the depression between it and the mainland. Between St Ives and Morvah the coast is crenulate. There are numerous picturesque and unspoiled coves. The cliffs slope rather gently, and only the lower parts show the result of marine action. Immediately within the coast is a bare plateau on the inner side of which rises the higher ground and tors. The bleak, wind-swept nature of the district is at once apparent, and this part of the Land's End peninsula is a distinct unit. Porthmeor Cove shows clearly the contact of the granite with the slates and greenstone. Along the northern and north-western margins of the granite mass several patches of killas[4] and greenstone are

[1] See p. 207: now called Gramscatho Beds.

[2] There are sand dunes in Carbis Bay.

[3] *Mems. Geol. Surv.* 'The Geology of the Land's End District', C. Reid and J.S. Flett, 1907.

[4] The name given in Cornwall to highly crushed sediments of uncertain age.

present. These are best seen between Pendeen and Cape Cornwall, where occur the finest examples of cliffs in metamorphosed rocks in the county. This stretch runs parallel to the strike, so that similar rocks constantly recur. 'In a general way, the sections may be described as showing a mass of fine-grained sediment, into which has been intruded a thick greenstone sill, both dipping towards the north-west at about the same angle as the underground surface of the granite, which here slopes in the same direction.'[1] A characteristic feature of this piece of coast is the number of clefts, or zawns as they are called locally, most of which are due to the vein material of tin or copper lodes having been removed on account of its relative softness. The zawns often end in vertical walls. The cliffs slope fairly gradually seawards, but the evidences of marine erosion are very clear. Unfortunately between Pendeen and Cape Cornwall the country just inside the cliff top is disfigured by numerous remains of former, and some present, mining.

The cliffs from Cape Cornwall as far as Mousehole are formed of granite. Their form depends a great deal on the nature and inclination of the bedding joints. Between Cape Cornwall and Land's End there is an attractive coast of high sloping cliffs which, near Sennen, are fronted by a good sandy beach, backed at its southern end by an accumulation of very large boulders. It is suggested that the sands in Whitesand Bay may result from the fact that the killas was once there, and because certain slaty rocks on the western side of the bay suggest that the floor of the bay is of slate or greenstone. Land's End is perhaps mainly noteworthy as the most westerly point of the country. The castellated nature of the granite is apparent, but this is even more characteristic of the truly magnificent coast between Land's End and Penberth. The granite cliffs are steep, there are numerous coves, and on the cliff top is a fine open heather-covered moorland reaching as far as St Levan. Eastwards of that place the cliffs are not so steep, but are still castellated in appearance. Towards Lamorna the sloping nature of the cliffs tends to increase. Lamorna Cove and valley are very typical and the valley is well wooded. Towards Mousehole the cliffs become more convex and less interesting (**53**).

(5) MOUNT'S BAY: LOE BAR

The coast of the inner part of Mount's Bay is quite different: at Newlyn the foreshore consists largely of flaggy hornfels, with greenstone possibly below. At Penzance the arrangement of the rocks is very confused with repetitions of slate, greenstone, and a hornblende-slate. Although less

[1] *Mems. Geol. Surv.* 1907, *op. cit.*

obscure in its structural details, the shore east of Marazion also shows alternations of slate and granite. Between Penzance and Marazion is a marshy, low-lying area where erosion is now serious, but is checked by the railway embankments. St Michael's Mount, a granite hill, was once part of the mainland, and when Diodorus Siculus wrote it is thought by Reid to have deserved its Cornish name which meant 'the hoar rock in the wood'.[1] At that time the swampy ground must have encircled it. These conditions probably lasted for some time, but no known record exists to show when it became an island joined to the land by its present low-water causeway (52).

Between Marazion and the outcrop of the Godolphin granite mass the coast shows many details of considerable interest, for the most part associated with the relative hardnesses of the various rock types. It is a much lower coast than most of those so far dealt with in Cornwall. North-west of Perranuthnoe the Head makes low and easily eroded cliffs. On the other hand, the greenstone is responsible for the long point called the Greeb. Between Perranuthnoe and Cudden Point the softer sedimentary rocks are worn back, and Cudden Point (greenstone) is very prominent. The local coastal scenery is here very fine: Piskie's Cove, Bessy's Cove, and Prussia Cove are very delightful, and compare with any of the better-known coves in wilder parts of the county. At Hoe Point the killas is very contorted. Prah Sands correspond closely with the seaward end of the metamorphic aureole of the Godolphin granite which forms good and typical cliffs especially at Trewavas Head. At Porthleven the killas is again contorted; the town is pleasantly built around its harbour. To the south-east the cliffs, fronted by sands, run on as far as Loe Bar. There are also many off-lying rocks in Mounts Bay: Cressars, Western Cressar, and Great Hogus are formed of greenstone, whilst the Long Rock and Ryeman are felsite (elvan) dykes.

A very different feature needs comment before describing the Lizard area: the bar across Loe Pool,[2] a fresh-water lake. The bar is about a quarter of a mile long, but a little less between the cliffs on either side and is about 600 ft. wide at its narrowest point. Like Prah Sands, it is formed mainly of flint shingle. The lake reaches up to the neighbourhood of Helston, and the lower part of that town was liable to be flooded. In the middle of the nineteenth century an adit was made through the rocks at the north-western end of the bar; when this in course of time became blocked, it was reopened and enlarged. Toy has discussed the formation of this bar, but his argument is not easy to follow. He appears to suggest that tidal action would succeed in moving sand and shingle from Porth-

[1] This is disputed by Rice Holmes. See p. 303 and footnote thereon.

[2] H.S. Toy, 'The Loe Bar...', *Geogr. Journ.* 83, 1934, 40.

leven towards the bar, but that strong south-easterly winds would cause a return motion. Before the bar was formed there was strong tidal action in the inlet, and he assumed that eddies set up by the outrush of water would cause the deposition of sand and fine material so that initial spits could have been formed on either side of the inlet. Later, the final closing is ascribed either to gradual processes, a 'tidal wave', or to sudden storms. A remarkable phenomenon was observed in January 1924, when the coastguard's log chronicled a 'tidal wave'. The wave was neither preceded nor followed by storms or unusual seas, but it did considerable local damage and, apart from destruction at Porthleven, it threw much sand and shingle on and over the bar. The precise cause of this so-called 'tidal wave' is not stated: but it is at least sure that it was not 'tidal'.[1] Since much of the shingle of the bar is high up and above ordinary high-water mark, it is reasonably certain that ordinary tidal action has had comparatively little to do with the formation of the bar, which is more likely to be the result of wave action. As in all such features, there have been considerable changes in the course of time. It is probable that in 1302 the Loe was inaccessible to sea-going ships; the bar seems to have been of considerable size in Henry VIII's time; Leland states that cuts through it to lower the level of the Loe were necessary in the sixteenth century, while Carew in 1602 commented that it was not a permanent feature. Borlase, in 1758, and Martyn, in 1784, both speak of a bar and pool. The bar and lake, partly surrounded by woods, add greatly to the variety and beauty of this part of the Cornish coast.

(6) THE ISLES OF SCILLY: LYONESSE

The Isles of Scilly[2] lie about thirty miles south-west of Land's End. They are flat-topped masses of granite, 140 in number, of which five are inhabited. Fig. 46 shows their general disposition. There is a deepening of the interior sea south-westwards towards a submerged valley. The present shape of the islands results largely from subaerial erosion before the downward movement took place which gave them their present general form. The submarine contours show clearly that the whole group is due to the erosion and drowning of a once continuous granite mass similar to those on the mainland. The killas that presumably once surrounded them has been eroded away except for a small patch on White Island off St Martin's. The present waste of the granite, mainly by marine action, produces much sand which is spread all over the floor of the interior sea, and it also adds

[1] Dr Chapman tells me that 'tidal wave' is a term used in this part of Cornwall in referring to a particularly big roller breaking on a flat or shallow shore. These waves are presumably the product of special conditions.

[2] *Mems. Geol. Surv.* 'The Geology of the Isles of Scilly', G. Barrow, 1906.

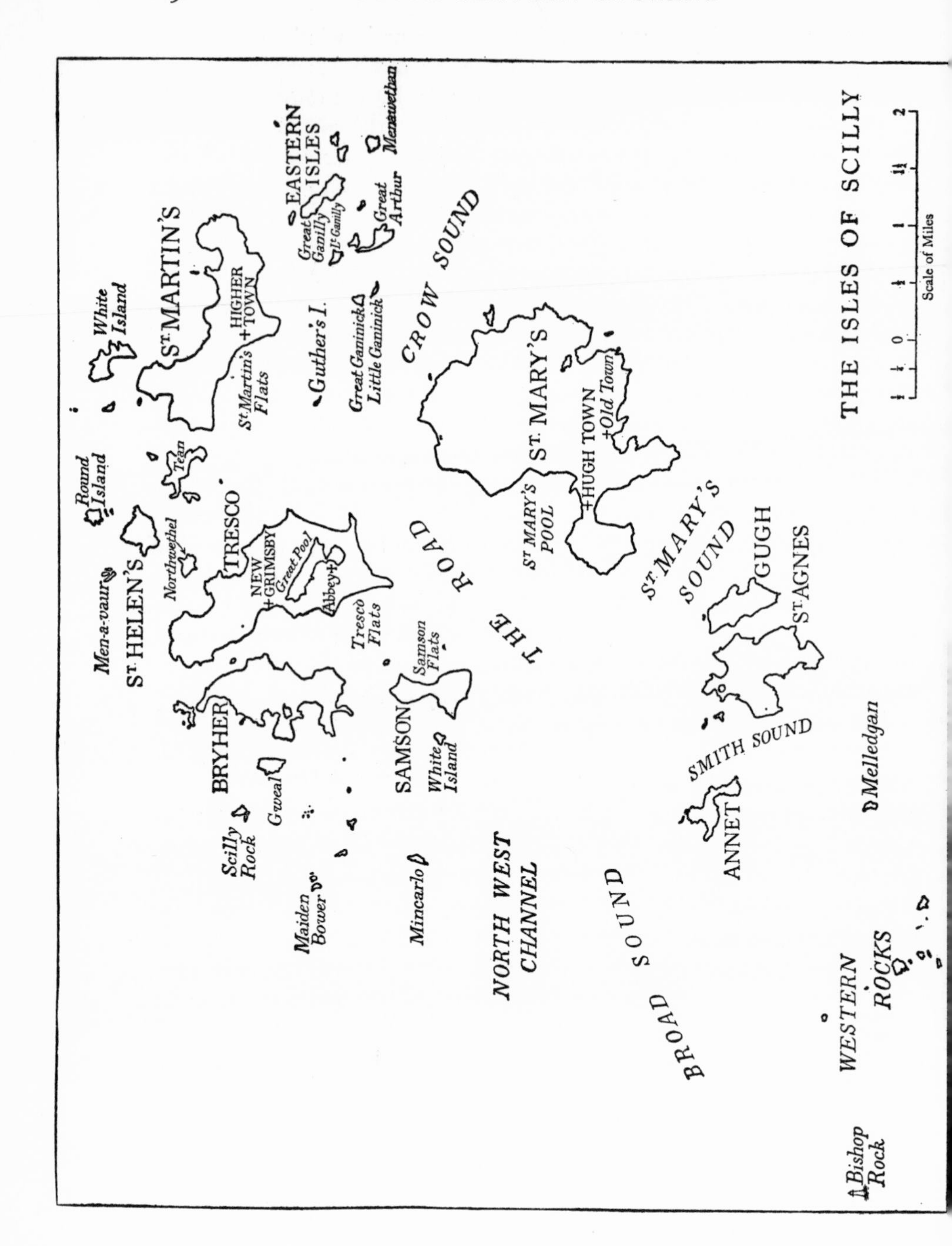
THE ISLES OF SCILLY
Scale of Miles
0
1
2
ST. MARTIN'S
HIGHER TOWN
St. Martin's Flats
White Island
EASTERN ISLES
Great Ganilly
Lt. Ganilly
Great Arthur
Menawethan
Guther's I.
Great Ganinick
Little Ganinick
CROW SOUND
ST. MARY'S
HUGH TOWN
Old Town
ST. MARY'S POOL
ST. MARY'S SOUND
GUGH
ST. AGNES
Round Island
Tean
TRESCO
NEW GRIMSBY
Great Pool
Abbey
Tresco Flats
Northwethel
Men-a-vaur
ST. HELEN'S
THE ROAD
BRYHER
Gweal
Scilly Rock
SAMSON
Samson Flats
White Island
Maiden Bower
Mincarlo
NORTH WEST CHANNEL
SMITH SOUND
ANNET
Melledgan
BROAD SOUND
WESTERN ROCKS
Bishop Rock

considerably to the actual land area when it is washed up and blown by the wind to form spits connecting formerly separate islands. Generally it may be said that coarse-grained porphyritic granite is dominant, and that it surrounds, with no sharp boundary, fine-grained and non-porphyritic granite. There are many vertical cracks in the rock, and, especially in Great Ganilly, they are all greisenized[1] and easily eroded into narrow depressions traceable across the island. Later than the greisening there were formed cracks or fissures tending slightly west of north, and best seen on the northern side of the islands. The waves have worked along them, and the evolution of the fissures bears in no small degree on the water-ways leading to the interior sea. Osman[2] discusses the formation of this sea, particularly that part farthest to the west called Broad Sound. He notes that the south-east side of the sound shows a relatively high shelf, whereas on the north-west it is much lower, and the junction of the two he calls a slip-shear fault of Caledonian age hading at 70° to the north-west. This fault can be traced also in other islands and increases towards the Bishop Rock. The older view, therefore, that Broad Sound is an ancient submerged river valley needs modification, especially if its relation to the North-West channel is considered. Broad Sound may indeed be the result of differential faulting, increasing in amount to the south-west, 'and...possibly this movement occurred during the post-Pliocene uplift. The same movement also accounts for the North-West channel, as it evidently indicates the termination of the low-lying shelf.'[3]

Around the coasts of the islands terrace features are common, and are mainly formed of Head: the true raised-beach deposits have nearly all disappeared, but the beach platform remains, and since the Head occupied much the same position as the beach, erosion has given it a terrace-like aspect. It is possible that there are two deposits of Head separated by a glacial deposit; in the last chalk flints occur which come mainly from a hard ferruginous sandy deposit. Barrow gives the full succession as follows:

Upper or Recent Head
Iron Cement or Glacial Deposit (Containing several different types of stones which must have been carried by floe-ice.)
Main or Lower Head
Old Beach
Sloping Surface of Granite

[1] 'Greisening is a process of metasomatic alteration due to the action of superheated steam and fluorine. In granite the felspars are attacked and converted into white mica, which is often lithium-bearing. Hence, there result aggregates of muscovite and quartz which are called greisen' (G.W. Tyrrell).

[2] C.W. Osman, 'The Granites of the Scilly Isles...', *Quart. Journ. Geol. Soc.* 84, 1928, 258.

[3] Osman, *op. cit.*

Barrow argues that the plane of marine denudation in the Scillies was of Eocene age and bases his view on the presence of flint and some chert in the gravel on St Martin's Head. Osman points out that flint occurs on most of the islands, and so it is more probable that the erosion surface was made in Pliocene times and that any Cretaceous remains were washed by waves into joints and hollows of the granite where they are now found.

It has already been remarked that there is much blown sand in the islands: on Tresco a great ridge, some 35 ft. high, impounds a lake of fresh water, and similar, but smaller, formations are found elsewhere. It is worth noting that St Mary's really consists of three islands united by sand bars; Samson and St Agnes' Islands show similar features, that is to say, in each case two granite knobs have been joined by sands. Much of the sand must have been formed when the sea level was lower by some some 25 ft. It is all granite sand, but now most comes directly from the Head (**51**, **50**, **49**).

This area, and perhaps more especially that part around the Seven Stones,[1] is associated with the Legend of Lyonesse. The nature of the legend is summed up by Crawford:[2] 'Once upon a time (so tradition says) a region of extreme fertility lay between the Scilly Isles and Cornwall. This land was called Lyonesse; and where now roll the waters of the Atlantic there once stood prosperous towns and no less than 140 churches. The rocks called the Seven Stones...are said to mark the site of a large city. This country was overwhelmed by the sea, and the sole survivor, one Trevilian, escaped destruction only by mounting on a swift horse and fleeing to the mainland.' It will be recalled that there are points in this reminiscent of Cantref-y-Gwaelod (see p. 148). There is, as usual, some truth in the legend, but it is almost certain that the lost land was in the Scillies, and not on the site of the Seven Stones. In general, historical as well as archaeological evidence supports the view. Between St Mary's and Samson Islands there is a long line of stones, and at low water it is possible to walk between certain islands. Crawford maintains that this line of stones represented a wall of some kind: some of the stones are still standing, and the structure is very similar to the present walls on the islands. Prehistoric walls of the same type occur on all the larger islands, and their courses are often interrupted by round cairns: they were clearly field walls. Samson Flats are in the interior sea, the floor of which is covered with sand. It is, therefore, possible that the now submerged walls were buried, and so partly protected, by the sand before the submergence.

[1] Granite rocks awash about three-quarter tides, and situated a little to the north of a line between the Scillies and Land's End.

[2] O.G.S. Crawford, 'Lyonesse', *Antiquity*, 1, 1927, 5.

Physiographically there is, of course, no doubt that the islands once stood higher, and that at no very remote date.

Direct historical evidence is also interesting, but it is hardly conclusive. Solinus, in A.D. 240, and Sulpicius Severus, in A.D. 400, speak of the Scillies in the singular, Siluram Insulam. Whatever the value of these statements, they are at least in harmony with other sources of information. Prehistoric studies, however, as far as they are available, suggest that neither the islands nor Cornwall were inhabited before the Bronze Age.[1] The legend of Lyonesse may then have a factual background, but Crawford's comment needs attention; he thinks that it is not a direct traditional record of the submergence, but that it may have arisen in later times as a result of observation by fishermen. If this view is correct, it follows that the Seven Stones must once have been a habitable island; whether it was connected with the mainland or not is a matter of pure speculation.

(7) MAINLAND FEATURES OF SOUTH-WESTERN CORNWALL: THE LIZARD AREA

Attention has already been drawn to the flints on the Scilly Isles, on Prah Sands, and the Loe Bar. They also occur elsewhere in Cornwall. It has been suggested by Milner[2] that they may have come from the east or north-east by a river which perhaps traversed the St Erth channel, the valley that separates the Land's End peninsula from the rest of Cornwall. Clement Reid[3] suggested that it was an Eocene valley; it is certainly not Pliocene, because both the mode of occurrence and elevation of the Pliocene St Erth deposits suggest a greater age. A return to the general question of these Pliocene deposits will be made later (p. 255).

In south-western Cornwall Hendriks[4] has noted the occurrence of various types of valleys: there are ordinary V-shaped valleys with straight slopes and interlocking spurs as in the Hayle River and west of Porthleven; others are of the incised type and grade to a point either above or below present sea level. The former sort are seen about Kennack Bay, and the latter in the Loe valley and part of the Helford River. Another type occurs east of Porthleven: it is characteristically wide and steep sided with a small stream meandering on a flat floor. Examples are the Cober, Porthleven, Gunwalloe, and Poldhu valleys. They are regarded as 'old'

[1] Crawford, *op. cit.*

[2] H.B. Milner, 'Origin of the Pliocene deposits of...Cornwall...', *Quart. Journ. Geol. Soc.* 78, 1922, 348.

[3] Quoted in H.B. Milner's paper.

[4] E.M.L. Hendriks, 'The Physiography of South-West Cornwall...', *Geol. Mag.* 60, 1923, 27.

features. Many parts of the valleys in this group are approximately parallel with one another, and they seem at different periods in their histories to have carried streams or tidal creeks. The inner valleys are probably early Pleistocene in age, since they grade with the raised beach.

Between Porthleven and Marazion hanging valleys are common, but this feature is characteristic mainly of the smaller streams. East of Porth-

Fig. 47. Geology and river systems of West Cornwall (after Hendriks, *op. cit.*)

leven there may be evidence of a superseded drainage system. The Cober, for example, rises near Carn Menellis and at first flows to the south-east; it then turns sharply to the south-west, but there is evidence that at one time it held its south-easterly course, and flowed to Gweek and the Helford River. Reference to Fig. 47 will make this point clearer, and it is worth noting that in Hendriks' view several streams flowed to what is now Crousa Common. These valleys all imply a former south-easterly slope and should possibly be associated with the post-Cretaceous peneplain. They are relics of the drainage system following the post-Cretaceous uplift and all may have carried Cretaceous detritus and flints. The gravels on Crousa Common may, on this view, represent the site of the combined mouths of some of these streams before the captures associated with the Helford River. In brief, these original south-east flowing streams were later intercepted by other streams flowing along belts of disturbance running in a south-westerly direction, not unlike river development in South Wales. Gullick[1] is of the opinion that the river system is not superimposed. He maintains that in the Lizard district most of the rivers originated along lines of weakness, particularly along the boundaries of different rocks and along faults. This applies to the Polurrian, Kynance, Cadgwith, Porthoustock, Porthallow, and Trelowarren streams, and also possibly to that part of the Gwenter River north of Gwenter. Hendriks, referring to the Land's End peninsula, notes that the rivers still pursue their south-easterly trend; she points out the significance of the Chalk flints in the raised beaches, that is to say, at the mouths of the north-flowing valleys. In her view the Hayle River is one of the youngest in Cornwall and is made up of three sections: the coastal pirate stream, a mid-portion, and the upper north-easterly-flowing part.

In the Lizard[2] area we find a completely different series of rocks whose general relations to the killas have been outlined above (p. 207). The oldest rocks, mica schists and bands of quartzite all much folded, are seen at Old Lizard Head, while the gneiss of the Man-o'-War Rocks and the serpentines were intruded later. Hornblende schists also occur at Porthallow and Porthoustock, in a few patches, including Mullion, on the west coast, and form the remainder of the Lizard Point east of the lighthouses. All three are penetrated by numerous dykes. The serpentine (see Fig. 38), the most characteristic rock, is probably a great laccolite; it reaches the sea at Kennack and Coverack, and for a great part of the coast from Mullion to Kynance, and from the Lizard to Cadgwith. It is cut through by numberless gabbro dykes of which a dozen or more may be seen in a single cove, for example, at Coverack. One large dyke forms the promontory of Carrick Luz. Later, a series of basic dykes penetrated the serpentine and are evident at Manacle Point, Coverack, and Kennack.

Taken as a whole, the Lizard groups of rocks are harder than those to the north, a feature which is plain at Porthallow and Polurrian. At both

[1] C.F.W.R. Gullick, 'A Physiographic Survey of West Cornwall', *Trans. Roy. Geol. Soc. Cornwall*, 16, 1936, 380.

[2] *Mems. Geol. Surv.* 'The Geology of Lizard and Meneage', J.S. Flett and J.B. Hill, 1912, and Sir J.S. Flett, *Summary of Progress*, Pt II, 1932, p. 1.

these places the junction is seen, and the Lizard rocks project farther seawards than the killas. The serpentine is relatively soft, and since it is cut back into the Lizard platform it gives rise to lofty cliffs. As it is but feebly jointed the cliffs are usually bold and rugged. At Predannock the coarse hornblende schist is more resistant to erosion, but the mica and green schists at the Lizard do not withstand it very effectively. The gabbro stands up well, but the banded gneisses, often with many joints, yield rapidly. Local erosion is aided in many places by landslips, the effects of which can be seen south of Rosemullion Head (a much-contorted projection in Gramscatho Beds) and on the cliffs between Mawnan and Toll Point. Where the serpentine is penetrated by dykes, water is apt to form channels, and thus again landslides may be initiated. The sea also occasionally cuts in along faults or dykes and forms coves like those at Mullion, Kynance, and Cadgwith. The roof of a large cave may collapse and lead to the formation of a more or less circular depression. These are most frequent where dykes or veins of banded gneiss occur in the serpentine, and the best-known cases are at Lavarnick Pit (west of Kynance), Holstrow, and Kildown Cove. In 1847 a cave roof collapsed at Housel Bay to form a great cauldron known as the Lions' Den. The cliffs on the western, exposed, side of the Lizard peninsula are very imposing and usually dark in colour. As the interior is a plateau, their tops are clearly defined. Deep cut coves are characteristic, and the finest scenery is undoubtedly at Mullion and Kynance where the off-lying stacks and islets afford interesting detail. At Mullion, in particular, the old harbour also adds much to the picture. After rounding the Lizard the type and profile of the cliffs gradually change. This change is almost entirely associated with the degree of exposure. In Housel Bay the cliffs are still steep, but at Landewednack Cove the slopes are less and grass covered. This is true of the cliffs as far as Black Head and becomes, as would be expected, even more noticeable nearer the Helford River and Carrick Roads. One interesting feature of some of the coves on the eastern side of the Lizard—Cadgwith, Poltesco, and Kennack—is that they are double, a small peninsula separating the inlet into two distinct coves. In some respects the eastern part of the Lizard coast, while far less grand, is more attractive than the western coast. There the very flat, barren, wind-swept plateau is so pronounced that although it cannot spoil the actual coast, it does perhaps detract from its setting. On the east the cliff scenery is far less imposing, but the many deep valleys, the woods, and the less emphatic nature of the plateau all combine to give a calmer and more restful landscape (**48** and **47**).

Coves are common at the mouths of streams, and are somewhat pronounced as a result of submergence. Those at Porthallow, Porthoustock,

Church, and Mullion all illustrate this point. Kynance Cove is in part formed from a valley which has been attacked on its western side by the sea, 'but the rock pinnacles and numerous islets have been separated from the land by the sea taking advantage of lines of fault and brecciation and of dykes of intrusive gneiss that were comparatively easily removed'.[1] Where faults reach the coast there is also a natural tendency to cove formation as at Porthallow and Porthoustock, although the effect of the faults may be only indirect.

The raised beach is easily traceable in this area, especially near Coverack and Lowland Point, and the rock platforms, sometimes covered with Head, are often continuous for considerable distances. On the western side the modern beach may be seen resting on the Pleistocene shelf. The old beaches are at different heights; the majority are only a few feet above high-water mark, but there are others at 25 and 35 ft. On Mullion the beach is 75 ft. high, and it may correspond to the 65-foot remnant at Penlee, south of Newlyn. The eastern part of Mullion is flat, while the western part rises to about 100 ft. Formerly the raised beach may have extended all round the island, but the western part has been eroded away leaving the present striking and rather curious outline (**46**).

There is little blown sand on the coast apart from small accumulations at Gunwalloe, Loe Pool, and Kennack Cove. At Church Cove blown sand, coming round and over the cliffs, has partly buried the church tower.

(8) THE LIZARD TO BOLT TAIL

Northwards and eastwards from the Lizard is an extensive tract of the Gramscatho and Mylor Beds, and the most striking physical features are, perhaps, the drowned valleys now forming the Helford River and Carrick Roads, the latter, certainly, covering an old alluvial plain. They are similar in general features to other submerged valleys of Devon and Cornwall, and are most conveniently dealt with in the section on recent vertical movements (p. 259). The description of distinctive features of this section of the coast is continued at once.

The general strike of the rocks is east-west, and there is much isoclinal folding.[2] Between Pendower and Zoze (Zone) Point the killas is much contorted: faults are numerous, and may occur locally every few feet. They may be normal or reversed, but their throw is small, seldom more than a foot or two; strike faults are mainly responsible for the large number of caves. The same tendency occurs west of the estuary near

[1] *Op. cit.* (1912).

[2] *Mems. Geol. Surv.* 'The Geology of Falmouth and Camborne', J.B. Hill and D.A. MacAlister, 1906.

Pendennis, where many gullies, locally called 'drangs', have been eroded along small faults. Similar features are present between Pennance Head and Maen Porth. The raised beaches resemble those of to-day, but may be so much cemented by oxide of iron as to offer great resistance to marine erosion. Occasionally caves cut during raised-beach times are to be seen above modern caves, the roofs of the latter consisting of the cemented floor of the former. Sand and shingle beaches of small extent are common, for example at Pendower, Porthcornick, Towan, Falmouth, and Maen Porth. At Swan Pool the waves have built a natural dam which holds back a fresh-water lake. At Gullyvase and Maen Porth rather similar features have been initiated, but swamps instead of lakes have been formed. There is not, indeed, much blown sand, but the conditions at Porth Farm, near Towan beach, are exceptional. There sand carried by the winds has choked the old channel which once existed near the end of the St Anthony peninsula.

Around Veryan and Mevagissey[1] the coast is rugged and rocky, but on account of the seaward slope of the plateau the cliffs seldom exceed 200 ft. in height. Nare Head is a great mass of igneous rock, and the Gull Rock, Middle Stone, and Outer Stone are stacks of the same material. To the west of the headland the trend of the coast cuts across the strata almost at right angles, so that alternations are rapid: Greeb Point is cut in beds of volcanic ash, but the material is poorly preserved. To the east of Nare Head the most interesting point is The Dodman, a uniform mass of pale grey, fine-grained, soft slate or glossy phyllite. It is well cleaved, and both dip and cleavage planes incline to the south-east at high angles. Hence, much slipping occurs in this direction, while on the western side the cliffs are extremely steep. The prominence of the headland can hardly be explained by the hardness of the rocks: in fact, they are not particularly resistant. But at the same time it projects out into deep water and consequently there is no beach or other material with which the waves can work. Both here and at Nare Head the submarine contours suggest that formerly both headlands stretched at one time perhaps a mile farther seawards. Black Head, bounding Mevagissey Bay on the north, is another greenstone promontory. Other capes on this stretch of coastline are usually due to harder beds, for example, Maenease Point. A harder band also forms the wall of Gorran Haven. Catasuent Cove lies between rocky ridges, the cove itself being cut in black shales, and Porthluney beach probably corresponds with a syncline in the comparatively soft slates. Resistant conglomerates hedge in Portloe Cove, while Kiberick

[1] *Mems. Geol. Surv.* 'The Geology of the Country around Mevagissey', C. Reid, 1907.

Cove lies between the igneous masses of Nare Head and the Blouth, and the Straythe between the Blouth and Manare Point. Jacka Point, Hartriza Point, and Caragloose Point are all formed of igneous rocks and, between Porthluney Cove and The Dodman, outcrops of igneous rocks of various types form minor prominences. In short, on this part of the coast the projections are nearly all of either igneous or hard sedimentary rocks, and re-entrants are in the softer sediments (**45**).

East of Par[1] the coast is rocky and beaches are few and small, mainly at valley mouths, like Polridmouth, Coombe Hawne, and Readymoney. They may also occur on parts of the raised-beach platform, for instance in Lantic Bay. Once again, the capes are mainly the result of harder bands of rock or contorted parts of the sediments: Gribbin Head is imposing.

From Mevagissey Bay as far as Plymouth Sound and beyond, the main coastal rocks are the Dartmouth Slates; these often show bright colours, especially in Talland Bay where Indian red and green tints predominate.[2] Green colours are very striking near Sharrow Point. The slates are by no means homogeneous in composition; they are often hard, glossy, and partly siliceous, but cleaved mudstones or slates, much more argillaceous in character, are also common. Rame Head is composed of grey slates with frequent inter-stratified beds of a fine, grey, quartz-veined grit which probably explains its prominence. Between Rame Head and Penlee Point (slate with some quartz) there are many intercalations of hard grit and many quartz veins. Throughout, bedding and cleavage are rather obscure. The Meadfoot Beds crop out between, approximately, Porthnadler and the east side of the bay: the bay in fact is cut in them, although Looe Island consists of a large stack of Dartmouth Beds. The latter are sometimes much contorted and form stacks and beach reefs. The succeeding grey slates form both the promontory at the eastern end of Millendraeth beach and Raven Rock. The same rocks also front the sea opposite Sheviock. A fine sandy beach in Whitsand Bay provides material which travels to the east.[3]

East of Plymouth Sound[4] the Dartmouth Beds form the coastline as

[1] *Mems. Geol. Surv.* 'The Geology of the Country around Bodmin and St Austell', W.A.E. Ussher, G. Barrow, D.A. MacAlister, 1909.

[2] *Mems. Geol. Surv.* 'The Geology of the Country around Plymouth and Liskeard', W.A.E. Ussher, 1907.

[3] Dr Chapman tells me that he is informed that drift seaweed in the bay comes from east of Plymouth. Whitesand Bay is so situated with reference to the prevailing winds and waves as to favour eastward beach-drifting. Incidentally, there are many shifting sands in the bay.

[4] *Mems. Geol. Surv.* 'The Geology of the Country around Ivybridge and Modbury', W.A.E. Ussher, 1912.

far as Ringmore, whence the Meadfoot Group continue to Hope. The strike is approximately west-south-west to east-north-east, whereas the cleavage is more nearly south-west and north-east. The Mew Stone and the cliffs near Wembury show the strike well. Staddon Point, in the grits of the same name, is prominent, and these beds form pronounced features inland as well as on the north coast of Cornwall. The Yealm, Erme, and Avon valleys have typical drowned mouths. Since the rocks are all relatively hard, waste of the coast is slow, especially where reefs help in its protection. Reefs also tend to hold shingle. Borough Island is a large stack consisting of the Meadfoot Beds. Near Thurlestone, which may take its name from a fine, isolated, natural arch, several small diabase dykes form minor features, and between South Huish and Hope the small headlands of Great Ledge, Beacon Point, and Workman Point are all in the Beeson Grits.

No purely physiographical or geological account can do proper justice to the fine and largely unspoiled coastline between the Lizard and Bolt Tail. In general the cliffs are less steep than in north Cornwall, and in many parts they fall to a lower platform which consists frequently of Head resting on the raised-beach erosion platform. The major headlands, Zone Point, Nare Head, The Dodman, Black Head, The Gribbin, and Rame Head, are all prominent, and are not only beautiful features in themselves, but, as it were, carry seawards the pleasant countryside of this part of Cornwall. The larger inlets, Carrick Roads, the Fowey River, Looe River, Plymouth Sound, Yealm Mouth, and Erme Mouth, are very picturesque and give great character to this coast. Their shores are often well wooded, and the surrounding country quite untouched. The remote Roseland peninsula, or rather peninsulas, gains greatly because of its pleasant interior. The smaller coves are equally attractive. It is impossible to mention all, but Kiberick, with its upper 'flat' ringed by 'cliffs', Porthluney, backed by Caerhays Castle, Mevagissey with its fine harbour and eighteenth-century buildings, and Polridmouth in the Menabilly valley and its almost sub-tropical flora, are outstanding. Farther east the mouths of the Yealm and Erme add greatly to the quality of the scenery (**44** and **43**).

On such a coast physical changes are usually slow, but every now and again storms sweep away beaches, and at Par the estuary penetrated much farther inland before the coming of the railway. Two old fish ponds still exist on the landward part of Par Sands.

(9) BOLT TAIL TO START POINT

The extreme south of Devon[1] is composed of highly metamorphosed schists of undetermined age. They were certainly involved in the Hercynian folding, and many steep plications were produced, so that the local strike is often at variance with the general strike. The rocks fall into two main divisions, the green schists and the mica schists; it is probable that the former are the newer and represent altered basic igneous rocks. They are most clearly exposed at Bolt Tail and Prawle Point. The magnificent cliffs from Bolt Head to Bolt Tail (only excluding Bolt Tail itself and Whitechurch) are in mica schist, which also reappears for a short distance at Hope village. Mica schist also forms the little 'look-out' headland which separates the two coves at Hope. On the eastern coastline mica schists form the craggy[2] arête or crest running out to Start Point. Both types of schist are penetrated by numerous veins of quartz which strengthen the rocks and enable them as a whole to offer great resistance to marine erosion. The extent to which they project farther seawards at Bolt Tail and Start Point than the killas on their northern flank is very noticeable (42).

Salcombe Harbour, partly restricted by a bar, runs inland as far as Kingsbridge. The upper reaches are very muddy at low water, but Waterhead and Southport creeks are most attractive. The raised-beach platform is often very well developed near Prawle Point. The coastline of that time was a little landward of the present beach, and great masses of Head have gathered upon it. Erosion in the Head occasionally produces interesting features, such as the pinnacles at Matchcombe.

(10) START BAY

The view northwards from Start Point resembles in a most striking way a textbook diagram of a submerged coast in a mature stage of development. The alternation of cliffed headland and valley bordered by a continuous shingle beach is extremely clear. At Hallsands[3] is an area of

[1] T.G. Bonney, 'On the Geology of the South Devon Coast...', *Quart. Journ. Geol. Soc.* 40, 1884, 1; W.A.E. Ussher, 'Excursion to the Start, Prawle and Bolt Districts', *Proc. Geol. Assoc.* 17, 1901, 119; *Mems. Geol. Surv.* 'The Geology of the Country around Kingsbridge and Salcombe', W.A.E. Ussher, 1904.

[2] The craggy or spiky features resulting from weathering are conspicuous in many parts of these schist cliffs.

[3] R.H. Worth, 'Hallsands and Start Bay', *Rept. and Trans. Devon Assoc.* (1) 36, 1904, 302, (2) 46, 1909, 301, (3) 55, 1923, 131; R.H. Worth, 'Statement on the Geological Conditions affecting the Coastline from Exmouth to Plymouth', Royal Comm. Coast Erosion, vol. 1, Pt II, p. 177, Appendix No. 14. Erosion at the north end of the village was still severe in 1943.

severe erosion, one of the few in Devon and Cornwall. Erosion here has undoubtedly been much aided by the removal by man of shingle during the last century, and it has been estimated that something like 650,000 tons were carted away from above low-water mark. A reef of rocks at Tinsey Head, just to the south, excludes the possibility of any renewal of supply, and the beach level at Hallsands fell about 12 ft. This naturally allowed wave action to attack the coast more easily, and the village of Hallsands was severely damaged. Since the removal of shingle was prohibited, a certain amount has gathered at Greenstraight, but noticeably not at Hallsands, where occasional severe storms still do great damage; that on 26 January 1917 was conspicuous. Near Beesands a broad has been dammed by shingle. North of the faulted schist boundary on the eastern coast the Meadfoot Beds continue as far as Strete Gate. Generally they are pale, or lilac-red, glossy slates with intercalations and harder beds of grit or quartzite, and with occasional igneous rocks, like those at Dun Point. Beacon Point is in hard slates, and Long Stone promontory in red slates (**41**).

At Torcross is the interesting shingle bar enclosing Slapton Ley. The shingle contains many flints, a good deal of quartz and material from Dartmoor, but very little from purely local rocks. Worth argues that dredging in Start Bay shows that this beach, or any other beaches in the bay, cannot receive any appreciable amount of material from offshore. He also maintains that at present the drift of beach material is nicely balanced, otherwise the bar would have disappeared long ago as a result of a pronounced one-way direction of drift. The presence of flints is also evidence for ascribing a considerable age to the Start Bay beaches, since there are no near outcrops with indigenous flints, and perhaps the physical influences working on them again support this view. The land on the inner side of the ley slopes down gradually, and consists of reddish and grey slates. Ussher had in mind the existence of a westward drift in raised-beach times, an argument apparently derived from the distribution of flints. If this be admitted (and it is by no means conclusive), he thought that the origin of the ley was due to the deflection of the Gara by a spit of sand and shingle. This southerly drift might also explain, as a fluviatile deposit, the spread of gravel on the north side of Slapton Bridge. There seems to be no very conclusive argument put forward for the origin of the beach and no observations appear to have been made on its foundations. In a minor way it has some resemblance to the Chesil Beach and it is perhaps worth asking whether it rests on a solid foundation at a comparatively small depth. The rounded shingle in Start Bay is only found between Start Point and the river Dart. At the present time flint sand is almost absent from the northern end of Slapton beach, and flints were much commoner in the past than they are now at Hallsands. Hunt maintains that the great spread of shingle in the Slapton beach means a decided excess in the past of supply over loss. Occasionally the bar is breached in storms. On a much smaller scale a somewhat similar beach at Blackpool has formed across the mouth of a low-lying valley, but the beach material is smaller than that at Hallsands. Gales from time to time drive in the shingle and expose a submerged forest.

The Dart flows to the sea through a fine drowned mouth. Borings have shown that the drowned valley is trough-like in form and had not reached base level, thus indicating a short and quick elevation of the land relative to the sea. There is no bar at the mouth. The cliffs between Blackpool and Sharkham Point are steep and often inaccessible,[1] and differential erosion is well shown. They are mainly made of Lower Devonian (Dartmouth Series) grits and hard cleaved shales with strengthening intercalations of igneous rocks which form several points on the coastline. In some respects the finest part of this coast, with several stacks, lies between Stoke Fleming and the Dart, but that immediately eastwards of the Dart is more remote and less spoiled. The igneous material, mainly diabase, is conspicuous near Stoke Fleming and on the coast between Matthew's Point and Redlap Cove. Igneous rocks are mainly responsible for the promontories at Matthew's Point, Redlap Cove, Combe Point, Compass Cove, and the Inner and Outer Froward Points. Sharkham Point is formed of limestone with intercalations of tuff. Between it and Berry Head, Mudstone Bay is eroded in Mid-Devonian slates and shales (40).

(11) TOR BAY

Tor Bay, with its two enclosing promontories Berry Head and Hope's Nose, is most suitably treated as a unit, and local details of coastal forms may be discussed separately (Fig. 48). Jukes-Browne[2] has investigated the origin of the bay. Both the promontories have hard and resistant cores of Devonian rocks, whereas the bay is mainly cut in the softer Permo-Triassic deposits. An anticline near Paignton brings up another mass of Devonian, probably part of a pericline which seems to die out some three miles to the eastward. There is another Lower Devonian area near Meadfoot and Kilmorie which also formerly continued some way seawards. If in past times the Torquay anticline in its southern prolongation had a

[1] *Mems. Geol. Surv.* 'The Geology of the Country around Torquay', W.A.E. Ussher, 2nd ed., W. Lloyd, 1933.

[2] A.J. Jukes-Browne, 'The Making of Tor Bay', *Rept. and Trans. Devon Assoc.* 44, 1912, 718.

structure similar to the existing northern part, then it included a more or less continuous band of limestone on its western side. The southern part of the bay was occupied by shales, tuffs, and limestones running eastwards from Stoke Gabriel; a reconstruction inferred from the tract of limestone extending from Galmpton to Berry Head. The northern limit of this belt was on the northern side of Saltern Cove where a fault brings up the

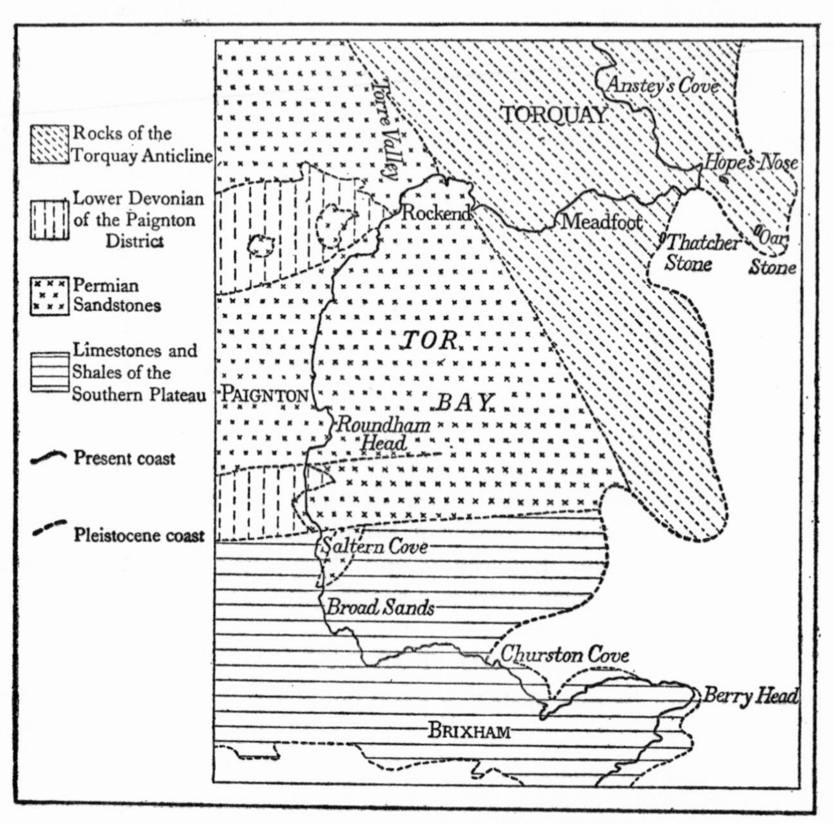

Fig. 48. Tor Bay in Early Pleistocene time (after Jukes-Browne, *op. cit.*)

Staddon Grits. To-day most of the southern part of the bay consists of mud and sand, except for 'The Ridge', which is of limestone. The Permian is unconformable and occurs in two troughs, one at Paignton and the other in the Torre valley: at Paignton the base of the Permian falls to 164 ft. below sea level, but to a rather lesser depth in the Torre. This suggests that the two tracts probably met eastwards to form a continuous floor to part of the bay. The Saltern Cove fault is the southern boundary of this. In the north-west of the bay the Permian-Devonian junction is a line of faulting which, in all probability, continues some distance to the south-east. It may thus be inferred that a triangular area of Permian sandstone and conglomerate exists beneath the bay, but that it does not extend beyond the bay mouth because it is cut off by faults to south and east. The map (Fig. 48) shows these features and also the general position when the Pliocene land was first invaded by the Pleistocene Sea.

In Pliocene times the Torquay anticline formed the highest ground, and, as may be inferred from the present valleys and their former probable continuations, the whole north-western part of the bay was an undulating area, with valleys trending generally eastwards; they also opened out in this direction and probably drained into a major north-west and south-east valley. In the southern part of the bay was a plateau-like area, somewhat hilly to the north-east and undulating in the north-west. The streams draining it may have met in the middle part of what is now Tor Bay, to unite in a stream running to the south-east as suggested by the submarine contours. 'Hence, I think we may infer that the Torre Valley was the main valley, and that in Pliocene times it ran obliquely across the Torbay area, passing out of it a little north of Berry Head, and being itself the tributary of a larger river which we may regard as a prolongation of the Exe.'[1] At this time the English Channel did not exist, and this proto-Exe may have drained eventually into the 'English Channel River'.

Subsidence began about the end of Pliocene times, and an early stage in this is represented in the raised beaches at Hope's Nose (20 ft.), Thatcher Rock (15 ft.), and elsewhere.[2] Raised beaches are absent from the western side of the bay, which may mean that only a small part of the present bay was then invaded. In any case, later erosion would presumably have destroyed any traces which may have existed. The period of emergence represented by the submerged forests implies that the bay once again became a land area, and that the then coastline could not have been far from the present 100-foot submarine contour. All the streams would necessarily have been rejuvenated and have deepened their valleys, and one result of this was that the Brixham caves ceased to be water-ways. It was at this time, also, that the many small gloups in the limestones were formed. Probably the old main valleys were also deepened.

Once again subsidence ensued: the sea advanced up these valleys which by degrees became small estuaries. The Tor Bay River itself was by that time a wide estuary; thus it allowed the ingress of the sea to the western part of the bay, and marine erosion of the Permian strata began. In the same way the proto-Exe valley was drowned, and erosion began within

[1] Jukes-Browne, *op. cit.*
[2] See also Chapter XII: this suggestion is not in conformity with Arkell's views.

it. But as the former extension of the Hope's Nose promontory was made of hard Devonian rocks, it would have resisted erosion far more effectively. It follows, therefore, that for some considerable time Tor Bay was a much narrower inlet than it is now. But as submergence continued the rocks were attacked more and more fiercely and the southern part of the bay, the plateau area, was eventually covered by water so that with the combination of increasing subsidence and erosion the bay began to assume its present form.

The floor of the bay now shelves gently seawards to the 50-foot contour, and the shore of the bay proper is fairly smooth. This is due mainly to the fact that erosion is slowly being checked by the relative calmness of the waters in the bay and that, whereas minor capes are naturally still subject to a certain amount of wear and tear, inlets in the softer rocks are either being silted up or protected by beach material. The building of sea walls has also helped in this process of protection. But even in the last 150 years or so there have been interesting, if minor, changes. In the early part of the nineteenth century there were cottages and gardens on the site of the present Livermead Sands, and at Corbon's Head there was a natural arch similar to the one now called London Bridge. Quite recently, indeed, the shingle ridges at Oddicombe and Babbacombe (to the north of Torquay) have been lowered by several feet, either on account of the cessation of quarrying, or because the beaches, when higher, were actually in an abnormal condition (**39**).

Around Brixham there are steep limestone cliffs. Broad Sands are cut in Upper Devonian slates, and the now drained lagoon and pebble bar are comparatively late formations. The Staddon Grits and Shales form the cliffs between Goodrington Sands and a small cove just north of Saltern Cove, except in those places where little pieces of Permian have been infaulted or lie unconformably on the older rocks. Roundham Head, for instance, is formed of Permian conglomerate. Daddy Hole, Thatcher Rock, and the Oar Stone are in line, and if the one-time limestone masses were restored, they would form the south side of the extended Ilsham valley. The erosion of this limestone has allowed the sea to work vigorously in the slates, and so to truncate the stream. The whole of the Torquay promontory is described by Shannon[1] as 'an anticline converted into a fault breccia on a large scale with thrust planes and slides as the dominating features'. Faults influence the cutting of Smugglers' Cove where dolerite is thrown against Staddon Grit, and at Oddicombe two faults let in the

[1] W.G. Shannon, 'The Geology of the Torquay District', *Proc. Geol. Assoc.* 39, 1928, 103, and also 'Erosion in the Torquay Promontory', *Rept. and Trans. Devon Assoc.* 55, 1923, 148.

Permian. South of the promontory the Thatcher Rock-Daddy Hole thrust-plane is very important. As suggested above, the islands now represent the south side of the formerly extended Ilsham valley.

The arch of London Bridge has been eroded along cleavage planes, and Pig Tor results from the sea cutting along a crush belt at intersecting faults. The minor headlands between the Ilsham valley and Hope's Nose result from the comparative resistance of recumbent folds, whereas the re-entrants are often cut along shatter-belts. 'Hope's Nose is due to a remnant of limestone or [on?] a lower thrust-plane....Armorican folding with volcanic tuffs is found on the lower platforms; this is cut by a thrust-plane from east-south-east; on this plane folding is from the east-south-east, in turn cut by a secondary thrust-plane.'[1] In Shennell Cove the reefs are formed on Staddon Grits, and a major thrust has led to erosion cutting out Hope Cove. Black Head is mainly a mass of dolerite. In Anstey's Cove erosion cut along the slates once the limestone cover had been removed, so that the slates had no protection from the dolerite: the cove is immediately north of the end of the dolerite sill. The glen at Babbacombe is cut in a shatter belt, and the inlet at Redgate beach is explained by faulting parallel to the coast. The base of Babbacombe cliffs is partly protected by a sill of igneous rock. Fault gullies occur at either end of Oddicombe beach, and Permian conglomerates have been let in between them. Petit Tor Cove is cut in Upper Devonian slates; north of this inlet erosion is rapid and the Permian cliffs are steep. Hanging valleys are common.[2]

(12) BABBACOMBE BAY TO THE MOUTH OF THE EXE

From this point to another about two miles east of Exmouth the coast is fringed by the Permian rocks which give rise to fine, high and brightly coloured red cliffs.[3] The oldest Permian rocks are the Watcombe Clays which are overlain by coarse breccio-conglomerates, so conspicuous in Watcombe and Petit Tor Crags. North of the Teign estuary, which is far more open and fringed by less wooded slopes than those farther west, the boulder breccias with finer intercalations of hard breccias appear in the cliffs, but towards Dawlish a more sandy matrix is introduced. From Horse Cove to Dawlish thick intercalations of a paler-coloured sand rock

[1] W.G. Shannon, *op. cit.* 1928.

[2] See also *Mems. Geol. Surv.* 'The Geology of the Country around Torquay', W.A.E. Ussher, 2nd ed., W. Lloyd, 1933; and L.G. Annis, 'The Geology of the Saltern Cove Area', *Quart. Journ. Geol. Soc.* 83, 1927, 492.

[3] *Mems. Geol. Surv.* 'The Geology of the Country around Newton Abbot', W.A.E. Ussher, 1913.

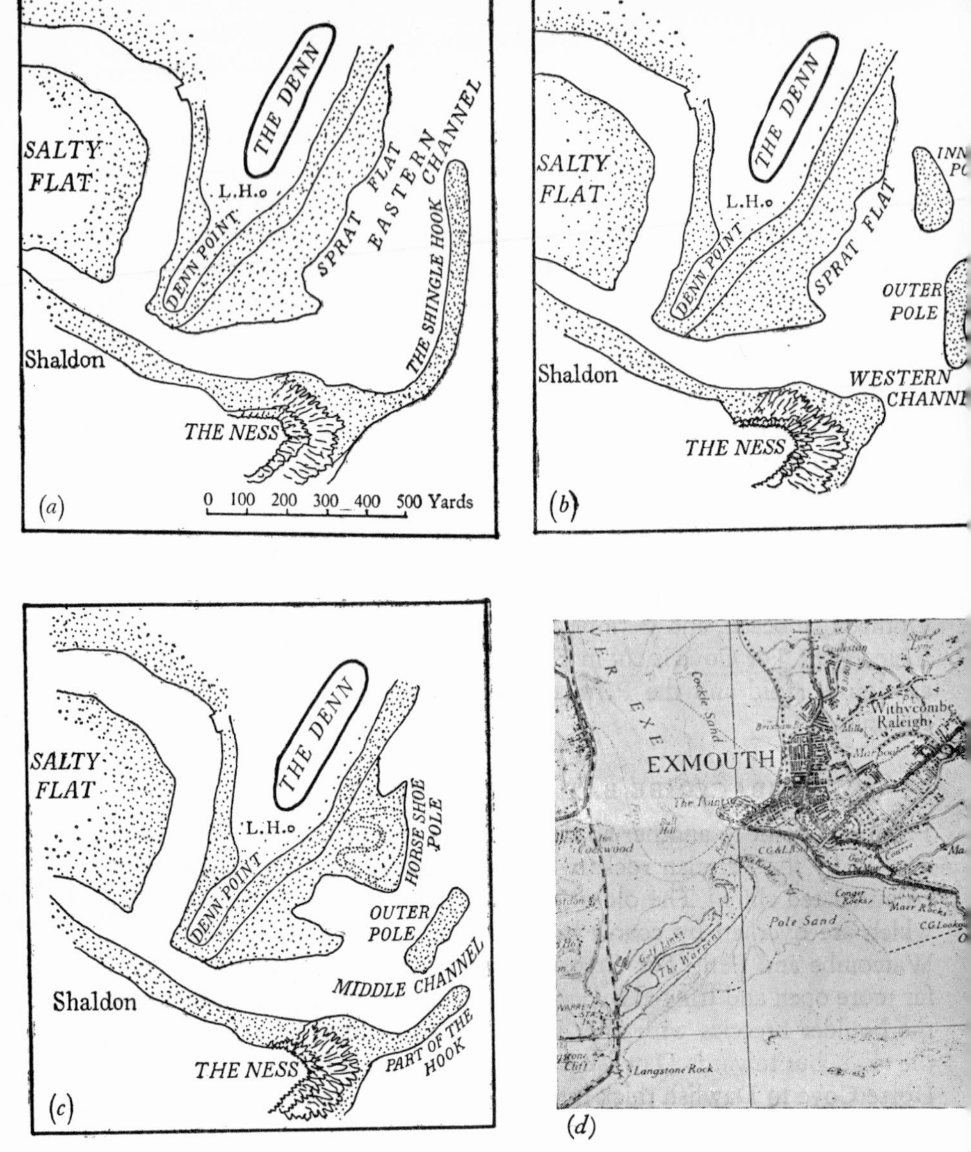

Figs. 49*a*, 49*b*, 49*c*, 49*d*. *a*, *b*, *c*, The evolution of Teignmouth Bar (after Spratt, *op. cit.*); *d*, mouth of the River Exe (reproduced from the one-inch map, 5th edition, of the Ordnance Surv

help to show up numerous faults. Thence to the mouth of the Exe sand rock with breccia bands is dominant. Langstone Point is formed of breccia. The breccia reappears east of the river, but towards Littleham Cove (on the east side of Straight Point) red marls with greenish mottling are seen in the cliffs which again are affected by many faults. At Littleham Cove the sandstones and overlying marls are faulted against the main mass of the Lower Marls. Coryton's Cove (south of Dawlish) is associated with a fault. Littleham Cove, as far as the West Down Pebble Beds, is eroded in marls. The whole of this coast is subject to considerable erosion, and this is plain from the disappearance or lessening of stacks or heads such as the Parson and Clerk. Now many parts of the coast are protected by sea walls built by the Great Western Railway Company. Near Budleigh Salterton the Triassic rocks form prominent cliffs, and contain the far-famed Pebble Bed.[1] The coastal type, however, does not differ markedly from that of the Permian rocks.

Two features of considerable interest on this stretch of coast remain to be discussed—the bars at the mouths of the Teign and Exe (**38**). At Teignmouth[2] the changes appear to be cyclical. The prevailing wind is approximately from the south-west, and if the outfall of the river is strong there is a good deal of commotion at the mouth. Spratt made a series of interesting observations on the bar, and estimated that a complete cycle of changes occupied about seven years, but that it varied necessarily on account of the incidence of winds and storms. Sir John Rennie[3] wrote in 1838:

> The southernmost part of the Den had been considerably reduced in length and height, and was beaten down by the violence of the waves, so that the waters of the Teign, instead of issuing in one combined and uniform mass, so as to enable them to act in the most effective manner in resisting the opposite and more powerful action of the sea and wind from without, are dispersed over a large area, and thus their force was considerably diminished, and became incapable of maintaining a deep and direct channel to seaward, and are necessarily compelled to run parallel to the shore until they had acquired sufficient force to discharge themselves, and in this manner the bar may be said to be in a continual state of change. When the easterly winds prevail there it extends a considerable distance along the shore from the Ness in front of the Den; when the westerly winds prevail, then the Teign, being protected at the mouth, naturally takes the most direct course seawards, the southern point of the Den accumulates both in length and height, and forms, as it were, a pier or breakwater to quiet the waters rising from the river in the most beneficial direction, and the bar is then

[1] Falls from this bed still contribute many easily recognized pebbles in the beaches east of Budleigh.

[2] See J.C. Inglis in discussion on 'The Sanding-up of Tidal Harbours', A.E. Carey, *Mins. Proc. Inst. Civ. Engs.* Pt II, 1903–4, 146, 215.

[3] Quoted in above reference.

considered to be in the best state, the entrance being close to the Ness, so that depending upon either of the above causes the bar and entrance is confined to a range of about three quarters of a mile.

Spratt's views[1] are best summed up in the three sketches in Fig. 49. In the first sketch the long hook of coarse sand or shingle is developed: and the Denn Point is then less high: in the second the Hook may break up and leave the Outer and Inner Pole Sands. Finally, the Inner Pole is driven onshore to form the Horse Shoe Pole. The fluctuations in the actual mouth of the channel are clearly shown. The Hook is liable to be broken near the Ness point by gales, and when a gale has weakened it, the river water may first flow seawards over a weak part of the bar, and later will cut a channel. At the same time the outlet near the end of the bar tends to be blocked, especially as a gale from the eastwards naturally turns its distal end shorewards. The effects of an easterly gale are clear; a westerly gale has similar results, although indirect because it drives water up-channel, which later, as it were, expands laterally (i.e. approaches the Hook so as to drive it ashore). The growth of the Hook is naturally irregular, and although the cycle of changes suggested by Spratt seems to take place, it does not do so at exact intervals. The Denn Point suffers greatly from strong westerly and north-westerly winds acting with the ebb directly down the estuary, but easterly gales are even more effective, especially if coincident with a high spring tide. Spratt's views on the growth of the Hook are different from Rennie's:

> By the contraction of the channel off the Denn Point, the stream abreast of it is... strong....Now as the ebb...runs with force directly down the Shaldon shore, and strikes against the extremity of the Denn Point..., much of the matter...will be swept across the stream...(so that)...material removed from the Denn Point...will eventually reach the Ness. But, as the ebb there meets with the along-shore tide, or with slack water, its strength and direction is suddenly checked, so that a deposition of the heavier material brought out will necessarily take place there.
>
> Thus, a spit or tongue of gravel and sand gradually grows out from the Ness Point, curving slightly at first, till the along-shore and river tides balance each other.

The spit across the mouth of the Exe is an interesting feature, and has been discussed by Martin (Fig. 49 *d*).[2] He refers to several early maps and also shows that the one-time belief that the spit 'grew' from the eastern side of the estuary rests on a confusion of Langstone Point with Starcross. The earliest available map (1611) shows a spit running eastwards, while a

[1] 'An investigation of the Movements of Teignmouth Bar.' London, John Weale, 1856. The paper is an Appendix to a 'Report on the Present State of the Navigation of the River Mersey', by the same author, published in London in 1884.

[2] J.M. Martin, 'Exmouth Haven and its Threatened Destruction', *Rept. and Trans. Devon Assoc.* 5, 1872, 89; 'The Changes at Exmouth Haven', *ibid.* 8, 1876, 453, and 25, 1893, 406.

later map (1637) shows an island nearly in the middle of the estuary. Martin, however, does not place much faith in either of them. The first authentic map, by Lang of Leyland, is dated 1787, and shows the Manor of Kenton. Most unfortunately it does not include Exmouth, but it is accurate as far as it extends. A glance at the fifth edition of the 1-inch Ordnance Survey map shows that there is an outer and an inner warren separated by a low. On the 1787 map there is marked on the river side of the Inner Warren, at a point about half a mile from the Light, an old salt works; moreover, close to the point where the Inner Warren joins the mainland is marked the Great Intake, which Martin interprets as an intake from the flooded lands for embanking. This suggests the Warren was low in this part, and that at an earlier time the existence of a high-water channel hereabouts may have left the Inner Warren either as a sand island or even as a peninsula running westwards from the Exmouth side, although such reconstructions are doubtful. The Outer Warren then, as now,[1] seems to have been a long line of sand dunes. Hence, Martin suggests that communication between the sea and the river must have been between the two warrens and through the tidal channel at the western end of the Inner Warren. He is uncertain whether this was navigable or not. As the map does not show Exmouth, it seems at least possible that the real channel was there, and that the gap in the Inner Warren could be explained by a ponding-back of the river water by a storm, so that it overflowed through some lower part of the Inner Warren. This is guess-work, and it is difficult to picture the conditions suggested by Martin, especially when dealing with a river of the magnitude of the Exe.

There seems to be no doubt, however, that the Outer Warren, in 1772, was a continuous line of sand hills, which were both high and wide, and that at their eastern extremity there was a sand hill or hills some 50 acres in extent and 20–25 ft. high. Martin also states that in 1787 there was a gap between this bluff and Exmouth about 430 yd. wide: by 1809 it had increased to about 520 yd., and the bluff was replaced by three small sandy islets, and now is gone altogether. The width of the channel continued to increase, attaining a breadth of 970 yd. in 85 years. (It is not quite clear what dates are in question, but presumably the 85 years previous to the writing of Martin's first paper which appeared in 1872.[2]) In 1872 there was a gap in the Outer Warren about 600 yd. from its western end, and the sea reached the lagoon through it; the lagoon certainly existed in 1787. This gap appears to have been made in 1859, close to an earlier gap of 1824. The 1859 gap widened, reaching 240 yd. by 1872, and so became a serious menace; it was artificially dammed in 1872.

[1] In January 1944 it was discontinuous and eroded.

[2] Or from 1787, Lang's map, to 1872.

The Ordnance Survey map of 1809, if compared with earlier maps, shows that the Outer Warren was being reduced in width as a result of degrading of its outer side, but that in 1809 there was no gap in it except close to its eastern end: this left an island to the east of it. In fact the inner part of the Outer Warren and the north side of Inner Warren, which was less regular in 1809 than in 1789, were divided into three elongated ridges which were in existence in 1876 (the date of Martin's second paper). Of particular interest is the fact that measurements made from the 1787 and 1839 maps show clearly that the Warren was wasting before the railway was built. Other small-scale changes were observable at the distal end of the Warren, for example, the continuous loss of land was again noted in Martin's third paper in 1893. I am not aware of any further detailed work, but there is a clear instance of the evolution of an interesting sand spit. An unusual feature is the formation of two spits—an inner and an outer—which is, as yet, unexplained. It is a problem that might repay further study. The Inner Warren is lower than the Outer; it is reasonably stable, and associated with mud flats and ooze. Martin is undoubtedly right in ascribing the general origin of both spits to the prevalent winds and waves carrying material to the north-east. The ebb from the Exe would maintain a channel between them and Exmouth. Furthermore, it is clear that the building of the railway and its embankment has largely deprived the Warren of new supplies of material, but this loss also preceded the building of the railway. It is noteworthy that the Warren is set back from Langstone Point, and may be compared, in this respect at least, with Ro Wen enclosing the Mawddach estuary in Cardigan Bay (p. 142). In January 1944 the eastern end of the Warren was an island at mid-tide.

(13) THE HIGH-LEVEL PLATFORMS IN DEVON AND CORNWALL

In both Devon and Cornwall, but particularly in Cornwall, are several high-level platforms, some of which are remarkably well developed. These features raise many acute problems: the personal factor seems to play a part not only in interpretation but also in actual observation. For instance, in Dewey's summary account, as given in *British Regional Geology*, 'South-West England', only three of the platforms are noticed. They stand at altitudes of 1000 ft., between 750 and 800 ft., and at about 430 ft. The highest is necessarily of limited extent and is clearest around Moretonhampstead: Dewey regards it as a general base-level plain of Upper Oligocene age, and the Tertiary (? Pliocene) sediments in the Bovey Tracey Basin result from its erosion. It is also possible, in his view, however, that the plateau may be a Miocene feature. The middle platform is most clearly seen in Dartmoor, but occurs elsewhere, for example around

Lydford: this is probably of Miocene age. The 430-foot (Pliocene) platform needs no immediate comment. Gullick[1] has traced the 750-foot level in western Cornwall in the moors of Land's End and Carn Menellis, and in north Cornwall Balchin[2] refers to it as the Treswallock platform. He states that it varies in height between 750 and 800 ft., and that farther west it may fall to 650 or even 600 ft. Individual fragments of it are not extensive, and are usually limited to hilltops or to the interfluves separating valleys cut into it, such as occur south of Polrunny. It is clearest near Camelford. Balchin suggests it is later Miocene in age, while at a height of about 850 to 920 ft. he recognized another level referred to as the Condolen phase. It is represented only as a cliff fragment: that, and not the platform, is the conspicuous feature, and at best it seems to represent but a minor still-stand. Gullick notes traces of a 600-foot surface on the moors of Land's End and Carn Menellis, and traces of this can be found as far east as the Dart, but there is apparently no evidence of it in the Padstow area (Balchin).

The Pliocene level of 430 ft. is most important. (430 ft. is the height to which it is referred in the *Regional Memoir*; other workers add subsidiary levels.) Its upper boundary is a degraded cliff which cuts indiscriminately across any geological boundary or across rocks of varying hardness. Pliocene deposits are associated with this level, and it is convenient to say something of them before dealing with the actual erosion surfaces. On and around St Agnes Beacon they occur between 420 and 350 ft., but their upper and lower limits are obscured by Head. At St Keverne (Crousa Common) Pliocene deposits are found on the Lizard plateau; their outcrop is well defined and shown clearly by the vegetation. At Polcrebo (480 ft.) there is a pebble bed of more doubtful nature. The classic occurrence is at St Erth. Here, too, the boundaries are obscured by Head, but the deposits are at a lower level (170–50 ft.), and in the bottom of a wide, open valley. The full sequence of beds is:

Vegetable Soil
Head
Yellow Sand
'Growder' (very coarse ferruginous sand)
Yellow Sand
Blue Clay with fossils
Quartz Pebbles
Fine Quartzose Sand
'Growder'

[1] C.F.W.R. Gullick, 'A Physiographic Survey of West Cornwall', *Trans. Roy. Geol. Soc. Cornwall*, 16, 1936, 80.

[2] W.G.V. Balchin, 'The Erosion Surfaces of North Cornwall', *Geogr. Journ.* 90, 1937, 52.

Dewey[1] writes: 'If the depth at which the clay with fossils was deposited were only 90 ft., the elevated adjacent land must then have been dry land; in fact, the shoreline must have been coincident with the 190 ft. contour....The lithological character of the clay points to deposition in fairly still water at probably not less than 50 fathoms. If this depth is correct, the land would then have been about 400 ft. lower than to-day, and this figure approximates with the amount of uplift of the plain of erosion.' As Milner[2] has pointed out, however, the mode of occurrence of these beds and their lower elevation suggest a greater age, and they are possibly Eocene.

The composition of all the deposits indicates a common origin for all of them, and Milner suggests that they are the product of an erosion period. Later erosion removed most of this material, and what remains is explained by favourable local conditions. Mineralogical as well as topographical evidence supports the idea of contemporaneity. There was certainly a general subsidence of this part of England at least in early Pliocene times, and the lower level of the St Erth deposits may imply that they were laid down to begin with in a hollow and were later modified by the erosion of the platform. 'Thus the St Erth valley, from its initiation as an early Tertiary river-course, became transformed at the close of Miocene times into a strait separating the Land's End area from that of the main Cornish land mass; and it is noteworthy that, even with the present configuration of the land, a subsidence of only about 150 ft. would be required to re-establish such conditions.'[3]

As all these deposits (with the modifications suggested in the case of the St Erth beds) rest on the so-called 400-foot platform, it is safe to regard this platform as definitely Pliocene in age, an opinion borne out by gradients in some of the river valleys. Gullick, however, in referring to the Polcrebo gravels (480 ft.), suggests that they are possibly not of quite the same age, and that they are perhaps to be correlated with the transgression just below the 600-foot stage. The 400-foot platform is not by any means a simple feature. It begins as a steep descent (an old cliff) from a bluff at 430 ft., but at about 300 ft. ends in a marked curve-over the precise origin of which is uncertain. Reid and Dewey suggest that it may either be a newer Pliocene shoreline or due merely to a rounding-off of the cliff edge which originated under an Arctic climate. It is not developing at the present time. Balchin, who calls it the Trevena platform, states that in north Cornwall it is plainest between Tregardock and Boscastle, in which area are also found the finest examples of V-shaped valleys. Tin-

[1] *British Regional Geology*, 'South-West England'.

[2] *Quart. Journ. Geol. Soc.* 78, 1922, 348.

[3] Milner, *op. cit.*

tagel Island shows the platform to advantage, where it is cut regardless of structure. Balchin also puts its lower termination at about 300 ft. The river Camel follows the Trevena coast for some miles before it turns to follow the slope of the platform, and it is hard to see how an age later than the cutting of the platform can be ascribed to this river. In this district the curve-over at 300 ft. may be as much related to the strike of the rocks as to the two factors mentioned by Reid and Dewey. In Balchin's district the curve-over only shows where the coast runs parallel to the strike: where it is at right angles, vertical cliffs occur. There is no level, however, at 180 ft. in this part of Cornwall; instead, at 240 ft. there is a platform (the Rosken platform) very clear on several headlands near Padstow, including Port Isaac and Park Head. But the rise between this and the Trevena level is not outstanding.

In south Cornwall there are many examples of flattenings at 180 ft., apparent in Carrick Du, Pedn Olva, Cudden, Perranuthnoe, Pendeen Point, Gurnard's Head, and the actual point of Land's End. It is found again on the western side of the St Ives-Mount's Bay valley. River gradients, especially in the Gwenter, Marazion, Cober, and Kennell also emphasize this level. Like that of the 400-foot level it ends in a curve-over.

> The bulk of the evidence in Cornwall does, indeed, appear as a whole, to favour a marine origin for the...platforms. Firstly, there seems to be no reason for regarding the river-system as superimposed; secondly, the gravel deposits...indicate a marine origin for the 400-foot platform—and possibly even for the 600-foot platform; thirdly, occasionally, and especially with regard to the 600-foot level, the well planed lower platform is succeeded, above a small bluff, by a further plain suggestive of a peneplain; fourthly...the granite-killas contact is, in the 400-foot platform, frequently crossed without any surface indication of contact, as it is also in the 600-foot platform at Troon and, occasionally, in the 750-foot platform as in the Camelford area of North Cornwall.[1]

(14) RAISED BEACHES AND DROWNED VALLEYS

At lower levels, around the foot of the cliffs and in re-entrants, are numerous traces of raised beaches, more especially of the rock platforms on which the actual beach deposits formerly rested. Naturally, these remains vary somewhat in height just as does the height of the present beach, if its full vertical range is taken into account. The beaches or platforms occur at various heights: about 65 ft. at Penlee, 75 ft. on Mullion Island, and between 15 and 40 ft. on Plymouth Hoe. However, the beach most frequently seen is that commonly referred to as the 'pre-Glacial' beach. Recent work by Arkell has suggested that it is inter-Glacial in age, and in any case is probably best called the *Patella* beach.[2]

[1] Gullick, *op. cit.*

[2] *Proc. Geol. Assoc.* 54, 1943, 141.

Many descriptions have been written on various occurrences of the *Patella* beach in Cornwall. In the literature existing up to the war, it would not be unfair to say that, in general, the following sequence of deposits (not necessarily complete at any one place) was usually associated with it:

Blown Sand.
Boulder Bed = Boulder Clay in North Devon and South Wales.
Head.
Current-Bedded Sand with a temperate fauna.
Boulders and Gravel.
Raised-Beach Platform.

East of the Lizard the platform is continuous for long distances: it is often Head-covered, and only a few feet above high-water mark. It is extremely well seen between Prawle Point and Start Point, especially near Lannacombe and Matchcombe. At the latter place, stacks have been cut in the overlying Head. The coast road from Penzance to Newlyn follows the platform, and in the low ground between Penzance and Marazion the old cliff is sometimes half a mile inland. It is often also well preserved in sheltered inlets.

Our knowledge of this beach has been greatly extended by Arkell who examined the Pleistocene rocks at Trebetherick Point near Padstow. The generalized sequence of deposits is given in Chapter XII where also will be found a discussion of the age of this beach and of others on the south-west and south coasts of England and Wales. It is remarkable that these admirable exposures were not discussed before 1942.

Below present sea level there are numerous traces of submerged forests,[1] well known in Cornwall on account of the workings for stream tin. The Pentuan (Pentewan) section is classic; the lowest significant deposit is there an ancient soil with oak stools *in situ* at 65 ft. below sea level, thus implying a change from the present of at least 85 or 90 ft. In other places borings have shown that alluvium as much as 120 ft. thick occurs in estuaries. The submerged forests are usually limited to the upper parts of estuary waters. Many deposits have actually been taken away from these waters in the search for tin, but in the Fal low-water deposits accumulate quickly. Comparison of charts suggests that between 1698 and 1855 the bottom rose 12 ft. at Tolvern Point and 18 ft. at Tregothnan boathouse. The quickest sedimentation was in Durran Creek. Thomas calculated that the silt carried into the harbour would, if uniformly dis-

[1] See the several *Geological Survey Memoirs* cited in previous footnotes, and also Chapter XII of this book.

tributed in all parts, cause a rise of the bottom of 1 foot in about 43 years.[1] The general nature of the submerged-forest deposits are dealt with in Chapter XII, and only discussion of the question of the age of the valleys and the formation of the drowned estuaries (rias) of Devon and Cornwall is necessary here.

The stream-tin deposits are probably glacial in age in the sense that they corresponded in time with the glaciation of that part of England and Wales which was ice-covered. All valley deposits resting on the stream tin are recent. The relation of the valleys to the Pliocene platform suggests that they are later Pliocene in age. Hendriks (*supra*) appears to put them earlier, following the post-Cretaceous uplift. The sequence of the platforms appears to mean a more or less long-continued negative movement, punctuated by still-stands during which the platforms themselves were cut. The lower raised beach is the *Patella* beach and its nature and extent imply a considerable time for its formation. As it is quite close to the present beach, we may reasonably suppose that the river system and valleys were then similar to what we see to-day. During the Pleistocene there is no doubt that the rivers were much more powerful and that they enlarged their valleys. This is stressed in the Falmouth memoir. The Head, according to Arkell, was probably formed in the Riss Glacial period. In Carrick Roads both the Head and any traces of the raised beach on which it rested were removed during the torrential conditions that led to the stream-tin deposits. It is also certain that during the glacial period the sea level fluctuated considerably with reference to the land. This would lead to the deepening of the valleys, some of which are known to have very narrow floors, and at that time must have resembled gorges. Worth[2] has investigated Plymouth Sound: 'To sum up, the general conclusions to which we can safely arrive are, that the rock-bed of the Hamoaze and associated estuaries forms a channel varying from 60 ft. to 150 ft. in depth, steadily growing deeper seawards, and with ever-increasing rapidity, presenting constant alternations of wide-spreading basins and much contracted straits. So far as our information extends, these conclusions apply also to Cattewater.' Worth supposes that ice action produced these forms: this is not in agreement with modern views. When the climate ameliorated forests began to grow on the lowland and the rise of the sea can be traced in their several beds. With the return of the sea to its present

[1] *Mems. Geol. Surv.* 'The Geology of Falmouth and Camborne', J.B. Hill and D.A. MacAlister, 1906, and *Trans. Roy. Inst. Cornwall*, 7, 1881, 12.

[2] *Ann. Repts and Trans., Plymouth Inst.* 11, 1890–1, 65. The paper contains a map showing areas reclaimed in the estuary, and diagrams showing the amount of silt in several branches of the estuary.

level the lower ends of these valleys were drowned and converted into the estuaries (rias) of to-day.

Note. In addition to papers cited in the footnotes, general reference may also be made to:

C. Reid, 'Memorandum on the Geological Conditions affecting the Coast of... Cornwall', Appendix 12 (B), vol. 1, pt II, R. Comm. Coast Erosion, 1908.

J. Prestwich, 'The Raised Beaches and "Head" or Rubble Drift of the South of England...', *Quart. Journ. Geol. Soc.* 48, 1892, 262.

M.A. Arber, 'Outline of South-West England in Relation to Wave Attack', *Nature*, 146, 1940, 27.

P. Macar, 'Quelques remarques sur la Géomorphologie des Cornouailles et du Sud du Devonshire', *Ann. Soc. Géol. Belg.* 60, 1936–7, 152.

(See also *Geogr. Journ.* 91, 1938, 89, note by C.F.W.R. Gullick.)

J.F.N. Green, 'The High Platforms of East Devon', *Proc. Geol. Assoc.* 42, 1941, 36.

H. Fox, 'Some Coast Sections in the Parish of St Minver', *Trans. Roy. Geol. Soc. Cornwall*, 12, 1896–1904, 649, and 'Further Notes on Devonian Rocks and Fossils in the Parish of St Minver', *ibid.* 13, 1904, 33.

* (See page 234.) "Mr Geoffrey Grigson has suggested in a letter that, e.g., in Samson, the two hills have been joined, before the subsidence, by deposits over the granite which were eroded away, and later the neck built up again by wave action on this foundation."

Chapter VIII

THE CHANNEL COAST AND THE STRAITS OF DOVER

(1) INTRODUCTION AND BRIEF OUTLINE OF THE STRUCTURE OF THE REGION[1]

The coastline of the part of southern England considered in this chapter is almost wholly in the Mesozoic and Tertiary rocks, and contrasts strongly with the older and usually harder rocks of Devon and Cornwall. The newer and more easily eroded rocks have allowed the coast to assume a more mature aspect: long sweeping curves are characteristic. The headlands are still of harder rocks, but harder in a relative sense only. The drowned valleys of the south-west find their counterpart here also, but in the newer rocks silting has gone on to a greater extent, and the protecting headlands have been cut back. At first sight there is not much resemblance between, for example, Salcombe Harbour and the lower valley of the Sussex Ouse. But if we were to remove the alluvium from the latter and allow for the differences in rock hardness, a strong resemblance would be seen, except that, in general, the country east of the Dorset border is lower than that of the Salcombe district.

Southern England shows the effects of Tertiary mountain building extremely well. The violent movements of the Alps gave place to gentle folds and occasional overthrusts in this region. The folds, although by no means regular, are in a general way all directed east and west. We are not here concerned with structural problems except in so far as they touch the coast. It may be stated that the contrasts between the types of folding shown in Wessex, the Weald, and the London Basin depend largely on the depth of the underlying Palaeozoic floor, and the ease with which its cover of newer rocks has yielded.

The three chief structural units are the London Basin, the Hampshire Basin, and the Wealden anticlinorium. Each contains many minor folds, and it is important to realize that the Weald is not by any means a simple anticline, but a domed-up area with many minor anticlines arranged *en echelon*. The age of the folding over most of southern England is Miocene, as is seen so well in the Isle of Wight where the Oligocene beds are sharply

[1] There are numerous papers on the structure and geology of this area. For the present purpose the *British Regional Geology Memoirs*: 'The Wealden District', F.H. Edmunds, 1935, and 'The Hampshire Basin and Adjoining Areas', C.P. Chatwin (no date)—are admirable summaries.

upturned. Near Weymouth, however, the folding began somewhat earlier; this is shown by the fact that the Jurassic and Wealden Beds were tilted and eroded before the Upper Greensand and Chalk were laid down.

The rocks range in age from the Permian to Recent (Fig. 50). The Permian, Trias, and Lias only occur in west Dorset and east Devon. In the Isle of Purbeck, which is described more fully below, the Jurassic rocks are of considerable importance. Throughout most of the region, the Cretaceous rocks are dominant except in the Solent and Spithead district. The Chalk is the most widespread, and if we follow its main outcrops we can see its profound influence on some of the most striking coastal features. In the east it forms the North and South Downs running out to the South Foreland and Beachy Head. (The Isle of Thanet with the North

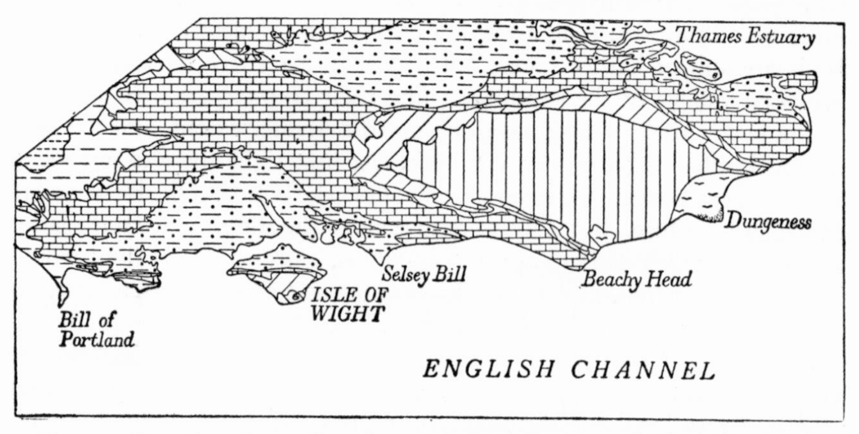

Fig. 50. General geology of south-east England (based on Geological Survey) (See Fig. 11 for explanation)

Foreland is a separate mass.) Between these two headlands the older beds of the Weald crop out in turn; the Gault at Folkestone and Eastbourne (where it is mainly hidden by alluvium), the Lower Greensand near Hythe and beneath Pevensey Levels; the Weald Clay scarcely reaches the sea, but is seen behind Romney Marsh and Pevensey Levels. Between Bexhill, Hastings, and Fairlight are the oldest Wealden Beds forming a famous cliff section of the Fairlight Clays, Ashdown Sands, and Wadhurst Clay. From Beachy Head to near Brighton the Chalk reaches the coast, but westward of Brighton and continuing into Hampshire is the Sussex coastal plain formed of recent deposits, of Coombe Rock, brickearth, and raised-beach material. Outcrops of the underlying Tertiary rocks are small, but are seen at Bognor, Bracklesham Bay, and elsewhere. The Tertiary beds form both shores of the Spithead and Solent and extend inland to the line of the South Downs. In the Isle of Wight they form all that part of the island north of the central Chalk ridge. They also form the coast from Hurst Castle spit to Poole Harbour. South of Poole Harbour the strata are turned up on end by the strong fold and thrust running east and west through the Isle of Purbeck. This corresponds with the Sandown anticline of the Isle of Wight, whose structure approximates to that of the Weald. The Chalk, standing almost vertically between Culver Cliff and the Needles, reappears in St Boniface and St Catherine's Downs, but, unlike the South Downs, it does not reach sea level. Between the two Chalk masses the Wealden and Lower Greensand beds are exposed as in Kent. In Purbeck the folding is similar, but erosion has removed all the Chalk except that of the ridge from Ballard Down to Worbarrow Bay and beyond Lulworth Cove. At the same time older rocks, down to and including the Kimmeridge Clay, are exposed in Purbeck. The central lowland of Wealden Beds forms a striking feature as compared with the Chalk to the north and the plateau of Portland and Purbeck rocks to the south. In the Isle of Wight the true Wealden outcrops are limited to Brighstone Bay and Sandown Bay, but the Lower Greensand, which there covers it, forms a similar belt of low ground. The structure between Weymouth Bay and Lyme Bay is also anticlinal. Portland Island is all that now remains of the southern limb of this fold.

East of Lyme Regis the beds of the Lias are seen to dip gently eastwards, and they are capped by horizontal masses of Upper Greensand which is resistant. These beds form Stonebarrow Hill, Golden Cap, and Thorncombe Beacon. Just west of Bridport, outcrops of the Fuller's Earth and Forest Marble which have been let in by faults interrupt the regular succession. The details are discussed below: all that needs stating here is that the Chesil Beach may be regarded as beginning at Bridport, and that from there to Portland, especially where the beach encloses the Fleet between it and the mainland, the Jurassic beds, striking east and west, are cut off from the coast. Westwards from Lyme Regis the Lower Lias and Keuper beds are nearly horizontal, or only gently folded. They are, however, often broken by minor faults. With the exception of that part of the coast approximately between White Cliff and Beer Head, they form the lower part of the cliffs to the mouth of the Otter. The upper parts of the cliffs are all in the Cretaceous rocks, Gault, Upper Greensand, and Chalk.

Roughly speaking, therefore, east of the Chesil Beach there is a very strong connection between tectonic structure, rock hardness, and coastal detail. To the west this connection is not so evident except in local detail since we are passing out of the area where the Tertiary folds have any particular influence on the coast. The accompanying map showing the main features of the geology will make a fuller discussion unnecessary.

(2) THE OTTER TO THE CHESIL BEACH: THE DEVON-DORSET LANDSLIPS

Eastwards from the mouth of the Otter Permian and Triassic rocks form the foundation of the area.[1] Otterton Point itself is formed of red sandstones, which at the cliff base are rather conglomeratic. Brandy Head is somewhat similar, and northwards therefrom the cliffs rise in height to about 200 ft. The sandstones are nearly horizontal from Otterton Point to Ladram Bay, but from High Peak to Peak Hill the beds dip slightly in the same direction, and the cliff is much hidden by slips and debris. The Red Marls of Sidmouth and Branscombe are less resistant than the sandstones at and near Otterton Point, but nevertheless they stand up boldly, and are often fronted by rocky platforms. There are several small bays, often picturesque and diversified by stacks and arches, which bear a close relation to the effects of erosion along joint planes and the less hardened parts of the rocks. There is a more or less continuous shingle beach from Peak Hill to Branscombe: the pebbles are mainly of chert and flint. Sand is exposed at low water. Hutchinson estimated that the loss of land through erosion averaged about 8 ft. a century along this tract. In Hooken Cliff, between Branscombe and Beer Head, we meet the first of the great landslip areas of Devon and Dorset. At sea level the contact between sandy Cretaceous beds and the clays of the Trias forms a plane of weakness, and a major slip occurred in 1790. Previously, South Down Common was bounded by a sheer cliff, but a fissure opened at its top in the winter of 1789–90, and in March 1790 in one night seven to ten acres moved suddenly seawards, and dropped 200–260 ft. vertically. The whole mass, instead of being completely shattered, broke up into columns and pinnacles, and pushed seawards some 200 yd. At the same time the sea floor was raised up into an offshore reef (**37**).

Between Beer village and Seaton a fault brings the Red Marls alongside the Greensand, and so has been mainly responsible for the erosion that has produced Seaton Bay as we now see it. Hard Chalk (of the *Rhynchonella Cuvieri* zone) forms the cave buttresses near Pound's Pool. The river Axe, which in the last mile or so of its course, keeps close to the eastern margin of its valley, has also been deflected slightly eastwards by a spit of shingle. Like the Otter it must at one time have formed a long narrow arm of the sea very similar to the Sussex river valleys (**35** and **34**).

[1] *Mems. Geol. Surv.* 'The Geology of the country near Sidmouth and Lyme Regis', 2nd ed. 1911, H.B. Woodward and W.A.E. Ussher.

From the Axe to Lyme Regis the general cliff structure is fairly simple. The lower part of the cliffs consists of Trias and Lower Lias, and for considerable stretches the Rhaetic is visible below. These beds are thrown into gentle folds and broken by faults. The Gault, Upper Greensand, and Chalk rest on them unconformably. This is the region of the best-known landslips.[1] For some four miles eastwards from Culverhole Point the undercliffs form a rough and broken tract about a quarter of a mile wide between the sea and the inland cliff. There are large pinnacles of displaced rock, and at Dowlands and Bindon a great chasm separates a large isolated mass from the main cliff. The ravines in the cliffs are called Goyles. It is not known when slipping began in these cliffs: presumably it is very old.

To understand the nature and causes of the slipping the following table will be useful; it shows the general rock succession:

Clay with Flints	
....	
Chalk	Middle Lower
Upper Greensand	Calcareous sandstones and chert beds Foxmould (sands) Cowstones (lenticular concretions of calcareous sandstone in sands)
Gault (sandy clay)	
....	
Lower (Blue) Lias (limestones and shales)	
Rhaetic (limestones, shales, marls)	
Keuper Marls	

Briefly, the cause depends mainly on the dip, on the sub-Cretaceous unconformity, and on the relation of both to sea level. If the junction plane between the Foxmould and underlying clay occurs above sea level, and if it also slopes seawards, erosion, by cutting into the cliff face, removes the outward support of the beds, so that the upper layers slide over the lower. This happened at Hooken, west of Beer Head, and between Axmouth and Lyme Regis. At Beer Head and in Whitecliff the unconformity is almost all submerged, and so the cliffs are wholly in the Cretaceous, and falls of Chalk drop directly on to the shore (**36**).

The undercliff gradually develops on the underlying clays, and then the factors influencing slipping become more complicated. Water is held up, and there soon accumulate great masses of talus and unstable material. But the actual forms of the undercliff depend mainly on the low angle of dip, and the coherence of the upper beds which are really let down by the

[1] *Ibid.* and M.A. Arber, 'The Coastal Landslips of South-East Devon', *Proc. Geol. Assoc.* 51, 1940, 257.

foundering of sand layers. The Chalk and cherty sandstones do not easily disintegrate, although they may break: they can and do slide *en masse*, but as the dip is not high their movement is steady and so they remain largely undamaged. It is this factor which has given the peculiar character to these landslips.

Between Culverhole Point and Humble Point the inland cliffs are of Chalk and Upper Greensand; the undercliffs are of fallen Cretaceous material very much disturbed. They slope downwards eventually to rest on the Lias shore reefs. Dowland's Chasm is the most striking feature: it was formed in 1839. Since June 1838 there had been a very wet period, and also one of strong gales. Fissures and cracks began to appear on the cliff top shortly before Christmas 1839, and on 23 December one of the cottages began to subside, at first in no very alarming manner. But by 5 p.m. the cottage was settling rapidly, and later other cottages were 'upheaved and twisted'. The great landslip occurred on Christmas night. 'During December 26th the land that had been cut off by the fissures in the cliff-top gradually subsided seawards, and by the evening had reached a position of equilibrium in the undercliff. A new inland cliff, 210 feet high in its central portion and sinking to east and west, had thus been exposed, backing a chasm into which some twenty acres of land had subsided. The length of the chasm was about half a mile, while its breadth increased from 200 feet on the west to 400 feet on the east.'[1] In all some eight million tons of earth foundered. The movement also caused a ridge of Upper Greensand (Foxmould and Cowstones) to rise in the sea near the beach; it was about three quarters of a mile long, and reached 40 ft. above sea level at high water. The beds were much broken, and the mid-part of the ridge was connected to the mainland by shingle. The reef very soon disappeared, but the main chasm of the slip remains much as it was, except that most of the many pinnacles have naturally gone.

There was a big subsidence in Whitlands cliffs in 1765, probably in part due to the wetness of the previous year. There was another slip in February 1840. The main movement was on the undercliff, and one result was to push Humble Point farther out to sea. As at Dowlands, reefs were formed, but the outer one was of shingle. It is worth mentioning here that reefs were also formed at Folkestone Warren in 1937, but (see below) the causes were quite different. At Pinhay the Chalk shows in the inland cliff: Chapel Rock in front of it is only the slipped face of the inland cliff. Chapel Rock is said to derive its name from the secret worship practised there during the Marian and later periods of religious troubles. The Great Cleft opened in 1886. The undercliffs at Pinhay are much disturbed, and

[1] Arber, *op. cit.* 1940.

there are three ridges and gullies parallel with the coast, between the main cliff and the sea. At Ware the Lower Lias forms the seaward cliffs, the Upper Greensand the inland cliffs. The Upper Lias forms a rough tract of ground with much Cretaceous debris on it. Both cliffs are much over-

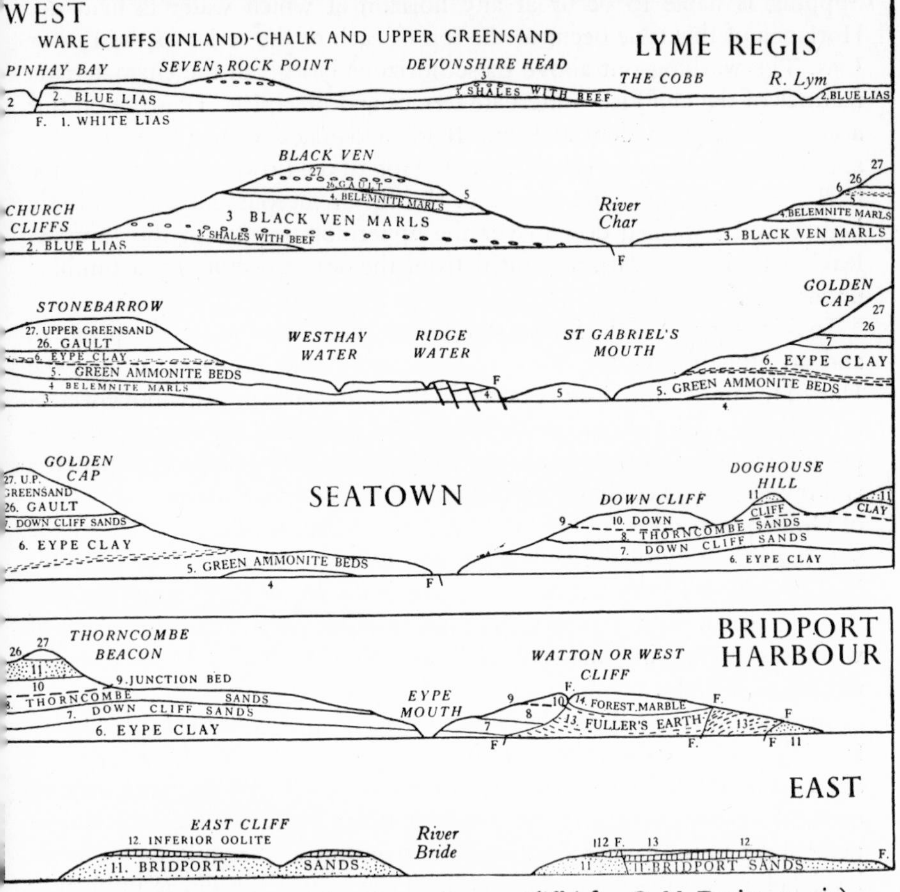

Fig. 51. Cliff sections: Pinhay Bay to Burton Cliff (after G. M. Davies, *op. cit.*)

grown, and a pinnacle of calcareous sandstone, the Chimney Rock, is the only prominent feature.

Apart from some settling each winter there have been no big falls in the last century. This may mean that a state of general equilibrium has been reached. Only minor changes occur at all frequently, but they are largely superficial.

Between Black Ven and Watton Cliff (both inclusive) there has also been much land slipping. Reference to Fig. 51 will show the general nature of the cliff section, which is largely composed of Lower and Middle Lias. The beds are clays, shales, marls, thin limestones, and sands; they dip to the east-south-east and are capped by the Gault, consisting of loams and clays, the Foxmould and the chert beds of the Upper Greensand. Slipping is liable to occur at any horizon at which water is held up. Horizons of this type occur at the top of each of the series of the Lower Lias. The washing-out above these horizons has given the characteristic features of the cliff face—alternate terrace and precipice. Damage on this account is often evident at Lyme Regis and Black Ven. On Black Ven and Stonebarrow water is held up at the top of the Lower Lias and in some horizons of the Gault and Foxmould. Hence 'Cretaceous and Middle Liassic beds may be washed down over the uppermost terrace of the Jurassic, leaving an inland scarp separated from the seaward cliffs by a tumbled undercliff'.[1]

In this way part of the former coast road over Black Ven has disappeared, and on Stonebarrow the Middle Lias sometimes slides *en masse* into the Fairy Dell undercliff. Lower down in the cliffs, sands or drift above water-bearing clays are worn back and the upper surface of the clay is thus exposed. Later it becomes boggy and then may surge seawards as avalanches and mudflows. There has been but little slipping on the higher sandy beds of Golden Cap, Down Cliff, or Doghouse Hill. Thus in west Dorset the varying lithology of the beds gives rise to water-bearing horizons. This leads to slipping which grades from simple mudflow to a large volume of moving strata. In south-east Devon the cause lay in one major plane of weakness underlying massive beds and so allowing slipping of vast, unbroken, masses.

In and near Lyme Regis the cliffs are formed mainly of the Lower Lias.[2] In Pinhay Bay a small fault has been followed by a stream which has cut a gully. Near Lyme Regis the limestones on the shore were formerly quarried, so that the rate of erosion was the greater. All the beds have a general dip to the east or south-east, so that they are brought down to sea level before we reach the Cobb. But even where the dip is uniformly seaward there are minor inequalities in the coast which give the impression of changes in the dip. A bay cut back into the lower beds suggests an anticline, and the curved beds at Seven Rock Point indicate a small anticlinal fold. At Lyme Regis the thin alternating bands of shales and ar-

[1] M.A. Arber, 'The Coastal Landslips of West Dorset', *Proc. Geol. Assoc.* 52, 1941, 273 (280).

[2] *Mems. Geol. Surv., op. cit.* 1911, and G.M. Davies, *The Dorset Coast, A Geological Guide*, 1935.

gillaceous limestones are very conspicuous. The Cobb itself stands on the main limestone beds on the foreshore, and inspection of the several ledges which run out to sea in curved forms clearly indicates a synclinal structure. Some of these have been removed artificially, a process which has increased erosion in Church cliffs, which like the reefs below consist of the higher beds of the Blue Lias.

The Lias cliffs continue to Charmouth, but they are there topped by Cretaceous beds. The dip of the Lias is best seen in the numerous reefs exposed at low water: it is mainly east-south-east. There are minor folds and faults. Mouth Rocks form part of these reefs covered with shingle. According to the Geological Survey the Lower Lias of this area is divided into four series: at the base the Lyme Regis or Blue Lias, which contains most of the limestone beds; then the Black Ven Bed or Black Marl which forms the two lower precipices on Black Ven and most of the western part of Stonebarrow Cliff. It is composed chiefly of blue-black clays and shales. Above it is the third series consisting of the Stonebarrow Beds or *Belemnite* Marl: this forms the highest precipice on Black Ven, and the seaward cliff on Stonebarrow. The fourth series is called the Wear Cliff Beds,[1] the whole of which are seen in Stonebarrow, but only the lower half in Black Ven. The Cretaceous beds overstep the Lias from east to west and are seen only on the tops of the hills. Locally, apart from chert beds, they are usually rather loamy and glauconitic sands. The hill-top deposits are post-Cretaceous and not fully understood; in general they consist of jagged fragments of chert in a sandy matrix at the base, followed above by flints and cherts in a loamy matrix (**32** and **33**).

The river Char tends, like other streams on the borders of Devon and Dorset, to follow the left bank of its valley, probably as a result of working in that direction down the dip-slope. Lang has investigated the evolution of the Char and neighbouring streams, and maintains that originally they had nearly meridional courses consequent on an east-west post-Eocene fold north of the present Vale of Marshwood. Jukes-Browne (see Fig. 52) supposed that the Winniford (=Chid) had been beheaded by the Char, and Lang supposes 'that this was but one in a progressive series of similar incidents, beginning in the west with the Wootton stream, or even with the Buddle, and ending in the east with the Simene....'[2] The mouth of the Char is deflected to the east by a bar of shingle; the smaller streams merely soak through the shingle beach, but may possess temporary mouths after

[1] The Green Ammonite Beds of W.D. Lang.

[2] W.D. Lang, 'The Submerged Forest at the Mouth of the River Char and the History of that River', *Proc. Geol. Assoc.* 37, 1926, 197, and see also W.D. Lang, 'The Geology of Charmouth Cliffs, Beach, and Foreshore', *ibid.* 25, 1914, 293.

storms. It appears that in the early part of last century the Char had, for a short time at any rate, a mouth on the western side of its flood plain. Soon after 1900 an attempt was made to give it a permanent mouth: the river ignored this. There is a small exposure of what appears to be a submerged forest[1] at the mouth. The beach is of mixed sand and shingle, mainly of

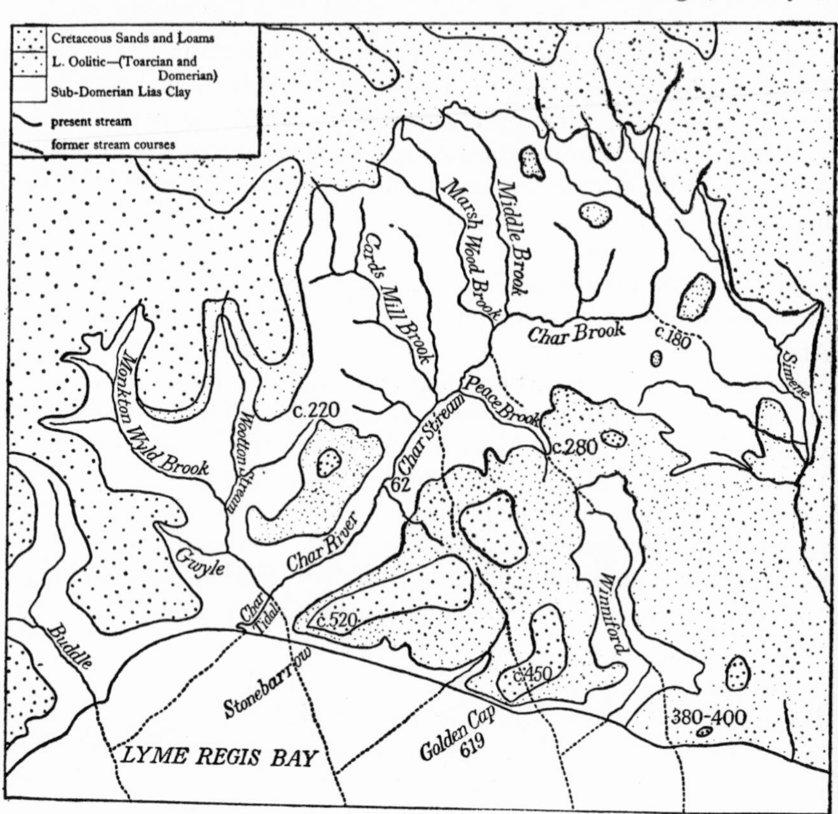

Fig. 52. Geological map of the Charmouth area (after W. D. Lang, *op. cit.*)

local origin: the large pebbles are of Lias limestone. Others consist of chert and Chalk flints. The amount of sand is variable, but is usually more noticeable just east of the mouth where the quantity of shingle is small. But eastwards again shingle increases, and at Westhay Water and St Gabriel's Mouth there is a steep pebble beach.

Westhay Water and Ridge Water make small cliff waterfalls, and erosion is rapid along Wear cliffs. The Three Tiers (Middle Lias) in these cliffs

[1] Dr H.D. Thomas tells me it is really a small accumulation of drift.

form ledges and waterfalls. Golden Cap is topped by the Upper Greensand, and owes its name to the sandy outcrops. The cliffs drop rapidly in height to Seatown where a fault coincides with the valley (Chid). From Seatown to Eype Mouth the beds in the cliffs undulate slightly: Thorncombe Beacon is high enough to have a capping of Upper Greensand, but otherwise Cretaceous beds are absent. The same beds are found just east of Eype Mouth, but very soon the cliffs recede and a fault occurs, trending about east and west, and lets down the Fuller's Earth and Forest Marble against the Middle and Upper Lias. Just before Bridport Harbour two other smaller faults are seen, and their effect is to bring in the Bridport Sands. The main part of West Cliff, therefore, is let down by trough-faulting (Fig. 51).

Bridport Harbour (West Bay) is held in place by jetties; some small shingle accumulates on the eastern side of these, but in general shingle tends to be swept away from the west side. Hence groynes have been built to try to collect shingle and so protect the esplanade. The cliff on the eastward side of the harbour is very conspicuous: it is sheer, and is made of yellow Bridport Sands capped by Inferior Oolite. Between it and Burton Cliff is the mouth of the Bride or Bredy, a small strike stream. It is slightly deflected to the east, and at the mouth is usually rather too deep to be forded dry-shod. Burton Cliff generally resembles East Cliff: all its lower part is formed of Bridport Sands, but a small fault brings in the whole thickness (only about 12 ft.) of the Inferior Oolite and some of the Fuller's Earth (**31**).

(3) THE CHESIL BEACH

Eastwards from Bridport there is a continuous beach of shingle which passes into the Chesil Beach proper (Fig. 53). In England and Wales there are many examples of shingle structures, but the Chesil is unique. It is a single main ridge of shingle running in a nearly straight line from Bridport to Portland: from Abbotsbury to Portland it is separated from the mainland by the Fleet. The most significant feature is perhaps the remarkable grading of this shingle: at the north-western end it is fine, from a pea to a walnut in size; at the south-eastern end it is coarse, pebbles of 2–3 in. or more in diameter being prevalent. There are no recurved ridges running back from it such as are found on nearly all other structures of comparable size. Not only are the individual pebbles smaller near Bridport, but the actual beach is smaller there: it is less steep, less broad, and less high than it is at Portland, and the whole intermediate portion shows that it increases with remarkable regularity in all these ways as it is traced south-eastwards. By some writers the Chesil is limited to that

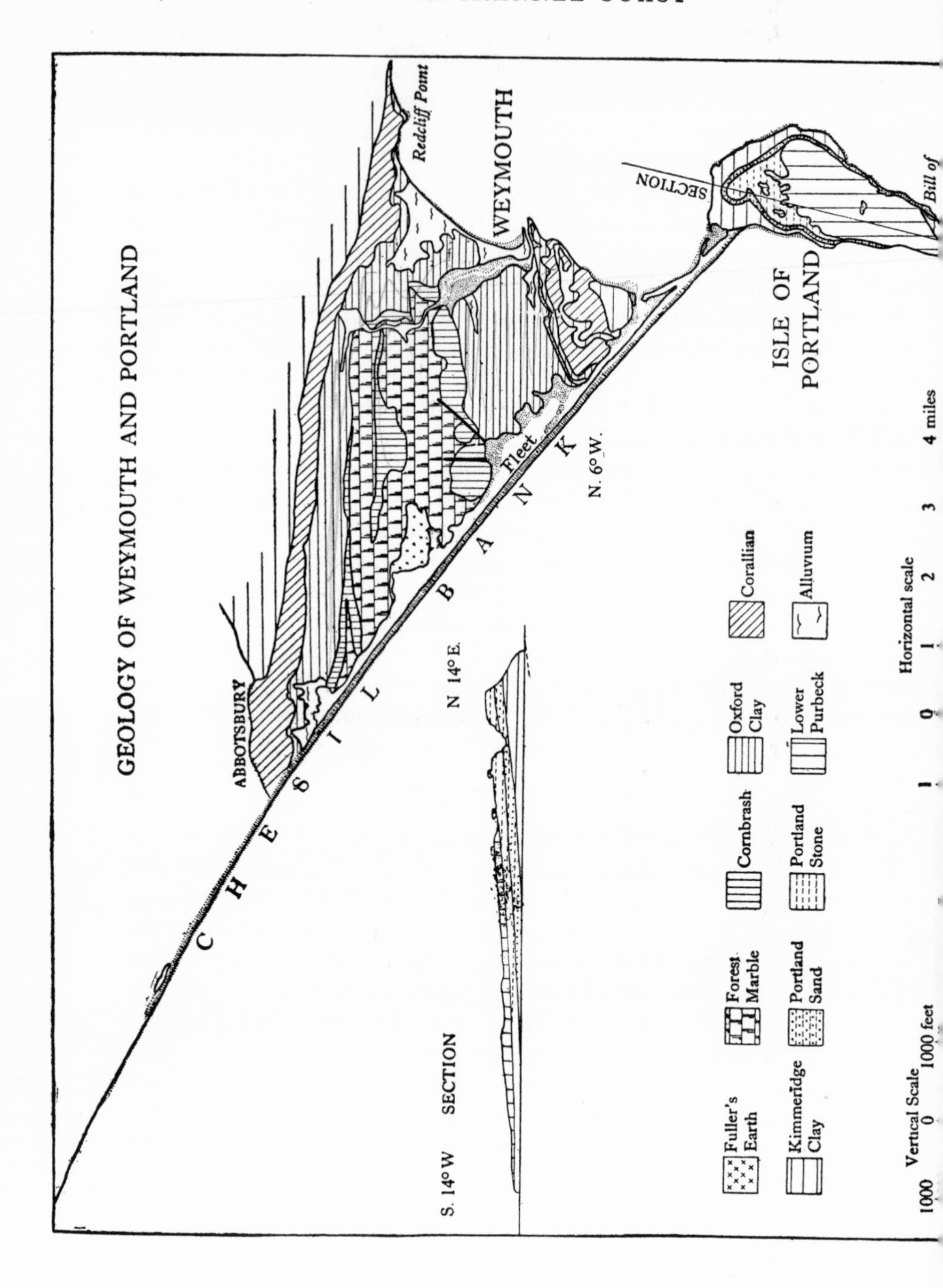
GEOLOGY OF WEYMOUTH AND PORTLAND
ABBOTSBURY
CHESIL BANK
Fleet
N. 6° W.
WEYMOUTH
Redcliff Point
SECTION
ISLE OF PORTLAND
Bill of
S. 14° W SECTION N 14° E.
Fuller's Earth
Forest Marble
Cornbrash
Oxford Clay
Corallian
Kimmeridge Clay
Portland Sand
Portland Stone
Lower Purbeck
Alluvium
Vertical Scale
1000 0 1000 feet
Horizontal scale
1 0 1 2 3 4 miles

part between Abbotsbury and Portland, but as there is no break, real or apparent, between Portland and Bridport, it is better to take the greater stretch. The mere fact that the Fleet ends at the Swannery at Abbotsbury is unimportant (30).

Before discussing the mode of origin of this beach, the following facts may properly be stated. From Bridport to Portland its length is eighteen miles. For six miles it is in contact with the coast. The Fleet varies in width from 200 to 1,000 yd. Two miles north of Portland the beach runs direct to the sea to join the island to the mainland. It is about 170 yd. wide at Abbotsbury and 200 yd. at Portland. Its height at the same two places, measured from high-water mark, is nearly 23 ft. and nearly 43 ft. The gradient of its crestline averages 1 in 8,450 from Abbotsbury to Wyke, and 1 in 880 from Wyke to Chiswell. At Abbotsbury the shingle extends downwards below low water springs some six fathoms, and eight at Portland. The stones above sea level increase in size with marked regularity to the south-east: those below water are said to decrease in the same direction. On its landward side the beach rests on a clay bench 3 or 4 ft. above low water springs: this clay is said to have been exposed on the seaward side after storms. Probably, then, the beach stands on a clay bench, which for part of its length only is Kimmeridge Clay. The Fleet is tidal: the tide gains access only at its south-eastern end, and at Abbotsbury the range is only a few inches even at springs. At times, especially in storms, a certain amount of water leaks through the loose upper shingle, and in this way it is probable that the cans or hollows on the leeward side of the bank have originated. The eastern bank of the Fleet is deeply indented and is not cliffed: there is no doubt that it has not been subjected to wave attack. The harder rocks, of the Cornbrash and Corallian as well as the hard bands in the Forest Marble, project farther than the softer clays. These headlands have been nipped by the small waves that arise in the Fleet. Although the Fleet ends at Abbotsbury, there are small hollows, wet and dry, farther on: Burton Mere is the most conspicuous, but there is also a dry hollow half a mile east of Abbotsbury coastguard station.

It is unnecessary to review the whole of the considerable literature concerning the beach, but certain papers are outstanding. In 1853 Sir John Coode[1] published the first, and very important, serious scientific paper on the Chesil. He argued that the source of the shingle lay between Lyme Regis and Sidmouth, that the shingle was transported by wind-waves and not by tidal currents, and that the varying dimensions of the bank are related mainly to the degree of exposure and depth of water offshore. He also maintained that large pebbles are more readily moved than are small

[1] *Mins. Proc. Inst. Civ. Engs.* 12, 1852–3, 520.

stones, and that consequently the largest shingle is found to leeward, that is, at Portland. The isolation of the beach, he thought, was due to the clay bench mentioned above. He noted also the commonly observed fact that heavy onshore winds scour the beach, whereas offshore winds usually result in accumulation. But his main point was that the beach grew to the south-east as a result of wind-waves carrying the shingle in that direction from the western part of Lyme Bay. Bristow and Walker (1869) and Fisher (1873) were more concerned with the Fleet and need not detain us here.

In 1875 Sir Joseph Prestwich[1] put forward views which ran counter to those of Coode. He attempted to show 'that the Chesil Beach is formed by the accumulation of shingle derived from the southern end of Portland and from the sea-bed westward thereof, and that the movement of the beach is northward'. It will probably be clearest and simplest to quote his main conclusions verbatim:

1. That the shingle of the Chesil Beach is chiefly derived from the materials of the raised beach, of which a remnant still exists *in situ* on the Bill of Portland, and partly from the harder beds of the Portland and Purbeck formations of that island.

2. That the storm-waves, in conjunction with the tidal current, drive the shingle of this old beach from the bed of the Channel on to the southern end of the Chesil Beach, whence it travels by the agency of wind-waves in a north-westerly direction towards Bridport Harbour, and on the other side of which shingle travels in the opposite direction....

3. That the growth of the Chesil Beach has been from the south-east under the influence of the two above-named forces.

4. That the shingle of the raised-beach itself was formed of materials which had travelled from the coasts of Devonshire and the adjacent parts of Dorset eastward to Portland.

5. That the sea for a time passed between Portland and Weymouth, and that the Fleet is merely a portion of the old shore-line dammed out by the growth of the Chesil Beach.

6. That the existing beaches are formed not only of the debris of the present coast, but also from shingle derived from old beaches, which, together with gravel of a former land-surface, is now scattered over various parts of the Channel bed, whence it is,—under certain conditions of proximity, wind, and tides,—thrown up on to the present shores.

This paper, like that of Coode, provoked a lengthy discussion. Prestwich based his argument about the direction of drift of the shingle on the assumption that the mean direction of wind approaching the beach was south-south-west, or even a point farther south. This wind is not quite at right angles to the general trend of the beach, but makes a greater angle with the beach to the north-west than it does to the south-east. Hence

[1] *Mins. Proc. Inst. Civ. Engs.* 40, 1874–5, 61.

he argued the waves would approach the beach somewhat obliquely and cause a north-westward movement of beach material. This, he maintained, was in agreement with (*a*) the smaller size of the pebbles towards Bridport, (*b*) the smaller dimensions of the bank in that same direction, (*c*) the decrease in depth of the beach below water, and (*d*) the presence of Portland rock debris, which gradually disappears north-westwards from Portland Island. At and near Bridport he thought that there were local and possibly oscillatory movements.

In 1898 Vaughan Cornish[1] reviewed the evidence once again. He noted that since the beach is from time to time raked over by the sea, it follows that the present arrangement of stones is at any rate conformable to present conditions of wind, wave, and tide, 'whatever may have been the original mode of supply of the shingle'. He was concerned largely with the remarkable grading of the shingle on the beach. If it is regarded as stretching from Bridport to Portland, then it is to some extent fed at both ends, but the material coming to the westward end is fine, whereas that at the other end is coarse, partly because of the nature of the local rock (Portland Stone) and partly because much of it is quarry waste. He concluded that, whereas the main drift of the water is to the east, much of the fine shingle which at first is carried eastwards from Bridport is later brought back by waves from the east, and that near Chesilton the strong outset carries out to sea any fine material which may come from Portland, so leaving only the coarser stones.

In 1930 Baden-Powell[2] argued that the logical view of the origin of the beach was that it formed the last stage in the post-Tertiary history of the area. Although flints make up well over 90 per cent of the pebbles, there are also stones from the Permian, Trias, and Tertiary deposits of Devon and Dorset, and from the older rocks of Cornwall. All these were first assembled in the Portland raised beach (at its full extent, whatever that may have been). These raised-beach pebbles became mixed with those of the extinct Fleet River which was drowned and whose right bank was captured by marine erosion. This would place the formation of the Chesil late in Quaternary time, and probably after the formation of the Head,[3] because the Head reaches sea level along the shores of the present Fleet. The pebbles are now sorted by the waves, the softer limestone ones yield fairly quickly to erosion. The raised beach and the drowned valley of the Fleet imply that this part of Dorset is typical both of emergence and submergence, and as the Chesil does not appear to be added to appreciably to-day we may regard it as a mature marine deposit.

[1] *Geogr. Journ.* 11, 1898, 528 (see p. 628 for the Chesil Beach).

[2] *Geol. Mag.* 67, 1930, 499.

[3] See p. 476.

Strahan, in the *Geological Survey Memoir* (1898), concludes that the pebbles travel from west to east, and states that 'A large proportion of the gales that visit that coast, even if they commence in the south-east or south, veer to the west and north-west, so that the direction of S.S.W. assigned to the prevalent wind-waves (by Prestwich) is not strictly accurate'. Wreckage, too, seems definitely to travel to the south-east. He also agrees with Coode who maintained that the larger pebbles travel faster than the small.

The most recent writer to discuss the Chesil Beach is Arkell.[1] He rightly notes that it cannot be regarded as having reached stability since big storms throw some pebbles over its crest. The lack of vegetation at the Portland isthmus suggests that this process goes on more quickly there than elsewhere. He refers to Richardson's[2] experiments which showed that pebbles, larger than the average for a given stretch of the beach, move quickly to the south-east until they reach a part of the beach where they conform to the average size. The fact that the pebbles seem to be graded in an opposite direction below water level Arkell ascribes to tidal influences. For 18 out of every 24 hr. a current flows eastward along the coast of West Bay. Such a current would be better able to move fine material, which, in consequence, gathers at the eastern end of the bay. Further investigations on this matter seem desirable.

Arkell's views on the origin of the pebbles are interesting. He admits that floating ice would help, but ordinary shore migration is probably sufficient. The point raised by Strahan and others that material from the west could not pass such headlands as Otterton Point, Beer Head, Golden Cap, and Thorncombe Beacon is, in Arkell's opinion, illusory. Such headlands are only transient features of a retreating coast and really depend on local accidents. In the past other headlands, now destroyed, would have had similar effects and would only temporarily have held up the supply.

Of the origin of the beach in general Arkell supposes that at the beginning of the 'Neolithic' subsidence it was a bay-mouth bar running from an extended Portland Bill to some lost headland farther west. Pebbles drifted to it as they do to-day. Behind the beach was a low tract of land, largely composed of Jurassic clays, and gradually drowned during the subsidence. The beach slowly retreated under the action of Atlantic rollers, which also excavated the submarine cliff at its foot. He agrees with Lewis's views that the present orientation of the beach is conformable with the direction from which the main waves approach.

[1] In his unpublished *Memoir of the Geological Survey* on Weymouth and Swanage.

[2] *Proc. Dorset Nat. Hist. and Antiq. Field Club*, 23, 1902, 123.

The Fleet is not thought to be a former river valley because it runs obliquely across all the drainage lines of the district.

We are, then, faced with very divergent views on the origin of the beach, and the time has not yet come when we can afford to be dogmatic. But it is important to note that, e.g., hard pebbles of Portland limestones travel north-westwards, and that they also get smaller in this direction and are found as far as Abbotsbury (Prestwich). Whilst I do not think that the regular grading has yet been properly explained, it is relevant to point out that it is quite common to find large pebbles at the leeward ends of beaches. This is well seen at Scolt and Blakeney (see pp. 358, 348). On the other hand, neither of these two formations shows any attempt at grading. My own observations on a good many shingle beaches all strongly suggest that wind-waves are the main transporting agent. If Prestwich was right about the mean wind approach, then I think a north-westward movement would result, but I am not convinced that his statement is either absolutely accurate or sufficiently comprehensive. Lewis, basing his views to some extent on Cornish,[1] has suggested that the bigger waves approaching from the Atlantic break on the Chesil with sufficient obliquity to give a slight easterly component, whereas the local Channel waves from east and south, because they are smaller, drive the lesser shingle to the westward along the beach. With this I am inclined to agree, although I do not think that either Cornish or Lewis has carried his reasoning sufficiently far to explain the very even grading of the shingle of the beach. We need simultaneous observations of wind direction and strength, of wave incidence and type, and of shingle movement at three or four selected places along the whole beach and extending over a considerable period of time in order to take into account the usual types of weather and storm acting upon the beach, before we have a final answer to this difficult problem.

(4) THE PORTLAND AND WEYMOUTH BEACHES

It will be convenient here to speak of two other shingle beaches before commenting on the Isle of Portland. Portland Beach runs north-westwards from Portland Castle as far as Small Mouth, and is quite distinct from the Chesil although it actually touches the Chesil for a part of its length. It is about a mile and a half long, and at the Portland end encloses the Mere. It is built mainly of Portland and Purbeck rocks. These must have come from the north and east coasts of the Isle as a result of waves set up by easterly winds, and also, be it noted, by south-westerly and even

[1] 'On the Grading of the Chesil Beach Shingle', *Proc. Dorset Nat. Hist. and Antiq. Field Club*, 19, 1898, 113.

westerly winds, because waves caused by these winds are deflected from their normal course by the Isle and may run in from the eastward on this part of the shore. Pebbles of Portland and Chesil Beaches do not mingle, and the little valley between them is a sharp geological boundary. The building of Portland Breakwater has naturally influenced this beach considerably, and has practically stopped its growth.

The third beach is that at Weymouth and Lodmoor: it runs to the north-east for about two miles and protects the low-lying areas within. The first area, Weymouth Back Water (Radipole Lake), has recently been decreased in area by the reclamation of that part between the railway and Radipole Park Drive. The lake is really the estuary of the bay 'and was fully open to the tide until a dam was carried across it to prevent the laying bare of the mud at low tide'.[1] Separating the estuary from Lodmoor is a low tract of Oxford Clay, and Lodmoor, a drowned valley, is faced by a continuation of Weymouth Beach which here is higher and narrower, but it is a true storm beach of large pebbles. Sorting action on the pebbles shows itself in that the size of the pebbles decreases towards Weymouth, where the beach is mainly sand, and towards Bowleaze. The beach has suffered seriously since the building of Portland Breakwater, and is prevented by a wall from being thrown back on to the low ground of Lodmoor.

(5) PORTLAND AND WEYMOUTH[2]

Structurally, the Weymouth-Portland area is an anticline, the axis running through Lodmoor. The southern limb, which dips but very gently, has suffered very much from erosion, and the Weymouth peninsula and Portland Isle are all that remain. We have already noticed the Fleet which by some is regarded as a river valley,[3] and is widest where the rocks are softest—in the Oxford Clay and Fuller's Earth. Fig. 53 shows an approximately north-to-south section through Portland. The undermining by the sea of the Kimmeridge Clay causes great falls of Portland Stone which for a time act as breakwaters. The cliffs were[4] admirable all round the island, even in the south where they are much lower. There is a strong contrast between the cliffs on the eastern and western sides of the Isle. Those on the eastern side are much affected by landslips produced by a falling outwards and forwards of great blocks of limestone. This is helped by the master joints in the rock. As a matter of fact the cliff face is often

[1] A. Strahan, *Mems. Geol. Surv.* 'Isle of Purbeck and Weymouth', 1898.

[2] For this and others parts of the Dorset coast, see G.M. Davies, *op. cit.* 1935.

[3] See Arkell's views *supra.*

[4] In many places they are greatly spoiled by quarrying.

a joint face. The joints run roughly north-north-east to south-south-west. The overlying limestone seems to squeeze out the underlying Portland Sand, and perhaps also the Kimmeridge Clay. As a result large masses of the limestone fall down. On the western side squeezing has less, and marine agencies appear to have greater, importance. The grain of the country is opposed to slipping along joint planes, and, instead, produces a spalling effect.[1] Arkell[2] notes that in the northern part of the Isle the

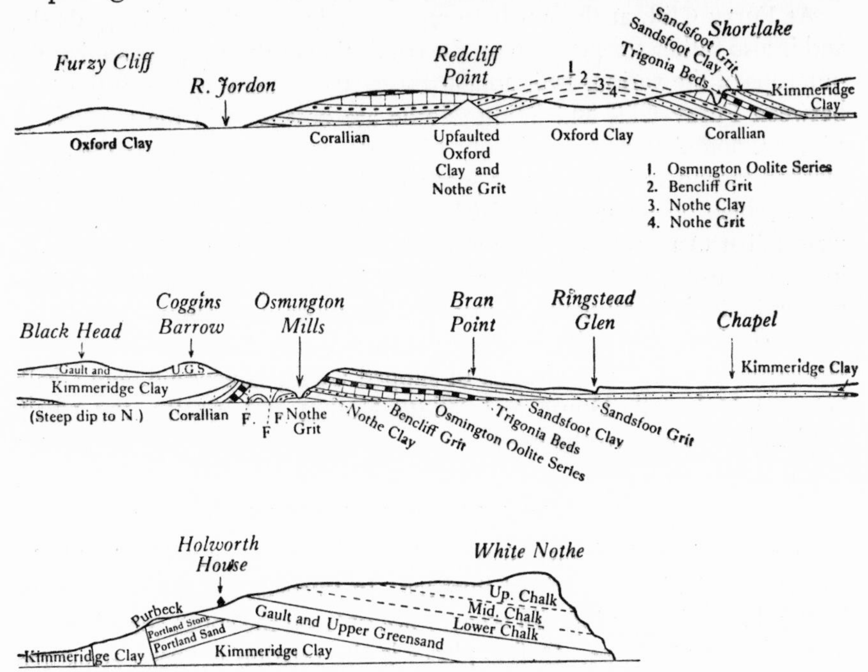

Fig. 54. The cliffs between Weymouth and White Nothe (after G. M. Davies, *op. cit.*)

landslides are intermediate between normal cliff slides and landslips. Some are very old. The greatest in modern times occurred in 1792, when a mass one and a quarter miles long by 600 yd. broad slipped down 50 ft. on the north-east coast. Another big slip occurred in 1858 on the west coast, and forced the underlying beach upwards on a bulge of Kimmeridge Clay. At the Bill is the famous raised beach at about 65 ft. O.D. The wave-cut platform in Purbeck and Portland Beds, on which the beach rests, slopes down from 50 to 20 ft. O.D.

Between Weymouth and Small Mouth are some little coves and valleys

[1] I am indebted to Mr F.H. Edmunds for this information.

[2] *Op. cit.*

in the Sandsfoot and Nothe Clays (e.g. Bincleave): the limestones form the higher ground. Sandsfoot Castle was built in 1539, and its present position on the edge of the cliff indicates considerable erosion since that time. The actual harbour (i.e. the estuary of the Wey) at Weymouth owes its existence to the soft Oxford Clay in which Weymouth Bay is mostly eroded. But, where the dip brings the overlying Nothe Grit to sea level, erosion is checked, as on the Nothe promenade.

At Furzy Cliff, at the northern end of Lodmoor, the dip is northerly, and is also rather steeper than on the south side of the anticline so that the outcrops of the various beds are narrower. The general nature of the cliffs between Furzy Cliff and White Nothe are best seen in the sections of Fig. 54, and comment need only be made on a few matters: the well-marked meanders in the lower Jordon valley; the small fault at Redcliff Point which brings up Oxford Clay with a little cap of Nothe Grit; the waterfall at Osmington Mills[1] on the Nothe and Preston Grits; the Burning Cliff west of Holworth House where, in 1826, rapid oxidation of iron pyrites ignited the oil shales so that they smouldered for about four years;[2] the marked unconformity between the Jurassic and Cretaceous beds near Holworth House; the undercliff of slipped Gault, Greensand, and Chalk between Holworth House and White Nothe; and, finally, the Chalk headland of White Nothe itself.

(6) THE ISLE OF PURBECK[3]

From approximately White Nothe to Studland Bay we must deal with the coast as a unit: it is the Isle of Purbeck. In Purbeck we find some of the best and most interesting cliff scenery not only on the south coast but also in England and Wales. In few places is the relation between structure, rock type, and erosion so well seen. It is, in some ways, an equivalent of the south Pembrokeshire coast: there we had to deal with Hercynian folding, here with Alpine folding. It will, therefore, be best first to describe the main traits of the structure of Purbeck before discussing the more interesting points of its coastal detail.

The northern boundary of the Isle is the Frome valley, but for the present purpose we may conveniently take the Chalk ridge running from

[1] The 'mud-glaciers' at Osmington Mills are a special type of landslip in very greasy materials with low dips. The material flows rather than slips.

[2] This burning is intermittent: it probably occurred before 1826, and again about 1916. It seems to occur whenever rapid oxidation of iron pyrites is brought about.

[3] *Mems. Geol. Surv.*, *ibid.*; 'Guide to the Geological Model of the Isle of Purbeck', A. Strahan, 1906; S.H. Reynolds, 'The Isle of Purbeck', *Geogl. Teacher*, 13, 1926, 433; G.M. Davies, *op. cit.* 1911.

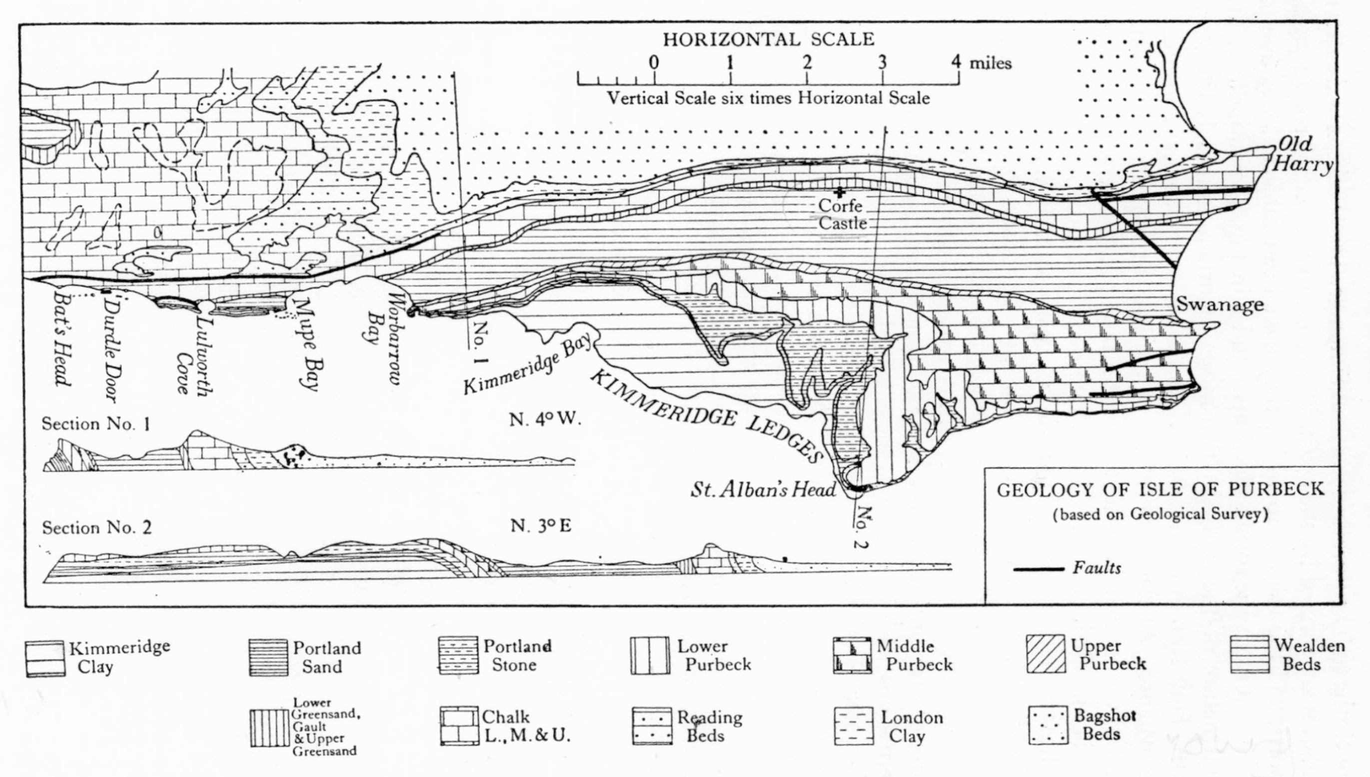

Fig. 55. Geology of the Isle of Purbeck (based on Geological Survey)

White Nothe to The Foreland north of Swanage. The southern and eastern boundaries are formed by the coast. The strike of the beds is roughly east and west, and the folding to which they have been subjected was due to a movement from the south. Sections 1 and 2 in Fig. 55 will indicate the general nature of the folding. The most important feature is the great reversed fault seen in Ballard Point. As shown on the map, it runs inland some distance. At the western end of the Isle it reappears behind Arish Mell, Lulworth, and Durdle Door. Throughout the whole length the Chalk is upended, and the Chalk ridge now forms the backbone of the Isle. In the southern part of the Isle, especially east of Worbarrow Bay, the dip flattens and the Kimmeridge Clay, Portland, and Purbeck Beds are nearly horizontal. It will also be noticed that the outcrops of all beds narrow considerably westwards, and partly for this reason and partly because of erosion the Jurassic beds west of Worbarrow Tout form but a narrow coastal fringe with a steep inland dip: all the plateau area of south-eastern Purbeck has disappeared. In the north the Tertiary beds of the Frome valley are all soft and form low ground. South of the Chalk ridge the topography is closely related to rock hardness. The broad outcrop of the Wealden Beds in Swanage Bay forms a belt of low ground which runs continuously to Worbarrow Bay to reappear west of Mupe Bay and at Lulworth and Stair Hole. Farther west the Chalk comes to the foreshore. Where softer beds reach the coast bays are formed, and where the Kimmeridge Clay occurs in the cliffs on the south coast undercutting is usually rapid. If these major points are borne in mind and reference is made to the map and sections in Fig. 55, we may now turn to the consideration of the coastal details (see **29**, **28**, **27**, **26**).

The Chalk headland of White Nothe[1] reaches 500 ft. Eastwards therefrom the best view of the actual cliffs is from a boat, although there is a shingle beach as far as Bat's Head. The beds are at first almost horizontal, but soon an easterly dip appears, at first gentle and then more steep towards Middle Bottom which marks a syncline. Bat's Head, like Black Rock, Cover Hole, and Ballard Point, is formed of hard Chalk of the zone of *Holaster planus*, while Man-o'-War Point is in the *Rhynchonella Cuvieri* zone.[2] Bat's Head is penetrated to form an arch by Bat's Hole. The Cow and Calf are isolated portions of the nearly vertical wall of Portland Stone which is again visible at Durdle Door. Just beyond is Durdle Cove where a slide plane in the Chalk and near the cliff foot forms a weak belt which has been worked on by the sea in such a way as to form a series of small

[1] See footnote, p. 285.

[2] A.W. Rowe, 'The Zones of the White Chalk of the English Coast, II. Dorset', *Proc. Geol. Assoc.* 17, 1901–2, 1.

caves. The rocks, Blind Cow off Swyre Head, and the Bull off Scratchy Bottom, are remnants of the Portland Stone wall. The Door at Durdle is a natural arch cut in the vertical Portland Stone which at this particular place may have been more jointed and fractured than usual. The Man-o'-War rock is mainly formed of Portland Stone with a little Lower Purbeck still attached to it; this rock, together with several smaller fragments, still marks clearly the outcrop of the Stone between Durdle Door and Dungy Head. At Man-o'-War Head the crushing of the rocks, due to folding, is very intense (**60**).

Between Dungy Head and Lulworth Cove the sea has already cut through the Portland Stone in three places; it is now eating into the broken Purbeck Beds, and once they are demolished erosion in the soft Wealden Beds will be rapid. Stair Hole shows this extremely well. Strahan regards Stair Hole, Lulworth Cove, and Mupe and Worbarrow Bays as representing three stages. In the first the sea has as yet only bored holes through the Portland-Purbeck wall, and the softer Upper Purbeck and Wealden Beds are just beginning to be scoured out rapidly. Lulworth illustrates the second stage: the wall has been broadly breached, and the sea within has reached the Chalk on the northern side, but the cove is mainly extending east and west in those parts where the Wealden Beds form its shores. In Mupe and Worbarrow Bays the wall has gone altogether, except for Mupe Rocks. The Chalk now forms the inner cliff of the bay. Generally the Chalk is fairly resistant, but is itself yielding at Arish Mell, and, as noted above, at Bat's Head. We shall return to the nature of the erosion in the Chalk when we reach The Foreland.

Before leaving Lulworth Cove, however, a paper by Burton needs consideration.[1] He somewhat disagrees with Strahan's views, and maintains that Lulworth has no real counterpart elsewhere on this coast. His view is that the cove is not primarily due to marine action, as that action does not, in his opinion, fully account either for its form or for its actual position. On the hypothesis of simple marine erosion it is rather difficult to see why the process in these very similar places is so much greater in one than the other two. In short, Burton holds that Lulworth Cove is due in the first place to physical geography and drainage, and secondly to marine erosion. Originally, when the valley was formed, the land stood higher, and the re-excavation of the older and higher valley took place following changes in the coastline in Pleistocene times. The older, wider, and higher valley, and the deeper inner valley are well seen just above the cove. Even at Stair Hole there are several springs issuing from the junction

[1] E. St J. Burton, 'The Origin of Lulworth Cove', *Geol. Mag.* 74, 1937, 377.

of the Purbeck and Wealden Beds, and one of them is causing much wastage and leading to cliff falls.

Although the ground immediately north of the coastal escarpment forms a long slope down to the main valley, the chief line of drainage follows the strike of the rocks, and independently at least of the northern slope of Lulworth Cove. Subterranean water seems to be held up by the Wealden Clays, and emergence is effected between them and the limestones of Purbeck age. In other words, it makes progress between the upturned edges of the strata. Such a constitution of local drainage has been in operation a long time, that is before the development of Stair Hole or of Lulworth Cove. Most of the drainage, a mere remnant of an earlier southern system, eventually crosses the strike whenever opportunity is best provided by the relatively lower ground, and the exposures of steeply-inclined surfaces of the Jurassic rocks above Little Bench, in the vicinity of West Point, show marked results of subaerial denudation, which at Stair Hole, by saturation and solution, has largely determined the initiation and enlargement of this impressive feature.[1]

Arkell[2] regards Lulworth Cove, Chapman's Pool, Winspit, Seacombe, and many small 'gwyles' elsewhere as the heads of much longer valleys which formerly extended southwards to the 'Channel' river. The surviving heads are too high to have been affected by the submergence of the

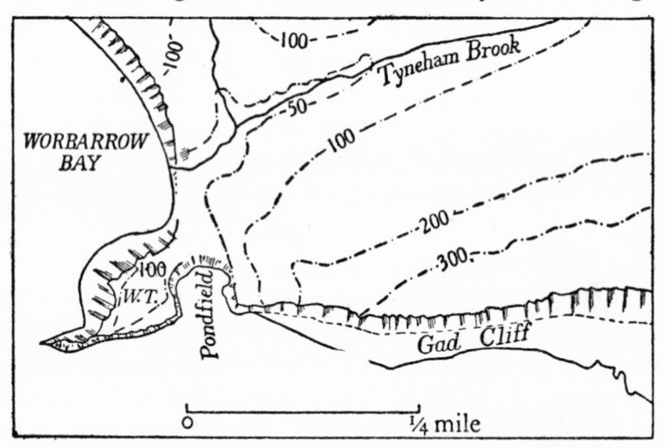

Fig. 56. Worbarrow Tout (W.T.) (after H. Bury, *op. cit.*)

lower parts of the streams, and thus they now afford a measure of the amount of coast erosion that has taken place.

Worbarrow Tout,[3] at the eastern end of Worbarrow Bay, presents some local but interesting problems (Fig. 56). The Tout is separated by a

[1] Burton, *op. cit.*

[2] In *Mems. Geol. Surv.* 'Isle of Purbeck and Weymouth' (unpublished at time of writing).

[3] H. Bury, 'Some Anomalous River Features in the Isle of Purbeck', *Proc. Geol. Assoc.* 47, 1936, 1.

narrow neck of land, about 35 ft. high at the head of Pondfield, from the main mass of Gad Cliff. A little to the north the Tyneham brook runs almost due west to Worbarrow Bay. This brook is only about a mile long, but has cut a small ravine inside a wider valley, indicating recent and probably constant rejuvenation and advance of the sea. There are at least two explanations for the low and narrow neck of land. A stream may have flowed to the *north* from Pondfield and joined the Tyneham brook, or the Tyneham brook may have turned sharply southwards and flowed through this neck. Since the neck is definitely graded to the *south*, and because there is no trace of any ravine on its northern side, it is held by Bury that the Tyneham brook once flowed this way. Consequently the present geography is due to the encroachment of the sea on the east side of Worbarrow Bay which resulted in the capture, at the angle of junction, of the combined Pondfield and Tyneham streams.[1]

From a point a little east of Pondfield, to St Alban's (Aldhelm's) Head, the Kimmeridge Clay alone reaches the coast. The cliffs are nearly vertical, and in many places the limestone bands in the clay run out to sea as the Kimmeridge ledges. In Kimmeridge Bay itself the cliffs, quite vertical, do not resemble normal clay cliffs: they are formed of alternating shales and stone bands. Several small faults are visible, and on the flat slabs of rock on the bay floor they can easily be seen in plan at low water. Along the western stretches of the Kimmeridge shore there are wild and tumbled undercliffs covered with slipped masses of overlying beds. The Portland and Purbeck Beds, a mile or so inland, enclose Kimmeridge Bay in a kind of inner cliff, which is an erosion escarpment.

Between St Alban's Head[2] and Durlston Head the beds are nearly horizontal. The Portland Beds are at sea level and capped by Purbecks. The Portland Stone throughout most of this stretch affords much resistance to erosion so that the cliffs are vertical, and erosion is mainly limited to joint planes. There are many coves usually associated with

[1] Arkell, *op. cit.*, remarks upon an interesting point concerning the grading of shingle in enclosed bays on Purbeck, such as Lulworth Cove and Worbarrow Bay. The largest shingle is usually found in the centre of the bay, where the longest waves strike head-on. This is very well seen in Worbarrow Bay where, on account of rock structure, opposite ends of the bay contribute different types of pebbles, the larger of which travel to the centre of the bay. 'When (the larger pebbles) have remained long enough to be worn down to smaller sizes they come under the influence of smaller waves which strike the beach obliquely, and they begin their journey back, arriving finally as coarse sand at one end of the bay or other.'

[2] At this headland, as well as Hounstout and White Nothe, landslips have often occurred. 'A wild undercliff of chalk or limestone boulders encumbers the slopes and often remains stable for many years while a special flora takes possession' (Arkell, *op. cit.*).

joints or small faults. Where the chert beds of the Portland formation rise above sea level, rocky ledges are produced as at Dancing Ledge. Where the Portland Sand is brought up, erosion is much quicker, and large blocks of the overlying stone are undercut. Tilly Whim Caves are artificial, having originated as quarries. In Durlston Bay is the best section of Purbeck rocks seen in this country: nearly the whole succession is repeated by two strike faults. Peveril Point is formed of the Upper Purbeck Beds.

Swanage Bay is cut in the soft Wealden Beds and, lying as it does between the resistant Purbeck strata of Peveril Point and the Chalk of Ballard Point and The Foreland, it illustrates remarkably well the effect of differential erosion. At Ballard Point the Chalk is vertical, and about 120 yd. north of the point the bedding planes are cut by the magnificent curve of the overthrust fault which, at sea level, is marked by a small cave. Nearer The Foreland are first the Pinnacles, detached columns of Chalk, then the large cave called Parson's Barn, and finally the fine Chalk stacks of Old Harry and what remains of Old Harry's Wife, most of which collapsed in 1896. The mass known as The Foreland (Handfast Point) was joined to the mainland by a narrow ridge up to the early years of this century. In fact, in this part of the coast there are many good examples of all stages in the formation of caves, arches, and stacks (**25**).

(7) STUDLAND BAY AND POOLE HARBOUR

North of the Foreland the coast is cut back again into Studland Bay: at the south-eastern corner there are low cliffs of Reading Beds, London Clay, and Bagshot Beds. These, however, soon give place to the sandy area known as South Haven peninsula, which, together with the Sandbanks peninsula, encloses Poole Harbour. South Haven peninsula has been studied in considerable detail by Diver,[1] Good,[2] and others. A traverse across it from east to west shows the following main features. (1) A wide sand beach with marram-covered dunes forming the first ridge so called by Diver in his paper (this was actually the last formed, apart from incipient new dunes). (2) In the more northerly part is a belt of saltings, characterized especially by *Juncus maritimus*. This belt runs southwards first into Eastern Lake, then Eastern Marsh, and finally Southern Heath. (3) In the north is the main mass of dunes, the product

[1] C.A. Diver, 'The Physiography of South Haven Peninsula...', *Geogr. Journ.* 81, 1933, 404.

[2] R. Good, 'Contributions towards a Survey of the Plants and Animals, of South Haven Peninsula.... II. General Ecology...', *Journ. Ecology*, 23, 1935, 361.

of recent erosion in Shell Bay—southwards these run into the second ridge, on which *Calluna* predominates. (4) Behind the second ridge is a belt of lakes in the southern part. Little Sea drains artificially to South Haven Point, a grassy area of considerable botanical interest. Little Sea covers

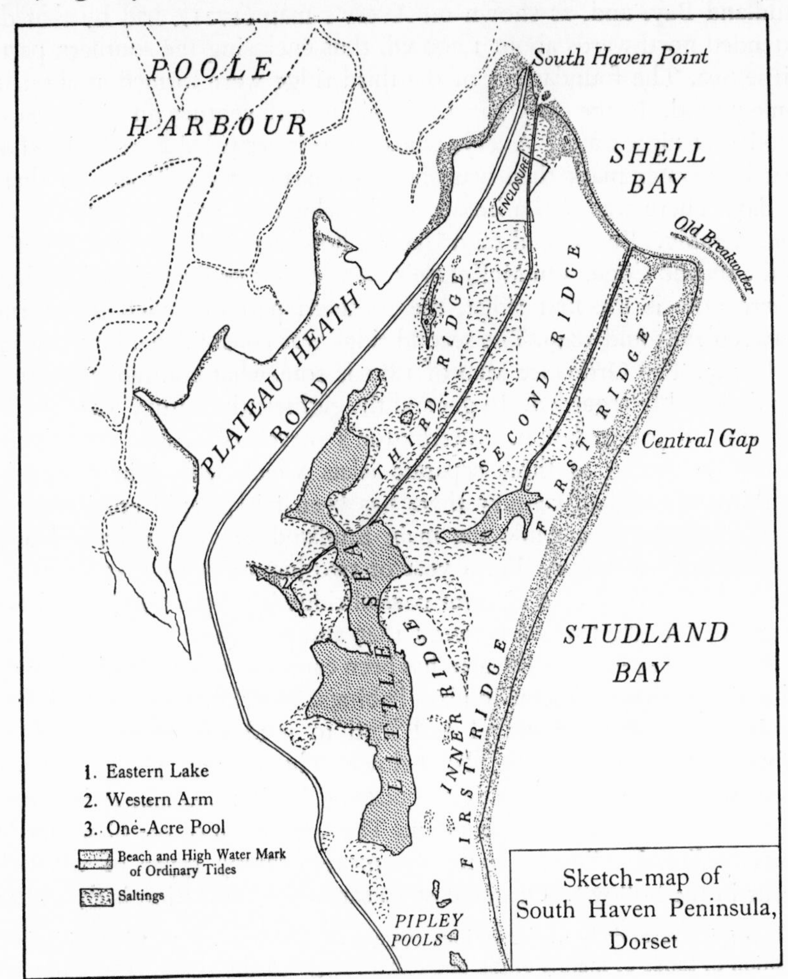

Fig. 57. South Haven Peninsula, Dorset (based on C. Diver, *op. cit.*)

some 70 acres and has a maximum depth of 6 ft. Eastern Lake is only about $1\frac{1}{2}$ ft. deep and 6 acres in extent. It fluctuates a good deal. The Central Marsh runs north from Little Sea, and on its western side is (5) the third sand ridge. Finally (6) there are the gravels, clays, and ferruginous sands of the Lower Bagshot Beds forming heathland (Fig. 57).

Diver[1] has also made an interesting study of the historical evolution of the district. From a critical examination of old maps and records it is clear that Little Sea first appeared in the seventeenth century. At that time Southern Heath was growing on what was then the sandy floor of Studland Bay, and, as shown on Avery's map (1721), had by that date extended northwards about 1,000 yd. thus enclosing the southern part of Little Sea. The foundations of the third ridge were formed at about the same period. In the early part of the eighteenth century all Little Sea was flooded at high water. Avery, whose map is regarded as reliable, shows that low-water mark then was about a quarter of a mile *east* of that of to-day. There was also a broad beach, which by 1785 had been eroded considerably. Four small sand islands, however, indicate the beginnings of the second ridge. On McKenzie's map of 1785 ordinary beach is shown where now is the first ridge, the southern part of which was formed between 1785 and 1849. The second ridge was consolidated between 1721 and 1789. The Ordnance Map of 1811 is somewhat contradictory and is not favourably regarded. In general the eighteenth century was therefore a period of progradation. Sherrington's survey of 1849 shows that the first ridge began as three separate islets which later joined together. Within the last half-century there has been fairly rapid erosion on the north-eastern coast, while the main Studland Bay shore has advanced eastwards about 80 yd. For a year or two just prior to 1933 there were signs of the building of a newer coastal ridge just opposite Central Gap. Eastern Lake was formed between 1849 and 1900 (**24**).

The plant ecology of the area has been carefully investigated by Good[2], who distinguishes three main plant habitats. The shores of Poole Harbour grade outwards from gravel and sand to mud and show an edaphic climax[3] of vegetation (see also below). The western plateau carries a heath vegetation characteristic of podsolized soils; *Calluna*, *Erica cinerea*, *E. tetralix*, and *Pteris* are abundant. The vegetation of the plateau represents a climatic climax. Naturally, more swampy types of vegetation occur in the hollows. On the eastern sands the vegetation is dynamic:

> It consists essentially of the various stages of a xerosere commencing with the deposition of sands of marine origin and culminating in a closed heath formation, complicated at almost all stages by the retarding or diverting effects of fresh- or salt-water flooding and by secondary invasion of blown sand. The general result is to give an extremely complex mosaic of successional vegetation states. The gradual emergence of blown sand from below sea-level gives rise to three initial plant associations according to the details of the process. If the sand accumulation is rapid, resulting in dunes,

[1] *Op. cit.* [2] *Op. cit.*

[3] 'A climax community of which the existence is determined by some property of the soil.'

the first colonisers are strand plants, very soon giving place to dune-grasses. If, however, the accumulation is gradual and at first inter-tidal, the initial vegetation is salt marsh dominated by *Juncus maritimus*. A third condition is found in the slacks or levels amongst the dunes and out of direct influence of the tide. Here the first species are some of those characteristic of the damp sand communities, and before long this type of vegetation becomes well developed.[1]

There is no need to analyse the vegetation in further detail. That of the shore proper, with which we are primarily concerned, differs in no essential aspects from that of similar places elsewhere, and a glance at Chapters XIII and XIV will show how the plants affect the topography and lead to the growth of dunes and other features.

Unfortunately the causes of the evolution of the South Haven peninsula and Sandbanks have not been investigated. We may, perhaps, compare them in a very general way with Portland Beach and Weymouth Beach, but the precise effects of wave incidence and direction of travel of beach material are uncertain. The two spits have clearly grown in opposite directions, and the southern one has prograded considerably and is stepped forward about half a mile relative to the northern one. In this western part of Poole Bay the effective waves are those generated in the Channel, as the area is protected from the Atlantic waves by the Isle of Purbeck. Lewis[2] argues that the length of fetch and wind strength are approximately equal from any direction between south and south-east, and that the important factor seems to be the position of the enclosing headlands, especially Ballard Down and The Foreland. In short, waves approaching Poole Bay from an easterly direction would cause beach material to move westwards from Bournemouth and also from The Foreland. Whilst I am inclined to agree with this as a general statement, it is clear that more detailed work is necessary in order to account for the very different forms of the two spits, especially the broadening of the southern one and the marked narrowness of the northern.[3]

Poole Harbour owes its main outline to submergence converting what was formerly a low moorland area with sandy knolls into a broad lagoon or estuary. The former knolls now form the islands, Green Island, Furzey Island, and Brownsea Island. The numerous bays and inlets of its margins represent the former dales and dingles that might reasonably be expected to have characterized a lowland heathy area traversed by a big

[1] Good, *op. cit.*

[2] W.V. Lewis, *Proc. Geol. Assoc.* 49, 1938, 107.

[3] There is no clear reason why there is so much sand in Studland Bay. As Arkell (*op. cit.*) suggests, it may be a secondary effect of the great erosion along Bournemouth and Christchurch cliffs in historic times.

river. Unfortunately the area has not yet been investigated as a unit. Green[1] and his pupils have analysed the regime of the tides, the salinity, and the bottom deposits, which are almost entirely formed of quartz sand. They also found that the main channel tends to migrate eastwards, especially where its curvature is greatest. It seems to maintain its depth, but it narrows because the western edge (the eastern part of Middle Ground) migrates more rapidly than the eastern bank. One point, in some way as yet unknown, must bear upon the changes of the enclosing sandspits: there is no conclusive evidence that more sand enters the harbour than passes out of it. As the ebb streams are generally stronger than the flood, an outward migration of material is suggested quite apart from the effect of strong inshore winds and storms. The harbour is also interesting because of the great development of *Spartina Townsendii*[2] which first appeared in 1899 and now covers many square miles. A full botanical study of the harbour would be of great interest.

(8) POOLE BAY AND CHRISTCHURCH BAY

From Canford cliffs[3] to the Solent and beyond, the coastal rocks all belong to the Eocene and Oligocene. Around Bournemouth and Christchurch we have the Bagshot Sands, Bracklesham Beds, and the Barton Clay, all of which yield easily to erosion. In the neighbourhood of Bournemouth the chines form a noteworthy feature. Dursley and Branksome chines each show two distinct valleys of different ages, and the newer and steeper inner valley alone carries water. The upper and wider valley also extends farther inland. The same feature is evident in other chines and also in the Christchurch 'Bunnies'. It does not occur in the 'Bottoms' which run down to the Stour and Avon. Fig. 58 illustrates diagrammatically the nature of these chines. The word 'chine' is strictly applicable to the inner valleys. Bury,[4] who shows conclusively that the views of Lyell and Gardner are extremely improbable, considers that the older, broader, and longer valleys were formed in a cold period when the ground was impervious owing to frost, and when the rocks in which they are cut would have been effectively harder and more resistant than now. This cold period probably coincided with the formation of the Coombe Rock, and

[1] F.H.W. Green, *Poole Harbour: a Hydrographic Survey*, 1938–9. London: Geographical Publications, 1940.

[2] A.E. Carey and F.W. Oliver, *Tidal Lands*, 1918, and (F.W. Oliver), *Rice Grass*, Min. of Agric. and Fisheries, Misc. Pubns., No. 61, 1929.

[3] Esplanades now protect large parts of these cliffs. The Sandbanks peninsula is largely built over.

[4] 'The Chines and Cliffs of Bournemouth', *Geol. Mag.* 57, 1920, 71.

the valleys may have been cut by floods of melt water from local snow melts. The valley-in-valley structure is associated only with those chines that run to the sea, and even apart from any slight change of sea level which may or may not have had an influence, the actual encroachment of the sea is sufficient to explain the feature. As we shall see later, Bournemouth Bay was once traversed by the Solent River which flowed a mile or two out from the present coast. The chine valley streams originally were short tributaries of that river. When it was breached (see p. 301) the broad, open chines were left. Reference to Fig. 58 shows that without

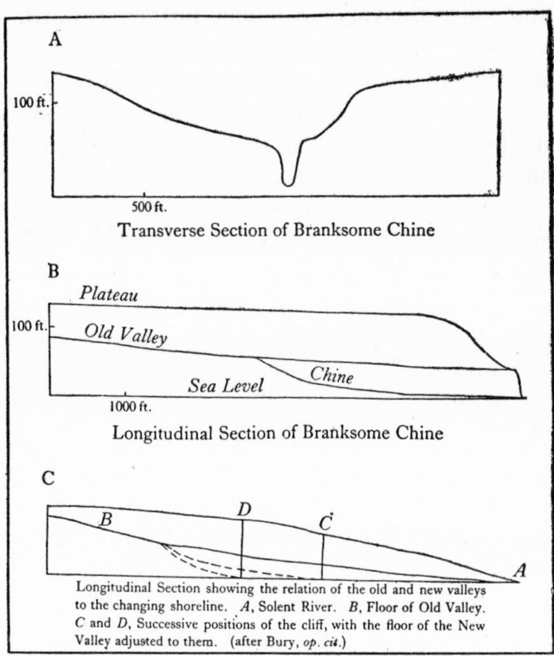

Fig. 58. Transverse and longitudinal sections of Branksome Chine, Bournemouth

any change of sea level in quite recent times the remains of the old valley floors would increase in height in proportion to the retreat of the cliff by marine erosion. The smaller inner valleys are the result of present climatic conditions and merely represent the power of the present tiny streams in cutting a small **V**-shaped valley in contrast to the wide, open valleys cut by the greater volume of water in the Coombe Rock period. Marine erosion is more powerful than that of the chine streams. It follows, therefore that, as the cliffs are cut back, a time will come when the only traces of the chines, provided conditions remain constant, will be notches in the line of the cliffs.

Bury also comments on the view that the Bournemouth cliffs are steeper now than they were about fifty years ago. He disagrees with the over-elaborate views of Ord, and adopts the simpler and much more reasonable view that at certain points the cliffs were more easily scaled than now.[1] This fact is easily explicable by local erosion rather than assuming, as Ord did, a former wide stretch of raised beach for which there is no evidence whatever.

The Bourne valley is not usually regarded as a chine: it is a simple V-shape in cross-section. The peaty alluvium or submerged forest probably marks a time when the valley extended farther seaward. This deposit is seldom seen nowadays.

Between Southbourne and Hengistbury Head there is a strip of low ground which some people think may represent a former outlet of the Stour. If that is the case, it is possible that the ridge of St Catherine's Hill formerly extended farther south-eastwards and joined the high ground of Hengistbury. The Stour and Avon, working on either side of a low and relatively inconspicuous ridge, may conceivably have breached it, and the two rivers then united. They may then have flowed south through this gap. But White[2] justly questioned this view. First, Christchurch Harbour is in a more direct line with the Stour which would have had to make a somewhat abrupt turn to pass through the Southbourne gap, and in any case the gap is a good deal narrower than the Stour vale. This view also suggests that the slope of the gap was southward, whereas it is actually to the north and grades naturally to the present river. There is also a dry watercourse just west of Double Dykes which runs through the gap downwards to the present harbour. In fact, it is far more likely that instead of the gap being a former outlet of the Stour, it is part of a valley which arose southward of the present coast when the land extended farther in that direction, and that it drained northwards to the Stour.

Hengistbury Head is built of the sandy Upper Bracklesham Beds. It is now almost an island, being connected with the mainland by the low isthmus noted above. Running north-eastwards from it is a fluctuating spit of sand and shingle enclosing Christchurch Harbour.[3] The more permanent part of the spit extends about three-quarters of a mile from Hengistbury Head, and in its widest part measures some 260 yd. Maps dated about 1654 show that it ended opposite Mudeford ferry; Bowen's

[1] In the last two or three decades the Bournemouth cliffs have been 'preserved' artificially and are now largely planted.

[2] *Mems. Geol. Surv.* 'The Geology of the Country around Bournemouth', H.J.O. White, 1917.

[3] E. St J. Burton, 'Periodic Changes in the position of the Run at Mudeford...', *Proc. Geol. Assoc.* 42, 1931, 157.

map (1717), Ellis's (1768), and the original Admiralty Chart (1785) all agree reasonably well on this point, apart from quite minor variations. The Ordnance Survey (1811) shows the end near Mudeford, but by 1870 (also Ordnance Survey) the spit reached nearly to Cliff End: it was even longer in 1880. In 1893 the end had retreated to Steamer Lodge, and the spit broke in 1895 or 1896. In 1911 it once again extended beyond Cliff End, but in November of that year it broke opposite Gundimore. In September 1930 it extended nearly to Steamer Lodge.[1] On the fixed part near Hengistbury there are dunes, the most mature of which are about

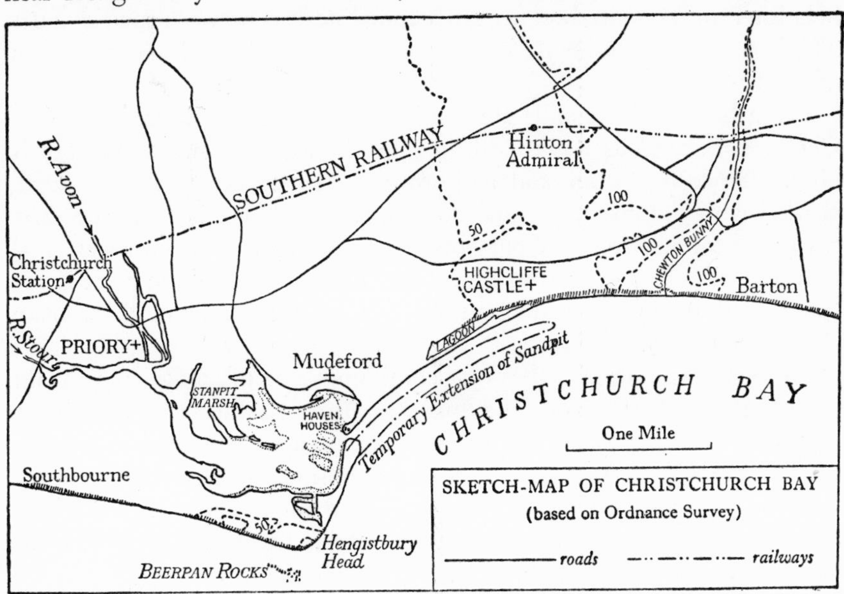

Fig. 59. Christchurch Bay

20 ft. high. The older dunes carry *Calluna* and *Ammophila arenaria* on their outer side, and gorse (*Ulex Europaeus*), bracken (*Pteris aquilina*), and a few brambles (*Rubus* spp.) within. Many of the smaller dunes show signs of not infrequent over-running by the waves. There are flints up to 8 in. long on the outer beach, and these are probably derived from the Bracklesham Beds of Bournemouth Bay. Small flints are more frequent near the distal end (Fig. 59) (**23**).

Local observers have held that the growth after the middle of the last century was related to the removal of ironstone from between tidemarks on the headland (Beerpan Rocks). As Burton points out, this may have

[1] Another breach was made during a severe storm and high tide in the early morning of 25 February 1935 (*Bournemouth Daily Echo*, 25 and 26 February 1935).

been merely a contributory cause, because many large flints came from the Boscombe Sands and passed Hengistbury long before the reef was removed. Much material is derived during onshore gales from the cliffs of Christchurch Bay, and this material is later drifted westward by easterly winds. In general, it appears that more material is brought into this bay by south or south-easterly winds than by south-westerly winds.

The changes in the length of the spit are related to weather conditions. A breach near Mudeford is most likely to occur during a combination of a heavy gale from the south-east with spring tides. Because such a combination is infrequent, there are fairly long intervals between such changes. The lagoon near Highcliffe Castle represents part of a former extension of the Run (i.e. channel).

From Cliff End the cliffs are continuous except for minor breaks (e.g. Chewton Bunny) as far as Milford. At Barton they are about 115 ft. high. Between Barton and Milford the Danes stream runs roughly parallel with the coast, and the deep notches in the cliffs indicate those places where erosion has cut them back so that they intersect small southern branches of this stream. Since the material of the cliffs is soft (sands and clays), they are frequently much tumbled and show many signs of landslips.

Milford beach is really part of a continuous line of shingle which runs out into, and nearly across, the Solent as Hurst Castle spit. The spit first runs to the south-east for about one mile, then changes direction to the east, and finally turns abruptly to run north-north-west. It is entirely a shingle formation, apart from some sand on the foreshore. An underlying clay bench is exposed near high-water mark. Lyell first drew scientific attention to the spit, but Lewis[1] has done the main work on it. He has applied to it his views on the approach of dominant waves to the direction taken up by shingle structures. The clay bench, probably an extension of the marsh clay within, has clearly aided its south-easterly extension from the mainland. On this part of the spit the normal direction of wave approach is directly onshore, and so drift of material along it is fairly slow. At the turn to an eastward direction the water deepens: the shingle which has travelled along the spit passes round this corner and comparatively large supplies gather there to be thrown up on the shore by waves, especially those coming down the Solent from the north-east. In this way the earliest recurves would form. If we also assume, as we well may, that this east-west part of the spit more or less marks the boundary between the former shoal water and the deeper part of the Solent, it seems reason-

[1] W.V. Lewis, 'The Effect of Wave Incidence on the Configuration of a Shingle Beach', *Geogr. Journ.* 78, 1931, 129 (and unpublished work).

able to suppose also that the shingle which continued to travel along the main south-eastern part of the spit would turn here and form the succession of recurves such as we see to-day. The direction taken by the recurves would be largely influenced by the north-easterly waves travelling down the Solent. This modifies the views set down by Lewis in 1931, but it is closely in accordance with his later, and yet unpublished, work (**17**).

(9) THE MAINLAND COAST OF THE SOLENT AND SPITHEAD AND THE INLETS NEAR PORTSMOUTH[1]

Material continues to travel eastwards along the Hampshire and Isle of Wight coasts of the Solent. This is evident from the small spits at the mouth of the Beaulieu River, at Yarmouth, at Newtown Bay, and particularly at Calshot (**22**). The discussion that follows shows that the Solent and Spithead are really part of the ancient Frome. In the Lymington and Portsmouth areas alike, the river valleys are gently graded to the Solent or Spithead. Southampton Water, in common with smaller rivers like the Lymington, Beaulieu, and Medina, is a drowned valley. In Southampton Water alluvial deposits fill this submerged valley. When they began to form, the land stood *at least* 40 ft. higher than now, and at that time Southampton Water was merely a non-tidal river which flowed in a narrow channel bordered by lowlands and marshes.[2] Dock excavations revealed these as submerged forests and peat beds. Mud still continues to accumulate and is very thick off Fawley. There are extensive banks also at the mouth of the Itchen, opposite Hythe, and at Hamble mouth, as well as inside Hurst Castle spit, at Calshot, and in the Beaulieu River. A good deal of reclamation work has been carried out, especially during dock building at Southampton; there are also enclosures at Lepe, Titchfield, Hook, Dibden, Farlington, and other places.[3] The rapid spread of *Spartina Townsendii* over much of the mud flats of the whole area is noteworthy, and is also of economic importance since indiscriminate growth would sooner or later lead to hindrances to navigation. The new docks at Southampton are excavated in alluvial beds and the underlying Bracklesham Beds (**16**, **15**, **14**).[4]

[1] For general reference to the coast from Lymington to Portsmouth, see *Mems. Geol. Surv.* 'The Geology of the Country near Lymington and Portsmouth', H.J.O. White, 1918.

[2] *Mems. Geol. Surv.* 'The Geology of the Country around Southampton', C. Reid, 1902; K.P. Oakley, *Proc. Prehist. Soc.* 9, 1943, 56.

[3] T.W. Shore, 'Hampshire Mudlands...', *Papers and Proc. Hants Field Clubs*, 2, 1894, 181.

[4] F.W. Anderson, 'The New Docks Excavations at Southampton', *ibid.* 12, 1934, 169; H. Godwin, *New Phytologist*, 39, 1940, 303.

It is difficult to be quite systematic in our treatment of the coast because the study of the Isle of Wight breaks the sequence of observations. It seems best to continue the comments on the mainland as far as Chichester Harbour before speaking of that island and of its separation from the mainland. All this area lies on the low ground formed by the Eocene and Oligocene beds. The dips of the rocks are generally low, and, on the mainland, inclined to the south. The Upper Chalk is seen in the northern parts of Portsea, Hayling, and Thorney Islands. The outcrop of the Reading Beds can be seen in these same islands by the ruddy tint of the mottled

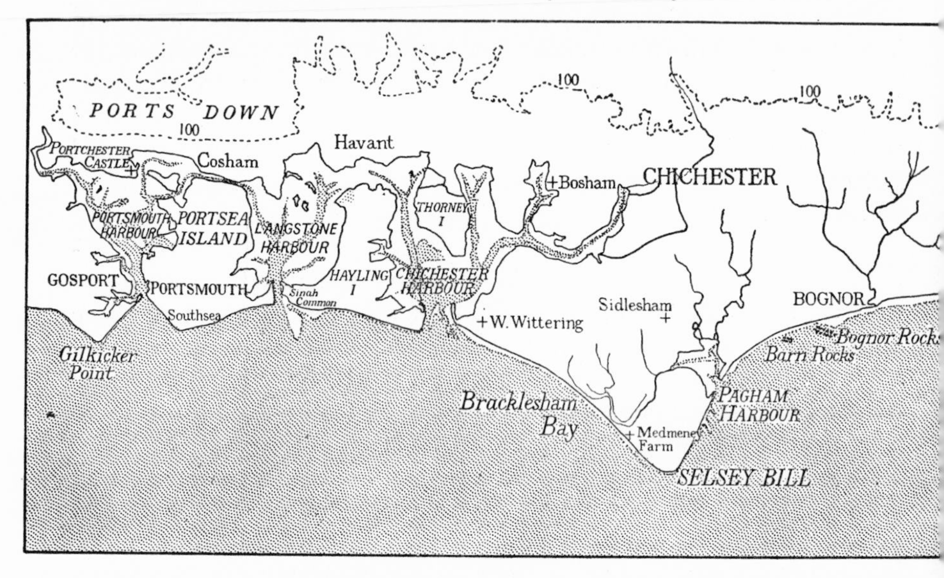

Fig. 60. Sketch-map of the coast between Portsmouth and Bognor (based on Ordnance Survey)

clays giving place on either hand to the Chalk and London Clay. The Barton Beds are under the Solent and Spithead, but emerge west of Southampton Water. The very indented coastline between Gosport and Chichester has not perhaps received the attention it merits. It must be remembered, as White points out, that at the end of the Pleistocene period the bigger rivers had entrenched themselves in the low coastal plain and at that time were 30 ft. or so below present high-water mark. The coast (see below) was then considerably to the south of its present position, and it is a justifiable inference that the rivers flowed down to it in shallow broadly open valleys which were cut in the gravel and brickearth-covered flats of which islands such as Portsea and Hayling are remnants. Later, subsidence drowned these lower valleys and produced the beginnings of the inlets we see to-day. Their present intricate outlines depend on several factors—marsh growth and reclamations, artificial alterations due to dock buildings, especially at Portsmouth, some aggradation by the streams flowing into them, and erosion. Hence, although slowly shallowing, they are at the same time often growing in size.[1] Their shores are all low, and wave erosion is affecting them in many places. This, owing to the prevalent westerly winds, is usually more noticeable on the eastern banks and near Longmere Point, at the southern extremity of Thorney Island, where the gravel and London Clay cliffs show little coves and stacks (**13**).

The shingle of the modern beaches is mainly derived from the Plateau Gravels and Coombe Rock, and consequently is decidedly subangular. It forms a fairly continuous belt near Portsmouth, and at Sinah Common, on Hayling Island, is piled up into fulls. The general drift is to the east, but naturally this is often modified by the in-and-out currents into the harbours. There are also strips of sub-angular shingle around the harbour shores.

(10) THE ISLE OF WIGHT

The Tertiary beds described in the previous section as having a low southerly dip on the mainland pass under the Solent and Spithead and reappear in the Isle of Wight; this, together with the Isle of Purbeck, forms the southern side of the Hampshire Basin. The general structure[2] of the Isle of Wight is dominated by the main Sandown anticline, the trend of which corresponds fairly closely to the Chalk ridge running from the Needles to Culver Cliff. Here the Chalk dips steeply to the north: to the south, as in Purbeck, the dip is gentle, although minor folds occur along the Sandown syncline and Brixton anticline. The Chalk reappears in the south and forms the high ground between Blackgang and Luccombe chines. Between the two Chalk outcrops is the lower country of the Lower Greensand and Wealden Beds. Reference to the geological sketch map (Fig. 61) will make the general structure quite clear. The quadrilateral form of the island depends almost entirely on the structure and the relative hardness of the rocks: the dominance of the central ridge of Chalk is evident. Through it cut three rivers, the Eastern and Western Yar and the Medina. Their origin dates back to the time when the Chalk extended over the intervening ground between its two present outcrops. Rivers rose on this and cut down to form the ancestors of the present valleys, in

[1] *Mems. Geol. Surv.* 'The Geology of the Country near Fareham and Havant', H.J.O. White, 1913.

[2] *Mems. Geol. Surv.* 'A Short Account of the Geology of the Isle of Wight', H.J.O. White, 1921.

course of time much of the Chalk was removed, and river erosion in the softer underlying Lower Greensand was comparatively rapid. Once the Greensand was exposed, the Eastern Yar in particular was enabled to develop important tributaries running along the strike, and by this process probably captured some of the headwaters of the Medina. Possibly the same thing happened with the Western Yar, but erosion has removed much of the south-western part of the island and one can only speculate. It would require but a small submergence to make three islands of Wight because the gaps at Freshwater and Brading are not much above sea level. We may also note that the Spithead coast is shelving, while that of the Solent is bolder, especially towards the west, and nearly all the rest of the

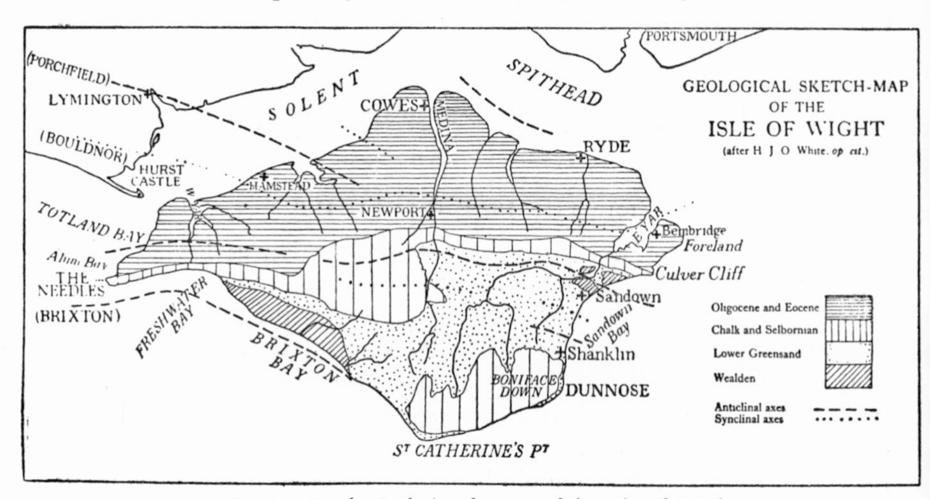

Fig. 61. Geological sketch-map of the Isle of Wight

coastline is bounded by cliffs cut by deep ravines, called chines, like those in Shanklin Bay (**20, 19, 18**).

The Wealden Beds form the cliffs on the west coast between Compton Bay and Atherfield Point. Between Sedmore Point and Brook chine there has been a good deal of slipping, and an undercliff called Roughlands has been formed. At Hanover Point is the so-called Pine Raft: it is of Wealden age and rests on marls, but is hardly a true submerged forest as the trees have drifted there from a distance. There is a corresponding but smaller outcrop of Wealden Beds on the east coast in Sandown Bay. These beds consist mainly of marls and shales with bands of sandstone and limestone. The Lower Greensand is best seen in Sandown Bay: the rock at its base runs out as a reef north of the Red Cliff. It is succeeded by the Atherfield Clay, seen best at its western outcrop near the village from which it takes its name. The succeeding sandstone forms Red Cliff, and reappears, on the southern side of the anticline near Shanklin. This sandstone is part of the Ferruginous Sands. South of Shanklin the cliff is prominent: the upper part is the Sandrock; at the base are clay bands. These lead to slipping and give rise to the undercliffs so common on the island. Near Luccombe chine a slip of some magnitude took place in 1910. The area called the Landslip near Bonchurch and nearly all the stretch from Ventnor to Niton have been similarly affected. At Atherfield the hard bands of the Lower Greensand run out to sea and form dangerous reefs. The northern outcrop on the west coast is very narrow.

The Upper Greensand, resting on the Gault, often forms very striking features, but they are not entirely coastal. Between Bonchurch and Niton is the innercliff which stands back from the actual coastal cliff. The cliff wall is of Upper Greensand, and large masses have slipped forward on the underlying Gault Clay, but whereas the main slips are very old others are quite recent. The Upper Greensand comes to the shore in Compton Bay and at the Culvers, and again near Ventnor where the lower cliff consists largely of masses which have fallen down. The Chalk outcrops on the coast, although narrow, give rise to characteristic scenery. The Needles are wedge-shaped masses of *Mucronata* Chalk with a steep dip and are very resistant to both subaerial and marine erosion. At Scratchell's Bay and the Needles the Chalk is extremely hard, and this is the reason for the survival of the Needles.[1] The Grand Arch in Scratchell's Bay is entirely in *Cor-anguinum* Chalk, the structure of which seems to be favourable to recesses. There is what Rowe calls 'an incipient grand-arch' in the same zone at Culver cliffs, and White notes that the early stages of such recesses can be seen in quarries (**21**).

North of the Chalk ridge is the lower ground of the Tertiaries. Whitecliff Bay shows the succession very clearly: the cliffs are much lower than the Chalk, and the bay has a fine sand beach. The rocks consist of varicoloured sands and clays standing nearly vertically. Towards the northern end of the bay the Bembridge Limestone, at first vertical, flattens out and runs to sea as Bembridge Ledge. North of this limestone the strata are nearly horizontal. The corresponding section is seen in Alum Bay on the west coast; here the coloured sands are well known (see Figs. 62 and 63). The northern part of the island is formed by flat or gently undulating Oligocene strata:[2] they include, in ascending sequence, the Headon Beds, the Osborne and St Helen's Beds, the Bembridge Beds

[1] A.W. Rowe, 'The Zones of the White Chalk of the English Coast, V. The Isle of Wight', *Proc. Geol. Assoc.* 20, 1907–8, 209.

[2] The Tertiary beds are liable to much slipping, especially near Hamstead.

(including the Bembridge Limestone), and the Hamstead Beds, and all are apt to vary much in short distances. The Osborne Beds are exposed on the low shore between Cowes and Ryde, and also between Seaview and St Helen's. They are mostly red and green clays and do not form any conspicuous features. On the north shore of the island the strata rise a little and there are also other minor undulations. The Bembridge Lime-

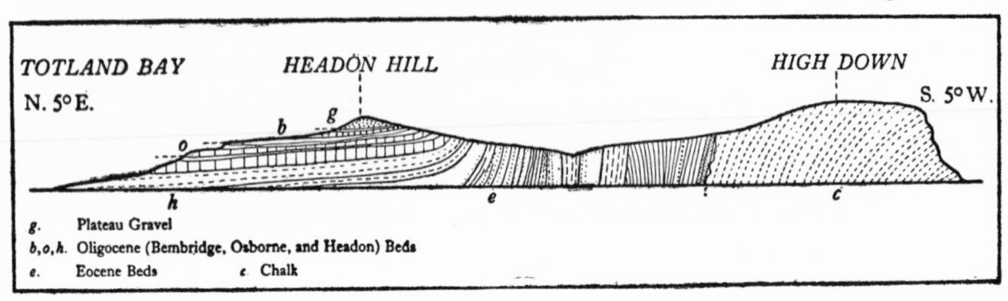

Fig. 62. Section from Totland Bay to High Down, Isle of Wight (after H.J.O. White, *op. cit.*)

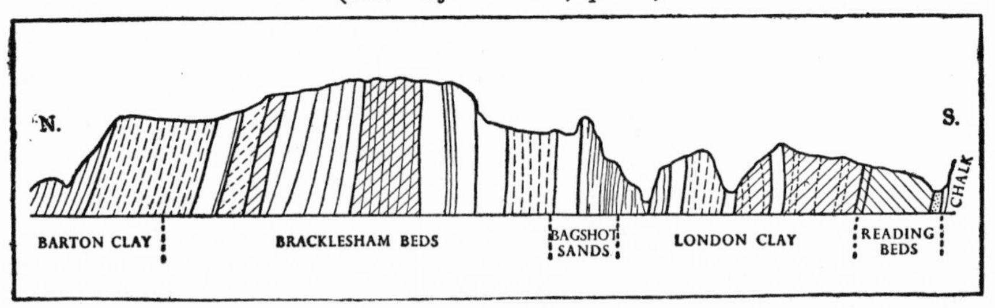

Fig. 63. Section in Alum Bay, Isle of Wight (based on Prestwich, *Quart. Journ. Geol. Soc.* 2, 1846, Pl. IX; and on Fig. 18, *British Regional Geology*. The Hampshire Basin (no date)).

stone is thus brought up to form prominent ledges as at Hamstead, Thorness Bay, and Gurnard Ledge. Bembridge Foreland consists of a thick bed of flint gravel resting on Bembridge Marls; these same marls also appear over the three ledges just mentioned. The Hamstead Beds are essentially marls; they only occur on the coast at Hamstead itself where they form most of the cliff which is rather over 200 ft. high. The Headon Beds scarcely affect the coast, but are visible at Whitecliff Bay.

(11) THE SEPARATION OF THE ISLE OF WIGHT FROM THE MAINLAND[1]

An examination of a geological map of southern England shows that the order and arrangement of the beds in the Isle of Wight are similar to, and

[1] See *Mems. Geol. Surv.* H.J.O. White, *op. cit.* 1921; *Rept. Brit. Assoc. Adv. Sci.* 1911, 'Portsmouth', 384 (a joint discussion of Sections C and E on the former connection of the Isle of Wight with the mainland).

in many respects the replica of, the beds in the Isle of Purbeck. The Needles are the direct counterpart of the Foreland and Old Harry, and to north and south the disposition of the Tertiary and older Cretaceous beds is the same. The Portland and Purbeck Beds are not seen in Wight, but presumably occur beneath the southern part of the island. We may, therefore, have little hesitation in joining in imagination these similar rocks, reconstructing a former land area in Bournemouth Bay, and thus completing the southern side of the Hampshire syncline. This former land connection existed long before the time of the submerged forests, and the whole area stood higher above sea level than it does to-day. At that time the Solent and Spithead were not arms of the sea, and the belt of low-lying Tertiary rocks was traversed from east to west by an extended river Frome. Precisely at what stage the Purbeck-Wight barrier was breached is unknown. Fig. 64, after Reid,[1] shows the probable extent of the former land. The sea was attacking its southern side, and there is no reason for thinking that the barrier was very wide. The Chalk almost certainly formed its backbone, and once this was breached the sea would have had an easy task in the softer Tertiary beds. We may also suppose that, as in Purbeck and Wight now, this Chalk ridge was cut through by rivers, and these gaps, once the sea had reached the Chalk, would help its work of erosion. But we must also bear in mind that the several submerged-forest beds indicate a considerable lowering of the land relative to sea level, possibly not far short of 100 ft.[2] Once the barrier had been partially breached by the waves, complete removal would not take long when submergence began. What is now happening at Mupe Rocks in Worbarrow Bay (see p. 283) may well illustrate, on a small scale, the preliminary breaching and erosion of the Purbeck-Wight ridge.

The submergence not only finally destroyed the ridge, but it converted the Solent, Spithead, Southampton Water, and all the smaller valleys into the drowned valleys and estuaries which we now see. It also helps us to reconstruct the former extent of some of the Isle of Wight streams. Inspection of a map suggests that the lower valleys of the Eastern and Western Yars are out of proportion to their upper parts, a fact particularly noticeable in the Western Yar. As Fig. 64 suggests, since the destruction of the connecting ridge there has also been a great deal of erosion on the west and east coasts of the Isle of Wight which, among other things, has resulted in the loss of the upper parts of the two Yars. It is no far stretch of the imagination to suppose that the Compton Farm stream formerly

[1] *Mems. Geol. Surv.* 'The Geology of the Country around Ringwood', C. Reid, 1902.

[2] See also p. 493.

joined the Western Yar somewhere in what is now Freshwater Bay. The small streams in Shanklin and Luccombe chines may similarly have joined the main stream of the Eastern Yar in what is now Sandown Bay. It must be remembered that the main stream of this river is a strike stream in the Lower Greensand Beds, and that this strike section probably originated as a subsequent tributary to the main dip stream which flowed northwards

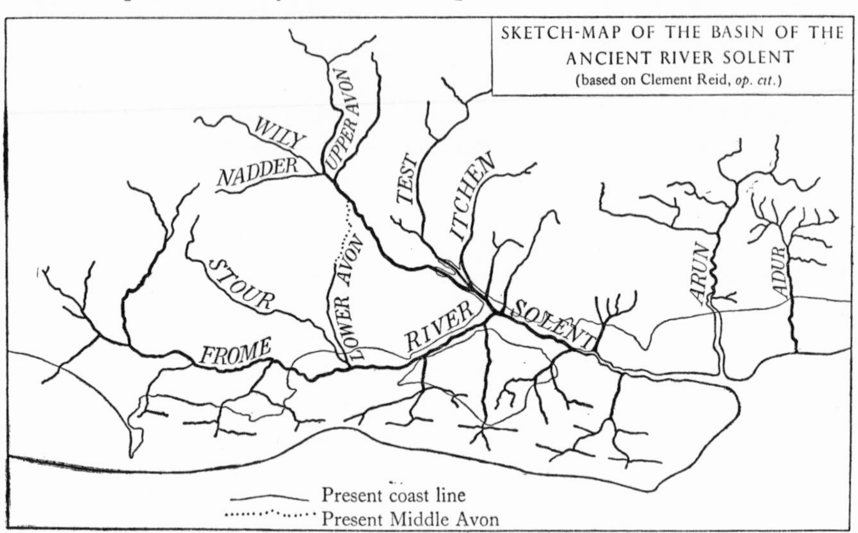

Fig. 64. Basin of the ancient river Solent (based on Clement Reid, *op. cit.*)

through Brading Gap. The Western Yar and the Medina still possess typical drowned mouths: that of the Eastern Yar has been considerably modified by reclamation. It is also worth mentioning here that the Lower Avon, instead of having to turn into the Solent River and flow to the sea somewhere near Portsmouth, was much shortened and revivified by the final disappearance of the former south 'bank' of the Solent River. As a result it was able to cut back and capture the Upper Avon waters together with those of the Wily and Nadder.

(12) ICTIS AND THE TIN TRADE

It is relevant at this point to call attention to the problem of the Cassiterides and the Phoenician trade in tin. Diodorus Siculus, in speaking of this matter, said that the tin for export was carried 'to a certain island lying off Britain called Ictis. During the ebb of the tide the intervening space is left dry....'[1] When the names Mictis, Ictis, or Vectis are used by the

[1] In C. Reid, 'The Island of Ictis', *Archaeologia*, 59, 1905, 281.

ancient writers they all appear to refer to the same island, and Vectis was the name given to the Isle of Wight in Roman times. The low-tide causeway implied in D. Siculus's statement is usually taken to refer to that between St Michael's Mount and Cornwall. Reid, however, argues that at that time St Michael's Mount was nothing more than an isolated mount rising out of a swampy wood, and that it was not until considerably later that erosion had cut back the Penzance coast to its present form. He then attempts to show that, since, in his opinion, St Michael's Mount is out of the question, it is necessary to find another island that will fit D. Siculus's description. This he maintains was the Isle of Wight because, as shown below, he sees reason to think that in those days a causeway probably existed between the western side of the Western Yar and Pennington marshes in Hampshire. This argument is purely geological. Immediately east of Yarmouth the strata form a syncline underlain by the Bembridge Limestone. The syncline is now incomplete because the western part is cut into by the Solent, leaving ledges at Yarmouth (Black Rock) and Hamstead. Black Rock is believed by Reid to be the last remnant of the old causeway. Reid drew a reconstruction of the outline of that part of the island, and allowed for erosion. (Hurst Castle spit was presumably formed much later, and before its formation erosion in the western Solent would have been quicker than now.) The limestone ledge was postulated across the Solent and then, after turning east, under the Lymington marshes.

Other writers had assumed a ford between Stour and Gurnard Bay, supposing that at that time (say 1100 B.C.) the island was still joined to the mainland. Reid, however, argued that even if the water had been sufficiently shoal it was extremely improbable that carts could have crossed as the bottom must have been clay or loose sand. Only on the limestone causeway could wheeled traffic have made its way. It may be stated that there is really no sound archaeological support for Reid's thesis. In the first place there is no convincing reason for supposing St Michael's Mount was then incorporated in the mainland. Secondly, his main argument for this choice of route, followed by the tin traders from England to the Continent, rests on the idea that the harbours west of Vectis were too much exposed to wind and weather. On the latter point it is more than pertinent to use Rice Holmes's[1] evidence that (1) there is ample shelter in the numerous deep inlets of Cornwall and Devon, and (2) it is most improbable that the business-like Phoenicians would bring the tin overland as far as this causeway for transhipment to the Continent when they

[1] See p. 230 and T. Rice Holmes, 'The Cassiterides and Ictis', in *Ancient Britain and the Invasions of Julius Caesar*, 1907.

could make the journey direct from Cornwall. That they were capable of making long voyages is amply proved.

(13) SELSEY BILL AND PAGHAM HARBOUR

It is suitable to resume here the study of the mainland coast at Selsey Bill which, with Pagham Harbour, is really of the same general nature as the islands and harbours around Portsmouth (Fig. 60). Selsey is built mainly of the sandy Bracklesham Beds, but the surface of the island shows superficial deposits almost exclusively. The island is virtually cut off from the mainland by a narrow and marshy depression, partly below sea level, running from Pagham Harbour to Bracklesham Bay. Before the reclamation in Pagham Harbour this severance was even more complete. The whole coastline is subject to considerable loss: opposite Bracklesham Farm it has been estimated at 6–8 ft. a year, and as much as 10–13 ft. off Cookham Manor Farm. Heron-Allen[1] estimates that a strip from 110 to 300 yd. wide disappeared from between Chichester Harbour and the Bill in the years 1778 to 1896. He also notes that land water generally aids marine erosion. There is a good deal of shingle; that on the south-western coast is apt to be washed away if strong westerly winds coincide with high tides (**12**).

Perhaps the main feature of interest from the coastal point of view is Pagham Harbour. In A.D. 681 Bede described Selsey as an island except for a connecting shingle tongue on its north-western side at Medmeney. Breaches of this tongue are recorded from time to time, and it is worth noticing that here, too, the shingle travels to the east. A great deal of misinformation about Pagham Harbour can be traced to a statement by Dallaway who wrote: 'The Nonae Roll, in 1345, bears indubitable testimony that the whole of Pagham Harbour was occasioned by a sudden irruption of the sea not many years prior to that date.'[2] Cavis-Brown, who knew the area intimately, maintained that this really referred to the destruction of sea walls between Pagham and Sidlesham which led to the flooding of already reclaimed land. Incidentally Pagham Harbour, as a name, was first used in the eighteenth century: previously the place was known as Selsey or Sidlesham Harbour.

There have been many reclamations in the harbour: indeed, one scheme, which did not materialize, was put forward as early as 1664. By 1672 there was a ferry, and its site was fordable at low water by a path called the Wadeway, which in 1805–9 was replaced by a strong bank. In

[1] *Selsey Bill, Historic and Prehistoric.* E. Heron-Allen, London, 1911.

[2] 'Selsey or Pagham Harbour', *Sussex Arch. Colls.* 53, 1910, 26. See also H.R. Mill, *Geogr. Journ.* 15, 1900, 205.

those years another 312 acres were inned. It is not our purpose, however, to list the various banks as they were built, but rather to call attention to the shingle bank enclosing the entrance to the harbour. Cavis-Brown gives the following figures of its growth from Selsey: 1672, 1445 yd.; 1774, 1540 yd.; 1823, 1766 yd.; 1852, 2260 yd. Unfortunately no information is given about the smaller spit of land running south-westwards from Pagham. On a smaller scale the two spits resemble those enclosing Poole Harbour, and, allowing for local differences, we may probably suppose that they are similar in origin.[1] There is no doubt that the general movement of beach material on the east side of Selsey is eastwards or north-eastwards; how far there may be a local counter-movement on Pagham Beach Estate is not clear.

(14) CHICHESTER TO BRIGHTON AND THE SUSSEX COASTAL PLAIN

In treating the coastal area between Chichester Harbour and Brighton it will be helpful first to study the present coast, and then to say something of the Sussex coastal plain and raised beaches. This is a somewhat unsatisfactory division of what is really a single unit, but is useful in that it separates present and past events which in most of this strip are not confused geographically.

The Bognor district is drift covered, and the only outcrops of solid rocks are between the tide marks. The foreshore rocks are mainly clays (London Clay), but outcrops of harder material appear at Barn rocks and Bognor rocks.[2] The former are a true sandstone outcrop, running as a double reef to the south-east. They are rather softer than the Bognor rocks which are beds of arenaceous limestone. Here again the reef is double, with water between, and the stone is hard. At both places the dip is low, about 3°, and inclined to the north-east. The rocks are for the most part only accessible at low water. At Felpham[3] the land is low, and as elsewhere in this area, forms part of the old marine plain reaching inland to the Downs. There are low cliffs, about 8 ft. high, of brickearth, which may sometimes be seen to rest on Coombe Rock. The mouth of the Aldingbourne may at one time have been farther west, and Venables suggests that erosion has probably removed a considerable amount of land over which the stream may have had a westward course. In other words, local knowledge does not suggest that the easterly trend near the mouth is simply the work of beach-drifting. Erosion is serious along all this coast, and there are numerous groynes between Bognor and Brighton.

[1] See Lewis, 1938. [2] E.M. Venables, *Proc. Geol. Assoc.* 40, 1929, 41.
[3] E.M. Venables, *ibid.* 42, 1931, 362.

The Arun shows the usual eastward deflection clearly. It must be remembered that when its valley, like those of the other Sussex streams, was cut, the land stood higher. Since that time there has been a subsidence, and at one stage the Arun must have resembled a fiord. It is now silted up. Reid[1] compares this ancient and now silted harbour with that at Chichester, which still remains open, the difference being due to the fact that no stream flows into the latter. At Arundel, Chalk was reached under 100 ft. of marsh clay: even under 150 ft. at the Station Inn. It is, of course, possible that these figures may include a certain thickness of the Reading Beds. But at Arundel causeway 84 ft. of marsh clay have been proved, and 117 ft. at Warningcamp. These figures give some idea of the valley before silting brought it to its present state. This arm of the sea formerly reached as far inland as Pulborough. It is also said[2] that high water once reached Angmering church, and that the village of Ford takes its name from the ford across Binsted Brook to Tortington, and not from a passage of the Arun. It may be mentioned here that the course of the small Ferring Rife stream, like that of other minor rivers, is determined merely by slight irregularities in the drift and is not concerned with the underlying geological structure.[3] Near Angmering Chalk appears on the foreshore. Between Littlehampton and Worthing there are low cliffs of gravelly loam often buried beneath sand and shingle. The material of the beach travels eastwards, but not much passes across the mouth of the Arun. Up to this point the coast is still considerably affected by the shelter given by the Isle of Wight.

In the past erosion has been more serious at and near Worthing than it is at present. A map of Worthing Manor (1748) shows that a common existed for about one furlong to the south of the present beach. This has disappeared. Since early in the last century erosion has been checked, mainly by the building of groynes. This liability to erosion appears to be less farther west according to the following statement made to the Royal Commission on Coast Erosion: 'The seaboard under the jurisdiction of the Commissioners of Sewers for the Rape of Arundel is not liable to erosion or accretion to any appreciable extent.' Eastwards from Worthing shingle increases in abundance, and there is an enormous quantity held up at Shoreham Harbour. The Adur, whose early history generally resembled that of the Arun, is still deflected for just over a mile to the east (Fig. 65). In earlier times there appears to have been a wide strip south of the river on which the village of Pende was situated. At that time the

[1] *Mems. Geol. Surv.* 'The Geology of the Country around Chichester', 1903.

[2] A. Ballard, *Sussex Arch. Colls.* 53.

[3] E.C. Martin, *Proc. Geol. Assoc.* 45, 1934, 427.

river was navigable as far up as Bramber, but when (probably in the fifteenth century),[1] this southern arm was destroyed, the harbour entrance fell back to a point opposite Shoreham. There then began that part of the deposition of shingle which we see to-day. In the seventeenth century the harbour mouth was in much the same place as it is now; later, approximately between 1698 and 1760, the spit was presumably cut through, and the river debouched opposite New Shoreham (i.e. the present town: Old Shoreham seems to have been a little northward or north-eastward

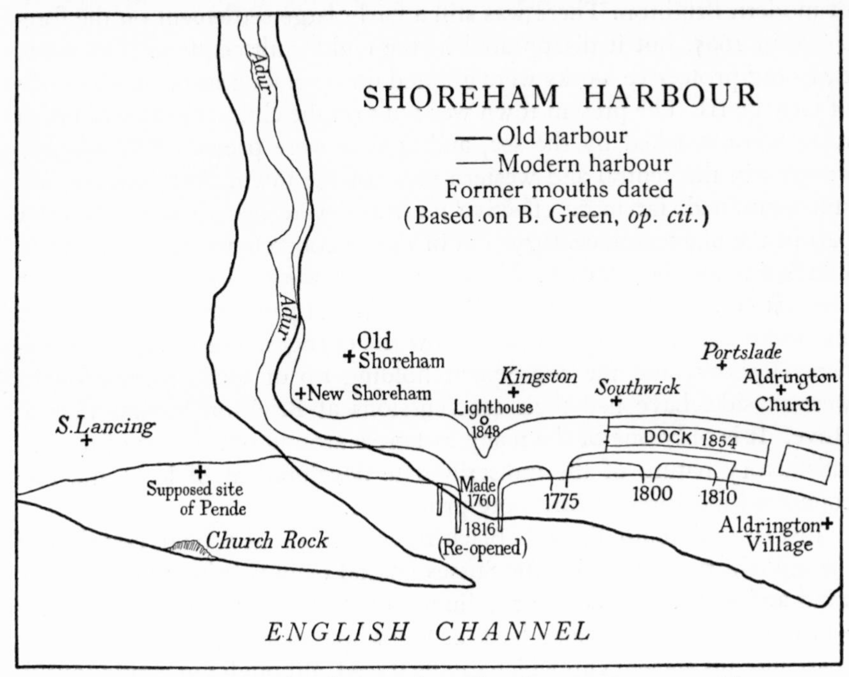

Fig. 65. Shoreham Harbour (based on B. Green, *op. cit.*)

of the present town). In 1760 a new opening was made opposite Kingston. But from this time onwards there set in a rapid lengthening of the spit so that in 1810 it ended opposite Portslade, but temporary openings had been made in 1775 and 1800 opposite Southwick. The outlet of 1760 was re-opened in 1816, and in 1821 harbour works were built to render this opening permanent. The shingle is piled up into fulls, and is partly stabilized by the planting of tamarisk. As in other similar places, the shingle is liable to be rolled landwards in high storms, and White states that near Lancing this inward movement between 1875 and 1891 amounted to distances varying from 70 to 320 ft.

[1] B. Green, *Sussex Arch. Colls.* 27, 1877, 69 and following footnote.

The harbour works at Shoreham considerably influenced conditions farther east, and not until groynes were built at Hove in 1885 was there local accretion at that place. At Brighton there have been much greater changes, and erosion has been serious since early historical times. The old town was built on the foreshore beneath the cliffs, and was known as Brighthelmstone.[1] It was a fishing port of note in Norman days. This Norman settlement stood farther south than its late mediaeval successor which grew up around the Pool valley and which was the direct ancestor of modern Brighton. There was still a fairly large settlement on the foreshore in 1665, but it disappeared in the eighteenth century. Powers to carry out protective works were granted on several occasions in the reign of George III. The present town was built on the cliffs, and these, in their turn, were attacked by the sea, and at first were protected by wooden groynes in the central and western parts of the town. But since erosion still continued, far more effective measures were necessary in the early part of the nineteenth century, and in 1803 a concrete wall was built from Old Steine to Kemp Town. Wooden groynes were still usual, but in 1867 the first concrete one appeared. This system of protection has continued and improved, but we need not follow the changes. The fixing of Shoreham Harbour, and the consequent holding up of much shingle on its western side, have had their repercussions at Brighton no less than at Hove. It is only one of the many instances around our shores of the unsatisfactory nature of the indiscriminate development of local and disconnected measures of foreshore protection.

Just east of Kemp Town is Black Rock, and about 200 yd. east thereof the old cliff running inside the Sussex coastal plain is being uncovered by wind and wave. But before continuing our survey of the actual coast, it will now be best to glance back at the whole area between the present coast and this ancient cliff which can be traced, although not continuously, westward to approximately the Hampshire border.

At Brighton, Portslade, and Worthing there is a raised beach[2] resting on a platform in the Chalk and Eocene beds. The platform rises northwards until it is about 30 ft. above O.D. and at that height meets an old cliff at the inner margin of the coastal plain. The cliff is usually buried beneath Coombe Rock, and, as noted above, the Black Rock is the only place where its face is visible. Elsewhere a change in gradient is usually the only means of clearly identifying this ancient cliff. At Worthing[3] the

[1] *Mems. Geol. Surv.* 'The Geology of the Country near Brighton and Worthing', H.J.O. White, 1924.

[2] *Ibid.*

[3] E.C. Martin, *Proc. Geol. Assoc.* 48, 1937, 48.

raised beach is seen at about 15 ft., and consists of sands and pebbles with numerous shells of Recent age. Its base is some 5–10 ft. above O.D., and rests on Eocene beds: at West Worthing it is on the Chalk. The beach can still be traced at Goring, and even as far as Ferring.

The 100-foot beach first appears about one mile west of Arundel. The true beach is composed largely of flints interstratified in false-bedded sands. Fowler[1] notes that it is associated with a bed of marine sand which might readily be confused with a Tertiary deposit as it is so massive and persistent. Between Arundel and Chichester the beach is on the dip-slope of the Chalk between the 70-foot and 130-foot contours. Fowler has not found the old sea cliff in this area: it may be buried beneath Coombe Rock. The bluff separating the higher and lower raised beaches makes its first appearance west of Arundel near the 50-foot contour; it seems to mark the inner limit of the lower beach. The lower beach may not necessarily be absent west of the Arun, it is possibly obscured beneath the Coombe Rock and brickearth. It was during the 15-foot beach stage that, according to Martin,[2] the Chalk downs between Worthing and the Arun were removed, except for Highdown Hill.

The higher beach can be traced to Chichester and beyond. Although called the 100-foot beach, its deposits vary from 80 to 130 ft. above O.D. Oakley and Curwen[3] interpret recent archaeological evidence as suggesting that this range of deposits implies not one, but two distinct levels: a 90-foot beach and a 135-foot beach. Together they certainly mark a period of rising sea level which led to aggradation in all the southern rivers during Clactonian and mid-Acheulian times. This submergence was almost certainly inter-Glacial, but the later fall in sea level coincided with solifluxion deposits and a cold period. It is likely that the first period of solifluxion happened before the sea fell below 120 ft. In the Portsmouth area Palmer and Cooke[4] recognize three terraces at 15, 50, and 100 ft. respectively, each being marked by a bluff or degraded platform. The 100-foot gravels yield a warm fauna, occasional erratics, and Acheulian implements. The two lower gravels, on the other hand, possess a cold fauna, many erratics, and Mousterian implements. Palmer and Cooke believe also that the estuarine and marine beds at Selsey, Wittering, and Stone were contemporaneous with the 100-foot gravels.

[1] Rev. J. Fowler, *Quart. Journ. Geol. Soc.* 88, 1932, 84.

[2] E.C. Martin, *Proc. Geol. Assoc.* 49, 1938, 198.

[3] K.P. Oakley and E.C. Curwen, *ibid.* 48, 1937, 317.

[4] L.S. Palmer and Lt.-Col. J.H. Cooke, *ibid.* 34, 1923, 253. See also C. Reid, 'The Pleistocene Deposits of the Sussex Coast...', *Quart. Journ. Geol. Soc.* 48, 1892, 344, and F.H. Edmunds, 'The Coombe Rock of the Hampshire and Sussex Coast', *Summary of Progress, Geol. Surv.* (1929), pt 2, 63, 1930.

The results...also lead to the following general conclusions with regard to oscillations of level in this district, namely, that at a given time the land was at least 100 feet lower with respect to sea-level, and that an elevation occurred at, or very soon after, this time, resulting in the deposition of the 15-foot estuarine deposits and lower fluviatile gravels. Subsequently the land was again depressed, but only by about 50 feet. This was followed by a rise, which took place in two movements at least. During the first stationary period the 15-foot beach was formed, and during the second stationary period the land was elevated some 70 feet above its present level causing the formation of the now sunken cliffs and buried river channels of the Arun, etc. The presence of the Neolithic forests beneath the sea shows that this recent elevation was followed by a depression to the present level, which depression is probably still in progress.[1]

The 100-foot level is now covered by three layers of Coombe Rock, the intermediate level by two, and the lowest by one layer. Each layer indicates a cold period. In western Sussex and eastern Hampshire the Coombe Rock is spread out south of Ports Down for an unknown distance, and disappears beneath the brickearths of Portchester and Farlington marshes.

Thus the Sussex coastal plain extends westward from Brighton, and in so doing it also runs farther inland, behind Worthing, to Goodwood Park, Chichester, and Ports Down. Its inner edge is closely coincident with the southern margin of the South Downs, but this margin is more often than not hidden by Coombe Rock and other superficial deposits. Thus the plain widens westwards and is on the Chalk and Eocene strata which were planed down to a level only a few feet higher than modern sea level. It is now mainly covered by rubble and raised-beach deposits, and east of Brighton modern erosion is cutting into these incoherent materials. The details are not yet clear, and a closer correlation between the levels recognized by Oakley and Curwen in Sussex, and Palmer and Cooke in Hampshire, is at least worth further investigation.[2]

(15) BRIGHTON TO EASTBOURNE

A mile or so to the east of Brighton the Chalk comes to the shore and forms cliffs. Between Brighton and Rottingdean there has been a good deal of erosion,[3] and this, as the Royal Commission noted, is due very largely to groynes holding the material at Hove and Brighton. In storms

[1] Palmer and Cooke, *op. cit.*

[2] See also in this connection and in other matters, C. Reid, 'Mem. on Geol. Conditions....The South Coast Generally', First Rept. Royal Comm. Coast Erosion, vol. 1, 1907, Appendix No. 12 (B).

[3] A sea wall now extends from Black Rock to Rottingdean: a wave-cut platform is well seen in front of the cliffs. Coombe Rock occurs in little valleys in the Chalk at Rottingdean.

the direct effect of wave erosion on the cliffs is also noteworthy, and this is helped by the fact that surface drainage water is (or was) carried over the cliff edge. The Ouse outlet is now regulated at Newhaven; at one time shingle had deflected the mouth as far east as Bishopstone. The Ouse has a wide mouth, but a few miles upstream it is contracted, although it widens out below Lewes and near Glynde into a large basin. Much of Lewes itself is actually below high-water mark, and the tide flows to Hamsey Place Farm on the main river, and to one and a half miles above Glynde bridge. Just as in the other Sussex rivers at the end of the Pleistocene the Ouse bed near Lewes[1] was at least 40 ft. lower than it is now. This was followed by a rather quick depression which led to the flooding of the lower ground and also to active sedimentation, so that during (about) the Bronze Age there were extensive tidal flats which were later reclaimed. The Rises are salient features in this low-lying area. The general evolution of the Cuckmere was similar: high water would have flooded as far as Litlington or even Alfriston. The fine meanders and river cliffs near its mouth deserve to be better known.[2] The river is also deflected a little eastwards by a shingle spit, and, in point of fact, much of the Cuckmere drainage has been diverted to the Ouse. Elsden[3] explains the break in the continuity of the cliff between Seaford Head and Newhaven by the fact that the coastline here intersects the trough of a basin: 'Seaford Head is the last remnant of a limb of the uniclinal fold, the trend of which is lost westwards in the Channel. The Newhaven cliffs are on the north margin of the basin, and the apparent horizontality of the beds there is due to the fact that the cliff approximately follows the direction of the strike.' The line of cliffs, known as Seven Sisters, runs parallel to the strike of the Chalk. Originally, the Chalk extended farther to the south, and was drained by valleys running down its dip-slope. Marine erosion has cut away much of this former land, and so the present cliffs show a profile section of the ground parallel to the strike and at right angles to the valleys. Since there are few open joints in the rock, and since the Chalk is very homogeneous, the cliffs are cut cleanly and nearly vertically (**11**).

Beachy Head, where the cliffs are more than 500 ft. high, is one of the prominent Chalk headlands of our coasts. The highest part faces south and is exposed to south-westerly gales; consequently it is undercut. The rate of undercutting is sufficiently fast to maintain a vertical cliff face. At

[1] *Mems. Geol. Surv.* 'The Geology of the Country near Lewes', H.J.O. White, 1926.

[2] At Cuckmere Haven and near Brighton and along many parts of the East Sussex coast there is a very fine wave-cut platform in Chalk.

[3] J.V. Elsden, *Quart. Journ. Geol. Soc.* 65, 1909, 442.

the Head itself the cliffs are perpendicular and formed of Upper, Middle, and Lower Chalk: the dip is to the west. Erosion is most active along joint planes, and in this way produces chimneys. Madock[1] states that formerly there were seven towers of Chalk standing out from the cliff, and called the Seven Charleses: the last fell in 1853. The nature and appearance of the cliffs between the Head and Eastbourne are very different. The vertical cliffs give place to steep curves, and there are many quite complicated structures in the shore platform which is made up of Lower Chalk, Upper Greensand, and Gault Clay. The westerly dip gives an exposure of Upper Greensand called Head Ledge which curves towards the cliffs in a manner suggesting a pitching anticline, the lower part is cut off by a thrust-plane. The Upper Greensand reappears at Cow Gap. Nearly all types of fault are represented on this platform. 'Whatever may be the cause, an inclined sheet of Upper Greensand has been broken into pieces and these forced under one another, whilst the Gault and softer Chalk flowed round them as the movement occurred.'[2] The recession at Holywell results largely from springs in the *Plenus* marls which flow out on to the beach. Eastbourne itself is mainly built on Coombe Rock which covers the Upper Greensand and Lower Chalk. Local groyning has resulted in the accumulation of a considerable area of beach.

(16) LANGNEY (LANGLEY) POINT AND PEVENSEY LEVELS

The interesting shingle foreland called Langney or Langley Point (Fig. 66) begins at Eastbourne. It consists of series of shingle fulls trending approximately north-east and south-west with a more eastward trend near Pevensey. Many fulls are more than a mile long. The pebbles are mainly of flint, chert from the Upper Greensand, and fragments of hard Chalk. The non-flint pebbles, in particular, get smaller towards the north-east. There are also pebbles of more distant origin, quartzites and quartz-tourmaline rocks probably derived from the Pebble Bed at Budleigh Salterton. There are also some grit pebbles and debris of igneous rocks. The whole foreland has grown to the east and north-east. It probably began as a spit or bar across a re-entrant in the coast when the sea flooded the low ground within. The foreland rests on Gault and Weald Clays, and these formed the foundation on which the primary spit began to form. The southern ends of the present fulls are all cut short, and the actual point has probably migrated along the spit rather in the manner of the points of Orford Ness

[1] Rev. H.E. Madock, 'Changes in the Coastline, especially between Beachy Head and Hastings', *Eastbourne Nat. Hist. Soc.* 19 February 1875.

[2] H.B. Milner and A.J. Bull, 'The Geology of the Eastbourne-Hastings Coastline', *Proc. Geol. Assoc.* 36, 1925, 291, and *ibid.* 48, 1937, 329.

and Dungeness. The district was carefully mapped by Desmaretz in 1736, and a comparison of his map with the Ordnance Map of 1844 shows that in 1736 the line of high-water mark was a good quarter of a mile to the south-east of Wish Tower. This is taken by Madock[1] to mean that, unless an old sea beach (of which there is no record) existed here as at Old Brighton, the cliff must have extended about a quarter of a mile to the south-east of its present position.

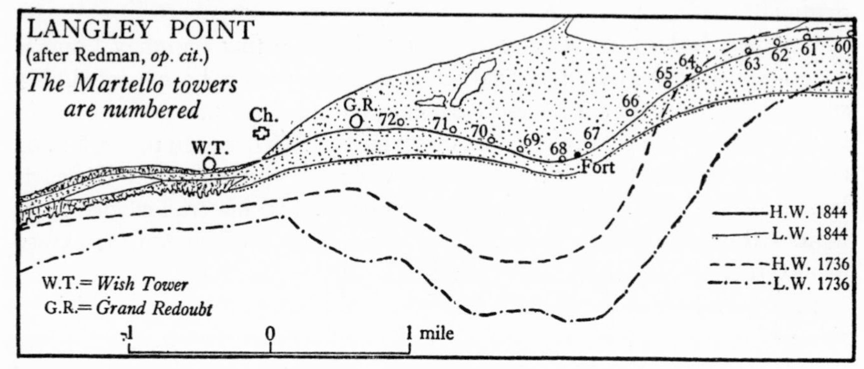

Fig. 66. Langley Point

On the other hand, Nowell's map of 1576 shows nothing whatever resembling Langney Point. This suggests several possibilities, but the most likely interpretation is either that there was very little shingle at that time, or that the ness had not begun to form.[2] Budgen's map of 1724 shows Langney Point projecting about one and three-eighths' miles, and in 1736 (Desmaretz) this had increased to a mile and a half. The idea has often been put forward that the destruction of Brighton beach between about 1665 and 1705 and the growth of Langney Point may be interconnected. It is doubtful if any positive proof can be found for this view, but it is certainly possible. It is not inconsistent with the later history of Langney: if, only for the sake of argument, we continue to accept as reasonably accurate the cartographical evidence, we may note that in 1778 (Yeakell and Gardner) the ness had shortened to one and a quarter miles, and the Admiralty survey of 1844 shows it as only seven furlongs in length. Erosion has continued since that time, and a certain number of precise measurements are obtainable from the situation of the Martello Towers

[1] *Op. cit.*; see also Milner and Bull (1925), *op. cit.*; also J.B. Redman, on the 'Alluvial Formations...of the South Coast...', *Mins. Proc. Inst. Civ. Engs.* 11, 1851–2, 162.

[2] Collins's plan does not show the ness, and it is not mentioned by Camden. These two facts are at any rate consistent with what is known of its positive history.

which were built in 1806. Writing in 1875, Madock notes that in 1806 towers numbers 72, 71, and 70 were (from their centres) 143, 184, and 177 ft. from high-water mark. No. 72 disappeared some long time before 1875; No. 71 became untenable in 1860 and was destroyed; No. 70 was below high-water mark; No. 69 was breached. On the other hand, new shingle has accumulated in front of Nos. 62–59. Farther east, between Pevensey and St Leonards all towers had been washed away or condemned.

On the whole, the evidence appears to indicate that Langney Point is a comparatively modern formation which, as would be expected, has suffered most erosion at its southern end, and has also prograded slightly in the east. In 1925 there were six or seven beaches between tower No. 62 and the actual beach. It is nearly a mile wide at its broadest, and there has been no change of level since it began to form. The material eroded from it now travels along the coast and eventually feeds Dungeness: except locally there is not much accumulation between the two nesses.

Pevensey Levels form the site of a former inlet, partly divided by a ridge of higher ground running from Polegate to Pevensey. There were also other islands, as at Horsey and Chilley. At a later date, two main streams, the Ashburn (= Waller's Haven) and Hurst Haven, together with numerous tributaries, drained the area to Pevensey Haven, where they united to form Pevensey Harbour. Later, the Ashburn was diverted to its sluice outlet at Northeye. All this marshland area averages about 8 ft. above mean sea level, and, in general, the lowest ground is now found around its margins. Salzman[1] has described the inning of the levels, but notes that a good deal must depend upon circumstantial evidence. Quite early in the process of eastward beach-drifting, the Bowen Level stream was blocked at Langney, and was diverted to the common outlet at Pevensey.

There is no evidence of any draining by the Romans. There are three main groups of levels: Willingdon, Pevensey, and Hooe. These at one time drained separately to the sea, at Langney, Pevensey, and Northeye respectively; the first two now find exit at Pevensey.

Anticipating the results of an examination of the documentary evidence, we may say that the main changes in the methods of draining the levels were as follows: (1) From the first inning down to the end of the fourteenth century all the levels drained out at a point on the borders of Pevensey and Westham, due south of the Castle. (2) In 1396 a large cut was made from Fence Bridge to Wallsend, to replace the former outlet. (3) In 1402 the greater part of the Ashburn, draining Hooe Level, was diverted to the Sluice. (4) The diversion of the Hooe drainage was completed in 1455 by a new

[1] L.F. Salzman, *Sussex Arch. Colls.* 53, 1910, 32.

ditch in Northeye. (5) As a result of the diminished volume and decreased scour of the water the outlet at Pevensey silted up, and the mouth of the Haven was forced eastward until it eventually reached the sluice, though on a number of occasions during the sixteenth and seventeenth centuries its mouth was re-opened at different points between Wallsend and the Sluice.[1]

The evidence of salt pans suggests that the mouth was flooded at the time of the Domesday assessment, but a good deal of inning had been done by 1180. Most of Mountney Level was reclaimed by the middle of the thirteenth century, and after 1300 embanking was fairly quick. Thus, as more and more land was saved, the volume of water lessened and so did the scour in the harbour. The Winchelsea storm of 1287 was apparently severe in this area also. In 1402 Waller's Haven was diverted eastwards to a point near the present sluice: the new channel to the sluice dates from 1455. There were incursions of the sea in 1428, 1469, 1481, and 1542. About the time of the Armada a careful survey was made of the Sussex coast; from it we may deduce that the haven mouth had been forced eastwards a considerable distance, and also that much of the water which formerly ran out of the haven now did so at Northeye sluice. By 1609 the blocking of the haven had become very common, and 'within living memory' its eastward deflection amounted to half a mile or more. In short the sluice really became the port and replaced the old haven. In about 1580 the export of iron began from the haven (it was stored close to Pevensey Bridge), and in this way the old port for a short time recovered activity. In 1633 a survey mentions 50 acres of salt marsh between the sluice stream and the haven: there are similar records in 1663 and 1696. In 1698 the Harbour Commissioners made a careful map which showed that the harbour was then about one and a quarter miles south-east from Pevensey Castle. The old outlet was marked close to the site of the present hotel at Wallsend. The same Commission also stated that up to about 1690 vessels of 50 or 60 tons loaded at the town bridge, but at the time they reported they added that a 14-ton vessel could only with difficulty enter the haven's mouth. Thus, by about 1700 the state of the levels was in the main similar to that of the present day.

Turner[2] calls attention to the lost towns of Northeye and Hydneye. The termination -eye probably signified an islet in the marshes. Northeye was in Bexhill marshes, and at the time it was built it would seem that they were flooded. So far as can be estimated the town stood near the marsh edge under the shelter of Barnhorne Hill. It was probably overwhelmed by the sea at much the same time as Winchelsea (1287). It may have stood

[1] L.F. Salzman, *Sussex Arch. Colls.* 53, 1910, 32.

[2] Rev. E. Turner, *Sussex Arch. Colls.* 19, 1867, 1.

at the mouth of Hooe Haven which was open up to 1748. The position of Hydneye is very vague, and is usually described as between Pevensey and Eastbourne. Probably it was an eye in Pevensey Marsh. Turner supposes it to have been in what is now Eastbourne, which he thinks was a small port in early times. The mouth of its port was the lowland close to the Wish. Both Northeye and Hydneye were limbs of the Cinque Ports.

(17) BEXHILL AND HASTINGS

Hooe Level is on the Weald Clay. The low cliffs at Cooden are in the shales, clay, and sandstone of the Wadhurst Clay. They are obscured too much by bungalows to allow of their structure being fully investigated. It is, however, very probable that the cliffs are disturbed. A little farther east the Ashdown Sand forms a cliff 20–60 ft. high for about one mile. Bexhill is mainly built on this formation, but the sea wall obscures detail on the coast itself. Natural exposures are seen at low water. At Galley Hill there has been a good deal of erosion and land-slipping: the isolated Ashdown Sand stack near Glyne Gap Halt has decreased by erosion in recent years. The Ashdown Sand continues through Bulverhythe, West Marina, and St Leonards (Goat Ledge). The next eastward ledge, however, is of Tunbridge Wells Sand which forms all the reefs as far as Hastings pier.

There have been several interesting changes in historical times at Hastings, and these have been made the subject of a careful study by Ward.[1] Where the Forest Ridge reaches the sea there are several small valleys: at Hastings is the Priory valley, farther east are the Bourne, Ecclesbourne, and Fairlight valleys (Fig. 67). The alluvial tract of the Priory valley is now converted into the Alexandra Park. The present coast is fairly straight and there is throughout a good beach of sand and shingle. Clearly, all the cliffs formerly extended farther seawards: they are usually fairly steep, and at Fairlight recede about a yard a year. The parades naturally protect the cliffs behind them. Hastings existed in Saxon times, and the eleventh-century settlement probably stood on low ground on the west side of the mouth of the Priory valley. All this has now been lost to the sea. The White Rock, which has also disappeared, was then a bold headland east of the town. West of this rock there were low cliffs (now partly visible behind houses), small bays, and undulations, the upper parts of which are now covered by the streets of St Leonards. Here, or a little farther west, was the Asten valley. It is likely that the

[1] E.M. Ward, 'The Evolution of the Hastings Coastline', *Geogr. Journ.* 56, 1920, 107.

Bourne valley was not an arm of the sea, as the town that later grew there had no haven of its own.

The coast near the Bourne valley, on the other hand, has not been greatly modified. Eastern Hill is now protected by groynes and a good deal of shingle has gathered. The wall of the town, which almost reached the beach, appears to have suffered no loss through erosion, and 'to-day a wide bank of shingle...extends in front of the town, which itself has expanded seawards from the line of its own sea wall'. In course of time the Saxon town on the western side of the Priory valley declined, partly

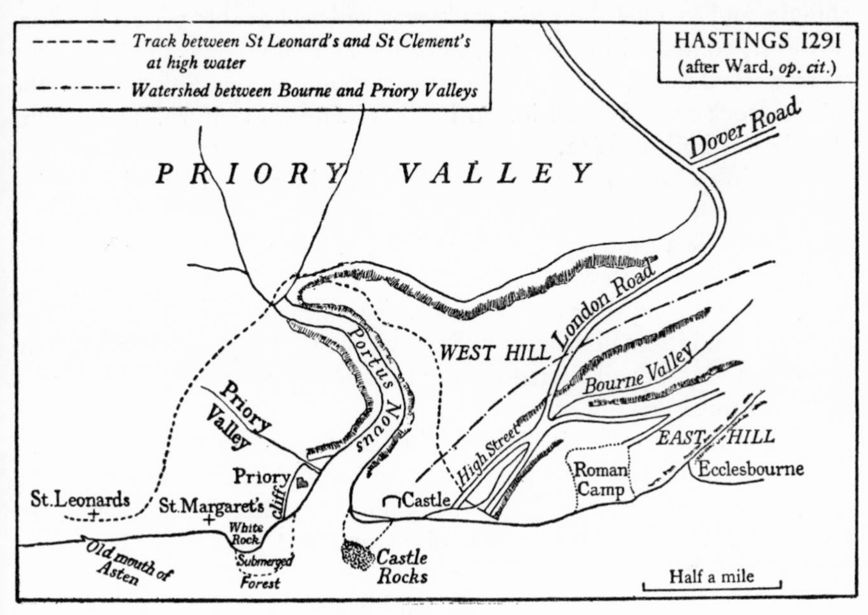

Fig. 67. Hastings in 1291

owing to erosion. Records suggest that it depended upon the protection of sea walls. The Priory buildings had to be evacuated by 1412, and Ward interprets the available evidence as meaning that by this date the low-lying region at the valley's mouth, and extending inwards to the modern railway station, had either been washed away or had become too swampy for habitation. By 1440 erosion was considerable at St Leonards and St Margaret's, and the Castle Hill was receding. By 1380 the eastward drift of shingle had deflected the Priory brook to a course below Castle Hill and a new settlement grew up on the western part of this shingle, but nevertheless from this time onwards the Priory valley itself remained uninhabited, until, in the mid-seventeenth century additions to the old town of the Bourne valley began to spread in that direction. The westward increase, however, was very slow, and the Priory valley eventually silted up more or less completely.[1] A small tidal inlet, it is true, still existed in the sixteenth century, and a haven appears on a map of 1746, while at the end of the century tidal water penetrated for a mile or so up the valley. This was still a waste area in 1746, but after that time, as silting continued, and as the shingle belt widened, reclamations took place, and the present streets were either laid out or planned by 1850. In 1883 the uneroded part of the White Rock was artificially removed: its site is marked by the curve in the parade at the eastern end of the baths. Groynes continued to trap shingle, and in 1878 the promenade was widened at this place. Erosion, however, continued in the west, since the Martello Towers built in 1805 were washed away, and the main road through St Leonards west of the White Rock was diverted inland in the nineteenth century. The Marine Parade was built in 1812.

The cliffs at Hastings are of pale yellow Ashdown Sand, and the massive sandstones of the Castle Hill and elsewhere have made them famous. There may be a capping of Wadhurst Clay. The development of strong joints trending north-eastwards has played no small part in their evolution, especially in their influence on the formation of caves. Ecclesbourne Glen is a small hanging valley over sandstone ledges. At Fairlight Glen are the clays and alternating sandstones of the Fairlight Clay group. They are much overgrown and slipped. The glen itself also hangs to seaward, and the small fall is discharged over sandstone ledges in the clays. The cliffs to the east show a fine section 'of rapidly alternating mottled clay, white, yellow, and deeply ferruginous sandstones, grey and carbonaceous clays, together with lignite seams and pockets'.[2] There also appears to be a fault parallel with the cliffs, because the reefs of sandstone and clay are discordant in their relationships with the cliffs. At Cliff End the massively and vertically jointed sandstones reappear. Toot (or Tout) Rock is a remnant of Ashdown Sand formerly washed by the sea and a little modified by ancient cliffing on its south-eastern side. The submerged forest at Cliff End rests on Fairlight Clay and Ashdown Sand (see **10**, **9**, 8).

(18) ROMNEY MARSH AND DUNGENESS

At Cliff End the cliffs run inland, and can be traced, apart from river gaps, all round the inside of Walland and Romney Marshes to Sandgate. This great marshland area with the huge spread of shingle at Dungeness is a

[1] It seems to have become nearly useless for navigation as early as the twelfth century.

[2] Milner and Bull, *op. cit.* 1925.

unit. It has been the subject of numerous papers, and provides many problems, not all of which are by any means satisfactorily settled, partly because the approach made to these problems has been almost exclusively historical, geological, or that of the engineer; only in recent years has there been some attempt at a more comprehensive view. In this kind of region, natural changes are going on in several ways at the same time. They precede the artificial manipulations of man, and continue alongside human activities. In order to try to achieve a clearer picture Dungeness, *sensu stricto*, is treated separately from the marshes, but it must always be remembered that these areas have closely interrelated features. The problems common to both, which include the alterations in the river courses, demand studies and policies simultaneous and inclusive in scope. Dungeness is the greatest shingle structure in these islands.

Fig. 68 shows the main points in the distribution of the shingle ridges. There is first the simple beach running north-eastwards from Cliff End towards Rye. This divides into three main groups of ridges, on the innermost of which is Camber Castle, built by command of Henry VIII in 1538/9. On the eastern side of Rye Harbour the ridges lie at first in quite distinct groups—the Midrips, the Wicks, Holmstone Beach, Lydd Beach, and the smaller unnamed group east of the inn. All these are comparatively low, and their general trend is shown on the maps. (The significance of high and low sets of ridges is discussed below.) Denge Beach proper begins at the point *K*. Both the inner and outer groups are high, but roughly in the middle of this wide tract is a set of lower fulls. In all it will be noticed that they are cut off abruptly by the sea at their southern ends and that many swing round to run approximately parallel to the eastern shore. There are local complications, especially near the ness point. Great Stone Point is the northern limit of these ridges. There is another large mass of fulls at and near Hythe. These show numerous recurved ends running north and north-west, towards the old cliff. Between Hythe and New Romney is the Dymchurch Wall which also rests on ridges; most of these are no longer visible, but they reappear south of Dymchurch. The angle at which the ridge groups between the Midrips and the point *K* meet the shore shows considerable alteration. Since at the time of its formation each individual ridge was the outer shore of the ness, we have in the trend and disposition of the ridges a means of reconstructing in part the earlier conformations of the ness.

The marshlands are all within the ridges. Pett Level and Rye Marshes are west of the Rother. Walland Marsh is enclosed by the Holmstone Beach-Lydd ridges and the Rhee Wall, and Romney Marsh lies wholly north of the Rhee Wall. Denge Marsh is the newest and is shut in between

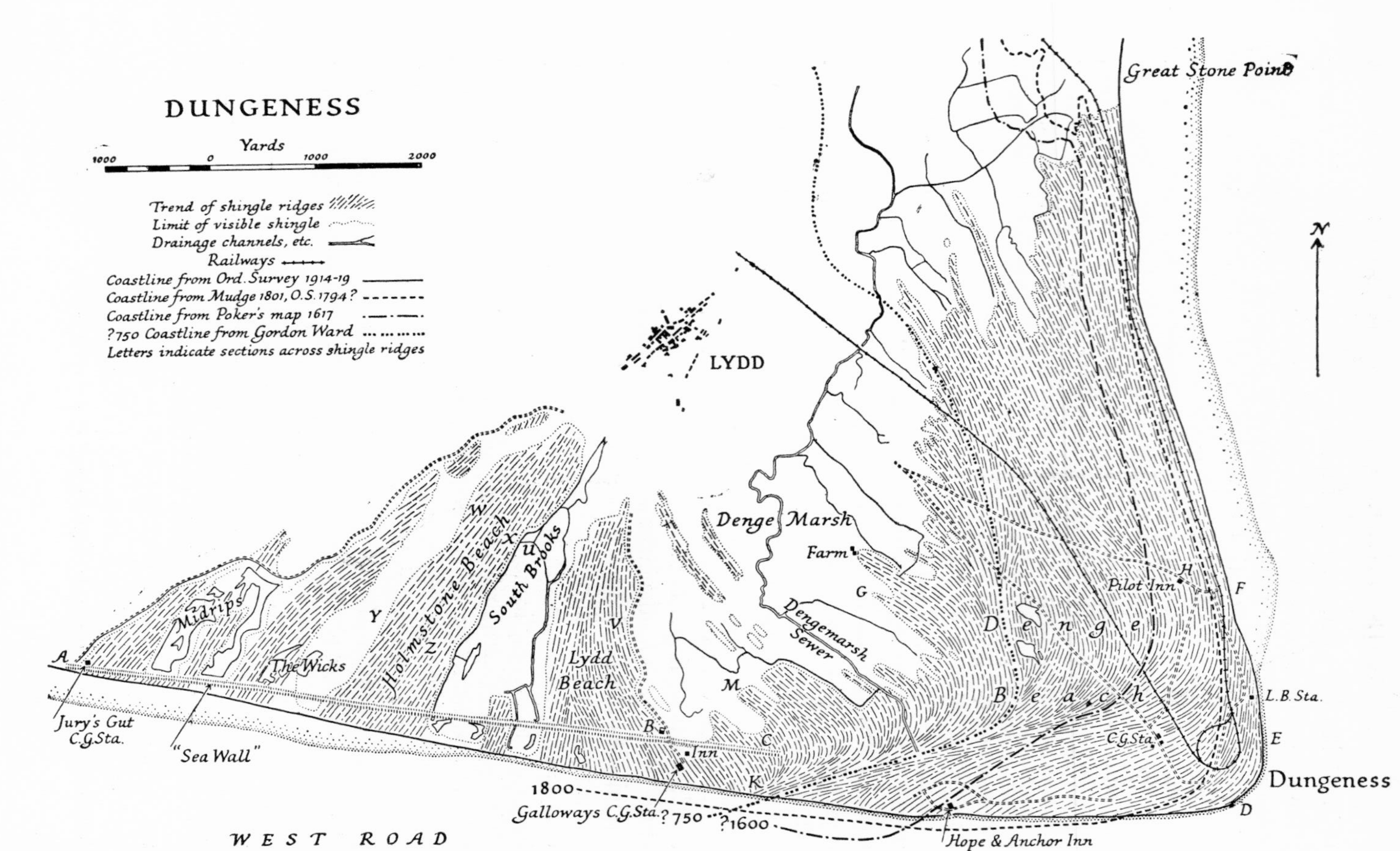
DUNGENESS
Yards
1000
0
1000
2000
Trend of shingle ridges
Limit of visible shingle
Drainage channels, etc.
Railways
Coastline from Ord. Survey 1914-19
Coastline from Mudge 1801, O.S. 1794?
Coastline from Poker's map 1617
?750 Coastline from Gordon Ward
Letters indicate sections across shingle ridges
Great Stone Point
N
LYDD
Denge Marsh
Farm
Dengemarsh Sewer
Denge Beach
Pilot Inn
L.B. Sta.
C.G.Sta.
Dungeness
Hope & Anchor Inn
Galloways C.G.Sta.
1800
?750
?1600
Inn
Lydd Beach
South Brooks
Holmstone Beach
Midrips
The Wicks
Jury's Gut C.G.Sta.
"Sea Wall"
WEST ROAD
A
B
C
D
E
F
G
H
K
M
U
V
W
X
Y
Z

Lydd, the ness and Great Stone Point. The rivers Rother, Brede, and Tillingham break through the old line of cliffs, and long arms of alluvium run up their valleys. Oxney Island lies between two arms of the Rother. The level of the marshes varies within small limits: usually those reclaimed earliest are the lowest. Appledore Dowels, artificially drained, is the lowest part of all. Part of the controversy about the nature of the evolution of the marshes, a physiographic problem in some ways distinct from that of the shingle ridges, turns on the possible existence of a former island at Lydd and also on the much-debated matter of the river Limen. Did this river flow out to the sea near Lympne at the old Portus Lemanis? The general trend of modern physiographical and engineering work is opposed to such a view, but Walker has adduced certain historical arguments in its favour **(6)**.

It is convenient first to consider the shingle ridges, and the general formation of Dungeness. Earlier writers, such as Drew,[1] suggested the meeting of the tides of the Channel and the North Sea as accounting for the throwing-up of the ridge. Redman,[2] who described it at some length, really offered no explanation of the ness. Topley, in the famous memoir on the Weald, gave no reason for it. Gulliver[3] ascribed it to tidal currents. As his views are given considerable prominence in the official *Geological Memoir*[4] of the district, some brief comment on them is necessary. He supposed that the present cycle of shore erosion started after what he called the Neolithic submergence. The inner cliffs preserve this coast inside the marshes, but in reconstructing the cliffs at Fairlight and Folkestone it is generally conceded that Gulliver imagined them extending too far seawards; he therefore assumed that far too much has been lost by erosion. Between the enclosing headlands of the old bay he supposed a bay bar formed. Gulliver thought that it was pushed back in its subsequent evolution and eroded particularly at its southern end (keeping pace with the erosion of the cliffs at Fairlight), and that in its mid-portion it became more and more cuspate (see Fig. 69). His views concerning its northern ends are even more open to question, and their improbability is clear from a study of the map. The origin of the ness he puts down to tidal currents. Wheeler also invokes tidal eddies. The relative competence of tidal currents and waves in the formation of shingle forelands has already been discussed (Chapter III). All that need be said here is that modern work on Dungeness strongly suggests the importance of wave action, but there

[1] *Mems. Geol. Surv.* 'Folkstone and Rye...', 1864. [2] *Op. cit.* 1851–2.

[3] F.P. Gulliver, 'Dungeness Foreland', *Geogr. Journ.* 9, 1887, 536.

[4] *Mems. Geol. Surv.* 'The Geology of the Country near Hastings and Dungeness', H.J.O. White, 1928.

is no idea of overlooking the significance of tidal *currents* in the movement of shingle below water. There is not the slightest ground for considering Dungeness to be the meeting-place of North Sea and Channel tides, nor for supposing its formation to be the work of tidal eddy currents. Lewin,[1] Montague Burrows,[2] Drew,[3] Burrows,[4] and Elliott[5] all suggest that there

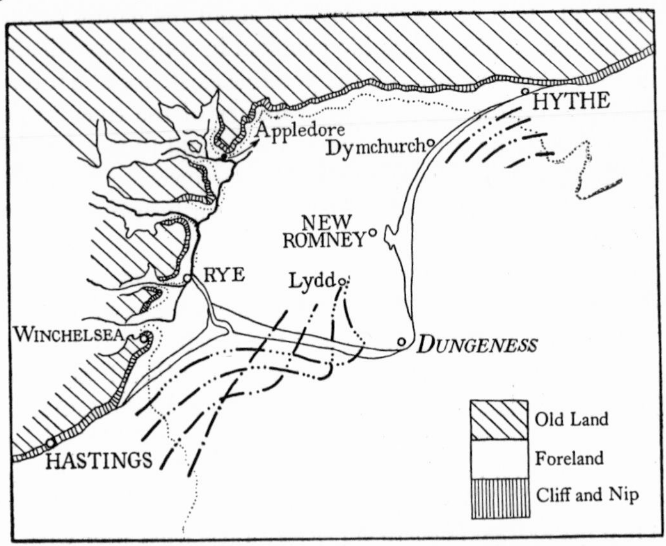

Fig. 69. The evolution of Dungeness (after Gulliver, *op. cit.*)

was originally a continuous line of shingle from Fairlight to Folkestone. To-day this can be traced through Lydd to the Dymchurch Wall. Lewis,[6] who has provided the most recent and comprehensive treatment of the shingle problems, also concurs in this, but is not necessarily in agreement with certain writers as to the date at which the line of shingle was formed. In discussing the arrangement of the shingle fulls, it must be emphasized that Lewis and Gilbert[7] also consider the likelihood of lesser but significant

[1] T. Lewin, *The Invasion of Britain by Julius Caesar*, 2nd ed. 1862.

[2] M. Burrows, *The Cinque Ports*.

[3] F. Drew, *Mems. Geol. Surv.* 'Folkestone and Rye', 1864.

[4] A.J. Burrows, 'Romney Marsh, Past and Present...', *Trans. Surveyors' Inst.* 17, 1884–5, 335.

[5] J. Elliott, 'Account of the Dymchurch Wall...', *Mins. Proc. Inst. Civ. Engs.* 6, 1847, 466.

[6] W.V. Lewis, 'The Formation of Dungeness Foreland', *Geogr. Journ.* 80, 1932, 309.

[7] C.J. Gilbert, 'Land Oscillations during the closing stages of the Neolithic Depression', *Second Rept. Comm. Pliocene and Pleistocene Terraces, Int. Geogr. Union*, 1930.

vertical movements since the so-called Neolithic depression. Gilbert's views can be summarized in a simple diagram.

1 = Neolithic Depression
2 = Uplift for —
3 = Forest Bed and Neolithic occupation period
4 = Depression
5 = Uplift for —
6 = Era of Roman occupation
7 = Depression
8 = Present sea level

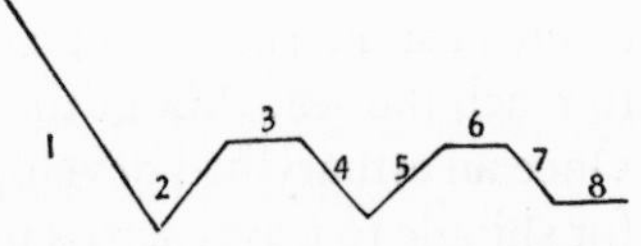

Fig. 70. Diagram illustrating earth movements in Romney Marsh

Lewis, having in mind recent work by Jolly, Godwin, and others, sees in the average heights of ridge-groups, evidence for still later, post-Roman, vertical movements.

In Neolithic times, as now, the main supply of shingle came from the west, and it is reasonable to suppose that the earliest ridges began near Fairlight, when that headland extended rather farther seawards, and the shingle travelled north-eastwards. These early ridges no longer exist, but if we assume that they evolved on lines similar to the much later Camber ridges, there are grounds for showing them diagrammatically as *AB*, *AC*, and *ADE* in Fig. 71. The ridges marked *DE* remain to-day just west of the Midrips, and these are the oldest in existence. Rather later is the line marked by Lewis *AFG*. This is the long and important shoreline, or bay bar, quoted by the early authorities. In Lewis's view it followed on the first uplift suggested by Gilbert, and it is noteworthy that the *DE* ridges are some 8 ft. lower than those on the present ness foreshore. When the full extent of this period of uplift was felt, it would not have been difficult for the waves to build a continuous ridge right across the bay on its shallow or even partly exposed sandy floor. On the other hand, it may have formed as a spit constantly extending from the south-west; this is more probable since the shingle was supplied mainly from that direction. What is not known is the measure of its extension towards Hythe before the land returned to its present level. The height of the recurves near *G* suggests that they are later, and that they were built when the sea was at its present level. If so, there was evidently a very long, and presumably not uninterrupted period, between the beginning and end of this shore, especially since Oldham[1] and others are of the opinion that the return to the present level did not take place until the thirteenth century.

Before dealing with the Hythe problems Lewis's views on Dungeness Foreland need examination. It is wise to stress again his contention that

[1] R.D. Oldham in discussion *Geog. Journ.* 69, 1927, 46 and see also *Quart. Journ. Geol. Soc.* 86, 1930, 64.

shingle ridges show a tendency to turn at right angles to the dominant winds and waves. As the sea rose again it probably breached this old shore near the site of New Romney, which became the outlet of the Rother, and again near Fairlight; the second gap enabled the Tillingham and Brede to reach the sea. About the same time the bend marked *K* began to form. Once an estuary had developed near Fairlight it would have been less easy for shingle to travel across it, and so the shore to the east of the river would have been starved, and would have retreated more readily under wave action. Some shingle, however, would still have come, and so parallel ridges could have been built. Since this shingle was no longer attached to the headland at Fairlight, it would have tended to swing round to face the dominant waves, and to have adopted a form such as *HKL*. Further driving back would have accentuated the bend at *K*, and gradually the shape would have become more and more like that which we see to-day. But the sharper the bend became, the less the amount of pebbles that would have reached its lee side; a fact that can be demonstrated to-day on the ness point. From this summary the lines in Fig. 71 showing the probable subsequent evolution are now clear.

The sharpness of the point is explained by the sudden change of exposure: large waves can reach it from the south or from the east, so that the two shores face these directions. The proximity of the coast of France prevents waves of any size reaching the ness from a south-easterly direction, hence its angular nature. As the point grew out into deeper and deeper water its extension became slower.

After the presumed break-through of the Brede and Tillingham (probably in the thirteenth century) less shingle travelled to the ness, but more accumulated west of the breach. This was built up into the Camber Castle ridges. The position of the *old* town of Winchelsea is not definitely known. Homan and Ward are not in entire agreement in detail, but both suggest that it was on the old shingle bar formerly joining Fairlight and Lydd, and possibly where the Wythbourne reached the sea. It may even have stood on an island at an earlier time. Behind it there was certainly marshland, and so Ward[1] argues that it may first have been an eye or island surrounded by marsh except for the shingle on the south. Then, at some time unknown, the shingle was breached and the inhabitants realized that the cut through the shingle and the flooding of the marsh within had made their town a fair haven. Old Winchelsea, however, was finally destroyed in 1287 by a storm. Lewis, having the probable thirteenth-century change of sea level in mind, supposed that this played a large part

[1] G. Ward, 'The Little Brooks of Old Winchelsea', *Sussex Arch. Colls.* 75, 1934, 192.

in the disappearance of the town. Such a change, however, could not have been sudden, and perhaps we may safely assume that it was the great storminess of that century combined with the gradual rise in sea level which progressively weakened the site of the old town and made possible its final destruction in a single violent outburst (**7**).

Before discussing the problems more directly connected with the marshes and rivers, we must turn again to the different heights of the groups of shingle ridges and to the interpretations of this phenomenon given by Lewis and Balchin.[1] Their work depended on careful levelling, and no valid criticism can be made on that score. But as the discussion on their paper showed, it is by no means certain that the comparatively slight differences of height in structures such as shingle ridges afford absolutely certain data for establishing fluctuations of sea level. Bearing this in mind, however, we may quote their conclusion: 'If we are justified both in relating the average heights to past sea-levels and in our attempts at dating, then the following changes of sea-level relative to the land seem to have taken place in historic times at Dungeness: a rise in level of approximately a foot a century from about the fifteenth century to the present day; approximately the same level as to-day about the thirteenth century, and a foot or so lower in the eighth century; five or six feet below the present level in Roman times.' These figures, read in conjunction with the map (Fig. 68), suggest the parts of the ness formed, in the views of the two authors, at the times mentioned.

The marshes have clearly grown up inside the protecting shingle. They are nearly all reclaimed, but unfortunately we do not possess any figures of the rate of accumulation of marsh mud. A great deal depends upon the date of construction of the Rhee Wall. By most authorities this is ascribed to the Romans, although a few suppose the Belgae to have been responsible. The wall enclosed the whole of Romney Marsh proper, but the eastern side of these marshes, whatever the actual date of construction, must already have been protected by shingle in the space now occupied by the modern Dymchurch Wall. Elliott[2] maintained that this shingle was the old bar running from Fairlight to Hythe, and this is in agreement with Lewis's more recent work. Elliott did not, of course, take account of the probable historical changes of level. It is fair to state, however, that when allowance is made for such changes the evidence on which Elliott relied is strengthened. Elliott argues that at this time the site of Lydd was bounded on the north by an arm of the sea: this is corroborated in an

[1] W.V. Lewis and W.G.V. Balchin, 'Past Sea Levels at Dungeness', *Geogr. Journ.* 96, 1940, 258.

[2] Quoted at length by Lewin, *op. cit.* 1862.

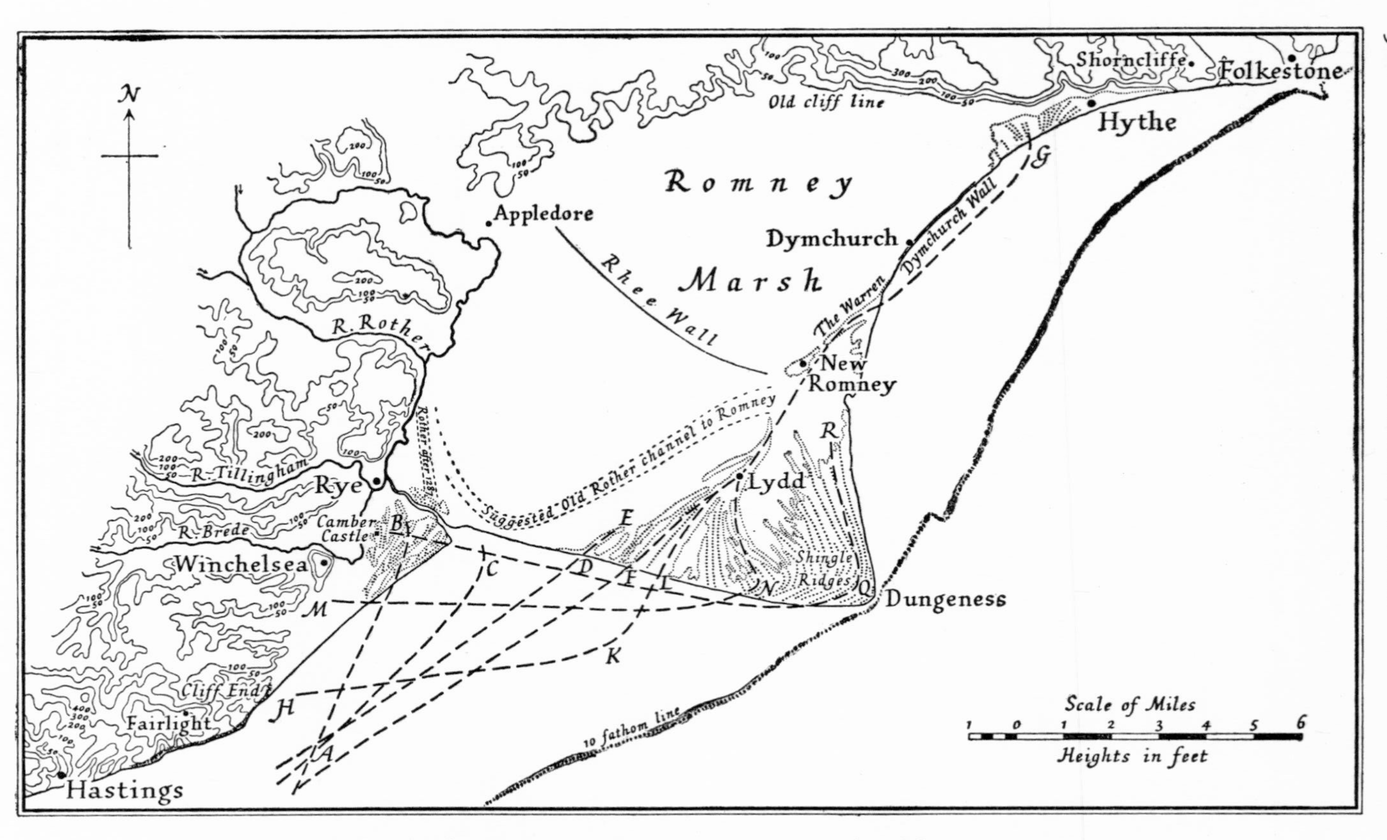

Fig. 71. The evolution of Dungeness (W.V. Lewis, *op. cit.* reproduced from *Geogr. Journ.* 80, 1932, 310)

account of a grant of land in A.D. 774 by King Offa to Archbishop Janibert. This seems to imply an opening between the positions of Lydd and Romney, and the configuration of the land seems to bear it out. Romney at that time must have been an island at high water, with a wide strip attached to it, barely above high-water level, extending towards Hope.

By 1479 nearly all of Walland Marsh was enclosed, except for that part drained by Wainway Watering and Wainway Creek. This creek was enclosed in 1661, so finally inning all Romney, Walland, Denge, and Guldeford Marshes. The details of the individual innings do not really come within the scope of this study, and a consideration of the three maps (Figs. 72 *a*, 72 *b*, 72 *c*) will serve better than a long description. If Elliott's views are correct, it will be clear from the first of these maps that the supposed river outlet of the Portus Lemanis was closed many years before the sea was shut off from the marsh itself by the building of the Rhee Wall. Elliott did not think that the Limen ever reached the sea near Hythe, but several other writers disagreed with him. There seems little doubt that the shingle at Hythe and Romney was connected: the digging of the foundations of the Dymchurch Wall showed ridges which clearly implied this. The many close recurves near Hythe at first sight definitely suggest a former outlet there, but it does not follow that it was the outlet of a river. It may have been merely a tidal inlet enclosed by shingle. Certainly, a great deal of marsh water drains off after high water from its exit here: the slope of the marsh alone indicates this. Elliott noted that about halfway between Hythe and Appledore there is a tract of land higher than the general marsh level. This he contended was due to the inrush of the tide into Hythe Haven losing its force about this part and so eventually depositing a bar. It thus caused the silting-up of the so-called river Limen. Silting continued until even Hythe Haven itself disappeared, and this, Elliott believed, took place in pre-Roman times. The fact that there is a depression along the foot of the hills is, however, no proof of the former existence of a river. Moreover, as Rice Holmes has noted, many superficial writers have misquoted Elliott about this estuary, because Elliott finally concluded that in Caesar's time there was no haven of any sort at Lympne. Rice Holmes[1] himself discussed the whole problem very fully and critically: 'The conclusions which we have now reached are, first, that the Rother did not, in the time of Caesar, enter the sea at Lympne, but debouched into the estuary near Appledore; secondly, that the marsh was closed at West Hythe Oaks, and therefore, that there was no harbour at Lympne; thirdly, that the Rhee Wall had not then been built, and therefore

[1] T. Rice Holmes, *Ancient Britain and the Invasions of Julius Caesar*, 1907 (Appendix dealing with Romney Marsh).

that the marsh was still flooded at spring tides by the inrush of the sea between Romney and Lydd; fourthly, that the Portus Lemanis was a pool harbour extending from West Hythe to a point nearly opposite Shorncliffe; and lastly, that the Rhee Wall was built in Roman times.'

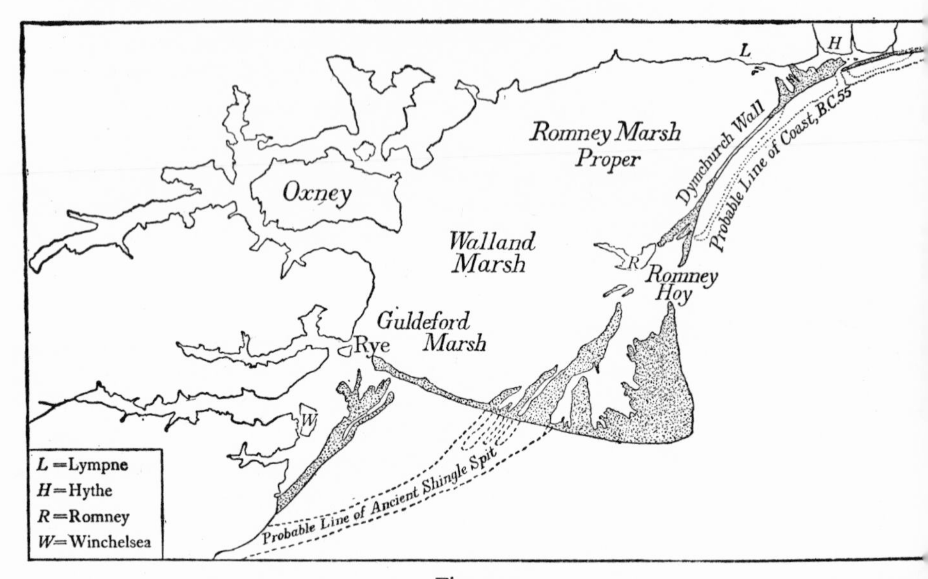

Fig. 72 *a*

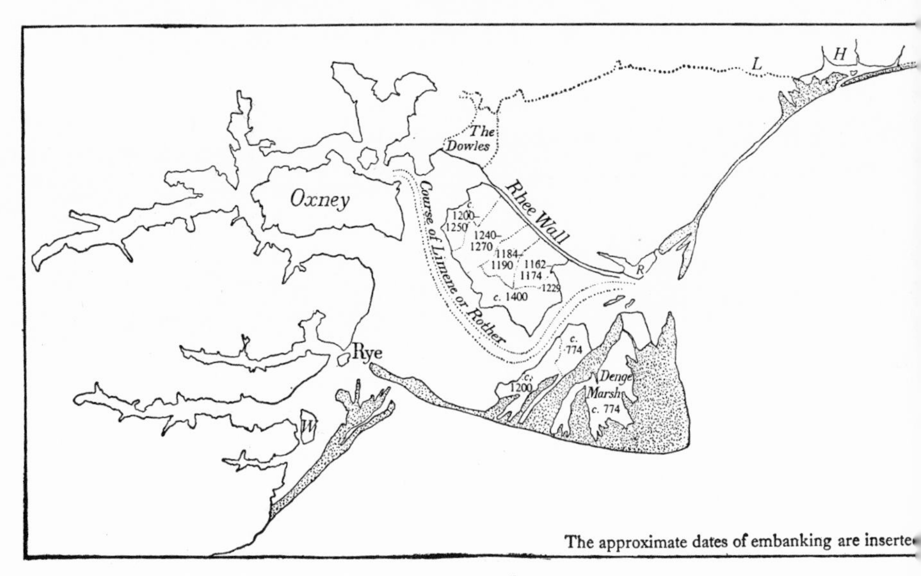

Fig. 72 *b*

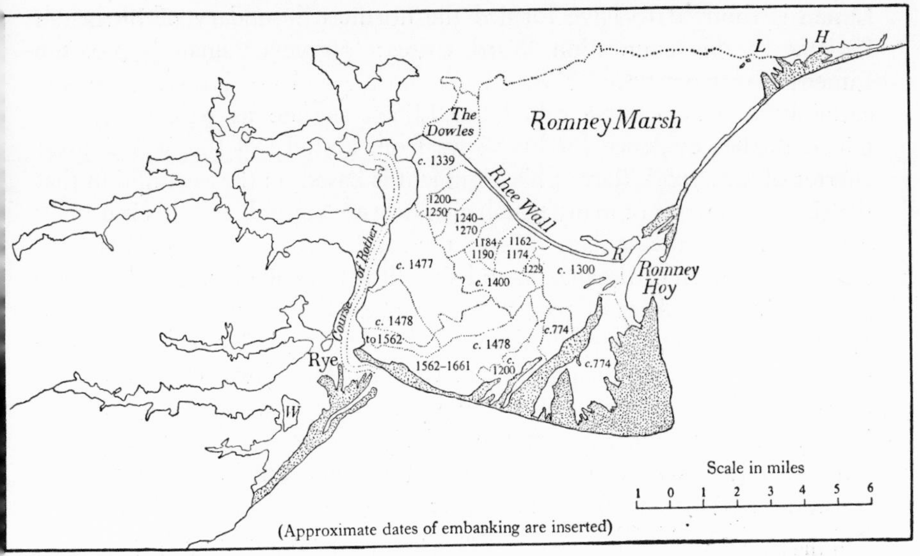

Fig. 72*c*

Figs. 72*a*, 72*b*, 72*c*. Outline maps of Romney Marsh and Dungeness (after T. Lewin, *The Invasion of Britain by Julius Caesar*)

This rather lengthy discussion is necessary in view of recent papers by Ward. Relying mainly on the evidence of charters, he seeks to reopen the whole matter of the course of the Limen, and definitely thinks that the Portus Lemanis was the outlet of the river. In a charter of A.D. 732 there is a reference to Sandtun,[1] and at that time the river is said to have run on its southern side; it is possible that the Willop Sewer marks its old course. In discussing Lydd[2] in the eighth century Ward concludes that by that time it had ceased to be an island (if it ever was one) and that there was an extensive estuary on its north-eastern side. This, of course, was the Rother outlet and is not disputed. In 1933 he discussed the significance of other early writings which, in his view, indicate that in A.D. 724 the Limen ran through the south part of Ruckinge: 'That the West Hythe or Willop section of the River Limen of 732 was continuous with the Ruckinge or Sedbrook section of 724 can hardly be doubted. It follows that any theory of the formation of the marsh which can merit acceptance must allow for the fact that between 724 and 732 the river Limen ran along the north-west side of Romney Marsh to an exit somewhere to the south of West Hythe.'[3] Again on the evidence of a charter of *c.* A.D. 850 the

[1] G. Ward, 'Sand Tunes Boc', *Arch. Cant.* 43, 1931, 39.
[2] G. Ward, 'Saxon Lydd', *ibid.* 43, 1931, 29.
[3] G. Ward, 'The River Limen at Ruckinge', *ibid.* 45, 1933, 129.

Limen is thought to have formed the northern boundary of Burmarsh Parish. In this connection Ward writes: 'However small it [i.e. the Limen] may have been or however estuarine and un-riverly, there was certainly a water-course called the Limen as late as 946.'[1] He obtained further evidence for his views from a study of the Wilmington charter of A.D. 700.[2] Part of his argument is based on the assumption that the sinuous courses of many marsh channels are natural, and this is usually conceded. In that case there follows the deduction that Sellinge Farm was a natural division of the marsh, and since it is not called an island or even a marsh in the charter of 700, Ward deduces 'that its normal features of 1,200 years ago were very similar to those of to-day'. He then goes on to develop the argument, suggests that Romney Marsh was old and settled country and good agricultural land when the Saxons first came to this island, and that Elliott's maps are entirely wrong and have misled later writers. His views are shortly summed up in his contribution to the discussion on Lewis and Balchin's paper:

> If the [Rhee] wall was Roman we ought at least to find that the land on the seaward side of it should be in some way distinguished from that to the north. It might well be flooded at high tide. It should at least be marsh and not good arable land. We find in the Saxon charters no difference at all. No wall is ever mentioned. Even the very oldest place-names, those which end in *-ham*, are scattered impartially on either side of the line of the wall. The line of a river or estuary running from Romney towards Appledore is mentioned as the Rumenea in 895, and as the Genlida in 830. But it does not approach the Rhee Wall. It encloses a large area to the south of the wall in which there are named settlements.... To this we may add that there is ample evidence of the northern branch of the Limen (later called the Rother), which must have run right through the wall, if it were there at that time, to reach its outlet below Lympne.

This difference of opinion is interesting and to some extent is due to the fact that the area has been worked by specialists rather than by one who has taken the whole into full consideration. There are grounds for thinking that, except for details, the physiographical studies are sound as far as they have gone. On the historical side it is difficult to express an opinion. It is reasonable to ask whether Ward has not perhaps interpreted the charters to mean a little more than they actually state, a query that is rather encouraged by his reference to the 'un-riverly' behaviour of his Limen. Is there not room for interpreting his evidence as referring to marsh creeks rather than to a true river? One can, however, only ask this question quite tentatively. We must await a careful survey, including borings, of the marshes before reaching a final conclusion.

If doubts exist about the Limen running out at Hythe, there is none about the more recent history of the Rother. Whatever may have taken

[1] G. Ward, 'The Saxon Charters of Burmarsh', *Arch. Cant.* 45, 1933, 129.

[2] G. Ward, 'The Wilmington Charter of A.D. 700', *ibid.* 48, 1936, 11.

place in Roman and Saxon times, from the eleventh to the mid-sixteenth century the Rother flowed out at Romney. After that time it adopted its present outlet near Rye.

The ness to-day lengthens slowly by shingle being taken away from the southern shore between Rye Harbour and the point, and, after rounding the point, being thrown up into local ridges. Temporary additions usually develop in this way, and the new, and quite local ridges, are low. They indicate the prevalence of south-westerly winds for some little while. The incidence of strong easterly winds piles some of this material higher up the new point, but also removes and redistributes the shingle, some of it to the westward. That shingle which succeeds in travelling right round the ness point will sooner or later be built up by waves from a north-easterly quarter. Hence, in general, the ridges are all truncated on the southern shore, and run roughly parallel to one another on the eastern side. As Dungeness, as a whole, does not appear to be wasting, new shingle must come to it from the westward. Precisely how this travels across Rye Bay is not fully known.

Some 98 per cent of the shingle of Dungeness is flint. There are also pebbles of the cherty sandstones of the Upper Greensand, fine-grained sandstones from the Hastings Beds, red and grey quartzites, dark quartz-tourmaline grit, and liver-coloured quartzites resembling those found at Budleigh Salterton.[1] At Camber, Lydd, and Great Stone there are small dunes. The exposures of mud just below high-water mark, especially on the southern shore, are nearly always outcrops of the marsh mud over which the shingle has either been pushed inland or from which the protecting shingle has been removed. The following figures illustrate briefly the rate of growth of the ness[2] (see Lewis and Balchin, *op. cit.* 1940):

Date	Distance from Lydd	Authority	Average annnal growth
Elizabethan	3 miles	—	—
1617	$3\frac{1}{8}$ „	Cole	5 yards
1689	$3\frac{1}{4}$ „	Collins	3 „
1794	$3\frac{3}{4}$ „	O. Survey	$8\frac{1}{4}$ „
1809	$3\frac{3}{4}$ „	Graeme Spence	None
1844	$3\frac{7}{8}$ „	Redman (?)	$4\frac{1}{2}$ yards

[1] Some of the far-travelled material is probably ballast.

[2] Other relevant papers re Romney Marsh and Dungeness: W. MacL. Homan, 'The Marshes between Hythe and Pett...', *Sussex Arch. Colls.* 79, 1938, 199; R. Furley, 'An Outline of the History of Romney Marsh', *Arch. Cant.* 12, 1880, 178; W.A.S. Robertson, 'Romney Old and New', *ibid.* 12, 1880, 349; C.J. Gilbert, *ibid.* 45, 1933, 246, and also *Quart. Journ. Geol. Soc.* 86, 1930, 94.

(19) SANDGATE TO DEAL: THANET CLIFFS

At Sandgate the Lower Greensand reaches the shore and forms cliffs, and at Folkestone it is succeeded by the Gault. Folkestone Warren is the second large area of landslips on the south coast. About three miles of coast are affected. There has been, and perhaps still is, a considerable difference of opinion as to the actual cause of the slips. The old idea that the Chalk slides forward on a slipping surface of Gault is, as Osman[1] demonstrated, false. The dip is really inland. Osman recognizes several types of movement: (1) Subsidence of the high cliffs. For these to be stable, when the Gault outcrop is *above* sea level, there must be a sufficient bulk of talus to act as a buffer against the tendency of the Gault to flow or bulge under the great weight of the Chalk above. If the talus is absent or removed, movement may take place if conditions encourage it. If the Gault outcrop is *below* sea level, the cliffs will only be stable provided the resistance of the Gault to compression, or flow, plus the resistance of the overlying Chalk to shear and fracture, are equivalent to the weight of the cliffs above sea level. (2) Settlements in the Warren. These are those usually associated with clay made plastic by water. There are two types, the first resulting from very heavy rains on normally dry Gault areas, and the second due to such rains on places where the surface of the Gault is usually wet. Slips, therefore, tend to occur after heavy rains, or just after low-water spring tides when the shore no longer possesses the support of a large volume of tidal water (e.g. December 1915). (3) Movements on the shore. These depend mainly upon how far the Gault is below sea level, and are always controlled 'by the change in stress due to the difference in weight of the height of the undercliff and to the low level of the shore'. (4) Falls of Chalk from the high cliffs. The falls seem to depend mainly on the jointing of the nodular Chalk. Falls may result from long rainfall, or merely because of wind and weather wearing away the yellow band beneath the nodular Chalk which consequently overhangs (**5**).

Osman believes that the Warren is an old feature initiated when the North Downs were breached. There is no regularity in the time of the falls, the chief of which have taken place in the following years: 1765 (a big slip), 1800, Dec. 1839, Feb. 1877, Mar. 1881, Sept. 1885, Jan. 1886 (a big slip), Nov. 1892, Dec. 1896, and Dec. 1915. The last two were of considerable size.

At one time there was a tiny haven at Folkestone, and in Caesar's time a natural harbour at Dover, the Portus Dubris of Antonine. The harbour

[1] C.W. Osman, 'The Landslips of Folkestone Warren...', *Proc. Geol. Assoc.* 28, 1917, 59. See also note on p. 344.

is also mentioned in Domesday Book, and even as late as 1582 there seem to have been neither banks nor shelves in front of it.[1] The estuary is said to have been navigable as far as Crabble, but this is probably an exaggeration. The present harbour dates from the time of Henry VII when a pier and two forts were built by John Clerk on the site now occupied by the harbour station. We are not concerned with details of later artificial changes: it is sufficient to state that the problem has always been to stop the inrush of shingle.[2]

The Chalk cliffs between Dover and the South Foreland have been long subject to erosion, but its amount is apt to be exaggerated. There are several hanging valleys (valleuses), some of which are now artificially graded to the shore. Burrows supposed that erosion had widened the Straits of Dover by two miles since the time of Caesar's invasion, but this is unlikely. It is true that Caesar described the cliffs as 'precipitous heights', but that proves nothing.[3]

The normal faulting of the Chalk, trending in a north-westerly direction, may have determined the line of the Dour valley.[4] The dip of the beds in the Dover-Walmer cliffs is north-north-east, at about one degree only, thus giving a section from the zone of *Terebratulina lata* to that of *Micraster cor-anguinum*. The cliffs of St Margaret's Bay and thence to Walmer are wholly in the latter zone. The Chalk reappears as a result of a fold in Thanet, and the coastal exposures show good examples of stacks, especially in Birchington Bay. In Thanet the Chalk is affected by numerous joints and small faults, which are all lines of weakness. The sea attacks these places, and has often cut channels more or less at right angles to the coastline. Thus blocks of Chalk have been separated from the mainland. There is also a good wave-cut platform in front of the Chalk cliffs. On the western side of Pegwell Bay the Thanet Beds appear, dipping generally to the south-west. They consist of sandy clays, marls, and sandstones and are often obscured by cliff falls. At Richborough they are overlain by the Woolwich Beds. In the Chalk coombs which reach the coast there is often some valley gravel, and near Ramsgate brickearth covers the Chalk, and narrow strips of it follow the coombs to Ramsgate Harbour. The slopes of the Dour valley are also banked with talus. Elsewhere Chalk Head with sandy loam occurs on the cliffs as at St Mary's Bay, Kingsdown, Pegwell, and Broadstairs (**4** and **3**).

[1] Rice Holmes, *op. cit.*

[2] See, e.g. A. Macdonald, 'Plans of Dover Harbour in the 16th Century', *Arch. Cant.* 49, 1939, 108.

[3] Rice Holmes, *op. cit.*

[4] See *Mems. Geol. Surv.* 'The Geology of the Country near Ramsgate and Dover', H.J.O. White, 1928.

(20) SANDWICH BAY, THE RIVER STOUR, AND THE WANTSUM STRAIT

The two most interesting features between Deal and Herne Bay are, however, the shingle formations of Sandwich Bay and the river Stour, and the ancient Wantsum strait between Thanet and the mainland (Fig. 73).[1] The Stour shows two great deflections, first to the south to Sandwich then, doubling right back on its course, northward to Richborough, near which town it enters the sea. Inside the great southern loop is a spit of shingle, arranged in fulls and rising only a few feet above the alluvium, at the southern end of which the former port of Stonar was situated. The shingle is well rounded and the ridge is about 100 yd. wide. This bank grew southwards from Thanet. The *Geological Survey Memoir* explains it by an eddy-drift, whatever that may mean. Eastwards of the river is another shingle and sand spit which has deflected[2] the Stour mouth some three miles to the north of its outlet near Old Haven in the sixteenth century. This shingle is really continuous with the beach stretching north from Deal.[3] At Deal there have been several fluctuations of the beach, so that at times Deal and Walmer Castles have, at least in their outworks, been attacked by the sea. Between 1741 and 1884 the shingle widened by about 380 ft. opposite Walmer, and lost about 200 ft. off Sandown; the ruins of Sandown Castle have now been washed by the sea for many years. The proportion of sand and shell increases northwards to Shell Ness. There are also some dunes, seldom exceeding 20 ft. in height, between Deal and Pegwell Bay.

The chief problem for the physiographer lies in the explanation of the southerly drift of the Stonar shingle and the northerly drift of the outer ridges in Sandown Bay. There appears to be no clear reason for assuming, as does the *Survey Memoir*, first an eddy-drift to the south, and then, presumably an end of this drift, for reasons quite unexplained, finally giving place to the normal drift of to-day. Hardman and Stebbing view the matter very differently. They think that the Stonar ridge was formed

[1] Consult *Mems. Geol. Surv.* 'Ramsgate and Dover', 1928, *op. cit.*; G.P. Walker, 'The Lost Wantsum Channel...', *Arch. Cant.* 39, 1927, 91. F.W. Hardman and W.P.D. Stebbing, 'Stonar and the Wantsum Channel (Pt I, Physiographical)', *ibid.* 53, 1940, 62 '(Pt II, Historical)', *ibid.* 54, 1941, 41.

[2] Hardman and Stebbing state that the Stour waters passed along a slight hollow between the Stonar shingle and a bank of Thanet Sand eastward of it. This was practically a continuation of the Lydden valley, and one of several small parallel valleys leading to the Wantsum. In other words, the northward deflection of the Stour is not, in their opinion, directly due to the shingle drift.

[3] F.W. Hardman, 'The Sea Valley of Deal', *Arch. Cant.* 50, 1938, 50.

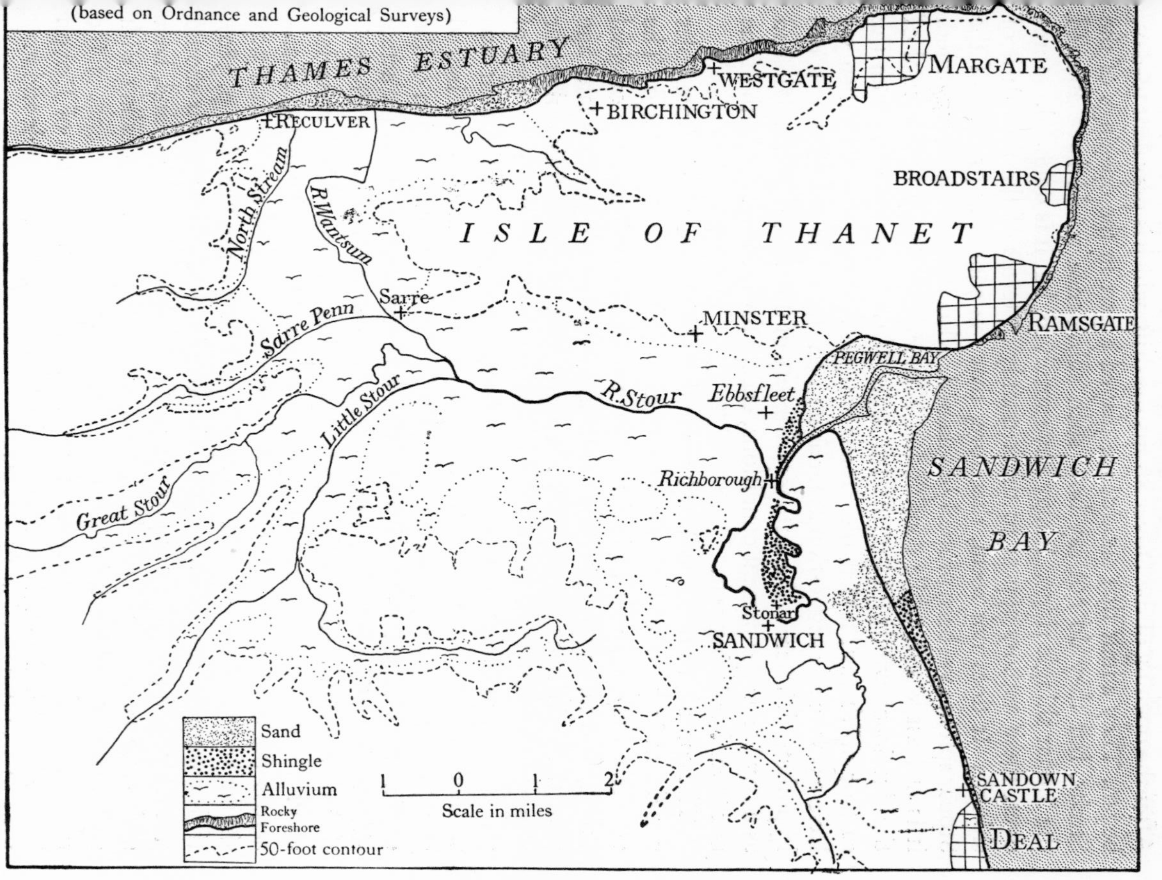

Fig. 73. Isle of Thanet and the Wantsum Channel

before the Straits of Dover were open and assume that the main drift was to the south. The reasons for postulating the southerly drift are not stated. After the opening of the Dover Straits the drift, which is prevalent to-day, began. They admit that subsidence played a part in the openings of the Straits, but they do not say how such subsidence would have affected the Stonar shingle ridge. It may have survived, but it is doubtful. Moreover, these authors appear to think mainly in terms of the tidal drift of shingle. It does not seem, therefore, that the southerly growth of the Stonar spit has yet been properly explained. Also, if it is as old as suggested, it is at least worth while considering Harmer's views on the reversal of the dominant winds during the Ice Age. If these winds came from the east or north-east, and if the Stonar shingle is sufficiently old, then we have a reasonable suggestion as far as it goes. But all this conjecture is too hypothetical. There is the further possibility of the spit's having originated as a bay bar under present conditions, except for the recent growth of the northward-pointing spit.

The comments made by Hardman and Stebbing[1] on certain bore holes put down in parts of the Stonar shingle are, again, unsatisfactory: 'The northern borings seem to indicate that much denudation of the original shingle had taken place, and that the shingle here was of a much later age than the main bank to the south. From this it seems to follow that for a long time after the drift of shingle had begun the northern entrance of the channel at Ebbsfleet remained open and that much of the drift was washed away by tidal action.' This is an equivocal statement, and indeed strengthens the view that there is as yet no convincing explanation of the origin of the Stonar spit. It is a pity that no detailed mapping of the actual shingle forms is available, since a careful map showing the exact disposition of the shingle fulls before they were disturbed would help considerably. Finally, one might suggest that the Stonar shingle formed as a bank across the eastern, and formerly wide, outlet of the Wantsum in a way similar to the formation of the original bank across the Yare (see p. 380). If that were the case, however, there is no very clear reason why it should not later have become joined to Thanet by local beach-drifting, and, later still, have been overlapped by the existing outer bank. In dealing with such features as these it is rash to accept any vertical movements of the magnitude of those responsible for the opening of the Straits of Dover. Those of the order envisaged by Lewis at Dungeness are possible and may have played their part at Stonar also.

The Wantsum strait is a channel partly evolved in a shallow syncline in the Chalk. Between Minster and Ebbsfleet are remnants of Thanet

[1] *Op. cit.* 1940.

Sand. Its northern entrance was controlled by the Roman fort of Regulbium (Reculver), and about a mile and a half from its eastern entrance there was an island on which the Portus Rutupis (Richborough) was built. Stonar, situated on the shingle bank, came into prominence in Roman times and prospered during the early Middle Ages, but was destroyed by the French in 1385. The town of Sandwich arose later than Stonar, and was on the south shore of the channel. There seems little doubt that Stonar beach had formed by Roman times, but Walker[1] thinks it was not of prehistoric origin. When all the present alluvial land was open water, Canterbury had direct access to an arm of the Wantsum channel. It remains to be proved whether this arm was constricted naturally or artificially. The problem is bound up with the line of the Watling Street. Originally, of course, the eastern entrance of the Wantsum channel stretched from Thanet to Deal, but this was narrowed first by the Stonar bank, and later by the outer shingle. At what period the outer shingle, travelling northward from Deal, first began seriously to contract the entrance, is uncertain; according to Walker it may have been as early as Saxon times. The eastern entrance seems to have silted first, and later a dam was built at Sarre (= the Sarre Wall). It was at this point that the tides entering either mouth of the channel met. In the north, Chambers' Wall joined Bartlett's to the Thames, and then followed the coast to Birchington cliffs. At the end of the eighteenth century the northern mouth was finally closed by walls joining Reculver to Chambers' Wall.

The average width of the original channel is estimated at two miles, and its depth, as indicated by boreholes, was roughly 40 ft. Bede, in A.D. 730, wrote that it was three furlongs wide, and Hasted noted that it was navigable for small ships up to the end of the fifteenth century. Sommer recorded that boats still carried coal to Canterbury up to 1699. The Stour, Little Stour, and other less important streams ran into the channel: all were sluggish. Boys (1775) stated that 'The River Stour from Fordwich Bridge to the sea runs in a narrow Channel and winding Course through a marshy Plain about one mile and a half broad at a Medium then called the Levels or Valleys or Meadows indifferently'.[2] The marshy plain really includes a good deal of the Stour estuary (**2** and **1**).

(21) THE GOODWIN SANDS

The actual coast must now be left for a discussion of two important matters—the Goodwin Sands and the formation of the Straits of Dover.

[1] *Op. cit.* 1927.

[2] In Hardman and Stebbing, *op. cit.* 1940.

The origin of the Goodwins is uncertain.[1] There is an abundance of speculation on the matter, but a paucity of fact. Traditionally, they are said to mark the site of an island called Loomea, once part of Earl Goodwin's estates, which was destroyed by storms in A.D. 1099. Lyell certainly regarded them as the relic of a land, and not merely as an accumulation of sea sand. He was of the opinion that the storms of 1099 may have swept away the last traces of this island. Borings have shown blue clay under 15 ft. of sand at the Lighthouse borehole, and there are records of clay having been found under 7, 57, and 75 ft. of sand elsewhere. The age of this clay is not established. 'The occurrence of Eocene strata in the sea bed beneath the northern half of the Goodwins is to be expected from the trend of the Richborough syncline. The extent of the blue clay is equally uncertain: it would appear from the sequel that the sands in places rest directly on the Chalk.'[2] At the present time the sands are subject to change. Holmes states that South Sand Head was a mile to the south-west of its present position in the time of Elizabeth. Brake Sand moved some 700 yd. to the west in the first half of the nineteenth century, and it is also possible that the whole mass expanded, or moved, towards the coast between 1895 and 1896. Possibly also the area left dry at low water was increasing. The maximum amount of sand occurs in Trinity Bay. White argues that this present shape implies a drift to the south-south-west, but the Channel Pilot says that 'the tidal stream sets with considerable strength north-westward and over them [i.e. the Goodwin Sands] at times'. This is an apparent contradiction of White's views, unless there is a south-westward movement of some magnitude induced by wind waves. Parenthetically, one might suggest that the formation of the Stonar bank and the movement of the Goodwins are not entirely unrelated phenomena. The historical evidence, especially that referring to Caesar's landing in Britain, has been best summarized by Rice Holmes[3]:

> The reader has doubtless already concluded that it is impossible to affirm either that the Goodwin Sands existed in the time of Caesar, or that they had not then accumulated to such a degree as to attract attention, or that their place was occupied by an island. If the silence of Domesday Book and, as it should seem, the absence of any other positive testimony constitutes an argument against the hypothesis of Sir Charles Lyell, the same argument may be advanced to show that before the Norman Conquest the sands had not begun to appear. Yet,..., there is some reason to believe that either sands or an island were there when Caesar invaded Britain. Tradition, vague as it is, combined with Lyell's authority, disposes me to accept tentatively the latter alternative.

[1] See *Mems. Geol. Surv.* 'Ramsgate and Dover' 1928, *op. cit.*; T. Rice Holmes, *op. cit.*; and The Channel Pilot.

[2] H.J.O. White, *Mems. Geol. Surv.* 1928.

[3] *Op. cit.*

In this unsatisfactory state the matter remains for the moment. One cannot help feeling, however, that any full treatment of the coastal problems of eastern Kent should not be too circumscribed; it should take into account all features and factors, even though they may seem at first to have no, or at best only remote, bearings upon each other.

(22) THE STRAITS OF DOVER

The origin and date of the formation of the Straits of Dover are questions about which more than one opinion is held. Briquet[1] points out that the raised beach with its associated cliffs and pebble ridge at Menchecourt implies a sea level 5 metres above the present one. A corresponding beach is found at Brighton and elsewhere on both sides of the Channel: it is the *Patella* beach of the south of England. Briquet's view is that the transgression connected with this beach explains the contemporary opening of the Straits. He deduces from the stratigraphical evidence in the Low Countries a continuous lowering of the floor of the North Sea Basin in the Tertiary period. This lowering enabled the sea to transgress on the Dover-Calais isthmus. At that time the Channel was a gulf, a depressed part of the 200-metre peneplain of southern England and northern France. There is no suggestion that on the actual site of the Straits the Chalk was dislocated in any way: to-day the cliffs show corresponding features on either side. In brief, the floors of both the North Sea and the Channel rose gradually to the isthmus which was a divide between them. The isthmus was attacked by subaerial denudation, and general reasoning suggests that it sloped more gradually to the north, whilst to the south it turned an escarpment face which was really a continuation of that of the North Downs. When, later, the sea level rose the isthmus was soon penetrated, and this took place when the ancient cliff at Sandgatte was formed, that is, at the same time as the Menchecourt transgression. This means that the sea was then attacking effectively from the north. But to allow the penetration to be made easily, Briquet supposes that the crest of the isthmus must have been breached in one or more places by rivers, just as the Downs are now breached by the Stour, Medway, and Darent. Granting this, we may suppose that the thalweg was not far above sea level, and that it afforded an easy means for the sea to break through the ridge.

Sea level fell after the Menchecourt episode. Quite possibly the isthmus returned, and even parts of the floors of the North Sea and Channel may have been exposed once more. This emergence, according to Briquet, would lead to the formation of the deep and buried channels of our rivers,

[1] A. Briquet, 'Sur l'Origine du Pas-de-Calais', *Ann. Soc. Géol. du Nord*, 44, 1921, 141.

and of the submerged forests. It also helps to explain the identity of the Quaternary fauna of England and the Continent. At this time, too, the re-exposed isthmus would have suffered subaerial denudation, and it is reasonable to think that the features now revealed in soundings were then formed: the Ridens, Colbart, and Varne ridges, and the low area of the Creux de Lobourg. Further deepening of the Creux de Lobourg may easily have resulted from tidal scour in the early stages of the next upward movement of sea level which gave rise to present conditions, and, on the French side, to deposits the level of which is virtually the same as those of a more ancient date.

Jukes-Browne supposed that there was a Calais-Dover ridge in Glacial times: Clement Reid, on the other hand, agreed with Prestwich that in early Glacial times there was a sea valley of some kind, but that later this became dry and England was once again joined to the Continent. Gregory[1] maintained that the Straits were an ancient valley; they were certainly closed in the Miocene, but he claimed that they were open in the early Pliocene in order that the warm fauna of the Coralline Crag could reach what is now East Anglia. This is not in agreement with the views of Stamp and other writers who consider that this fauna reached Suffolk through a strait at the western end of the London Basin. Since the Red Crag fauna is clearly evidence of colder conditions, Gregory supposed that the Straits were closed at that time; he also considered them closed much later because, in his view, the possibility, based on zoological evidence, that the Thames once flowed round Kent and through the Straits, is strong in comparison with that suggesting that the Thames and Rhine formerly joined, or at least flowed in roughly parallel courses, northwards over the exposed North Sea bed. In fact, he gives reasons (not very convincing) for assuming that the broad rise between Happisburgh Banks, Terschelling Bank, and Heligoland separated north and south basins in the North Sea, prevented Scandinavian glacial erratics from reaching Essex, and directed the Thames to the Dover Straits. It is not, however, clear from his account just when the Straits, as apart from a river valley, were reopened after Red Crag times.

Stamp[2] rightly points out the need for distinguishing between events before the maximum stage of glaciation, and after the retreat of the ice. At its maximum the ice covered all the Wash and Humber rivers, and the

[1] J.W. Gregory, 'The Relations of the Thames and the Rhine, and the Age of the Straits of Dover', *Geogr. Journ.* 70, 1927, 52.

[2] L.D. Stamp, 'The Thames Drainage System and the Age of the Straits of Dover', *ibid.* 70, 1927, 386, and 'The Geographical Evolution of the North Sea Basin', *Journ. Conseil Internat. l'Exploration de la Mer*, 11, 1936, 135.

zoological evidence based on the presence of similar fish (burbot and white bream), which do not occur in the Thames, but are found in the Wash and Humber rivers as well as the Rhine, suggests a connection of these streams after the maximum glaciation. Woodward and Kennard insisted strongly that the great similarity of the fresh-water mollusca of the Thames and Rhine indicated a former junction of the two rivers. As these hardy forms survived the Glacial period in the south of England, Stamp pertinently asks whether the junction of the Thames and Rhine that they suggest was not a pre-Glacial feature. If this were the case, it certainly explains an awkward problem.

In general, Stamp follows Briquet in tracing the probable development of rivers on the isthmus: he suggests that the Creux de Lobourg originated in much the same way as the Stour gap through the North Downs. It is very probable that the Diestian (the equivalent of the Lenham Beds) peneplain stretched across the Straits, and the early Thames flowed over it. If so, the present north Kent and isthmus rivers joined it as southern tributaries. Similarly, a southern Diestian sea on the site of the Channel probably reached as far east as Beachy Head: streams most probably drained to it also from the isthmus. If so, it is reasonable to suppose that a south-flowing stream cut backwards through the isthmus and captured the Thames. This is in general accordance with the point noted by Briquet that the south face of the isthmus was an escarpment, and so presumably steeper than the north slope. This would account for (*a*) the severance of Thames and Rhine, and (*b*) the existence in the Straits of a big river at the end of the Pliocene. As has been noted, there is a great similarity between English and Continental faunas and floras in Quaternary times: this perhaps limits the width of the river valley in the Straits, even if it does not negative such an idea altogether. On the other hand, there still remains the very interesting possibility that when the ice advanced on the North Sea Basin, the water would have been ponded in the south, and would have risen until it overtopped the lowest part of the Calais-Dover ridge through which it would have flowed and cut a valley. In any case, there is apparently no evidence for an open strait before the Sandgatte raised-beach episode, and there is no conclusive reasoning to prove that the Straits were open in early Glacial times.

This last comment appears to agree with W. B. Wright's view that if the *Leda myalis* bed of Norfolk (see p. 404) is the equivalent of the pre-Glacial[1] beach of the Channel coast, and if, as is commonly accepted, there was a general fall in sea level after its formation, then as the ice advanced

[1] W.B. Wright, *The Quaternary Ice Age*, 1937. (In the sense that Wright spoke of the pre-Glacial beach.)

(which coincided with the fall in sea level) in the North Sea the waters of that sea would have been ponded back, and the Straits would not have been open. As Stamp notes, only when the water either rose as a whole, or when in a local impounding it rose above the level of the lowest col, would the excavation of the gap have begun. In any case probably several changes of sea level in the Quaternary have to be considered, and it is at least possible that the Straits were alternately open and closed. This point is also stressed by Wright who, in discussing Quaternary man, shows that it is necessary to assume a connection with the Continent as late as the Magdalenian: 'Whether this connexion was continuous throughout the Palaeolithic Period it is impossible to say. The Straits of Dover may not have been cut until then, but, on the other hand, the connexion may have been only an intermittent one due to changes of sea-level.'[1]

Thus, for the actual origin of the Straits the views of Briquet and Stamp may well be followed. Their opinions are somewhat similar, except for the ice-impounding theory. The differences are those concerning time rather than cause. Moreover, if the oscillations of sea level in the Quaternary and the gradual widening of a river valley to form the Straits of to-day are borne in mind, there is plainly room for a variety of opinions concerning detail rather than the general nature of the evolution of the Straits.

(23) BRIEF SUMMARY OF PLATFORMS AND RAISED BEACHES IN SOUTHERN ENGLAND

There have been various references in the preceding pages to the raised beaches and platforms of the south coast. In this section only a summarized statement of the more important facts will be attempted.

In Sussex and Hampshire remains of beaches are traceable at *c.* 100+ ft., and *c.* 15–30 or even 50 ft. The higher beach is preserved inside the Sussex coastal plain. In the area around Portsmouth each of the three beaches is marked by a bluff or degraded platform. Offshore muds associated with the 100-foot level indicate, by their fossils, a warm climate. The 50-foot beach was formed later, in colder climatic conditions. The Coombe Rock spread over the 100-foot beach as well as the 50-foot beach. As the land continued to rise, there appears to have been a resumption of cold conditions forming a second Coombe Rock over both levels, and a beach at 15 ft. was formed. With the uplift of this, a third layer first of brickearth and then of Coombe Rock formed, so that there are three deposits of Coombe Rock on the highest beach, two on the next, and one on the third. The Portland beach rises from 25 to 50 ft.; it contains marine shells and

[1] W.B. Wright, *The Quaternary Ice Age*, 1937, p. 295.

shingle, which are overlain by loam with non-marine shells, and finally by Head or Coombe Rock. The lowest of the south coast beaches is the so-called 10-foot pre-Glacial beach of W. B. Wright: its height is distinctly variable. It is possible that the beach at Portland Bill is part of it. The beach is pre-Glacial in the sense that it is earlier than any drift in contact with it. Its probable position in the time scale is discussed in Chapter XII, where it is called the *Patella* beach.

The higher platforms of Devon and Cornwall are also traceable into this part of England. In east Devon and near the Dorset boundary, Green finds many traces of the Bodmin Moor platform at *c.* 690 ft. This probably corresponds with the Diestian and Lenham Beds. In the same area are fragments at *c.* 595 ft., probably the equivalent of the Netley Heath platform, and Scaldisian or Poedilian in age. There are also traces at *c.* 530, *c.* 505, and *c.* 440 ft. Green[1] thinks that the Bodmin Moor terrace can be followed on and off from western Cornwall to the Chalk escarpment and that it is always at much the same altitude. Judging from this similarity in level of the highest platform, he feels it safe to correlate any lower terraces found in the same area.

In the Wealden district there is a well-marked terrace at *c.* 200 ft. seen in several river gaps and also in the London Basin. Green[1] has named it the Ambersham terrace, and it is always associated with a sharp-edged gravel. It is found in several rivers farther west, including the Dorset Stour, Frome, Axe, Teign, Bovey, and Dart. The terrace is approximately horizontal where it approaches the sea, a fact which is in itself presumptive of stability. It appears to correlate with a sea level about 170 ft. above the present. Below it are two well-marked terraces at *c.* 100 and *c.* 50 ft.—the Boyn Hill and Taplow terraces. In the Mole valley the Taplow appears to be duplicated, there being two well-marked levels at 70 and 50 ft. Bull has also found spur levels in the South Downs at 130 and 80 ft., and similar levels can be traced in the Stour, the Hampshire Avon, and the Dorset Stour. Furthermore, the Boyn Hill and Upper and Lower Taplow terraces are present farther west in the Axe, Otter, Dart, and Teign valleys.

Finally, there are the buried river channels and submerged forests which were seen in dock excavations at Southampton and elsewhere, the well-developed drowned valleys of this coast, and the records of boreholes in various valleys, including the Sussex Ouse. Possible correlations are discussed in Chapter XII, but probably all the platforms from the Bodmin Moor to the Taplow represent not only a descending series, but also an age series, the oldest being the highest. They seem to have given

[1] J.F.N. Green, *Quart. Journ. Geol. Soc.* 92, 1936, lxviii.

way to the two (or three) main raised-beach levels, and the drowned valleys follow after the *Patella*, and lowest, raised beach, but at an unknown interval. Later still occurred the minor fluctuations noted so far in particular reference to Dungeness. It is probable that a slow downward movement is still in progress.

(*Note.*) There are numerous papers directly or indirectly concerned with the south coast which are not mentioned in footnotes. A few to which general reference is made include:

J.B. Redman and others, on part of the coast, *Rept. Brit. Assoc. Adv. Sci.* Aberdeen, 1885, p. 407.

D.L. Linton, 'The Origin of the Wessex Rivers', *Scot. Geogr. Mag.* 48, 1932, 149.

J.F. Kirkaldy and A.J. Bull, 'Geomorphology of the Rivers of the Southern Weald', *Proc. Geol. Assoc.* 51, 1940, 115.

A.W. Rowe, 'The Zones of the White Chalk of the English Coast, I. Kent and Sussex', *ibid.* 16, 1899–1900, 289.

[NOTE TO P. 332 FOOTNOTE.] In the *Geogr. Journ.* 105, 1945, 170, W.H. Ward discusses the landslips at Folkestone Warren. Sections show that they are primarily caused by erosion of the Gault 'toe' by the sea, and that they are rotational in character. '...the slip plane penetrates the full thickness of the Gault and there is no suggestion that the Chalk is sliding down the surface of the Gault.' After a slip the rotating mass builds up at the toe and forms islands, which are soon destroyed by the sea. Ward also maintains that the Axmouth landslip was rotational and very similar to those at Folkestone.

Chapter IX

HUNSTANTON TO RECULVER

(1) INTRODUCTORY

In treating the coastal features of East Anglia and Essex it might at first be taken for granted that the study should begin at the Wash and end at the Thames. For reasons that will be apparent later this regional division would be rather illogical and on this account the start in this chapter will be made at Sheringham. Thence there will be a study of the coast westwards, and secondly eastwards and southwards to the Thames. To avoid repetition, the beaches will have attention first, and then the cliffs, which between Weybourne and Happisburgh must be treated as a unit. Sheringham is chosen as a starting place because there, or very close to that place, the direction of travel of beach material changes.[1] The groynes at the town itself show by the shingle which is piled up on their western sides that the movement is to the east. But west of Sheringham there is an unbroken line of beach to Weybourne and Blakeney Point, and along the main beach of Blakeney the movement of beach material is definitely to the west. The reason for this sudden change is not quite clear. Waves coming in from the sector between north and somewhat south of east would, on the whole, cause westward-directed beach-drifting on the north Norfolk coast. Similarly, winds from roughly north-west round to somewhat north of east would set up waves to drive material southwards along the coast east of Cromer or Mundesley. Opposite winds cause local reverse movements. However, in general, the change of direction seems to take effect near Sheringham, and it is for that reason the coast lends itself to description west and east of that place.

From Sheringham to Weybourne there is a good shingle beach lying under the cliffs (see below). The Chalk is but just below sea level at and near Sheringham; an abrasion platform is cut in it, and the immediate offshore zone is shallow. There is comparatively deep water close inshore at Weybourne, but at no other point on the north coast of Norfolk. At Weybourne also the cliffs begin to recede from the present coast and they run inside the marshlands: they reach the sea again at Hunstanton. These distinguishing features make the coast between Weybourne and Hunstanton a unit, and it is an area which affords the finest example of coastal marshes in these islands. The marshes and offshore bars have all grown

[1] J.A. Steers, 'The East Anglian Coast', *Geogr. Journ.* 69, 1927, 24.

up on the extensive sand flats. It is not uncommon in places to see from half a mile to well over a mile of sand exposed, while farther seawards some banks may also dry out. In point of fact the flats extend well out to sea, especially near the Wash. At high water the sea not only floods the outer beaches, but also covers the marshes, so that there are really two shorelines, the outer formed by the offshore bars, the inner corresponding fairly well with the old cliff line, although on account of changes, both natural and artificial, the sea does not quite reach to the foot of these ancient cliffs.

Along the whole coast of eastern England the general movement of material under the influence of both beach-drifting and tidal currents is directed to the south. The Norfolk coast between the Wash and Sheringham acts as a great groyne so that material from Lincolnshire and farther north, to say nothing of that raised from the shallow floor of the sea, becomes banked up against the coast. This function of the coast probably explains the extensive sand flats while the marshes are merely a local growth on them. If this conclusion be correct, it also accounts satisfactorily for another anomaly. East Anglia, like all southern England, suffered from the rise of sea level, relative to the land, that took place in post-Glacial times. Its effects are clearly seen in the wide alluvial valleys in which lie the Broads. But in north Norfolk the results are not very obvious, partly because there are but few and unimportant streams, and partly because it is often assumed that offshore bars such as Scolt Head Island and Blakeney Point are, or should be, characteristic of emerged and not submerged shores. But if we grant that after the submergence of this already low coast with a shallow and very gently shelving offshore zone further effective shallowing continued as a result of the accumulation of material travelling from the north, it provides at once a complete explanation of the features of to-day, namely offshore bars enclosing marshes and tidal lagoons. It is an interesting illustration of the difficulty of classifying shores in detail.

On the shallow sand flats the larger waves naturally break well away from high-water mark since roughly speaking a wave breaks when it passes into water the depth of which is half the length of the wave. Thus only very small waves reach the inner edge of the flats. Bigger waves due to storms or stronger winds break farther out, and in doing so erode the sea floor, and sooner or later throw up a submarine bank. This may or may not grow upwards, and it certainly will not do so regularly throughout its length. Sooner or later, however, parts of it will be built up beyond the level of high water. In all these early stages the bank will necessarily be very unstable and will probably change its form with every tide. A new storm may dissipate it completely, so that the first stage of formation must be repeated all over again. Given the proper circumstances, however, some of these banks become relatively stable and form offshore bars. Once this stage has been reached seeds of strand plants probably soon reach it. Some of these take root, and vegetation of the bank begins. If the plants get a fair hold, they also begin to help in its stabilization. The original bank may be of any size or shape: the probability is that it will be long compared to its width, and will be built more or less parallel to the old shoreline.

Let us suppose such a bank has attained a fair degree of stability. Waves break on its outer shoreline, and if these approach it at an oblique angle the lateral drift of beach material begins: the bank, in fact, starts to grow in length. This process is not uniform: occasional winds from contrary directions, for example, are disturbing factors, so that there is every likelihood of recurved ends forming. There has so far been mention only of sand and sand flats, but a good deal of coarser material also forms shingle. There is, with the exception of the Sheringham-Weybourne cliffs, no shore source whence this shingle can reach places such as Scolt Head Island to-day. On the other hand, during the Ice Age vast quantities of stones must have been scattered in boulder-clay, sand, and gravel deposits over the floor of the North Sea. In all probability it is largely this source which gives the shingle to the offshore bars and spits of this coast. On any mixed sand and shingle beach it is plain that the shingle is pushed to the top, and accordingly when the original offshore bar begins to emerge there is likely to be a crest line of shingle on it. At a later stage, when lateral drifting has begun, it is the shingle which *shows* the major effects reflected for instance in the recurved ends. If shingle is wanting, the same kind of effect may be given by shells: the miniature ridges off Stiffkey illustrate this.

All stages of the evolutionary process can be seen on this coast, and many details are well worth examination as they occur. Before that, however, the area landward of the ridges needs attention. Here are areas of quiet water at high tide, and, at first, bare expanses of sand at low tide. The tides, in addition to any land water that may drain to them, will carry into these quiet spaces much fine material, a large part of which will be deposited in sheltered spots at slack water when the tide turns. In this way mud banks begin to form. Seeds of halophytic plants reach these places: some germinate and produce plants which disturb the run of the water immediately around them and so help to bring about deposition. At this, and also at an earlier stage, seaweeds have the same effect. Thus an original sand flat becomes covered by considerable and increasing areas of mud, and these, in their turn, tend to be covered by plants. In this way the growth of the marshes begins. As time goes on, the amount of mud increases, and the tidal water in the marsh area becomes more and more restricted to channels or creeks. The intimate connection between plant growth, the nature of the muddy or sandy substratum, and the amount of tidal inundation are discussed in detail in Chapter XIV, but a summary such as that given above is necessary by way of introduction to the study of this coast.[1]

(2) BLAKENEY POINT

The first major feature is Blakeney Point;[2] this for the purposes of this chapter may be defined as beginning at Weybourne, or even as far east as Sheringham. Where the beach runs seawards from the cliffs at Weybourne it is for some considerable distance merely one major ridge of shingle showing the usual minor ridges due to particular storms or tides on its seaward face. It is not a high ridge, and in severe storms it is overtopped by the waves, and occasionally breached. In this way the two or three large fans of shingle on its inner side have been formed. From Weybourne to Cley Wall the marshes have been reclaimed, although in recent years[3] large areas have been spoilt by the sea breaking through. At Cley the river Glaven runs out to the bank which has deflected the river westwards for nearly four miles. Between Cley and Blakeney (see Fig. 74) there is another large reclaimed area which extends as far north as the river, but not to the shingle. Along the whole length of the ridge from Weybourne to the Marrams, the occasional storm waves which overtop it are gradually pushing it inland. Within living memory there has been a good deal of salt marsh on the north side of the river near Cley, but this has now gone as a result of the bank moving inland over it.

At the Marrams occur the first recurved ends. There are six in all running back almost at right angles to the main ridge. Originally they probably curved off from it as, for example, the newer ridges at the Headland, but the over-rolling motion has pushed the main beach so far back that it rests athwart the laterals. The laterals nearly all bend round eastwards at their unattached ends, the result of waves inside the main ridge. Marshes have grown within the laterals, and it will also be noticed

[1] For general accounts of East Anglia, including the coast line, see *British Regional Geology*, 'East Anglia', C.P. Chatwin, 1937; and J.A. Steers, 'The Physiography of East Anglia', *Trans. Norfolk and Norwich Nat. Soc.* 15, 1941, 231.

[2] See Blakeney Point Publications No. 4 (F.W. Oliver) and No. 7 (F.W. Oliver and E.J. Salisbury): also Chapter XII, *Tidal Lands*, A.E. Carey and F.W. Oliver, 1918.

[3] And again in April, 1943.

that the entrance to them between the ridges is narrow. In the early stages of their formation the narrow entrances favour rapid upward growth, because the water within the ridges is quiet and drops its load. Once grown upwards the marshes are only covered by the highest tides, and their further upward growth automatically nearly ceases. The vegetation on them is zoned, and from below upwards the characteristic plants of each of the three zones are *Suaeda fruticosa*, *Festuca rubra*, and *Statice binervosa*.

Nearly half a mile farther west of the Marrams is the Hood which consists of one long recurve with two smaller branches, enclosing a small marsh covered only at very high tides. Unlike all the others, the major recurve is bent westwards near its distal end, due presumably to east and south-east winds. A *Salicornia* marsh is developing near the end of this ridge, and where the ridge joins the sea beach there is a small dune complex. This long lateral is similar to those farther west at the Headland where there are several ridges approximately parallel to one another. Dunes have grown on each of these, and between the ridges are lows, some of which are still flooded at spring tides. Fig. 74 will suffice to draw attention to the general disposition of these ridges. In recent years the sea has cut into the Headland and partly destroyed the dunes, but at the same time a long new projection has grown out to the south-west. This is mainly a shingle beach, which is gradually being covered with *Agropyrum* and *Ammophila*: it is still low and liable to be overrun by storms. Towards its end it divides into several recurves, some of which have subsidiary branches, and dunes have grown up on this part, which is higher than the stretch between it and the Headland. In the sheltered area within this new projection we see the first stages of marsh development. A little mud is being deposited on the sand flats and some pioneer plants, *Salicornia* spp. and *Suaeda maritima*, have appeared. Within the older ridges near the old life-boat house the mouths of the lows have often been cut off by the growth of new dunes, and the best development of a *Salicornia* marsh is in the great low between the Long Hill and the Headland.

Since the whole structure has grown westwards and has sent out a series of recurved ends, it follows that the marshes within these successive recurves should show a graded series, the oldest being near Cley. In a general way this is true, as is shown by a comparison of the Marrams marshes with those of the Headland and Far Point, but the sequence is not so clear as that at Scolt Head Island (*q.v.*). This is due in part to the different arrangement of the recurves at Blakeney which are on the whole shorter than at Scolt. Moreover, the whole area between the main beach and the main channel is not arranged in such definite 'regions' as at Scolt Head Island. Hence, there are large stretches inside Blakeney ridge which are not yet marshes in the full sense of the word. To the south of the channel, however, and apart from reclaimed areas, there is a typical development of saltings, especially west of Blakeney golf course. These are discussed below.

Experiments made on the outer beach between the Hood and the Headland showed clearly that the main direction of travel of beach material is westwards as a result of waves approaching the beach obliquely from the east and north-east (see p. 55).[1] This is a movement in opposition to the flood current which runs to the east. The effects of this current are possibly seen in the sands below the shingle. Blakeney Harbour channel, after having been deflected several miles westwards by the shingle ridge, turns round, and, although subject to considerable fluctuations, nearly always shows a marked deflection to the east in the sand flats. This appears to be the result of the tidal current carrying in this direction the sand churned up by the waves. There is nothing inconsistent in these two opposed tendencies: the shingle lying high on the beach is only worked on by breaking waves at and near high water and, for reasons given elsewhere, the dominant waves here come from a north-easterly direction. The sand is affected over a much longer time interval. It is disturbed by waves traversing it and is carried alongshore by the dominant current, in this case to the east by the flood current.

(3) STIFFKEY AND WELLS FORESHORES

Before describing the reclamation within Blakeney Point an account of the foreshore between the Headland at Blakeney and the eastern side of Wells Harbour is in place: it is an interesting stretch that illustrates well the early stages in the development of offshore bars.[2] The horizontal distance in places between high- and low-water marks at springs is nearly two miles. On the outer shore, outside Stiffkey Meols, there can be seen low irregular ridges composed mainly of sand and shells with very little shingle. These are miniature but unstable offshore bars, which may remain more or less in one place for a considerable time before being partly or wholly demolished in a storm. During a quiet period there is time for a thin film of mud to be deposited within them, and for *Salicornia*, the annual *Suaeda*, or *Glyceria* to grow. An embryonic marsh thus evolves. Separating the old and well-developed marshes off Stiffkey from this outer beach are fairly continuous ridges of shingle and sand on which

[1] See Steers, *op. cit.* 1927.
[2] 'Scolt Head Island Report', *Trans. Norfolk and Norwich Nat. Soc.* 14, 1937, 210.

small dunes occasionally rest. These are the Stiffkey Meols and represent simply a more developed form of offshore bar which has become relatively stable. They are not so precise in form as Blakeney Point or Scolt Head Island, and they do not carry recurves in any full sense of the term. On account of the wide extent of sand flats in front of them, they are not worked upon effectively except in storms or at high tides.

The eastern headland of Wells Harbour is a far more elaborate formation (Fig. 75). It consists of shingle and sand ridges bearing dunes, and arranged in triangular form, one apex of the triangle pointing north, and the two exposed sides facing north-east and north-west. Erosion is particularly noticeable on parts of the north-western side. The early stages of its history are unknown, but in the recent past there have developed new outer ridges on the north-eastern side, which enclose new marshes. The extreme south-western tip shows typical recurves. It seems probable that we have here a normal offshore bar which, largely because of its local position, became bent; once this had taken place, a certain amount of beach-drifting to the south-east and south-west resulted, mainly from north-westerly winds. Probably the wanderings of the harbour channel have helped in the erosion on the western side, but in April 1937, when the area was examined, erosion of the dunes on that side had no apparent relation to the position of the channel which was some distance away. It is also curious that the north-eastern side should show a seaward growth of new ridges. Conjecture alone is possible, but they may be related to the wide expanse of very shallow water which waves from the north-east or east would have to traverse before reaching the ridge. The precise effect of local details might well repay detailed study.

The inner marshes between Blakeney and Wells are typical of this coast; brief mention of them will suffice, since the fuller descriptions of the Scolt Head Island marshes will apply to all on this coast in so far as their general development is concerned. The Blakeney and Wells marshes are in an advanced stage of development and the usual plant zones are present (see below) (**154**). At Stiffkey the sea lavender (*Limonium humile*) is prolific, and gives a glorious show of colour in late June or July. The sea purslane (*Obione portulacoides*) covers large areas to the almost total exclusion of other plants. *Juncus* spp. occurs at the inner margin, whereas the outer, and less developed areas are characterized by *Glyceria maritima*, *Salicornia* spp., and *Suaeda maritima*. The older marshes are enclosed within Stiffkey Meols and the dunes and shingle ridges off Wells. The creek system of the marshes is similar to that elsewhere on this coast, and consequently a description of the creeks and their evolution at Scolt Head Island will suffice and prevent unnecessary repetition. It should, however, be noted that there are two or three permanent channels on the outer flats, one of which is that of the Stiffkey River. Further, the meols off Stiffkey are continued by similar ridges to Morston, where, to the west of the harbour creek there are extensive sand and shingle flats very little colonized by plants.

The inner line of the marshes is marked by a line of low cliffs except to the east of the mouth of the Stiffkey River. Here there is a great shingle bank running as far east as Morston. Solomon regards it as a raised beach. There is much to be said for this view, but the bank is not necessarily the 25-foot beach. It is comparable to the present main Blakeney ridge and, although higher, it might have been formed under present conditions before Blakeney Point and the marshes accumulated in front of it. At its western end it is partly covered by a deposit similar to the brown boulder clay of Hunstanton and, if the correlation between the shingle bank and the boulder clay be accepted, it implies that the bank is at any rate earlier than the last glaciation of East Anglia, and it strengthens the view that it was not formed at present level. The Stiffkey River finds its exit through a sluice, and the lower mile or so of its valley was reclaimed about 1750. Before the reclamation salt water must nearly have reached Stiffkey village.

(4) BLAKENEY POINT: HISTORICAL

It is convenient to call attention here to the growth of Blakeney Point and to the more important enclosures within it, although it means some retracing of the path.[1] Up to about the middle of the seventeenth century no important enclosure had been made, and the marshes and flats were in their natural state. Even in 1823 all tides flowed up the Glaven as far as Glandford; this explains the rather straggling nature of that place and of Wiveton, since they were both built with a view to access to a water front. In the marshes there were five small 'eyes', which originally were islands, and are possibly of glacial origin (see Fig. 76). The most conspicuous is that at the north-west corner of Cley Marsh, known as Thornham's Eye. Little Eye, Flat Eye, and Gramborough Hill remain, but the main beach has been pushed back on to them. There was a sixth 'eye' which was situated near the northern end of the present east bank, but which was destroyed by the sea.

In late mediaeval times, not only was the Cley channel wide and open, but also another running eastwards to Salthouse and Kelling. This is still traceable if viewed from high ground when these marshes are partly flooded. Embanking began in the seventeenth century; there are records of the Dutchman, van Hasedunck, being at work at Salthouse in 1637,

[1] B. Cozens-Hardy, *Trans. Norfolk and Norwich Nat. Soc.* 12, 1924–9, 354.

and building the bank which joins Little Eye and Great Eye to the mainland. This bank gave rise to a good deal of litigation on account of its obstructing the drainage from Kelling and Weybourne. It was followed in 1649 by a bank a little farther west and running due north to the main beach. This bank seems to have disappeared entirely. For about the next two centuries Salthouse main channel was flooded at each tide, and the name Salthouse Broads dates from that period. Calthorpe's Bank was built in 1627: this allowed ships to reach Cley quay, but obstructed navigation higher up the Glaven. This also aggravated the inhabitants of Wiveton, and the bank was demolished, so that once again Wiveton was on tidal water. Later, Calthorpe appears to have enclosed Blakeney marshes and the site of the present golf course, but the precise dates of this

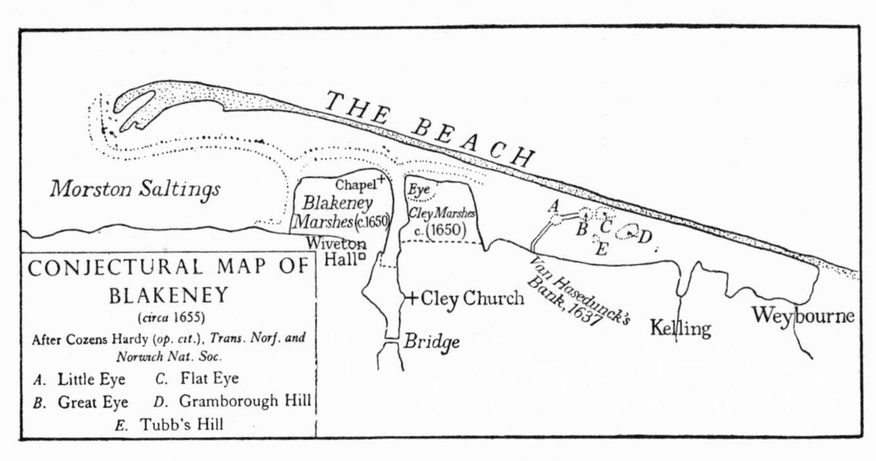

Fig. 76. Conjectural map of Blakeney

work are uncertain. Cley marshes were probably inned about 1650 by Simon Britiffe, who naturally made use of the eye at their north-western corner. Thus the marshes were left for about two centuries, but all this time the sea was encroaching. By 1850 Salthouse channel was nearly blocked by shingle which prevented the outflow of water from Salthouse and Kelling marshes and thus led to their becoming waterlogged. In 1855 a wall was built all the way along the coast from Weybourne, and an outlet was made for the fresh water of the Salthouse marshes into Cley channel. Unfortunately the 1855 bank broke in 1861, and Salthouse marshes again became sodden. This waterlogging increased later since the silting of Cley channel prevented a free fall of water into it from the marshes. It was not remedied by a new cut until 1924. The storms of 1861 and 1897 broke the north-eastern corner of Cley marshes, and the sea wall has also been breached in several places: in 1921 the struggle to exclude the sea was given up. In 1823 an Inclosure Act allowed the making of the present Cley-Blakeney road, which caused further silting of the Cley channel. The road bank broke in 1897 when there was a rough sea in Wiveton valley, but the breach was soon repaired. Before leaving this topic, it is of interest to note that Cozens-Hardy, from a careful study of documents and maps, estimates that between 1649 and 1924 the main beach off Salthouse travelled landwards about 275 yd.; it now partly overwhelms Flat Eye and Gramborough Hill which in 1649 were well inside the main beach.

Unfortunately it is not possible to give figures of any value for the westward growth of Blakeney Point. There is an interesting map of 1586[1] which shows the Far Point of that day almost due north of Blakeney Staithe: the haven channel was farther west as it is to-day, and is shown east of a line due north from Morston church. It is impossible to say how accurate this map is, but the positions of the Far Point and harbour entrance are at any rate consistent with the known prosperity of Cley and Blakeney in those days. There does not appear to be any large-scale later map until that made by Palmer in 1835.[1] This shows the life-boat house and the harbour channel which then ran seawards some three cables westward of it. The existing maps are those of the Ordnance Survey of 1907; they do not show the long ridge which has developed from the Headland and runs to the Far Point of to-day. A full and up-to-date physiographical survey of the Point is very much wanted.

(5) WELLS MARSHES AND HARBOUR: HOLKHAM

This brief account of the embankments inside Blakeney beach has shown some of the difficulties that beset enclosure schemes: advantage to one landowner or place is often a considerable liability to those responsible for adjoining land. This problem is still better illustrated at Wells. The foreshore features to that harbour have already been traced, but before touching on the reclamations there are one or two other coastal features which need mention. Immediately west of the harbour channel are the dunes known as Holkham Meols. These run south-westwards to Holkham Gap, and are there succeeded by another ridge bearing the same name. These ridges are similar to the headland east of the harbour: they are offshore bars, formed mainly of sand with some shingle, and later rendered prominent by dune growth. They are now stabilized by conifers which were planted between 1853 and 1891. These trees give the coast an artificial

[1] Both are reproduced in colour in the Second Report of the Tidal Harbours Commission, 1846.

appearance, which is heightened by the fact that all the marshes behind them are now reclaimed, and because of the wall built the whole way along the western side of the channel leading to Wells Harbour. The nature of the area before this extensive inning is seen on the map (Fig. 77) (**153**).

The Tidal Harbours Report of 1846 makes interesting reading about Wells, because two very different views were taken about the effects of bank building.[1] The harbour is a true marsh harbour, and in many ways similar to that at Blakeney, except that there is no deflecting shingle ridge, and no stream corresponding to the Glaven. The marshes grew up behind the protecting shingle and dune ridges (meols), and a creek system developed, the main creek being the harbour channel. It is important to note the long south-eastward projection of the old marshland along Warham Slade. The earliest noteworthy reclamations were made on the west side of the harbour (1 and 2, on Fig. 77) and the smaller Church Marsh (6) was inned about the same time. Any reclamation must mean that water is excluded from an area it once covered, and consequently there must be less to drain away at the ebb. This implies that the scouring effect of the ebb current in time grows less, and in harbours of this type a strong scour is essential. In 1758 Sir John Turner's bank across Warham Slade was built, and the inning of this area and of Church Marsh gave rise to a most important discussion about reclamation. In a report in 1782, Hodskinson, Grundy, Howard, and Nickalls argued very strongly that to cut off the highest waters was fatal. They contended that those waters, once the marsh was covered, could not get away until the lower marshes were largely uncovered, and that then the upper waters emptied themselves rapidly and so produced the necessary strong scour in the harbour channel. Smeaton's view of the matter is entirely different. He outlines, in a very modern way, the development of marshes on a sand flat, and discusses their drainage and especially the function and nature of creeks. He notes that the scour is not very effective until the water on the marshes has fallen to the point of retreating into the creeks. Further, as long as the marshes have not, in their upward growth, exceeded the height of neap tides, there is a scour developing at every falling tide. Where the marshes are above neap-tide level, the scour, in its total effect, lessens and creeks begin to be choked. He supposes that the extremities of channels will be first affected in this way. Smeaton's opinions are not borne out by observations at Scolt Head Island. He certainly admits that the Warham Slade bank would limit the scour, but he differs from the others in his estimate of the quantitative effects of scouring, and also in thinking

[1] Second Report of the Tidal Harbours Commission: Appendix B, No. 225 *et seq.* p. 437.

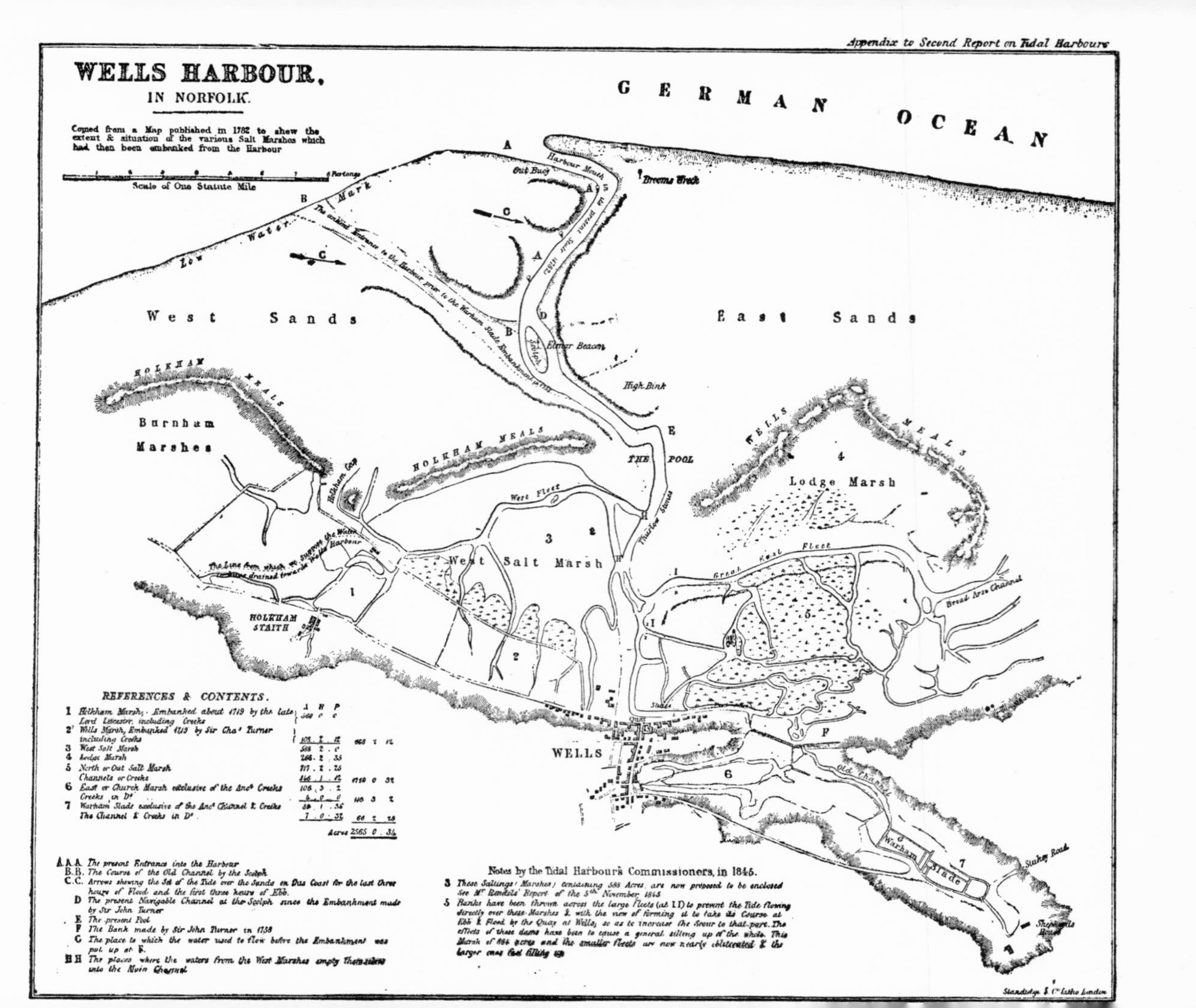
Appendix to Second Report on Tidal Harbours
WELLS HARBOUR,
IN NORFOLK.
Copied from a Map published in 1782 to shew the extent & situation of the various Salt Marshes which had then been embanked from the Harbour
Scale of One Statute Mile
GERMAN OCEAN
West Sands
East Sands
Burnham Marshes
HOLKHAM MEALS
WELLS MEALS
Lodge Marsh
West Salt Marsh
THE POOL
High Bink
West Fleet
Great East Fleet
Broad Arse Channel
HOLKHAM STAITH
WELLS
Old Channel
Warham Slade
Stukey Road
REFERENCES & CONTENTS.
Notes by the Tidal Harbour's Commissioners, in 1845.
Standidge & Co. Litho London

that the main silting was in the upper part of the harbour. The complaints about the depth of the Pool (see Fig. 77) all seem to come after 1758, but this trouble, in Smeaton's view, is but a part of the general problem outlined above, the practical effects of which were beginning to be clear. To other difficulties must be added, according to Smeaton, sand carried up the channel, and blown from Holkham Meols to the Pool. The Pool, in his view, had been shrinking for some time, but the inhabitants became aware of it only when boats could not easily swing in it. Since increasing shallowness coincided fairly well with the building of the Warham bank, it was not difficult to assume that the one was the cause of the other. This opinion Smeaton opposed: 'I must, therefore, conclude..., that whoever would find a cause for the alteration of the course of the out channel, for the filling up of the pool, for the landing up of the harbour, channel, or creek, and in general for the decaying state of the harbour of Wells, must seek some cause widely different from the embankment of the Slade marshes in the year 1758....' There is no need to follow the further reports: suffice it to say that Smeaton did not convince people, but since the banks were built, that ended the matter.

Lodge Marsh was once a farm, but the bank protecting it was destroyed, and in 1905 a new bank was begun across Lodge Gap, but as the material used was sand it was unable to stand up against an abnormal tide in the October of that year. In 1859 the main bank along the western side of the harbour channel was built, and 649 acres were inned.[1] The reclaimed marshes extend westwards from Wells as far as the sea wall running from Overy Staithe to Gun Hill. The foreshore consists of the same extensive sand flats backed by the dunes, and there are numerous traces of the old natural marsh creeks in the reclaimed area. Near Gun Hill several old shingle ridges can be seen in the hollows of the dunes, while Gun Hill itself has extended westwards of late years. The essential point to bear in mind is that, despite its artificial appearance to-day, apart from the constant variation of minor detail, the stretch between Wells and Overy Staithe is the counterpart of the coast to the east of Wells and to the west of Overy, except that west of Burnham Harbour there is more shingle which forms the foundation of the true island of Scolt Head.

[1] For details of minor embanking at Wells and Holkham see Appendix to Second Report of the Commissioners appointed to inquire into Tidal Harbours, 1846, and Royal Commission on Coast Erosion, vol. I, pt II, 1907, Appendix 23, p. 243.

(6) SCOLT HEAD ISLAND COMPLEX

Scolt Head Island, like Blakeney, belongs to the National Trust. The island is irregular, and about four miles long measured along the main beach. Its width varies with the tide, and only the highest tides reach the inner foot of the main dune ridge. At low water, Norton Creek, which separates the island from the mainland, dries in places. There are some shingle ridges (the Ramsey ridges) on the south side, the origin and significance of which is discussed below. The landward marshes, like those elsewhere on this coast, run up to the old cliff line, while the river Burn flows into the marshes at Overy Staithe and reaches the sea at Burnham Harbour. The shingle ridges and dunes of Brancaster golf course, although separated from the island, really form part of the whole complex (Fig. 78). Like the Ramsey ridges they have grown to the east, in a direction opposite to that of the island itself. It follows, therefore, that structures on the island are newer at its western end. The island consists of a main shingle beach with dunes running parallel to the sea, and a series of lateral ridges, partially dune covered, running landwards from the main ridge. These laterals first trend south and west, and then turn to a direction between south and east. These alignments are very different from those of the laterals of the golf course and Ramsey ridges.[1]

Any shingle ridge of the whole complex is composed almost entirely of flint pebbles mixed with a large proportion of sand. Stones, other than flint, form not more than about one-half per cent of the whole. The flints are usually well rounded, but more angular pebbles are seen on shingle flats such as that opposite Beach Point on the south side of Norton Creek. The ridges are typically long and narrow, but some terminate in spreads of shingle which on the laterals is now stable. Dunes have accumulated on them, arising partly because the ridges themselves form an obstacle, but mainly because tufts of *Psamma arenaria* and *Agropyrum junceum*, and sometimes other plants, form definite nuclei around which the embryo dune can grow. All stages of dune development can be seen: new ones form at the Ternery, and they sooner or later unite and form a continuous ridge. When a new shingle ridge forms in front of an old one, the sand supply to the latter is depleted. Some laterals such as those beneath House Hills and Long Hills carry high and well-developed dunes which must have taken a long time to form. This implies that there was little or no tendency for newer ridges to develop in front of them and so cut off their sand supply, but it is not possible to say precisely why this was so. In

[1] For full accounts of Physiography, Ecology, etc. see *Scolt Head Island*, edited by J.A. Steers, 1934.

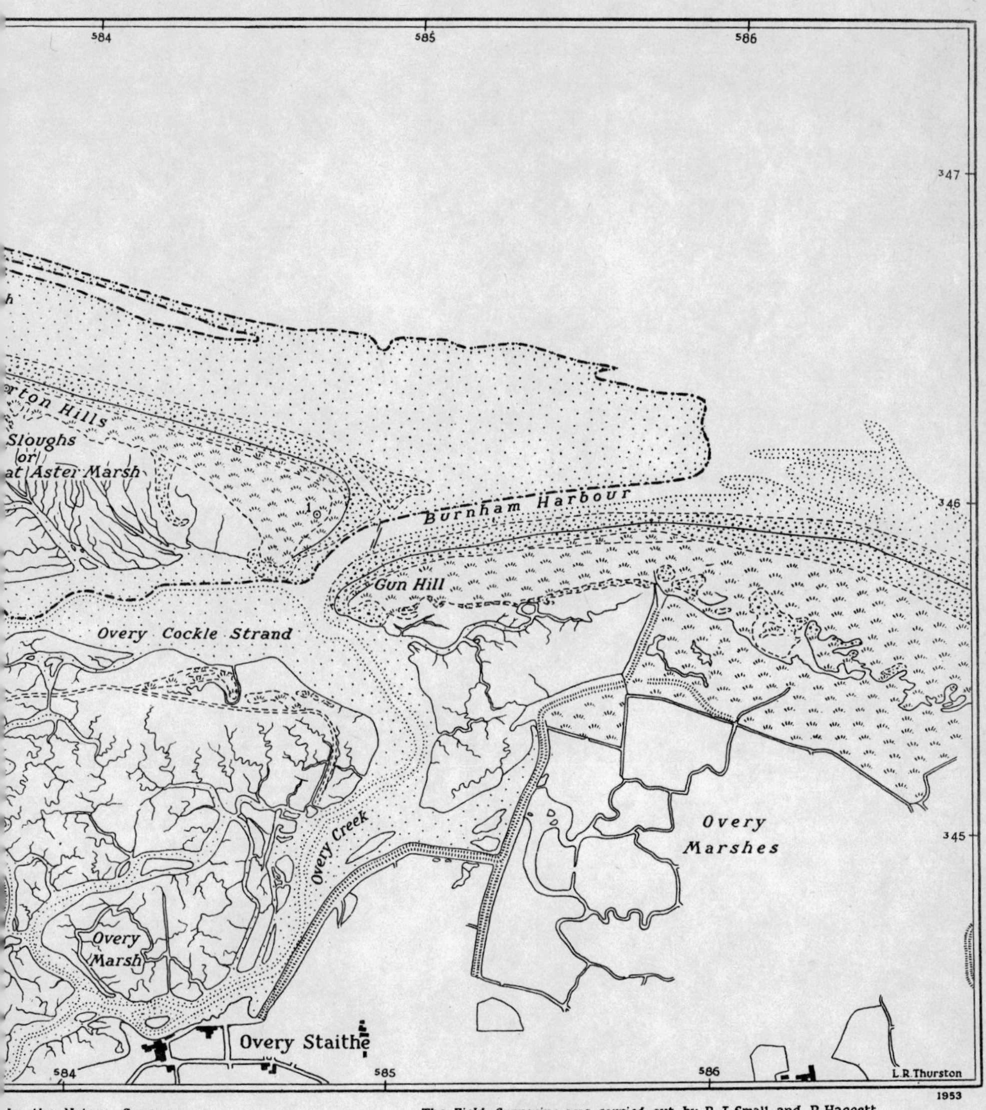

ave Square,
ondon, S.W. 1.

by the Nature Conservancy,

The Field Surveying was carried out by R.J. Small and P. Haggett under the direction of Professor J. A. Steers and A.T. Grove

such formations as Scolt Head Island there is a continual struggle between the forces which extend the island longitudinally and those which lead to recurving and lateral ridges. At certain times one set of forces will prevail conspicuously over the other. Since about 1880, for example, the westward extensions of the island were rapid: of late years there has been some loss. But before 1880 there must have been a considerable period when the Long Hills ridge formed the westernmost lateral, and sand flats lay to the west of it, affording an ample supply of sand to form dunes. The extensive sand flats exposed at low water supply the sand for the dunes on and near the foreshore.

If the sand supply ceases, dunes are liable to suffer erosion. *Psamma arenaria* never forms a very close turf, and it is characteristic of the plant that when the supply of blown sand ceases, it thrives less well; thus old dunes often carry rather dead-looking *Psamma*. Other plants have usually colonized the dunes at this stage, but patches of bare sand are always exposed and may be attacked by the wind. There is no theoretical limit to such an attack until the dune is cut through or removed. The roots of *Psamma* and other plants restrict it to some extent in practice and the normal feature of erosion is a bell-shaped hollow or blow-out. On the island these blow-outs usually face the prevalent westerly winds or the dominant north-easterly winds.

The shingle and dune ridges divide up the island into segments. The parts between the ridges are filled with salt marshes, the foundation of which is sand, and the marshes gradually increase in height from west to east, or from younger to older. This can be clearly seen from a vantage point when a big tide is flooding the marshes; the western marshes are inundated before most of the older ones. The creeks of the individual marshes all drain eventually to Norton Creek. Hence it might seem that the lowest part of any marsh should be near this creek, but this is not always the case. The fact that *Obione portulacoides* so frequently grows along creek margins leads to greater deposition just near the creeks, so that the inter-creek parts remain rather lower. On some of the older marshes there is a thick belt of *Obione* reaching down to Norton Creek. The other marshes are all similar. Plover Marsh, for example, is hemmed in by two long low laterals. The shingle of these ridges extends some way under the marsh, and where marsh and shingle meet on the surface there is a wide belt of *Suaeda fruticosa*. To the north, the shingle ridge is partly obscured by dunes, but the *Suaeda* zone is still present. The vegetation is that normal to a high marsh—*Glyceria maritima*, *Obione portulacoides*, *Artemisia maritima*, *Limonium vulgare*, *Plantago maritima*, *Spergularia media*, *Triglochin maritimum*. In pans and creeks at lower levels there are found *Salicornia* spp., *Suaeda maritima*, and *Aster tripolium*. The drainage is intricate, the numerous small creeks sooner or later converging into one large creek which runs to Norton Creek. The nature of the changes that may take place in a creek is discussed in Chapter XIV.

In 1933 ninety-eight shallow bores were made in the marshes. The two most important conclusions reached from this investigation were (1) that sand, sometimes black in colour, seems to be the real foundation of the island; and (2) that in some places there are buried pockets of soft mud which seem to represent old scour holes filled up with fine material. In several bores alternations of sand and mud were found, and in order to see how these occur a visit to the Ternery is instructive. During the last few years the dunes of the main ridge have been cut through and a large fan of shingle and sand has been pushed over one of the newer marshes within. This fan is encroaching on a pocket of soft mud and may in time bury it. Supposing, however, the breached ridge to remain fairly stable, the fan may settle a little and a new layer of mud, brought in by the tide, will be deposited on it, just as at present, thin and discontinuous layers of mud are gathering over the sandy floors of the newer marshes. Similarly, thin sheets of sand are often blown on to marshes, while sand may also be redistributed by tidal action. In all of these ways irregularities of distribution may take place.

The island has grown westwards, and beach-drifting is undoubtedly the main factor at work. Many experiments with marked pebbles have been made on the foreshore, and it is clear that waves coming in from any direction from between north and a point south of east cause a westward drift of material. Even quite small waves from the north show the same effect in a minor way, but the trend of the beach is important. West of the Head (i.e. Scolt Head) the trend is a little south of west, and so the effect of such waves is normally greater there than elsewhere. Winds and waves from a north-westerly direction have the reverse effect, and if they persist for a few days may drive material eastwards. There is, however, no doubt that the dominant direction of travel is to the west, and this can go on no matter what the direction or strength of the tidal current which here (unlike Blakeney) runs westwards up to two or three hours before high water (Fig. 8). Because the golf-course ridges are completely sheltered by the island, north-easterly winds have little or no effect. But north-westerly and westerly winds are important, and are mainly responsible for the eastward growth of these ridges. In the past similar conditions led to the growth of the Ramsey ridges.

There is no completely satisfactory explanation of the formation of the lateral ridges, here or elsewhere. A shingle ridge at the western end of the island will have its free end turned landwards by the waves. Once such a deflection is made, waves, because of their tendency to break more or less parallel with the shore, will swing round on to the recurved end and force it farther back. Waves approaching directly from a westerly direction will intensify this effect. There is no accounting, as yet, for the angle at which a lateral ridge first leaves the main beach, nor for the change of course about halfway along the length common to so many of the Scolt Head Island laterals. As at Blakeney, the extreme ends of several lateral ridges are strongly bent: this is due to wave action within Brancaster channel.

Old maps of the island are scarce. That of L. J. Waghenaer, 1585, is interesting only in that the island seems to be a spit deflecting the Burn. The next maps of interest are those of Faden (1797), the first edition of the 1-inch Ordnance Survey (1824), that of Bryant (1826), and a special survey made under the direction of J. Dugmore in connection with the Burnham Enclosure Award of 1825. The last shows the main ridge as continuous; House Hills and Butcher's Beach are much as they are to-day, and the Long Hills are also shown. The general positions of Burnham and Brancaster channels are on the whole identical with those of the present, except for the sweeps in the outer sands. A comparison of this map with the Ordnance Map of 1886 shows no great westward growth. Since 1886, or a few years previous, the long ridge between the Head and the Ternery was built. It is comparable to that between Blakeney Headland and the Far Point.

The island has grown westward, and each lateral is a former Far Point. There has also been another movement, a general but not extensive landward retreat of the foreshore. Where waves and scour have removed a good deal of beach, outcrops of marsh mud are seen on the foreshore of the eastern parts of the island. This clearly implies the inward movement.

The evolution of the island is shown in Fig. 79. The Far Point of any time has always had a tendency to recurve; sometimes conditions have favoured the building of a long lateral, at other times of only a short one. The figure also shows that the present length of a lateral is not an exact criterion of its original length, for, on account of the inrolling of the main beach, the present laterals are not necessarily as long as they were at first. The diagram also helps to explain the irregular spacing of the laterals. If conditions over a considerable period of time favoured a relatively rapid western growth, then only short laterals were probably formed. Conversely, the development of long lateral ridges implies a slow westward growth; compare, for example, the rapid growth and comparatively short laterals since 1886 and the Long Hills ridge.

We must now turn to the problem of Burnham Harbour and the Great Aster Marsh at the eastern end of the island. The harbour used to be regarded as a break-through of a continuous dune ridge from Holkham or Wells. The break was ascribed to erosion by a storm coinciding with a high tide within the Scolt 'spit', so that once the sea had made the breach low enough, the impounded waters emptied and scoured out a hollow

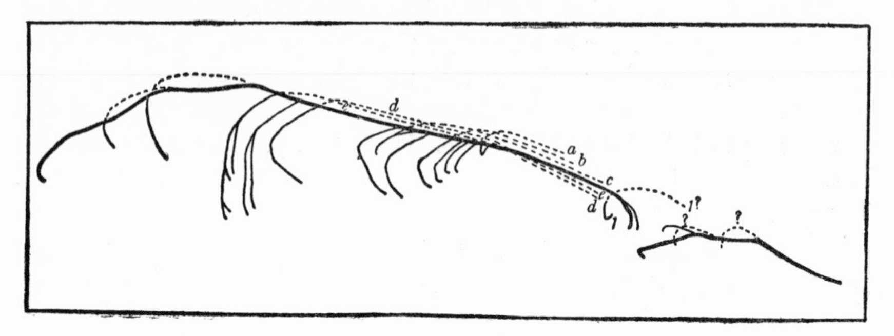

Fig. 79. Evolution of Scolt Head Island

later to become Burnham Harbour. The area now covered by Great Aster Marsh was thus assumed to have been rejuvenated by the greater influx of tidal waters through the newly formed harbour. Indeed, unless such a rejuvenation is granted, the break-through theory of Aster Marsh presents a difficulty, because the marsh should have been in a stage of development characterized by plants like those on the higher marshes just west of it.

A reconsideration of the problem, however, has led to the conclusion that the harbour is not new: (1) the bores put down in Aster Marsh showed no direct evidence of rejuvenation—it appears to be in a young stage of development; (2) the marshes east and south of the harbour seem to have passed through quite normal evolutionary stages; (3) north and east of Aster Marsh is Plantago Marsh, which is aberrant from an ecological viewpoint, but the bores in it lent no support to the rejuvenation hypothesis. The drainage of the whole marsh is toward Burnham Harbour: if a break-through had occurred, the drainage might be expected to resemble that of its neighbours and flow to Norton Creek and the west. The dunes on both sides of the harbour are growing towards it and constricting it, and are consequently favourable to the development of a strong scour which helps the harbour to maintain itself in its present position. Furthermore, granting the break-through hypothesis, a breach would tend to heal itself, and especially so if, as seems reasonable, a fair development of dunes as well as high marsh inside this spit is assumed. In fact, the evidence all seems to be against a *recent* break-through, whatever may have happened in the *early* stages of the island's history. If it be granted that the harbour is a permanent opening, we are far better able to explain the Ramsey ridges. These really form one main ridge with several laterals running south and east from it. They are old ridges and have grown eastwards like those of the golf course. At the time of their formation there must have been open water to the north, and even slightly to the north-east of them, and they are probably earlier than that part of the island west of Norton or House Hills. If we suppose a time when Burnham Harbour was wider than it is now, we may probably think of the Ramsey ridges forming under conditions similar to those now acting on the golf-course shore. Then the tide had easier access to Overy and the marshes near that place which then were in all likelihood bare sand and mud flats. The golf-course ridges were definitely built before the long western tongue of the island was formed. The Ramsey ridges probably grew in a similar situation; a little of the eastern end of the island had formed, and the Overy channel was more open. If such an explanation be rejected, the Ramsey ridges must be assumed to have developed on an open foreshore along which the drift of material was to the east and this is highly improbable (**151** and **152**).

To sum up, the island probably originated as follows. First there was an extensive sandy foreshore with some shingle. The waves separated shingle from sand, and sooner or later piled up the shingle into ridges near high-water mark, so forming an offshore bar. The early ridges were unstable and mobile. Eventually one became more stable, and dunes began to form on it; it gradually extended westwards, sending out a succession of recurved ends which now exist as laterals. At this stage the main ridge was probably increased by newer ridges formed by the waves and pushed back on to the main ridge, a process which can still be seen in action. A walk along the foreshore to-day usually shows one or more long, low, and relatively broad sand ridges which have a channel, draining westwards, between them and the main ridge; these are sometimes pushed back by the waves to form eventually an integral part of the main ridge. Ridges, often largely composed of shingle, are formed from time to time near the Ternery and are pushed on to the Ternery ridge. The material thus added is sooner or later swept round to the Far Point ridge and strengthens it. At other times, under certain conditions of wind and tide, these new ridges are of sufficient size to form a new and permanent addition to the island, and the old Far Point ridge is not increased but remains as the last recurved end, the newer ridge taking its place as the terminal of the island.

This mode of origin and growth is quite consistent with the existence of the Ramsey and golf-course ridges. Each of these groups is due to eastward directed beach-drifting; the golf-course ridges are still growing; the Ramsey ridges originated when the island had extended for only part of its length, but probably as far as House Hills. The long underlying shingle bars of these hills and the hills themselves both suggest a considerable halt in the westward growth, and conditions for the formation of the Ramsey ridges would have been very similar to those controlling the golf-course ridges' development to-day.

Although Scolt Head Island and Blakeney Point are similar, there are also interesting differences. There is nothing at the former to rival the great outer shingle beach at Blakeney. Along much of the outer beach at Scolt Head Island there is comparatively little shingle, which is relatively far more abundant at the Far Point and on the laterals. The spacing of the laterals is more regular at Scolt Head Island than at Blakeney Point, and they are generally somewhat longer. Hence, the intervening marshes are more distinct, and there is no exact counterpart to the almost bare sand flats within Blakeney Point. The most obvious difference, however, is that Scolt Head is an island whereas Blakeney is joined to Weybourne cliffs. In this sense it is fair to regard the first as an offshore bar and the second as a spit. Yet in physiographical essentials they are the same. The western part of Blakeney Point is in shallow water and comparable in every way to Scolt Head Island. Conditions are certainly favourable for the travel of material westwards from Weybourne cliffs, and there seems little doubt of the early forming of a spit there. Farther west, however, in the shallow water, conditions would favour the growth of an offshore bar. Both processes could easily have been concurrent and the spit could have lengthened into a bar. Although there is no proof, it may well have been that there grew first a spit and an independent offshore bar farther west, and that a time came when these joined. If this happened, the later inrolling of the whole formation could easily have obliterated the point of junction and any associated spread of shingle. This, however, is pure conjecture. The point needing emphasis is that all along this coast offshore bars are characteristic. The much greater amount of shingle at Blakeney and the fact that it is joined to the mainland should not be taken automatically as arguments that Blakeney Point is different in structure from Scolt Head Island, Wells Headland, or the smaller bars off Thornham, and elsewhere. The maps in Fig. 80 show the way in which the Far Point at Scolt Head Island has changed in recent years.

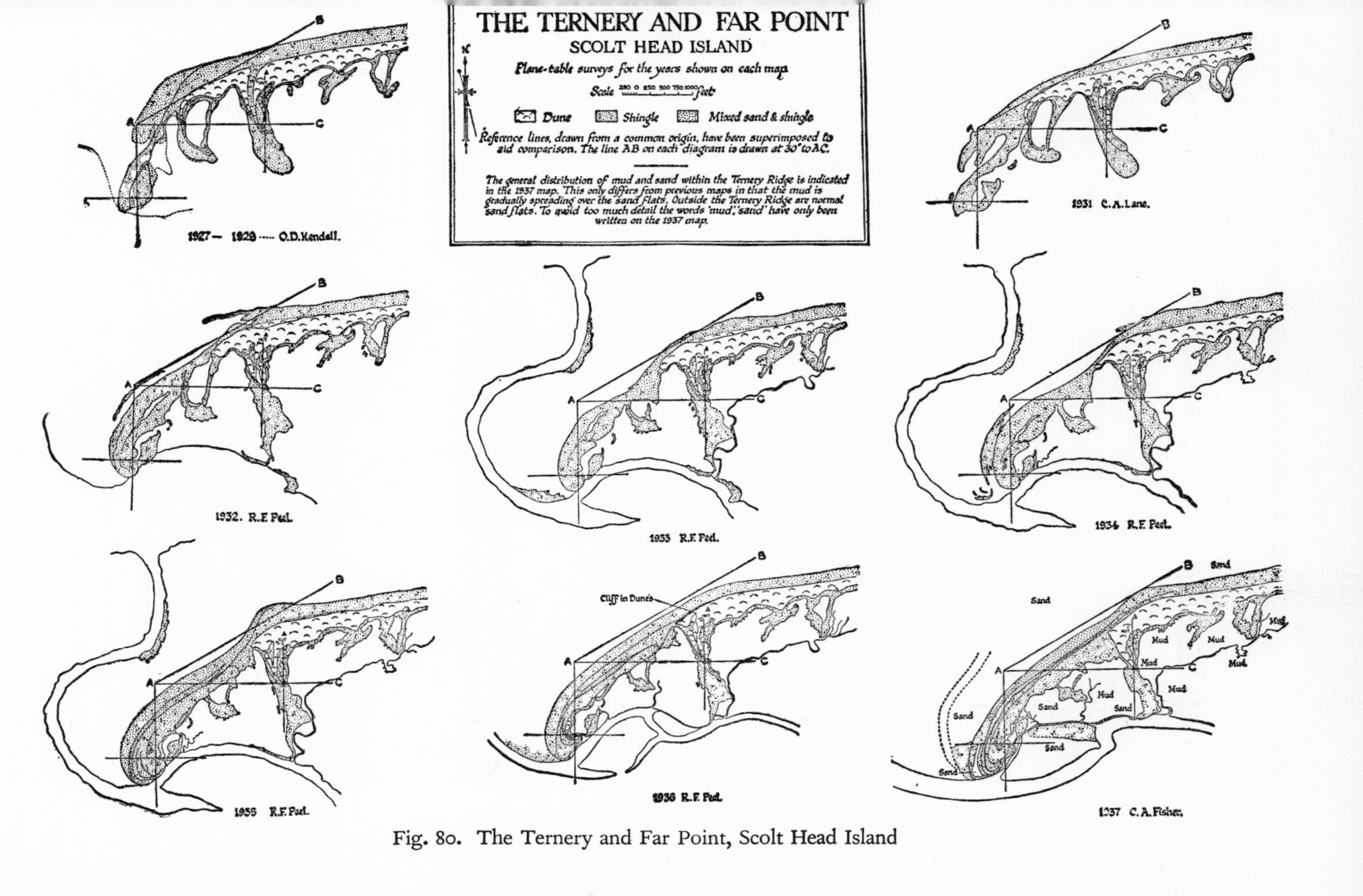

Fig. 80. The Ternery and Far Point, Scolt Head Island

(7) BRANCASTER TO HOLME[1]

The ridges forming Brancaster golf links, when traced westwards, give place merely to a sandy and shingly foreshore off Titchwell, where there is another embanked marsh. In these reclaimed marshes there are also traces of old ridges and dunes that can be followed into open natural marsh off Thornham (Fig. 81 *a*). Thornham Harbour is the counterpart of Wells, although on a smaller scale: it is a true marsh haven. The channel winds about in the sand flats, and higher on the beach, on either side of the harbour, are small sand and shingle ridges which form islands at high water. The island to the west of the harbour is interesting and has been carefully mapped several times. It is a miniature edition of Scolt Head Island: it has at least one good lateral, and dunes occur on its main crest especially at the eastern and western ends. Although its growth is restricted owing to the enclosure wall of Holme marshes, it has clearly grown a little to the west. It represents an intermediate stage between the immature ridges at Stiffkey, and the highly developed ones of Scolt Head Island. The ridges east of Thornham Harbour are much the same: all are small, and storms can, and do, cause great and rapid changes. Behind the ridges there is natural salt marsh. The major creek sends one branch up to the old Staithe, and the other runs eastward to the wall of Titchwell marshes through which there is a sluice. The same creek can be traced through Titchwell and is practically continuous with Mow Creek inside Brancaster golf course. Hence, it follows that the inner and older ridges, as well as the outer ridges of the present shore, bear to this creek much the same relation, although in one sense duplicated, that Scolt Head Island does to Norton Creek. In other words the development of offshore bars persists, but they are here less conspicuous partly because immature and partly because the sea walls of Titchwell reach them.

West of Thornham is another reclaimed area at Holme and Old Hunstanton. Although the foreshore dunes appear quite natural, they rest in places upon a sea wall. Broadwater within it is merely part of the old creek system expanded to form a long and narrow lake. Erosion is serious at the eastern end at Gore Point where the drainage of the natural Holme Marsh runs out. From Gore Point the older dunes stretch south-westwards to Hunstanton cliffs, but a new series of shingle ridges and dunes has grown out in front of the older dunes, thus forming the golf links and Holme Marsh.[2] The most interesting point about these newer ridges is

[1] J.A. Steers, 'Some Notes on the North Norfolk Coast', *Geogr. Journ.* 87, 1936, 35.

[2] A.S. Marsh, 'The Maritime Ecology of Holme-next-the-Sea', *Journ. Ecology*, 3, 1915, 65.

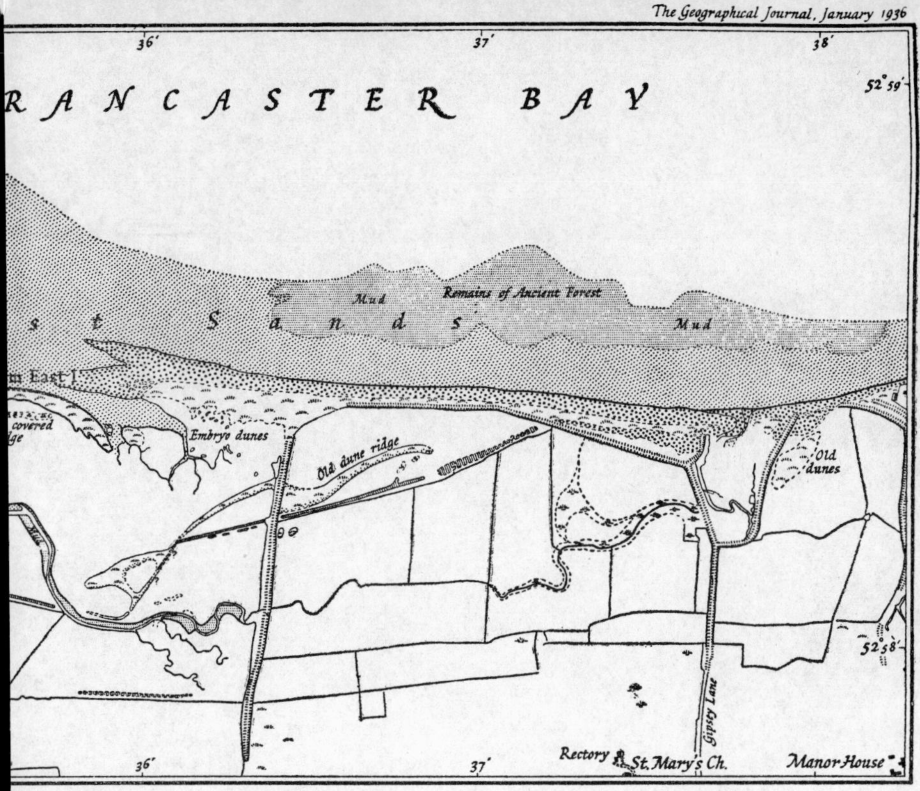

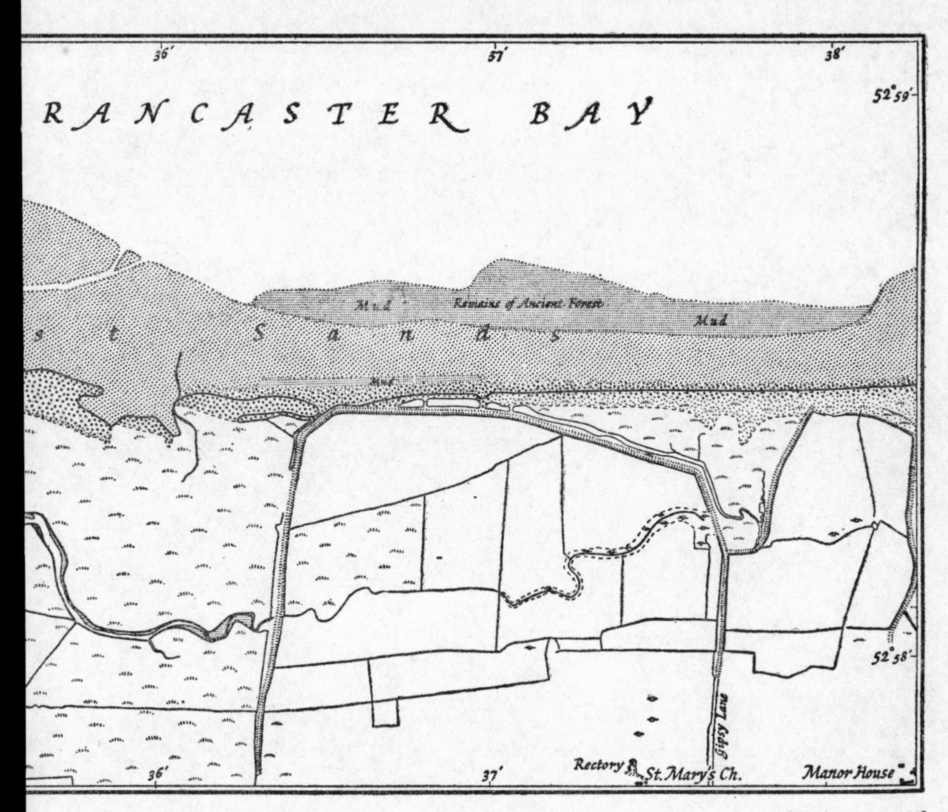

[To face page 366]

that they have extended eastwards or north-eastwards, in opposition to the prevailing movement on this coast. There does not appear to be any entirely satisfactory explanation of this process, but it is almost certainly related to the fact that the ridges are at the mouth of the Wash; they lie opposite to a fair stretch of open water over which winds from a westerly quarter can have a considerable effect. As in the case of the eastward movement of material at Wells (eastern) Headland, this feature shows the difficulty of applying general rules to a long stretch of shore. In general, a westerly movement of beach material certainly prevails from Sheringham to Hunstanton, but in these two places it does not.[1]

The channel of Blakeney Harbour is deflected eastwards on the sand flats. This change in direction also obtains in the channels of Burnham, Brancaster, and Thornham Harbours. It is less noticeable at Wells. All are subject to considerable changes in storms or at big tides, but the reasons given for the deflection at Blakeney apply, *mutatis mutandis*, at the other havens also.

Between Brancaster and Holme the most extensive exposures of submerged forest probably belong to the same series as that seen occasionally on the Suffolk coast. At Brancaster it is the middle bed of three: the lowest is usually invisible, but appears now and then near Brancaster Staithe; the highest is a very fragmentary and sandy bed, about 4 or 5 in. thick, that can be seen just west of the club house on Brancaster links. The nature and significance of these beds is discussed on p. 402.

(8) BRIEF SUMMARY OF THE VEGETATION OF THE NORFOLK MARSHES

Little has been said of the nature of the marsh vegetation, which is uniform all along this coast. The subject is treated more fully in Chapter XIV, but a brief summary is relevant here.[2] The Algae, whether rooted or lying loose on the sand flats, are the first plants to induce the accumulation of mud, and mud is also deposited directly on bare sand in sheltered places. If the substratum of mud and sand is firm, the first plants to thrive are *Salicornia* spp. and *Suaeda maritima*. On loose sloppy mud, *Zostera* spp. is first. *Glyceria maritima* is often contemporaneous with these, but is commonest on sandy areas. *Aster tripolium* follows closely, and can often be seen invading spreads of *Salicornia*: it is characteristic of wet, low,

[1] The general nature of the coast between Brancaster and the Wash can be seen on Figs. 78 and 81.

[2] See V.J. Chapman in *Scolt Head Island*, 1934: a brief account will be found in *Proc. Geol. Assoc.* 40, 1929, 341.

muddy marshes that are intersected by numerous incipient creeks. As the level rises, all these plants, except *Zostera*, are found; they gradually give place, however, to *Spergularia media*, *Triglochin maritimum*, *Armeria maritima*, and to *Obione portulacoides* which nearly always begins its growth on creek banks and thence spreads rapidly over great areas of marsh, completely smothering and obliterating all in its path. *Limonium humile* gives brilliant masses of colour on these middle-high marshes in July. At higher levels *Plantago maritima*, *Juncus* spp. and *Artemisia* are often prolific. Over all except the highest marsh levels it is common to find a thick spread of the unrooted alga, *Pelvetia canaliculata*, which is also abundant in the pans. The development of pans, creeks, and other physical features of the marshes is better dealt with in Chapter XIV.

(9) THE DECAY OF THE PORTS OF NORTH NORFOLK

The marshland coast of Norfolk is now commercially almost derelict, but in the past it was very different. In the fourteenth century, for example, the harbour of Cley-Blakeney was thriving.[1] At that time Norfolk produced goods required by the Low Countries, and a sheltered harbour, suitable for the small vessels of the period, was all-important, especially for the export of wool and cloth. Many early records concerning Cley deal with the compulsory purchase of supplies for the army of the Black Prince in France: the inhabitants of Cley, for instance, were frequently required to produce salt fish. In 1523 Wolsey again ordered Cley to supply provisions for military purposes. A consideration of the two short tables below will serve effectively to show the high importance of this haven in the sixteenth century:

Muster Roll, 1570, numbers of mariners and of all ships above 30 tons

Port	Ships	No. of mariners
Lynn	24	141
Holme	2	9
Wells	10	69
{Blakeney	11	52
{Wiveton	8	43
{Cley	13	65
Salthouse	—	38
Kelling	—	5
Weybourne	4	17

[1] B. Cozens-Hardy, *Trans. Norfolk and Norwich Nat. Soc.* 12, 1924–9, 354.

Another return in 1582 brings out very clearly the size of the ships using Cley at that time:

Port	No. of ships of 100 tons or over
Cley	2
Wiveton	4
Blakeney	1
Lynn	2
Yarmouth	4
Wells	3

During the early Middle Ages, the trade of this and of other north Norfolk ports was chiefly in the export of wool, and in the import of manufactured goods from the Continent. In the later Middle Ages, that is from the thirteenth to the fifteenth centuries inclusive, the export of different kinds of cloth became more important than that of raw material. Probably also there was intercourse with Iceland in connection with whale fishing. By the seventeenth century local coasting trade had become more important than foreign: grain was sent to the north, and coal was brought back. In 1728 we read of Cley 'from whence great quantities of malt and all sorts of corn and grains and divers goods and merchandize have been from time to time exported, and great quantities of Coals, Iron, Fir, and other Timber, Deal, Tiles, Stone, Salt and other merchandize imported'.

Up to the middle of last century at least, long lines of farm waggons filled with corn waiting to unload at the quay were a common sight. The same was true of Thornham, but there the harbour was never as important as that of Cley. Some trade still survives at Wells; its glory indeed has departed, but there is not the complete decay characteristic of the other ports. The building of the railways finally confirmed their decline although silting and sanding had long been growing and insoluble problems. The increase in the size of ships was also an important factor in diverting shipping from the small Norfolk harbours.

In ancient history and in the Dark Ages this coast must also have been of importance. Brancaster (Branodunum) was a Roman fort[1] the site of which lies about half a mile east of the church and on the higher ground immediately adjoining the marshes. Recent excavations and finds suggest that it belongs to the third and fourth centuries, and when it was built there may have been some kind of navigable channel running to it. Farther east, on Holkham marshes, the Ordnance Map marks a 'Danish' camp.

[1] J.K. St Joseph, 'The Roman Fort at Brancaster', *Antiq. Journ.* 16, 1936, 444.

The Brancaster fort is about five miles east of the place where the Peddar's Way reaches the coast. In Lincolnshire there is another Roman road running to the coast south of Skegness. The possibility of a Roman ferry service having existed across the Wash was first suggested by Canon Tatham in 1921. Phillips[1] notes that the line of the Peddar's Way now marked on the map between Ringstead and Holme is not necessarily authentic: all that is certain is that the Way reached the coast somewhere between Hunstanton and Holme. It is more difficult to find the end of the Lincolnshire road, but a ferry is a possible explanation of these two roads ending abruptly on the coast, and the Wash may have been easier to cross in those times than it is now.

(10) HUNSTANTON CLIFFS

At Hunstanton is the well-known cliff section in the Cretaceous beds. With the exception of a little Chalk in the cliffs at Weybourne, and the abrasion platform in that rock off Sheringham and Cromer, it is the only outcrop of solid rock on the East Anglian coast north of the London Clay at Bawdsey except for minor occurrences of Crag near Caister and on the Suffolk coast. The base of the Hunstanton cliffs is formed of the Carstone (Lower Greensand), the Red Chalk follows, and capping it is the Lower Chalk. The Lower Greensand is well bedded and jointed, and the lowest visible beds form a shore platform, which shows clearly two sets of rectangular joints. In the cliffs the vertical jointing helps erosion, and small niches and pillars have been formed, but the softness of the rock prevents their development on a large scale. In detail the erosion of the box-stones is worth noticing. The Red Chalk does not show any striking form (except for its bright colour), although the Dirt Bed is easily picked out by the waves where they can reach it. The White Chalk is gradually undercut by the wearing away of the underlying rocks and falls in massive blocks. The balance between the rate of marine erosion, the nature of the bedding, and the characteristics of the rocks gives almost vertical cliffs. The northward component of the dip brings the Chalk to sea level at Old Hunstanton where the cliffs end[2] (150).

The old cliff line can easily be traced the whole way from Weybourne to Hunstanton and it often corresponds fairly closely with the main coast road. The marshlands have grown out in front of it. Southwards from Hunstanton the cliffs again recede behind marshland. The Heacham River is deflected a mile or two to the south by shingle and sand drifting in from

[1] C.W. Phillips, 'The Roman Ferry across the Wash', *Antiquity*, 6, 1932, 342.

[2] See *Mems. Geol. Surv.* 'The Geology of the Borders of the Wash', W. Whitaker and A.J. Jukes Browne, 1899.

the north. The only point which needs attention is the occurrence near Wolferton and Dersingham of deep, narrow, and short valleys in the old cliffs, here formed of Sandringham Sands or partly of glacial material. They open out into wide flats which extend to the Wash, are quite dry, and were almost certainly formed under conditions that do not now exist. The mode of their formation is unknown, but, locally, they form a rather striking feature.

(11) THE CLIFFS BETWEEN WEYBOURNE AND HAPPISBURGH

It is now convenient to return to Sheringham and to consider the features of the coast eastwards from that place. As a preliminary, however, the nature of the cliff section between Weybourne and Happisburgh needs comment, and secondly, the foreshore characteristics and the changes at Cromer in historic times. It is, perhaps, rather awkward not to work systematically in one direction; but the arguments already presented show that, not only is there a great contrast in the type of coast east and west of Weybourne, but also that, because of the different directions of movement of beach material, it is more logical to make Sheringham a starting-point

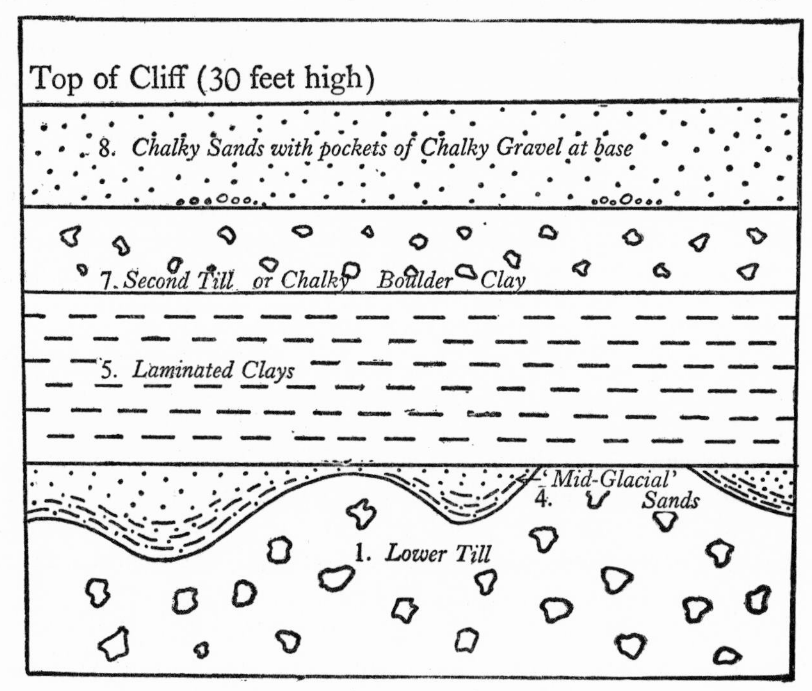

Fig. 82. Section north-west of Happisburgh Gap (after Solomon, *op. cit.*)

for coastal studies in each direction. Between Weybourne and Happisburgh the cliffs are all of glacial material: they form probably the most significant section of Quaternary deposits, not only in these islands but in the whole world. Since the structure of the cliffs is simpler towards their eastern end, a description of them beginning at the eastern end is the most satisfactory. They rise from under sand dunes, and about 100 yd. east of Happisburgh Gap the general sequence of the deposits is that shown in Fig. 82. The Lower Till is regarded by Solomon[1] as a deposit of the first glaciation, that of the North Sea ice, but in order to make this account clear it is best to give the complete sequence recognized by Solomon in tabular form:

Glaciation		Local Deposits in cliffs	Terms used by earlier writers
(Hessle)	H.	12. Brown Boulder Clay	=Hessle Clay of Woodward
(Little Eastern)	L.	11. Cromer Ridge Gravels	=‘Cannon Shot Gravels’ of earlier writers (in part)
		10. Sandy Brickearths, etc.	
	i.	9. Bacton Valley Gravel	=Bacton Valley Gravel of earlier writers
(Chalky or Great Eastern)	C.	8. Chalky Outwash Sand and Gravel	=‘Mid-Glacial Sands’ of Wood and Harmer
		7. Chalky Boulder Clay	=‘Marly Drift’ and ‘Contorted Drift’ of Reid. Also ‘2nd Till’ of Reid at Happisburgh
		6. Sands	
		5. Laminated Clays	=‘Intermediate Beds’ of Reid
(North Sea)	N.	4. Sands and Clays	
		3. Upper Till	=‘Stony Loam’ of Reid at Mundesley
		2. Mundesley Sands	
		1. Lower Till	=‘1st Till’ of Reid at Happisburgh; ‘2nd Till’ at Mundesley

It will be appreciated that nowhere are all these deposits visible at one place. Moreover, although generalizations are apt to be misleading, it is safe to say that the newer deposits come in to the west, and that some, including the Lower Till, occur in places in many parts of the entire section. Solomon's work followed the earlier mapping of the cliffs by Reid. The section is often much tumbled, and deposits vary a good deal: hence it is

[1] J.D. Solomon, ‘The Glacial Succession on the North Norfolk Coast’, *Proc. Geol. Assoc.* 43, 1932, 241.

not surprising that even now the sequence is open to criticism. However, since the nature of the cliffs, and not the great difficulties of precise recognition and dating of the deposits, is our primary concern, Solomon's scheme and nomenclature will serve very well (**156**).

The Happisburgh Gap section is simple: thence to Walcot there is not much variation, and between Walcot and Bacton the cliffs are often buried in sand. About half a mile north-west of Bacton Gap, the Bacton Valley Gravels (i. 9) appear, and a little beyond is the first typical basin of Chalky Outwash Sands (C 8) overlying a very Chalky Grey Till. About a mile from the gap the Till (N 1, N 3) is duplicated and separated by the Mundesley Sands (N 2) which thicken rapidly. Roughly 300 yd. from Mundesley parade the section is:

C 8 or C 6	Chalky Outwash Sand and Gravel with occasional patches of reconstructed Chalk; (?) Boulder Clay	25 ft.	
N 3	Upper Till: Brown, stratified	6 „	'Stony Loam' of Reid
	Upper Till: Blue, poorly stratified	14 „	
N 2	Mundesley Sands	45 „	
N 1	Lower Till	0–10 „	'2nd Till' of Reid
	Cromer Forest Bed Sands and Clays		

The big contortions, due to ice pressure, which are so characteristic of the western part of these cliffs, begin near the Clarence Hotel, and the section is almost undecipherable, but beds C 5 and C 7 appear to be present. Near Trimingham there is an undisturbed section:

C 8	Chalky Outwash Sands and Gravels	20–30 ft. to top of cliff
C 7	Chalky Boulder Clay	to 10 ft.
C 6	Gravelly and Chalky Yellow Sands	„ 12 „
C 5	Laminated Stoneless and Chalky Clay	„ 20 „
N 4	Clean Yellow Sands	10 ft.
N 3	Upper Till	40 „
N 2	Mundesley Sands	15 „
N 1	Lower Till	10 „ to beach

This is the most complete succession of the North Sea Drift and Great Eastern deposits.

Landslips and contortions obscure the section to the west; apart from minor variations there is no great change as far as Kirby Hill, which is made up of a 'basin-shaped mass of chalky outwash material, some 100 ft. in thickness, with a trace of Ridge Gravel above'.[1] Lighthouse Hill

[1] Solomon, *op. cit.* 254.

(Cromer) is similar, but the Ridge Gravel is thicker. West of Kirby Hill the sub-divisions of the North Sea Drift (N) are not clear, and Chalky Boulder Clay increases towards Cromer, where just to the west of the promenade there appear

C 7	Chalky Drift, streaky with patches of sand and grey till	60 ft.
N 3 / N 1	Till, brownish and streaky above, grey and homogeneous below	20 ,, to beach

There is no substantial change at Cromer and the Runtons, but the great erratics of Chalk first show about 300 yd. west of East Runton Gap. The Till (N 1) beneath them is curiously undisturbed: the Till (N 3) overlies the erratics, suggesting that they were transported during the North Sea Drift. They were later affected by the Great Eastern ice. At the western end of the erratics the succession is:

C 8 / C 7	Chalky Boulder Clay, sand and gravel, contorted	40 ft.
N 3 / N 1	Till	to 6 ft.
	Leda myalis Sands, etc.	20 ft. to beach

There is not much difference at Sheringham, except that sand basins occur at the cliff top. Beeston Hill is built up of basins of coarse Chalky outwash gravel contorted with Chalky Boulder Clay. Skelding Hill is similar, but here the glacial disturbances have also affected the lower glacial deposits. Near the life-boat house the contortions disappear. The Till gradually thins out towards Weybourne, where the Chalky Boulder Clay falls below beach level through the Chalk which forms the base of the cliffs.

The whole line of cliffs is formed of soft and incoherent materials. Clay or loam seams often occur in the gravels, and they, as well as the Till, are often contorted so that the limbs of the folds frequently slope seawards.[1] The falls are nearly all due to land water rather than to direct sea erosion, but the waves, after a fall has taken place, remove the debris, and so help to prepare for another one. The inclined clay or loam beds, when saturated with land water, form conspicuous sliding surfaces. Wind erosion of sand beds in the cliffs also produces falls, but usually on a smaller scale. These layers are hollowed out by the wind, and collapse of the overlying material follows. Instances of this kind of fall are often obvious between Cromer and Sheringham. At Beeston Hill, the loose and complicated structure causes an almost continuous steady fall of sand and pebbles, which is accentuated in high winds. West of East Runton Gap a great Chalk erratic reaches nearly to the cliff top. It is part of the Till, and con-

[1] See *Geogr. Journ.* 93, 1939, 399.

ditions are most unstable. The constant falls, caused by the undercutting through wave action of the sandy and clayey beds, yield vertical cliff faces. It is this part that shows curved and folded structures well, and when the 'dip' is seawards cliff falls are frequent and often serious. In 1938 a seam of sand was visible in the cliffs at Walcot at about high-water level. As this is washed out, the overlying Till collapses and allows fairly rapid erosion of the low cliff, but although loss has been continuous, there are no reliable measurements. It may be added that the stratigraphically important outcrops at Cromer of the Crag, Forest Bed, and *Leda myalis* beds (see below, p. 404) are of less significance to this study since they react like the glacial beds. From Weybourne to a point beyond Sheringham the cliffs are fronted by the abrasion platform cut in the Chalk. The flints produced by this erosion gather to form the Cromer Stone Bed.

The cliffs are, after all, but part of the seaward side of the moraine district of North Norfolk. The whole is a most attractive region which stands in strong contrast to all parts adjacent to it. Eastwards from Mundesley the coast is spoiled by poor development, but the natural scenery is also less striking than it is around Overstrand, Cromer, and Sheringham.

(12) CHANGES IN HISTORICAL TIMES NEAR CROMER

That erosion has been serious along these cliffs for many centuries is evident from the disappearance of Shipden, the predecessor in settlement of modern Cromer.[1] Shipden was first mentioned in the Domesday Survey, and Cromer, as a name, appeared first in 1262. In 1452 the form 'Shipden-alias-Cromer' occurs, and after 1483 Shipden falls out of use. By 1337 the sea had certainly made serious inroads at Shipden, for at that date its church had for twenty years been washed by the sea, although the old town was still inhabited. Shortly before 1391 erosion had become so serious that a jetty had to be built for the defence of the local fishing boats. The erosion could not be stopped. Camden says that another attempt to build a jetty was made just prior to 1580: presumably the earlier one had disappeared. A third attempt, also fruitless, was made in 1731. There were many other similar struggles and failures, and the present facing walls and jetty date from 1845. Shipden lay below the cliffs, and its site is now completely washed away. Its former church is held by many to lie about 400 yd. out to sea beyond the jetty, where Church Rock survives, a mass of squared flints and mortar. Others, however, claim that the site is half a mile out to sea where there are also blocks of masonry. Such incursions of the sea are, unfortunately, only too common in the history

[1] W. Rye, *Cromer Past and Present*, 1889; W. Rye, 'Incursion of the sea in Norfolk', East Anglian Handbook, 1881, p. 153; A.C. Savin, *Cromer, A Modern History*, 1937.

of Norfolk. The village of Snitterly, for instance, is said to have lain off Blakeney, and certainly at one time the term 'Blakeney-alias-Snitterly' was used. Sheringham was formerly of greater importance. Overstrand suffered severely even before Richard II's time. Severe breaches have occurred at Mundesley, near which place there appears to have been a village of Eidesthorp, now vanished. Keswick had a fine church in 1382, of which only the site remains. The church at Happisburgh would certainly not have been built in its present position, if the strength and persistence of erosion could have been foreseen. Whimpwell, which used to lie between Happisburgh and Eccles, has also disappeared (**157**).

(13) HAPPISBURGH TO FLEGG; THE HORSEY BREACH; ECCLES CHURCH

Beyond Happisburgh, and as far as the former island of Flegg, the coast is low. It is backed only by a thin line of dunes and the flat land behind them is often below high-water mark. This stretch is in some ways like the marshland coast; there are, indeed, no marshes, but the offshore bars are represented by a continuous line of beach, with some shingle and dunes separating the sea from the low land within. In their natural state dunes are not formed regularly: some grow higher than others; wind gullies develop, are cut lower and become lines of danger to sea attack. Hence these dunes have been to some extent artificially controlled; hollows are filled and the crest-line is kept reasonably level: *Psamma*, in particular, is planted to help guide their growth. But whilst this support may be of some value, it cannot overcome the main difficulty. If the coast is to be even reasonably safe from erosion, there must be a constant supply of new material brought to it to replenish loss. The drift is here to the east, and it might be argued that the waste of the cliffs east of Cromer should help to build up the beaches off Palling and Horsey, but unfortunately it does so only to a very limited extent. Carruthers thinks that most of this material goes instead to offshore banks. If this be so, and if it should continue, the stretch of coast just described will remain a region threatened by erosion, and man, even with the planting of faggots to retain a beach, and with the training of dunes (processes of somewhat dubious value), will be fighting a losing battle against the sea. How acute is the struggle, was made clear only a few years ago by the serious breach made at Horsey.

But before discussing the effects of this breach and the features of the existing coast, the peculiar configuration of north-eastern Norfolk deserves comment.[1] The cliffs disappear at Happisburgh, but reappear near Winterton, and apart from small gaps continue to Caister. Then follows

[1] Steers, *Trans. Norfolk and Norwich Nat. Soc.* 15, 1941, 231.

the wide gap of the Bure-Waveney outlet, and south of that are cliffs as far as Lowestoft Harbour. The coast from Winterton to Caister is the eastern side of the former island of Flegg, really a group of islands. The flat land behind the coast from Happisburgh to Winterton runs inland, and includes the whole of the Broads area, and much of the Bure valley up to Coltishall at least. This lowland narrows between Thurne and Acle, to widen once again from Acle to Reedham; between these two towns and the sea there is only Yarmouth (*q.v.*) and the northern part of the Isle of Lothingland. A long arm of the sea formerly ran up the Yare valley as far as Norwich, while a second arm extended behind Lothingland to join the Waveney valley and its wide expanse of low ground to beyond Bungay. Although Lowestoft Harbour is apparently mainly artificial, Lothingland was formerly an island, or at most joined to Suffolk by a narrow neck of low ground occasionally overridden by the sea. Behind the Horsey sand hills, Flegg, and Yarmouth, there are more than 56,000 acres of marshland all below 10 ft. high, and often below Newlyn datum. Removal of the protecting dunes and but a very slight downward movement of the land would convert all these acres, and many more in the Yare and Waveney valleys, into wide open estuaries. Such an outline description gives greater significance to comments on the effects of the Horsey breach, and later to those upon the Broads and on modern coastal features.

On 12 February 1938 the dunes at Horsey[1] were breached by waves during a north-westerly gale: the width of the breach reached 517 yd., and salt water pouring through this gap covered 7,469 acres (Fig. 83). The breach was repaired by building a barrier of sandbags high enough to keep out any ordinary tide. But on 3 April there was another north-westerly gale, and nearly all the new defences were washed inland, although the flooding within was not so extensive as that in February. The effects of this flooding on the fauna and flora, both natural and artificial, have been studied in some detail. The full results, while of much interest, cannot be discussed in this place, but a short description of the character of the flooded land affected is interesting: 752 acres arable, 3,459 acres grazing ground, 47 acres osiers, 174 acres woodland, 1,660 acres rough ground, 86 acres building sites, gardens, etc. Broads, reeds, and ronds, 1,291 acres.

The breach was made where the Hundred Stream reaches the coast. The early history of this river is almost unknown, but it was probably

[1] J.E.G. Mosby, 'The Horsey Flood, 1938', *Geogr. Journ.* 93, 1939, 413. For full details see J.E. Sainty, J.E.G. Mosby, A. Buxton and E.A. Ellis, *Trans. Norfolk and Norwich Nat. Soc.* 14, 1938, 334–91; A. Buxton and E.A. Ellis, *ibid.* 15, 1939, 22–41; A. Buxton, *ibid.* 15, 1940, 150–60; 15, 1941, 259–68; 15, 1942, 332–42; 15, 1943, 410–20.

embanked in the twelfth century when its mouth was open. Later the mouth silted up, and the gap in the sand hills was filled in artificially, possibly about two centuries ago, but the actual date is lost. Faden's map of 1792 shows that the mouth was blocked by that year. After 1812 the old waterway which joined Horsey Mere with the Hundred Stream was filled in, and both the mere and Horsey Level were diverted to the Thurne. The Hundred Stream then fell into disuse, and its bed is now normally almost dry. There are records of breaches in these same sand hills as far

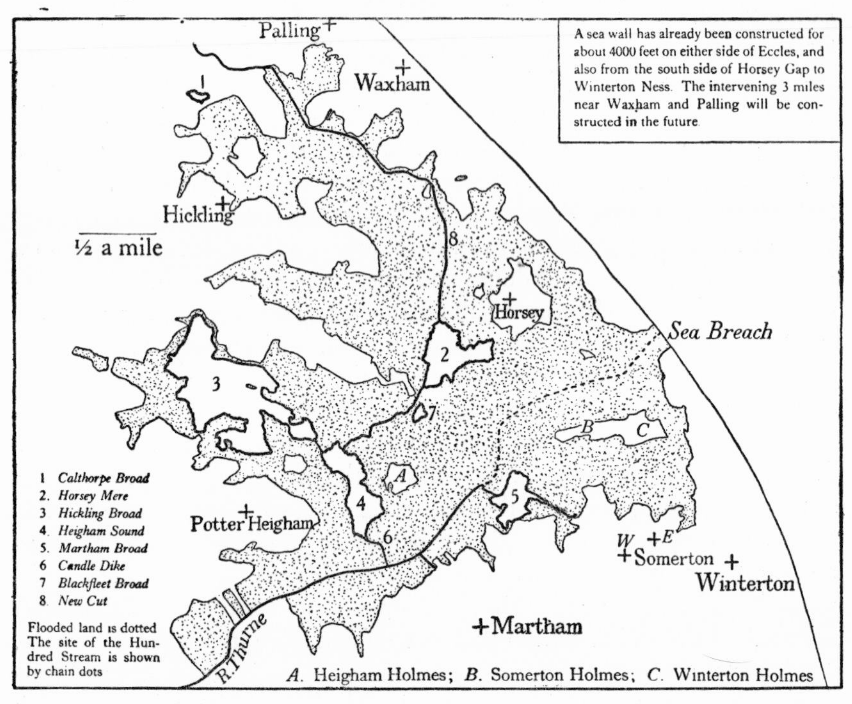

Fig. 83. The Horsey Floods, 1938 (after Mosby, *op. cit.*)

back as 1287; those of 1608 and 1655 are also noteworthy. Nine small gaps were made shortly before 1792, and the dunes were broken again in 1805. William Smith, the geologist, was called in to give advice on their repair. 'He proposed to make all the new artificial embankments as like as possible to the natural embankments thrown up by the sea (and wind) on the same coast, to make them of the same materials, and to give them such directions as might best shelter the new work by the old.'[1] Another serious breach, however, was made by the storm of 1897 which did great damage all along the East Anglian coast, and ten years later a storm broke

[1] Quoted in Mosby, *Geogr. Journ.* 93, 1939, 416.

through again. The boulder clay of the Cromer-Happisburgh cliffs can be traced to Waxham, but not opposite the Hundred Stream, although clay reappears about a quarter of a mile to the south. The absence of any foundation at, or even just below, sea level is probably the reason for the repeated breaks, and the mouth of the Hundred Stream itself was probably filled in with weak materials such as peat and sand. From 1924 to 1934 there were no serious north-westerly gales, and in those years much sand was blown up on to the cliffs and dunes. In December 1936 a gale did much damage and removed a good deal of the sand: with the help of ordinary storms in the next year it prepared the way to the 1938 breach.[1]

That erosion has long been serious between Happisburgh and Winterton is well known, and its effects are most spectacular in Eccles. There the dunes have passed completely over the church. In 1858 the tower stood at the edge of the dunes, but in 1862 strong gales and tides washed away the sand and exposed the original level of the village. Palmer[2] quotes Captain King: 'On referring to my notes made on 27th December 1862, I find the following entry: "To the north of the church considerable remains of cottages are laid out: the very roads and ditches are visible....The old tower now stands clear of the sandhill in which it was embedded....On the 1st May, 1869, I made a careful study of the church. At that time the foundations were perfect....On the night of 19th October 1869, occurred a north-easterly gale and high tide. On that occasion the eastern foundation of the Church was undermined and turned over on its side."' Destruction, however, continued, so that in 1894 the tower was strengthened to try to preserve it, but it was beaten down by a storm on 23 January 1895. Only its foundations can now be seen if the condition of the beach is favourable. A similar fate seems to have overtaken Waxham church, and Palling also has suffered severely. There also appears to have been a place near Eccles called Markesthorpe which has disappeared completely.[3]

Less than a mile south of the Hundred Stream is Winterton Ness: here accretion has taken place, dune ridges having grown out for a quarter of a mile in front of the old cliff. There does not appear to be a satisfactory explanation for this sudden change, but the change itself may help to explain why the mouth of the Hundred Stream shows no deviation to the

[1] Since 1938 two parts of what is to be a continuous sea wall have been built (see Fig. 83).

[2] F.D. Palmer, 'Eccles by the Sea', *Norfolk Archaeology*, 12, 1895, 304.

[3] The natural beauty of the coast at Bacton, Happisburgh, Waxham, Palling, and Eccles has been largely ruined by the indiscriminate erection of bungalows and shacks.

south as would be anticipated on this coast. Mosby[1] (see Fig. 83) calls attention to the 'islands' of Somerton and Winterton Holmes, and suggests the possibility that another existed near Winterton Ness. There is no lighthouse now at Winterton, although one was built before 1588, and there seem to have been two in 1843 (**158**). Dunes run on south to Hemsby and Scratby, but erosion is serious at and north of Caister, where the sea is eating into the soft cliffs of boulder clay and gravels. Crag appears occasionally on the foreshore at Caister.

(14) YARMOUTH AND LOWESTOFT

Yarmouth Denes and Lowestoft Ness (or Denes) must be considered in relation to one another.[2] According to tradition, Yarmouth originated as a small fishing encampment on a sand bank which formed across the Bure-Waveney estuary in about A.D. 1000. The so-called 'Hutch' map picturesquely attempts to show the coast as it then was, but, of course, no reliance can be placed on it. We need not doubt, however, that a sand bank might form in the estuary. It is certain that there were originally two entrances to the river, one lay to the north of the bank and was called Grub's Haven or Cockle Haven, and the other to the south which evolved into the present mouth. Cockle Haven soon became choked with shingle travelling from the north, and thus the original sand bank became part of a continuous spit running south from Flegg. The spit as a whole continued to grow southwards, and attained its maximum about 1347, when the mouth of the river lay between Gunton and Corton. A long spit of this nature was a great hindrance to navigation, and the burghers of Yarmouth made several cuts through it to try to bring the mouth near their town. Apart from minor oscillations, the present haven, actually the eighth, dates from 1566 (**159**).

The growth of the Yarmouth spit almost certainly had a considerable effect on Lowestoft, where the cliffs are now fronted by a flat expanse of sand and shingle called the Denes or Ness. It has been stated, without any supporting evidence, by Gillingwater that the cliffs behind the ness were washed by the sea in Roman times. This is by no means unlikely on physiographical grounds, since what little is known of the ness itself suggests that it is a fairly late formation. The first known mention of it appears in the Hundred Rolls, *temp*. Ed. II.[3] To-day, the trend of the cliffs from Gorleston to Corton is continued in the *outer* line of the ness, and near Corton the cliffs themselves change direction and run more to

[1] *Op. cit.* 1939. [2] J.A. Steers, *Geogr. Journ.* 69, 1927, 24.

[3] Information from Mr V.B. Redstone: see J.A. Steers, *Suffolk Inst. Arch. and Nat. Hist.* 19, 1925, 1.

the south. Reference to Fig. 84 will explain the possibility that a continuation of the supposed former line of the Yarmouth spit is seen in the cliffs inside Lowestoft Ness. A haven was cut through the spit in 1392, and there were four more cuts before 1566. The part of the spit south of these cuts would become 'dead', and could easily be washed southwards, and soon after the sea would begin to erode the now undefended cliffs between Gorleston and Corton. Erosion, especially after 1566, would probably be at a maximum near Gorleston, because there no beach material from the north would be forthcoming to act as even a minor protection.

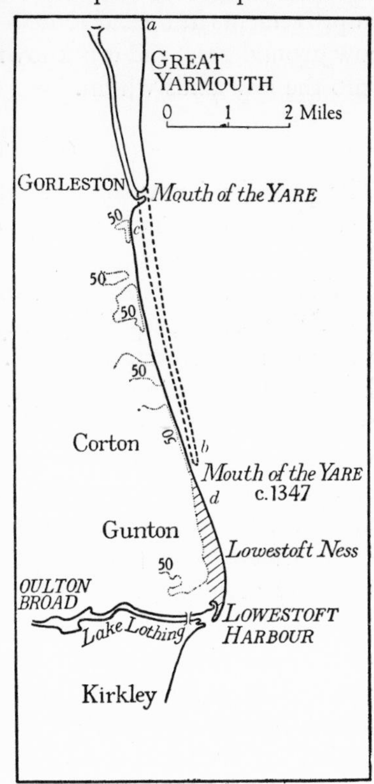

Fig. 84. Diagram of the Yare Spit

The sea was, during much of this time, probably attacking the cliffs at Lowestoft also, but the material cut off from the Yarmouth spit would very probably be driven ashore here much in the same manner as that from the end of Orford Ness is driven onshore at Shingle Street (see p. 389). If this did happen, then the shingle would soon have begun to protect the cliffs at Lowestoft. Working on this assumption, we may imagine the sea gradually forming the widely open bay between Yarmouth and Corton, and the southward movement of beach and offshore material working towards Lowestoft. In other words, the direction of drift of material gradually changed from the line *a...b* to the line *c...d* (Fig. 84). When this material reached Corton, any shingle already at the foot of the cliffs would have diverted the newer material a little seaward, leading to further deposition, and eventually to the formation of the ness. Now, however, the ness is suffering erosion and this has been the case for many years past; hence the Lowestoft authorities have to make expensive provisions in the form of sea walls. Just at the harbour piers there is some accretion, but this is very local.[1] In fact, whilst Yarmouth gains all the time because the beach is held up by the piers at Gorleston, Lowestoft and the neighbourhood to the north suffer because so little

[1] In recent years erosion has been locally very severe just near Pakefield church. Extension of the sea wall will probably stop this, but may lead to erosion farther south.

material travels past Gorleston. It is an example of the difficulties that arise through local control of foreshore problems. The harbour at Lowestoft does not, in all probability, mark a natural outlet of the Waveney. Between Lake Lothing and the sea two ridges of glacial material were found when improvements in the harbour were being carried out. There certainly was low ground here, and it is known that in storms and high tides the sea ran into the low lands within.

(15) THE BROADS

The Broadland next deserves study.[1] We have already drawn attention to the low ground in the valleys of the Thurne, Bure, Yare, and Waveney. In the subsidence or submergence associated with the post-Glacial period all this area was under water, and long arms of the sea penetrated far inland. The Broads, especially Breydon Water, are the last traces of the open water of these former estuaries. Seeing that the Broads are perhaps the best-known lakes in this country, it is odd that we have not yet had a full treatment of the many physiographical problems which they present. Gregory[2] regarded them as the unfilled remnants of former arms of the sea. Beach-drifting dammed back their seaward ends, Breydon Water being the last vestige of an open estuary. Around Wroxham, and in the upper waters of the other Broads rivers, Gregory supposed that the rivers, formerly more powerful and heavily laden, brought down quantities of material and built out deltas, thereby separating areas of water which remain as broads in so far as they are not yet filled with vegetation. The smaller broads to the east of the Waveney between Lowestoft and Yarmouth were ascribed to the damming back of small valleys with silt deposited by the main river. No complete explanation was given for the fact that in several of the broads, instead of the rivers flowing through them, a rond occurs between river and broad. Moreover, when Gregory wrote little was known about the local deposits, and nothing about the modern methods of pollen analysis in the dating of deposits. It is hardly necessary here to review the Robberds-Taylor controversy of last century which is fully described in the *Memoirs of the Geological Survey*.

In 1939 Mr J. N. Jennings began to investigate the problem, but the war curtailed his work. In a private note he writes:

Several lines of hand-bores were made across the Ant. If the deepest parts of these bores are plotted as a continuous section along the river, they show that underneath all

[1] For general summary of older information on the Broadland, including the Robberds-Taylor controversy, see *Mems. Geol. Surv.* 'Yarmouth and Lowestoft', J.H. Blake, 1890.

[2] J.W. Gregory, 'The Broads', *Nat. Sci.* 1, 1892, 347.

there is a considerable thickness of brushwood peat resting on sand and gravel. From Ludham Bridge as far up the river as Barton Broad there is a clay deposit, and throughout the section the upper layer is formed of *Phragmites* peat. My view is that Barton Broad is predominantly natural in origin and, in part at least, owes its formation to the incursion of the estuarine clay which does not seem to have entered the valleys in which Crome's and Alderfen Broads lie. Miss Pallis's map of the deposits in the Yare valley suggests a similar story. The clay in the Yare valley may well have been responsible for ponding back the remains of Buckenham Broads, but seems to disappear well before Surlingham Broad is reached. Dr G. Erdtman published a section of the Bure valley in Woodbastwick fen which shows similar features to the Ant sections. Some of the ronds may prove to be formed of silt of landward origin, and not be similar to the roddons in the Fens (see p. 432), but it is more than likely that ronds and roddons will grade into one another.

There, unfortunately, the matter rests for the time being, but it is to be hoped that circumstances will soon permit Jennings to continue his researches. There follows on p. 403 a comment on the possible relationships of the deposits to be found in the Ant to those elsewhere on the coast and in the Fens.

(16) LOWESTOFT TO ALDEBURGH

The cliff sections from Gorleston to Aldeburgh are all similar physiographically in that they are formed for the most part of sands and gravels and yield easily to marine erosion. Stratigraphically many details have yet to be worked out, but for the present purpose it is enough to state that Crag beds are often found in the lower parts of the cliffs, whilst the remainder are of glacial sands and boulder clay. As the following comments suggest, the individual sections are all short and are separated by tracts of alluvium along which flow small streams; the less important trickle through the shingle. The Blyth is controlled at Southwold Harbour, and the Alde is deflected by the great spit of Orford Ness.

A good deal of light on past conditions of the coast from Lowestoft to Aldeburgh is given by the Butley Cartulary which deals with the apportionment of wreck of the sea. The relevant portion may well be quoted:[1]

Inquisitio capta apud Donewicum, die Mercurii prox. post festum Sancti Gregorii, Papae, anno regni regis Henrici, filii regis Johannis, XXI, pro wrecco maris, et alius diversis Domino Regi tangentibus, coram Roberto de Laxinton...qui dicunt quod Henricus de Colville et Thomas Batun habent wreccum maris in villa de Pakefield, et Kessingland, quo warranto ignorant. Item Simon Perpond habet wreccum maris in villa de Benacre, viz., a portu de Kessingland, usque ad portum de Benacre. Item, Ballivus de Blything habet wreccum maris nomine regis in tota villa de Northaling, a

[1] Taken from vol. II of Suckling's *History of Suffolk*: the original document is in the Bodleian Library, Oxford.

dicto portu de Benacre usque le Southmere. Item Thomas Bavent capit wreccum maris in villa de Easton, viz., a Southmere usque Eston-Stone. Item Comes Gloucestriae capit wreccum maris in villa de Southwold a Eston-Stone usque partem australem de Eycliff. Item Domina Margeria Cressy capit wreccum maris in villa de Blithburgh, et Walberswick, viz., a Eycliff usque portum Donewych. Item Burgenses Donewici habent wreccum maris in villa Donewico, viz., a portu Donewici usque ad limitem de Westleton, abutt. super altum mare Cachecliff. Item, quod Willimus Hardyll capit wreccum maris in villa de Westleton, viz., a predicto limite usque portum de Menesmere. Item Abbas de Leyston capit wreccum maris in villa de Thorp, viz., a portu de Menesmere usque Almouthe. Item Prior de Snape capit wreccum maris in villa de Aldeburgh, viz., a Almouthe usque le Ness de Orford. Item Ballivi de Orford habent wreccum maris nomine Domini Regis in tota villa de Orford, viz., a le Ness usque le Newmore. Item Comes Marescallus capit wreccum maris a Newmore usque portum de Handford in comitatu Essex, et appropriat. et portum de Orwell, et Gosford, qui pertinent Domino Regi (160 and 161).

If we may take this document literally, it means that there were small havens at Kessingland, Benacre, Dunwich, Minsmere, and the Hundred River at Aldeburgh.[1] All except Dunwich are now dammed by shingle. Covehithe Ness, a fine expanse of shingle ridges and dunes, originated as a spit across the Kessingland river, while the water from the stream draining Benacre Broad is carried northwards by the New Cut to Kessingland sluice. Southmere is probably the part between Benacre and Covehithe Broads, and Easton Stone was probably the eastern limit of the present Easton cliffs, which still form one of the most attractive parts of the Suffolk coast. Easton Broad is now dammed by shingle. Eycliff was probably the southern end of Southwold cliff, Cachecliffe seems to have been at the south-eastern end of Dunwich common, and Almouthe must have been the outlet of the Hundred Stream. The small ness at Thorpeness, very similar to that at Covehithe, has developed here, and Thorpe lake (The Meare) is partly natural, partly artificial. The greatest changes have taken place at Dunwich, which was an important mediaeval city and a port, situated at the head of an inlet enclosed by a shingle bar. The Dunwich River joined the Blyth, and they shared a common but variable mouth near the present position of Dunwich. The history of Dunwich is the history of many east coast ports—that of a haven constantly blocked by storms, and liable to shift. There was also in the Middle Ages, as now, great erosion of the cliffs. Prior to 1328 the haven mouth was near Dunwich, but a great storm in January of that year blocked it up, and litigation with Southwold and Walberswick ensued before a new cut was made. Redman suggested that the succeeding haven was Buss Creek, to the north of Southwold. This is unlikely, and is probably the result of

[1] Steers, *op. cit.* 1925.

reading 'leagues' instead of 'miles' for the difficult word 'Leucas'. Originally Buss Creek was open to the sea, but it seems to have been closed before the thirteenth century. Briefly, the history of the Dunwich coastline seems to have been as follows. In the first place a spit developed which deflected the river Blyth to the south. In course of time this spit attached itself to the cliffs at Dunwich and the waters within, either naturally or artificially, found egress where the haven stood prior to the storm of 1328. It is impossible to say what happened before that date, but it is doubtful if this early fourteenth-century haven were the first. Certainly many later cuts were made through the shingle, none of which lasted long. In the eighteenth century the shingle seems to have been thrown on to the marshes, and subsequently erosion has removed almost all the site of the old city. The existing cliffs, however, are high and beautiful, and afford the best viewpoint on the Suffolk coast.

Before leaving this stretch of coast a word may be said about the extent of erosion. Exaggerated figures are often quoted without authority, and it is a pity that so few precise measurements are available. The *Geological Survey Memoir*[1] states that between 1878 (August) and 1882 (May) 130 ft. were lost at Covehithe, while between 1878 and 1887, 172 ft. disappeared there. At Dunwich over a period of 108 years the average annual loss was 18½ in. Other measurements are available, and it is reasonable to assume, for some parts of the coast, an annual loss of between 1 and 2 yd. It is at any rate possible that there has been a loss of anything up to two miles as a result of erosion. But this is far from justifying the assumption that all parts of the Suffolk coast have suffered as much. North of Lowestoft, a small hamlet called Newton, near Corton, seems to have disappeared. This is probably to be related to the 'retreat' of the protecting Yarmouth spit.

(17) ORFORD NESS

Southwards from Aldeburgh is the largest of the east coast shingle spreads, Orford Ness,[2] about eleven miles long (Fig. 85). Orford Ness is the name given to the apex of the spit, while the southern end is called North Weir Point. Sand is visible only at very low tides. The shingle is arranged in fulls, and the hollows between the ridges are called swales or slashes. As the spit grew southwards it reached deeper water, and as far as the ness it ran away from the land, which trends to the west of

[1] 'The Geology of Southwold and of the Suffolk Coast', W. Whitaker, 1887; J. Spiller, *Geol. Mag.* 3, 1896, 23; Royal Commission on Coast Erosion. See also *Essex Naturalist*, 12, 221 for notes on erosion in Suffolk and Essex.

[2] J.A. Steers, 'Orford Ness...', *Proc. Geol. Assoc.* 37, 1926, 306.

south, and is now quite plain as a line of high ground, a former cliff, within the marshes. The growth of the spit can be traced from ancient maps rather better than is usual. The oldest available, part of a coastal chart extending from the Orwell to the Bure, is that of the time of Henry VIII (*c.* 1530). On it the mouth of the Alde is shown nearly opposite to Orford, but the map is rough and unreliable. An Elizabethan chart suggests that the haven entrance was then about midway between Boyton Hall and

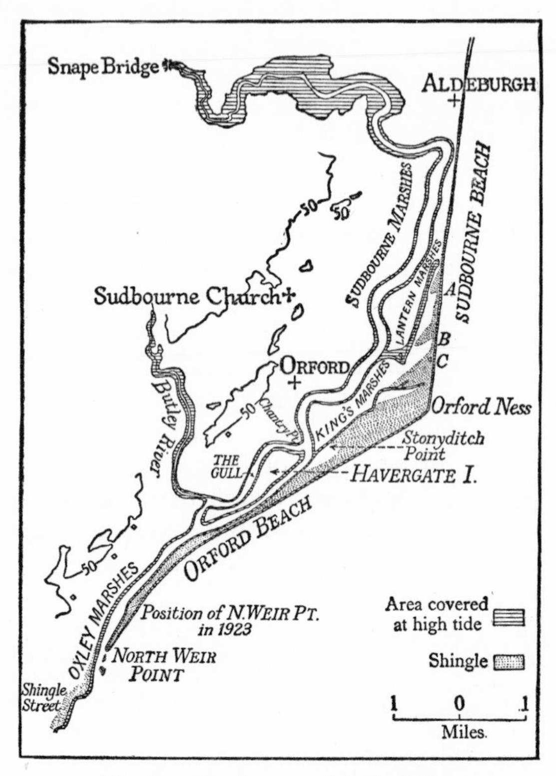

Fig. 85. Sketch-map of Orford Ness

Caldwell Hall, and in this agrees quite well with Appleton's map, also Elizabethan, of 1588. Norden, in 1601, made a detailed and accurate survey of the area: he shows the harbour mouth in the same position as that on the earlier and less accurate Elizabethan map. Saxton's and Speed's maps of 1575 and 1610 are diagrammatic and not very informative. Later maps show a progressive southward growth. The Ordnance maps of 1805 and 1839 are consistent with Bryant's map of 1824–5. From this accumulation of evidence it is clear that the spit's maximum length was reached in 1897, but that by 1902 it had been considerably shortened. In 1923, however, it was extending once again.

Between 1601 and 1897 the growth amounted to nearly two and a half miles, an average of some 15 yd. a year. Taking the Henry VIII and Elizabethan maps at their face value, we should have to reckon with a growth about four times as fast as this before 1601. Such a process is, however, unlikely. Orford probably was at the height of its prosperity as a port when the spit had advanced far enough south to protect Orford's haven, and not menace it as it does to-day. The castle was built in 1165, and we may reasonably assume that the town was then thriving. It is significant that, according to the physiographical argument, this was the time when the end of the spit had reached the position suggested by the Henry VIII map. Since this map is clearly very rough, there are grounds for thinking that it represents fairly well, even if quite by chance, the general condition of the spit two or three centuries earlier. It is also important to note that this position corresponds very closely with Stonyditch Point as far as calculation is possible. Here many fulls end, and imply a long stand-still period. If this argument be valid, it means a growth in the spit of about five and a half miles in 700 years (1165–1897). This development, however, was not necessarily continuous, and sometimes there may have been substantial retrogression (**162** and **163**).

The evolution of the spit can be traced by studying the disposition of the fulls, each of which was once the sea line of the spit. The older ones are naturally less distinct, but their trends are obvious. From Thorpe Ness to Aldeburgh and Orford Ness there is one major shingle ridge on which minor ridges are superimposed. This main ridge cuts off obliquely the older ridges marked *A*, *B*, and *C*, on Fig. 86. They are all probably earlier than the twelfth century. The figure shows that in each of these groups the general trend becomes more westerly as they are followed to the south. In each separate group the southerly ridges run more to the west than those in the north of the same group. The greatest development of ridges, often more than forty in number, occurs near the lighthouse. The following groups may be distinguished: (1) the present beach ridge, (2) the ridges enclosed between this and the lighthouse, and (3) the oldest ridges which run landwards of the lighthouse, and are truncated north of it by those of group (2). These old ridges are also cut into by the newer ridges near Stonyditch Point (*P* on figure). To the south of it the ridges all run in long, open curves approximately parallel to one another. Near North Weir Point local complications are common because large masses may be cut off and the spit may grow forward again: this is an oft-repeated process.

The spit owes its formation to the combined action of waves and the longshore drift which here runs to the south. It therefore works in conjunction with waves coming from the quarter between north and east, and from across the greatest fetch of open water relative to the beach. These waves cause a southward directed beach-drifting, and this is undoubtedly the dominant factor in the growth of the spit.

Orford Ness began as a simple shingle spit across the mouth of the Alde, and there is every reason for thinking that it recurved landwards and grew forward as a series of hooks or recurves. But as it extended southwards it also worked outwards from the land (see Fig. 86). At the

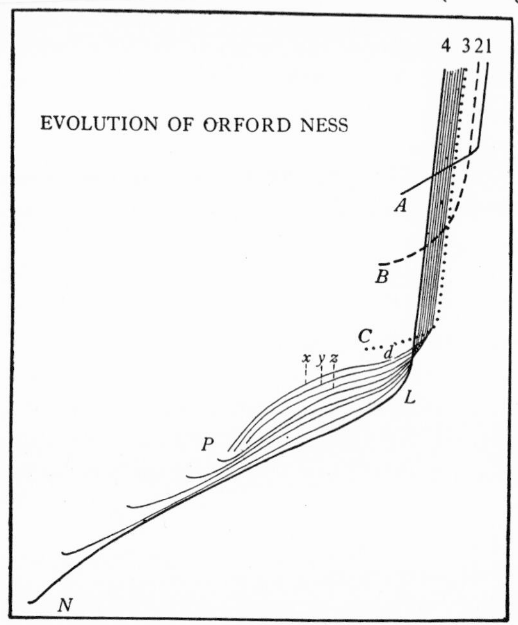

Fig. 86. Evolution of Orford Ness

same time there would have been a general landward motion of the whole spit due to occasional overcombings in storms. The result of these processes is shown in Fig. 86.

Suppose the continuous line 1 to represent the spit at an early period in its evolution. The recurved ridge marked *A* corresponds to those marked *A* on the map. In course of time the spit lengthened and at the same time was driven landwards, as shown by the chain-dotted line 2. Corresponding to this are the recurved ridges *B*. Repetition of the processes of lengthening and retreat landwards led to the formation shown by the dotted line 3 and the recurved ridges *C* which correspond to those a little north of the present lighthouse. The next succeeding ridges to be formed were those marked x, y, and z. When the earliest of these was formed there was a bay-like outline (see line dxP). As newer ridges formed, this outline gradually disappeared, so that now the outer line is shown as 4 LN on Fig. 86. Near P the x, y, and z bars are truncated by newer ones; this is the truncation already mentioned near Stonyditch Point. The spit continued to grow southwards, but also decreased in width, while the distal end curved landwards. At the same time fresh material coming from the north helped to lengthen the spit and the combined effect of these two processes produced the long tapering part of it south of Stonyditch Point.

The change in the trend of the beach at the ness implies that the part under the lee of the ness is a little sheltered from north-east winds, but this no longer holds good towards North Weir Point. Here the spit is thin and unstable, and in the great storm of 1897 a mile or more was cut off. This shingle was piled up in quantities at Shingle Street, the effect of the 1897 storm being to add to that already existing there and at the same time to form the lagoon between the old and the new shingle. The same material continues to the south and protects the marshlands as far as Bawdsey Cliff (London Clay and Red Crag). It still moves southwards and forms a small and fluctuating bar across the Deben.

The marshes within Orford Ness are mostly reclaimed, and the river runs between stout walls. In the natural parts the plants are in all ways similar to those of the Norfolk marshes and call for no special comment. It will be seen that the little Butley River has also been deflected towards the south.

(18) THE DEBEN TO THE COLNE

Between the Deben and the Orwell is a cliff of Red Crag resting on London Clay and, since it is protected, largely grown over. At the Orwell mouth is another shingle foreland, Landguard Point. This area is now much built over, and there are other alterations, since it partly encloses the important base of Harwich. Redman described it at some length. Seaward of the fort it consists of an accumulation of sand and pebbles, partly grassed over. The shingle is arranged in fulls. Redman pointed out also that the end of the spit is of sand and small shingle, and in this respect unlike Dungeness or Orford Ness. Between the fort and the lighthouse the parallel fulls indicate a westward growth. 'These meet the present margin in so rectangular a manner, as to show the way in which wasting has proceeded eastward, and the increase, westward; the latter in effect the resultant of the former....The rounded grass-grown "fulls", due south of the Fort, in a north and south direction, at right angles, nearly, to the immediate sea margin of the point on the south side, show that the spit stretched much farther into the sea in that direction at an early period.'[1] Before Redman wrote there had been a good deal of erosion. The fort was built in 1718, and there is also erosion on its west side helped, in the past, by taking shingle for ballasting. Thus the point wasted on both sides, and this narrowing was accompanied by a lengthening to the south-west. The extreme point altered much with varying winds and other local influences. Landward of the fort the point is all shingle arranged in fulls produced under the influence of north-easterly winds, like those at Orford. The growing spit enclosed Walton Creek, possibly the former outlet of the Orwell, and each new full overlapped the earlier ones, and can be traced westward south of the creek. The north-south ridges, mentioned above, are marginal to those older ridges, and on their eastern side. The growth to the west was estimated in 1848 to average 10 yd. a year. Leslie[2] states that the point may perhaps have been at times a high-water island: Bowen's map of 1750 seems to suggest that the fort was on a sand bank seaward of the point, but a map of 1790 shows it clearly as part of the mainland. Leslie quotes Silas Taylor,[3] who maintained, as did Redman, that the river Orwell once flowed out nearer Felixstowe, and this may mean, as Leslie says, 'that there is considerable evidence of the former existence of a tidal entrance to the Rivers Stour and Orwell between Landguard Point and the foot of the Felixstowe-Walton line of cliffs' **(164)**.

The cliffs at Harwich, Walton, and Frinton are in London Clay and Crag. They are all soft and yield readily to erosion except where protected. Hamford Water is a large inlet containing several islands, seven of which are formed of London Clay. They and the surrounding shores only rise but slightly above the level of the alluvium which is largely covered by marsh plants, especially *Aster tripolium* and *Spartina stricta*. A noteworthy feature is the very soft, sloppy, nature of these marshes. Hamford Water does not appear to have been the outlet of a river, and probably owes its origin to the drowning of a low, rolling, and possibly ill-drained area of London Clay. North of the Naze, at Stone Point, there is some shingle and blown sand. The small patch of Crag, resting on London Clay, is interesting as it is probably the oldest fragment of Red

[1] J.B. Redman, 'The East Coast between the Thames and the Wash', *Mins. Proc. Inst. Civ. Engs.* vol. 23, 1863–4.

[2] J.H. Leslie, *The History of Landguard Fort in Suffolk*, 1898.

[3] S. Taylor, *The History and Antiquities of Harwich and Dovercourt*, First Collected by Silas Taylor; edited by Samuel Dale, 1730. See also G. Harwood, *Mins. Proc. Inst. Civ. Engs.* 20, 1860–1, 4.

R.Cam
R.Stour
R.Orwell
R Deben
FELIXSTOWE
MANNINGTREE
HARWICH
Hamford Water
THE NAZE
Walton
Frinton
R.Pant
R.Colne
COLCHESTER
BRAINTREE
Roman River
Wivenhoe
BISHOP STORTFORD
R.Sturt
Pod's Brook
R.Chelmer
Brightlingsea
MERSEA I.
St Osyth
CLACTON
R.Blackwater
Tollesbury
Goldhanger
R. Blackwater
Othona
GUNFLEET
CHELMSFORD
MALDON
Bradwell
DENGIE FLATS
R.Roding
Ramsey I.
BUXEY
R.Wid
RAY SANDS
SUNK SAND
R.Crouch
Burnham
Canewdon
FOULNESS SANDS
BARROW
Hullbridge
R.Roach
FOULNESS
ROCHFORD
Pitsea
Prittlewell
Wakering
MAPLIN SANDS
Leigh
SOUTHEND
CANVEY I.
THE SHINGLES
R.THAMES
Mar Dyke
Purfleet
Grays Thurrock
Tilbury
Cooling Marshes
Cliffe
MARGATE SAND
Dartford
GRAVESEND
SHEERNESS
R.Medway
Minster
Warden Pt
SHEPPEY
ROCHESTER
CHATHAM
Gillingham
Herne Bay
Reculver
Birchington
Westgate
MARGATE
The Swale
Whitstable
FAVERSHAM
NEAR FOULNESS
1. Wallasea Island
2. Potton Island
3. Rushley Island
4. Havengore Island
5. New England
---- The Broomway
NEAR SHEPPEY
1. Isle of Grain
2. Elmley Island
3. Isle of Harty
NEAR MALDON
A. Northey Island
B. Osea Island
OUTLINE MAP OF THE COAST OF ESSEX
AND OF THE THAMES ESTUARY
(Based on Ordnance Survey)
4 3 2 1 0 4 8 12 16
Scale of Miles

Fig. 87

Crag in the country. South of Frinton, near Little Holland, the shore is of sand and shingle, with an outcrop of London Clay to the west. Presumably this blocks a former arm of the sea running towards St Osyth. The marshes to the south of this village are bounded by a broad belt of sand and shingle, which, according to Dalton,[1] was increasing in width towards the end of the last century, although much was used for road metal and ballast. It ends at Sandy Point to the west, and eastwards runs into the dunes and sandy beach at the foot of the London Clay cliffs at Clacton.[2] There are similar isolated, but shorter, ridges elsewhere in the marshes. Mersea Island is formed of London Clay with glacial gravels in patches on it. The flats around it are all tidal mud (see Fig. 87).

(19) THE COLNE TO THE THAMES

The coast of Essex south of the Colne is largely a study in the history of the rivers and embanking.[3] Some of the possible changes of the river courses are discussed below, but details of the embanking will be omitted; in the first place many dates of construction are unknown, since undoubtedly there was a good deal of more or less private inning; secondly a detailed study, although of considerable local interest, would not be of much general value. There are, however, certain coastal features that call for special mention; and on the Essex coast and in the Thames estuary there is abundant evidence of a post-Roman subsidence **(165)**.

At the mouth of the Blackwater is the Roman station of Bradwell-juxta-Mare, built in the period of the later Empire, and called Othona. Its precise position was established in 1864 during the digging by the Essex Estuary Company for embanking work. Erosion has been active at Bradwell. The estuary of the Colne is in many places fringed by salt marshes, the higher parts of which form small cliffs overlooking the lower and soft, sloppy, mud reaching to and below low water; there is no shingle. Near Brightlingsea at least 700 acres were reclaimed between 1700 and 1800, although there is no record of exact dates. When the Crouch reached the sea near the site of Burnham in comparatively recent historic times, there must then have been a wide, open, and shallow bay lying between Bradwell and Shoebury. This bay is now filled with alluvial material, including that forming Foulness Island. The alluvial area, to-day a maze of

[1] *Mems. Geol. Surv.* 'The Geology of the Neighbourhood of Colchester', W.H. Dalton, 1880; see also, *ibid.* 'The Geology of the Eastern End of Essex...', W. Whitaker, 1877.

[2] The natural beauty of the marshes and dunes between Clacton, St Osyth, and Brightlingsea has departed, In its place are endless huts and bungalows.

[3] For erosion on the Essex Coast, see *Essex Naturalist*, 12, 221.

islands and watercourses, most of which follow the lines of former natural marsh creeks, is converted into good arable and pasture land. There are lines of shingle in the marshes, and a long one running north from Holliwell Farm is shown on geological maps. Unfortunately there has been no detailed mapping of shingle and dunes, and consequently there can be but little reliance on the trend and nature of old shingle ridges to help in studying the evolution of the coast. At Wallasea Ness the Crouch is joined by the Roach, whose eastern bank consists of Foulness and the islands, or former islands, south of it. Dalton[1] regards these as in the nature of deltaic deposits of the Thames. Other material is carried farther north by currents to build the Maplin and Foulness Sands. In the enclosed part of Foulness field boundaries are often original marsh creeks. At Church End, the highest part of the island, a section in the churchyard showed 2 ft. of mould resting on 2½ ft. of clay, and below that 6 in. of sand. All below 5 ft. was waterlogged. At (?) Old Hall the London Clay was found at 77 ft. below the surface. During the formation of the island the Roach and Crouch changed courses a good deal, and Foulness is really composed of several islands, the creeks between them having silted up. Old walls are not necessarily ancient boundaries of one of these islands, but may represent the increase of such an island by the taking in of more marsh. All the eastern part shows the most recent general enclosures. The surface is nearly flat, except for depressions marking the sites of former large creeks. South of Foulness are four other islands, Potton, Rushley, New England (Little Wakering), and Havengore. The two last are joined to Foulness at low water, and it is interesting that the creeks separating the several islands disappear into mere trickles when they reach the Maplin Sands.

Sea walls line the foreshore from Bradwell to Burnham, and on the south of the Crouch they border the river and all the islands. The natural marshes outside the sea walls cover a great acreage. They usually show two distinct levels. The higher flats, the saltings, fringe the sea walls (usually built of mud and often in a state of disrepair) and extend from the high-water mark of spring tides to about 3 ft. below that level. They carry the vegetation common to this coast: *Glyceria maritima*, *Spartina stricta*, *Aster tripolium*, *Statice* spp., *Suaeda maritima*, *Armeria maritima*, and *Salicornia* spp. The abundance of a particular plant depends primarily on the height of the marsh. The outer edge of these saltings is much broken by creeks, and is usually a sharp cliff 3 or 4 ft. high. Beyond this lie the mud flats with a vegetation of seaweeds and *Zostera*. The saltings help to protect the sea walls, but erosion is often serious and the indigenous

[1] W.H. Dalton, *Essex Naturalist*, vol. 3, 'Foulness', and vol. 14, 'Wells on Foulness Island'.

plants are not sufficient to check the process at exposed places. Hence *Spartina Townsendii* was introduced experimentally in 1925.[1] At first progress was satisfactory at Northey Island where the sea wall was broken and the surface covered with a large spread of *Aster*. Later experiments were made at Goldhanger, Tollesbury, Mersea, and Bradwell. The Bradwell site was slightly higher than that at Northey, and the experiment succeeded. Those at Tollesbury and Mersea were failures. In each case the low *Zostera* flats were planted which were covered by 5 or 6 ft. of water at high tide. 'Assuming the plants to have been quite sound, it is difficult to assign a reason for these failures, unless they have been due to strong currents and heavy wash.'[2] Conditions were serious at Goldhanger since in places erosion was exposing the base of the wall. At first the experiment was not at all promising, but in 1928 good progress was reported. Plantations were also made at Holbrook on the Suffolk side of the Stour, and at Mundon in Essex. At Holbrook the difficulty was to maintain the plants on the sloping face of the wall: this was overcome by making a small shelf at the wall foot and holding it in place by alder stakes which stick up above the level of the shelf, forming a low palisade (**166** and **167**).

Outside the sea walls the surface of the natural marshes is as much as 8 ft. higher. Within the sea walls, surface level is almost exactly that of half-tide, and the average height of the sea walls is about 15 ft. The clay soil of the enclosed land seldom exceeds 5 or 6 ft. in depth, and rests on sand saturated with salt water. It is this formation that accounts for the absence of trees. There are similar walls along most of the Blackwater estuary, but no wide expanse of seaward marsh like that east of Southminster and Burnham. Maldon is built on glacial drift resting on London Clay and is well above the part of the river which is liable to considerable lateral shifting. The channel of the Blackwater was once south of Northey Island and north of Osea Island, and the river may once have flowed to the south of Ramsey Island. These all have London Clay as a basis and are much more akin to Mersea Island than to Foulness and its neighbours.

On the north of the Crouch the mud is as much as 80 ft. deep, and in Foulness creeks it reaches 30 or 40 ft. These facts are evidence of a time when the river was flowing at a considerably higher level. But before discussing the earlier courses of the rivers, it is interesting to note the way

[1] The Shelford Creek channel, the southern boundary of Foulness Island, was dammed and not bridged by the modern road. *Spartina* (I am not sure whether naturally or artificially introduced) has now (1943) nearly filled the old creek as far down as the sea. This has taken place since about 1926 or 1927.

[2] 'Rice Grass', *Min. Agric. and Fisheries*, *Misc. Publications*, No. 66, 1929. Information about the Essex marshes is also largely drawn from this source.

in which the sea channels all lead in a north-easterly direction. It is worth noting that Foulness Island existed and was inhabited before the Roman invasion. Christy[1] calls attention to the Roman Road from Prittlewell to the Thames at Wakering Stairs where begins the Broomway running very straight for several miles along the Maplin Sands and about a quarter of a mile seaward from Foulness. He suggests that this too may be of Roman origin.

The account given of the coasts of Norfolk and Suffolk contains several brief references to the severe storm of 29 November 1897.[2] This also did great damage in Essex. The wall from Parkeston to Dovercourt became a great cataract: that around Walton Hall Marsh was breached in thirty places: only about one-tenth of Horsey Island escaped flooding and the sea walls gave in between Frinton and Little Holland. In the Colne estuary there was great destruction at St Osyth; the Stroodway was impassable for several hours, while Brightlingsea became virtually an island; there were sea breaches at Burnham, Wallasea, and elsewhere. Serious floods occurred at Foulness and adjacent islands, and several walls along the Crouch gave way. Canvey Island was half submerged, and the amount of land flooded in Essex was estimated at nearly 30,000 acres. It is important to realize the extent of the damage done in such storms of which that of November 1897 and that at Horsey (Norfolk) in 1938 are outstanding examples. They are, fortunately, infrequent, but even if such storms themselves are rare and short lived, their effects exceed those of decades of more normal conditions. In general, the worst results of storms on this coast occur when, following a period of vigorous westerly and north-westerly winds, a strong gale from a north-easterly quarter coincides with a high spring tide. These three conditions produce abnormally high water on the coast. If, as in the Thames floods of 1928 and the Fen floods of 1937 these, or similar conditions, coincide with heavy flooding in the rivers, the river water cannot get away even at low tide and so is dammed back, whilst at the same time it is steadily swollen by more and more land water. Thus, the river flood banks are overtopped and wash away in weak places so that large areas are flooded.

It is difficult to say where the study of the coast ends and that of the river Thames begins. There are, however, one or two matters that may properly be touched upon here. At Shoeburyness and Southend there is now very little change; some protective works have stopped the slight

[1] M. Christy, 'On Roman Roads in Essex', *Trans. Essex Arch. Soc.* 17, 1923–5, 226, and W.A. Mepham, 'Great Wakering', *Trans. Southend-on-Sea and Dist. Antiq. Hist. Soc.* 2, No. 3, 106.

[2] See *Essex Naturalist*, vol. 10.

erosion at Southend. About half-way between Southend and Leigh there was once Milton Hamlet, and Milton shore is now nearly coincident with Southend front.[1] Salmon asserted in 1740 that within living memory the remains of a chapel had been seen at low-water mark, but unfortunately this statement is not verified. In 1327 there was a severe storm which inundated about 110 acres of land. Canvey Island was reclaimed from 1622 onwards. Norden (1594), Stent (1602), and Speed (1610) all show five islands between Hole Haven and Southchurch, of which three correspond to the present Canvey Island, one possibly with Leigh Marsh, but the fifth is untraceable. Nicholls suggests that this seventeenth-century island may have been part of a salt marsh which before 1327 probably carried the coastline well south of its present position.

At Prittlewell, Southchurch, and Shoebury there are three minor valleys reaching the Thames, and that at Southchurch has been carefully investigated. Up to about the second century A.D. it appears to have been a fresh-water valley: later, probably as a result of subsidence, it became a salt-water creek.[2] This development probably dates from roughly the third century. Some time after the thirteenth century a barrier was built to exclude the sea, and the area became a marsh.

Nearly all the coast of Essex is flat and protected by sea walls. It is nevertheless picturesque, and the sense of desolation which belongs to some of it is to many people a distinct scenic asset. There are low cliffs in parts of Mersea Island, and in the Thames estuary the old river cliffs between Leigh and Benfleet, crowned by Hadleigh Castle and overlooking Leigh Marsh and Canvey Island, form a prominent feature. Canvey Island, like Jay Wick south of Clacton, is now largely covered by a bungalow town. Above Canvey Island there are extensive marshlands off Coryton and East Tilbury, but near the former place large oil storage tanks stand on them. Down river from Leigh the north bank is built up as far as Shoeburyness.

(20) THE RED HILLS AND OTHER MOUNDS

Another set of interesting features, peculiar to Essex and parts of Kent, is found in the Red Hills which occur amongst other places at Langenhoe, Goldhanger, Canewdon, and Canvey Island. They are low, flat, mounds of variable outline, and contain compact masses of burnt earth and pieces

[1] J.F. Nicholls, 'New Light on the History of Milton Hamlet', *Trans. Southend-on-Sea and Dist. Antiq. Hist. Soc.* 1, pt III, 172.

[2] A.G. Francis, 'On Subsidence of the Thames Estuary...at Southchurch', *Essex Nat.* 23, 151, and *Trans. Southend-on-Sea and Dist. Antiq. Hist. Soc.* 2, No. 1, 49.

of clay which have been shaped. They are all near the inner edge of the alluvium, that is to say, near the original tide marks. 'The character of the burnt earth and associated pottery, and the presence at the site of deposits of sandy silt suitable for pottery manufacture indicate, in all probability, that the burnt earth is domestic and industrial waste material left behind by the Romano-British occupiers of the site, who were doubtless attracted thither by the deposits of serviceable raw material located close to tidal waters.'[1] Moreover, the mounds on Canvey Island appear to have been reoccupied later. Assuming that the hearths at Canvey are domestic, they must have stood 4 or 5 ft. above high-water mark, or 12 or 13 ft. above their present level. In these mounds occurs, then, one of the several lines of evidence pointing to a post-Roman submergence. The Red Hills must be distinguished from other mounds such as are seen in the marshes near Hullbridge, and formerly near Maldon. Christy and Dalton maintain that the latter are incidental: they are thought to be associated with mediaeval diggings for salt-making and might be the dumps of excavated material.[2]

(21) BRIEF NOTES ON THE RIVERS OF ESSEX AND SUFFOLK

Before discussing the submergence which gave the typical drowned mouths of the Essex and East Anglian coasts, brief reference must be made to one or two points in the river development of Essex. The whole subject is very complicated and it is outline rather than detail which is helpful in this chapter. The Blackwater estuary is clearly out of proportion to the small Chelmer-Blackwater river which now drains into it, but there is significantly a broad belt of low ground running across Essex from Romford to below Malden. North of this line high ground is more or less continuous: south of the line it is more broken. It is now generally held that the belt was a former line of the Thames, dating from a time before the Winter Hill stage of Wooldridge and Linton, and possibly belonging to the Lower Gravel Train stage. 'This (Blackwater) exit cannot have remained open during the Great Eastern Glaciation and we may conclude that the position of the present mouth of the Thames no less than that of its present course through London reflects the effects of the great glacial diversion. It is not without significance that for a long succeeding period, during the accumulation of its "100-foot", "50-foot", and later alluvia

[1] E. Linder, 'Red Hill Mounds of Canvey Island...', *Proc. Geol. Assoc.* 51, 1940, 283; see also Red Hill Exploration Committee, Report, by F.W. Reader, *Proc. Soc. Antiq.* 23, 1909–11, 66; *ibid.* 22, 1907–9, 164; and R.A. Smith, *ibid.* 30, 1917–18, 36.

[2] M. Christy and W.H. Dalton, *Trans. Essex Arch. Soc.* 18, 1925–8, 27.

the Thames swung northwards in the vicinity of the present coastline towards the region of its old mouth.'[1]

The line of low ground followed by the ancient course of the Thames ran via Upminster and Wickford, and the Thames was at a higher level than now. It is also likely that an extended Crouch drained part of the trough in later times.[2] In any case, the Crouch is of more recent origin than the Romford-Maldon-Blackwater line of the Thames. Wooldridge and Linton, in their synthesis on south-eastern England, do not agree with Gregory's reconstructions of the ancient courses of the Thames. The Roach is but little more than a delta distributary and, for most of its short course it lies within the delta islands of Foulness, Wallasea, and Potton. It rises in the Rayleigh Hills, and from there to Rochford is only a small brook.[3]

Boswell[4] has discussed the evolution of the rivers in northern Essex and southern Suffolk. These are mainly dip streams, and it is possible that the present Upper Stour from its source as far downstream as Wixoe was formerly the headwater of the Colne. The strike portion of the Stour above Long Melford may have cut back and captured the Upper Colne and diverted it to the Stour. Farther north the courses of the Deben and Butley rivers are suggestive, and the Upper Deben may have flowed into the Butley. Unfortunately proof of these and other possible diversions is not obtainable because boulder clay obscures detail. But such changes are of interest in that they throw light to some small extent on the incongruous sizes of the present estuaries.

Before leaving this matter the course of the Waveney needs comment.[5] Did it ever flow direct to the sea? Certainly not in historical times, but it may have done so earlier. At Kessingland the cliffs show a deposit in a shallow valley cut in the Chillesford Clay and Lower Glacial Sands. Not much is known of this deposit inland, but it may have been laid down in a former direct continuation of the Waveney to the coast, and consequently implies that the northward deflection to Yarmouth occurred in late Glacial times. This north-south reach of the Waveney is cut in the 'Lower' Glacial beds.

[1] S.W. Wooldridge and D.L. Linton, 'Structure and Surface Drainage in South-East England', *Inst. of British Geographers*, No. 10, 1939.

[2] B.R. Saner and S.W. Wooldridge, 'River Development in Essex', *Essex Nat.* 22, 244.

[3] See T.V. Holmes, 'Notes on the Ancient Physiography of South Essex', *Essex Nat.* 9, 193, and J.W. Gregory, *Evolution of the Essex Rivers and of the Lower Thames* (Colchester) 1922.

[4] P.G.H. Boswell, *Quart. Journ. Geol. Soc.* 69, 1913, 581, and (River Stour) *Journ. Ipswich and Dist. Nat. Hist. Soc.* 1, 1925, 7.

[5] J.A. Steers, *op. cit.* 1941.

The estuaries of Essex and southern Suffolk are similar to those of the Broads valleys of Norfolk, and also, allowing for the differences in size, to the smaller and shingle dammed valleys between Lowestoft and Aldeburgh. The evidence of submerged forests and post-Roman depression is outlined below. Here we may remark that the difference in appearance of the valley mouths north and south of the Orwell-Stour mouth is due mainly to present coastal conditions. There is an abundance of shingle as far south as Landguard Point. Hence, the northern rivers Bure, Blyth, Alde, and many minor streams, are dammed back by spits and the coastal outlines are simple. Farther south, except for the stretch near St Osyth and one or two minor areas, there is little shingle and so the rivers remain open, and presumably partly for this reason have not silted up as much as the Norfolk rivers in their lower courses. All the country drained by these rivers is low, especially near the coast. Furthermore the rocks are all soft and often incoherent sands and gravels. The valleys would, therefore, be wide and open, and consequently when drowned would easily assume their present appearance. The special influences in the Blackwater estuary have been already mentioned.

(22) SHORE FEATURES IN THE THAMES ESTUARY

It may seem illogical to describe certain features in the Thames estuary, especially on the Kentish coast, in the section on East Anglia and Essex. The reason is that it seems proper to have treated south-eastern Essex (i.e. south of the Colne-Blackwater mouth) with the rest of that county and with Suffolk and Norfolk, but this part of Essex is also virtually part of the Thames. Hence it is not unreasonable to continue along the south shore as far as perhaps the Wantsum channel. Thanet is clearly a distinct unit, but the marsh area from Gravesend to Whitstable, enclosing as it does the higher London Clay country, is similar to Essex, and as the London Clay continues to just beyond Herne Bay, the Wantsum channel forms a convenient end of this section. It may be said that since there is London Clay in the cliffs in Pegwell Bay, the section should end there. However, Reculver, the Wantsum channel, and Thanet are, for reasons given elsewhere, so clearly connected in their coastal development in historic times with Sandwich and Richborough, that the stratigraphical fact of London Clay occurring in Thanet is of less significance than the physiographical point mentioned above. The same kind of argument might be made for having a separate section on the Thames estuary coast beginning at the Colne-Blackwater mouth, but it would be an unnatural break in other ways.

Neither the structural history of the London Basin nor the vicissitudes of the Lower Thames from Pliocene times onwards is really relevant material for this chapter. These form a long study in themselves, but do not bear in any material way on the present problem. The Romans are said to have embanked the Lower Thames in the first place, but Spurrell states that below Purfleet there are no traceable remains of Roman banks, and above that place only small banks were needed. If the Romans indeed built any banks, their constructions have vanished. There may still be some Saxon banks below Gravesend, but above that place, apart from the Littlebrook Walls, there are no banks earlier than those of the thirteenth century.[1] The river is now tidal up to Teddington, but the ordinary marine tide reaches approximately only to Richmond. Moreover, in Roman times it seems that the tide only reached to London Bridge.[2] All along both banks of the river there are numerous salt marshes.[3] Some are now almost obliterated by buildings and other artificial changes, while some are still open, and it is these marshes that have been embanked. There has undoubtedly been a subsidence since Roman times and this is evident from the numerous places where abundant Roman material has been found just below marsh level. But the subsidence dates back much earlier and in fact the alternate silt and peat bands that have been demonstrated so effectively in dock sections imply a recovery in stages from the time when the North Sea floor was exposed. The earlier stages need not be considered here, and in any case they are but the concluding episodes of a long and tangled story dating back to Pliocene times; but the superposition of Roman on Bronze and Iron Age sites at Brentford gives us a good measure of the prehistoric and historic part of this movement which is still in progress.

The change of level since Roman times probably amounts to some 15 ft., and at that time parts of Central London were aits or eyots similar

[1] 'Many writers are impressed with the "mighty", "stupendous", or "vast" embankments which keep out the water of the river, while Dugdale and Wren seem to have thought that because they were so great, none but the Romans could have raised them. There is no need for such expressions. If embankments were needed in the Roman and early times, they were of minor importance as engineering works in the upper part of the estuary near London. The height to which we see them now rise is the gradual increase from slighter banks which costs but little exertion, although regular attention.... The most difficult place for embanking in the Thames is the Swale marshes...' (Spurrell, *Arch. Journ.* 42, 1885, 286).

[2] See p. 14, London (Roman), *R. Comm. Hist. Monuments*, 1928.

[3] Details of the vegetation of the salt marshes are unnecessary. At Canvey Island, which may be taken as typical, we find the usual East Anglian plants: bare mud with *Salicornia* passing into (1) an *Aster-Salicornia* zone, (2) an *Obione* zone, (3) an *Aster-Glyceria* zone, and (4) an upper *Glyceria* zone (see N. Carter, *Journn. Ecology*, 20, 1932, 352).

to the Eel Pie Eyot and Chiswick Eyot of to-day.[1] We need only mention Putney, Chelsea (Chesil or Shingle Island), Battersea (St Peter's Island) and Westminster (Thorney Island). The Isle of Dogs was then, and for some centuries later, a more or less isolated mud bank. Frog Island, now a tract of marsh, lay between Hornchurch and Rainham marshes. Before embanking took place the tide washed the higher ground at Woolwich, Barking, Purfleet, and Cliffe, just as it does at Gravesend to-day.

The earliest embanked marshes are usually the lowest.[2] The Roman level is well known on the eastern part of Canvey, around the isles of Grain and Sheppey, opposite Gillingham, and at Southend. To-day there are no marshes east of Sheppey, but Spurrell does not doubt their former existence. Erosion of the existing saltings is also serious hereabouts, and along the Harty-Whitstable shore Roman remains are often thrown up by the waves. Sheppey is but one of a series of similar islands, whose number was formerly greater. To-day we have the isles of Queenborough, Elmley, Harty, and others, and they grade downwards into mere tiny mounds but just above marsh level. They are all fragments of a former spread of London Clay. The Kentish Flats and Pan Sand lie several miles to the east of Sheppey. The Pudding Pan rock (so-called) is now never dry, but near it many specimens of Samian ware have been found. This rock seems to have been largely a mass of brickwork. Spurrell[3] suggests that the Pan Sand in Roman times was an island like Harty to-day.

Sheppey[4] seems to have been surrounded by walls, possibly until quite recent times. In 1780 Minster is said to have been about the centre of the island. Speed's map of 1608 contains a suggestion that a great deal of marsh extending eastwards from Shell Ness was capable of reclamation, but unfortunately his cartography on other parts of the coast does not inspire all physiographers with confidence. There has been, however, very great erosion on the northern shore of Sheppey in the cliffs of London

[1] See the several *Memoirs of the Geological Survey* covering the Lower Thames, especially 'The Geology of the London District', H.B. Woodward, 2nd ed. by C.E.N. Bromehead, 1922.

[2] There are many instances of breaches of the embankments. Dagenham Creek was ordered to be inned in 13 Elizabeth. In 1621 the bank broke and flooded the whole marsh. Another breach occurred in 1707 and did great damage, and was not repaired until 1724. The first embankment must have been long before Elizabeth's time as there is a record of a breach as early as 1376. The break in 1707 gave rise to great expense and litigation. There are several papers in the *Essex Naturalist* concerning Dagenham Breach, e.g. T.V. Holmes, 6, 142, and W. Crouch, 6, 155.

[3] *Op. cit.*

[4] See A.G. Davis, 'The London Clay at Sheppey', *Proc. Geol. Assoc.* 47, 1936, 328, and W. Topley (Coast Erosion), *Rept. Brit. Assoc.* 1885, 404; W. Topley, *Mems. Geol. Surv.* (District Memoirs), 'The Weald', 1875.

Clay eastwards from Scrapsgate. In the parish of Warden loss has been continuous and serious including the disappearance of the church and churchyard.[1] The storm of 1897 also did great damage at Sheppey, and the erosion is accentuated by land water. At the other end of the island, Sheerness is protected by walls: much of the area is below high-water mark. On the marshes between the Medway and Shellness there are up to two hundred small mounds, the Kentish counterpart of the Red Hills of Essex.

The London Clay rises up from the marshes at Seasalter, and from there to Reculver there is an almost continuous section of London Clay cliffs with the Oldhaven and Woolwich Beds coming in towards the east. There is a steady waste of this coast, and on the highest cliffs just east of Herne Bay mud streams are not infrequent. Erosion is much aided by land water.

(23) GENERAL NATURE AND RESULTS OF RECENT VERTICAL MOVEMENTS

There are no traces of raised beaches[2] between the Wash and the Thames, but at the same time there is a good deal of evidence of recent submergence. Physiographically it is clearly shown in the drowned valleys, but of late years some precise work has been done on submerged peat beds and Romano-British land surfaces. There has been some reference to three separate peat beds at or near Scolt Head Island. The highest is thin and fragmentary and occurs just west of the club house on Brancaster links. The next bed is nearly continuous from Holme to Brancaster, and is possibly also that seen between the tide marks elsewhere on this coast. The lowest bed is only known inside Scolt, and forms the floor of a creek at the landing place called Judy Hard. It has been bored and is 8 ft. thick. Godwin[3] has examined all three and has analysed them for pollen grains. Although the available evidence is not very complete, he has been able to date the Judy Hard bed as Boreal and Early Atlantic and the highest bed as probably Hallstadt. The extensive middle beds can only be regarded as intermediate between the other two in age.

On the Essex coast in several localities Warren[4] has shown that there

[1] The rate of erosion at Warden Point is 9·7 feet. Between Scrapsgate and Warden Point jetty (6½ miles) 103 acres were lost between 1865 and 1906.

[2] Except the Morston shingle bank, see p. 352.

[3] In *Scolt Head Island*, 1934.

[4] S.H. Warren and others, 'Archaeology of the submerged Land Surface of the Essex Coast', *Proc. Prehist. Soc.* No. 9, 1936. This contains full references to earlier papers.

is a buried land surface underneath the present salt marshes. The following table gives the sequence of deposits, but the full sequence is never found at any one place:

9. Present Salting Surface.
8. Tidal Silt or *Scrobicularia* Clay, and Red Hill Briquetage.
7. Peat.
6. Buried Prehistoric Surface.
5. Grey Marsh Clay. } Probably contemporary.
4. Rain Wash, Flint Implements and Pottery }
3. Pleistocene (?) Brickearth and some Erratics.
2. Layer of shattered Septaria.
1. Grey Marsh Clay with *Elephas primigenius.*
London Clay.

The buried surface (6) is usually seen exposed below the peat at 2–5 ft. above low-water mark. At times, it rises higher than this, merging into the dry land of to-day, and it also sinks considerably lower. It is seen at Hullbridge-on-Crouch, Walton-on-the-Naze, Clacton, Dovercourt, and elsewhere. Warren has called it the Lyonesse surface. The sites were originally slightly inland, because not only has submergence taken place but also erosion. The earliest traces of human settlement belong to the Lower Halstow (Mesolithic iii) Culture, and the latest remains that can be dated are a few Neolithic B beaker sherds. The occupation of the sites seems to have ended with a submergence which brought in salt-marsh conditions in places producing a thin peat bed. The *Scrobicularia* clay indicates a submergence of at least 10 ft.

Jennings[1] has noted in the Ant broads a brushwood peat at the bottom of all the bores. He thinks this implies a slow submergence. The overlying clay indicates that the rate of submergence quickened, and the top layer of *Phragmites* peat suggests either a stand-still period or possibly a slight elevation. Unfortunately pollen analyses of these deposits have not yet been made, but it is probable that the Lower Peat is the equivalent of the Lyonesse surface of Essex.

It is relevant here to anticipate one point: in the Fens (p. 430) there are underground peats separated by the Buttery Clay which is the equivalent of the clay deposits succeeding the Essex land surface. The Walton land surface equates with the upper part of the Lower Peat (– 13 to – 22½ ft. O.D.) at Shippea Hill in the south-eastern Fens. Furthermore, Godwin thinks that the extensive forest between Holme and Brancaster corresponds to the Lyonesse surface. It is also probable that the topmost Brancaster peat is of the same age as the top of the Upper Fen Peat.

[1] See p. 382 *supra.*

Although there are no raised beaches on this coast, it is just possible that in the *Leda myalis* bed of Norfolk we have the equivalent of the 'pre-Glacial' beach of the south coast. This was suggested by Wright.[1] The full sequence of the uppermost Pliocene and lowest Pleistocene beds seen in the Cromer cliffs is:

Base of Pleistocene	Arctic Freshwater Bed with *Salix polaris*, *Betula nana*, etc. (5)		
Pliocene	*Leda myalis* Bed (classed provisionally with the Pliocene (4))		
	Forest Bed Series	Upper Freshwater Bed	(3)
		Estuarine Bed	(2)
		Lower Freshwater Bed	(1)

No. (1) is seldom preserved, and the relation between it and No. (2) 'seems to be somewhat similar to that of the recent submerged forests in our estuaries to the deposits now forming in the same localities, in part from their destruction'.[2] The upper surface of (2) is weathered into a soil, and (3) is a lacustrine deposit. 'It will be seen that though a land surface does occur in the pre-Glacial deposits, it does not correspond with the horizon to which the name "Forest-bed" has been more especially applied. The main reason why this "Forest-bed" was for a time accepted as a land surface was the occurrence in it of many large stools of trees, often in an upright position, and occasionally with interlocking roots. That these have merely drifted into place is evident,...'[2] The *Leda myalis* bed is of fine, false-bedded, loamy sand, with but few fossils. *Leda myalis* and its associate, *Astarte borealis*, are arctic, and are in the position of life. Bed 5 is a flood loam with arctic species, and clearly indicates the incoming of arctic *land* species. Solomon considers the *Leda myalis* bed to underlie the North Sea Drift (i.e. the drift of the first glaciation), and does not agree with Reid's view that it is overlaid by (5) which cannot now be seen in the cliffs. Wright, arguing from its horizontality and constant height of 15 ft. above the present shore, equates it with the 'pre-Glacial' beach of southern England.[3]

The submerged forest, or peat bed, some 30 ft. below O.D. in the Orwell below Ipswich,[4] is much later and entirely post-Glacial. It precedes the other peat beds already described, and probably corresponds with beds of similar depth in southern England. In early post-Glacial times the

[1] *The Quaternary Ice Age*, 1937. [2] *Ibid.* p. 106.

[3] See also Chapter XII. It is clear that Wright fully recognized the possibility that the 'pre-Glacial' beach was really inter-Glacial—and so not to be equated with the *Leda myalis* bed.

[4] Boswell, *op. cit.* 1913.

North Sea bed was dry, at least in its southern part, and the Dogger Bank was a low flat island. This matter is discussed more fully in Chapter XII, but in general it means that as the water level rose in stages successive peat beds were formed. Thus the earliest of the post-Glacial beds are the lowest, and the uppermost Brancaster bed is the newest. The now submerged Romano-British surface implies a still later subsidence. This is proved not only by the evidence already given, but also by the relative levels of embankments, marshes, and high water in the Thames. Much of Southwark is below Trinity High Water Mark which is 12½ ft. above O.D., and Roman relics occur 9 or 10 ft. lower. At Tilbury, Romano-British huts are about 13 ft. below Trinity High Water Mark. Yew tree stumps on both banks of the river occur at about this depth, and as the yew is very sensitive to salt water the trees must have grown when the land stood at least 13 or 15 ft. higher. The same subsidence is proved at Southend.

The movement (see p. 496) is probably still going on. 'On the night of January 6–7, 1928, there occurred disastrous floods in London, accompanied by small loss of life. The occurrence of such disasters at long intervals of time is just the mode in which a gradual land subsidence may be expected to make itself felt. The accidental piling up of the tide through meteorological causes, taking place perhaps only once in a generation, sooner or later overtops the flood defences.'[1] It is interesting to compare this episode with that of the Horsey floods of 1938.

[1] H.L.P. Jolly, 'A Supposed Land Subsidence in the South of England', *C.R. Cong. Int. Géog.* Amsterdam, 1938, vol. 2, Sect. 2B, Oceanography.

Chapter X

HOLDERNESS, LINCOLNSHIRE, AND THE FENLAND

(1) INTRODUCTORY

From Sewerby, on the southern side of Flamborough Head, to the north-eastern corner of the Wash, near Hunstanton, the shoreline is in many ways very different from that in any other part of England and Wales. The present shore is formed almost wholly of boulder clay and recent deposits, mostly alluvium. A study of modern conditions shows an area of rapid erosion in Holderness, counterbalanced by the accretion inside Spurn Head and the Humber. Farther south there is a good deal of erosion, not always obvious to a casual observer, along parts of the east Lincolnshire coast, and another area of rapid accretion in the Wash.

But we must go back a little in geological times to appreciate the full amount of interest of this long stretch of coast. Before the Ice Age there was no Holderness and a much smaller Lincolnshire: furthermore the fens of the Humber, Trent, Ancholme, Don, and Idle were probably areas of open water. The Fenland itself was formerly a great bay, with the sea reaching nearly to Cambridge, and the East Midland rivers, Great Ouse, Nene, Welland, and others flowed into this old bay which, in many ways, must have resembled the undrained Zuyder Zee.

The Humber estuary and the Wash are somewhat similar. Their origin needs fuller discussion later on (p. 438), but both, it should be noted, breach the Chalk; and to-day, moreover, both are silting up. Material travels consistently southwards along all this coastline, and it is thus not only logical but also convenient to trace the evolution of the present coast from north to south (Fig. 88).

(2) HOLDERNESS BAY

Holderness marks the site of the northern part of a great bay, the southern part of which is to be found in north Lincolnshire. Its origin is comparatively simple.[1] The Lower Chalk of both Yorkshire and Lincolnshire is hard, the Middle Chalk is both hard and full of flints, while the Upper Chalk is softer and free from flints. A great synclinal fold causes the Upper Chalk to underlie Holderness, and in a former period this soft material was easily worn away by marine erosion. The harder Middle

[1] *Mems. Geol. Surv.* 'The Geology of Holderness', C. Reid, 1885.

and Lower Chalk, however, was left projecting as the bold headland now known as Flamborough Head. Reid has cogently argued that this erosion was aided by the absence of flints in the Upper Chalk, and also because at that time the ice had not advanced on to these islands and left its masses of clay with innumerable boulders. A Chalk cliff undefended by a beach

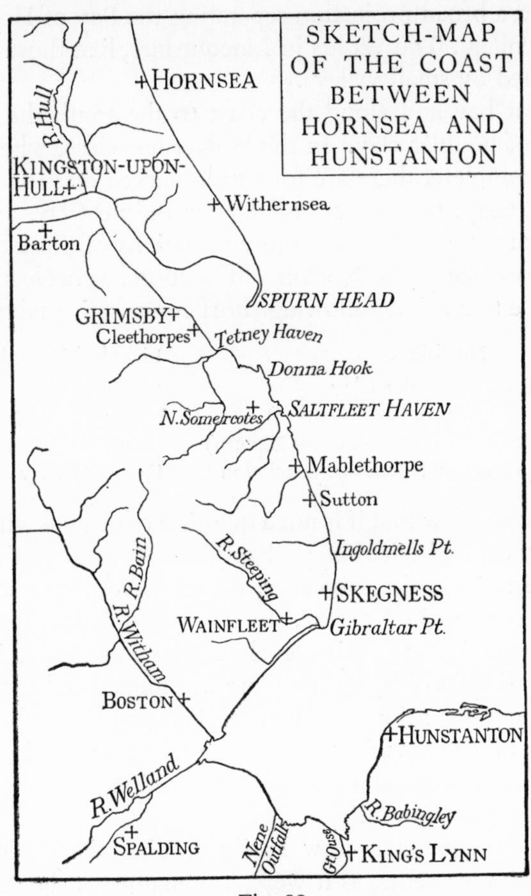

Fig. 88

of flint pebbles would waste far more quickly, and hence the bay was thus cut to follow fairly closely the curved strike of the beds. The available evidence suggests that this was in the Pliocene period. The bay was cut back to the line of the present Yorkshire Wolds, and it is clear that at one time the sea ran up the valleys which drained them. The valleys must have been later in age than the Pliocene because gravels are often found in them overlying an older boulder clay. In other words, these valleys are either inter-Glacial or possibly pre-Glacial in age. The outline of the former bay is indicated approximately in Fig. 89 by the 200-ft. contour line. To the west of the Wolds no trace of old beaches has been found. Flamborough Head projected farther seawards than it now does. Somewhat similar conditions obtain in north Lincolnshire but, there, south of Grimsby, the Wolds formed a broad projection separating the Bay of Holderness from that of the Fenland. The Wolds in Lincolnshire, like those of Yorkshire, were intersected by small valleys.

In this great bay and along the coast to the south the ice deposited great masses of boulder clay. In the Holderness cliffs, where the succession is most complete, there are four well-marked divisions of this drift: the Basement Clay, the Lower and Upper Purple Clay, and the Hessle Clay. It is not necessary to attempt correlations of glacial deposits but, since the importance of the Norfolk cliff section has been stressed, it may be worth while to add the following short comparative table:

Norfolk	Lincolnshire and Yorkshire
Hunstanton Brown Boulder Clay	Hessle Boulder Clay
Upper Chalky Drift (i.e. Little Eastern)	Upper Purple Boulder Clay
Great Chalky Boulder Clay (i.e. Great Eastern)	Lower Purple Boulder Clay
Norwich Brickearth or North Sea Glaciation	Basement Clay

This will serve to show that it is not a question of dealing with the boulder clay of merely one episode, but probably with four glaciations like those in Norfolk. In Yorkshire, however, details have not been worked out to the same extent as in Norfolk. In the inter-Glacial episodes changes naturally occurred in the relative levels of land and sea and in the erosion of the surrounding Wolds, but they are not immediately relevant to this chapter. At the present time the most southerly outcrop of boulder clay in Holderness is at Kilnsea: it reappears at Cleethorpes, in Lincolnshire, where there is only about half a mile of cliff section, the only cliff between Holderness and Hunstanton.[1]

In Holderness the sea is now busily eroding the boulder clay, and Bridlington Bay represents, as it were, a miniature of the old pre-Glacial bay. Near Bridlington the old Chalk cliff is being re-exposed, and at Sewerby (see p. 477) occurs the well-known section of the so-called pre-Glacial raised beach. Between Bridlington and Hessle on the Humber, the deposits banked against this old cliff are unknown, but at Hessle, the Chalk is clearly exposed once again. It must be assumed, as has been suggested above, that there were considerable changes in the relative levels of land and sea over the whole area of this great bay, and in post-Glacial times

[1] The Cleethorpes cliff is now inside the promenade!

there is clear evidence here, as elsewhere, of a former low sea level, shown especially in the submerged forest[1] exposed during the building of docks at Hull. This relative subsidence of the land was accompanied and followed by the deposition of a considerable thickness of estuarine warp. The small post-Glacial coastal valleys also became salt-water creeks, and Holderness at that period must in many ways have resembled the Broads of Norfolk. Most of the former meres have either silted up or their sites have been eroded away by the sea, while artificial drainage has also been partly responsible for their disappearance. The last surviving is that at Hornsea. Imagining for a moment the time immediately following the final retreat of the ice, Holderness might be pictured as a boulder-clay district of rolling and hummocky country with many hollows in which water could have collected. With the growing amelioration of the climate vegetation would soon have followed, and subsequently the now submerged forests would have filled much of the Humber estuary; land would have extended some way out into the North Sea, probably far enough to have included the Dogger Bank. With the following rise of sea level the coast would have begun to take the shape with which we are familiar today, but numerous meres would still have existed. Evidence of these meres occurs in place-names and in the occasional outcrops, mainly on the coast, of swampy ground or lacustrine deposits. A brief summary of recent changes along the coast between Bridlington and Spurn enables a reckoning of the rate of loss of land in historic time. In the course of it also a rough estimate of the position of parts of the coastline during the centuries of the Roman occupation is calculable and valuable for the purpose of deducing alterations since the fifth century A.D. For any earlier stages of coast development the evidence is insufficient.

(3) EROSION IN HOLDERNESS

Between Flamborough Head and Sewerby the rate of loss is small because both foreshore and cliff are made of hard Chalk, and the sea is still laying bare the old pre-Glacial cliff. The piers at Bridlington hold up material in its southward travel, and appear to be effective enough to affect the coast as far north as Sewerby, although at one time, before protective structures were built, erosion was very severe both north and south of the pier. Southwards from Bridlington and as far as the Humber erosion is continuously very serious, although there are naturally places where the rate is either above or below the average for the whole coast. The subject has been fully studied by Sheppard.[2] The map (Fig. 89) will show

[1] W.C. Crofts, *Proc Yorks. Geol. Poly. Soc.* N.S. 14, 1901, 245.

[2] Especially, *The Lost Towns of the Yorkshire Coast*, 1912.

better than any verbal description the number of villages and small towns which have disappeared into the sea. The Roman coastline is also indicated, but the demarcation must be considered approximate only. Sheppard's reconstruction of it is based on good evidence, however, and may be taken as reasonably correct.

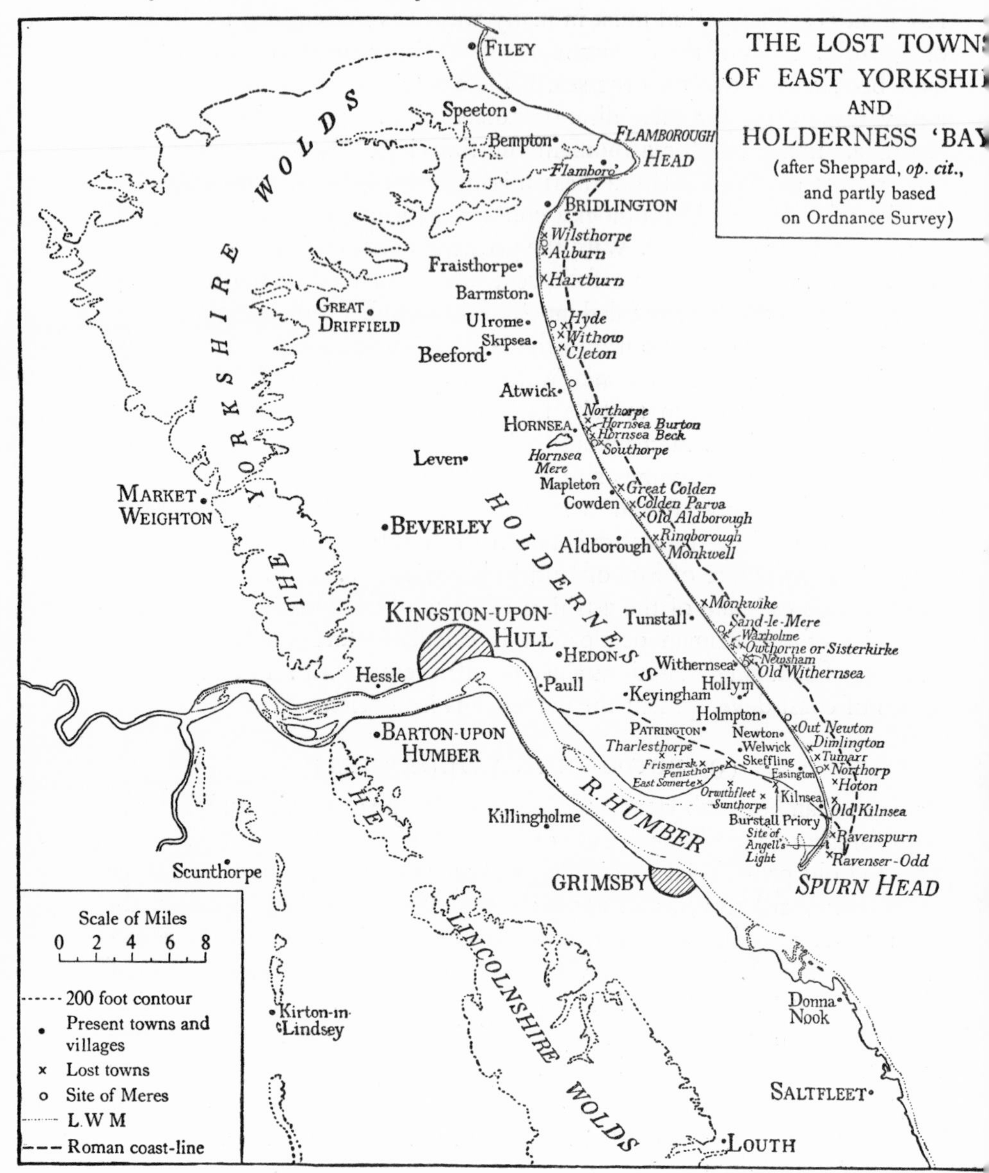

Fig. 89

It is unnecessary to describe the loss at each place marked on Fig. 89: the tale is much the same for all. It is sufficient to comment on some of the more interesting and important features (144).

Skipsea still exists, but to the south-east of it there was a place called Cleton, Cleeton, or Clayton, which has now vanished. The name 'Cleeton Lands' still applies to part of Skipsea. In the Domesday Survey, Skipsea was included under Clayton, which suggests that in the past Clayton was the more important of the two villages. The rate of loss here is now of the order of 2 yd. a year. There was formerly a good section of a lacustrine deposit to be seen in the cliffs, clear evidence of an old lake or mere,[1] so that the village must once have resembled Hornsea, Withernsea, and Owthorne. At Hornsea the present rate of erosion varies somewhat, but about 2 yd. a year seems a fair average; strong defence works of recent construction have checked the loss, otherwise the mere would in time have been reached by the sea. Formerly this mere, or another smaller one, extended farther east, but the fossils clearly show that it, like the present mere, was always a fresh-water lake. Holinshed speaks of Hornsea as a haven, and there are several references, probably belonging to the fourteenth century, which make it clear that vessels were safely harboured there. At Withernsea groynes were built in 1870 and have helped in preserving the sea front,[2] but just south of the town the rate of erosion has at times been severe: indeed between 1852 and 1885 it reached an average of 3·3 yd. a year. At the town itself it amounted to as much as 6 yd. a year in the twenty-four years before 1876. There is evidence that in earlier times there was a navigable creek at Withernsea like that at Hornsea: it is, for example, mentioned by Holinshed.[3] The finding of the pre-Viking Roos Car Images about 6 ft. below the surface in blue clay is another significant piece of evidence since they were discovered in clearing out a dyke. Lord Burleigh's chart of *temp.* Henry VIII also indicates a small creek. At Dimlington the cliffs are unusually high (about 100 ft.), and consequently the rate of erosion is less, although it often reaches 2 yd. a year. At Easington there are records of a haven in the fourteenth century, but as Holinshed does not refer to it, it may perhaps have disappeared by his time. Erosion has always been serious here, amounting at times to between 3 and as much as 5 yd. a year. The conditions at Kilnsea seem always to have been difficult because, whilst serious depletion takes place on the seaward side, there is also loss on the Humber coast. Carefully measured distances from the Blue Bell Inn showed that 334 yd. of land

[1] This was partly visible in July 1943.

[2] There is now a sea wall at Withernsea.

[3] Holinshed died *c.* 1580. His *Chronicles of England, Scotland and Ireland* appeared in 1573.

were lost in sixty years (1847–1908). At Kilnsea there was once a creek or haven. The cliffs now are low, but there is reason for thinking that formerly they may have reached 50 ft. on ground to-day eroded by the sea. In 1905–6 much of the low land of Kilnsea Warren was flooded by the sea breaking through the artificial coast banks.

In 1895 the British Association Committee appointed to enquire into the rate of coast erosion reported at the Ipswich meeting, and Capt. A. H. Kennedy gave a careful survey of the loss of land in Holderness based on the two 6-inch Ordnance Surveys of 1852 and 1889. Five sheets of the 6-inch map cover most of this coast, and in summary form the loss can be easily seen in the following table:

Sheet 197 (Hornsea)	Average loss	182 ft.
„ 213 (Aldborough)	„	237 „
„ 228 (Hilston)	„	139 „
„ 243 (Withernsea)	„	245 „
„ 257 (Holmpton)	„	273 „

Average for all sheets = 215 ft.
Average erosion each year = 5 ft. 10 in. (nearly).
Period covered 1852–1889 = 37 years.

The loss in four townships between 1086 and 1800 has been estimated by Sheppard on the evidence of the Domesday Survey and other documents:

	1086	1800
Easington	2,400 acres	1,300 acres
Holmpton	1,280 „	900 „
Tunstall	1,280 „	800 „
Colden	1,920 „	1,100 „

'The coast of Holderness, from Barmston to Kilnsea, is 34½ miles long. Assuming the land has been carried away equally throughout the distance, there has been a strip of land 1809 yards...wide, washed away, at the average rate of 7 feet 1 inch per annum. If the denudation since Roman times has been at the same rate, 53,318 acres, or about 83 square miles, have been lost. This is equal to a strip of land 2½ miles in width for the whole of the distance.'[1] Reid, writing in 1885, estimated an average annual loss of 2¼ yd., or, over the whole coast, about 34 acres. He added, however, that in the thirty years before he wrote the rate had increased as a result of the building of groynes and the carrying away of shingle.

The larger stones from the boulder clay collect in the form of the Stone Banks which are common at short distances from the shore. The beach itself, according to Reid, is composed of sand, which predominates, and

[1] Sheppard, *op. cit.* p. 42.

of stones less than 1 ft. in diameter. The material continually moves to the south. There seems to be a rapid change in the material as few stones are really well rounded, and there is also a tendency for the larger ones to work down below the tide marks. This accounts for a rather strange deficiency of material at some points where there is no artificial removal and no exceptional southward travel. Reid also remarks that there is much less beach along Holderness than one might expect, especially considering the large number of stones washed out from the boulder clay. 'What beach there is consists principally of shifting banks of sand, which, unless fixed by groynes, are of little use in protecting the coast during storms.'[1] As suggested earlier, the boulder clay finally disappears at Kilnsea Warren where, for about three miles, only a narrow bank of sand and shingle divides the North Sea from the Humber.

The whole sweep of the Holderness cliffs is somewhat monotonous. Yet the varying heights of the cliffs, from just over 100 ft. at Dimlington to sea level in other places, the pleasant, rolling, and usually well-wooded agricultural land within, and the wide expanse of sea and sky give the area a charm all its own. Not only in its geology are there resemblances to parts of Norfolk: the villages and countryside are often reminiscent of some in that county. Unfortunately an open coast of this kind is easily spoiled, and building development between the wars was often very bad. How pleasant the genuinely natural setting can be is seen well at Grimston and south of Withernsea. Usually erosion is greatest where the cliffs are lowest, but it is interesting to note that just to the south of Tunstall there is a break in the cliffs and a former small mere is dammed by sand dunes which are scarcely set back from the line of low cliffs on either side. Whilst one may agree in general with Reid's statement concerning the composition of the beach, it is nevertheless true that there is a good sandy beach all along this coast.

(4) SPURN HEAD

The material travelling southwards along the coast accumulates as Spurn Head,[2] the maintenance of which depends largely on the upkeep of groynes and the abundance of marram grass. Its general form indicates a balance between the effects of the North Sea and the Humber influences, and the lighthouse near the extremity has been shifted many times to conform with the changes in the position of the point. The first beacon seems to have been in use as early as 1428 when R. Reedbarowe, the hermit of Ravensporne chapel, obtained a grant for toll and a light. The second record is of Angell's light in 1676, in connection with accounts of a sand

[1] Reid, *op. cit.*

[2] See C. Reid and T. Sheppard, *op. cit.*

bank thrown up a few months previously in the Humber mouth, and apparently joined to the Yorkshire bank. Certainly, in 1684, there is further mention of the point's continued increase in length. In 1771 Smeaton, whose *small* light was built in 1771, 280 yd. east of the high light, reported that Spurn Head had extended 280 yd. since 1766. Another light was built 70 yd. farther west in 1816; a third still 30 yd. farther west in 1830, a fourth another 50 yd. to the west in 1831, and by 1863 the sea had reached the high light. Shelford estimated that the southerly extension of the point was 2,530 yd. between 1676 and 1851, and from 1851 to 1888 the high-water line shifted a further 200 yd. to the south. At this time the average westerly movement of the point was 8 ft. a year on the sea side, and 17 ft. on the Humber side, amounting thus to an annual increase in width of 9 ft. (**145**).

Callis, writing in the early part of the seventeenth century, remarks that 'of late years parcel of the Spurnhead in Yorkshire, which before did adhere to the continent, was torn therefrom by the sea, and is now in the nature of an island'.[1] Shelford also suggested that at an earlier time Spurn Head became an island, and that because of this the inhabitants of Grimsby, in 1289, claimed privileges there. This is quite possible, since the point has indeed been alternately island and peninsula on several occasions. The Trinity House chart of 1820 shows two breaks in Spurn, and as late as 1849 a large breach occurred on the neck, although not sufficient to sever the point; the Admiralty Chart of 1875–7 shows a long, narrow head and a broader neck. Shelford and Pickwell give the following figures for the southward growth:

1676–1766	1,800 yd.	or approx.	20 yd.	a year
1766–1771	280	”	56	”
1771–1786	150	”	10	”
1786–1851	300	”	4·6	”
1851–1864	113	”	8·7	”
1864–1875	60	”	5·4	”

Thus, for 200 years the average rate has been $13\frac{1}{2}$ yd. a year. Any such figure, or series of figures, must be taken with caution, but nevertheless they imply the possibility of the whole spit's forming in about 400 years. This rapidity in itself is in keeping with the accounts of occasional breaks and subsequent regrowths.

Undoubtedly Spurn does not represent more than a very small part of the waste from Holderness, and on the other side of the Humber there are large masses of sand at Donna Nook. These are unlikely to have been formed entirely by the waste of the Lincolnshire coast, because there the

[1] Quoted in Reid, *op. cit.*

native rocks are mostly warp and clay without stones.[1] Reid pointed out that there is little evidence of anything coarser than sand crossing the Humber estuary. But there, again, the fact that Spurn Head has at times become detached helps to solve the problem, because once such a break has occurred the island must of necessity work its way southwards towards Lincolnshire. There appear to be no records of this process in action, but it seems to be the most reasonable explanation for the known fluctuations of Spurn Head and the beach formations at Donna Nook.

On various old charts (e.g. *temp.* Henry VIII and G. Collins, 1684) islands are shown in the Humber mouth, and it is just possible that the former town of Ravenser existed on such an island. Ravenser is usually assumed to have been to the west of Spurn Head, but a map of the time of Henry VIII certainly shows an island to the east of the point. Sheppard notes that this map was not known to earlier writers on the matter. The town is supposed to have been of Danish origin and in 1305 returned two members to Parliament. There was serious erosion at Ravenser in 1355 and 1361, and by 1391 it had nearly disappeared. In 1538 Leland gives the last known reference to it. As Sheppard also suggests, Ravenser may well have been west of Spurn Head in the fourteenth century, but east of its present position. The settlement of Ravenser-Odd seems to have originated in the thirteenth century not long after Ravenser rose to importance. It, too, seems to have been on an island, but it had disappeared by 1360. Reference to Fig. 89 will show that other places have also been washed away in this locality, including Orwithfleet, Sunthorpe, and Burstall.

(5) THE HUMBER RECLAMATIONS AND FENLANDS

The Humber is probably the muddiest of our rivers, and there have been many discussions in the past about the origin of the silt. It is now generally agreed that it is brought in by the tide, and that the rivers add little if any material.[2] At the same time, in the estuary itself there is much local redistribution of silt. The original source of the material is undoubtedly Holderness. The mud and silt naturally first settle in sheltered places, and sooner or later rise high enough for salt-marsh plants to grow. The sequence of vegetation is generally similar to that in Norfolk and other parts

[1] Prof. Swinnerton calls my attention to the boulder clay so often exposed at low spring tides: in all probability this yields much material for the present beaches. Moreover, as the post-Glacial sea rose in level, boulder clay to the east of that now exposed would have yielded plenty of stones.

[2] T. Sheppard, *op. cit.* Chapter XXV; C. Reid, *op. cit.*; P.F. Kendall and H.E. Wroot, *The Geology of Yorkshire*, 1924, Chapter XXVII.

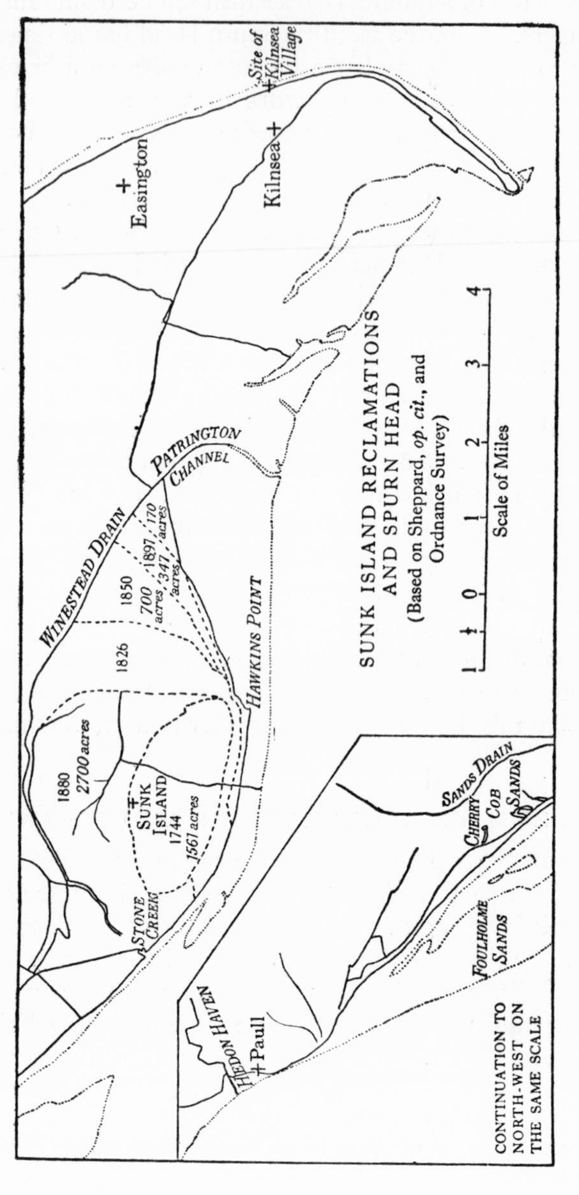

SUNK ISLAND RECLAMATIONS AND SPURN HEAD (Based on Sheppard, *op. cit.*, and Ordnance Survey)

Fig. 90

of the east coast: samphire gives place to the lavenders, sea purslane, and sea blite, while marram or other grasses appear on dunes at higher levels. This growth of marsh has led to the reclamation of much land: for instance Sunk Island is now part of the mainland south-west of Patrington, although the earliest account (*temp.* Charles I) describes it as an island of some 9 acres. Its subsequent physiographical evolution can be easily appreciated from a study of Fig. 90. Just north-west of Sunk Island there is another large reclaimed area known as Cherry Cob Sands, while some miles above Hull and west of Brough the former Broomfleet Island has been incorporated in the mainland. The last-named island began as a sand bank in the early nineteenth century; by 1820 it had reached high-water mark, and by 1846 130 acres had been enclosed. Up to 1912 some 600 acres had been added to the country in this locality. Reed's Island began to form much about the same time as Broomfleet, and enclosure started in 1840. In 1886 the total area of the island was 491 acres, but since then growth seems to have ceased, and between 1886 and 1901 67 acres were lost by erosion. Nevertheless, the gain in the Humber has on the whole been considerable. There is unfortunately little record of the dates of the earliest embankings of the mainland marshes, but Reid estimated that about 290 square miles had been gained by embanking since the Roman period.

Before resuming the survey of the present coastline it is worth examining shortly the Humber-Ouse valley inland beyond Goole and also the south-bank valleys of the Don, Idle, Trent, and, farther east, the Ancholme. It is a matter for regret that this former fenland area has not yet been investigated by the modern methods worked out by Godwin in the East Anglian Fens; hence, no proper correlations are possible. Phillips,[1] referring to the Ancholme valley, near Brigg, gives the following somewhat generalized sequence of deposits from the surface downwards: humus, peat, upper forest bed, brown alluvial clay, blue-grey alluvial clay, lower forest-bed, glacial drift. It is, perhaps, right to assume that a somewhat similar series of deposits characterizes much of the low-lying area on either side of the Isle of Axholme, and in the valleys of the Don, Idle, and Ouse. The *Geological Survey Memoirs* covering this area are of little use since they were written long before the modern technique of investigating recent deposits was invented.

Presumably all these deposits represent successive stages in the rise of the sea from the low level recorded by the deepest submerged forest. At

[1] C.W. Phillips, 'The Present State of Archaeology in Lincolnshire', *Arch. Journ.* 90, 1934, 106. See also *Mems. Geol. Surv.* 'The Geology of parts of North Lincolnshire and South Yorkshire', W.A.E. Ussher, 1890.

Hull the lowest forest bed is about 50 ft. below high water, while in 1877 Parsons described a forest bed traceable at Long Drax, Goole, Thorne, and Eastoft. Near Goole the tree stumps were seen in the river bed between tide marks, and a lower bed of peat, 12 ft. below the upper, was also found. During this, and somewhat earlier times, the Humber estuary must have been far more extensive and branching than it now is, but unfortunately it is not possible to draw a line of any accuracy to represent the boundary. The general evolution of the present scenery must have resembled that of the Fens and Broads. The old tidal inlets were reduced by accretion, and a comparatively narrow river-way was left.

The dates of the first river embankments, like those of marsh reclamations, are not known. Kendall and Wroot[1] give evidence suggesting that some rivers were embanked by 1315, but up to the seventeenth century there is little in the way of definite record. In 1626 Vermuyden began work on the marshes of the Idle and Don. The Idle formerly meandered northwards across Hatfield Chase, but was diverted eastwards to the Trent. The Don in the early seventeenth century had two mouths; one opened to the Aire at Snaith, and the other passed by Crowle to the Humber at Adlingfleet. The nearly straight and artificial-looking reach of the Don between Thorne and East Cowick (near Snaith) was in existence before Vermuyden began his work, and so may be mediaeval or even Roman. The Dutch engineer, in closing the Thorne-Crowle-Adlingfleet branch of the Don, certainly enabled the reclamation of Hatfield and Thorne marshlands, but was responsible also for the severe flooding of farm lands around Snaith because the Snaith channel was too limited in capacity. Consequently Vermuyden dug a new channel from Newbridge to Goole, which is still called the Dutch River and still followed by the Don. The Snaith channel silted up in 1877.

Prior to this seventeenth-century draining work, Axholme was truly an island of Keuper Marls, and marshy conditions prevailed in the Ancholme valley. In all these low-lying areas the process of warping has been practised from the eighteenth century onwards. This consists in letting the flood waters of the tide into enclosed areas to drop their silt, and is merely an adaptation of the natural process by which salt marshes are formed. These warped lands cover large areas and obscure natural details.

On the southern shore of the Humber there is a small boulder-clay cliff at South Ferriby, and the shingle near Barton is interesting. This is marked on the geological map as Glacial, but Ussher comments that it has something of the character of a raised estuarine beach which can be followed,

[1] *The Geology of Yorkshire*, Chapter XLVII.

on and off, round the Chalk escarpment to a point beyond Horkstrow. It does not rise more than 50 ft. above the sea and forms a terrace feature at the foot of the Wolds. There has been a good deal of reclamation on the southern shore of the Humber, but not to the extent of that practised in Yorkshire. The development of Immingham Docks, a few miles north-west of Grimsby and near deep water, is a comparatively new feature.

(6) NORTH-EASTERN LINCOLNSHIRE

The boulder-clay cliff at Cleethorpes[1] wasted at much the same rate as the similar cliffs of Holderness. Reid[2] argued that from the contour of the ground (in 1885), the cliff must have come into existence within the previous 200 or 300 years: before then salt marshes would have extended in front of it. It has, indeed, already been suggested that the beach material at Donna Nook comes from Spurn.[3] Inside Donna Nook, and a little south of it, is a large spread of shingle and blown sand between Somercotes and Saltfleet, usually referred to as the Somercotes beach. On the geological map this line is more or less continuous via Marsh Chapel and North Coates to Cleethorpes, but the ¼-inch geological map refers to the material as 'Valley Loams and Gravels of various ages'. Reid, however, called attention to the extensive alluvial tract in the Tetney district, lying within this belt of gravels. At first sight it suggests a former higher sea level, but the alluvium has only yielded fresh-water and land shells. It is thus probably of fluviatile and not estuarine origin, and therefore quite different from the present marshes which often intersect it. It is separated from the boulder clay by two metres of lower alluvium, and there is no obvious explanation to-day for water having once been held up in this area. Reid consequently suggested that the line of the Somercotes beach and dunes was formerly continuous as far as Cleethorpes, and that it formed a barrier to Tetney and other havens. Behind this obstacle alluvium could have accumulated, and it was not until some later time, when the shingle and dune barrier was breached by the sea, that the water drained away and allowed the cutting up of the district and its penetration by dykes. The formation may also be closely associated with the supply of material from Spurn, which is admittedly irregular.

Accretion in this part of Lincolnshire has enabled reclamation to take place, and the Royal Commission of 1907[4] noted that in Wragholme, North and South Somercotes, Skidbrooke-cum-Saltfleet, and Saltfleetby, between 800 and 900 acres were gained fifty to sixty years previously.

[1] The cliff is now inside the promenade.

[2] *Op. cit.*

[3] But see note, p. 415.

[4] See Appendix 23 of R. Comm. Reports: 'Lincolnshire', p. 235.

Accretion is, however, replaced by erosion, often fairly severe, at Mablethorpe, Trusthorpe, and Sutton, and even as far as Ingoldmells[1] (see below): it ceases south of Skegness.

(7) EASTERN LINCOLNSHIRE

East Lincolnshire, at first sight somewhat unattractive, presents in fact many points of interest, and of recent years Swinnerton[2] has greatly advanced our knowledge of this region. The marsh area is nearly all below the 50-foot contour. Alongside the Wolds it is made up of well-wooded boulder-clay country which slopes eastwards down to sea level. It then gives place to the Outer or Bottom Marsh which is flat, treeless and intersected by a network of drains. This part is almost all below high-tide level. The sea once reached the Wolds here as in Holderness: probably not later than the middle of the Glacial period, although the chronology is very uncertain. There would then have been no marshland, but a broad wave-cut platform covered by the sea and overlooked by Chalk cliffs. The platform was ice scoured, and the late ice left a moraine belt between Mumby, Anderby, and Huttoft. Near the edge of the Wolds glacial overflow channels were cut, and near Alford the water escaped to the sea, probably forming the shingly deposits seen sometimes on the beach north of Ingoldmells. When the ice finally retreated it left a spread of boulder clay over all this platform, averaging about 80 ft. in thickness. This clay is the foundation of the present marshland, and it is often seen at about low-water level on the beach. Immediately after the retreat of the ice, Lincolnshire stood higher than now relative to sea level, and the boulder clay became forest covered, but later there followed the submergence of post-Glacial times. This slowed up the rivers; the climate became wetter, and peat bogs, now often seen in the submerged-forest areas between Mablethorpe and Skegness, began to form. The sea then flooded the lower boulder-clay areas, and silt was laid down on which salt-marsh plants began to grow. This process of quiet sedimentation implies a sheltered area, and Swinnerton therefore argues that some form of offshore barrier shielded the coast from rough water. He suggests that a broken line of morainic hills running from Spurn Head to the Norfolk coast may have existed. But at the end of the Bronze Age sinking ceased, and may even

[1] Erosion is not always obvious to a casual observer. Prof. Swinnerton tells me that in places its extent is indicated by the traces of old drain floors in the clays of the beach.

[2] H.H. Swinnerton, 'The Physical History of East Lincolnshire', *Trans. Lincs. Nat. Union*, 1936; also A.J. Jukes-Browne, *Mems. Geol. Surv.* 'The Geology of Part of East Lincolnshire', 1887.

have given place to a slight rise of the land. At any rate, the waters became fresh and the corresponding fresh-water plants and reeds gained a hold. Close to Ingoldmells there was an area of tree growth which now forms a thin peat bed. Salt works were established at Ingoldmells by Iron Age people, and later still there was a Roman station. The site of the salt works is now at mid-tide level, while the Roman remains are just above it. Hence, it is clear that there has been a renewal of sinking since or during the Roman occupation of the site. This sinking seems to have been too rapid to allow salt-marsh plants to grow, and on the lower marshland the episode is recorded by a layer of clay containing *Scrobicularia*. This again implies quiet conditions, and it may therefore be assumed that the protecting barrier offshore still existed during and after the Roman occupation. In the thirteenth and fourteenth centuries, however, there are records of severe marine floods, so it is clear that by that time the barrier had gone, and it had almost certainly been weakening for some considerable time. Now that there is no barrier the marshland has been cut back by the sea and sandy beaches and dunes have formed.

(8) INGOLDMELLS

Such, then, has been the general history of the marshlands of east Lincolnshire. At the risk of a little repetition, however, it is useful to follow Swinnerton[1] more closely in his important investigations at Ingoldmells, partly because they result in one of the earliest and best demonstrations of recent sea-level movements, and also because they relate very closely to the investigations carried out in the Fens farther south. It is, moreover, to be hoped that future work will link the findings of research both in the Fenland and at Ingoldmells to the Ancholme and Axholme areas.

The erosion between Mablethorpe and Skegness is often obscured by sand swept along the beach, but when the sand is scoured away erosion hollows in the underlying boulder clay are clearly seen. The glacial deposits are fairly resistant and make hummocks which form points at Addlethorpe, Ingoldmells, and elsewhere. Overlying them are the post-Glacial deposits, the lowest of which is a peat bed up to $2\frac{1}{2}$ ft. thick, and rising and falling with the inequalities in the boulder-clay surface. Tree stumps of alder, birch, oak, prunus, and yew project through this peat in which 'Neolithic' implements have been found. Then follow the mid-post-Glacial deposits, mainly stiff clays: they usually rest on the peat, but occasionally this has been removed by local erosion. The clays vary a good deal in thickness from place to place, that is to say from 3 to even

[1] 'Post-Glacial Deposits on the Lincolnshire Coast', *Quart. Journ. Geol. Soc.* 87, 1931, 360.

8 ft.: the colour varies, but usually the top and bottom layers are blue-grey, whereas the mid part is purple. In the middle of the purple clays vegetation occurs, but the species are doubtful, although *Salicornia* spp. is probably included. Above the purple clays there is a definite plant succession. The first layers contain *Juncus*; in the second *Triglochin* is added; in the next these two lessen considerably, and *Statice* and *Armeria* become abundant. There follows a well-marked change to *Phragmites communis* (indicating brackish to fresh water) just below the upper peat.

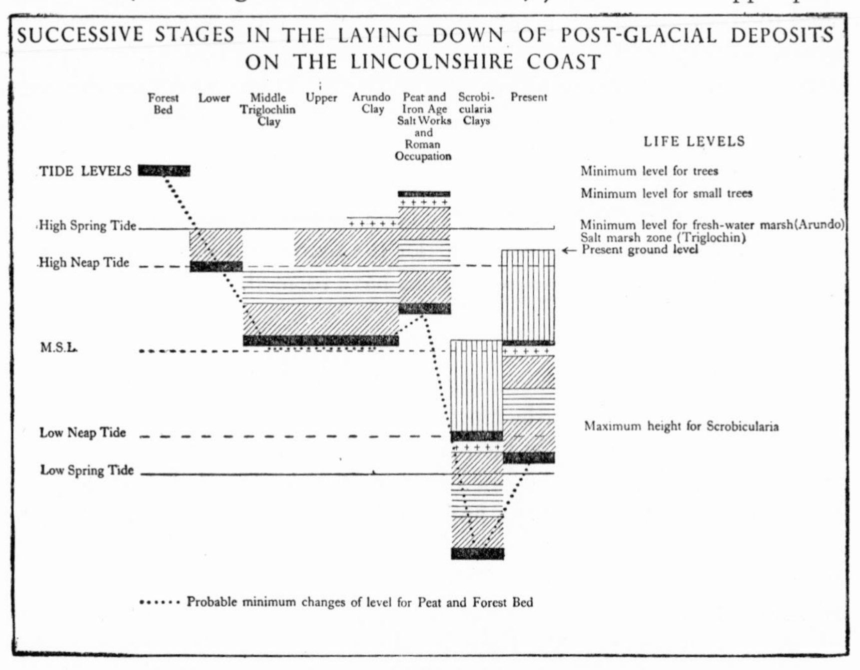

Fig. 91

In all this series there appears to be an entire absence of estuarine shells such as *Scrobicularia* and *Cardium*. The upper peat contains pottery, and was contemporaneous in its forming with the Hallstatt (Upper Bronze) Age. This layer forms the lower boundary of the upper-post-Glacial deposits, which consist of clays 7–17 ft. thick, purplish in colour, and often soft and sloppy. They are similar to clays now forming, and contain *Scrobicularia plana*, *Cardium edule*, and *Hydrobia ulvae*. Plant remains are by no means so abundant as in the lower clays[1] (see Fig. 91).

The tree stumps in the lower peat occur as far down as the lowest spring-tide level. If it be right to assume that the present tidal range of

[1] Fig. 91 was very kindly prepared by Prof. Swinnerton: he allows me to use it.

about 19 ft. also held at this time, it implies that, when the trees grew, sea level was, relative to the present, at least 25 ft. lower.[1] The trees also indicate a mild climate, while the succeeding peat indicates a change to greater dampness, possibly related to a slight rise in sea level. Such a rise certainly continued during the formation of the middle clays, and salt-marsh conditions obviously prevailed during much of that phase. Swinnerton argues that the absence of *Scrobicularia* and *Cardium* means that the surface was not below the level of mid tide.[2] The change from the *Triglochin* clays to the *Phragmites* layer is sudden and may be explained by the ordinary depositional processes during which coarser silts tend to gather at the seaward margin of wide mud flats. On the other hand, it may indicate a slight fall in sea level, since the ground was sometimes dry enough to allow the growth of small willow and yew trees[3]: there are also traces of a Bronze Age industry. The upper peat at the time of its formation must have been at the level of high spring tides. To-day it is at mid-tide level, indicating a subsequent rise of sea level of about 9 ft. The base of the Roman occupation level is but 4 ft. above the peat, and less than 4 ft. above the *Scrobicularia* clays. It is, therefore, argued that when these clays formed the Roman station was below high-tide level. In short, the sea rose and deposited the clays after the Roman occupation. The condition of the outer shore and the possible line of protecting banks have already been discussed.

Swinnerton[4] has also described several prehistoric pottery sites in this same district which, on a smaller scale, resemble the Red Mounds of Essex. Since no fragment of Roman pottery is found in any of the twelve sites examined, it seems clear that the workings had ceased before the Romans came into this district. All the available evidence points to their being of Bronze Age–Early Iron Age date.

The Lincolnshire coast between the Humber and Wainfleet is certainly flat, but it is by no means unattractive. In the past, accretion north of Mablethorpe has produced wide flats fronted by lines of dunes. Between Mablethorpe and Skegness, except where sea walls and promenades run, there is usually but a single line of dunes, sometimes diversified by high isolated hummocks. In the winter of 1942–3 erosion dealt severely with this coast, and broke through the dune belt in various places. The dunes

[1] Following Swinnerton.

[2] Godwin inclines to the view that the whole sequence is that of a natural transition of a salt-marsh succession to a fresh-water marsh.

[3] It is interesting to note that *Phragmites* flourishes at the inner side of the Brancaster marshes to-day.

[4] *Antiq. Journ.* 12, 1932, 239.

are usually well covered in vegetation, and the abundance of sea-buckthorn (*Hippophae rhamnoides*) is a marked characteristic. Unfortunately much of this coast has been, at any rate temporarily, spoiled by unsightly groups of huts and bungalows. South of Skegness, where accretion recurs, the dunes (in part a golf course) continue into the sand and marsh formation of Gibraltar Point: the whole stretch forms an interesting and attractive foreshore (**146** and **147**).

Just inside the dune belt there is agricultural land reaching as far as the line of the Wolds. On a clear day the view landwards shows up clearly the several sea walls (sometimes followed by roads) which have from time to time been built to reclaim the marsh. Seawards there are extensive sand flats, interspersed with ridges which will eventually develop into new foredunes. Where erosion takes place the underlying boulder clay is often exposed on the foreshore.

(9) THE BORDERS OF THE WASH

Gibraltar Point has been formed in part by the southerly drift of material along the Lincolnshire coast, and the salt marshes between it and the Witham have been described by Newman and Walworth.[1] As they are more or less typical of all those in the Wash, it is worth noting at once their main characteristics before discussing the intricate problems of the Fenland. The width of these marshes varies from 200 yd. to a mile, and they are all outside the sea wall built about seventy years ago.[2] On the bank zone, or sea wall, there is a settled type of pasture vegetation. At the base of the bank 10–20 ft. wide, *Agropyrum junceum* and *Spartina stricta* are dominants; below the bank on the marsh itself there is first a belt of *Festuca rubra* with much *Spergularia salina* and *Glaux maritima*, then an intermediate zone with many of the characteristics of the preceding one, but including *Obione portulacoides* and *Suaeda maritima*, next is a *Festuca-Agropyrum* zone followed by the most extensive plant area of all, an *Obione* zone with much *Suaeda maritima*. At about average high-tide level there is a belt of *Festuca* and *Salicornia*, much broken by channels and tidal action. Below it, the mud flats are first occupied by a nearly pure stand of *Salicornia*, while farther downwards and outwards *Zostera* and Algae occur: pans and channels are common. The unreclaimed marshlands[3] on the Norfolk side of the Wash are in the main like those of

[1] L.F. Newman and G. Walworth, *Journ. Ecology*, 7, 1919, 204.

[2] I.e. seventy years before the time Newman and Walworth published their paper.

[3] *Mems. Geol. Surv.* 'The Geology of the Borders of the Wash', W. Whitaker and A.J. Jukes-Browne, 1899.

Lincolnshire, although reclamation has been less. Reference to Fig. 96 will show the main innings(148).

The shores of the Wash certainly cannot be regarded as normal 'seaside'. On the other hand, to dismiss them as uninteresting or unattractive is wrong. On approaching the still unreclaimed parts of the inlet newer and newer sea walls and fields are traversed. The land is extremely fertile and the crops varied. The landscape resembles that of parts of Holland, but windmills are absent from the newly reclaimed land. The wide expanse of sky and cloud; the open country, broken by lines of trees, scattered farms and sea walls; the colours of the different crops; and the ever-changing level of the water on the natural marshes, the appearance of which alters greatly with the seasonal flowering of the plants, give to this part of the country a peculiar charm. It is true, however, that a visit during a north-east storm in winter will give a very different impression of the district from one made on a fine day in late summer.

The eastern side of the Wash is bordered by a belt of fairly high country which, at Hunstanton, is eroded to form the first cliffs south of Cleethorpes. Near North Wootton and Wolferton the heathy ground of the Sandringham Sands overlooks extensive reclaimed marshes: the edge of this higher ground is a distinct cliff notched by ravines, especially near Wolferton. That this cliff was formed a considerable time ago is evident from its present distance from the water: it may also have been formed during a period of slightly different sea level. The ravines, which are cut in soft and often rather incoherent sands, were probably formed during, or soon after, the Glacial period.

Along this part of the Norfolk coast beach material travels into the Wash: this is well exemplified by the considerable deflection of the Heacham River, and probably also by the large masses of shingle at Snettisham. Because shingle is absent, and the high ground is much farther inland on the Lincolnshire shore, it will be appreciated that there is a marked contrast between the two sides of the Wash.

(10) THE FENS

(*a*) *Introduction and Pollen Analysis*. The Wash, the origin of which is discussed on p. 438, is merely the unfilled-up remnant of the whole Fenland basin. The modern inner shoreline is subject to incessant change: it is only a matter of years before one set of embankments follows another and before new natural salt marsh grows. But to appreciate the present state of affairs we must go back to the beginning of post-Glacial times. The approximate boundary of the basin runs from Firsby through Tattershall, Lincoln, Billinghay, Peterborough, Earith, Cottenham, Burwell, and

Denver to King's Lynn: its area is just over 1,300 square miles. To-day over the Fenland 'The airman looks down upon a regular pattern of channels separating well-tilled fields. Rivers run directly to the sea through corn, fruit, potatoes and sugar beet. Straight lines dominate the scene. Small dykes divide the fields with mathematical regularity, and there are few hedgerows. Long straight roads are frequent, and they cross each other at right angles. The railway lines are also characterized by long straight stretches. Both river and railway are, more often than not, embanked above the level of the surrounding country. Not only have the island settlements increased in size, but houses have dispersed themselves over the fen as well—either in straight lines, or in groups, or as isolated farm-houses sheltered by trees.'[1] In the Middle Ages the scene was very different, and much earlier in geological time there was a great shallow bay, with a certain number of islands, such as Ely, March, and Chatteris.

In discussing the evolution of the Fenland the facts shown in Fig. 92 are of permanent importance. There is first of all the great distinction between the peat and silt areas, one indeed which becomes of greater significance every year. Secondly, the fen islands are outstanding features. The whole basin lies in a hollow in the Jurassic rocks, which in the Fenland are mainly composed of the Oxford and Kimmeridge Clays. If the shallow nature of the Fenland gulf be borne in mind, it will be readily appreciated that even small relative movements of land and sea will have far-reaching results. The seaward end of such a gulf will naturally tend to collect silt deposits, whereas peat will grow in the fresh water on the landward side. If silt and peat have thus formed, a slight rise of sea level implies that the silts will transgress over the peat. If, on the other hand, conditions favour the development of peat, then this formation may expand over the silt areas. It is thus to be expected that there will be alternate wedgings of the silt and peat deposits, especially around the margin of the basin, and this is certainly borne out in fact.

Before discussing the nature of the fen deposits and the information they give of past conditions, a slight digression is necessary to introduce very briefly the modern technique of pollen analysis.[2] Trees and plants distribute their pollen grains *widely* by the wind. These grains fall into various types of deposit and may or may not be preserved. Fortunately peat preserves them remarkably well, and so it is possible from a sample of peat to estimate purely by means of the preserved pollen grains the trees and plants growing at the time of its accumulation. Counting of the

[1] H.C. Darby, *The Draining of the Fens*, 1940, Chapter IV.

[2] H. Godwin, 'Pollen Analysis—an Outline of the Problems and Potentialities of the Method', *New Phytologist*, 33, 1934.

grains also affords a means of reckoning the relative frequency of the different tree genera. The information available from any one small

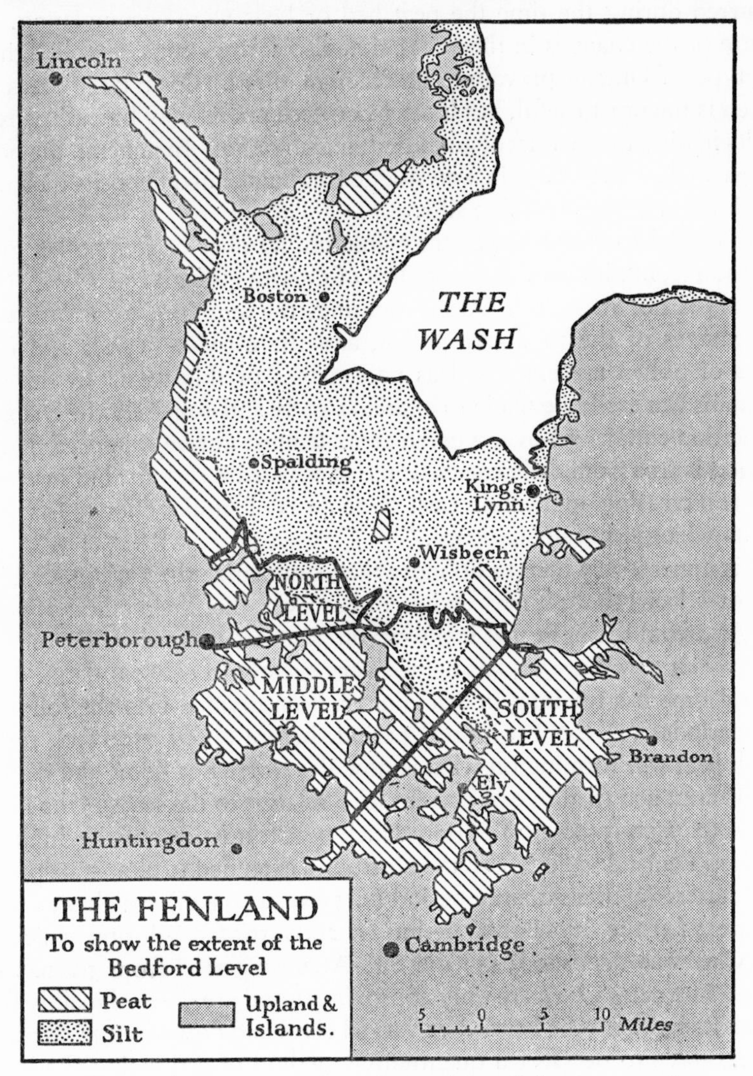

Fig. 92 (from H.C. Darby, *The Draining of the Fens*, p. 71)

sample of peat is, of course, very limited, but if samples be taken over a wide area and at frequent and regular intervals through the peat beds, then the deductions can become extremely valuable. Despite the accidents that may arise in the pollen distribution, such analyses give a fairly accurate picture of the types of vegetation flourishing at any one time over a large area, and also of the changes in the types of plants and trees that may have occurred during the time the peat bed or beds were accumulating. The nature of the changes in the vegetation also throws considerable light on the type of climate prevailing at different times. Over large areas tree pollen is the most useful, but locally non-tree pollen may be indispensable in elucidating the history of an old salt marsh or fen. In general, the pollen grains are best preserved in waterlogged and unaerated deposits including, besides peats, clays and fine silts.

It is usual to count 150 as the minimum number of free pollen grains in a sample, but in this total the pollen of *Corylus* (Hazel) and *Myrica* (Bog Myrtle) is not reckoned. After counting, the amounts are expressed as percentages of the total tree pollen, and the *Corylus-Myrica* and other types of pollen are expressed as percentages of this total. In this way diagrams can easily be made to express clearly and concisely the changing vegetation conditions over a given area. It must not be supposed that the method is free from deficiencies, some of which are serious, but nevertheless its increasing application over large areas of similar climate has given results which inspire confidence.

It is unnecessary here to do more than summarize the various divisions of post-Glacial time; it may be noted, however, that the method of pollen analysis, from the light that it throws on past vegetational conditions, *and therefore on past climates*, has helped very greatly to confirm them. Post-Glacial time has for long in Scandinavia been divided into the following five periods: (1) the Pre-Boreal period, before about 7500 B.C., during which peat was forming on what is now the North Sea floor, and the land stood about 200 ft. above its present level relative to the sea; (2) the Boreal period (*c.* 7500–5500 B.C.) during which peat formation was beginning in the deep Fenland valleys, and the North Sea extended to almost its present boundaries; (3) the Atlantic period (*c.* 5500–2000 B.C.) which was warm and wet; (4) the Sub-Boreal period (*c.* 2000–500 B.C.) during which the Fen Clay was deposited; (5) the Sub-Atlantic period from 500 B.C. onwards. The dates given are only approximate and are derived mainly from Scandinavia where the banded or varved clays laid down by the retreating glaciers gave to de Geer a quantitative method of estimating the length of post-Glacial time. In Scandinavia there is a close relationship between these clays and peat deposits and raised beaches. By pollen-analysis technique it has been possible to hazard an extension of this time scale beyond the Scandinavian peninsula.

With this very rough outline of the technique in mind it is easy to appreciate the significance of the following short table. (See also the appendix at the end of this chapter.)

	Fenland zones	Period
VIII	Alder-Oak-Elm-Birch (beech) zone	...Sub-Atlantic
VII–VIII	Transition	
VII	*d* Alder-Oak-Elm-Lime zone	...Sub-Boreal
	c, *b*, *a*	...Atlantic
VI	*c*, *b*, *a* Pine-Hazel zone	...Boreal
V	Pine zone	
IV	Birch-Pine zone	...Pre-Boreal

In order to examine the Fenland deposits numerous shallow bores and excavations have, of course, been necessary. The great mass of this work has been undertaken by Godwin whose industry and careful correlations have revolutionized our views on the whole area and superseded those of Skertchley[1] and earlier writers. His work has also been carefully related to archaeological finds at many places.

(*b*) *Sections at St Germans and Peacock's Farm.* Before discussing the general conclusions applicable to the whole area it will be valuable to examine two particular localities purely as examples of the nature and sequence of deposits. The first site in question is that at Wiggenhall St Germans, near King's Lynn,[2] and the section given in Fig. 93 will make clear the following notes. The section shows that the Fenland floor is here of Kimmeridge Clay. Later the land subsided relative to sea level, and bad drainage led to the formation of a thin peat bed. The sea

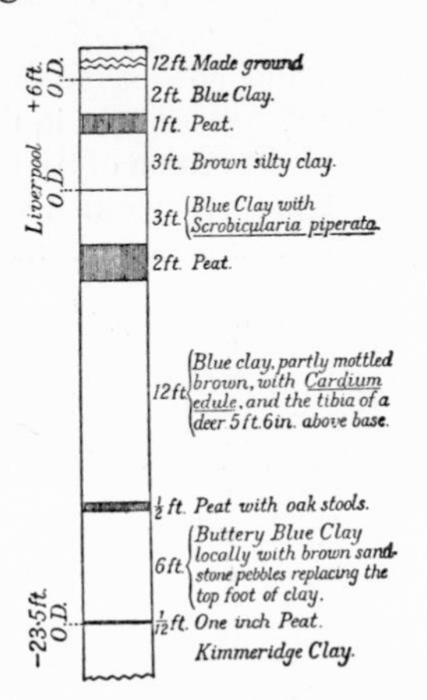

Fig. 93. Section of Fenland deposits at St Germans (after W.A. Macfadyen, *Geol. Mag.*, 70, 1933, 184)

then broke in, but at the same time the subsidence of the land either ceased or gave way to a slight elevation accounting for the development of the second peat bed. Further subsidence followed, to be succeeded by another stand still or elevation phase, during which a third peat bed formed, and there was later another repetition of this process. The four peat beds here are rather exceptional: as the Fenland floor graded upwards only two peat beds occur farther inland (near Ely), one is at the present surface and the other below the Buttery (Fen) Clay, which may be as much as 12 ft. thick.

[1] *Mems. Geol. Surv.* 'The Geology of the Fenland', 1877.

[2] H. and M.E. Godwin and F.H. Edmunds, *Geol. Mag.* 70, 1933, 168.

The second site is at Peacock's Farm,[1] about seven miles east-north-east of Ely (Fig. 94). Here the base of the Lower Peat is about $22\frac{1}{2}$ ft. *below* sea level, and was possibly some 10 ft. *above* sea level when the peat was formed. The *net* relative subsidence of $32\frac{1}{2}$ ft. holds good locally and may be indicative of like changes throughout the Fenland. The Upper Peat, with Early Bronze Age remains in its lower parts is now −6 ft. O.D., and at the time of its formation was about 16 ft. higher than at present.

Positive evidence of marine transgression is given by the returning wet conditions at the top of the lower peat and the semi-marine conditions of the fen silts and clays. It is difficult to postulate the causes which initiated peat-formation above the basal sand, but it is not unreasonable to attribute it to the influence of land subsidence and backing up of the river drainage. Increased general rainfall might, however, have similar effects. The continued formation of peat throughout the lower bed might represent conditions of stability or of movements of elevation or depression too slow to affect the process of peat accumulation.

The rooted willows in the upper peat seem to indicate stability or slight elevation after the deposition of the fen clay; the return of wet conditions later suggests renewed subsidence, a suggestion supported by the present level below O.D. It is, however, possible that here again the effect is due to increased rainfall or local drainage effect.[2]

The remains of three cultural periods in this section are of great interest, and unique in the Fenland. All are below present sea level. The Early Bronze Age appears at the base of the Upper Peat; Neolithic 'A' (=Windmill Hill) is recorded for the first time in this part of the Fens, and the Tardenoisian industry is found about 4 ft. deeper in the Lower Peat. Pollen analysis has shown that the oldest part of the Lower Peat is Boreal, and that the Tardenoisian level is near the Boreal-Atlantic transition. The pollen analysis of the Neolithic 'A' horizon has not proved valuable.

These two examples will serve to illustrate the nature and sequence of the deposits in the peat fens. The three main beds are always those of the Upper and Lower Peat and the Buttery Clay. There may be subsidiary peat beds like those at St Germans, usually for the reason given on p. 429. It is now profitable to turn to the general conditions in the Fenland during the several post-Glacial episodes, and in so doing to follow Godwin's methods very closely.

(*c*) *The Fenland in the Several Post-Glacial Periods.*[3] In the Pre-Boreal episode the Fenland was an area of open woodland, mainly pine and birch.

[1] J.G.D. Clark, H. and M.E. Godwin, and M.H. Clifford, 'Report on Recent excavations at Peacock's Farm...', *Antiq. Journ.* 15, 1935, 284.

[2] *Ibid.*

[3] H. Godwin, 'Studies of the Post-Glacial History of British Vegetation, III and IV', *Phil. Trans. Roy. Soc.* B, 230, 1940, 239. This, with its predecessor, *ibid.* 229, 1938, 323, is a paper of great importance.

The trees grew on the outcrops of the Jurassic clays, the Lower Greensand, and the Gault clay. It was a country of wide river valleys which were cut when the volume of water in them was greater than at present. The sea was far away, probably to the north of the Dogger Bank. In the Boreal the climate, as indicated by the trees, was warmer. The birch gave way to the pine, and the elm, oak, and hazel appeared, while at the end of the period, when the climate was at its best, came the aldor and linden. Mesolithic folk were living at Shippea Hill. But with the advent of the Atlantic period the climate grew wetter; the alder became very important, and with oak and lime began to predominate. The rivers still followed winding valleys, which often contained much sedge fen: in fact the Fenland became waterlogged and black peat gradually formed. The shore was probably not far from its present position. During Neolithic times the area was covered by sedge fens: there was not a great deal of open water. The chief trees were the alder, sallow, and birch, especially abundant at the close of the period and near the fen margins. Nearer the end of the Neolithic drier conditions returned so that fen woodland increased over the formerly wet surface. Oaks, pines, and yews grew in these woods, and the rivers increased the dryness of the country by cutting downwards in the peat. In the Sub-Boreal period there was an invasion of the sea, and it was then that the silts and clays spread far over the Fenland. The precise physical conditions which caused this inundation are not known, but it was possibly due to the breaking of a coastal bar, causing conditions not unlike those at Horsey in 1938 (p. 377). The evidence of the diatoms and foraminifera in the Buttery Clay implies that brackish-

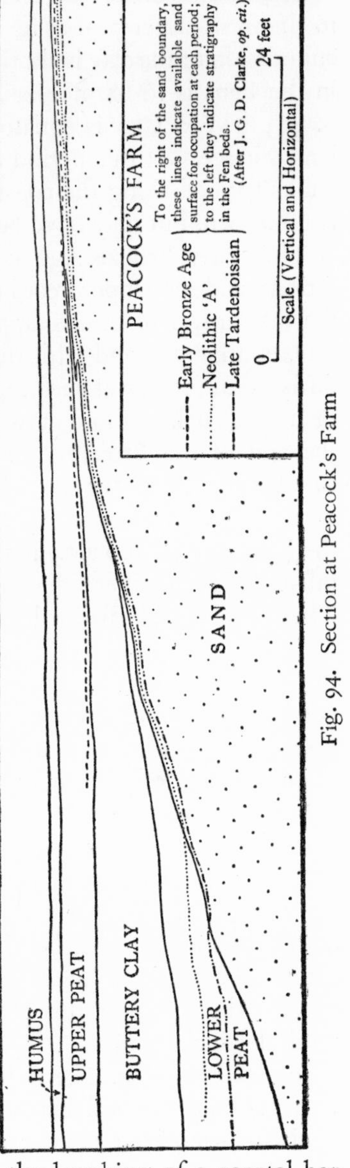

Fig. 94. Section at Peacock's Farm

water lagoon areas were reached by salt water at high tides, and that they were also maintained by rain and rivers. Marine influences are most noticeable at the base of the clay, and there is a fairly regular and gradual transition to the fresh-water peat at its upper surface. The time allowed for the accumulation of the clay is short, but this deduction must hold if the evidence in the Fens is to harmonize with that on the Essex coast. As noted on p. 403 the Fen Clay is equated with the clay above the Essex (Lyonesse) land surface. At this period also it is more than likely that sand and silt banks formed along the coast of that time. The main outlet of the rivers is unknown, but may have been north of the historic one at Wisbech.

The Early Bronze Age saw the re-establishment of fresh-water conditions, and the Upper Peat began to form by growing outwards from the fen margins. The climate appears to have been relatively dry, or, at any rate, continental, and with the marine recession thus caused, instead of a fairly continuous wet surface, open water persisted only in isolated basins or meres, such as those at Whittlesey, Yaxley, and in other places. Near the end of the Bronze Age the climatic and physical conditions were suitable for the rather extensive local growth of woods of pine and birch on the peat. Subsequently, however, the climate deteriorated, and during the Iron Age (*c.* 500 B.C.) it was colder and wetter. The Fens became uninhabitable, and there developed large open meres and much true fen. They were usually formed on the landward side of the Fens, beyond the margin of the Buttery Clay and the roddon silts (see p. 433), and some of them, including those at Ugg and Whittlesey, Soham and Streatham, persisted until the last century. In these lakes calcareous marl was deposited, and owing to the decrease in level of the peat fens (see below) these deposits of marl now stand up as low tables or mounds. About Roman times there was another, but smaller, marine invasion, which, instead of forming an inland lake, developed extensive coastal marshes and river levees, running far back into the peat areas. The beginning of this movement is difficult to date; it was perhaps rather earlier than the arrival of the Romans, but it certainly continued during Roman times. On these coastal salt marshes there was a good deal of Roman settlement and for the first time this part, at least, of the Fenland began to have a mildly artificial aspect.

(*d*) *The Roddons.* The roddons are common features in the landscape to-day. Their general nature was first studied by Fowler,[1] but it was Godwin[2] who later proved that they were formed of silt laid down as

[1] G. Fowler, 'Fenland Waterways, Past and Present', *Cambridge Antiq. Soc. Comm.* 33, 1933, 108 and *Geogr. Journ.* 83, 1934, 30.

[2] H. Godwin, 'The Origin of Roddons', *Geogr. Journ.* 91, 1938, 241.

natural levees to winding and meandering rivers in low-lying country. To-day they stand up above the general surface level because the wastage and shrinkage of the peat has brought it below the level of the roddons. On the other hand, since their formation the peat had had time to grow as high as their tops, and even to extend over them and hide them. The evidence of the fossils found in the roddon silts shows that the marine influence during their formation was more strongly marked than it was in the waters when the deposition of the Fen Clay took place. It is clear that strong tides flowed far up the rivers.

(*e*) *The Fenland in Historic Times*. The story of the Fenland in historic times is taken up by Darby. It has been noted that the silt areas, and also certain of the islands in the fens, were cultivated in Romano-British times, and the transition in human reaction from this land use to the almost complete abandonment of the district in the Dark Ages is not yet fully explained. Darby[1] suggests two answers, both of which may be correct. The Romans had practised artificial drainage and had built banks, but it is very likely that their work was allowed to deteriorate seriously after they had left the country. On the other hand, the turn for the worse may have been due to natural conditions, that is to say, the water-courses may have silted up, and there is evidence of a slight post-Roman subsidence. Certainly in Anglo-Saxon times the density of settlement was much less, and it is significant that the Fenland was a frontier region between East Anglia and Mercia. There is no record of any substantial drainage activity for many centuries after the Roman occupation. Little direct evidence exists for the Anglo-Danish period: Darby remarks that the available sources suggest a sparsely inhabited Fenland with a great deal of marsh and open water. The Domesday record shows that the early mediaeval villages were in the silt regions and the fen islands. There was no village settlement in the peat area, and the shoreline was not far beyond the line of siltland villages.

Up to the time of the great seventeenth-century drainage schemes there was not much change in the general appearance of the Fens. Darby, writing of the Middle Ages, says: 'A bird's eye view of the Fenland... would have revealed a country-side ranging in character from open pastures and meadows through reedy swamps to the pools of many meres connected by a confused network of channels. In dry summers, when the edges of the marsh dried up, the extent of the pasture was greater; while in winter, water might cover almost the entire face of the country. Around the islands, reclamation for cattle and tillage was gaining upon the marsh; but, in the main, the effects of this reclamation were comparatively un-

[1] *The Medieval Fenland*, 1940.

important in the peat lands.'[1] On the other hand, the changes in the river courses which have been made clear, particularly by Fowler, are of considerable interest. During the early part of the Middle Ages the rivers Nene, Great Ouse, Cam, Lark, Little Ouse, and Wissey, all with winding and tortuous courses, apparently entered the Wash in the neighbourhood of Wisbech. Only the Nar and Gay, with a few minor streams, reached the sea at Lynn. But before the end of the thirteenth century the outlet at Wisbech had become choked with silt and sand brought in by the tides. Towards the end of the century a part of the Nene and the western branch of the Great Ouse began to flow along Well Creek and so to Lynn and the estuary of the Nar and Gay (Fig. 95). It is also probable that the Little Ouse and Wissey changed likewise. The date of the diversion of the eastern branch of the Great Ouse from Littleport by an artificial channel to Lynn is not known, but it is clear that when the Ouse-Cam ceased to outfall at Wisbech a more direct course from Ely to Lynn was advantageous. Somewhat similar changes were taking place in the estuary of the Welland. The old Bicker Haven, to which the earliest reference is made in a ninth-century Crowland charter, seems gradually to have warped up and become marshland. It is uncertain when it was enclosed, but no embankment is shown in Blaeu's map of 1654. W. H. Wheeler[2] thinks that it was cut off from the sea in *c.* 1660, when the marshes of South Holland were reclaimed.

(*f*) *The Lowering of the Peat Surface*. There is no need to describe fully the intricate details of the drainage schemes of the seventeenth and later centuries. Whilst extremely interesting for the physical geographer they do not directly concern the present coast, except in so far as long outfall channels have been built at Lynn for the waters of the Ouse. It was not foreseen by Vermuyden and his successors that one result of the draining of the Fenlands would be the rapid lowering of the peat surface. This sinking was due mainly to two factors:[3] (1) the shrinkage of the peat as it dried, and (2) the wasting away of the drying peat as a result of bacterial action. There was some settling of the silt areas also, but to a far less extent. This differential movement had a great effect upon the rivers. The naturally flat gradients of the Fenland had always been a difficulty for engineers, and the rapid loss in the peat areas accentuated the problem. There is, however, no evidence that differential erosion is responsible for the silt areas now standing higher than the peat fens: they always were higher. It is, however, true that the beds of the outfall channels are now of much the same height as the lowest parts of the peat fen.

[1] *The Draining of the Fens*, 1940, Chapter IV.
[2] *A History of the Fens of South Lincolnshire*, 2nd ed. 1896.
[3] See Darby, *op. cit.* p. 71.

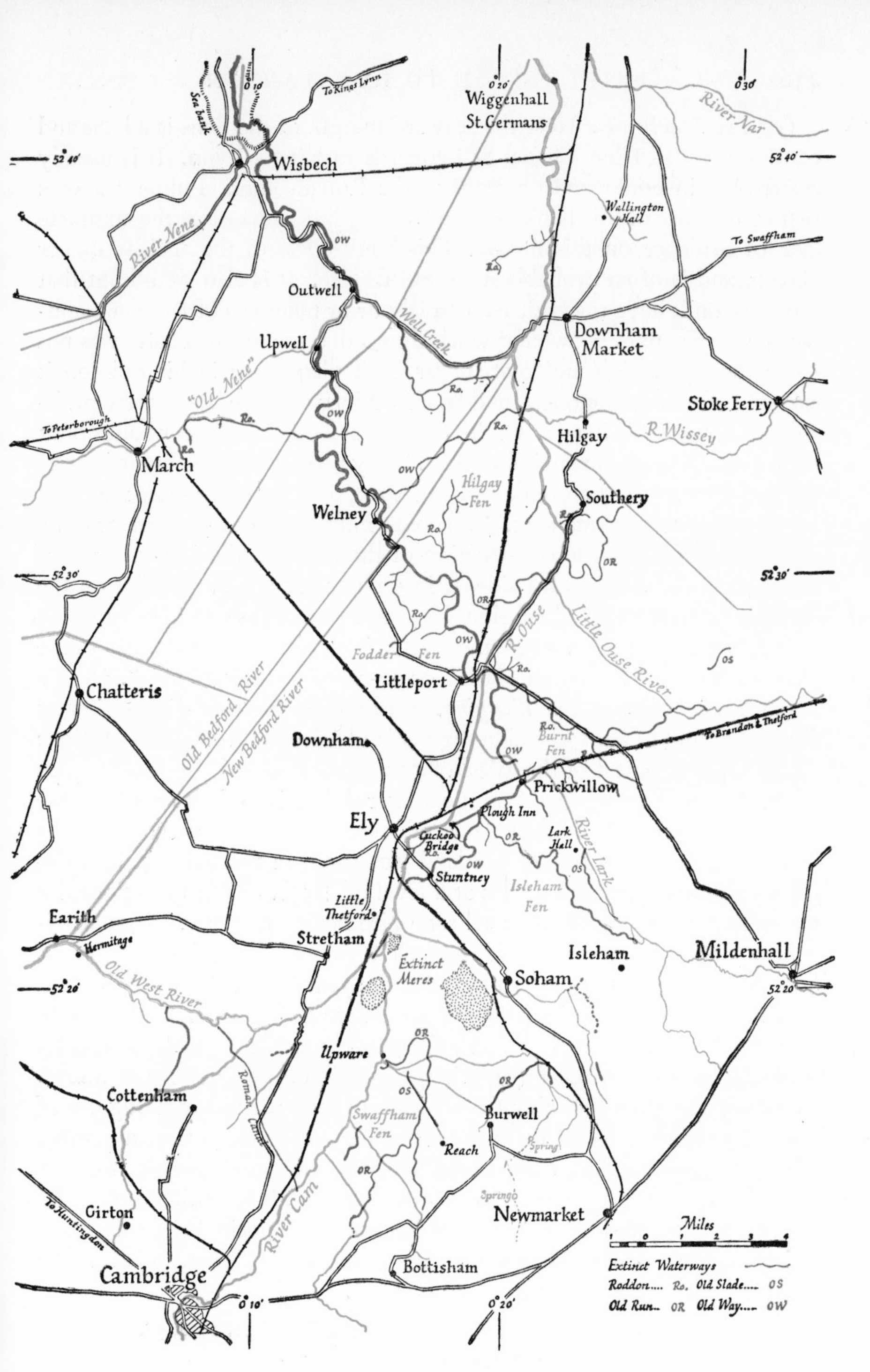

Fig. 95. The extinct waterways of the Fens (G. Fowler, reproduced from the *Geogr. Journ.* 83, 1934)

(*g*) *The Sea Walls.* Near the seaward margin of the Fens is a long and continuous wall from Chapel St Leonards to King's Lynn. It is usually referred to in books and on maps as the Roman Bank. Phillips[1] allows that some parts of this bank *may* be Roman, but emphasizes the complete lack of evidence on this matter. This bank encloses the ancient Bicker Haven, and is often probably a natural feature. It is also significant that nowhere on it have traces of Roman labour or materials of Roman manufacture been found. Fowler's[2] work shows that the original bank was not more than 3 ft. high and that it was built piecemeal: in his opinion it existed before the Conquest, and he also thinks that the banks were probably early monastic in age. On its seaward side there are other banks built from time to time to reclaim saltings. They usually enclose limited areas. The most recent structure dates from 1919, and the older ones are often levelled down to the marshes. 'Where as much as about 50 years or more separate the time of the construction of these banks, the general level of the ground on the seaward side is, in each case, higher than that on the landward one. Thus the land steps upward towards the sea till one reaches the present salt marshes. There the level...is considerably above that of the land behind the "Fossatum Maris" or so-called "Roman Bank".'[2]

(*h*) *The Enclosures and Warping.* Borer gives the following estimates for the amount of land gained in the Fenland since the Roman occupation: on the Norfolk coast of the Wash, 8,000 acres; in South Holland, 37,000 acres. It is now recognized that the longer the period of silt accretion, the better the reclaimed land, a sequence of events which was often overlooked in the nineteenth century, when land was taken in too quickly. The deposition of warp usually begins at a level of +5 ft. O.D.: neap tides rise to +8 ft., and at about this level samphire begins to grow. At approximately +10 ft. it is replaced by grasses. The foreshore, outside a new bank, usually becomes a grass marsh in about ten years, but it is best to leave it for another twenty or twenty-five years before enclosing it. In the Wash, as in Denmark and north-western Germany, accretion is encouraged by cutting a series of ditches (or grips) running up and down the marsh. These are given a fall of about 2 ft. to a mile, and the material dug out of them is normally placed on the side facing the flooding tide, a process which helps to produce calm water and deposition. Any erosion of the excavated material is washed on to the grassland. These ditches also aid the ebbing tide, and allow the consolidation of deposited silt before the advent of the next tide. The grips are about two chains apart, and are made when the marshes reach about +10 ft. O.D.

[1] C.W. Phillips, *Arch. Journ.* 90, 1934, 106, and 91, 1934, 98.

[2] Quoted in Godwin, *op. cit.* 1940, pp. 295–6.

The total amount of accretion outside the so-called Roman Bank is given by Borer[1] as follows (see Fig. 96):

	Before 1869 (acres)	Since 1869 (acres)
West Side or East Holland	6,336	—
Head of Wash and South Holland	35,163	985
Bicker Haven and Welland Marshes	10,464	—
Nene Marshes	9,536	1,177·6
Norfolk Coast	1,800	1,383
	63,299	3,545·6

(*i*) *Conditions in the Wash.* The Wash is merely the unfilled part of the Fenland Basin, and Skertchley's insistence that it is a bay and not an

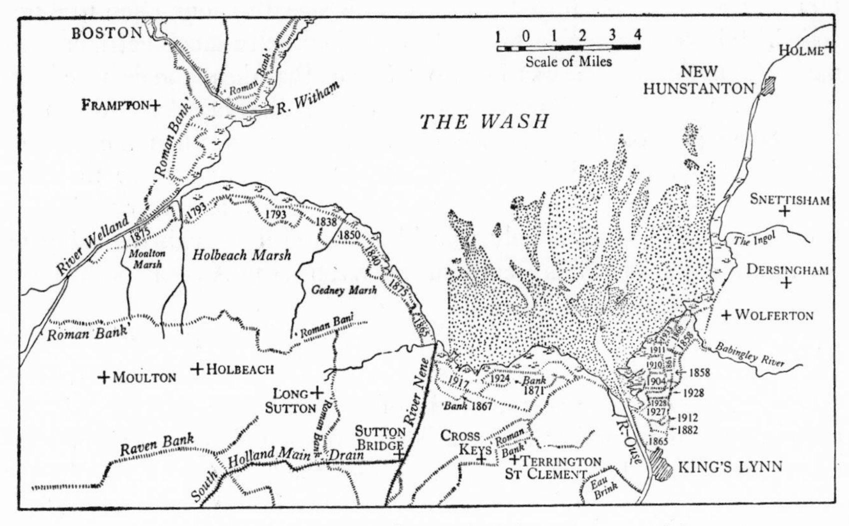

Fig. 96. Reclamations in the Wash (after Borer, *op. cit.*)

estuary has more than academic importance. It is generally shallow, that is to say, with average depths of five fathoms. In Lynn Deeps 15 to 26 fathoms occur near Lynn Well. Skertchley[2] ascribed the deep part around Lynn Well to a whirlpool-like action due to the run of the tidal streams. There is here, however, a long and rather narrow trough which may quite well represent part of the old valley of the united Fen rivers. This was the view expressed in the *Geological Survey Memoir*. The flood tide runs south-south-west down the Lincolnshire coast, and west-south-west along the Norfolk coast, whilst the directions taken by the ebb are nearly

[1] O. Borer, 'Changes in the Wash', *Geogr. Journ.* 93, 1939, 491.

[2] *Op. cit.* Chapter XVIII.

opposite. Accordingly high water is almost simultaneous at Boston and Lynn, and half an hour later at Wisbech. Skertchley argued that a circular movement was thus produced around Lynn Well which would consequently be scoured out. Even if this view no longer holds it is worth noting that deposition would be less likely here, and that silt carried in by the flood tends to gather round the head and sides of the Wash. The shoals in and near the Wash were extensive and liable to alteration. Unfortunately little precise information is available on this matter, but measurements made in the south-eastern part of the Wash by the Great Ouse Catchment Board showed that between 1871 and 1938 'generally the ends of the shoals have withdrawn towards the shore. A shoal like the Roaring Middle, which had an area of 140 acres, has now been reduced to zero, though the Inner Westmark Knock has increased...from 3,400 to 6,000 acres'.[1] It has not been possible to find out whether the shoals had increased upward, but 'there appears to be no evidence that these shoals have been completed up to any height, or that the material has been carried inshore to become part of the reclaimed land. It seems clear that the drift of material into the Wash has considerably slowed down since the days between Roman times and the seventeenth century, and information on this point would be extremely valuable when dealing with the fixing of the future channels in this section of the Wash' (**149**).[2]

THE ORIGIN OF THE WASH AND THE HUMBER

There remains for discussion at the end a point that might perhaps more logically have come first—the origin of the Wash.[3] In a sense it is not unlike that of the Humber, and a discussion of the two together is perhaps more convenient. Both of these inlets breach the Chalk, and undoubtedly at one time the Chalk of Norfolk and Lincolnshire was continuous. The rivers, before the wide breach was made, probably flowed, like those of the Weald to-day, through narrow valleys in the Chalk to a North Sea more distant than that of our time. Possibly, as Skertchley suggested, the united Welland, Nene, and Ouse followed the line of Lynn Deeps.[4] The estuaries so formed were probably enlarged both by erosion and subsidence, and so the original comparatively narrow river gaps through the Chalk were gradually widened, and the intervening parts of the barrier reduced to outliers. As these processes continued the Chalk finally dis-

[1] See Borer, *op. cit.* [2] Borer, *op. cit.*

[3] Skertchley, *op. cit.* and Kendall and Wroot, *op. cit.*, especially Chapters XXXIII and XXXVIII.

[4] *Mems. Geol. Surv.* 'The Geology of the Borders of the Wash', W. Whitaker and A.J. Jukes-Browne, 1899.

appeared, and the sea was brought directly on to the soft Jurassic clays underlying the Fenland. These were worn away into a broad and relatively shallow basin out of which rose islands like those on which stand Ely, and other places, formed of the parent rocks of the area. The surrounding higher ground to north and south was made of the Chalk, and to the west of the harder Jurassic rocks. After the Ice Age the surrounding highlands were all capped by boulder clay, but the basin character of the true Fens remained unaltered. The points discussed in this chapter show that the subsequent history of the basin was in the main a process of filling up with occasional but short-lived ingressions of the sea, and it is likely that the final destruction of the Chalk barrier and the hollowing-out of the basin took place in late Tertiary times. In Holderness and Lincolnshire the old shoreline ran along the foot of the Wolds, and the Chalk between Barton and Hessle is severed. Borings through the boulder clay which has filled the old and wide bay of Holderness and Lincolnshire show, indeed, that originally the Humber debouched into the sea by a trumpet-shaped mouth lying directly east of the Barton-Hessle gap. 'There is nothing in what is known of the form and depth of this mouth inconsistent with the idea that it owed its origin to the activity of such a river as the Humber operating at or near the present level.'[1] There is no need to discuss the different views on the origin and evolution of the Yorkshire rivers, beyond mentioning that the Humber, like the Mersey and Tyne, has a drift-filled valley above and deeper than the relatively shallow channel at the mouth. It is possible that a narrow trench in the lower river may yet be discovered, but it is unlikely, and the deeper upper valley is usually ascribed to the scouring action of ice. The river systems flowing into both Humber and Wash probably originated in the Miocene. They carved out their valleys, and in both cases cut through the Chalk like the Wealden rivers to-day. In the Wash the Chalk barrier is widely breached by the sea: in the Humber the gap is narrower, but traversed by salt water. In both areas changes of level, not necessarily identical, but probably kindred in nature, have taken place since the initiation of the river systems. In both, moreover, ice has played a great part in the formation of the present scenery, mainly by deposition, but partly also by erosion. Finally, in both regions extensive fens stretch away behind the Chalk barrier. In the Fenland proper there has now been pieced together after protracted and arduous research a long and connected story: in the south Yorkshire and north Lincolnshire fens there may be a similar physiographical history to relate, but no one has yet attempted with modern technique studies comparable in scale to those on the East Anglian Fens.

[1] Kendall and Wroot, *op. cit.*

APPENDIX

DIVISIONS OF POST-GLACIAL TIME

On p. 429 a short table of the Fenland zones and post-Glacial periods occurs with a brief explanation. It is, however, more convenient to complete the comments in an appendix. The main Fenland zones are indicated by Roman numerals: the sub-zones by small letters. No. IV is the earliest zone in this area, and its predecessors, I, II, and III need no explanation here. Evidence concerning No. IV is very scanty in the Fens, but is found in the 'Moorlog' recorded occasionally by trawlers from known depths in the North Sea. Pollen analysis shows that the Moorlog, from depths of 18 to 29 fathoms, was accumulating during zone IV in the Pre-Boreal period. Since this time it is right to assume a net subsidence at least of 174 ft. A Mesolithic harpoon was found on the Leman and Ower Banks at 19 to 20 fathoms. The peat in which it occurred belonged to the transition from zone V to zone VI, and since the peat was *in situ*, there is evidence of a subsidence of at least 114 ft. The peat at Judy Hard is mentioned on p. 402 and extends from zone IV to well into zone VII.

The sub-zones refer to minor divisions into which some of the major zones are conveniently divisible; for example, 'In the Fenland the formation of the Fen Clay very greatly affected the pollen diagrams by destruction of all but the most marginal fen woods, thus effectively setting a phase of wetness between two periods of dryness and extension of fen woods. It will thus prove possible over much of the Fenland to recognize, within zone VII, four sub-zones: VII *a*, in which open sedge fens prevailed; VII *b*, in which the diagrams show very strong local influences of alder and sometimes also of pine and birch as well; VII *c*, in which the transgression of the fen clay stage has caused local effects to be very small; and VII *d*, a phase of partial recovery from the wetness of the previous stage, marked by the development of fen woods of alder in all but the wettest places.'[1]

With this further comment in mind, there will be fuller appreciation of the correlation table given in Fig. 105 and also of the important diagram (Fig. 106) showing post-Glacial changes of relative land and sea level in the Fens. The localities mentioned by numbers or by numbers and letters are listed on the diagram with brief notes. This diagram also introduces a correlation between the Fenland and Ingoldmells, and it will be noted that only the Roman level and the *Triglochin* clay at that station fail to fit in precisely with the evidence in the Fens: the clay is too high by 10–15 ft.[2] The Roman station may possibly have been built behind protective walls, or 'Alternatively, we may put forward the view that the bulk of the upper silt in the Fenland had been deposited before the Romano-British colonization of the first century, that during the first to third centuries there was temporary marine regression, and that at the end of this time marine transgression, though not extensive, again became apparent.'[3]

[1] H. Godwin, *Phil. Trans. Roy. Soc.* 230, 1940, 246. I have made great use of this paper and its forerunner, *ibid.* 229, 1938, 323 (with M.H. Clifford) throughout. They are indispensable.

[2] In a personal note, Prof. Swinnerton tells me that *Triglochin* marks a vegetation belt the position of which shifts seaward as the process of silting progresses. If at the same time slow upward movement takes place, the silts deposited in the *Triglochin* belt in its later position will necessarily be of different level from those deposited on its earlier position. Hence, the difference of level referred to is perhaps not a real difficulty.

[3] H. Godwin (1940), p. 293.

Chapter XI

THE NORTH-EAST COAST: BERWICK-ON-TWEED TO SEWERBY

(1) INTRODUCTORY

In this chapter on the north-east coast from Berwick to Sewerby there are included from the geological point of view three entirely different coastal regions. In Northumberland, apart from one or two minor but interesting Permian outliers, only rocks of Carboniferous age reach the coast. In Durham the littoral strata are Permian and Triassic, and in north-eastern Yorkshire the Jurassic and Cretaceous rocks give a coast of particular beauty. These three parts are discussed separately, although the whole stretch is included in one chapter.

A great deal has been written on the stratigraphy and palaeontology of the Jurassic and Cretaceous rocks of Yorkshire, and there is also a considerable literature on the Permian of Durham. Along the Northumberland coast, however, there are several stretches of dunes (links), so that, although the bibliography of the Carboniferous rocks in that county is a long one, there is considerably less concerning the actual coastal geology than in the Durham and Yorkshire sources. Moreover, bearing the objects of the present volume in mind, it is not easy to give detailed descriptions of the Northumberland and Durham coasts because even a careful study of the literature yields comparatively little material dealing minutely with littoral topography. Furthermore, even in the Yorkshire literature, where material is abundant, it is largely a matter of gathering information from books and papers whose purpose is not primarily physiographic.

(*a*) *Structure of Northumberland and Durham.*[1] In Northumberland the structure of the rocks is concentric around both the Cheviot dome in the north and the smaller Bewcastle Fells in the south-west (Fig. 97). Between these two is the tectonic depression of Upper Redesdale, while to the south-east the outcrops run in a wide arc around both centres. Throughout most of the county the dips are fairly high, so that dip and escarpment slopes are characteristic. South of the Tyne the dips are small, and outcrops tend to run with the contours; consequently the main part

[1] E.J. Garwood, *Geology in the Field*, 'Northumberland and Durham', 1910; 'Contributions to the Geology of Northumberland and Durham' (G. Hickling and others), *Proc. Geol. Assoc.* 42, 1931, 217 (include also Summer Field Meeting, 1931); *British Regional Geology*, 'Northern England', T. Eastwood, 1935.

of Durham is of the nature of a dissected plateau tilted slightly eastwards. This brings in newer rocks—the Permian—on the coast, but both Northumberland and the greater part of Durham are formed mainly of Carboniferous rocks.

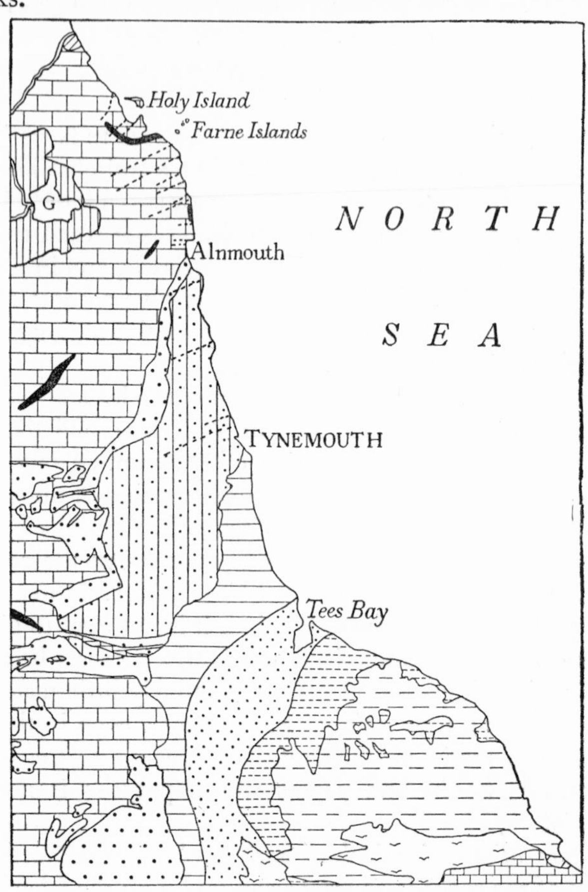

Fig. 97. General geology of the north-east coastal areas (see Fig. 11 for key) ----- = faults. (Trias is undivided)

Little need be said of the Cheviots since they do not directly affect the coast. The massif is a deeply eroded volcano of Lower Old Red Sandstone age, in the centre of which is a granitic mass. In this area the pre-Carboniferous surface is more than 3,000 ft. above sea level, whereas in the south of the county it is about 3,000 ft. below the base of the Coal Measures. The contrast between the structures of Northumberland and Durham is really due to the different geological histories of the two counties in Carboniferous times. Northumberland, especially in its southern parts, was then an area of rapid depression, whereas Durham was comparatively stable and was all the time an area of thin sedimentation. The rocks which accumulated in the Northumbrian trough, which was much enclosed by land, were often lagoonal in nature. In the north of the county the Cementstone group, the earliest Carboniferous deposit, consists of a thick series of shales, with numerous thin bands of fresh-water or estuarine cementstones. To the south-west the conditions are more marine. At this time Durham was still land, and so also was the Cheviot area. The Cementstone group is about 3,000 ft. in thickness: it is succeeded by 1,000 ft. of the Fell Sandstone, and then by the Scremerston[1] group. All these are missing in Durham. There is no doubt that this contrast results from the trough-like area of sedimentation between the Cheviots and Durham.

In the second half of Lower Carboniferous times, submergence was more widespread and sedimentation more uniform over the area now covered by both counties. This phase is represented by rhythmic repetitions of limestone, shale, sandstone, and coal. In the Tyne Gap district submergence began early, before that resulting in the Main Scar Limestone of the Pennines, and these lower beds seem to correspond to the Scremerston Coal group of the north part of Northumberland. Northwards from the Tyne another change appears 'first at the bottom, and then higher up, the marine limestone bands disappear, while the coal horizons in this part begin to include workable seams, which on the whole increase in number and importance as we proceed northwards, until in the Scremerston district eight or ten workable seams are present'.[2] In both counties the Upper Carboniferous rocks are more or less normal in character except that the Millstone Grit and the Lower Coal Measures are much attenuated as compared with their development farther south in Yorkshire.

The Cheviot dome, then, is the dominant feature in north Northumberland. Towards Berwick a shallow syncline opens towards the coast, so that the rocks dip away from the Cheviots and east and south of these hills the dip is away from them. At the same time there is a tendency towards the development of east-north-east to west-south-west folds nearer the coast, whilst there are several fault lines with the same trend which reach the coast. Their effects on the nature of the coast are discussed below (p. 454).

Over most of the south-eastern part of the county, outside the area affected by the east-north-east folds and fractures, the rocks dip, with

[1] Essentially the Cementstone type, but with frequent seams of inferior coal.

[2] G. Hickling, *Proc. Geol. Assoc.* 42, 1931, 222.

little variation, gently to the east-south-east. Nearer the Tyne the dip swings round more to the east, and becomes somewhat steeper, while farther inland, near the Tyne Gap, there is relatively intense folding. The tilted plateau of Durham is really a low dome the maximum elevation of which, both structural and topographic, is near Cross Fell. This chapter, however, relates more to the eastern part of the county, where '...the lower section of the Coal Measure strata rises again from below the upper around the mouth of the Tyne',[1] and 'where the dip is inland, instead of seaward. Between Newcastle and the sea, the rocks, consequently, lie in a trough, the axis of which runs S.S.E. with a marked pitch to the south....'[1]

The Permian is unconformable to the Carboniferous rocks, and there is a continuous belt of Permian rocks along the Durham coast—a feature unique in Britain. The series begins with the Yellow Sands, false bedded, possibly of aeolian origin, and reaching a maximum thickness of 170 ft. These are succeeded by 15 ft. of Marl Slate, and then comes the main mass of the Magnesian Limestone, divided into Lower, Middle, and Upper divisions, the total thickness of which is about 800 ft. The Lower Limestone is a blue-grey, evenly-bedded rock. The Middle, or Shell, Limestone group is particularly interesting on account of the nature of the brecciation which is of three types. 'The earliest seems to have been the "cellular breccia", where the fragments, and not the matrix, have often weathered completely away....The "massive brecciation" is clearly a later event, for amongst the rocks affected are the "cellular breccias" themselves. The blocks are in places incompletely separated, they may be 3 ft. or more across, and they are more calcitic than the matrix....The "gash breccias" ...are obviously a late phenomenon due to the collapse of the roofs of natural cavities or the walls of deep fissures made by the circulation of sub-soil water.'[2] The Upper Limestone includes the local north Durham phase of the 'Flexible Limestone', which is succeeded by the Concretionary or Cannon Ball Limestone. This is well known from the variety of odd forms taken by the concretions. Finally come the Hartlepool and Roker Dolomites, about 100 ft. thick, massive, and oolitic. The way in which these several formations of the Permian approach the coast is clear from the six sections in Fig. 98.

Triassic rocks outcrop near the Lower Tees. There appears to be a complete series from the Magnesian Limestone to the Lias, but there are few surface exposures. They form an area of low ground heavily drift covered and gradually giving place to the Tees estuary deposits.

[1] G. Hickling, *Proc. Geol. Assoc.* 42, 1931, 224.

[2] *British Regional Geology*, 'Northern England', T. Eastwood, 1935, 59.

With the interesting features of the inland scenery of the two counties we are not concerned. In the coalfield area much of the ground is covered by thick masses of drift, broken only by the harder sandstones of the Coal Measures. Much of the coastal area is low lying, and, along the bays, is fringed by belts of dunes. Near the Durham coast the Magnesian Limestone forms a marked escarpment beyond which small streams have cut deep gorges into its dip-slope towards the sea, forming the picturesque denes of the coast.[1]

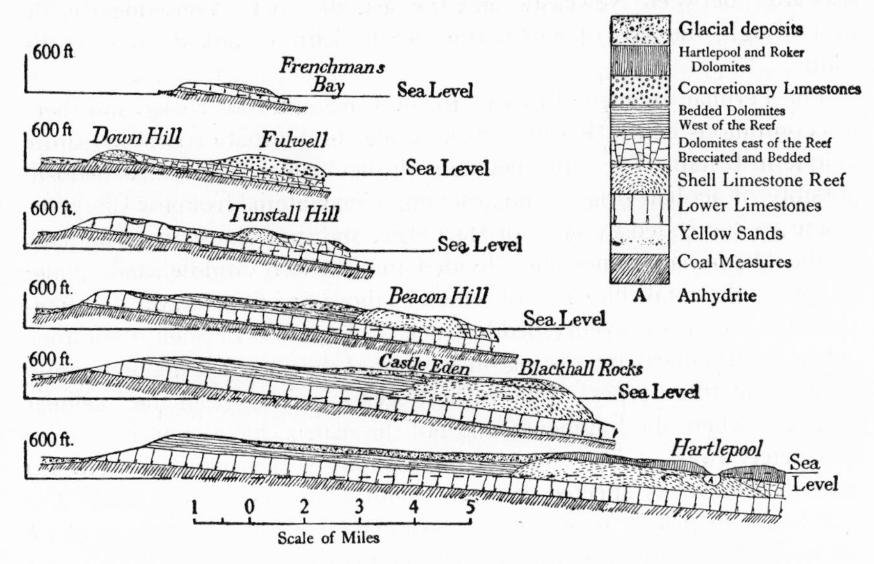

Fig. 98. Sections through the Permian Rocks to the Durham Coast (after C. T. Trechmann, *Proc. Geol. Assoc.* 70, 1925, 135)

The igneous intrusions of later Carboniferous and Tertiary times need comment in so far as they affect the coastal scenery. The great Whin Sill is the most striking of all. It begins in the Kyloe Hills, and runs east and south to Bamburgh and the Farne Islands. It reappears farther south at Snook Point (near Newton), and follows the coast from Dunstanburgh to Cullernose Point whence it retreats inland again. The sill is definitely transgressive, and in its most commonly developed form it is a dark blue-grey quartz-dolerite. It has locally metamorphosed the sedimentaries into which it is intruded, and associated with it are many quartz-dolerite dykes. Just north of the Kyloe Hills there are dykes arranged *en echelon* and extending from the Tweed to Holy Island, while another similar group

[1] In Dwerryhouse's view a glacial lake was ponded behind the limestone ridge, and its overflow led to the formation of the Wear mouth at Sunderland.

extends from the North Tyne to Boulmer. The general trend of all is roughly east-north-east. In addition there are the tholeiite dykes, considered to be of Tertiary age, which trend east-south-east to cut the coast from the neighbourhood north of Blyth to that south of the Tyne.

On many parts of the coastal districts there are thick deposits of drift, the vast majority of which are boulder clay interspersed with occasional patches of gravel and sand. Nearly all the coal area between the Coquet and the Tyne is so covered, and to a lesser extent the northern part of the county between the Cheviots and the coast. The Wear valley is similarly swathed, and south of Seaham Harbour, patches of Scandinavian[1] drift are found in hollows of the Magnesian Limestone, and underlie the main drift cover. The latter is often in two divisions, a lower clay, generally blue or purple, with (in Durham) numerous boulders of Pennine rocks and parted by sands and gravels from an upper clay often reddish coloured and relatively stoneless: this contains material from the Cheviots, the Tweed valley, and western Scotland. If the intervening sands are wanting, a marked unconformity separates the two clays when in juxtaposition. It is clear that the Scandinavian ice first impinged on the coast, and that the Pennine and Scottish ice came later.

> After a period sufficiently long to allow of some weathering of the Scandinavian Drift, and the formation of a loess-like material, the ice-streams from the Pennines made their way to the coast, and the blue clay was deposited over much of the area of Teesmouth, and in the lower Tyne valley and Weardale. In South Durham (...) this ice-stream brought erratics of the Shap granite that had crossed Stainmoor and mingled with the Teesdale ice. With the growth of the Scottish glaciers of the eastern valleys the ice congestion of the North Sea basin produced many modifications in the direction of flow of the inland ice. The Tweed and Forth ice, debarred from direct passage eastward, was deflected southward down the coast, passing southward as a buffer stream between the Scandinavian ice and the growing Pennine glaciers.[2]

(*b*) *Cleveland and Flamborough.*[3] From the Tees to south of Filey the coastal rocks are entirely Jurassic. The nature, thickness, and disposition of the beds vary much from place to place. The complete sequence is given in the table on p. 42. The immense shale beds of the Lias, with valuable ironstones near the top, are well seen on the northern and eastern sides of Cleveland. The lowest beds, calcareous sandstones, appear as scars in Robin Hood's Bay which stand out because the softer shales have been worn away. There are also other sandy facies in the Middle Lias, but

[1] The Scandinavian fragments *may* correspond to the Basement Clay of Yorkshire.

[2] A. Raistrick, *Proc. Geol. Assoc.* 42, 1931, 285.

[3] Details will be referred to later. For general use in this connection see P.F. Kendall and H.E. Wroot, *Geology of Yorkshire.*

generally, in Liassic times the sediments gathered in fairly deep water to form shales, clays, and impure limestones.

The Dogger is the lowest bed of the succeeding Oolites. This bed shows several distinct facies, but consists mainly of gravels and sands forming conglomerates and sandstones. The Estuarine Beds are composed of a succession of shales, sandstones, coal beds, and underclays, and indicate the muddy delta flats of a great river entering the Oolite sea, but the occasional limestones, the Ellerbeck Bed, Millepore Bed, and the Scarborough Limestone mean that there were invasions of marine water in the delta area. This Estuarine period ended with the deposition of the marine Cornbrash which is well developed on the Yorkshire coast; it is followed by the ferruginous sandstones of the Kellaways Rock which, inland, is of considerable physiographic significance. The Oxford Clay represents a deep-water phase, whereas the Corallian Beds, as a whole, indicate a return of shallow seas. Finally the Kimmeridge Clay and the Speeton Clay imply a closing phase of deep water.

The Lias and Oolites, forming the whole of the surface of north-eastern Yorkshire, are very gently folded into broad anticlines and synclines and also disturbed by faults. There is also a north-south transverse trough running approximately along the line of the Goathland Beck, thus giving a saddle shape to the whole mass and causing the peak of the hills to lie just west of Robin Hood's Bay. The North Cleveland anticline is the most northerly of the east-west ridges, and runs from Roseberry Topping to the sea. Its northern flank forms an escarpment overlooking the Lower Tees. Its south flank merges into a shallow syncline, but this is a structural and not a physiographical feature. The main Cleveland anticline follows the line from Stony Ridge to White Cross, Loose Howe, Pike Hill, and Three Howes, and so to Robin Hood's Bay. Eight to ten miles to the south the Jurassic outcrop is interrupted by the alluvial plain of the Vale of Pickering. Bores show that the Kimmeridge Clay here underlies the superficial deposits, and that the vale is really a faulted synclinal trough. Some miles farther south is the Market Weighton axis. The movements which produced these features were all mainly post-Jurassic and part of the great series of Tertiary movements, although some admittedly took place concurrently with the deposition of the rocks. The point is hardly relevant here, except for its bearing on the accumulation of the great thickness of Lias in Robin Hood's Bay.

Faulting has also influenced the Jurassic area to no small extent, especially in the east and south. There are several important north-south faults, some of which are presumably associable with the formation of the North Sea basin in this region. 'One of these faults dictates in large measure the line of Runswick Bay; another is reputed to give rise to the harbour of Whitby, though the structure here has been interpreted by Mr S. W. Hughes as an unfaulted monocline. But the most important is that which is known as the Peak Fault, which breaks off abruptly the Cleveland main anticline.'[1] Thus Robin Hood's Bay is carved out of the end of this ridge, and but for the hard sandstones of the *Ammonites margaritatus* zone which occurs at sea level (see p. 469) at the North Cheek and again, owing to the fault, at the South Cheek, erosion would have been far more extensive than is actually the case.

There is probably no better summing up of the tectonic structure as it affects the coast than that of Kendall and Wroot:[2]

> The hillsides from Guisborough to the moors of Newton Mulgrave, near Hinderwell, and thence northward to the sea present the geologist with a much dissected dip-slope of the North Cleveland anticline. The coast between Skinningrove and Port Mulgrave is the northern anticline cut obliquely to the axis so that the curve is very flat. At Runswick we begin to get into the trough of the syncline, though this fact is difficult to trace because of the disappearance of the solid rocks from the cliff—which is there of boulder-clay, the infilling of an old river-course. From Kettle Ness to Sandsend we are still in the syncline, but the section is very oblique so that the dip seen at the coast is everywhere slightly southward. There is almost no upward rise to complete the trough, but a great fault or fold at Whitby harbour, which has an upthrow of 200 feet to the east, gives the rocks beneath Whitby Abbey the elevation they would have had if the syncline had been normal. About Hawsker we begin to perceive in the dip of the rocks the effects of the main Cleveland anticline, and this great arch is cut squarely across at Robin Hood's Bay. From the Peak there is a continuous dip to the south—following the south flank of the main anticline—to Filey Bay, where the Kimmeridge Clay succeeds the Corallians and passes under the Cretaceous rocks.
>
> Some of the prominent 'nabs' on the coast—for example, the headland crowned by Scarborough Castle, and Osgodby Nab—interrupt the regular sequence by the interposition of faulted masses.

Near the end of Jurassic times, the calcareous oolitic rocks of various types gave place, as a result of the slow sinking of the sea floor, to the deposition of the mud of the Kimmeridge Clay. This repeats the muddy conditions of the Lias at the beginning of Jurassic times. The Kimmeridge Clay in Yorkshire is succeeded by the Speeton Clay, about 300 ft. thick. In turn this gives way to a calcareous, iron-stained, deposit—the Red Chalk—succeeded by the White or Grey Chalk with pink bands, and finally by the Chalk itself. The dip of the Cretaceous rocks is to the south in the coastal area, as a result of Tertiary movements which again emphasized the north-south Pennine axis, and the east-west Cleveland axis. This folding together with erosion has removed the Chalk from all except the

[1] P.F. Kendall and H.E. Wroot, *Geology of Yorkshire*, 1924, p. 319.
[2] *Op. cit.* p. 327.

south-eastern part of Yorkshire. The most conspicuous coastal feature formed by the Chalk is Flamborough Head, where the great pressures due to folding have hardened the rock. On the other hand, the almost undisturbed beds of much of the Yorkshire Chalk seem to have obtained their unusual density as a result of the percolation of waters highly charged with calcium carbonate redeposited in the pores of the rock.

(c) *Summary*. This brief and cursory survey[1] of the rocks adjacent to the coast between Berwick and Sewerby will at least have shown that they are great in number and type. The alternating thin shales and limestones of much of Northumberland, folded and faulted, and often hidden by dunes, give interesting sections which, on a small scale, show differential erosion. The same may be said of the Coal Measures. In Durham the great development of the Magnesian Limestone, and of the breccias associated with it, occurs in its various forms. In north-eastern Yorkshire the beauty of the coast results from the variety of the Jurassic rocks with their folding and faulting. Near Flamborough the Chalk is the dominant rock. In Northumberland and Durham additional detail is given to the coast by the igneous rocks, and throughout the whole stretch from Berwick to Sewerby boulder clays play a large part in shaping the scenery. They occur in almost every bay and are subject to more rapid erosion than the solid rocks.

From the points of view of human and economic geography, the coastline of Northumberland and Durham is sharply divided by the Coquet mouth into two strongly contrasted parts—the rural north and the industrial south. Between Coquet and Tees it is rare to find a coastal view unspoiled either by collieries and all the untidiness that accompanies them, or by built-up areas. In the north the open country and sweeping views give a most effective setting to the coast *sensu stricto*. But even here there can be found many examples of ill-conceived and misplaced houses and bungalows. On the other hand, the extensive dunes and flats near Holy Island, the fine sweeps of dunes in Beadnell, Newton, and Embleton Bays, and the peculiar interest of the Whin Sill between Dunstanburgh Castle and Cullernose Point give to this bleak and wind-swept coast a great natural charm and distinguish it from any other part of the coastline of England and Wales.

[1] For general information on the coastal features see, *The Sea Coast*, W.H. Wheeler, 1902, and Royal Commission on Coast Erosion, vol. 1, 1907, Appendix 23.

(2) THE BORDER AND BERWICK: HOLY ISLAND AND THE FARNE ISLANDS

From the Border to the Tweed the coast is formed by fairly high cliffs of Lower Carboniferous rocks: Marshall Meadows Bay is the main feature. The Tweed is deeply incised and southwards from its mouth[1] the cliffs and foreshore platforms are all in the Carboniferous rocks, the general dip of which is to the south, so that the highest strata outcrop near Goswick. The rocks in the cliff section are first the Scremerston Coal group, mainly shales and sandstones with thin coal seams, and, south of Doupster Burn, shales and sandstones of the Lower Limestone group, including thin bands of limestone. Small faults occasionally break the sequence at Spittal and Seahouse. The Black Rocks of Cheswick are in the Dryburn Limestone and behind them is a narrow belt of sand hills reaching a maximum height of about 50 ft. (**125**).

From Cheswick to a point near Bamburgh there is a different type of coast. Holy Island (Lindisfarne) lies offshore and encloses the Holy Island Sands and Fenham Flats, although the latter are more within the point of Old Law, a northern extension of Ross Links (Fig. 99). Holy Island is some 1,050 acres in extent,[2] and the solid rock of its foundation appears mainly near the northern, eastern, and southern sides of the main mass of the island. The eastern side presents an almost continuous section, and the rocks are principally sandstones, lying nearly flat or with but a slight easterly dip. The eastern beach is encumbered with large boulders, and the remainder of the island is covered with superficial deposits, blown sand, boulder clay, and alluvium. The long western-pointing arm, called the Snook, is built of dunes up to 30 or 40 ft. high, resting on a base of boulder clay. The sand is arranged in ridges enclosing damp hollows containing such plants as *Samolus valerandi*, *Erythraea littoralis*, *Anagallis tenella*, *Parnassia palustris*, and *Gentiana amorella*.[3] At the south-eastern corner of the island the castle stands on a prominent crag of quartz dolerite (whin) which extends to St Cuthbert's Island. It dates (probably) from 1539, and was built as a defence to the haven, formerly called the Ooze (Ouse). It is interesting to note that shingle beaches which have been built on Holy Island have really joined up what were originally three or more separate islets, on one of which the castle stands. The sea formerly

[1] *Mems. Geol. Surv.* 'The Geology of Berwick-on-Tweed...', A. Fowler, 1926.

[2] *Ibid.* 'The Geology of Belford, Holy Island, and the Farne Islands', 2nd ed. R.G. Carruthers and others, 1927.

[3] J.S. Hull, 'Holy Island', *The Vasculum*, 14, 1928, 95; see also G. Johnston, *Proc. Berwickshire Nat. Club*, 7, 1876, 27.

filled the valley that runs northwards from the Ooze. The present direction of beach-drifting (to the south) is well seen in the small but markedly recurved shingle ridge at Castle Point. At low water a great expanse of marine alluvium is laid bare between Holy Island and the mainland. The northern part, mainly Holy Island Sands, is of very fine sand, but Fenham Flats are formed chiefly of silt and mud.

Ross Links,[1] with their northern extension of Old Law, an island at the highest tides, enclose Fenham Flats. The main body of these links is a

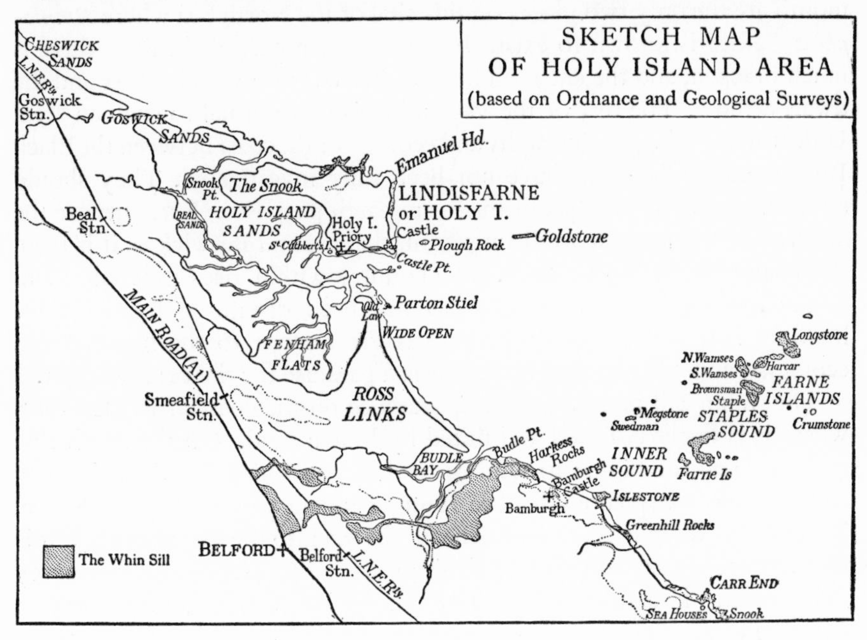

Fig. 99

flat moorland fringed seawards by a broad belt of dunes. The low-lying and rather basin-shaped inner part of the links is easily inundated after heavy rains because it is composed of boulder clay, and is only about 12 ft. above sea level. Seawards from the dunes the slope of the beach is very gentle, and this, combined with the set of the flooding tide, has earned the locality an unpleasant reputation. Ships wrecked on the Plough, Goldstone, or Guzzard banks usually drive up later to the foot of the dunes, which run in marked north-easterly trending ridges reaching 60 ft. or more in height. The amount of shingle on the beach is small. Parts of the links are marked on the geological maps as 'raised beach', but reference

[1] J.E. Hull, 'Ross Links', *The Vasculum*, 14, 1928, 121.

to the *Geological Survey Memoir* shows that this is by no means a definite interpretation, because nowhere can be seen a rock shelf or flat of indisputable raised-beach type. On the other (western) side of Fenham Flats, as far north as Beal Point, there are narrow storm beaches of sand, loam, and shingle which are wider opposite the mouths of streams. A southern arm of Ross Links almost encloses Budle Bay, in which there is much silt and clay of an estuarine nature. Bronze Age pottery has been found under 8 ft. of surface sand on Ross Links.[1] Incidentally, here and at Great Yarmouth are the only two places on this side of the North Sea where *Ammophila Baltica* is known to exist. In the southern part of the links there is much bracken, and the enclosed hollows are overrun with heather and the dwarf willow. The rest of the vegetation is that normal to such an area. Unfortunately the physiography and ecology of the coast between the Black Rocks and Budle Point have not been fully investigated. They should afford many interesting problems for research purposes. Furthermore the area forms a unit, quite as distinct from that around Berwick as it is from the whinstone of Budle Point. The country within rises gradually, and fortunately the natural scenery has not been disfigured by unsightly buildings. It is a locality which gives great scope to the botanist and ecologist, and with the Farne Islands is a paradise for the ornithologist.

The Whin Sill makes its first appearance on the coast at Budle Point, where it is finely polished and fluted by blown sand. On the south side of Budle Bay the higher ground formed by the whinstone rises abruptly and forms a conspicuous feature running inland past Belford to Kyloe. Because the Whin is continued beneath the sea into the Farne Islands[2] it is relevant here to say something of this group before continuing the survey of the coast. The actual number of islets in the Farnes depends largely upon the state of the tide, and also upon whatever difference the observer may care to make between islets and mere rocks. Tate stated that there were between fifteen and twenty-five. They fall into two groups separated by Ox Scar Road. Farne or House Island is the largest, and its name is now applied to the whole group. They are all parts of the Whin Sill, and stratified rocks are known only in four places, two on the Brownsman, one on Nameless Rock, and the remaining and largest exposure at the Bridges, where it can only be examined at low water. All these sedimentary fragments are merely inclusions in the Whin and are much metamorphosed limestones of Carboniferous age. Farne Island itself has an area of about 16 acres at low water, but of these 11 are bare rock. The south and west

[1] P. Brewis and F. Buckley, *Arch. Aeliana*, 4th series, 5, 1928, 13.

[2] G. Tate, 'The Farne Islands...', *Proc. Berwickshire Nat. Clubs*, 1850–6, p. 222; J.E. Hull, 'Among the Farnes', *The Vasculum*, 14, 1927, 1.

sides of the island are precipitous and cut in rudely columnar basaltic cliffs. To the east of the Farne, and separated from it by a shallow channel, are the Wedums or Wide Opens, which with the Noxes really form one island at low water, as they are joined by a shingle ridge called the Bridges. The Swedman is a bare rock; the Brownsman (Fosseland) is mainly rocky, reaching 40 ft. high; the Stapel derives its name from three (originally four) pillars of basalt standing rather apart from the island; the Wamses are two fairly large islands with, like Farne itself, a scanty vegetation and a peaty soil; the two Harcars become a single islet at low water; Northern Hares and Crumstone are little more than rocks; Longstone (Grace Darling's island) is only a rock about 4 ft. high, but *Glyceria* is found on it and on several other islands in the group.

In general, the Whin is rudely columnar, and well fissured. The more prominent fissures run more or less north and south, and have been converted into deep chasms, thus forming the Churn and St Cuthbert's Gut on Farne Island. On Stapel the trend of the fissures is more nearly north-west. In all the islands the more prominent cliffs face south and west since the inclination of the Whin is usually downwards to the north-east. There is some boulder clay on Farne, Stapel, and on one or two of the other larger islands. The most abundant of the sixty-two species of plants noted by Tate are *Silene maritima* and *Glaux maritima*, but there are no trees. St Cuthbert made the islands famous: he retired there in A.D. 676.

(3) BUDLE POINT TO DUNSTANBURGH

At Budle Point there are altered shales on the Whin, and between that cape and Bamburgh[1] there is a considerable amount of blown sand, often reaching more than 100 ft. high, but resting on rock or drift. The links are continuous south of the Harkess Rocks to the neighbourhood of Shoreston Hall. The Harkess Rocks[2] are low lying, and are formed of the Whin Sill together with some sediments. They outcrop on a bleak and wind-swept coast on which the sand is constantly shifting in such a way as to reveal new exposures. The sediments, which appear to have been folded before the intrusion of the Whin, are much broken, and large slabs, often with abrupt and almost vertical sides, have been enveloped in the magma in a way which gives the effect of shale flats bounded by tiny basalt cliffs. This is particularly noticeable near Budle. Wind-polished whin surfaces are also seen at Monkhouse Rocks.

From a point slightly north of Seahouses (North Sunderland) to Ebba's Snook the sandstones, shales, and limestones of the Carboniferous series

[1] The castle is built on the Whin Sill.

[2] J.A. Smythe, 'The Harkess Rocks', *The Vasculum*, 16, 1930, 9.

are arranged in an anticline. This is broken by an important fault which runs out to sea almost in the mouth of the Annstead Burn. Lebour, in his *Geology of Northumberland and Durham*, called attention to the east-west faults. 'Several of these great faults, however, are not visible on the coast-line, but reach the shore where it is flat and sandy, and so common is this arrangement that it may be said that almost any sandy bay between the Tweed and the Aln marks the line of a fault.'[1] Annstead Bay[2] is a good example. Reference to Figs. 100*a* and 100*b* will show clearly the differential nature of the erosion in the various beds outcropping on the shore, and the marked re-entrant where the fault occurs. Detailed comment on the map is unnecessary, but it is plain that the thin limestone beds usually form minor 'headlands', especially to the south of the fault, whereas the sandstones are more prominent to the north. Beadnell Haven, Beadnell Dyke, and Lady's Hole are noteworthy features (**126**, **127**, **128**).

Beadnell Bay is fringed by sand dunes, often reaching about 50 ft. in height: part of them is known as Newton Links, which are cut through by the Tughall Burn. Garwood suggests that there is a fault in this bay comparable to that in Annstead Bay. Snook Point, at the southern end of the bay, is a whin point and so also are Whittingham Carr and Robin Hood's Rock. A limestone overlies the whin at Snook Point, and this rock forms the northern margin of Football Hole. 'The succeeding sand dunes mark the trough of a synclinal, possibly faulted, so that, after passing an outcrop of false-bedded sandstone, the limestone is again encountered resting on the whin as before, but dipping in the opposite direction. A small fault then brings in the whin again, dipping in the usual direction to the south-east at eight degrees, and overlain by the Great Limestone, with which it forms Newton Point.'[3] Southwards from this point shales and shaley sandstones outcrop on the coast, and another fault occurs at Embleton Bay. The off-lying rocks at the northern end of the bay are all formed of limestone, except Out Carr and the eastern part of Embleton, which are formed of whin. The vegetation of the Embleton and Newton sand hills is somewhat unusual.[4] Bracken is the dominant plant, and in the spring bluebells and cowslips are conspicuous. Bruce knows of no other comparable association on Northumbrian dunes, and the occurrence is the more interesting as there is no woodland near. The dunes are low, seldom

[1] G.A. Lebour, 'On a Great Fault at Annstead...', *Trans. N. England Mining and Mech. Engs.* 33, 1883–84, 69; also see J. Hardy, *Proc. Berwickshire Nat. Soc.* 12, 1890, 502, and E. Tate, *ibid.* 4, 1858, 96.

[2] Lebour, *op. cit.* p. 133.

[3] E. Bateson, *A History of Northumberland*, vol. II, 1895, Embleton Parish.

[4] E.M. Bruce, 'The Vegetation of the Sand Dunes between Embleton and Newton', *The Vasculum*, 17, 1931, 94.

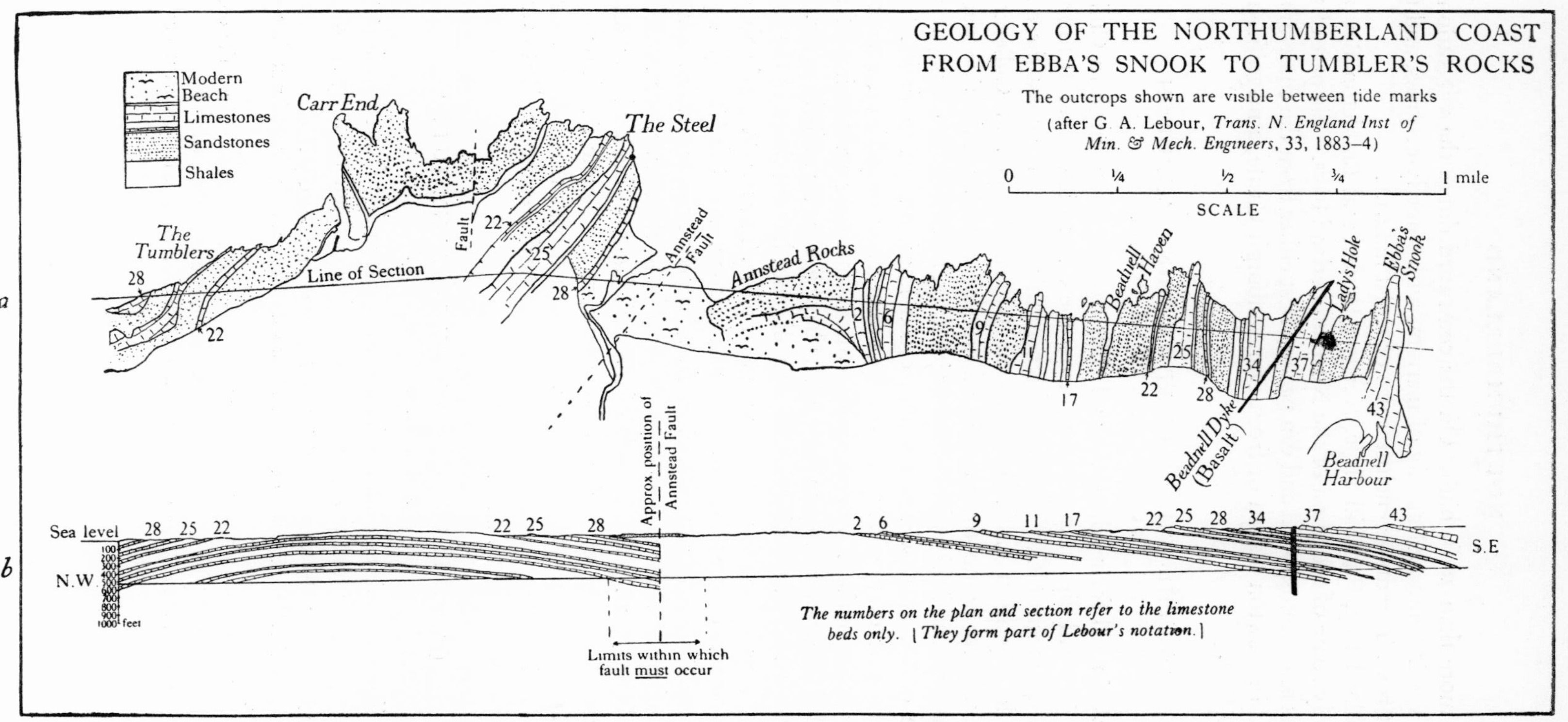

Figs. 100 *a*, 100 *b*

much more than 20 ft. high. On the coastward dunes the vegetation is open, and is composed mainly of marram, with some lyme grass, and the burnet rose. The dunes enclose rows of moist hollows.

Beadnell Bay, Football Hole, Newton Haven, and Embleton Bay are certainly some of the finest of the Northumberland bays. This part of the coast is almost unspoiled and can be seen best either from the coast-guard station in Newton Haven or from Dunstanburgh Castle, a magnificently placed ruin.

(4) THE WHIN SILL FROM DUNSTANBURGH TO CULLERNOSE POINT

The Whin Sill forms the coast and cliffs from Dunstanburgh to Cullernose.[1] The sill dips seawards just below the Great Limestone. It forms a bold escarpment facing inland, and where this reaches the coast there are quite locally high cliffs, especially at Dunstanburgh and Cullernose. Between these points there are broad dip-slopes, between the tide marks, which are much broken by columnar jointing, and resemble crazy paving. There occur, also, widened joints, crush lines, and small faults. This seaward part of the Whin is usually swept clear of pebbles, except at Queen Margaret's Cove and Cushat Stiel, where the weathering of the included sediments and two small faults have given rise to little coves. At Craster[2] a stream, coinciding with a small fault, has cut through the rock, and has thus led to the small haven at that place. The two 'Carrs' offshore are fragments of the overlying Great Limestone. At Gull Crag (the cliff north of Dunstanburgh Castle) 50 ft. of rudely columnar whin rest on 40 ft. of sandy shale and sandstones, which, in their turn, are based on 6 ft. of sandstone. From the top of the Gull Crag the buckled-up mass of limestone, known, on account of its shape, as the Saddle Rock, or Grey Mare, is a feature of the foreshore. The beach near the Saddle Rock is covered with huge rounded masses of whinstone decreasing in size towards Embleton Sands[3] (**129** and **130**).

(5) CULLERNOSE POINT TO ALNMOUTH[4] AND THE COQUET

At Cullernose Point the Whin leaves the coast and is faulted against the Four-Fathom Limestone. The fault itself is occupied by a dyke which

[1] J.A. Smythe, 'The Saddle Rock', *The Vasculum*, 15, 1929, 56.

[2] J.A. Smythe, 'The Erosion of Rocks', *The Vasculum*, 3, 1917, 73, and next-but-one footnote.

[3] One almost regrets that the Whin Sill escarpment, facing west, is not the coast. It is a good feature. The seaward inclined whin gives a very tame effect.

[4] *Mems. Geol. Surv.* 'The Geology of the Alnwick District', R.G. Carruthers and others, 1930.

reaches a width of 12 yd. In Swine Den large blocks of sedimentary rocks are seen embedded in the Whin while shaley sandstones come in to the south and are intersected by a dyke now weathered out into a wall-like ridge. These beds, together with a limestone which for a short distance rises above sea level in an anticline, continue to the south until new beds are brought in at the Howick fault. This is a complex disturbance of 400 ft. or more, strikingly seen on the coast. In Howick Bay the conspicuous north-south strike of the shaley sandstones forms reef-like islands at low tide.[1] There are also minor disturbances in the cliffs. 'From Iron Scars massive false-bedded sandstones (of Millstone Grit) passing in places into coarse red grits, extend to Seaton Point. These are interrupted, north of Boulmer, by a fault running due east through Long Houghton which brings in a thin limestone near the cliff, and by a vertical whin dyke, 100 ft. wide, running east and west, and immediately north of the lifeboat station at the Torrs.'[2] Boulmer Haven, wholly in Millstone Grit, situated between tide marks, also corresponds with a small fault. The small bay south of Seaton Point is sandy and, once again, appears to correspond with a fault. Marden Rocks, just at sea level, are also part of the Millstone Grit.

Alnmouth is pleasantly situated. The links form a flat and attractive space in front of former cliffs. The mouth of the Aln has changed in comparatively recent years. The river, which is still tidal to Lesbury Bridge, used to flow to the south of Church Hill, which was then joined to Cheese Hill by a low ridge. But in 1806 the river broke through this ridge, thus making Church Hill, which is formed of glacial deposits, into a high-tide island. It was not, however, until some years later that this new channel became the main outlet of the river. Church Hill is now joined to Buston Links by a line of sand hills. The harbour was of importance in early times, and there are several references to it in fourteenth-century records. In 1730 Mark wrote: 'It [is] a very good harbour for ships, and is the only flourishing place of trade and shipping except Blythes-Nook between Newcastle and Berwick.'[3] It dwindled in importance, however, after the French wars. South of the river there is a narrow dune belt backing on to the Millstone Grit which outcrops as far south as Warkworth and the Coquet mouth. The lower reaches of the Coquet valley are of post-Glacial

[1] Dr Carruthers tells me that off the mouth of the Howick Burn there have been of late years frequent exposures of the submerged forest so notable farther south at, e.g., Amble, Newbiggin, and Sunderland.

[2] E. Bateson, *op. cit.* 2, 1895, 328, 'The Geology of Horwick, Long Houghton, and Lesbury'.

[3] G. Tate, 'An Account of Lesbury Parish, Northumberland', *Proc. Berwickshire Nat. Club*, 8, 1879, 238; see also 'Alnwick District', *Mems. Geol. Surv.*, 1930.

date: in several places it cuts through solid rock and eventually reaches the sea at Amble after a meandering course over rock and boulder clay. In pre-Glacial times the mouth may have been about four miles farther south in Druridge Bay. The present mouth has been deflected somewhat to the south by the beaches and dunes. It is relevant here to point out that between Amble and Sunderland there are several widely open bays, all of which are sandy. In each of them boulder clay reaches the sea, and they also mark the outlet of pre-Glacial valleys or the lower parts of the rock surface.[1] The more important bays of this class are Druridge, Whitley, Whitburn, the bay south of Blyth, and those at the end of the Sleekburn and Tyne valleys. Between the bays the coast is formed of harder rocks, usually thinly capped by boulder clay. The former valley which drained to Druridge Bay appears to have been broad and shallow. Two borings prove the underlying rock surface to be at 45 ft. and more than 47 ft. below sea level, thus indicating a pre-Glacial valley emptying in the middle of the bay (**131**).

(6) THE COQUET TO THE TYNE

Southwards from the mouth of the Coquet the Coal Measures reach the coast, and form off-lying islands or rocks, such as Coquet Island and Hadstone Carrs, and the Scars opposite Cresswell. Dunes fringe the coast in the bays as already noted, and boulder-clay sections occur often, for example, between Hadstone Carrs and the Chevington Burn in Druridge Bay. Lyne Sands, completely spoiled by mining, at the mouth of the Lyne River extend for nearly a mile between Lyne Skears and the rocks near Beacon Point. Nearer Newbiggin[2] and between that place and the Wansbeck mouth there are several promontories running roughly parallel with one another forming Newbiggin Point, Spital Point, and Hawks' Cliff: they shelter small bays and coves and are all in the Coal Measures. The rocks are hard and form precipitous cliffs, but become softer on rounding Hawks' Cliff. The mouth of the Wansbeck was at one time picturesque. The spit of shingle and sand on its northern side illustrates the prevailing southward drift of beach material on this coast. South of the river are Cambois Links which form a long spit running to the south-east and continued into reefs covered at high water. The river Blyth is thus deflected to the south, and the modern town of Blyth is situated on its south shore. This shore, however, has suffered change in recent times.

[1] D. Woolacott, 'The Superficial Deposits and Pre-Glacial Valleys...', *Quart. Journ Geol. Soc.* 61, 1905, 64, and see *Geogr. Journ.* 30, 1907, 36.

[2] R.G.A. Bullerwell, 'A Section of the Cliffs near Newbiggin-by-the-Sea...', *Trans. Nat. Hist. Soc. Northumberland, Durham and Newcastle-on-Tyne*, 4, 1909–13, 61.

The name Blyth really applies to the eastern part of the town standing on what was formerly a small peninsula which was severed from Cowpen and almost entirely cut off from Newsham by the Blyth Gote slake.[1] A sand hill, Stob Hill, dominated the south end of this peninsula, and Cowpen Quay was then tidal mud. The Snook is a name that has been applied both to the spit north of the river as well as to the southern peninsula, to which it really belongs: it is first mentioned in 1208. The modern Wansbeck, Sleekburn, and Blyth are all post-Glacial streams, and have cut their ways through rock or boulder clay since the Ice Age. The depth of the pre-Glacial Sleekburn valley is some 93 ft. below sea level, and is thus midway between that of the pre-Glacial Druridge valley (47 ft.) and the pre-Glacial Tyne (141 ft. below sea level).

Another dune belt extends from the Blyth to Seaton Sluice. The dunes average about 240 yd. wide, and their seaward edge is well defined.[2] The sea runs up the beach, but seldom reaches the actual dunes, the spacing and height of which are irregular. The succession of plants is generally typical of most Northumberland links. Sea purslane (*Arenaria peploides*), with couch and lyme grass, gives place to a pure *Ammophila* association, which, in the older dunes, is mixed with ragwort, wild carrot, and hogweed. Mosses soon appear, and also much bloody cranes bill and dune meadow rue. Finally, there are ordinary land grasses with burnet rose and gorse. Skinner[3] analysed the dune soils south of Meggie's Burn, and found a fairly uniform mixture of a large proportion of coarse sand with a smaller amount of fine clay: this is to be associated with an outcrop of boulder clay on the shore and below low-water mark. The humus content is low, but increases in the flats, especially those containing *Salix repens*. In this more southern part the dunes show signs of denudation. Sandy Island, at the mouth of Seaton Sluice, is really artificial, being mainly composed of Kentish Chalk dumped from ballast.

Seaton was an entirely natural harbour up to the middle of the seventeenth century, and between it and the Brierdean Burn there is a good deal of variety in the coast. Alternations of rock and creek occur near Seaton itself; St Mary's Island, more correctly named Hartley Bates (Baits), is a rock isolated at high water; from Curry's Point (rock) to the Brierdean is a boulder-clay coast fronted by sand and shingle. From Seaton to the Tyne all the headlands consist of massive beds of sandstone, often con-

[1] E. Bateson, *op. cit.* vol. IX, 'Blyth'.

[2] P.G. Fothergill, 'The Blyth-Seaton Sluice Sand Dunes', *The Vasculum*, 20, 1934, 23.

[3] E. Skinner, 'A Survey of the Dunes between Meggie's Burn and Seaton Sluice', *ibid.* 20, 1934, 122.

tinued farther seawards as reefs.[1] Seaton Sluice Point, Charley's Garden, Crag Point, St Mary's Island, Table Rocks, Brown's Point, and Sharpness Point are all of this nature. The Brierdean valley coincides with a fault, invisible at the surface, and there are also faults at Seaton Sluice and Crag Point, where in addition there is a whin dyke. The 90-Fathom Dyke (fault) runs in a north-west to south-east direction through the southern side of Cullercoats Bay,[2] and intersects the southern headland of the bay, which is cut in an outlier of Yellow Sands (Permian) resting unconformably on the Coal Measures. In Collywell Bay, which really marks the position of a big trough fault, there are outcrops of Coal Measures capped by Permian and Glacial beds. From Smugglers' Cove, at the southern end of Cullercoats Bay, as far as Sharpness Point, the rocks are hidden beneath boulder clay and blown sand. Permian outliers rest on the Coal Measures at Tynemouth, but between the life-boat station and Lowlights the cliffs are of boulder clay, much subject to slips, and erosion is rapid.[3] The Tyne again exemplifies post-Glacial down-cutting, especially east of Newcastle, but it is noteworthy that, as with the Tweed at Berwick, the pre-Glacial outlet must have been in a relatively narrow rock gorge, now deeply buried (140 ft. or more).[4]

The coast from the Coquet to the Tyne is almost entirely industrial. Its natural beauty is disfigured by collieries, railways, and houses. Nowhere is this better exemplified than at Cambois. The fine beach and dunes give place immediately to railway sidings and industrial squalor. South of St Mary's Island there is almost continuous town as far as the Tyne. The general contrast, therefore, between the Northumberland coast north and south of the Coquet is clearly due to the fact that the southern area corresponds with the Coal Measures.

The general direction of beach-drifting along the Northumberland coast is to the south. This is seen, for example, in the small shingle ends deflecting the Aln, Coquet, Lyne, and Wansbeck.

[1] R.G. Absalom and W. Hopkins, 'The Geological Relations of the Coast Sections between Tynemouth and Seaton Sluice', *Proc. Univ. Durham Phil. Soc.* 7, 1923–7, 142. (For dune plants at Seaton, see A.W. Bartlett, *The Vasculum*, 16, 1930, 129.)

[2] S.R. Haselhurst, *Proc. Univ. Durham Phil. Soc.* 4, 1911–12, 15.

[3] R.M. Tate, 'On the Erosion and Destruction of the Coastline...', *Trans. Nat. Hist. Soc. Northumberland, Durham and Newcastle-on-Tyne*, 11, 1894, 187.

[4] See E. Merrick, 'Superficial Deposits around Newcastle', *Proc. Univ. Durham Phil. Soc.* 3, 1906–10, 141; S.R. Haselhurst, 'Some Features of the Glacial Deposits at the Tyne Entrance', *ibid.* 5, 1912–15, 147, and 'On the Rate of Recession at Tynemouth', *ibid.* 2, 1900–6, 121; D. Woolacott, 'The Geological History of the Tyne, Wear, and Associated Streams', *ibid.* 2, 1900–6, 121.

(7) THE DURHAM COAST

South of the Tyne the Coal Measures are no longer on the coast and from South Shields to Hartlepool the Permian rocks form a particularly interesting coastal section unique in Britain. Although much has been written on the stratigraphy of these rocks, little has been said of the resulting coastal scenery, and a more detailed investigation on the relation of coastal forms to erosion and rock structure would be of considerable interest. Between South Shields and Marsden the breccias of the Magnesian Limestone are very evident, and the limestone is much disturbed. The vertical fissures seen in the cliffs, the breccia-filled gashes, the fantastic and endlessly diversified cavities, geodes, cellular and other structures, so well seen at Marsden Rocks and elsewhere, have with much reason been ascribed to the hydration and subsequent leaching-out of inter-bedded anhydrite layers.[1] The stacks in Marsden Bay are very fine. The large one is itself riddled with tunnels and arches, and affords a textbook example of the work of marine erosion on a particular kind of rock.

Woolacott[2] in several papers referred to certain gravels, between 100 and 150 ft., on the Cleadon and Fulwell Hills as raised beaches, a view not in agreement with later workers. There is, however, much diversity of opinion on the matter, and it must suffice to draw the reader's attention to these sections which deserve prolonged study. There is a great deal of boulder clay along the Durham coast, especially capping the cliffs (see Fig. 98), and drainage changes have taken place since Glacial times. The Wear now enters the sea at Sunderland. (According to Woolacott it formerly flowed along the valley of the Wash to the Tyne, but later carved a way through the Magnesian Limestone and boulder clay to Sunderland.[3]) There have been many local changes in the harbour at that place, and an interesting series of plans is given by Murray.[4] Other noteworthy coastal details are the submerged forest at Roker (Sunderland), the stack of the Parson's Rock which shows the concretionary limestone in one of its characteristic forms—the cannon ball—and the caverning of the Magnesian Limestone at Hole Rock, near Ryhope (**132** and **133**).

Between Sunderland and Hartlepool the coast is intersected by several deep denes, all of which are post-Glacial in date. Excellent examples occur at Hendon, Ryhope, Seaham, Hawthorn, Foxhole, Castle Eden, Hesleden,

[1] G. Hickling and A. Holmes, *Proc. Geol. Assoc.* 42, 1931, 252.

[2] E.g. *Proc. Geol. Assoc.* 24, 1913, 87; *Proc. Univ. Durham Phil. Soc.* 1, 1896–1900, 247.

[3] D. Woolacott, *Proc. Univ. Durham Phil. Soc.* 1, 1896–1900, 247.

[4] J. Murray, *Mins. Proc. Inst. Civ. Engs.* 6, 1847, 256.

and Crimden. These are nearly all cut through boulder clay, but in one or two cases, including Hawthorn Dene, they also cut through the underlying Magnesian Limestone rocks, forming extremely picturesque coastal features. Crimden Dene, where a spit of boulder clay forms a low cliff capped by blown sand, marks a somewhat abrupt change in the nature of the coast; to the north lies an austere and rocky section, with high cliffs and foreshore scars in the Magnesian Limestone, often topped by boulder clay, and broken here and there by the denes; to the south is a long stretch of sands backed by a belt of dunes and occasional low cliffs. There is also much ballast on the beaches between Crimden and Hartlepool. A Mesolithic site of later Boreal age has been found near Crimden Dene,[1] and Neolithic settlements are also known elsewhere on the Durham coast. They afford some evidence as to the amount of erosion that has taken place. The site of Horden,[2] about a mile north of Castle Eden Dene, suggests that there has been very little coastal recession here since Neolithic times: the storm beach and sands form adequate protection. Between Horden and the mouth of the Wear, and even as far as the Tyne, Neolithic remains occur only sporadically, and the coast is less protected by dunes and shingle. Moreover the limestone is softer, and there is less boulder clay, and for these reasons erosion has been greater.

South of Seaham Harbour patches of boulder clay containing Scandinavian pebbles are found in fissures and hollows of the Magnesian Limestone. This clay is probably the equivalent of the Basement Clays of the Yorkshire coast. It is well exposed at Hawthorn Dene, Hesleden Dene, and Warren House Gill.[3] The blue or purple clay with Carboniferous and Cheviot rocks overlies this, and is laid bare over long stretches of the coast. Between this and the Upper Clay are sands, gravels, and laminated muds. The Upper Clay itself is more variegated in colour and contains Cheviot, Tweeddale, Lake District, and west Scottish material. Sands, gravels, and morainic deposits of the retreat stages of the ice sheet succeed to it.

In one way the Durham coast is the most disappointing in this country: it is so completely spoiled by industrialization, yet it is clear that in its natural state it was very beautiful.

[1] A. Raistrick and G. Coupland, *Trans. North Nats. Union*, 1, 1936, 207.

[2] C.T. Trechmann, *Trans. Nat. Hist. Soc. Northumberland, Durham and Newcastle-on-Tyne*, 4, 1909–13, 67.

[3] C.T. Trechmann, 'The Scandinavian Drift of the Durham Coast...', *Quart. Journ. Geol. Soc.* 71, 1915, 53. (For other papers partly connected with coastal matters, see D. Woolacott, 'The Stratigraphy and Tectonics of the Permian of Durham', *Proc. Univ. Durham Phil. Soc.* 4, 1911–12, 241; G.A. Lebour, 'Breccia-Gashes, Durham Coast', *Trans. N. England Inst. Min. Mech. Engs.* 33, 1883–4, 165.)

The coastal details, especially in the Magnesian Limestone, are interesting and varied. Marsden Bay with its stacks, caves, and arches is especially worthy of comment: at its southern end industry closes in almost to the cliff top. Cliffs and stacks are still found, but the actual cliff scenery falls off towards Sunderland. Whitburn Bay has good sands. South of Sunderland the cliff scenery is again of a high order, but inland from the cliff edge the skyline is ruined. There is a good deal of local erosion to the south of Sunderland, and again to the south of Seaham Harbour. At and near the mouth of Hawthorn Dene is the best remaining part of this coast. The lower parts of the cliffs are steep; here, as often elsewhere, the upper parts, formed of boulder clay, slope at a much gentler angle. The cliffs remain good almost to Crimden Dene, but in two places colliery waste is dumped below high-water mark, and at Easington the cliff face in part is colliery waste which has been shot over the existing cliffs. All this waste material sooner or later travels south—one reason for the coal dust on the beaches. The denes, deeply cut and well wooded, are not only in themselves, but also by contrast to the industrialism of the area, features of great beauty, and with the cliffs (particularly if seen from beach level) still give a real charm to this coast. In earlier days the whole effect must have been most pleasing. Industrialism has also taken possession of the southern half of the dunes running to Hartlepool. The Tees marshes, except for the works near the mouth of Greatham Creek, are mainly open and unspoiled. The old square at Seaton Carew also makes a pleasant break from the industrial towns.

(8) THE TEES MOUTH AND MARSHES

Near Hartlepool the nature of the coast alters. Hartlepool itself is situated on the Heugh, an outlying mass of Magnesian Limestone which has been joined to the mainland by the spit of sand and dunes running from Crimden Dene. Between the dunes and the higher ground within is an area of flat alluvial land on a part of which lie the docks. It seems that if a rocky platform exists under the dunes of North Sands, it is certainly below sea level. Hence, in all probability the spit has grown from the north-west and united Heugh Island to the mainland. The Heugh was formerly subject to erosion, but is now enclosed in a sea wall. It is generally fronted by limestone scars, but between the lighthouse and the pier is a bay with sand, shingle, and large boulders of limestone. Hartlepool Bay, containing remains of a submerged forest, is enclosed between the Heugh and Long Scar, a projecting mass of red Triassic sandstone. Between Hartlepool Bay and the Tees the coast is formed of low cliffs of sandstone of Triassic age, but these soon give way to the extensive area of Tees Bay. There has been much reclamation here, and the nature of most of the northern part of the area has been fully described by Harrison[1] who surveyed it ecologically before industrialization began to alter the appearance of the district. All this part is of post-Glacial origin. At the close of the Ice Age the Tees and several minor streams ran into a broad and complex open estuary which extended well inland and also up the Billingham Beck. The southward travel of beach and offshore material along the coast, coupled with the somewhat sluggish nature of the Tees, led to the deposition in the estuary of enormous amounts of silt and warp. These deposits tended, like those at the southern end of the Wash, to impede the land drainage, so that the arms and upper parts of the main estuary were filled by alluvial material, composed mainly of sandy clays impregnated with carbonates derived from the Magnesian Limestones. These clays, and associated boulder clays, are often about 100 ft. thick. Probably this infilling was neither a continuous nor a simple process, because borings near Middlesbrough have passed through peat beds, and also sub-fossil oaks have been found.

The mud flats are known locally as Slems. The first major reclamation, in Saltholme and Cowpen marshes, was made in 1740 by the building of a great earth wall skirting Greatham Creek, supported by a second wall running south-west from it. Rather more than a century later a massive slag wall was completed, and 26,000 acres were secured. This is now partly grazing land, and partly the site of iron and salt works: relics of fourteenth- and fifteenth-century salt works occur on all the marshes as irregular mounds. The innings also led to the development of lakes and fleets. In Billingham Bottoms (i.e. the alluvial land in Billingham Beck) the marsh is mainly a Juncetum. The name, Saltholme Marsh, refers, in Harrison's paper, to all the reclaimed marsh outside the earliest earth wall, and so includes Saltholme, Cowpen, and Fore marshes as depicted on the maps of the Ordnance Survey. There are large fleets in this stretch, all of which, when in a natural condition, are kept open by tidal scour. They are usually isolated by walls, but some, including Swallow, Mucky, and Todlers, had almost silted up by 1921. Holme Fleet is still open. Harrison stated that they were linked up and drained by a deep channel which still follows the sea wall and empties by locks into Greatham Creek.

The natural salt marsh at Greatham even about 1920 was but a fragment of its former self. It consisted mainly of two big tracts bordering Greatham Creek which was deep and tidal. The minor channels meander

[1] J.W.H. Harrison, 'A Survey of the Lower Tees Marshes...', *Trans. Nat. Hist. Soc. Northumberland, Durham and Newcastle-on-Tyne*, 5, 1918–21, 89; see also J. Taylor, 'River Tees...', *Mins. Proc. Inst. Civ. Engs.* 24, 1865, 62.

in typical fashion. The general nature of the vegetation is much the same as on other east coast marshes. At the lowest levels green algae, especially *Rhizoclonium*, *Vaucheria*, and *Chaetomorpha*, thrive, while at higher levels *Salicornia*, and possibly some *Glyceria*, are found. The *Salicornia* is also accompanied by *Suaeda maritima*, and upwards gives way to abundant *Glyceria*, which merges into an Armerietum. Where the surface is muddier, *Aster* usually succeeds *Glyceria* and *Salicornia*, and is closely followed by *Statice limonium*. Within the Armerietum, *Obione portulacoides* and *Artemisia vulgaris* occur.

There are also extensive flats on the south side of the main stream, but Coatham Marsh as early as 1888 was threatened with obliteration by slag tips. Barrow[1] commented on a number of small flat-topped hillocks there, and despite earlier and often fanciful explanations of them he concluded that they were remnants of an older and higher alluvium, most of which has been eroded away. West of the railway skirting Coatham Marsh, and between the tide marks, is a thick deposit of sand, and dunes become prominent nearer to Redcar. The Tees mouth is constricted by Coatham Sands running out towards South Gare Breakwater. The present river entrance has been altered much by man, and industrialism has spread over much of the sand spit running north-westwards from Redcar.

(9) THE CLEVELAND COAST AND FILEY

Near Redcar the boulder clay rises above sea level to form small but pleasant cliffs often capped by blown sand and fronted by a good beach. These cliffs extend to Saltburn, but no solid rock appears in this stretch. The so-called raised beach[2] at Saltburn is probably an erosion terrace of the Skelton Beck before the ravine was cut down to its present level, and also when its mouth was somewhat seaward of its present position. To the east of the town the highest beds of the Lower Lias, capped by Middle Lias, are conspicuous in the prominent and flat-topped Hunt Cliff. Although the shales of the Lower Lias are soft, they are nevertheless more resistant than the boulder-clay cliffs west of Saltburn, and so Hunt Cliff forms one of many examples on the north Yorkshire coast of a promontory standing forth as a result of differential erosion. The Lower Lias shales also form extensive scars at the cliff foot.

The small beck with its mouth at Skinningrove is one of several on this coast (see Fig. 101) the courses of which have been somewhat altered in post-Glacial time. The general nature of these changes is best discussed with the study of the Mickleby and Bick Head becks at Sandsend (see

[1] *Mems. Geol. Surv.* 'The Geology of North Cleveland', G. Barrow, 1888.

[2] P.K. Kendall and H.E. Wroot, *The Geology of Yorkshire*, 1925, pp. 733 et seq.

p. 467): at Skinningrove no noticeable change has taken place in the position of the mouth. The dale itself is very picturesque except near the sea where it is completely ruined by the iron works and ugly appearance of Skinningrove itself. Boulby Cliff,[1] between Skinningrove and Staithes, was at one time the highest cliff in England (660 ft.), but its form was greatly altered by the enormous excavations made between 1615 and 1871 for the Boulby and Loftus alum works. This excavation was cut into the cliff, and its waste was tipped into the sea. The Lower Lias appears in the scars at the cliff foot, the Middle Lias in the lower part of the cliff, the

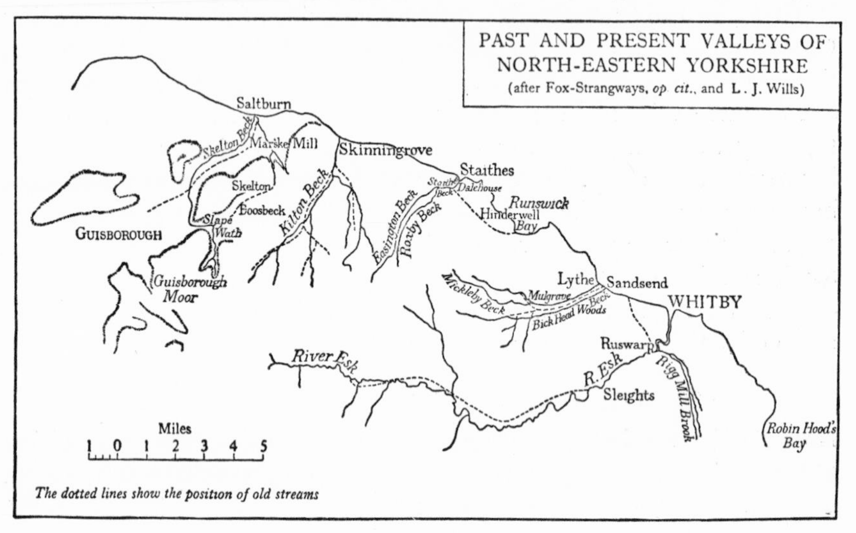

Fig. 101

Upper Lias, including the Alum Shale, in the works themselves, while above is the Dogger with a capping of Lower Estuarine Sandstones (**134**).

The Easington and Roxby becks unite about a mile inland, and run out to sea at Staithes as the Staithes Beck, but this channel was not the original mouth (see Fig. 101). At Dalehouse there is an old drift-filled valley which used to carry the waters of these streams through the parish of Hinderwell to the sea in Runswick Bay, and it is this drift which forms the boggy hollow close to Runswick village. In the gully above the village there is a fault along which several landslips have taken place, and these have been responsible for the terrace-like appearance of the village. North of Runswick the Upper Lias forms the cliffs to Port Mulgrave, which harbour was made to convey to Middlesbrough the Cleveland ironstone worked on the cliff side. The coast between Skinningrove and

[1] Kendall and Wroot, *op. cit.*

Runswick Bay is magnificent, and the unspoilt nature of the country within gives a fine setting to the beauty of the cliffs. The old part of Staithes around the harbour, and shut in by high cliffs, is picturesque and adds to the attraction of the coast. The fine sweep of Runswick Bay is somewhat marred by unsuitable shacks (**135, 136, 137**).

Two small streams in Runswick Bay, quite incapable of eroding it to its present form, occupy the old drift-filled valley of the Staithes Beck. At Kettle Ness the extremity of the point is composed of the Ironstone series which makes first a low cliff and then falls to form reefs. The ironstone is displaced by many small faults. These Lower Lias beds are hard and are the cause of the projecting headland. The higher parts of the cliffs at Kettle Ness are Middle and Upper Lias with the Dogger and some of the Estuarine series. The point itself is almost barren of vegetation and is, in parts, very subject to mudflows. South of the ness it is not easy to explore the foreshore; the cliffs are tumbled and broken and subject to slips. In parts the railway runs on the lower cliff. It will be clear from Fig. 101 that the Mickleby and Bick Head becks which reach the sea at Sandsend flow in a valley formerly occupied by one stream the course of which followed the dotted line.[1] Although this is the only case where the two streams remain separate right down to the sea, Fig. 101 also makes it clear that it is not uncommon on this coast to find two streams in what was formerly a single valley. The drainage from the two sides of these old valleys has been separated by boulder clay, and post-Glacial erosion has not yet been long enough at work to remove the barrier. It will be noticed that in several places the streams have cut down and into solid rock. East of Sandsend and past Raithwaite Gill the Jurassic rocks give place to boulder clay which forms lower cliffs, now partly inside the road. It has been suggested that Raithwaite Gill marks the former outlet of the Esk. Near Upgang Beck the Estuarine Sandstones reappear to form the small rocky cape called Lector Nab about half a mile from Whitby pier.

Whitby Harbour channel, a gorge-like feature, is of pre-Glacial age and situated on a line of faulting. The western side is down-thrown some 200 ft. relative to the eastern side, and the effect of this fault does not cease at the water line. Fox-Strangways[2] regarded the line of breakers extending from north of the harbour to Saltwick Nab as marking the outcrop of the Jet Rock[3] which is abruptly truncated by the fault. Im-

[1] C. Fox-Strangways, 'The Valleys of North-East Yorkshire and their Mode of Formation', *Trans. Leicester Lit. Phil. Soc.* 3, 1892–5, 333.

[2] *Mems. Geol. Surv.* U.K., 'The Jurassic Rocks of Britain', vol. 1, Yorkshire, 1892.

[3] The *best* jet occurs in lens-shaped masses in the top 10 ft. of the Jet Rock. Jet is wood that has suffered a peculiar form of decomposition.

mediately to the western end of this line is deep water, and the trench, known as the Swatchway, which is cut through the outcrop, is possibly an old lake-overflow channel formed during the Ice Age. Beyond Whitby east pier the Upper Lias is exposed on the foreshore, and the neighbouring cliff gives an excellent section of the Alum Shale, Dogger, and Lower Estuarine Sandstone. It is noteworthy that the great oblique joints on this cliff simulate bedding, and small caves are often cut in them. Saltwick Nab, a bare promontory much lower than the actual cliffs, is formed of Alum Shale, but the hollow at Saltwick is artificial: the works associated with it were shut down in 1821. Black Nab is an island at high water. Little comment is necessary on the geology of the cliffs and scars between Saltwick and Robin Hood's Bay: they generally resemble other similar features on this coast, and their nature can readily be seen from the sections in Fig. 102. But no section can bring out the beauty of this range of cliffs, which are as fine as any in Yorkshire; they are often very high with little talus below; good farm land exists on their tops and stretches far inland (**138**).

Robin Hood's Bay is one of the finest features of this coast.[1] It is very picturesque and the cliffs are lofty and steep. The view of Robin Hood's Bay from Ravenscar is admirable. The bay is backed by the open moorland of Fylingdales, and the enclosing headlands appear far more prominent on the ground than on the map, so that the whole forms a circumscribed area of great beauty. Fig. 102 *c* shows the arrangement of the rocks which form scars along the foot of the bay. On account of the obvious anticlinal structure in the bay, which is situated on the Cleveland axis, the harder beds of the Middle Lias reach the sea at the bounding headlands of North Cheek and South Cheek, which are consequently protected. The softer shales of the Lower Lias have been worn back to form the bay. Differential erosion, however, is seen in the numerous scars where the harder beds of calcareous sandstone or limestone offer greater resistance to the sea. As will be shown later there is a good deal of boulder clay in this and other bays, and in Robin Hood's Bay, where there have been several fairly recent slips,[2] both Upper and Lower Boulder Clays occur. The Peak cliffs near the south end of the bay reach about 600 ft., and at Peak a great fault brings in between its two branches hard sandstones of the Middle Lias, and against the point of the cliff hard bituminous shales. The downthrow of the fault is to the south and measures 400 ft. Thence southwards for some three miles there is a considerable development of undercliff due to falls and slips. The upper part of the cliffs for

[1] Kendall and Wroot, *op. cit.*; Fox-Strangways, *op. cit.*

[2] L. Walmsley, *The Naturalist*, 1913, p. 280.

a distance of over two miles has given way and now forms a lower terrace. This feature is interrupted at Hayburn Wyke where there is a shingly beach at the mouth of a deep and well-wooded ravine. The local scenery is somewhat reminiscent of that near Lynton in north Devon. The undercliff reappears to the south: its formation possibly began early in the seventeenth century. Cloughton Wyke is another small inlet comparable to Hayburn Wyke, although rather less enclosed. Southwards

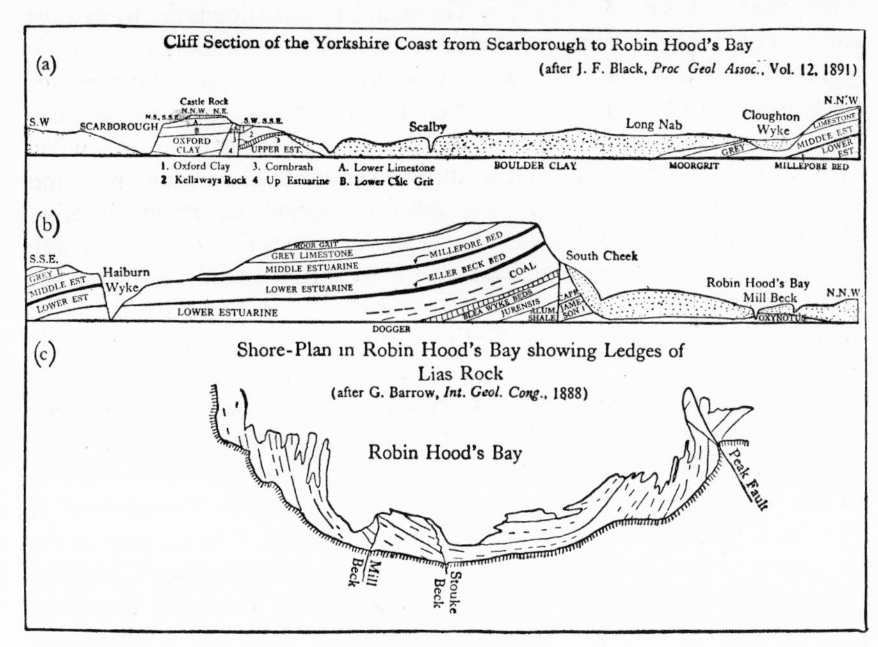

Fig. 102 *a*, *b*, *c*

from it the cliffs, whilst still fine, are not so spectacular as those to the north (**139**).

The whole coast from Peak to Scarborough trends approximately north-north-west to south-south-east, and is much influenced, not to say determined, by a series of faults which run in a roughly north and south direction.[1] The Lower Oolites are throughout the dominant rock group, but they are in places much obscured by drift, especially south of Long Nab, from which place as far as Scarborough North Bay the cliffs are almost wholly in drift and fronted by scars of sandstone. At both Blea Wyke and Cloughton there are fine cliff sections showing beds upwards from the Dogger to the Moor Grit, the base of the Upper Estuarine

[1] Kendall and Wroot, *op. cit.*

series. The north-south faults are traceable at Scarborough, but the conspicuous promontory is due mainly to the resistance of a thick bed of sandstone. This ridge consists of Oxford Clay, with Calcareous Grit and Limestone topped by Passage Beds. The Oxford Clay outcrops to the north of the ridge, and in this soft formation and in drift the North Bay has been cut. Two faults converge on the north side of the Scarborough ridge, and the great and ancient defensive ditch which runs north to south across the ridge is cut along them. These faults throw down the Kellaways Rock to the level of the marine drive at the harbour entrance.

The South Bay also shows some interesting features. The mass of coarse gravel overlying the boulder clay near Spa Gardens forms a hummock, a feature not infrequent at the mouths of valleys once obstructed by ice.[1] Indeed, a rather similar feature appears at Scalby about a mile and a half north of Scarborough. Near the bathing pool the strong jointing of the Upper Estuarine Sandstone aids the erosive action of the sea. The Black Rocks, the colour of which is due merely to seaweed, are formed of calcareous shales and sandstones, the so-called Scarborough Limestone. White Nab is in the Upper Estuarine series, and there is much pisolitic iron ore thereabouts. In the reefs off White Nab the faults cutting the Scarborough promontory can be traced, but whereas at Scarborough the downthrow is to the east, it is here to the west, so that Kellaways Rock and Oxford Clay are brought alongside Middle Estuarine Sands resting on the Millepore Limestone, which is admirably seen in Cornelian Bay. The notch left by the fault in Osgodby Nab affords an entrance to Cayton Bay which otherwise is often unapproachable from the north. Cayton Bay is almost entirely the product of faults which have carried the Millepore Bed to water level at the two ends of the bay—and this bed is the hardest on this part of the coast. The bay contains a fine sandy beach and is shut in between picturesque headlands; unfortunately it is ruined by huts and shacks. On Yons Nab the Millepore Bed is found at the seaward extremity overlying Lower Estuarine Sandstones. Then follow, in upward succession, Middle Estuarines, Scarborough Limestone, and Upper Estuarines, all of which are on the foreshore, whilst the cliff shows the Cornbrash, Kellaways Rock, Oxford Clay, and a capping of Lower Calcareous Grit (140).

The dip carries the Middle Jurassic rocks below sea level in Gristhorpe Bay, and between there and Filey the cliffs reach about 250 ft. in height and are steep. The land, all given to agriculture, slopes inland towards Filey. The Kellaways Rock and Cornbrash are near water level, and the cliffs are cut in Oxford Clay, Lower Calcareous Grit, and boulder clay. The grits are vertical and overhanging, whereas the slopes on the clays are more gentle; in addition, the jointed nature of the grits has led to the

[1] Kendall and Wroot, *op. cit.*

formation of buttress-like projections. The island-rock which formerly stood near the northern end of the bay is now represented only by its sandstone base. The most conspicuous feature, however, between Scarborough and Flamborough is Carr Naze with its extension of Filey Brigg.[1] The Brigg is composed of Lower Calcareous Grit, and shows the same general strike of the rocks that can be traced from the west of Scarborough. The rocks of the promontory dip to the south, and are broken, by erosion, into steps and holes. The surface is 200–500 ft. wide, and the scarped edge faces the north and most exposed side. This partly accounts for the numerous coves. The Doodles, which are features of Carr Naze rather than of the Brigg proper, are also cove-like features. Approximately at water level there is a hard grit, and the Ball Bed lies above it. The balls are hard siliceous masses, possibly concretions, resting in a much softer sandy bed. The waves eat out these balls, especially if they occur near the top of the bed, and once they are freed, they are rolled about, thus forming the hollow or 'bath' near the centre. Another hard band extends as a roof over the hollow. To the south of the Brigg the cliffs are mainly cut in stiff purplish boulder clay, with the Oolites, never much exposed, at the base. The boulder clay is tenacious and hard, and forms vertical cliffs which show the furrowing action of rain extremely well, giving almost a 'bad-land' topography. When the basal Oolites are absent, slips are frequent. The great curve of the bay is very graceful, but despite the relative hardness of the drift deposits erosion is serious, and due, for the most part, to land springs and atmospheric causes. In this and other respects it is interesting to compare these cliffs with the glacial ones of north Norfolk. Several denes or gills run down to the bay. These are fairly well wooded and give considerable variety to the scenery. The beach all round the bay is formed of hard sand. Unfortunately, man has seriously spoiled the natural beauty of the bay by putting up so many inappropriate buildings (**134, 136, 137, 138, 141**).

(10) SPEETON AND FLAMBOROUGH

To the south the boulder clay abuts on to the Speeton Clay, but in this section of Filey Bay the beds are very confused and slipped. This is largely due to the weight of the overlying Chalk which seems to cause bulging of the cliff foot, but the irregularities of the bedding are also explained by the thrusting action of ice in the Glacial period. At Speeton Beck a fault is suspected, but its presence is so far unproved. Nevertheless the beck stands at a significant place: the Speeton Clay occurs to the north, and the White Chalk, here slipped and out of place, to the south. A little farther

[1] C. Fox-Strangways, 'Filey Bay and Brigg', *Proc. Yorks Geol. Poly. Soc.* 13, 1895–9, 338; Kendall and Wroot, *op. cit.*

south the Red Chalk outcrops, and also the lower beds of the White Chalk. The southerly dip still prevails, so that the top of the Lower Chalk falls to sea level between Buckton and Bempton. From that place on to Flamborough the hard, thick Chalk with Flints forms a magnificent range of almost vertical cliffs which show many features of interest. Scale Nab (Bempton) is perforated, while the cliff rocks are much contorted and Staple Nook is due to the comparative softness of the contorted beds. There are arches between Chatterthrow and Little Thornwick, and many ridges and gullies at Great and Little Thornwick (**141**).

Near the headland[1] itself is the ancient Bempton valley just inside the cliffs. The coast cuts with slight obliquity across the line of this valley, which is pre-Glacial and drift filled. Eastwards of Thornwick, cliff erosion accentuates this valley, since all the headlands show the solid rock to a considerable height above sea level, whereas the small inlets cut back into the drift. In the deeper bays the Chalk surface is only a little above high-water mark, and it is thus that the alternation of ridge and inlet is explained. The fissures and caves at Great Thornwick are cut along vertical joints, which often coincide with small faults showing throws measurable only in inches. The arch near the North Sea Landing is also cut along a joint or joint-fault, the first step in the formation of stacks so evident at this place. The King and Queen rocks are two good stacks at Breil Point, while the little bay at Selwicks results from a fault downthrowing to the north. The Adam and Eve pinnacles are striking stacks in this bay, and there are also arches at Kindle Scar: the High Stacks are at the eastern extremity of Flamborough Head (**142** and **143**).

Near this part of the headland there was once a pre-Glacial ravine now filled and hidden by drift. As at Thornwick, this valley is intersected by the cliff, and in one place the sea has cut out a cave in the Chalk near the valley's eastern margin, and, reaching the drift, has washed much of it away. This process has led to the collapse of the roof of the inner part of the cave, and so to the formation of a great hollow, which now functions as a blow-hole. The cave seems to have grown even in recent years and the neighbouring Pigeon Hole was formed in like manner. The whole headland was once drift covered, which is evident east of the South Landing, as well as in other places. On both sides of this inlet the Chalk is that of the *Micraster coranguinum* zone, and at the south end of the Danes' Dyke it is disturbed so that the dyke becomes a natural ravine of Glacial or pre-Glacial age. It is now partly filled with chalky and flinty gravel, and is traversed by a stream. There has also been, however, some artificial alteration. The Chalk cliffs run on to Sewerby where they disappear

[1] Kendall and Wroot, *op. cit.*; G.W. Lamplugh, 'The Geology of Flamborough Head', *Proc. Yorks Geol. Poly. Soc.* 13, 1895–9, 171.

under boulder clay. Finally, it is worth noting that the northern cliffs of Flamborough are all in flinty Chalk, whilst those on the south side are free from flints.[1]

The Jurassic cliffs of Yorkshire are undoubtedly some of the finest in this country. Their setting usually in good open agricultural land, and their height and often sheer faces give great character to them. On the other hand, they lack the infinite variety of gully and crag so impressive on the harder cliffed coasts facing the Atlantic. They are, of course, directly comparable to the Dorset cliffs. There, however, the detail of Lulworth, Stair Hole, Mupe, Worbarrow and other places adds something that the Yorkshire cliffs lack. On the other hand, the fine sweep of Robin Hood's Bay with Fylingdales Moor and the picturesque setting of Staithes have no counterpart in the south. Fine though they are scenically, and interesting as they are geologically, there is nevertheless something a trifle monotonous about these Yorkshire cliffs—although many readers may disagree with this opinion. The Chalk cliffs eastwards from Filey are equal to, or better than, cliffs of the same type elsewhere. Moreover, the details dependent on the boulder-clay covering of the Chalk at Flamborough add something which is necessarily absent from most of our Chalk cliffs. The natural setting of Thornwick and North Sea Landing is beautiful: they must be visited before the way in which man can spoil them is fully realized.

(11) BOULDER CLAY, RAISED BEACHES, AND SUBMERGED FORESTS

Along the Yorkshire coast boulder clay[2] is present in nearly all the bays. It occurs in Filey, Cayton, Scarborough (North and South) Bays, in that north of Scalby, in Robin Hood's, Dunsley, and Runswick Bays, in that between Staithes and Boulby, and in Skinningrove and Marske Bays. The clays are red or brown in colour, very tough, and full of boulders of all sizes. There are also intercalations of sand and gravel, while the nature of the boulders clearly indicates their northern origin. One result of particular significance is that the old valleys originally draining to this coast have been plugged up, so that there is now a marked absence of streams and rivers of any size. The Esk is the only one of importance, and even this stream had to cut a post-Glacial gorge to reach the sea, because its natural outlet in Dunsley Bay was blocked. Again, in Filey Bay the enormous amount of boulder clay has had the effect of turning the original

[1] It is deplorable to see the ruin of the scenery at Flamborough Head due to the completely unsuitable and misplaced bungalows and shacks.

[2] See, e.g. 'East Yorkshire' in Jubilee Volume of Geologists' Association, 1910, p. 592.

Derwent back along its old course. Similarly, the Upper Derwent could have flowed out into Scalby Bay, but instead has had to cut the Forge Valley gorge into the Vale of Pickering.

The bottoms of many of these pre-Glacial valleys are often below sea level, thus showing a depression of the land relative to the sea since that time. In north-eastern Yorkshire there are two boulder clays: the sequence is not so diversified and complex as it is south of Scarborough. The Lower is best seen in Robin Hood's Bay and west of Whitby. It is usually of a dark purplish colour, very compact, and full of stones and boulders, often of large size and well striated. The Upper Clay is redder in colour and more loamy, less calcareous and with a smaller admixture of stones. In north Cleveland it can be seen capping cliffs consisting entirely of clay, for example, in the bay west of Whitby, in Runswick Bay, and in the cliffs west of Saltburn.

There are no undoubted raised beaches between Berwick and Sewerby, although attention has been called to those features on Ross Links, the Cleadon and Fulwell Hills, and Saltburn which resemble raised beaches, and which in the past have been described as such. In each case there is much doubt and it is best not to include them in any general assessment (Chapter XII) of the problems concerning raised beaches.

Various examples of submerged forests are known to occur, as at Hartlepool, Sunderland, and Amble. Little detailed work has yet been attempted, and beyond alluding to the post-Glacial submergence, there is no point in further comment in this chapter. They must naturally fit in with the better-known examples in Lincolnshire, East Anglia, and the Fens, but the precise relationship is not yet clear.

Note. There are many books and papers on the Geology of the Yorkshire Coast, including the following not previously referred to:

J. Phillips, *The Geology of Yorkshire*, pt I, 'The Yorkshire Coast', 3rd ed. (R. Etheridge), 1875.

'A Synopsis of the Jurassic Rocks of Yorkshire' (with Summer Field Meeting), *Proc. Geol. Assoc.* 45, 1934, 247 (various authors).

R.S. Herries, 'The Geology of the Yorkshire Coast between Redcar and Robin Hood's Bay', *Proc. Geol. Assoc.* 19, 1906.

Mems. Geol. Surv. 'Whitby and Scarborough', 2nd ed. 1915, C. Fox-Strangways and G. Barrow.

J.W. Davis, 'Sections in Liassic and Oolitic Rocks', *Proc. Yorks Geol. Poly. Soc.* 12, 1891–4, 170.

W.L. Carter, 'Field Excursion, Bridlington and Filey', *ibid.* p. 421.

G.W. Lamplugh, 'Bridlington-Filey Coast', *ibid.* p. 424.

Rev. E.M. Cole, 'Carr Naze', *ibid.* p. 442.

C. Fox-Strangways, 'Redcar-Scarborough Coast', *ibid.* 13, 1895–9, 248, and 'Hayburn Wyke and Filey', *ibid.* p. 356.

See also J.F. Blake, *Proc. Geol. Assoc.* 12, 1891–2, 115, and *ibid.* p. 207.

Chapter XII

RECENT VERTICAL MOVEMENTS OF THE SHORELINE

(1) INTRODUCTORY

Chapter II contains a brief discussion of the probable effects of the Ice Age on changes of sea level. These changes are revealed in such phenomena as river terraces, raised beaches, and submerged forests, but, at present, a complete understanding of them is lacking. The whole subject is one of great difficulty, and much that has been written on it may mislead because of premature attempts to generalize and correlate which are, as yet, unjustified. Thus, in quite recent years new beaches have been found which are not easily placed in any sequence. For post-Glacial, and more particularly late post-Glacial, changes the technique of pollen analysis has helped greatly,[1] but it has not necessarily simplified the problem. In pre-Glacial times there is a still greater difficulty, namely, that of the many high-level surfaces of erosion. There have been occasional references to them in previous chapters, but even in this one only a summary of some of the views about them will be attempted. Indeed, it is not unfair to say that since investigation is still in an elementary state, personal interpretations are apt to play a somewhat conspicuous role.

Although this book deals specifically with the coasts of England and Wales, it is desirable, in this chapter, to consider raised beaches and related phenomena in Scotland and occasionally in Ireland. This is necessary in order to make the account logical and intelligible. If the current interpretations of high-level erosion surfaces be correct, it would be proper to discuss them first since they are older than any of the raised beaches. Because, however, they do not so directly affect our coasts, and because so much more data are required about them they will be treated in a separate section at the end of this chapter.

It is clear that two major processes have been in operation in Glacial and post-Glacial times: general rises and falls of sea level due to the waxing and waning of the ice sheets, and independent movements of the lands resulting primarily from their reaction to increasing or decreasing weight of ice. Since the ice in these islands was concentrated mainly in the hilly regions of the north and west, it is clear that for this reason alone there will be differences, in so far as vertical movements of the coasts are con-

[1] Pollen analysis will almost certainly be applied sooner or later to more ancient formations—inter-Glacial and possibly pre-Glacial.

cerned, between north and south; consequently correlations between the two regions may be difficult or impossible. Further, since these islands are on the margin of the European land mass, vertical movements in them must not be considered solely as the outcome of our own local ice caps and glaciers. Raised beaches and submerged forests thus result from the combined effects of sea-level (eustatic) movement, and those of the land masses (isostatic).

The treatment in the following pages is roughly chronological, but it would be rash to suppose that this method of approach implies a complete sequence. One other word of warning is necessary. Since land and sea moved independently, a beach of a certain age may not necessarily lie at the same height throughout. In this country we have a habit of speaking of the 100-foot or 25-foot beach, but such terms are often misleading because the heights mentioned only hold good for certain areas; therefore, if too much stress be placed on the 'height-name' it may prove disconcerting to find, for example, that towards the southern part of Britain the 25-foot beach is at or near sea level.

(2) THE *PATELLA* BEACH

In a recent paper Arkell[1] has thrown much new light on the beaches of southern England, and has evolved a more satisfactory interpretation of these features. It is well known that at intervals on the south coast between Sussex and Cornwall, around the Bristol Channel, in southern and western Ireland, and along the Channel coast of France, there is a rock-cut platform about 10 ft. above present high-water mark. This feature and the associated beach deposits have been known hitherto as the pre-Glacial or *Patella* beach. The latter term will be used in this chapter. It was generally assumed that 'pre-Glacial' implied that the beach was formed entirely before the Ice Age began. Before discussing this point the characteristics of the beach and its covering deposits, as described by Wright, will first be considered. In these islands the beach[2] is frequently covered with 'Head' or 'rubble-drift', which is an ancient scree or talus produced by the wearing away of the cliff behind the beach. But it differs from an

[1] W.J. Arkell, 'The Pleistocene Deposits at Trebetherick Point, North Cornwall: Their interpretation and correlation', *Proc. Geol. Assoc.* 54, 1943, 141.

[2] The word 'beach', strictly speaking, refers to the superficial deposits of sand, shingle, etc. resting on the cut platform. Naturally such loose deposits are easily removed. In this discussion, therefore, the term 'raised beach' includes the rock platform as well as any superficial deposits that may remain. The actual nature of the occurrence will, it is hoped, be clear from the context. In order to avoid misunderstanding, the name 'pre-Glacial beach' for the 10-foot platform of southern England, South Wales, and South Ireland will not be used. It will be called the *Patella* beach.

ordinary modern scree because it is washed out to a far greater distance than is usual from the cliff foot, and also because the contained fragments lie at a much lower angle than is found in a modern scree. The Head also contains locally streaks or wisps of loam which give a roughly stratified appearance, and there is no doubt that it was formed under far more severe, tundra-like, climatic conditions than those which prevail to-day (**46**).

This beach is absent from the east coast between Kent and just south of Flamborough Head. But at Sewerby (see p. 472) there is the famous

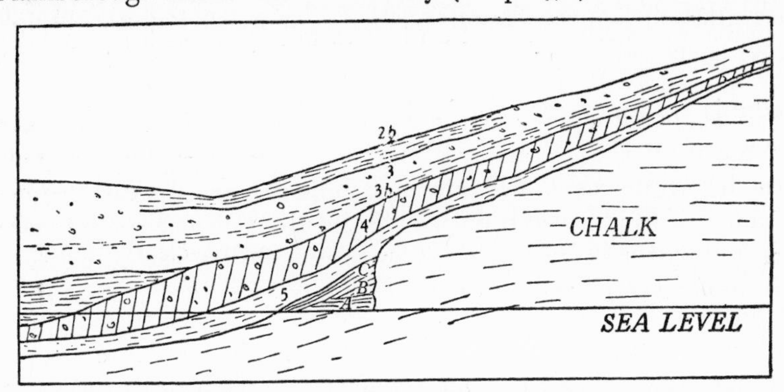

Fig. 103. The Sewerby Section. *A* = Old Beach, *B* = Land Wash, *C* = Blown Sand, 5 = Chalk Rubble, 4 = Basement Clay, 3 = Upper Clay, 3*b* = Intermediate Stratified Series, 2*b* = Sewerby Gravel (after W. B. Wright, *op. cit.*)

section first explained by Lamplugh[1] (see Fig. 103). Beach deposits are banked up against an old cliff, and overlying all are beds of clay and gravel: that marked 5 in the section apparently possesses all the characteristics of the Head of the south of England, and can doubtless be correlated with it. Since there is also no doubt of the glacial character of beds 3 and 4, which are subdivisible into an Upper Boulder Clay, an intermediate stratified series, and a Lower Boulder Clay, it follows that the beach is 'pre-Glacial' in the sense that it is earlier than the earliest glacial bed resting on it. This conclusion is confirmed by somewhat similar sections found in Gower and in southern Ireland, although there is no need to describe these in detail: it is enough to cite the papers listed in the footnote.[2] The beach

must obviously have been raised beyond the reach of the waves some time before the advent of the land-ice, or the lower head could not have accumulated on it. The oc-

[1] G.W. Lamplugh, 'The Drifts of Flamborough Head', *Quart. Journ. Geol. Soc.* 47, 1891, 384.

[2] W.B. Wright and H.B. Muff, *Proc. Roy. Dublin Soc.* 10, 1904, 250; R.H. Tiddeman, *Rept. Brit. Assoc.* Bradford, 1900.

currence of blown sand without any great admixture of head close down upon the wave-cut rock platform indicates that this elevation commenced even before the conditions were suitable for the formation of head. The 'land-wash' at Sewerby points to a similar conclusion. Thus we see that between the elevation that put a stop to the cutting of the pre-Glacial shore platform and the invasion of the country by ice, there was time for the production of land-wash, blown sand, and the head. The first two deposits are of a trifling importance as an indication of a lapse of time. A year or two is ample to account for them. With regard to the head, however, one cannot be so sure. It is very hard to put a figure to it. It may represent a very long period of time, and on the other hand it may have been formed very rapidly. It certainly requires more than a couple of years, and I am inclined to suspect that it represents many hundreds, if not thousands.[1]

There is also another line of evidence which assuredly points to the formation of the beach during a cold climatic period. It contains in many places, of which Sewerby and Sandgatte are examples many far-travelled erratics that could only have arrived on these coasts as a result of incorporation in blocks of ice and icebergs. Their presence clearly implies that the Ice Age had begun, although the ice may not at this stage have stretched southwards to its greatest extent. The study of these erratics has, moreover, brought up another point of great interest. Their nature and places of origin clearly indicate that they were carried from east to west along the Channel, in the opposite direction, therefore, from any general drift of material at the present day. In other words, as Harmer[2] first showed, the prevailing winds in the Channel were then easterly and not westerly, a phenomenon which would be the probable result of the growing ice caps. The winds blowing out from them would have been northerly and easterly, and the westerly winds associated with the southern part of depressions would have been pushed farther southwards. Harmer originally evolved this theory to account for the occurrence of the Pliocene Crags of East Anglia. To-day relatively little shell sand is cast up on the shores of Norfolk and Suffolk, but vast masses accumulate on the coasts of Holland: perhaps, therefore, during the accumulation of the Pliocene Crags, ice caps existed to the north. As Wright remarks, these arguments are all rather speculative, and assume that the beach is definitely pre-Glacial. 'A preceding glaciation would perfectly well explain the distribution of erratics which is essentially the same as that of the overlying drift.'[3]

The last sentence suggests that Wright had in mind the possibility of an inter-Glacial date for the *Patella* beach rather than an entirely pre-Glacial one. It is at this point that Arkell's work may be introduced.

[1] W.B. Wright, *The Quaternary Ice Age*, 1937, p. 117.

[2] F.W. Harmer, *Quart. Journ. Geol. Soc.* 57, 1901, 405.

[3] Wright, *op. cit.* p. 119.

Arkell examined with great care the area at Trebetherick Point, at the mouth of the river Camel, and established the following sequence of deposits in that locality: (1) the *Patella* beach and 10-foot platform, the oldest Pleistocene deposit in the district; (2) false-bedded sand and sand rock, probably of aeolian origin, which show clearly that the period of formation of the raised beach was followed by recession of the sea and the blowing up of sands; (3) Head, which implies peri-Glacial conditions and which formed probably during a stage of low sea level; (4) the Trebetherick boulder gravel which is *not* covered by Head, and which is thought to be a beach built during a short period of high sea level; (5) a pebbly solifluction deposit, which is really a poorly developed Head previously unrecognized in Cornwall; and (6) blown sand of Holocene age, which contains *Neritoides* and which is probably associated with a low sea level. The submerged forest of Mesolithic date in Daymer Bay is probably a little older than the sand.

These beaches near Trebetherick Point have not been described by previous writers, and it is important at the outset to emphasize the very well preserved nature of the 'pre-Glacial' beach. Its freshness is extremely difficult to explain here and elsewhere if it be assumed that the beach was formed before any glaciation occurred in these islands. The oldest deposit resting on it is the Head of Upper Acheulian or Levalloisian age at latest, and which thus belongs to the Riss[1] glaciation. Both Riss and Würm glaciations have left widespread traces over the area, but there is not the least evidence of any older advance of the ice. Hence, it seems likely that the beach dates from some time, probably early, in the Mindel-Riss inter-Glacial. The meagreness of the associated fauna and the presense of erratics noted by earlier observers both imply a cold climate, and Arkell thinks it most likely that the conditions in which the beach was formed were those of 'waning' cold. As noted above, the blown sand containing shells which follows the beach suggests strongly a rise in temperature and a fall in sea level.

[1] W.J. Arkell, *op. cit.* p. 157. When Arkell read his paper the nomenclature Gunz, Mindel, Riss, and Würm was used. In the printed paper, *op. cit.*, the following terms are used except in fig. 15:

Berrocian or Northern Drift Glaciation,
Catuvellaunian or Eastern Drift Glaciation,
Cornovian or Irish Sea Glaciation,
Cymrian Glaciation, and
Cymrian II or Clwydian Glaciation.

The first four may be taken as roughly equivalent of Gunz, Mindel, Riss, and Würm, but 'For our present purposes it is essential to find a classification and nomenclature free from...ambiguities and independent of assumed but unproved correlations'.

Before the publication of Arkell's paper no satisfactory correlation of beaches at other levels on the Bristol Channel and English Channel coasts was available. In Barnstaple Bay there is a 10-foot beach, but it does not appear to be the *Patella* beach. Shingle associated with it rests on the 10-foot platform, but also extends upwards to 60 ft. O.D. It is likely that it belongs to the same inter-Glacial period, but 'the only explanation seems to be a second oscillation within the Boyn Hill interglacial, during which the materials of the old 10-foot beach which must have existed at Barnstaple Bay were reworked and redeposited with a warm fauna, and finally covered with fresh blown sand. The heavy erratics...remained on the platform.'[1] The remains of a beach at 65 ft. at Mousehole in Cornwall may also be associated with the rise of sea level to the Upper Boyn Hill stage.

In Sussex there are beaches at 120 ft. at Goodwood and at 120 and 135 ft. at Slindon, while certain terraces at about 150 ft. in the Stour, Avon, Test, and at St Helen's (I.O.W.) appear to correspond with the Goodwood beach. The higher beach at Slindon is regarded, on rather slight evidence, as being later than the one at Goodwood. The sea level obviously fell after these high beaches were built, but the exact sequence of events during the fall is somewhat uncertain. Arkell suggests that both the 100-foot terrace and the 30-foot raised beach at Brighton belong to this episode, the first being the earlier in time.

No explanation is given about the possibility that, if the Goodwood and Slindon beaches follow the *Patella* beach, the high sea level would probably have uncovered any deposits overlying the *Patella* beach.

There is in Islay, Colonsay, Mull, and the Treshnish Islands a 'pre-Glacial' shoreline between 90 and 120 feet. It is not at present possible to relate this to the other beaches, and it should be carefully distinguished from the 100-foot beach of Scotland and the *Patella* (often referred to as 'pre-Glacial') beach of Southern England and Wales.

(3) THE BEACHES IN THE GOWER PENINSULA

Before examining further the beaches on the south coast, it is worth turning for a moment to the Gower peninsula in South Wales, where George[2] suggested the sequence of events to be the following: the oldest is the *Patella* beach identical with the 'pre-Glacial' beach already discussed: in Gower it can be correlated with a cold period. In Minchin Hole this beach is overlain by an ossiferous breccia containing a warm fauna,

[1] W.J. Arkell, *op. cit.*

[2] T.N. George, 'The Quaternary Beaches of Gower', *Proc. Geol. Assoc.* 43, 1932, 292.

and immediately above the breccia is a beach deposit containing *Neritoides*. The deposit is only little more recent than the breccia, and *Rhinoceros hemitoechus* occurs both below and above the beach. George refers the subsidence which accounts for the beach to the genial episode in the early part of the Mousterian. If this be the case, it seems that the succeeding glacial gravels which belong to the older drift are of Middle or Upper Mousterian age, or even younger. Between the formation of the older and newer drift are the traces of human occupation of Paviland cave in Aurignacian and early Solutrean times, and this is strongly suggestive of the view that the area was ice free from the Mousterian until near the end of the Palaeolithic period. Recent work shows that the last major glaciation of southern Britain was approximately of Magdalenian age, and George considers that all the available evidence points to the older drift of Wales being the equivalent of the Upper Chalky Boulder Clay[1] of East Anglia. Next in order, in the Gower peninsula, lie the Heatherslade beach and platform, coinciding with the modern beach. George suggests that the cementing of the Heatherslade deposits occurred during the regression of which the submerged forest beds are a record. It may be Early Neolithic[2] in age. In the Late Neolithic period (and possibly still more recently) the submerged-forest beds were formed: these probably correlate with newer forest beds elsewhere. Finally, there is the modern beach platform. This summary omits a well-marked feature in certain parts of Gower—the 50-foot platform which is usually cut in glacial drift at Rhossili Bay, Eastern Slade, and Overton Mere. It is possible that the final form of this platform is due to marine erosion[3] (**78**).

The geological surveyors distinguished in Gower the newer Head resting on the older drift, which in turn overlies the older Head, itself the equivalent of the Cornish Head. George puts the older Head and older drift in the Middle and Upper Mousterian in order that the sequence shall agree in his scheme with the position of the *Neritoides* beach. His evidence, however, is based largely on the equivalence of two beds of blown sand;[4] he maintains, moreover, that the *Neritoides* beach never lies in direct contact with glacial drift, but is, in fact, often separated from the *Patella* beach by Head, which, in Arkell's view is the lower Head, and

[1] In the sense used by Boswell (*Proc. Geol. Assoc.* 42, 1931, 108) and equivalent to the third glaciation of East Anglia—the Little Eastern of Solomon.

[2] George here is presumably referring only to the forest beds of South Wales. Many forest beds are Mesolithic. *Neolithic is used as in George's paper.*

[3] It should be noted that even before the *Patella* beach was formed there was a period of severe cliff erosion. The present-day cliffs are mainly the result of this.

[4] In view of his own remarks in his paper, a correlation depending largely on this factor would appear to be open to question.

any Head *on* the *Neritoides* beach is the newer Head. Arguing from the close similarity of the section described by George at Caswell Bay to those at Trebetherick Point, Fistral Bay, and Godrevy, Arkell tends to disagree with George's view that there is a gap in the sequence of deposits between the *Patella* beach and the blown sand on it. Briefly, therefore, Arkell would place the *Neritoides* beach of Gower and the 25-foot beach found in the Channel Islands in the Boyn Hill (Mindel-Riss) inter-Glacial.

The boulder gravel at Trebetherick has been correlated by Dewey with clay beds and striated erratics (? boulder clay) at Croyde and Fremington: at the latter place they rest on the raised beach and together with the Head reach 65 ft. O.D. The boulder gravel is probably rather later than the Barnstaple stage and may be the equivalent of the 50-foot platform in Gower, which is more recent than the older drift: in the peninsula it is only found in drift and so presumably is evidence of a brief still-stand of sea level. The raised beaches of Portland, Tor Bay, and Hallsands also probably belong to this same Riss-Würm (Taplow) inter-Glacial. The Portland beach was clearly formed before a fall of sea level, since it is covered with a stratified loam containing snail shells, and clearly has not been submerged. It is possibly the equivalent of the 50-foot beach recognized by Palmer and Cooke in the Portsmouth district.

The upper solifluction pebble bed at Trebetherick was formed probably during the Würm glaciation, and is the Cornish equivalent of the newer drift of South Wales and of the upper Head of Gower. The amount of Head forming in Devon and Cornwall at this time was limited, but in the Chalk areas farther east, Coombe Rock was produced in greater abundance. Palmer and Cooke recognized three Coombe Rocks near Portsmouth, the oldest overlay the 100-foot beach, and of the two younger the one contained Mousterian implements and the other Aurignacian.

Arkell's findings hardly simplify the problems of the south- and west-coast beaches, but may make possible their arrangement into an orderly sequence. Together with their drift covers, they embrace two Glacial and two inter-Glacial periods, and the 10-foot *Patella* beach, the oldest formation, goes back to the final stages of the second or Mindel Ice Age. The position in which the *Patella* beach often lies strongly suggests that north-easterly winds were then prevalent, whereas the Goodwood, Tor Bay, Start, and Mousehole beaches indicate the south-westerly or south-easterly winds of an oncoming glaciation.

Hitherto Arkell's views have been followed in some detail, but it is opportune now to refer to the work of Movius.[1] He maintains that before

[1] *The Irish Stone Age*, Cambridge, 1942.

the advent of the Upper Chalky Boulder Clay[1] of East Anglia, the coastline of these islands was very much the same as it is to-day. This follows from his two assumptions (*a*) that the *Patella* beach dates from the third, or Riss-Würm, inter-Glacial period, and (*b*) that at this time the Mousterian culture had found its way into western Europe. That the age of the beach is third inter-Glacial 'is established by the fact that everywhere it is overlain by rubble head or ancient talus, resulting from the degradation of the old cliff, which contains derived implements of Middle Palaeolithic type, and the overlying deposits are composed of either drifts or solifluction layers referable to the period of the Würm ice-sheets of the Continent'. This conclusion, however, is based on the findings of Dewey and Wright, and not on first-hand field work: accordingly Arkell's estimation of the date of the *Patella* beach seems the more probable. Moreover, the opinions of W. B. Wright are also open to reinterpretation, and especially his contention that the *Leda myalis* beds of the Norfolk coast may be of the same age as the beach. If, as is generally conceded, these beds occur below the North Sea Drift, which is usually accepted as representing the first glaciation of East Anglia, it is clear that they must be much earlier than the *Patella* beach. In the Gower sequence Movius follows George very largely and writes: 'The *Neritoides* beach is overlain by Old Drift, which indicates that its age is probably Third inter-Glacial, although it may be older.'

(4) RIVER TERRACES AND RAISED BEACHES: A GENERAL DISCUSSION

Before discussing some of the problems associated with later, and often higher, beaches, some of the general points, not only of beach formation but also of the closely related subject of river terraces, demand examination. It is clear that, if the sea were to rise 100 ft. relative to the land, all that part of the land below the 100-foot contour would be flooded. Were the change of level maintained for any length of time, deposits of shallow-water clays, silts, and sands would form on this lower ground, and they would contain the remains of animals. When the sea level fell, these deposits would be eroded and only patches would eventually survive. But they would be enough to indicate to an investigator that such a movement of sea level relative to the land had taken place. In the estuary of a river, deposits would likewise accumulate. With the rise in water level, and with the resultant lowering of the gradient of the river, material would be deposited and, given sufficient time and supply, would eventually fill up the estuary to the new level. Should the sea level fall again later, the river

[1] Or Little Eastern Glaciation.

would easily excavate this material, and most of it, perhaps the whole, would be lost. But fragments in sheltered places, or for some local reason, would be left, and they would be sufficient to tell the physiographer of the oscillation. It is logical to suppose that if a series of such oscillations had taken place there might now be found traces of all. But it is also obvious that, unless each change in level were smaller than the preceding one, the whole series might be buried by a later major oscillation. Even this brief discussion will suffice to show that the study of terraces is extremely complicated, since only fragments are likely to remain at different heights and in different places. A complete 'staircase' of terraces at any place may be found, but such sequences are very rare. Higher up the river, above its tidal section, the gradual increase of gradient, the lateral as well as the vertical corrasion of the river itself, and the effects of hard and soft rocks add further complications; these, however, are outside the scope of this chapter.

On the shoreline itself other problems occur. On paper it is easy to state that a beach is so many feet above sea level, or high-tide mark, or Ordnance Datum, but in the field it is more difficult. In general there is much to be said for giving heights above Ordnance Datum, but for reasons that are all too obvious this suggestion is not put into practice even in this chapter. Should reckoning from sea level be used, investigators should be precise as to which level is meant. Mean sea level can be obtained carefully by long-term survey methods, and in our Ordnance maps lines are drawn representing the high and low water of ordinary spring tides. From a good deal of experience on flat coasts the writer can state with some emphasis that they are of little value, for the very sufficient reason that the surface of a beach is inconstant, if for no other. Hence, even if they can be guaranteed accurate in the first place (and this is doubtful), precision cannot be assumed months or years later. In actual practice observations in the field are inevitably made in varying conditions of weather and tide, and unless great care is taken it is extremely difficult either to discount or to give proper weight to wave action. Moreover, the investigator must decide whether to measure the height of a cut platform or that of the deposit on it. If the former, what part of the platform should be measured? It is likely to possess a lesser or greater slope seawards, and it will vary a little in height in different types of rocks, and for many other reasons. If beach deposits be measured, the observer must remember that they may have been thrown by storms to a considerable height and thus be quite unrepresentative of the true beach level. Exposure also plays a great part: deposits in exposed places are likely to be at a higher level than those in sheltered places. These difficulties met with in field work can easily be expanded: all that needs stress here is that precision should be aimed at, but it is extremely difficult to attain.[1]

Finally, the effects of subsequent tilting due to isostatic movements must be kept in mind, because they are often responsible for a given beach being at sea level in some places and 25 ft. or more higher in others.

(5) THE 100-FOOT BEACH OF SCOTLAND

Let us now turn to the 100-foot beach of north-western Scotland (Fig. 104).[2] The evidence for this submergence appears both in actual beach deposits and platform, and also in relatively deep-water clays. The last were laid down, as outlined above, on many parts of the lower ground. From what is known as yet, the beach extends over but a limited area which includes most of Scotland north of the Southern Uplands. Many of the shorelines have since been eroded away, and to-day the best remaining examples of the 100-foot beach are in the western islands and parts of the mainland coast north of Skye. There it may be represented by great shingle ridges, by a well-marked notch, or by both. Everything points to there having been an enormous amount of material available to the waves. There is one very interesting characteristic of the beach, namely that it is always absent from the upper part of the Highland lochs, whereas lower beaches may be present in such positions. The only valid conclusion to be drawn from this fact is that at that time the upper reaches of the lochs were occupied by local glaciers. This feature is exemplified in the lochs just to the north of Kyle of Lochalsh. In some places, for example at the narrows of Loch Carron, lie huge gravel embankments: these are not beaches but great fans of outwash gravels. But in the south, on the lower ground, the submergence is represented by clays still preserved at such places as Errol, Portobello, Paisley (where they are still worked for bricks), Renfrew, Kilchattan on Bute, and Benderloch in Argyll. The Paisley section has its surface about 40 ft. above present sea level. It consists of a finely laminated tenacious mud, containing many Arctic shells in the position of growth. Such features alone prove its Glacial age, and are consistent with the absence of the beach from the higher parts of the lochs. Further evidence of its Glacial age is found in great blocks of rock, often completely or partly covered with barnacles, which obviously dropped into the accumulating soft clays from rafts of floating ice when these had become too small to carry the weight of the

[1] For a general discussion concerning raised-beach and related problems see J.A. Steers, *The Unstable Earth*, 1937, Chapter v.

[2] See Wright, *op. cit.* Chapter xx; and also H.L. Movius, *The Irish Stone Age*, Appendix V, 1942.

rock masses any longer. The absence of the beach from the upper parts of the lochs presumably also means that at that time Scotland was undergoing

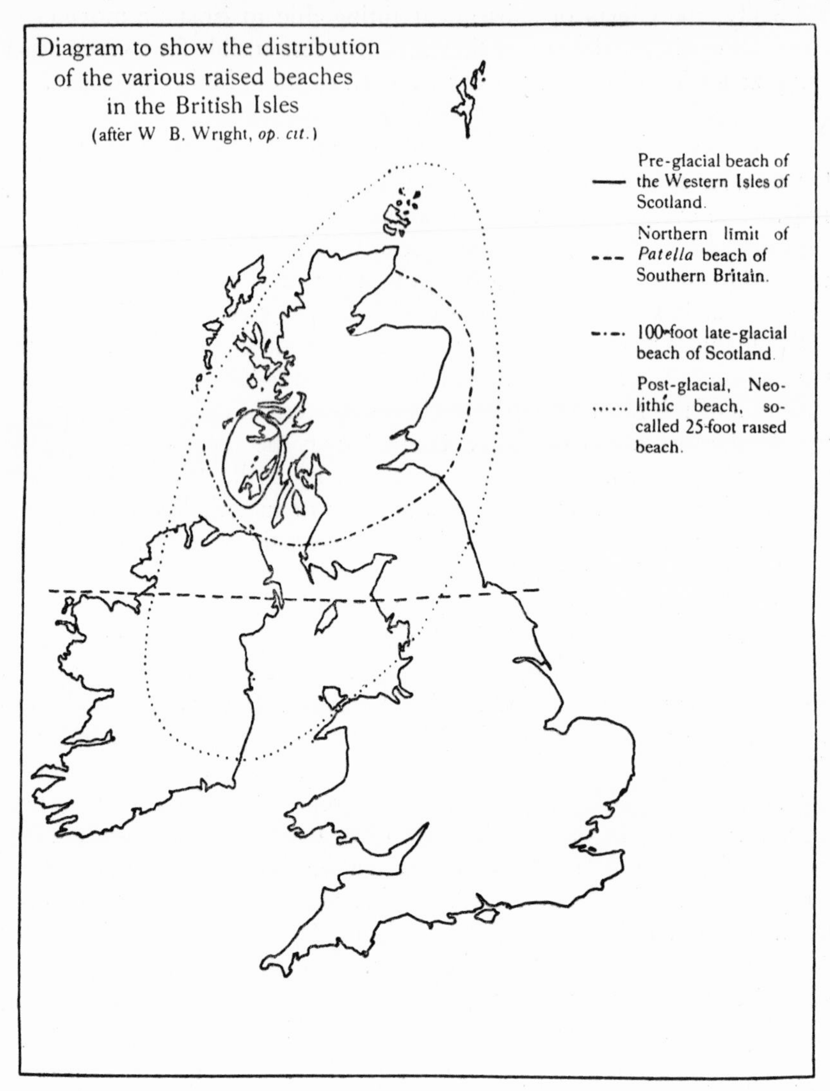

Fig. 104

what is often called the moraine glaciation. It is not quite clear whether the maximum extent of the existing glaciers was contemporaneous with the beach or not, but other evidence clearly shows that the sea had fallen lower than the 100-foot level before the glaciers finally retreated. In Mull, for example, the bottoms of various glens, quite low enough to have contained the beach, are covered by the moraines of local glaciers. In Glen Forsa there are many small eskers and gravel fans of the nature of outwash gravels. These are all below the 100-foot level, and seawards are covered by the deposits of the 25-foot beach: they are definitely subaerial formations. Clearly the presence of these and similar gravels in other glens means that the sea had fallen well below the 100-foot mark before the glaciers made their final retreat.

Incidentally the comparatively restricted development of the 100-foot beach and related deposits implies a local isostatic movement of the land rather than a general, eustatic, change in sea level. There are no traces of this beach in England, Wales, or Ireland.

(6) THE SO-CALLED '50-FOOT' BEACH

In Scotland there are many beaches at levels between about 60 and 35 ft. The average is about 50 ft., and the beach is often termed the 50-foot beach. It is not a precise name in any respect, nor is it by any means certain that what is usually called the 50-foot shoreline in various localities is always the same. This lower beach has been said to penetrate into the upper parts of certain lochs from which the 100-foot beach is absent.

> Now it is a difficult matter, without seeing all the evidence, to make anything like a definite statement on the question, but the writer[1] is in considerable doubt as to whether the terraces ascribed to the 50-foot beach in this inner area are raised beaches at all. They seem to have rather the character of glacial outwash fans, having gently sloping surfaces dropping from heights of 40, 50, or 60 feet down to varying levels sometimes as low as 12 or 15 feet above high-water mark. The lower limit is in all cases determined by the amount of subsequent cliff cutting mostly effected during the much later early Neolithic submergence....It seems desirable at present to reserve judgement on this question of the penetration of the 50-foot sea into areas from which the 100-foot is excluded.

However, the undoubted beaches associated with the 50-foot level are, like those of the 100-foot, also associated with deep-water clays containing the shells and remains of animals living in an Arctic climate. Whilst then we are still dealing with an event of late glacial time, the evidence available is not such as to let us say that the general series of 50-foot beaches represents a fall of sea level in stages from the 100-foot level, or that they are not due to a rise of sea level after it had fallen some way below present sea level after the formation of the 100-foot beach.[2]

[1] W.B. Wright, *First Rept. on Pliocene and Pleistocene Terraces*, Int. Geogl. Union, 1928, p. 102.

[2] Movius, *op. cit.*: 'Thus, as the ice receded from the fjords, beaches marking stages in the isostatic uplift of the land were formed at lower levels. These are later, as they

(7) THE SUBMERGED FORESTS AND PEAT BEDS

Whatever may be the true significance of the so-called 50-foot beach, still less is known of the conditions which followed it. At some later date, possibly considerably later, instead of raised beaches being formed, there was a low sea level and a temperate climate during which the submerged forests grew. Many accounts exist of these forests and of their time relations to the beaches. There are no longer grounds for thinking of them all as earlier than the 25-foot 'Neolithic' beach, and it is also necessary to start at a much lower level for the deepest beds than has usually been recognized. Before discussing them it will be helpful to insert a generalized table showing the probable dates of Mesolithic and later ages. It will not do to think of the 'Neolithic submergence' as the last major, and comparatively simple, movement of land relative to the sea. Moreover, it is as well to remember that the many forest beds at different levels probably mean halts or recessions in a general transgression of the sea at least up to the time when the 25-foot beach was formed: the beach itself is, of course, not necessarily everywhere of the same height or age, and it is only found in northern Britain.

Roman Conquest		A.D. 43
Iron Age	C. Belgae, Two waves	*c.* 75 and 50 B.C.
	B. Middle La Tène (Marnian overlords)	*c.* 300/250 B.C.
	A. Late Hallstadt-La Tène Groups	*c.* 500 B.C.
Bronze Age	Carp's-tongue Sword people (Millennium of integration)	*c.* 750 B.C.
	Wessex-Breton people	*c.* 1750 B.C.
	The Beaker Invasions	*c.* 1900/1800 B.C.
Neolithic	Baltic (Peterborough) (Windmill Hill) Western and Megalithic	*c.* 2500/2200 B.C.
Mesolithic	Baltic (Maglemosian)	
	Western (Tardenoisian)	
Upper Palaeolithic		

(after Grahame Clark, *Prehistoric England*)

occur below the others and penetrate the fjords, ice-free at this period, where the higher beaches are absent. *They are also later than the 50- and 75-foot beaches formed in the peripheral parts of the area, which seem to correspond to the highest beaches farther north.* (The italics are mine, J.A.S.) Below the 50-foot level data on the recovery of the land, as recorded by strand-lines, are lacking; this may indicate a sudden uplift or extensive Post-Glacial marine erosion.' This bears out Wright's point concerning the uncertainty of the age of many of the so-called 50-foot beaches in Scotland.

The submerged-forest beds are numerous around our shores. They vary a good deal in composition, but fen peats preponderate, and there occur often remains of plants which are now characteristic of salt marshes. These are sometimes followed by *Phragmites* peat, and that again by sedge or reed peat, while above all may lie a wood peat containing alder and willow stumps. Such a sequence implies a change from salt-marsh conditions to those of fen scrub. Other comparable sequences are found, and indeed one like that just outlined may be followed by peats which suggest a return to marine conditions. Knowledge of these beds has increased enormously of recent years on account of the application to them of the methods of peat investigation and of pollen analysis. In this chapter the treatment is inevitably limited mainly to the British Isles, but a full account of submerged peat beds and raised beaches should, needless to say, take full account of Scandinavia and especially of the excellent work accomplished by Scandinavian investigators.

The lowest deposits of which we have record come up in dredges from the bottom of the North Sea, many parts of which are covered with a hard brown fissile peat. It was Clement Reid who first proved that this peat had formed in fresh-water fen-like conditions during a temperate climate, and recent investigators[1] have applied the technique of pollen analysis to many specimens brought up from depths between 100 and 170 ft. below sea level. These lower samples contain only birch and pine pollen with a little hazel, and therefore presumably belong to zone IV or V (p. 440). Specimens from shallower water close to the British coasts contain pollen grains which imply alder and mixed oak forests, and are referable to zone VII. These higher beds may be continuous with those underlying the marshes fringing the North Sea. The deepest peat (−23·5 ft. O.D.) found in a boring at St Germans, near King's Lynn, certainly belongs to zone VII. Thus there are grounds for believing that the growth of the North Sea to its present extent took place between zones V and VII. Reference to Fig. 106 makes clear that this period includes Mesolithic, Neolithic, and Bronze Ages.

Unfortunately pollen analyses of peat beds are still limited in number, and many beds, especially those exposed from time to time in dock and other artificial sections, have never been studied. Some reference to them, however, is inevitable, if only to enable comment on their location, and

[1] Outstandingly, H. Godwin. In this account much use has been made of his writings, especially 'Studies of the Post-Glacial History of British Vegetation', I and II, *Phil. Trans. Roy. Soc.* B, 229, 1938, 323 (with M.H. Clifford); III and IV, *ibid.* 230, 1940, 239; and 'Pollen Analysis and Quaternary Geology', *Proc. Geol. Assoc.* 52, 1941, 328. See also, *Submerged Forests*, Clement Reid, 1913.

also an adequate survey of the general sequence of deposits, and, within the limits of particular sections, the minimum amount of change of level that has taken place. A well-known section was described by Austen[1] in the Pentewan valley stream-tin works in Cornwall. At 50 ft. below present high-tide level are stumps of trees rooted in subaerial beds, and resting on water-worn materials. Above them lie some 30 ft. of marine beds, covered again by 20 ft. of estuarine beds which contain the skeleton of a whale. Other sections of a like nature occur elsewhere in Cornwall, while dock sections illustrate the same kind of features. As another instance, the section described by Strahan[2] at Barry in South Wales is interesting:

12. Blown Sand.
11. *Scrobicularia* Clay.
10. Sand and Gravel—base of *Scrobicularia* Clay. Strong line of erosion.
9. Blue Silt with many sedges, 1–6 ft.
8. The Upper Peat Bed, 1–2 ft. thick, and 4 ft. below O.D.
7. Blue Silty Clay with many sedges.
6. The Second Peat, mainly sedges, 3–8 in. thick.

CHRONOLOGY (APPROX)	PHASES OF FEN HISTORY	ARCHAEOLOGICAL HORIZONS	MAJOR FEATURES OF THE PERIODS.	CONDITIONS AT SITES: MARGINAL	MIDDLE	SEAWARD	PREVALENT STATE	LAND LEVEL RELATIVE TO SEA	CLIMATIC EVIDENCE IN FENS.	CLIMATIC PERIODS	RECURRENCE HORIZONS (DRYNESS IN S. SWEDEN) GRANLUND.	FOREST ZONE
1900 1.500 1,000	DRAINAGE & MARSHLAND RECLAMATION	HISTORIC		PEAT	GROWTH	SILT LOCAL PEAT	WET (locally dry)	? LOW ? HIGH	WET	RECENT SUB-ATLANTIC (cold & wet)	X RY I	VIII
500 A.D. 0	UPPER SILT & MERES	ROMANO-BRITISH	MARINE TRANSGRESSION	? MERES	RODDON SILTS	UPPER SILTS	WET	LOW	WET (LAKE MARL)		X RY II	
B.C. 500 ?1,500	UPPER PEAT	IRON AGE (LA TÈNE, HALSTATT) BRONZE AGE EARLY BRONZE AGE Beaker 'A'	stagnation phase Brush wood on raised bogs – Marginal pine woods & tendency to form acidic (sphagnum) peat [Wet phase] [Dry phase · local woods]	OLD RUNS CUT PINE & YEW IN FEN-WOODS - SOME ACIDIC PEAT	RODDON-CHANNELS CUT FEN-WOOD ↑ BRUSH-WOOD ↑ FEN	FEN-WOOD SILT	(DRY) DRY	HIGH	WET (IRON AGE IN FENS) DRY (RAISED BOGS)	SUB-BOREAL (warm, dry continental)	WETTER ↑ X RY III ? RY IV	VII-VIII VII d
?1,500 ?2,000	FEN CLAY		shallow brackish lagoons EXTENSIVE BUT SHALLOW MARINE TRANSGRESSION	FEN-WOODS	FEN-CLAY	SILT	WET	LOW	WARM (TILIA)		DRIER ↑	VII c
?2,000 3.500	LOWER PEAT	(ESSEX) Beaker 'B' NEOLITHIC NEOLITHIC A	Dry Fen woods alder brush wood peat-formation becoming general in Fens peat forming only in river channels etc	EROSION CHANNELS CUT DRY OAK FEN-WOODS			DRY WET	HIGH ? LOW	WARM (MOLLUSCA) ? DRY (BLOWN SAND) INCREASING WETNESS (ALDER)	ATLANTIC (warm & wet)	? RY V ↓ WETTER	VII b VII a
5.500		LATE TARDENOISIAN MESOLITHIC							? DRY (BLOWN SAND) DRY (NO ALDER)	BOREAL (warm dry)		VI c VI b VI a
7.500			peat at JUDY HARD & on DOGGER & LEMAN & OWER BANKS – N SEA					V. HIGH SAY 200' (61 M.) HIGHER THAN NOW		PRE-BOREAL (cold)		V IV

Fig. 105. Correlation table of Fenland deposits (from Godwin, *Proc. Roy. Soc.*, 230, 1940, 284)

[1] R.A.C. Austen, 'Superficial Accumulations of the Coasts of the English Channel', *Quart. Journ. Geol. Soc.* 7, 1851, 118.

[2] A. Strahan, 'Submerged Land Surfaces at Barry...', *ibid.* 52, 1896, 474.

5. Blue Silty Clay.
4. The Third Peat: large logs and stools with roots in place, 8 in. thick and 20 ft. below O.D.
3. Blue Silty Clay with reeds, willow leaves, and fresh-water shells, 12 ft. thick.
2. The Fourth Peat, trees, and roots in place, land-shells, 3–4 in. thick.
1. Old soil, roots, and land shells, 35 ft. below O.D., resting on rock.

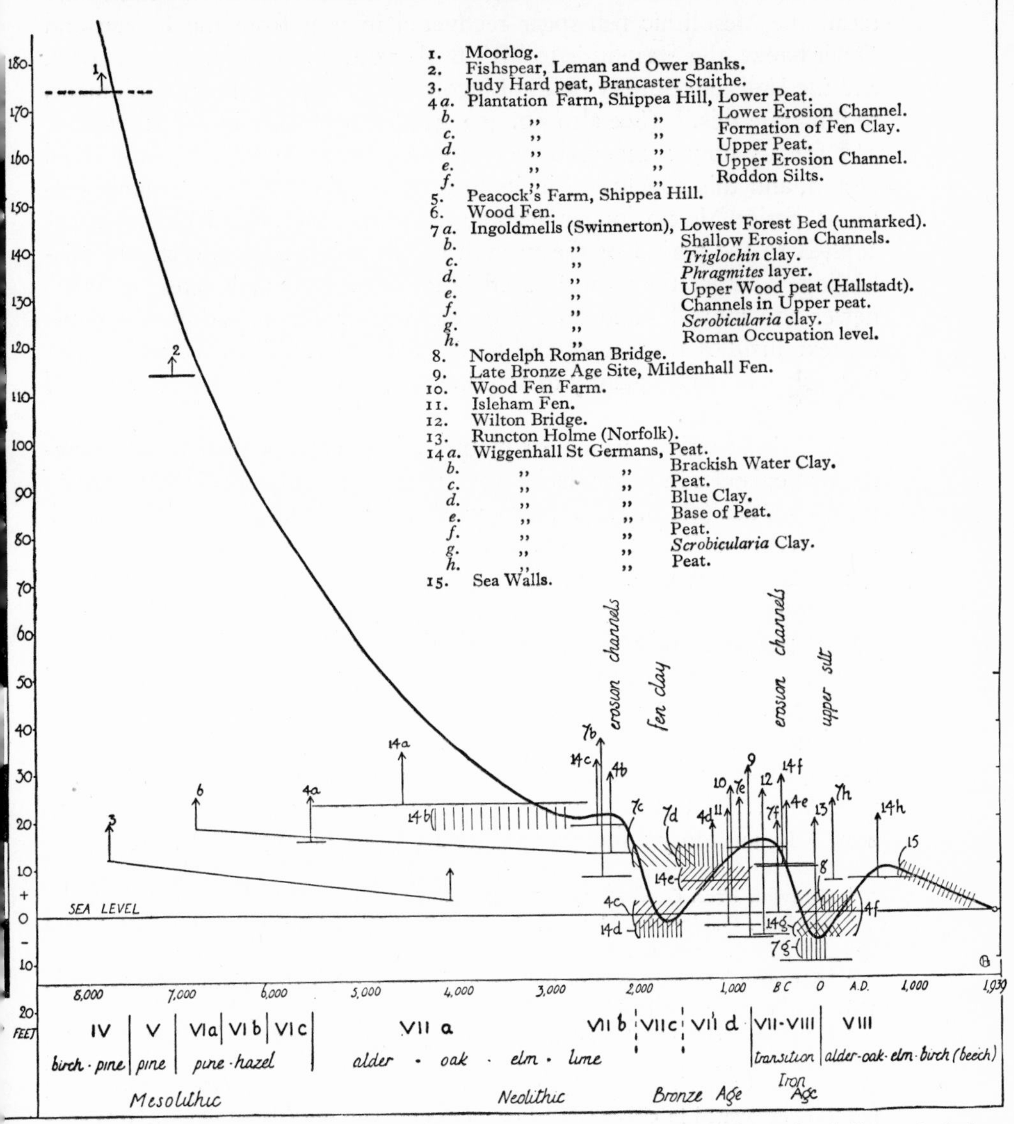

Fig. 106. Post-Glacial changes of relative land and sea level in the English Fenland (from Godwin, *Proc. Roy. Soc.*, 230, 1940, 287). [For explanation of symbols, see also p. 440]

This is typical of many such sections, and the occurrence of several beds of peat is interesting. Although these particular peats have not been re-investigated in recent years, it is probably true to include them and others in the statement 'the scattered outcrops of peat may belong to any part of the pollen zonation from IV onwards to zone VIII, and archaeological discoveries confirm the great range of age thus disclosed. We need only recall the Mesolithic fish-spear recovered in peat from the Leman and Ower banks, the Mesolithic to "B" Beaker occupation of the Essex coast, and the Hallstadt pottery on the Lincolnshire coastal peats at Ingoldmells, to illustrate this.'[1] (See also pp. 403, 421.) This clearly implies that peats and forests now submerged were formed throughout the post-Glacial period, and this conclusion in its turn means that the term 'submerged-forest period' is no longer accurate. Further the adjective 'Neolithic' applied to the period of the growth of the forest beds is certainly misleading, whether the term be used in its older and wider sense, or in its newer and more limited one. Hence, in this volume, except where the context justifies the use of the term 'Neolithic', this word which was formerly used so frequently to describe the submergence has been replaced by 'post-Glacial'.

Godwin has given us a clear picture of the various climatic and vegetational episodes in post-Glacial time in the Fenland and adjacent areas of the North Sea (see p. 428). Much work has yet to be done in places such as Morecambe Bay, Southampton Harbour,[2] the west Wales estuaries, the Thames estuary, and the fens adjacent to the Humber before a complete and consistent story can be told. In the meantime Godwin's[3] studies on the peat beds of South Wales and the Somerset Levels are extremely useful. In Swansea Bay, particularly in the old channel of the Tawe, peat beds are known at various depths down to 54 ft. They have all been carefully analysed for pollen, and the results show 'conclusively that those at −54 to −28 ft. O.D. belong to zone VI*b*, and those at −20 ft. O.D. to zone VI*c*....Peat beds at about −5 ft. O.D. on the coast are referable to zone VII, and only a few feet higher are peats belonging to zone VIII.'[4]

[1] Godwin, *Proc. Geol. Assoc.* 52, 1941, 350.

[2] See K.P. Oakley, 'A Note on the Late Post-Glacial Submergence of the Solent Margins', *Proc. Prehist. Soc.* 9, 1943, 56.

[3] 'A Boreal Transgression of the Sea in Swansea Bay', *New Phytologist*, 39, 1940, 308; and 'Correlations in the Somerset Levels', *ibid.* 40, 1941, 108.

[4] Godwin, *op. cit.* (1940). See also O.T. Jones, *Quart. Journ. Geol. Soc.* 98, 1942, 61. 'The relation of the deep channel in which these (alluvial) deposits lie to the glacial and other deposits in the vicinity is discussed and the conclusion is reached that the channel has been excavated in either Glacial or more probably pre-Glacial times....It is considered that before the depression set in the sea-level may have been about 200 feet lower than at present' (p. 85).

This sequence shows that there was a rapid transgression of the sea in the second part of zone VI at least, but that the rate fell off considerably at the end of zone VI*c*, and may have given place to retrogression. Such evidence as is available suggests that the two peat beds at Combwich in Somerset correlate with the higher peat beds of the Swansea district and probably belong to a zone not later than VII.

At present it is impossible to date all the peat beds known around our coasts. In general, it is safe to say that the highest are the newest, but it does not follow, especially in comparisons of northern and southern Britain, that beds at the same level, in terms of Ordnance Datum, are of exactly the same age. In view of the trend of the evidence from the Fenland and Bristol Channel areas, and assuming for the moment that it is typical of southern Britain at least, a great transgression of the sea in post-Glacial times seems certain, but more particularly in the Boreal period. It filled up the basin of the North Sea, and if the lowest sample of moorlog be taken as a datum, caused a rise of sea level relative to the land of at least 170 ft. This increases by nearly 100 ft. the usual estimates made in standard textbooks, since these figures were based on the lowest peat beds known at Pentewan and other places. The actual rise may well have been of the order of 200 ft. (see footnote 4 on previous page).

(8) THE 25-FOOT BEACH

The questions now arise: when did the rise in sea level cease, and how is the formation of the 25-foot beach related to it? They are not easy to answer. The 25-foot beach is usually thought to be later than the submerged forests, a view which derives from Jamieson's[1] excellent work in Scotland: he showed from several clear sections that the 25-foot beach or its offshore clays overlies the highest peat beds in the river Ythan and other places.

Before considering this question, however, something must be said of the beach itself. The name,[2] 25-foot beach, is not a good one, because in point of fact it occurs at all heights from sea level to 35 ft. It is the most clearly marked of all the raised beaches of Scotland, and platforms, terraces, and shingle ridges are all well preserved. In Wright's[3] opinion, the traces of the beach indicate more erosion and deposition than occur on the modern shore. The limits of the occurrence of the beach are shown in Fig. 104. The fauna contained in the beach and associated clays is like

[1] T.F. Jamieson, 'On the History of the last Geological Changes in Scotland', *Quart. Journ. Geol. Soc.* 16, 1865, 161.

[2] Movius (*op. cit.*) calls it the Litorina (*sic*) beach because its deposits contain the periwinkle *Littorina littorea*, which is characteristic of its counterpart, the Littorina Sea of the Baltic area.

[3] *Quaternary Ice Age*, pp. 380 *et seq.*

that of the present day, and there is no indication whatever of Arctic conditions. Praeger, who considered the fauna as a whole, concluded that at the time of the formation of the beach the temperature was some degrees higher than that of the present day. It is worth noting that the associated estuarine clays now form the level carse lands at the heads of the Firths of Tay and Forth. The age of the beach, in some places at least, can be determined with fair precision: traces of Neolithic man are definitely found in it, and at Oban, Azilian culture layers occur in caves which were formed during the maximum of submergence. Since these layers are covered by storm shingle, it seems that the Azilians must have occupied the caves soon after the retrogression began, but also at a time when their debris could be still overwhelmed by a great storm. At Larne in north-eastern Ireland and in the Grune Point spit of Cumberland the beach contains Nöstvet or Campignian axes and other implements.[1] In both places the evidence implies that the Campignians were living on these coasts when the submergence was at a maximum. Since the Azilian culture is distinctly earlier than the Campignian, it is logical to conclude that in the Oban district the maximum of submergence occurred earlier than in north-eastern Ireland and Cumberland. Needless to say evidence may yet be forthcoming to suggest that nearer the periphery of the area in which the beach occurs the date of the maximum submergence was even later.

George has suggested tentatively that the Heatherslade beach of Gower is the equivalent of the 25-foot beach, which admittedly falls in height as it is traced southwards. But this proposition, as George frankly admits, presents considerable difficulties since 'It will scarcely be contended that there has been a sufficient interval between Neolithic times and the present day for the platform of the post-Glacial Heatherslade Beach to have been not merely planed, but also to have been elevated to a sufficient height (at least 30 or 35 ft.), so that cementation of its pebbly deposits might take place, and then to have subsided to its present level.'[2] In any case, there can be no precise correlation of this, or any other beach, to the submerged forests. Godwin has clearly shown that in South Wales there is no gap in the series representing progressive subsidence from early in the Mesolithic period probably into the Iron Age or even later.

If, then, a precise date is impossible, the tentative opinion that the 25-foot beach was formed somewhere near the maximum of the great

[1] W.B. Wright, 'The Raised Beaches of the British Isles', *First Rept. of the Comm. on Pliocene and Pleistocene Terraces, Int. Geogr. Union*, 1928, p. 99.

[2] T.N. George, *Proc. Geol. Assoc.* 43, 1932, 307. In view of Arkell's recent work and reinterpretation of the Gower sequence, the difficulty may be lessened or nullified.

post-Glacial transgression may be put forward. The beach occurs mainly in areas affected by considerable isostatic movement, and is not always at the same height; moreover, the archaeological evidence from Oban and Larne suggests that it is not uniformly of the same age. As Wright remarks, investigators may yet find it, or rather its associated clays, covered in places other than 'the coasts of Lancashire and Cheshire...[where]... an upper submerged forest [is] rooted in what seems to be the marine clays of the raised-beach period'.[1] The difficulties of correlation associated with the Heatherslade beach form, however, only a part of the problem. Since no occurrence of the true 25-foot beach can be definitely proved in southern Britain, it is natural to ask whether there be any way of correlating it with other events or features. At the present time the question cannot be answered because it is only in the Fenland and North Sea that there is a sufficiently comprehensive record of events in post-Glacial time. Further, the Fenland area was more affected by eustatic than by isostatic movements.[2] It may perhaps be justifiable to suggest cautiously that the 25-foot beach corresponds with the maximum of transgression shown in the Fenland curve at the end of the Neolithic period (Fig. 106).[3] This figure also makes it clear that in zones IV, V, and the early part of VI, the sea level stood low relative to the land in the Fens, as it did also in South Wales, and probably throughout southern Britain. (On the Continent the figure shows that contemporaneously the land was at an intermediate height in north-western Germany, and that in the southern Baltic it was not far from its present height.)

The transgression continued until well after Neolithic times. In northern Britain isostatic movements predominated and so caused the land to rise faster than the sea. Hence, also, in the north, at some stage after the formation of the beach, it was raised above sea level. This agrees also with the fact that the level of the 25-foot beach decreases southwards, and that in

[1] W.B. Wright, *First Rept. of the Comm. on Pliocene and Pleistocene Terraces, Int. Geogr. Union*, 1928, p. 100.

[2] It should be borne in mind throughout this discussion that eustatic movements affected the whole of north-western Europe. Isostatic movements were more localized. This difference explains the limitation of the Late Glacial 100-foot, so-called 50-foot, and 25-foot beaches to north Britain. The 'submerged forest' eustatic movement certainly affected Scotland, but the local isostatic movements were of greater import. Hence the correlation of events in post-Glacial times in southern and northern Britain can hardly be achieved with any great accuracy.

[3] Such a suggestion does not agree with Wright (*op. cit.* 1928), who speaks of the beach as 'Early Neolithic'. If, however, as shown by comparisons at Oban and Larne the beach becomes progressively older away from the central areas, it may be that the corresponding event in the Fenland would be as late as the end of the Neolithic.

southern Britain the eustatic rise of sea level continued in excess of any isostatic movements. These, indeed, were in fact probably negligible approximately south of the Mersey-Humber line. As Godwin says:

> In the period between the Neolithic and Romano-British times, for which our evidence is best (i.e. in the Fenland), there is clear indication of the continued progress of the general transgression. There is some evidence, which we have accepted, that the transgression was broken by two periods of retrogression, one in the Bronze Age and the other after the Iron Age. In each of these, erosion channels appear to have cut through peat beds and underlying deposits. Possibly a third retrogression was in progress about 2500 B.C. The evidence presented seems to require periods of emergence breaking a long trend of submergence.... It is, moreover, clear, in the author's view, that the combination of factors determining eustatic and isostatic control of relative land- and sea-movement is so complex that a simple resultant such as unbroken 'sinkings', or sinkings broken by periods of halt, is less likely than some combination of 'sinking' and 'elevation'. To be sure, the 'sinking' has much the upper hand in this part of the North Sea.[1]

(9) RECENT MOVEMENTS

Chapter IX contains an account of the sinking of the land since Romano-British times in Suffolk, Essex, and the Thames estuary. There seems no doubt from the numerous discoveries of Roman pottery that parts of Southwark now 5–6 ft. below Trinity high-water mark were freely inhabited in Roman times. Longfield cogently argues that the Romans would hardly have selected for their area of settlement in this district a low-lying tract liable to inundation. Moreover, the Thames banks are probably entirely post-Roman, and were built when the Southwark and similar tracts became far more liable to flooding. Perhaps then the floods of 1928 (see p. 405) are the most recent reminder that the transgression since post-Glacial times has not yet finally ceased.

It is unfortunate that we cannot base rigid deductions on the differences found between the two geodetic levellings of Great Britain completed in 1860 and 1921 respectively. If these changes of level be plotted on a map, they show that, whereas south and east England has sunk, the north and west of the country has risen a little, although not in proportion to the sinking of the south-east.

> Taken at their face value, these differences might indicate that the surface of England had actually tilted and that the south-eastern counties are now 1·5 to 2 feet lower than they were in 1850. This suggestion, however, could only be true if there was no error of such magnitude in the original levelling, whereas when these differences between the old and new levels were first brought to light they were considered to be entirely due to a large cumulative error in the original levelling. There was no definite proof

[1] *Phil. Trans. Roy. Soc. Lond.* B, No. 570, vol. 230, 1940, p. 297.

that such an error existed, but conversely it was equally difficult to prove that it did not exist, and as the second would be equivalent to a definite statement that the south-eastern counties were sinking into the sea at the rate of 2 feet per 60 years...perhaps the simpler of the two alternatives was accepted.[1]

In a discussion on a lecture by Longfield to the Geological Society[2] Bernard Smith made some interesting comments which bear on the subject matter of this chapter. In Chapter IV there is comment on the abundant evidence of raised beaches and terraces on the Solway and Cumberland coasts: these features suggest formation in stages. Longfield warned his audience that any figures for changes of level in the north-west were likely to be less accurate than those in the south-east and thus liable to mislead. But it is nevertheless tempting to follow Smith in suggesting that 'This rise (0·28 inch near Silloth, in sixty years), if more or less continuous throughout the greater part of Neolithic and modern times, would easily account for the "Neolithic raised beach" of the Silloth district, which descended nearer sea-level at Walney Island and could not be detected south of Morecambe Bay....' The whole truth about this post-Glacial movement of sea level may be difficult enough to ascertain: a tilt to the south-east of a line running through Lancaster and Hartlepool, however, would help to explain the long continuance of the transgression and the complete absence of post-Glacial raised beaches in the south and east of our country. At the same time, the tilt, if it can be substantiated, is consistent with the elevated coastal flats of Cumberland and the Solway.

That the downward movement of the land may still be continuing in the Thames estuary district has been proved. Tidal observations at Felixstowe covering a period of only fifteen years (1917–32) also seem to bear this out. It is yet to be proved whether the views of Lewis and Balchin on shingle-ridge heights at Dungeness have a direct bearing on this problem.

Before leaving the matter, the significance of the fact that peat beds and submerged forests are preserved at all is worth noting. They are all soft formations, and most of them are thin, and it is, thus, somewhat surprising that with a slow subsidence they were not completely worn away by wave erosion. Possibly the halt or even minor emergence, during which the beds formed, was terminated by an abrupt change of level so that they were quickly brought below sea level and covered with sand and other deposits. Such a suggestion should not, however, be taken as implying the existence of a full record, nor that every bed has escaped erosion.

[1] Capt. T.E. Longfield, 'The Subsidence of London', *Ordnance Survey Professional Paper*, N.S. No. 14, 1932, p. 4.

[2] *Abstracts Proc. Geol. Soc. Lond.* No. 1260, 3 March 1933, p. 75.

Note. In speaking of, for example, the 10-foot *Patella* (pre-Glacial) beach, it is not meant that sea level was then 10 ft. higher than now. The remains of the beach platform occur, on an average, at this height. Clearly the platform was cut by waves, and observations show that the upper limit of the raised-beach *deposits* in contact with the old cliff is about 50 ft. O.D. Assuming a tidal range at that time, comparable with the one obtaining at present, we may assume that mean sea level was about 30 ft. O.D. higher than now. Similar reasoning applies, of course, to other terraces and beaches. This would, in Green's opinion,[1] equate the *Patella* beach with the Muscliff or Staverton terraces. The actual platform of the beach may be the product of both episodes.

HIGH-LEVEL EROSION SURFACES

There is no need to treat this recent and difficult matter in great detail because it has no very direct bearing on the themes treated in this book. In other chapters there are allusions to the way in which these surfaces have affected the nature of cliffs, especially in South Wales and Cornwall. There is still much difference of opinion about the real origin and nature of high levels of erosion, and clearly also there is as yet no final agreement as to their number and height. It is, in fact, a topic at the stage in which the personal factor is bound to play a considerable part. But such a comment does not imply adverse criticism, since in the early stages of investigation it is only natural that there should be a good deal of speculative work.

There is, however, no doubt that these surfaces exist. In some places they are very clear, and observers, conservative and otherwise, have recognized them. Two of the main difficulties in their investigation are, first, that usually only fragments of platforms remain, and secondly, it may be the case that too much reliance tends to be placed on map evidence. Since any such surfaces are likely to have had a slope, and at times a considerable one, to the sea level prevailing at the time of their formation (this applies whether the surfaces are marine or subaerial in origin), as a result of later erosion any given platform, which may once have been of wide extent, is cut back irregularly. Consequently, its remaining traces are at different levels and, unless means can be found of interpreting these features in terms of a given sea level, it is easy for observers in different areas to give varying estimates of the number and levels of the possible platforms. A mathematical technique has been devised whereby the slope of a given part of a river valley, above a rejuvenation head, can be prolonged 'in the air' to its former mouth.[2] In some cases the reconstructions may well be correct, but the hypothetical character of such reckoning is undeniable. Further, if the difficulties in the way of reconstructing former

[1] *Proc. Geol. Assoc.* 54, 1943, 129.

[2] O.T. Jones, *Quart. Journ. Geol. Soc.* 80, 1924, 568.

river profiles are substantial, *a fortiori*, those in reconstructing a whole platform are still greater. Other troubles are likely to arise in some places through the possible confusion with erosion levels of more or less plane surfaces which derive mainly or entirely from rock structure.

These general points claim attention, not because any physiographer would wish to minimize the significance of the platforms or the importance of the work already done on them, but it is wise to emphasize that in any intricate and relatively new investigation differences of opinion and interpretation cannot but be noticeable.

There is no need to recapitulate systematically the descriptions given of these platforms in previous chapters: a short summary will suffice. The areas in which they are most conspicuous in England and Wales are the Lake District, a great part of Wales, Devon, and Cornwall; in rather a different fashion they occur notably in the London Basin and the Weald. The chief erosion levels occur at 430 ft., 730–800 ft. (it is possible that two distinct shelves are included in this range), 1000–1070 ft., and 2000 ft. There are subsidiary levels at about 320 ft., 550–570 ft., 900–920 ft., and 1130–1170 ft. In the Lake District, platforms are thought to occur at the 1470 and 1600 ft. levels and others between 1800 and 2600 ft. may in time be recognized. At lower elevations there are the 275-foot Menaian platforms of Greenly, and the 200-foot platform of South Wales and of the St Austell district in Cornwall. In the Scilly Isles there is evidence of an erosion level at 130–140 ft., and in the South Downs at 130 ft. These are but generalized figures for the levels. Authors will often give more precision to a level in a certain area, but as a rule it will fall, notwithstanding, into one or other of the categories just listed.

There is rightly much reticence in interpreting the significance of these platforms. Hollingworth, who has summed up the evidence for most areas, writes of the more pronounced examples:

> From their correspondence in altitude and in relative size and position, and from their relation to subsidiary shelves at intermediate levels, it is deduced that the platforms at any one level are contemporaneous and represent an unwarped shelf or notch formed during a halt in the emergence of the land. The order of formation appears to have been one of decreasing altitude, indicating progressive fall in base level interrupted by major and minor stand-still periods of considerable duration.... It is difficult to envisage a sub-aerial origin of so widespread a series of platforms of small breadth and limited vertical range, and their formation as coastal marine-cut platforms seems more probable. A diversified topography of sub-aerial origin, first submerged and then subjected to periodical uplift relative to the sea-level, would provide the most favourable circumstances for the formation of the platforms. The latter would then represent notches of limited depth cut during longer or shorter periods of stationary sea-level.
>
> The absence of warping over so large an area is more readily accounted for by the

hypothesis of a stable land and falling sea-level than by that of the uniform uplift of a land mass.[1]

Full agreement with these views is perhaps not likely at present, but some writers on specific areas, however, give the impression of accepting some such scheme even though their recognition may be implied rather than expressed. There are obvious difficulties concerned with it: if, as is usually assumed, the platforms are early Miocene and later in age, the

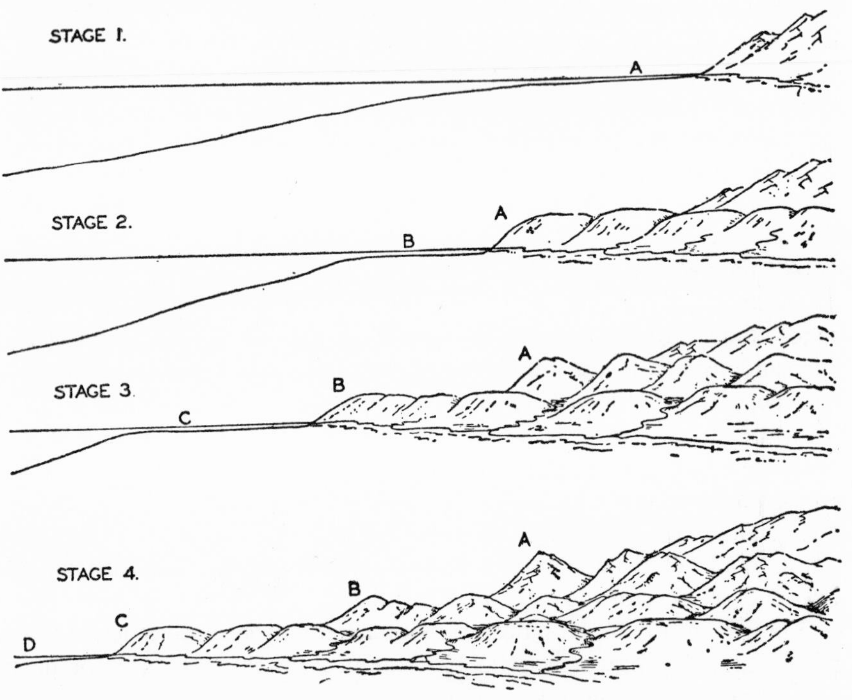

Fig. 107. Diagram illustrating the formation of Coastal Platforms (S. F. Hollingworth, reproduced from *Quart. Journ. Geol. Soc.* 94, 1938, 59)

features of East Anglia create a dilemma. In that region warping of some kind probably took place during Pliocene times: possibly this did not extend far to the west, but the occurrence of the Pliocene Crags is disconcerting. Other troubles arise if the question of river development is fully considered.

Whatever the difficulties may be in their final interpretation the existence of these platforms must be admitted. Their full extent and number are not yet known, but in some parts of the country they are so distinct

[1] *Quart. Journ. Geol. Soc.* 94, 1938, 75.

that they cannot possibly be overlooked. They also provide definite features in the appearance of cliffs and of river valleys and so they must be included in any study of our coasts. They are all older than the earliest *Patella* raised beach, and they may signify an interrupted fall of sea level between some part of the Miocene period and the Ice Age. During the Ice Age and after it, fluctuations of sea level are more easily correlated with the effects of the waxing and waning of the ice sheets and with isostasy. Should they be coastal platforms due to marine erosion of an intermittently rising land or, more precisely, to a falling sea level, the process of their evolution may be that shown in Fig. 107 which is given by Hollingworth.

Chapter XIII

COASTAL DUNES

(I) INTRODUCTORY

The many dune areas of our coasts have been described in the regional chapters of this book.[1] It may prove helpful here, however, to try to give a general account of dune formation and of the effects of vegetation on the different types of dune. It is essential to stress at the outset the importance of plant growth, since in this way, allowing naturally for differences of climate, scale, and of area covered, coastal dunes differ fundamentally from those in deserts.

The supply of sand to all coastal dunes comes from the flats and banks exposed at low water, particularly during spring tides. The wind blowing over these flats soon dries the sand and carries it inland. The nature of the resulting dunes thus depends to some extent on the incidence of prevalent or of occasional winds. On most of the western coasts of England and Wales the westerly winds blow steadily throughout much of the year and dune ridges form at right angles to them. On the eastern coasts, however, the prevalent westerly winds are offshore, and dune growth depends, according to locality, much more on the incidence of local or on shore winds, especially those blowing during stormy periods.[2] On a windy day the amount of sand moved is very great, and it can be felt as well as seen: it will be dropped when the force of the wind ceases, and also when the wind meets an obstacle. On an open beach after a period of strong winds small dunes are noticeable around large pebbles or other objects, and sand naturally collects on their lee sides. Such forms are, however, only very temporary. When the sand blows up on to a shingle ridge or beach a great deal is trapped between the stones. Much of this may be washed out again by waves at high water, but it is certainly one explanation of the large amount of sand in some beach ridges. Drift wood, boxes, or any other jetsam washed up on to shingle ridges out of the reach of normal waves, provides the nucleus of a dune which may grow to the height of a few feet. Where there is a plentiful supply of drift rubbish, low and immature dune ridges may be seen in process of formation.

[1] Various names are used for dunes in different parts of these islands, e.g. Meols or Meals in Norfolk, Towans and Burrows in Cornwall and the west country, and Links in Scotland.

[2] Probably the dominant winds.

But the really important factor is plant growth. On the top part of the foreshore, just out of reach of spring tides, various plants survive, while on the drift line itself the most common species are *Cakile maritima*, *Salsola kali*, *Arenaria peploides*, and *Atriplex hastata*. These are annuals, but during their period of growth they form excellent sand traps and lead to minor dune formation. Even in winter their indirect effect is still present, since, unless the embryo dune is washed away by storms or blown away in a gale, the hillock itself is an obstacle and collects more sand around it. As shown on p. 132 heaps of seaweed are often most effective agents of this kind.

(2) *AGROPYRUM JUNCEUM* AND *PSAMMA ARENARIA*

Certain grasses, however, are the most influential in dune growth, and the sea couch grass (*Agropyrum junceum*) is usually the pioneer. Its seeds take root on the summits of shingle ridges, and once the shoots push above the surface they begin to gather sand around them. The grass grows rapidly and forms small tufts, while neighbouring tufts spread, and sooner or later coalesce. In this way a fairly continuous growth develops and sand gathers all along it. Thus a small dune ridge, a true embryo foredune, comes into existence. Although not a halophyte, *Agropyrum* can withstand a certain amount of flooding by sea water and thus, even if these tufts are overrun in a storm, they do not die away. Indeed, their root systems seem to be rejuvenated and send up more tufts next season. *Agropyrum*, however, is a pioneer and accordingly seldom collects dunes of any size. Its place is soon taken by the marram grass, *Psamma* (formerly *Ammophila*) *arenaria*. This is by far the most important and valuable of the coastal dune-building plants. Marram grass spreads by seeds on the shingle ridges, but it has qualities of almost unlimited growth. Occasionally as a pioneer, but far more often as a successor to *Agropyrum*, it forms tufts and so collects sand around it. These tufts grow upwards and outwards rapidly, and very soon unite to form a long low dune ridge and a true foredune. If sand be constantly blown to this ridge, the marram grass grows upwards equally with the accumulation of sand, and the whole ridge increases in height and breadth. Needless to say, it is not likely to grow regularly: some parts in time overtop others and gullies form. These, in their turn, become valleys through which the wind with its load of sand passes, and in this way a certain amount of sand spreads in the rear of the dune line. Later, tufts of marram may take root and hold some or all of this shifted sand.[1]

[1] In certain parts of the country the lyme grass (*Elymus arenarius*) plays an equally important part and replaces the marram.

The root system of marram grass is very complex and ramifying. The seaward face of a foredune in normal weather is usually covered with fresh sand and grades downward to the beach in a steep concave curve. After a storm and high waves this face is eroded: much sand is washed away, and the ramifying root system is often very apparent on the new face exposed. Although it is clear that a marram dune cannot withstand a severe attack, it is equally obvious that without marram grass and its intertwining roots the dune would be much more at the mercy of the waves. As a rule the greater the supply of fresh sand the more active the growth of the grass. Even the most casual visitor to sea-coast dunes will notice the fresh appearance of this grass on the outer formations and its rather mangy appearance on the older ones deprived of a continuous renewal of sand. These older dunes are often referred to as grey dunes. *Carex arenaria* replaces *Psamma* to a considerable extent at this stage, and this is frequently accompanied by an abundant growth of lichens.

(3) DUNES FORMING ON SHINGLE RIDGES

It has been shown in Chapter IX and elsewhere that new shingle ridges often form in front of older ones on an extending spit or bar, and on a rapidly prograding spit several dune ridges are very usual. This process has a very noticeable effect on dune growth. If a new ridge be established, it will sooner or later get its share of *Agropyrum* and *Psamma*. Hence a new dune ridge comes into being and, since it is to the windward of the former ridge, the latter is deprived of most of its sand supply. Consequently its dunes either cease to grow or only increase very slowly. Should the new barrier accumulate quickly, the dunes on the older ridge may even begin to disappear as the result of erosion.

If the original ridge maintains its sand supply, the dunes will continue to grow upwards. There is no obvious or definite reason for the upward limit of growth, but in England and Wales 60 ft. is about the maximum height in most dune areas. In Scotland, the Culbin dunes on the Moray Firth reach just about 100 ft. These now have some marram grass on them, but much of it has been planted. Of recent years a good deal of reclamation has been carried out in the Culbins, but the greater dunes appear to have grown up as bare sand masses reminiscent of those in deserts. In other parts of the world much greater heights are recorded for coastal dunes, but in the first place some of these measurements require confirmation, and in the second it is not always clear whether the dunes reach down to sea level or not: they may rest on a solid platform.

(4) EROSION OF DUNES, PARABOLIC DUNES, REMANIÉ DUNES

All dunes are subject to erosion even in the initial stages, and once they attain some size the wind may at any time begin to cut in to them. This happens because the plant covering is never close, and thus bare sand is always exposed. Such patches may be made larger in many ways: by rabbits, for example, or by human beings digging or sliding on the dunes. Any bare place of this kind may be attacked by the wind and once erosion has begun it will increase, often rapidly. It is thus that blow-outs form, which may range in size from a mere scratch to a valley cut right through the dune. The upper edges are often nearly vertical or even slightly overhanging on account of the marram and other vegetation. The sand so shifted and later carried through the gap may merely be spread out on the marsh, flat, or other surface, or it may begin to gather as a new hummock. If marram seeds take root on this hummock, it will grow in the usual way, and it is no uncommon thing to see new dunes forming in this fashion.

In some cases where the sand supply is plentiful and where vegetation, particularly marram, has not gained a complete foothold, except on the lower ends of the dune ridge, the higher central part may continue to grow and move forward. Many dunes which are not fully anchored by vegetation move in this way. The sand blows up the gently sloping windward side, which may, of course, be steepened by wave erosion at high tides; it flies over the top and forms an abrupt lee side. The constant addition of new sand, partly from the front of the dune itself, may raise the whole formation, but it will also move over the dune and cause it to advance inwards. Should this happen, particularly on the central part of a dune ridge, the resulting form will be a parabolic dune, that is to say, an arc concave to the direction from which the wind comes. Minor examples often occur in any fair-sized dune area, but the most impressive ones in these islands are to be seen on the Culbin Sands, where at Maviston a parabolic dune has advanced over and destroyed part of a forest. A second and smaller type of parabolic dune can also develop in certain conditions from a blow-out. The parabolic shape is the direct opposite of the crescent dune or barchan so typical of the deserts, and can only develop as a result of the interaction of wind, dune advance, and vegetation. Minor examples of true crescent dunes may often be found on a foreshore, since those forming round jetsam or a marram tuft may take on a crescentic pattern as long as their growth is not hampered by a neighbouring formation. But whatever explanations may be given for the desert barchan they cannot wholly apply to coastal forms because of the presence of vegetation in the latter.

Once the supply of new sand is cut off from a dune area, the forces of erosion may become stronger at any moment for the reasons just given. Since no new sand is available, cuts or gashes cannot be healed, and, unless artificial means are taken to prevent erosion, the whole dune individuum may in time be blown away. By the time this threat occurs other plants besides *Agropyrum* and *Psamma* have usually colonized the dunes, but they do not form a close and protective turf. These older, grey, dunes, once erosion has begun, take on all manner of forms which are influenced in part by size, height, plant covering, relation to underlying shingle ridges, and other factors. They may be collectively described as remanié dunes. The substratum on which they formed is largely exposed, and the former shingle ridges are often brought to light once again. Incidentally this provides further proof of the observed fact that new dunes usually have their origin on such banks, and in many parts of Norfolk those which are re-exposed are now largely covered by a prolific growth of *Statice binervosa*.

(5) CLASSIFICATIONS OF COASTAL DUNES

Various classifications of coastal dunes have been attempted mainly from a physiographical point of view. Briquet[1] adopted a simple method which is given below:

(*a*) Accumulation forms: in which the vegetation gradually gains the ascendancy, and embryo dunes characteristic of prograding shores begin to form.

(*b*) Fixation forms: in which the inner and older dunes of a prograding shore are gradually deprived of their sand supply. These become liable to wind attack and so give way eventually to remanié forms.

(*c*) Remanié forms: these are subdivisible into (*a*) moderate forms: in these the sand supply ceases, while marram tends to disappear and is replaced by other plants; it is in this stage that blow-outs are perhaps most common; (*b*) intensified remanié forms which often result from a strong frontal attack by the sea or from the general decay of the vegetation. At first they will be very similar to the moderate forms, but blow-outs and cuts in time enlarge and join laterally, and the sand may be removed from a whole zone. In the rear of the wind-swept area the sand may gather again, and form a moving dune which may give some difficulty before it is finally held.

[1] A. Briquet, 'Les Dunes Littorales', *Ann. Géog.* 32, 1923, 385.

(*d*) Parabolic dunes, which have already been described. Sometimes the mid-part of the ridge advances until it is blown away altogether, or is

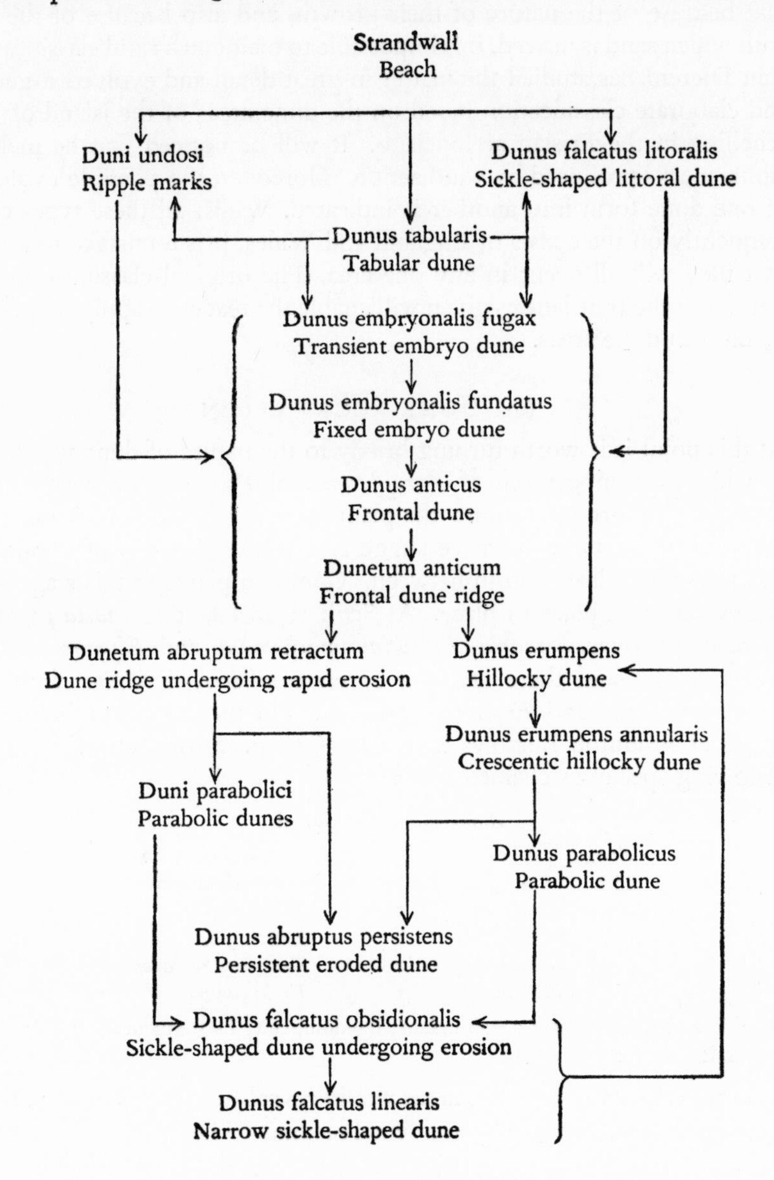

represented only by lines of dunes running generally parallel with the direction of the wind. These in time will be fixed by vegetation.

There is naturally every transition between different types of dunes.

There may, of course, be clear examples of a foredune, a fixed dune, or a parabolic dune on a foreshore, but the reader will readily understand that because of the nature of their growth, and also because of the ease with which sand is moved, it is impossible to maintain a rigid classification. Van Dieren[1] has studied the matter in great detail and evolved a genetic and elaborate classification based on the dune areas of the island of Terschelling in the Frisian archipelago. It will be noticed that he includes ripple marks produced by wind action. Moreover, the possible evolution of one dune form into another is indicated. Whilst all these types occur frequently on the coasts of England and Wales, it is a mistake to assume that they will all occur in any one area. The original classification is in Latin, but the translations given will enable the reader to apply the scheme to English dune areas.

(6) DUNE VEGETATION

At this point it is worth turning briefly to the nature of dune vegetation, in which the importance of *Agropyrum* and *Psamma* has already been stressed. On growing dunes the plant growth is usually very open and their characteristic appearance suggests a confused series of hummocks and tufts with clear sand in between. Common plants at this stage differ somewhat from place to place. At Scolt Head Island *Arenaria peploides*, *Festuca rubra* var. *arenaria*, *Eryngium maritimum*, and *Elymus arenarius* are common, and there are also *Crepis capillaris*, *Erodium cicutarium*, *Senecio vulgaris*, and *Euphorbia paralias*. On the corresponding newer dunes at Braunton Burrows in Devon (see p. 216) Watson[2] gives the following species as characteristic:

Ammophila arenaria d

Anagallis arvensis	o	*Bryum argentium*	r
Cynoglossum officinale	r	*Potentilla anserina*	r
Erodium cicutarium	r	*Sedum acre*	o
Euphorbia paralias	o	*Senecio Jacobaea*	o
Leontodon hispidus	r	*Teucrium Scorodonia*	r
Nepeta hederacea	r	*Viola canina*	r

(In this and other tables: a = abundant; d = dominant; f = frequent; l = local; o = occasional; r = rare; v = very.)

Examples drawn from other areas would reveal further differences, which Tansley[3] explains largely by accidental seeding. 'Their successful germina-

[1] J.W. van Dieren, *Organogene Dünenbildung*, 1934.

[2] W. Watson, 'Cryptogamic Vegetation of the Sand Dunes of the West Coast of England', *Journ. Ecology*, 6, 1918, 126.

[3] A.G. Tansley, *The British Islands and their Vegetation*, 1939, p. 851.

tion and establishment depend no doubt on periods of damp weather when the surface of the sand is temporarily stable.'

The dunes of older systems are much greyer in colour and the distinction is valid ecologically between the newer, yellow or white dunes and older, grey ones. On the latter the vegetation is as a rule more nearly closed, but it seldom resembles a continuous turf, and rabbits often expose considerable areas of sand. Although *Psamma* may still be plentiful, it is not usually very vigorous, and it is often replaced by *Carex arenaria* and *Festuca rubra*, for example, at Scolt Head Island. At the same time the ground between the bigger plants is carpeted by mosses and lichens the appearance of which really justifies the name 'grey' dunes, while many of the plants common on white dunes have been driven out by newcomers. In the early part of the year the sides of grey dunes, especially at Scolt Head Island, are covered with such ephemerals as

Valerianella olitoria	a	*Senecio vulgaris*	o
Stellaria boreana	la	*Veronica arvensis*	o
Erodium cicutarium	f	*Aira praecox*	o
Cerastium semidecandrum	f	*Aira caryophyllea*	o
Stellaria media	f	*Arenaria serpyllifolia*	o
Myosotis collina	f	*Phleum arenarium*	o
Cerastium tetrandum	l		

In addition, at this stage, the Scolt Head Island dunes, which are probably typical of north Norfolk in general, all carry

Senecio Jacobaea	a	*Cynoglossum officinale*	o
Viola canina	la	*Galium verum*	la
Cnicus arvensis	l	*Hypochœris glabra*	f
Anthriscus vulgaris	l	*Veronica chamaedrys*	o
Agrostis alba	f	*Leontodon autumnale*	o
Cnicus lanceolatus	f	*Leontodon nudicaulis*	o
Lotus corniculatus	la		

On the grey dunes, also, variations occur with locality. Pearsall[1] gives the following as characteristic of the fixed dunes at the northern end of Walney Island (p. 89):

Ammophila arenaria	a	*Lotus corniculatus*	f, la
Calystegia soldanella	f	*Ononis repens*	la
Festuca rubra var. *arenaria*	a	*Rosa spinosissima*	la
Galium verum	f, la	*Salix repens* (hollows)	la
Geranium sanguineum	f, la	*Thalictrum dunense*	la

Rosa spinosissima is also characteristic of large areas of the dunes on Morfa Harlech, and *Hippophae rhamnoides* occurs locally.

Grey dunes also often bear a considerable number of shrubs including

[1] W.H. Pearsall, *The Naturalist*, 1934, p. 201.

Rubus spp., *Ligustrum vulgare* (privet), *Sambucus nigra* (elder) and *Ilex aquifolium* (holly). In some places, for example, Walney Island and the Studland peninsula in Dorset, a Callunetum has developed. Mosses and lichens, on account of their very limited rooting systems, are unable to establish themselves on mobile sand and for this reason await the comparative shelter of rather older dunes. On a typical grey dune the lichens exceed the mosses in number of species and abundance. Of the mosses *Tortula ruraliformis* is the most important pioneer, while amongst the lichens the species of *Peltigera* and *Cladonia* are the most prominent.

(7) DUNE SOILS

In the grey dune stage the soils are very different from those in the white or yellow stage. The relatively stable soil of the grey dunes is enriched by the humus of plants, mosses, and lichens, and, at the same time, *Ammophila* decays noticeably. The reasons for its decline are uncertain. Tansley[1] suggests that the most potent factors are the 'increasingly severe competition for available water, and the effect of lack of oxygen, or of increased carbon dioxide or some other toxic product of the new vegetation carpet'. As the dunes increase in age the flora changes gradually into that associated with inland regions, but in this respect the nature of the sand composing the dunes is important. The Norfolk dunes are mainly composed of quartz sand, whereas others contain a large proportion of lime derived from minute fragments of shells, and calcareous algae. In the latter case the flora is naturally likely to contain a high proportion of calcicolous species as, for example, in the dunes of Dog's Bay, in County Galway, and Rosapenna, in County Donegal. Salisbury investigated the carbonate and humus contents of dunes at Southport and Blakeney Point.[2] He found that in the Southport dunes the percentage of $CaCO_3$ in embryo formations is about six, but that owing mainly to leaching it drops in the course of time, so that the dunes of about 150 years and more contain less than 1 per cent. The Blakeney dunes begin with a percentage of less than 1 and drop like those at Southport after some 250 years to about one-tenth of this quantity. On the other hand, the humus content increases with time. At both Southport and Blakeney the percentage in new dunes is just less than 1, but the rate of increase at Southport is far more rapid than at Blakeney, so that in dunes estimated to be about 300 years old the percentages are about 8 and $2\frac{1}{2}$ respectively. Salisbury maintains that in any given dune area these changes are regular, so that it is possible approximately to estimate the age of dunes in this way.

[1] *Op. cit.* p. 856.

[2] E.J. Salisbury, 'The Soils of Blakeney Point...', *Ann. Botany*, 36, 1922, 391, and 'Note on the Edaphic Succession in some Dune Soils...', *Journ. Ecology*, 13, 1925, 322.

(8) WATER CONTENT

An important factor both in dune habitat and physiography is their water content. Various experiments have been made on this aspect of dunes, and it is interesting to refer to some work by Chapman[1] at Scolt Head Island. He chose an area of old grey dunes known as House Hills, dug pits around the dunes, and then established the levels of the upper and lower surfaces of these in relation to Island Zero Level (see p. 527). The pits were prevented from collapsing by being revetted with wire netting. Within the dunes is a natural pond, and the mass of the dunes probably rests on two main lateral ridges and on the space intervening between them. The water movements in the pond were recorded over a given time and were clearly found to be related to the cycles of spring tides. Daily tidal fluctuations appeared to have no effect on the level of the pond, and, in fact, the tide had to flood the adjacent marsh before any record could be obtained. The rise of the water in the pond took place about 48 hours after the first flooding tide over this marsh, and the maximum height in the pond occurred also about 48 hours after the highest spring tide of the cycle. It is interesting to note that Oliver likewise noticed this lag of about 48 hours at Blakeney Point.

This rise and fall of the water level means that the fresh or brackish supply in the dunes is resting on the salt water below. When spring tides occur, the salt water-table is increased and raised, and consequently the fresh water in the dunes also rises. The lag of 48 hours is due to the slow percolation of the water through the dune sands. It is easy to understand that dune plants grow in a very rapidly drained medium. They have long roots, and are intolerant of sea water, using instead the fresh water on top. In dry summers this fresh water probably often sinks below the root zone, and were it not for the occasional liftings just described, the plants would be deprived of their supply for too long and would die. Capillary action also probably helps to supply the shorter rooted plants with water.[2]

(9) WANDERING DUNES: FIXATION OF DUNES

Reference has frequently been made to wandering dunes. These may occur on a coastline that is prograding, stationary, or in retreat. On a prograding shore they are perhaps less likely to be a serious menace since,

[1] V.J. Chapman, in *Scolt Head Island*, 1934, and 'Note on a Dune Drainage System', *Mems. and Proc. Manchester Lit. and Phil. Soc.* 81, 1936–7, 77.

[2] Mention may also be made of the suggestion that there may be an internal formation of dew in shingle ridges. In the dry summer of 1911 sheep could graze on the vegetation on the Chesil Beach. See T.G. Hill and J.A. Hanley, *Journ. Ecology*, 2, 35; also A.E. Carey and F.W. Oliver, *Tidal Lands*, 1918, p. 113.

owing to the forward growth of the shore, the new supplies of sand are trapped by the more seaward ridges. It is difficult, however, to be precise in detail: Morfa Harlech and Morfa Dyffryn are rightly considered in part as prograding, but in part also as undergoing erosion, and in each area there is a belt of inward-moving dunes (see pp. 131 and 139).

If the coastline be stationary, supplies of sand are constantly added to the same dune masses. There are impressive examples of these conditions on the Kurische Nehrung and the Frische Nehrung on the Baltic coast of Germany. The dunes reach 150 to 180 ft. in height and during the time when regulation was not practised advanced 15 to 20 ft. a year, overwhelming farms, villages, and cultivated land. The Culbin estate on the Moray Firth has been likewise invaded, but it is as well to point out the fictitious nature of the stories of a sudden invasion by sand which still survive.[1] Other cases of settlements being smothered by dunes have been noted in this book—for example, St Piran's in Cornwall and Kenfig and other places in South Wales.

If the coastline be suffering erosion, the dunes, as well as the platform on which they rest, are affected, and the sand is blown back on to the foreshore. It is also true that much of the dune sand may be carried away by marine action, but in each case the mutability of conditions with time needs emphasis. Careful acquaintance with a dune coast will enable any observer to realize how quickly erosion may give place to accretion or vice versa. Moreover, dune wandering may also start in old formations. If, for example, on account of severe burning of the vegetation, or because of depredations in a rabbit warren, much bare sand is exposed, the wind can easily work on this and build new dunes: should these be some little way behind the belt of thriving marram, they may not readily be colonized and by shifting may present a serious problem. Carey and Oliver call attention to an odd fact—that records of dune-wandering in Europe are all relatively modern. Certainly, if the phenomenon existed in earlier times, it is somewhat curious that writers did not call attention to it. They suggest that the wandering may be mainly due to man's negligence 'unless we have entered a new geologic era as far as dunes are concerned'.[2] Whatever the reason, the process became of considerable economic significance more than a century ago, and a problem formidable enough to provoke state action.

It may be of interest to outline shortly the means by which moving dunes, and dunes in general, are fixed.[3] Since the sand is derived from the

[1] J.A. Steers, 'The Culbin Sands', *Geogr. Journ.* 90, 1937, 498.

[2] A.E. Carey and F.W. Oliver, *Tidal Lands*, 1918, p. 69.

[3] *Ibid.* Chapter VI.

shore, in stabilizing dunes it is essential to distinguish between the newer and the older formations. If the older and inner dunes have for some reason started to wander, the new sand blown up from the beach could in part reach them and aggravate the movement. Hence, it pays to hold the new blown sand along the foreshore and to tackle the inner moving dunes independently. It is not difficult to trap the sand blowing up from the foreshore: indeed, the problem is one of upkeep rather than establishment once a dune ridge has formed. The new dune must stand far enough back to escape erosion by the waves: it should follow the general trend of the coast rather than its minor irregularities, and its crest must be kept reasonably level and restrained from undue height. Experience has shown that

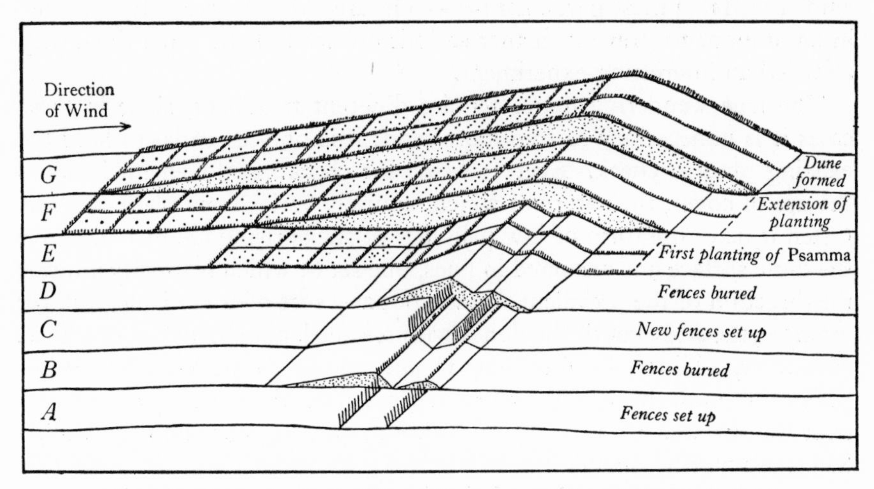

Fig. 108. Stages in the building of a new sand dune (after F. Solger)

a broad and low dune of this type will hold as much sand as a high and narrow one, and it is also much easier to maintain. The point to remember is that it is intended mainly, perhaps exclusively, to prevent sand reaching too far inland. Thus some means must first be found to trap the sand, and then to continue the process as the dune overtops the initial obstacle. Simple brushwood fences are quite effective and, should they become buried, can be simply and cheaply replaced by others at higher levels. When a fair bank of sand has accumulated, it is often planted with *Psamma* which grows up with the dune. This plant may be established as early as the end of the first year of dune growth, or at the beginning of the second year. It is important, however, to cause the seaward slope of the new dune to rise gently and uniformly to a broad and flat crest. This may be achieved by the judicious planting of the *Psamma* which should be spaced more widely on the lower parts of the slope than on the upper (see Fig. 108). Such a distribution prevents the development of hummocks, and if growth for some reason proves too irregular it can be checked by uprooting some of the *Psamma*. Should a part of the dune be damaged it ought to be quickly repaired by planting more *Psamma*, or possibly by the erection of new brushwood fences. Blow-outs easily form in neglected ridges and may soon become a serious menace: they should be checked at once. At Scolt Head Island some large blow-outs in older dunes have been replanted with *Psamma*, but since they are now more or less cut off from a direct supply of new sand, their recolonization by *Psamma* is a slow and only partially successful process. It should be emphasized that these particular blow-outs are old and were only replanted in an attempt to prevent further serious erosion: at the same time, they afforded an interesting experiment.

The true wandering dunes require different treatment, since in these cases it is unnecessary to trap more sand: it is only essential to hold the existing sand in check. Every scheme for checking movement includes the use of cover, the nature of which naturally varies from place to place. Often it takes the form of low palisades of sticks rising about 1 ft. above the dune surface and arranged in lines, one set of which is usually athwart the direction of the prevalent winds. But any sort of cover is helpful, for example, the strewing of the dunes with seaweed, pine branches, or refuse. Moreover, additional refinements in *any* dune area are possible by improving the soil through spreading salt-marsh earth or harbour dredgings over the surface. By way of illustration the methods used on the Culbin Sands are worth comment.

In 1921 the Forestry Commission began to acquire land at Culbin and so saved the Low Wood area (cut down in 1914–18), which would probably have soon reverted to loose sand once the trees were felled.[1] Low Wood was replanted, and since then the sand and the great hills to the east, that is to say the leeward, are gradually being conquered, partly by the planting of marram grass and partly by thatching. The latter process consists of laying out and pegging on the dunes branches of birch and broom which grow naturally in this district. This is a more effective method than planting, because, as the Commissioners have worked from west to east, the available sand supply becomes more and more restricted, so that *Psamma* does not thrive too well. The grass has, however, gained much ground in the mobile area, and will thus help in later reclamation work. Stable sweepings are also spread over the thatched areas, and numerous grasses and weeds are helping to bind the sand. Scots pine,

[1] Steers, *op. cit.*

Corsican pine, and the Lodge Pole pine are the species used in tree planting, and have grown most successfully. Small trees, two or three years old, from the nursery are planted 4–5 ft. apart on the fixed sand, and later these are thinned out, and the felled trees used mainly for pit props. The Corsican pine has proved the most suitable for the less stable dunes, while the Scots pine thrives well on shingle and fixed sand areas, and the Lodge Pole pine grows almost anywhere. The only places presenting real difficulty for planting are the low swampy areas. These are numerous, and may contain shallow lochs in winter, but one result of increased planting has been a marked drying of the whole area.

Other extensive reclamations of dune areas have been made at Southport and on Lord Leicester's estate at Holkham on the Norfolk coast. At Southport *Psamma* was chiefly used. At Holkham about 300 acres were planted between 1853 and 1891 with Corsican, Austrian, and Scots pine. These were put in behind the dunes which were also systematically watched and regulated. The Austrian pine was used on the more exposed, seaward places, and the others in more sheltered positions. The whole belt thrives and is rightly regarded as a most successful example of coastal planting. The regional chapters contain comments on various other reclamation schemes, to which further reference in this chapter is unnecessary.

Note. The references given in the footnotes do not by any means cover the enormous literature on coastal dunes. The chapter has been written in general terms, and many references to particular dune systems are given elsewhere in this volume. General observations by the writer have also been incorporated. For further details, see

A.G. Tansley, *The British Islands and Their Vegetation* (1939).

A.E. Carey and F.W. Oliver, *Tidal Lands* (1918).

Scolt Head Island, edited by J.A. Steers (1934).

D.W. Johnson, *Shore Processes and Shoreline Development* (1919).

J.W. van Dieren, *Organogene Dünenbildung* (1934).

All these contain many references to detailed papers. The Royal Commission Reports on Coast Erosion (1907–11) are also full and helpful.

Chapter XIV

SALT MARSHES

(I) INTRODUCTORY

In the foregoing chapters there have been frequent references to individual salt marshes and to the plants which characterize them. But, in addition, it is in place to describe shortly the development of marshes as a whole and the chief features of physiographical interest found in them.

Marshes tend, as a rule, to develop in two types of locality, that is to say, either behind shingle bars or similar protective features, or in the upper and sheltered reaches of gulfs, bays, or estuaries. There are excellent examples of the first type along the north coast of Norfolk where some of the finest marshes in this country have formed. They lie between the mainland and the offshore bars, of which Blakeney Point and Scolt Head Island are the best known. The marshes in the Dyfi (Dovey) and Mawddach estuaries also belong to this class since they have grown behind the shingle bars across the mouths of the estuaries. Those, natural or reclaimed, near Westward Ho!, at Dungeness, in the Tees mouth, and on the lee side of Walney Island are yet other instances. But the marshes at the south end of the Wash, in Southampton Water in part at any rate, those in Morecambe Bay, those at the heads of some of the lochs on the west coast of Scotland, and some in the Thames estuary belong to the second type. A rigid distinction between the two types is, however, neither necessary nor even desirable since they grade into one another.

Chapter III contains a discussion of the origin and nature of spits and offshore bars. These often show a series of recurved ends, each end having once been the termination of the spit or island. In consequence between these old ends there occurs a series of compartments which are ideal places for marsh development. Sometimes the openings between adjacent ends are wide, while at other times they are very narrow. In general, marshes grow most quickly when the entrance between the laterals is narrow. An offshore bar such as Scolt Head Island, for example, has grown to the west by throwing out a long series of lateral ridges; here it will be clear that, apart from any local or particular changes which are, for the moment, beside the point, an approach to the west leads to the newer structures and most recent marshes, and consequently these westernmost marshes will be in a stage considerably different from those farther east.

All marshes are due in the first place to silt and fine material being carried into some sheltered area by the tides. Although circumstances will differ somewhat from place to place, it is commonly found that the foundation on which a marsh originates is a sand flat, a part of the sea floor before the offshore bar was built. In the earliest stages the flats are nearly level, or but gently inclined to some main tidal channel. Therefore, at first, the tides ebb and flow over them as sheets of fairly tranquil water. On a fine and windless day, indeed, the water rises and falls as calmly as if a lake were being artificially raised or lowered. Nevertheless, near the turn of the tide there is a period of relative stagnation during which deposition may take place. The original sand flat is likely to be strewn here and there with patches of algae such as *Enteromorpha*. Moreover, around the inner edge of the flat is the shallowest water and here as well as on the flat itself sediment is likely to be dropped and held in position. Thus the original surface sooner or later becomes more irregular, while it is easy to see that, other things being equal, sedimentation will probably be quicker along the inner margin of the marsh. The patches of algae, however, are liable to shift in more stormy times, and any sediment they may have aggregated is then redistributed. In time, however, they become relatively or even absolutely stable.

This stabilization is greatly helped by the spread of vegetation. Not only do the tidal waters carry silt, but they also bear with them numerous seeds of halophytes derived from neighbouring areas, or even from a distance. These seeds, like the silt, are deposited, and if circumstances are favourable take root. Once this has happened true marsh development has begun. The algal masses and the quiet inner margins of the marsh form suitable places for this process, but the kind of plant growth will depend largely on the nature of the substratum. In sandy areas, for example, *Salicornia* spp. are common, but where sloppy mud is forming *Zostera* spp. are more frequent. The general result, however, is the same. The plants grow and therefore in their turn become effective trappers of silt, because they naturally upset the run of the tidal waters round about them and so bring about further deposition. Thus in the early stages the future marsh shows itself as a more or less discontinuous and narrow fringe along the inner margin of a flat, and as a few isolated and inconspicuous rounded hummocks gradually appearing over the rest of its surface. Although *Salicornia* spreads fairly rapidly, it is never a very good accumulator of silt because, since it is an annual, it dies each year, and the thin straggly stems that remain during the winter cannot bring about much deposition. In time, however, the colonies expand.

The continuation of this process implies the coming of a stage when the tidal waters, instead of flowing as a thin film over the whole area, will be divided by these islands. Deposition in favourable places continues all the time, and so, as the hummocks grow, and the inner fringe of embryo marsh grows outwards, probably with some irregularity, the tidal waters become slightly restricted to wide, open, flat channels. Moreover, the sand flat is unlikely to have been quite level in the first place. Therefore natural and original depressions easily tend to develop into more pronounced drainage channels, and, as will be seen later, these, in their turn, often become the major creeks of fully developed marshes.

As deposition goes on and as the vegetation spreads, other factors emerge which influence the development of the marsh. The early stages in fairly open surroundings probably take a long time to evolve, but the marsh surface rises gradually in level as well as extending horizontally. Accordingly, first the tidal waters become increasingly restricted to fairly definite lines which later form marsh creeks, and secondly as the vegetated patches gain in height, the number of times they are completely submerged by tidal waters steadily decreases. It is this process, together with the nature of the substrate (sand, firm mud, sloppy mud, etc.) which greatly influences the spread of vegetation. Certain plants, true halophytes, are better adapted than others to less frequent immersion. Hence with increase in height of the marsh, other plants sooner or later take root. These usually have a far greater effect on the rate of sedimentation because of their greater trapping powers, and also because they often maintain an effective leaf area throughout a part or the whole of the winter. The fact of their dying down in the cold season matters little, provided that their leaves remain intact during some or all of that period. On many sandy marshes *Glyceria maritima* may either accompany or follow *Salicornia*, or may now and then precede it. *Glyceria* grows rapidly in tufts and is a very effective trapper of silt. On muddier marshes *Aster tripolium* frequently follows *Salicornia* and both may be accompanied by *Suaeda maritima*, although the latter is usually a little later than *Salicornia*. Once this stage is reached sedimentation goes on apace: the mud, indeed, is seldom firm, but the spread of the plants is rapid and the true creek system of a mature marsh is definitely outlined. On the newer parts of *Aster* marshes in Norfolk there is usually a multitude of small creeks opening into a lesser number of major creeks. The little creeks may become permanent, but may also atrophy and become channel pans (see below).

Thenceforward, the development of the marsh is usually rapid. In Norfolk the *Aster* marshes in early summer afford a luscious green appearance. *Aster*, *Salicornia*, and *Suaeda maritima* are often equally abundant,[1] and are, so to speak, revetted together by algae, especially

[1] *Spartina stricta* also occurs in patches on some of the Norfolk marshes at this stage.

Pelvetia canaliculata var. *libera* and *Bostrychia scorpioides*, both of which are unrooted forms. Thus, apart from the pans and the creeks, the whole surface becomes covered with a close mat of vegetation which may reach an average height of 5 or 6 in. above the mud surface. The general level remains low, and spring tides at least cover the surface. It is clear that everything now favours the rapid accumulation of sediment. The figures given on pp. 527 *et seq.* show how this can be measured.

With further increase in height of the marsh new plants appear, including *Limonium vulgare*, *Spergularia marginata*, *Triglochin maritimum*, and *Glyceria maritima*, although the last may appear much earlier, especially on sandy marshes. *Armeria maritima* is also abundant, and in late spring gives a magnificent display of colour. One other plant, especially in eastern England, deserves special mention. Even in the almost pure *Aster* marshes *Obione portulacoides* is prominent, growing as a rule along the banks of creeks, presumably because of the congenial drainage conditions, and thence spreading rapidly outwards. It is a thick bushy plant and increases like a weed over the marshes often nearly or entirely obliterating all the other plants in the areas it covers. It is rare on most of the west-coast marshes.

It is clear that the marsh has now reached a stage when the number of tidal inundations becomes appreciably less on account of the general increase in height. It follows, therefore, that the rate of increase of height now drops, although accretion may continue for a long time. The plants already mentioned persist in the later stages of the east-coast marshes, although *Salicornia* spp., *Suaeda maritima*, and *Aster tripolium* gradually decline in importance. Occasional plants may, however, be found in any part of a marsh, for example when locally its surface is lower, or on account of some other circumstance such as growth on the lower banks of a creek. On the other hand, new plants make their appearance, especially *Artemisia maritima* and *Plantago maritima*. At the highest stages of marsh growth *Juncus maritimus* and *J. Gerardii*, with *Glaux maritima*, are common.

Thus far, the general evolution of the marshes and their vegetation inwards from the enclosing bar or spit alone have been discussed. Like processes naturally appertain to the outward spread from the old shoreline, and probably are the more important because they frequently begin in more sheltered places. However, it is sufficient to visualize the general growth of the marshes outwards from the old shore and inwards from the offshore bars. Bearing these facts in mind, it is obvious that growth results in an all-over rise of the surface and also in a continuous restriction of the tidal channels. From the *Aster* marsh stage onwards, true marsh creeks are noticeable. The bigger ones may be wide and open, and resemble, at high water, large rivers, while at high spring tides the whole marsh surface is flooded. At low springs all the upper creeks, and perhaps also the major ones as well, dry out, and during neap tides only the lower marshes and creeks are inundated. The higher marshes are thus exposed to the air for some days on end and dry up. The importance of this in the spread and control of vegetation is discussed on p. 540. Before leaving this matter, one other point deserves mention. Along parts of the Norfolk coast, particularly at Brancaster, there are excellent examples of the transition from a true salt marsh to a fresh-water swamp. The highest marshes on the mainland side show a profuse growth of *Phragmites*, a fresh-water plant associated with other brackish-water species.

If a marsh should begin to grow in a space between two laterals, and if the mouth between the ends of the laterals be narrow, the development is usually rapid. This follows mainly from the fact that the inflowing tide, carrying in its load of silt and debris, finds a calm space in which sedimentation takes place quickly. When the ebb begins there has been a fair time for the particles to settle, and thus, for the most part, they escape being washed out again, and are less liable to the irregular shiftings that are almost certain to take place on an open marsh. The individual marshes of Scolt Head illustrate these circumstances very well, but the Marrams at Blakeney Point may perhaps be regarded as the classic examples.

Marshes like those in the Dyfi estuary are formed in much the same way as those on the east coast, but, as suggested on pp. 533–4, the sedimentation brought about by direct tidal influence is equalled, and in some localities possibly exceeded, by the sediment brought down by the rivers. Generally the Welsh marshes are much sandier than those of the east coast and further differences in appearance arise from the cropped nature of the turf of which they are largely formed, and from the absence on them of plants like *Obione* and the sea lavenders which are so common in the east. The contrast is also accentuated by the almost complete absence of lateral shingle ridges in the Welsh examples.

(2) SALT PANS

One of the most conspicuous features of a salt marsh is the salt pan. Pans are of different origins. When the vegetation first starts to spread over the original surface, whatever this may be, plant growth is not regular. Sooner or later parts of the bare surface may become wholly or largely enclosed by vegetation without being covered by it, and this process forms an embryo and rather ill-defined pan. As the vegetation continues to spread and grow upwards it may soon form a true enclosure with no outlet channel. Should this happen, the pan is likely to persist for some considerable time because, mainly on account of the lack of drainage, the soil becomes quite unsuited for plant growth. Furthermore, the fact that the pan fills at spring tides means that the evaporation of this water leaves a high percentage of salt, which in its turn is probably inhibitive to seed

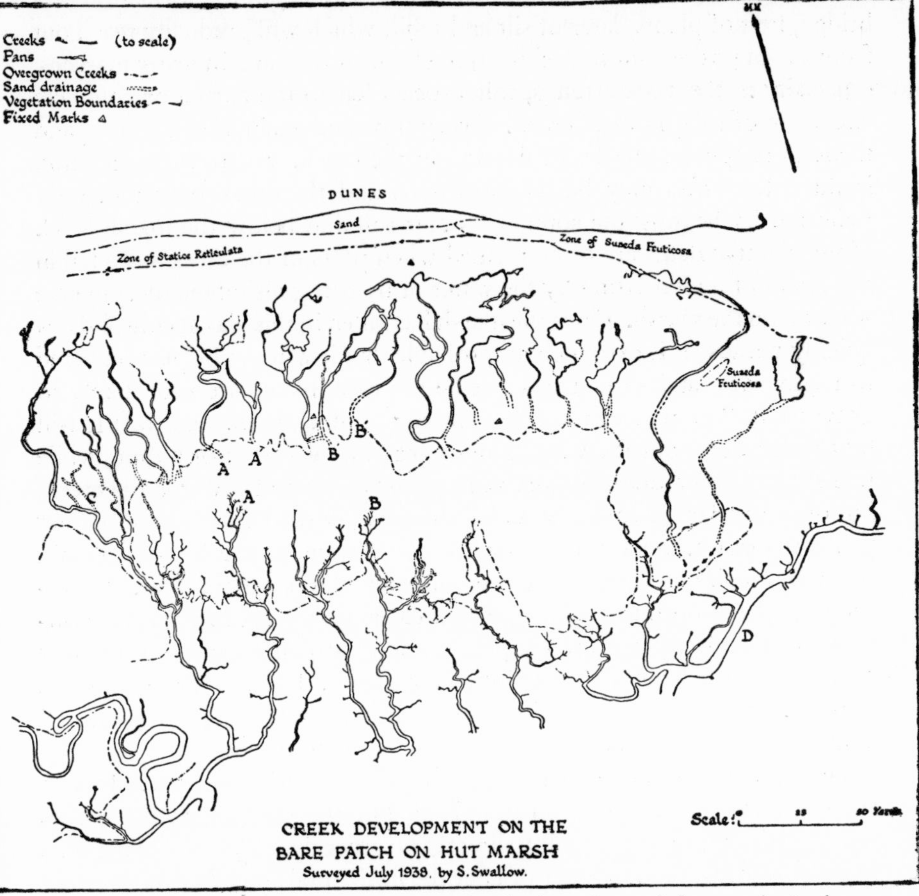

Fig. 109. The continuous dotted line marks the boundary of the uncolonized area in 1939. In uncolonized area—a scattered distribution of *Salicornia* spp. Mud in west, largely sand in east

germination. The conditions just described lead to the formation of what are called primary pans, obviously contemporaneous with the sward that bounds them. If, however, a creek cuts back (see p. 523) into a pan and so gives it normal drainage, vegetation rapidly spreads into it and the pan disappears.

Another type of pan is more elongated in outline, and on inspection can nearly always readily be related to an old channel or creek. The maintenance of a creek depends, on the one hand, upon the relative power of the ebbing and flowing tide to keep it clear, and, on the other, upon the rate of growth of the vegetation. If, as sometimes happens, the plants which overhang a narrow creek meet, they may begin the formation of a bridge, first of plants, later of silt and mud, which will gradually reach out from the opposite banks. A short tunnel thus forms, and in many marshes, especially in the Dyfi estuary, this process led to the partial or complete enclosure of long lines of creek. Every now and again a hole shows, and through it appears the water flowing in the creek. Where the vegetation is thick the holes may be inconspicuous and the creek only indirectly indicated by hearing the sound of water running below the marsh level. More effective dams are often formed when parts of the creek sides fall in as a result of undercutting by the water or by being dislodged by someone walking on the marsh. Once the marsh has fallen in this way it consolidates quickly because it is extremely improbable that it will not be plant covered to begin with, and should this not be the case, plants are soon likely to grow on it. The part of the creek above the block is thus cut off: it will fill at high tides, but the water cannot get away easily. For a time, while there is still some drainage, the vegetation will spread towards the creek, but once the dam is complete and the drainage totally interrupted, a long pan is the result. Such formations are very common, and it is often plain that a particular creek has thus been obstructed in several places. Nevertheless, its old outline usually remains clear, partly because the pans indicate it, and partly because the places where the dams started are often at a slightly lower level than the rest of the marsh. In eastern England it is noticeable that, once the initial stage is over, dams or tunnels are soon well defined as a result of the bushy nature and growth of *Obione*.

Secondary pans develop subsequently to the primary pans. In one sense channel pans can be so described, but the term probably applies most precisely to those forming directly on secondary marsh which grows in the following way. Sooner or later the horizontal spread of the primary marsh is arrested, either by erosion or tidal scour. Hence, a low cliff is left at the outer margin of the primary marsh and masses break away. The breaking process comes to a stop, however, because secondary or lower marsh soon develops at the foot of the small cliff. It is in this marsh that the secondary pans develop along lines very similar to those of the primary pans in primary marsh. It is likely, however, that the surface of the lower marsh will be somewhat irregular to begin with, since it grows on and around the blocks fallen from the higher marsh, and thus these initial hollows may sometimes form the pans.

(3) FURTHER DEVELOPMENT OF CREEKS

It has been shown that as the marsh grows both horizontally and vertically, the tidal waters become more and more confined in definite channels and creeks. But a casual visitor to a marsh might well suppose that the creeks, instead of being for the most part built up, were really erosion features. Erosion certainly takes place in creeks, but as a rule it is directed laterally, and, as described above, is one agent in the formation of channel pans. Creeks, however, are notoriously meandering, and a certain amount of erosion with some alteration in course thus occurs at bends for somewhat the same reasons as in an ordinary river. The difference lies in the fact that water flows up and down a creek, and for hours or even days on end the channel may be dry. An inspection of the banks of deep creeks usually reveals a series of layers or laminae of mud and silt, often with intercalations of blown sand, which indicate the stages by which the banks have grown upwards. The vegetation, especially *Obione*, growing on the marsh and on the creek edges, always tends to overhang the creek except where undercutting is so pronounced as to cause falls. Thus the banks are usually steep.

There are, it is true, certain erosion processes which help to form creeks, and in particular circumstances they may be very effective. On the largest marsh on Scolt Head Island the lower part (see Fig. 109) is a normal silt and sand formation, which has grown up in the usual way, and the inner edge of the whole marsh is fringed by a broad belt of dunes. Big tides cover the whole and often sap the dunes, thus causing the accumulation of what may be described as a beach of sand at the dune foot. Certain plants, especially *Statice binervosa*, *Obione portulacoides*, and *Suaeda fruticosa*, have colonized a good deal of this area and, by a growth rather similar to that on a normal marsh, have left small runnels or creeks down which the water drains when the tide is ebbing towards, but not as far as, the true lower marsh. Between this and the upper beach is an area as yet comparatively unvegetated. A few years ago there were far fewer plants on it than now and large stretches were almost bare sand, covered by a very thin film of mud. The creeks of the lower marsh extend upwards to this uncovered patch, while those at the dune foot reach down to it, and some now are continuous across it. A few years ago there were no 'through' creeks, but the reason for their recent development is clear: when the tide is ebbing from off the patch the water runs off as a whole, but when it comes near the ramifying heads of the lower creeks, it falls into them in minor cascades. The fact that there is a thin mud layer resting on sand often leads to a certain amount of actual undercutting due to the washing-out of the top layer of sand. The general result is, however, that the waterfalls recede and cause the creeks to lengthen headwards appreciably even in one tide. Since the water ebbing from the upper runnels near the dunes falls until the marsh is nearly drained, there is a tendency for it

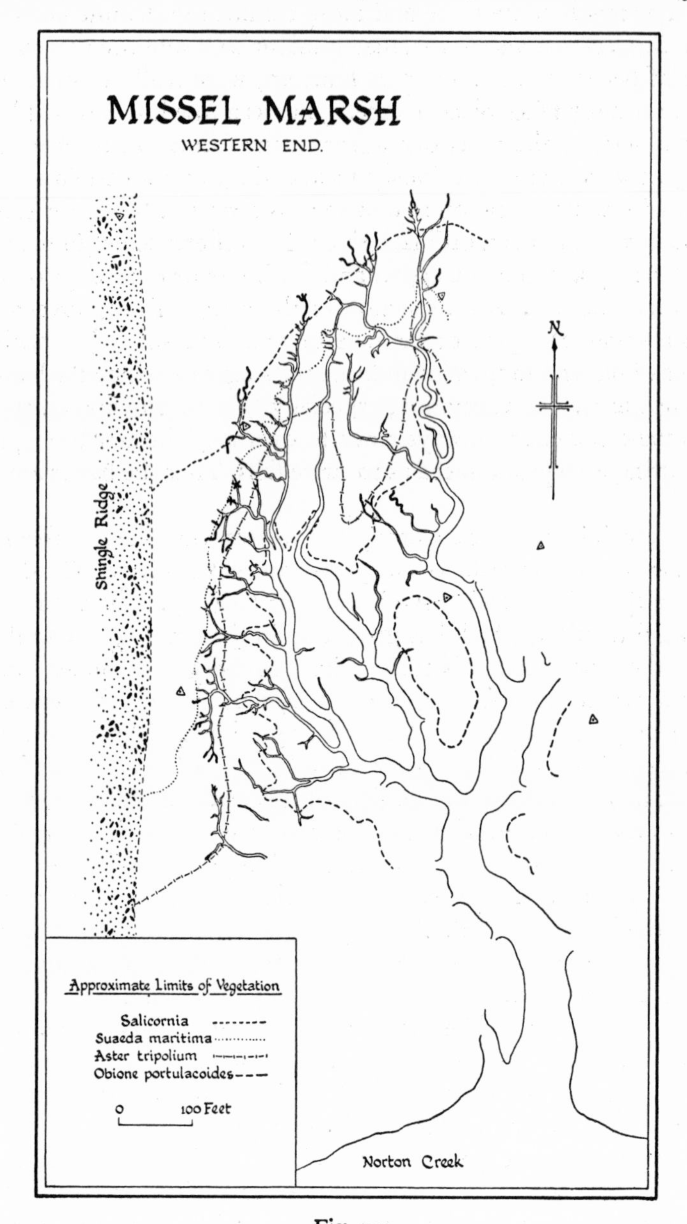

Fig. 110

to follow what were, at first, almost imperceptible lines of slightly lower level. Even now these are often scarcely noticeable, except that the upper sand spreads down them and so shows them up by a slight difference of colour and texture. It is clear enough that these drainage lines led in several instances to the heads of the lower creeks, and with the continuance of the headward erosion already mentioned, the creeks spread back along these lines and sooner or later cut through the less vegetated area to meet the runnels near the dunes. The process of headward erosion is illustrated in Fig. 110,[1] and precisely the same sequence of events is clear on other marshes on Scolt Head Island and elsewhere. Careful surveys to demonstrate the backward cutting of creeks were made of a part of Missel Marsh on Scolt Head Island in 1934 and 1937, and in those years several had extended headwards 30–40 ft. It is, however, in the opinion of the writer, a distinctly secondary process in creek formation as compared with the upward building on developing marshes. The creeks had also deepened somewhat due to the general upward building of the marsh.

Where marshes are very sandy, for example, like those at Talsarnau near Harlech, and in other parts of Cardigan Bay, the creeks are somewhat different in detail from those of muddy formations. At Talsarnau there is little if any mud, and the upper few inches of the marsh surface are bound together by roots of grasses and plants. Lateral erosion in creeks, which here, too, grew with the development of the marsh, is consequently easy, since the incoherent sand has been sapped away, and the top layer with roots accordingly has fallen down. Along a big creek the result is very striking in appearance, and in smaller creeks, where both sides may have been equally affected, the usual consequence has been the formation of a small inner channel. A glance at Plate **104** will show the effect of this process far better than a long description.

(4) UPWARD GROWTH OF MARSHES: MEASUREMENTS

In addition to the silt and mud trapped by plants on a marsh surface a good deal of blown sand is likewise captured, especially in those marshes partly enclosed by dunes. Strong winds carry the sand with them and in a storm a thin layer of it may be laid down all over the marsh. Lighter winds, however, and the sapping of the dunes by high tides induce a good deal of short-distance transport of sand. This naturally gets caught in the vegetation; in several places on the north Norfolk marshes, for example, *Obione portulacoides*, growing in patches in suitable areas, traps consider-

[1] J.A. Steers, *Trans. Norfolk and Norwich Nat. Soc.* 14, 1937, 212, and 1938, 393.

able quantities. The plants grow quickly and in this way quite large areas have risen several inches in a few years. Occasionally other plants have a similar effect, but much depends upon their actual position in relation to the available sand supply.

There are thus many factors controlling the rate of marsh growth. These include locality; supply of material—mud, silt, or sand; the nature of the vegetation; the rate of spread of certain plants; the height of the original surface on which the marsh forms, which will determine the amount of tidal inundation it receives; the strength and direction of the winds in carrying sand to it; and the effects of river-borne material. Any marsh must consequently be in a state of continuous evolution: it is simultaneously growing upwards, extending outwards, and also, in places, suffering erosion. It is worth turning, therefore, for a moment, to observations on the measurements of these processes.

The upward rate of marsh growth has been analysed fully in three places: in the Dyfi estuary by Richards, on the Skalling peninsula (Denmark) by Nielsen,[1] and at Scolt Head Island in Norfolk by the writer of this book. There are certainly plenty of records both in this country and in America of the rate of sedimentation at isolated spots, but not apparently of whole marsh areas. The method employed in the three cases has been the same, and owes its inception to Oliver who began to use it at Blakeney Point and who has not yet given a full account of any complete series of measurements. Since these marshes are formed mainly of mud and silt, a thin layer of any distinctive material laid down on them will form a basis for measuring the accretion. Distinctive coloured sands, such as those from Alum Bay in the Isle of Wight, have been used for this purpose, but, provided the localities are carefully marked, there is no sound reason why the ordinary shore sand of the area should not serve, although a distinctive sand is clearly less likely to give misleading impressions.

The sand is first of all scattered by hand in patches, perhaps 3–5 ft. in diameter, on the marsh surface and vegetation. If this be done at a period of spring tides, the next covering tide, or tides, will wash the vegetation clean and spread the sand fairly evenly over the marsh floor. A second visit is usually sufficient to smooth out any lumps with the hand or with a trowel. Once it has settled in and under the vegetation it will remain secure, and can be left for as many years as the investigator wishes. Should he aim at approximating with some accuracy the upward rate of growth of the whole marsh surface, he must, to begin with, decide upon a series of lines. When these have been surveyed the sand should be spread in

[1] N. Nielsen, 'The Marshes of the Skalling Peninsula...', *Kgl. Danske Selskab. Biologiske Medd.* 12, 4.

patches at intervals along the lines, depending on the details of the particular marsh. Since it is not easy to find such patches again unless they are clearly marked, it is well to set up at a distance from them of 3 ft. or more a stake driven firmly into the marsh. The sand patches must then be left for some years. Later, when the amount of sediment forming on them is to be measured, the investigator must take care not to trample over the sanded area and also not to spoil it for observation methods. There are plenty of ways of avoiding damage. The writer found that an ordinary bicycle pump with the nozzle end cut off and sharpened was most effective. The sharpened end could be carefully pushed into the marsh, and the sample brought up was cleanly cut and could be gently pushed out of the tube by the plunger. The layer of sand was always absolutely clear, and the amount of sediment on top of it was easily measurable to a millimetre. If four or five samples were taken from each patch and the results averaged, the final reading gave a very reliable indication of the amount of sediment which had accumulated since the sand was laid down. It was a simple matter then to find the annual rate of accretion.

On Scolt Head Island[1] several such lines of patches were prepared in September 1935. The lines were put down on different types of marshes, and also on marshes of different levels. The first investigation was made in June 1937, and the second in June 1939. Since great care was taken on each occasion in avoiding damage to the sanded areas, and since several samples were examined at each station, there is no reason for thinking that the observations of 1937 in any way upset the later series of measurements. Further, careful lines of level were run between not only the actual stations on any one line, but also between the different lines, at least as far as was practicable. The results were then related to Island Zero Level—a known height in relation to the local tides and Ordnance Datum. The results can be shown as on Figs. 111 and 112, but before studying these figures certain facts must be borne in mind:

(*a*) Station 1 on the Lower Hut Marsh line is 0·80 ft. lower than station 1 on the Upper Hut Marsh line.

(*b*) Station 1 on Missel Marsh line is 0·52 ft. lower than station 1 on Lower Hut Marsh.

(*c*) Station 1 on Missel Marsh is 1·32 ft. lower than station 1 on Upper Hut Marsh line.

In the sections the level of the marsh surface shown at each station is that for 1937, and the lines, that is to say the apparent surface of the marsh and

[1] J.A. Steers, *Geol. Mag.* 75, 1938, 1926, and *Trans. Norfolk and Norwich Nat. Soc.* 15, 1939, 41.

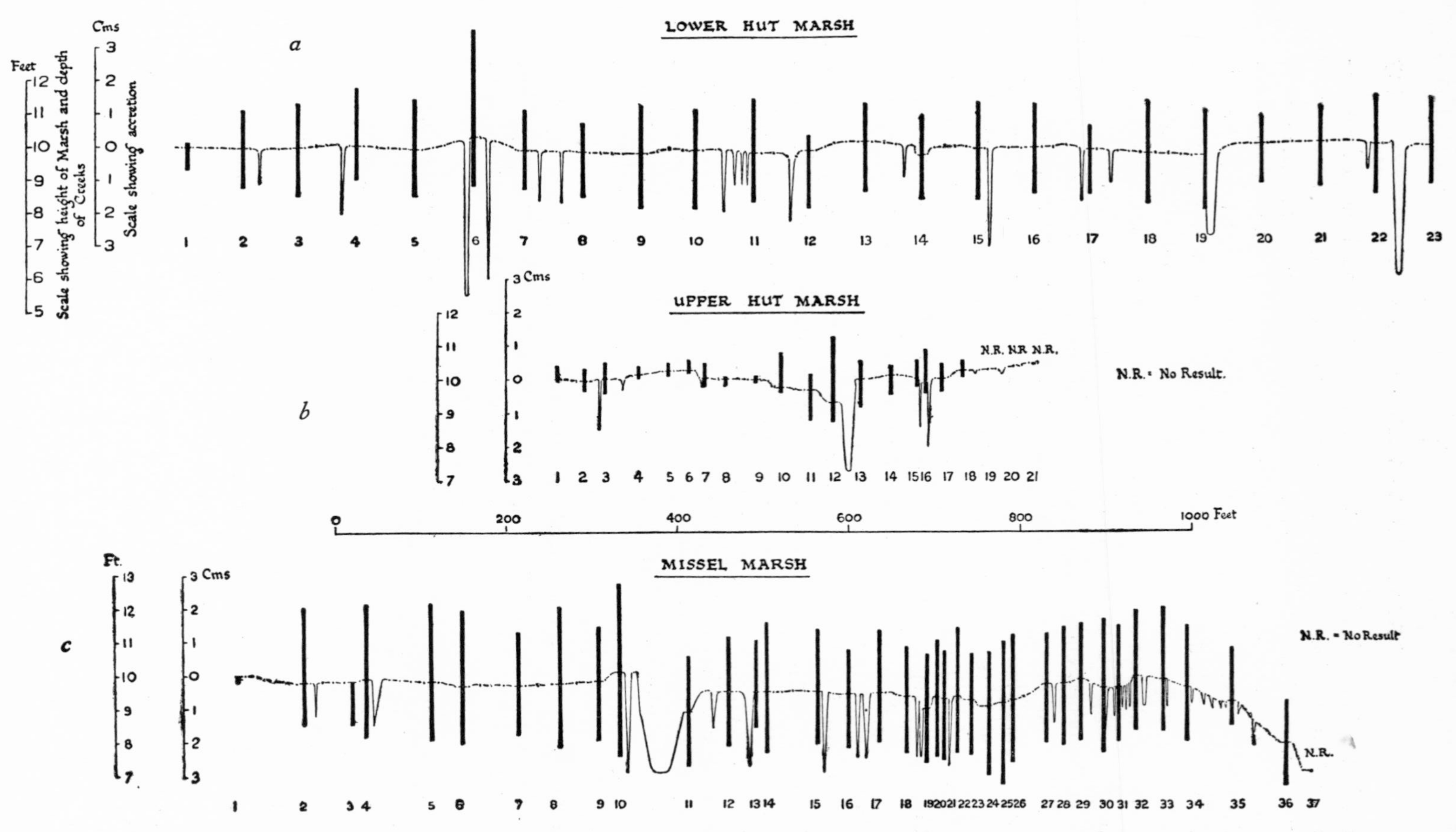

Fig. 111 *a*, *b*, *c*. Sections on Missel Marsh and Hut Marsh showing accumulation of sediment (see page 547)

sections of the creeks, joining these stations, are only approximate. It is impracticable to level every foot of the marsh surface. The vertical scale on which the marsh surface is shown is noted in Fig. 111; the creeks are represented on the same scale and their relation to the respective stations is also carefully indicated. The vertical columns showing the amount of sediment accumulated before and after 1937 are on a metric scale, and in

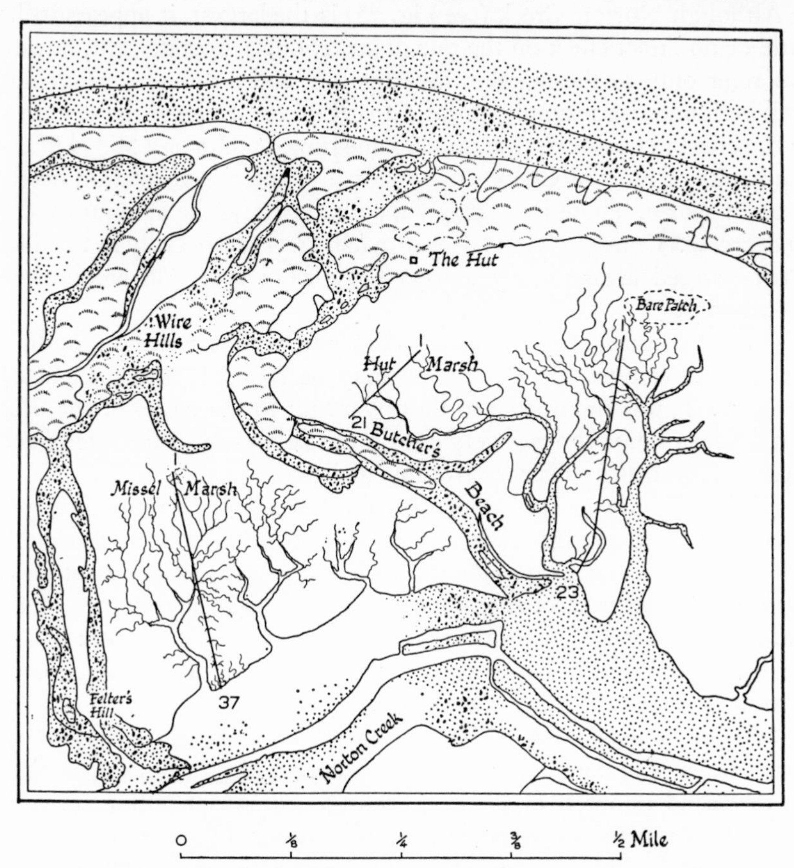

Fig. 112. Lines of sedimentation on Missel Marsh and Hut Marsh

reading the sections it is essential to bear in mind the two different vertical scales mentioned. The part of any vertical column *below* the marsh surface indicates the amount of accretion in the twenty-one months before June 1937, while the part above indicates the amount of accretion in the twenty-four months following that date.

The main features demonstrated by the sections are:

(1) That sedimentation rates are usually highest on the lowest marshes.

This is well illustrated by the readings for Missel Marsh, or by comparing those for Lower and Upper Hut Marsh.

(2) That the rate and amount of sedimentation is usually greater near big creeks, provided (*a*) that there is a thick spread of vegetation to trap the silt, and (*b*) that there is thick *Obione* bordering the creeks, particularly on the older marshes.

Although Norton Creek (see Fig. 78) is the largest, it appears to have little or no direct effect on the rate of accretion. This statement, however, can refer only to the marshes investigated; and a different result might well be found, for example, in the Great Obione fringe at the eastern end of the creek in question. But, at the same time, on Hut and Missel Marshes the banks of Norton Creek slope gradually, and the vegetation on them is often sparse. These two factors apparently cause the amount of silting to decrease. Conversely, the deeper marsh creeks, especially in the thickly plant-covered areas, nearly always show high accretion amounts near their banks.

(3) Golf Links Marsh is a closed marsh, with a narrow mouth, and sedimentation rates for this reason may be rather higher than in open marshes. It was impossible to tie this marsh to the levels of the others, but measurements showed very clearly the falling-off of the rate of accretion with increasing distance from its main drainage channel.

(4) On the Lower Hut Marsh line the amount of accretion, except for minor and probably insignificant variations, is fairly uniform throughout. There are, however, two noteworthy exceptions: (*a*) station 1 shows very little accumulation, probably because it is an area relatively bare of vegetation; (*b*) station 6, in contrast, shows an unusually high rate: it lies on a narrow peninsula between two deep creeks, and is also in thick *Obione*, two factors favouring rapid accretion.

(5) On Missel Marsh sedimentation is fairly uniform. Station 10 in some ways resembles station 6 on Lower Hut Marsh, but differs from it in that sand rather than silt is collecting. Stations 1 and 3 are curious because the 1939 measurements showed lesser amounts than those of 1937. Both lie in bare areas, except for loose *Pelvetia*, and so support the view that until a marsh is thickly covered with plants sedimentation is either slow or may be temporarily replaced by local erosion.

(6) Upper Hut Marsh stands at a higher level than the others, and for this reason alone shows a smaller rate of accretion. The sand put down at stations near the dunes was blown away and thus no record was available. In general, however, this line also illustrates the decrease of accretion away from the main creek.

(7) On Plantago (Aster) Marsh the slope of the surface was steeper,

and the highest values were recorded for the lowest stations, which were also closest to the main creek and in thick *Obione*.

(8) Slight and perhaps significant differences were reflected in the sedimentation on all the lines in the twenty-one months before June 1937, and the twenty-four months after that date. (*a*) On Lower Hut Marsh, except for station 6, the amounts in the second period were slightly less than those in the first. (*b*) Although there were exceptions, the converse generally applied to Missel Marsh. (*c*) On the Golf Links Marsh the amounts were generally higher during the second period. (*d*) Upper Hut Marsh afforded no satisfactory data. At all stations, except No. 12, the rates for both periods were low, but in general the accretion for the second period was very slightly higher than for the first. This seemed peculiar, considering the height of the marsh along this line. It appeared then that, with the exception of Upper Hut Marsh, the total amount of sedimentation showed a tendency to fall off on the higher marshes, but to increase on the lower and closed marshes. It did not necessarily follow, however, that the rate per month at any station showed an increase in the second period. Whilst these results were to be expected it must be emphasized that two periods of approximately two years each probably afford insufficient evidence upon which to base any definite statement. But whatever the limitations, the measurements as a whole suggest that (*a*) the rate of sedimentation is slow until a marsh is covered with a fairly thick vegetation and has already accumulated a spread of mud over it, and (*b*) after a marsh attains a certain height the amount of tidal flooding is less, and continues to decline with increasing height; consequently the rate of sedimentation falls off. There is, therefore, a period of some length during which the most rapid growth occurs, and it is probable that Missel Marsh is now nearing the end of this phase.

If the whole period of forty-five months be considered, and the rates of sedimentation for all stations showing records on each line be averaged, the mean amounts of accretion prove to be as follows:

Missel Marsh	*c*. 3·33	cm.
Lower Hut Marsh	*c*. 2·67	,,
Golf Links Marsh	*c*. 2·16	,,
Aster Marsh	*c*. 1·74	,, [1]
Upper Hut Marsh	*c*. 0·81	,,

The position of three of the lines is shown on Fig. 112. The main part of Hut Marsh is flooded by most ordinary spring tides, but is bare at neaps. It is seldom covered for more than one and a half to two hours at

[1] Not necessarily typical of the whole marsh.

any tide. Nearly all the creeks are fringed with *Obione*, and the other predominant plants along the longer line are *Aster tripolium*, *Spergularia marginata*, *Salicornia* spp., *Suaeda maritima*, *Limonium vulgare*, occasional patches of *Triglochin maritimum* and *Glyceria maritima*, and nearly everywhere a close mat of *Pelvetia canaliculata* var. *libera* and *Bostrychia scorpioides*. The upper part of this marsh (the Upper Line) is not covered by any but comparatively high spring tides, and then only for a short time.

Missel Marsh in its lower parts is covered by a close turf of *Aster tripolium*, *Salicornia* spp., *Suaeda maritima*, *Spergularia marginata*, and some *Triglochin maritimum* and *Glyceria maritima*. *Pelvetia* and *Bostrychia* are again abundant. Near the upper part of the line the vegetation is sparser, but *Salicornia* spp. is spreading. The plants on the Golf Links Marsh and Aster Marsh are similar, but *Plantago maritima* occurs abundantly on the latter, and the upper end of the line is in the *Juncus* and *Artemisia* belts.

Chapman, using the sedimentation rates obtained from the Scolt marshes between 1935 and 1937, reached some interesting, even if very tentative, conclusions on the rate of marsh development. Recognizing again the short duration of the sedimentation experiments, Chapman's[1] conclusions are worth quoting. On the Norfolk coast the primary succession proved roughly to be: Zosteretum → Salicornietum → Asteretum → Late Asteretum → Limonietum → Armerietum → Plantaginetum → Juncetum. If the sedimentation results be analysed under these headings, the following table emerges:

Vegetation	No. of sites	Range in height referred to Island Zero Level	Average rate of accretion in 21 months	Average rate of accretion per annum
Salicornietum	8	− ·90 ft. to + ·40 ft. = 1·30 ft.	1·19 cm.	0·68 cm.
Asteretum	22	+ ·40 ft. to + 1·30 ft. = 0·90 ft.	1·71 "	0·98 "
Late Asteretum	4	+ 1·30 ft. to + 1·90 ft. = 0·60 ft.	1·57 "	0·90 "
Limonietum	5	+ 1·90 ft. to + 2·30 ft. = 0·40 ft.	1·41 "	0·80 "
Armerietum	5	+ 2·30 ft. to + 2·90 ft. = 0·60 ft.	0·63 "	0·36 "
Plantaginetum	3	+ 2·90 ft. to + 3·30 ft. = 0·40 ft.	0·73 "	0·42 "

The rate of development of any one marsh thus passes through a number of phases, with accretion gradually becoming less as height increases and tidal submergences become less frequent. Moreover, bearing in mind the figures of the same table, a very rough estimate is possible of

[1] V.J. Chapman, 'Marsh Development in Norfolk', *Trans. Norfolk and Norwich Nat. Soc.* 14, 1938, 394.

the time taken under average conditions for a marsh to develop from the Salicornietum to the Juncetum stage:

Vegetation	Average max. depth of silt at conclusion of phase	Average time required for the accumulation of depth of silt found at conclusion of phase
Salicornietum	1·30 ft.	*c.* 58 years
Asteretum	2·20 „	*c.* 86 „
Late Asteretum	2·80 „	*c.* 106 „
Limonietum	3·20 „	*c.* 121 „
Armerietum	3·80 „	*c.* 172 „
Plantaginetum	4·20 „	*c.* 201 „

These figures suggest that the total time required for a marsh to pass from the beginning of the Salicornietum to the Juncetum is about two centuries. They would only be very slightly modified if the results of the sedimentation experiment up to 1939 were included. Clearly, exceptions such as any abnormal depression in a marsh area, or conditions in almost closed marshes, cannot be taken into account. In Denmark, and on similar reasoning, Nielsen has estimated that the Skalling marshes develop to maturity in about 100 years. These marshes, however, are sandier than those of the east coast of England and resemble the formations in Cardigan Bay.[1]

In the Dyfi marshes like experiments were made by Richards[2] who found 'that the rate of accretion was so dependent on the height of the point that it was impossible to trace clearly the connexion with other factors while this was unknown'. The strips of sand were laid down so as to cross the marsh vegetation zones from the Juncetum outwards, and the two following tables illustrate the nature of the results:

Line I: about 400 ft. long:

Association	Accretion (cm. in 100 lunar months)	Land height	No. of observations
Glycerietum	6·61	1·65 ft.	8
Transition Glycerietum-Armerietum	5·53	1·53	7
Armerietum	3·53	2·13	30
Transition Armerietum-Festucetum	2·00	2·85	1
Festucetum	1·75	3·24	4
Juncetum	2·03	2·69	3

[1] Stable conditions of land and sea are presumed in all these estimates.

[2] F.J. Richards, 'The Salt Marshes of the Dovey Estuary', *Ann. Botany*, 48, 1934, 225.

Line III: about 250 ft. long:

Association	Accretion (cm. in 54 lunar months)	Land height	No. of observations
Glycerietum	3·67	1·15 ft.	6
Glycerietum-Armerietum	6·93	1·64	2
Armerietum	4·09	1·67	15
Festucetum	3·38	2·31	6
Festucetum-Juncetum	4·60	2·33	1
Juncetum	2·25	3·00	1

The chief factors which governed the rate of accretion in any given place appeared to be (*a*) the density of the vegetation, (*b*) the height above mean sea level, and (*c*) the distance from the river front of the marsh. The third influence is somewhat more complex than appears at first sight. On line I accretion was high near the erosion cliff: then it fell off and then again increased. The greater rate near the cliff was traced first to the deposition by higher tides of coarser material near the creek banks, secondly to the amount, in places, of wind-borne sand, and thirdly to the general slope of the land. Where the slope is relatively steep, much silt laid down by the tides is later easily washed away by them and by heavy rain storms. This process naturally has a greater effect on those parts of the marshes where the vegetation is sparser.

The Dyfi marshes, it will be remembered, are in a very different environment from those in north Norfolk, and their general setting influences the supply of material. In the Dyfi this is definitely greater near the head of the estuary, because there the fine particles carried down by the river meet the salt water and deposition is rapid. The distance of the marshes from the river front or from a large channel is significant because it affects the supply of material in suspension and also that of wind-borne sand. These two factors are also important at Scolt Head Island. The others, including height above mean sea level, the presence of artificial obstacles such as groynes, the density and nature of the vegetation, and general topography need consideration in all marshes, but naturally their effects on marsh formation, whether single or collective, vary widely from place to place. It should be noted here that the technique of measurement just described does not seem to work at all effectively unless there is already a fair coating of mud on the marsh. Experiments made on a closed *Salicornia* marsh, just developing on sand at Scolt Head Island, were a failure because the new and distinctive sand was washed about, especially in winter when the thin stringy stems of the plants afforded no adequate protection at all.

Quite plainly as a result of the rapid upward growth of a marsh, the horizontal spread is also very considerable in a short time. This is easy to show by mapping a marsh before and after a period of years, or, even more quickly, by re-mapping a marsh on the existing 6-inch Ordnance Survey Map of the area (see Fig. 26: Talsarnau marshes). Just outside the mouth of the Wash is a small enclosed marsh at Holme, a region which has been carefully mapped at irregular intervals during the last one hundred and fifty years on account of reclamations made for grazing purposes. From a study of these maps, it seems clear that Holme Marsh has developed since the Enclosure maps of 1858, or even a little later. This finding, considered alongside the measured rates of accretion at Scolt Head Island, suggests that sufficient time has elapsed for Holme Marsh to develop into the lavender marsh of the present day: that is to say, the measured rates of accretion at Scolt Head Island and the historical evidence for the formation of Holme Marsh are corroborative. Anchor Marsh, a small formation almost encircled by a recurved shingle ridge at Scolt Head Island, has become completely plant covered since about 1920.

Whilst all marsh plants aid the accretion of silt, the most spectacular example of this activity is *Spartina Townsendii*. This is a hybrid of the introduced American species *S. alterniflora*, and the English species *S. stricta*, which is common on many marshes, and it was first reported from Southampton Water in 1870. It is a strongly growing perennial grass which thrives on deep mud often too mobile for the growth of *Salicornia*. Since 1870 it has spread remarkably quickly on our south coast, especially in Southampton Water and Poole Harbour. Holes Bay (Poole Harbour) in 1911 contained numerous clumps of the grass, but by 1924 it formed a nearly continuous and dense cover over large parts of the bay, and it has been planted in other marshes in order to help reclamation processes. *S. Townsendii* is a bigger plant than *S. stricta*. 'The stout stem bears stiff erect leaves: from the bases of the stems stolons radiate in all directions, binding the soft mud; and feeding roots, mostly horizontal in direction, ramify through the surface layers of the mud, while stouter anchoring roots extend vertically downwards. The leaves offer broad surfaces to the silt-bearing tidal water, and their points catch and hold fragments of seaweed and other flotsam. The thick forest of stems and leaves breaks up the tidal eddies, thus preventing the removal of mud which has once settled on the marsh....No other species of salt-marsh plant, in north-western Europe at least, has anything like so rapid and so great an influence in gaining land from the sea.'[1]

[1] A.G. Tansley, *The British Islands and their Vegetation*, 1939, p. 828.

(5) THE STRUCTURE OF THE MARSHES ON SCOLT HEAD ISLAND

The sub-structure of marshes naturally varies a good deal. On the west coast they are mainly sandy throughout, and, at most, only a thin layer of mud or silt may be present. Elsewhere, especially on the south coast, they are largely formed of very soft watery mud on which few plants, other than *Spartina Townsendii*, can take hold. This kind of mud is also found in the marshes in Hamford Water, Essex, where the main phanerogam is probably *Aster tripolium*. In part of the Hamford marshes death by drowning in this watery mud would be all too easy. The contrasts between different marsh sub-structures are difficult to explain, but a rapid upward growth of certain plants and a continuous supply of fine silt into a quiet backwater such as Hamford may produce soft marshes, especially where the underlying stratum of sand or of any other material is several feet below the present marsh surface. On the Norfolk coast there are restricted areas of mud much softer than that generally prevailing over the marshes. These patches are probably related to depressions in the substratum, a point which is discussed further on p. 537.

In order to obtain some idea of the geological structure of marshes the writer undertook an investigation of those at Scolt Head Island.[1] Shallow bores were put down in all the marshes, but some were much more fully examined than others. Fig. 113 shows, in the form of a stereogram, the nature of the deposits on Hut Marsh, the largest individual marsh on the island. It is largely shut in by dunes resting on shingle ridges, but the entrance to Norton Creek is a wide one, and there are two major creek systems. The western part of the marsh is distinctly higher than the rest, a fact that is particularly noticeable during the rise of a flooding tide. Twenty-eight shallow bores were made, and their positions are shown on the upper part of Fig. 113. The deposits encountered are marked below. Under the bare patch near the north-eastern corner of the marsh, there was evidence of a pocket of soft, inky-black mud, beneath a covering of sand and a very thin layer of surface mud (see also p. 521 and bores 11 to 17). The long line 1–28 shows that sand underlies the central part of the marsh, while the surface mud generally thickens southwards, that is, towards Norton Creek. The shingle seen in bores 21 and 22 represents the thinning-out of this material from the enclosing beach. The banks of the major creeks are built up largely of sand, and a great deal is blown on to the marsh from the surrounding dunes, which are also sapped by small waves at high spring tides.

[1] *Scolt Head Island*, 1934.

The marshes near the western end of the island are all in an early stage of development, and consequently show the gradation from an open sand flat to one covered by a few inches, or perhaps even a foot, of mud and silt. The structure of the older eastern marshes varies somewhat, since they usually show more surface mud, although sand is still a very important ingredient. In several there were found pockets of deep black mud like that under the bare patch on Hut Marsh. On the easternmost marshes (Plantago and Aster Marshes), the surface layer of mud near Norton Creek often proved more than 5 ft. thick and rested on sand. In the north-western part, near the main line of sand dunes, there were found to be a few inches of surface mud resting on 3–4 ft. of sand which was in its turn underlain in places by nearly 7 ft. of black mud.

On the marshes on the mainland side of Norton Creek the layer of surface mud was found to vary a good deal in thickness, but in the marshes, away from any dunes or shingle, it averaged roughly between 1 and 3 ft. Sand was usually discovered below, but lower down still, several bores struck the soft inky mud again. In one place this proved more than 7 ft. thick, and its whole extent could not be measured.

Clearly, the evidence of all the bores points to a sandy substratum for the island, although many contained alternations of sand and mud. A study of existing conditions easily explains this mixture, since near the western end of the island the sea has broken through the outer ridge and spread a thick layer of sand over the thin mud layer which had previously accumulated. This phenomenon is common to many marshes, for example, Holme Marsh where shingle also has invaded it. Shingle and sand shifted in this way can obviously overrun a pocket of the deep inky-black mud, and sand blown from the dunes has a similar effect. The deep pockets of black mud are less easily explained. They may fill scour holes. On the other hand, the pockets sometimes are partly enclosed between shingle ridges, or a lateral ridge may later extend and turn from its original course so as to enclose a small hollow. This deep mud at Scolt Head Island is always soft and stinking. It recalls somewhat the environment in which the soft marshes of Hamford Water are accumulating, although the patches on the island are much smaller.

Marshes in themselves need not indicate any change of level of land and sea. It has been shown that they form in sheltered waters on existing sand flats or other bottoms. In an area of considerable tidal range (that on the Norfolk marshes is nearly 21 ft.) it is clear that, given favourable circumstances, a pure mud marsh of at least equal thickness could form without involving any change of level. In depressed areas it could be even deeper. If, however, the marsh be everywhere consistently deeper than the tidal

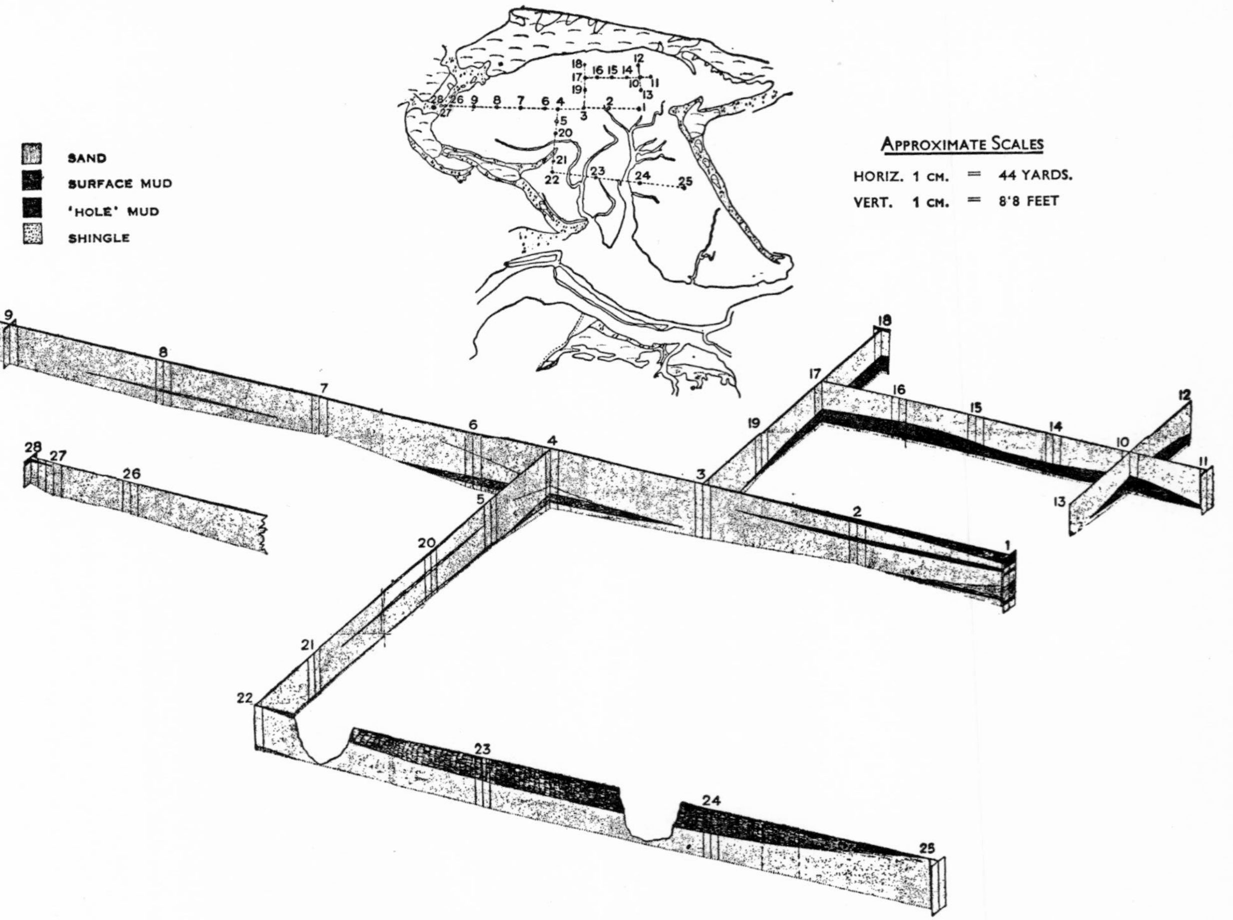

Fig. 113. Sketch and sections of the lines of bores on Hut Marsh, Scolt Head Island. No attempt has been made to show minor creeks

range, relative movement of land or sea is suggested. This phenomenon is the clearer if evidence of characteristically high marsh plants in the now lower levels of the mud is forthcoming. But if such plant remains appear within the limits of the tidal range, no such conclusion is allowable because breaks through the protecting shingle bars may lead to erosion of the enclosed marsh. Later, therefore, the break may be healed, and marsh growth begin again. Such a marsh may be termed 'rejuvenated', but provides no evidence of genuine change of level.[1]

(6) THE TIDES, WATER-TABLE, AND OTHER FACTORS OF MARSH ENVIRONMENT AND THEIR EFFECTS, ESPECIALLY ON THE PLANTS

Chapman, who began his studies on marshes at Scolt Head Island, and who later extended them to many other areas in this country and to the New England marshes of America, has analysed certain marsh phenomena more fully than any other investigator.[2] Moreover, although his studies were primarily ecological, much of his work is of great interest to the geographer. On the Norfolk marshes he cites ten main interacting and formative factors at work: Tides, Salinity, Drainage, Aeration, Water-table, Rainfall, Soil, Evaporation, Temperature, and Biota (that is, the inter-relations of plants, algae, and animals).

Table 1

	Periods of maximum submergence
Upper Marshes (above +1·10 ft. I.Z.L.)	(*a*) March–May (*b*) September–October
Lower Marshes (below +1·10 ft. I.Z.L.)	(*a*) December–January (*b*) April–May (*c*) July (*d*) October

The most important agent at work in his opinion, and certainly the most prominent and interesting from the geographical point of view, is that of the tides. Chapman developed a special technique for his work at Scolt Head Island. Careful levelling was carried out on the marshes, and a datum level, called Island Zero Level (I.Z.L.) was obtained, and found to be + 7 ft. O.D. The investigation covered the following matters: the number of submergences per annum; the periods of greatest submergence;

[1] Dr Chapman tells me that he is unable to agree with all the views outlined in this paragraph.

[2] V.J. Chapman, 'Studies in Salt Marsh Ecology, I, II, III', *Journ. Ecology*, 26, 1938, 144: see note at end of chapter.

the maximum period of non-tidal exposure—that is, two or more days during which no tide covers a part or the whole of a marsh; the hours of submergence each month; the hours of submergence in daylight. (Daylight was assumed to begin one hour before sunrise and end one hour after sunset.) Table 1 makes clear that marshes above and below a level of +1·10 ft. I.Z.L. show very considerable differences in tidal conditions.

On the upper marshes, therefore, submergence is most frequent at the spring equinox when the seedlings are just coming up, a point also noticed by Wiehe when working on the Dyfi marshes. Again, the autumnal submergences coincide with the equinoctial period which is, in addition, the fruiting season. These circumstances are important because the spread of many marsh plants depends on the tidal distribution of seeds. 'The rejuvenation of the desiccated marsh algae must also be secured by these tides and therefore the spring equinox tides will be of the utmost importance in perpetuating the algal species. Between these two periods there is a long period of non-tidal exposure in the summer (two or more days during which no tide covers the marsh) when considerable desiccation occurs. The whitened remains of algae strew the ground, but some filaments remain alive through being buried in the soil or under the dead plants until the incidence of the autumn tides restores them to vigorous growth. This long summer exposure probably plays a fundamental part in limiting the upper boundaries of some species.'[1] Chapman discussed the relation of the phanerogams to submergence in 1934. At that time he chose +1·30 ft. I.Z.L. as the significant level, which differs, of course, from the +1·10 ft. more recently adopted. However, from the present point of view, the difference is not of great significance, and the table for the Scolt Head Island marshes serves well to illustrate the manner in which the communities and species are related to submergence. Although it refers to one area only, it may well apply to the whole of East Anglia and even to other British coastal areas.

Amongst the algae, fifteen species occur almost wholly on the upper marshes, and of these, nine are Cyanophyceae, a group well able to tolerate adverse conditions. Incidentally, no species of the Cyanophyceae is wholly confined to the lower marshes, which have a period of maximum submergence in summer and therefore escape desiccation. Chapman is of the opinion that tidal submergence is of greater significance to the phanerogams than to the algae, because the distribution of the latter may be also determined by such factors as the effect of the higher plants, light, and space relations. It is also possible that the lower limit of the phanerogams is determined by the degree of submergence in daylight.

[1] Chapman, *op. cit.*

Thus only two plants are found below − 1·00 ft. I.Z.L., namely *Zostera nana* and *Salicornia herbacea*.

Marshes below +1·1 ft. I.Z.L.	Marshes above and below +1·1 ft. I.Z.L.	Marshes above +1·1 ft. I.Z.L.
	Communities	
Salicornietum	Late Asteretum	General Salt Marsh
Asteretum	Obionetum	Sea Meadow
Creek Asteretum	Suaedeto-Salicornietum	Juncetum
Zosteretum	Obioneto-Staticetum	Plantaginetum
		Sandy Obionetum
		Obioneto-Glycerietum
	Species	
Zostera nana	*Aster tripolium*	*Armeria maritima*
Salicornia perennis	*Salicornia herbacea*	*Triglochin maritimum*
Spartina stricta	*Obione portulacoides*	*Limonium vulgare*
	Suaeda fruticosa	*Spergularia media*
	S. maritima var. *flexilis*	*Glyceria maritima*
		Plantago maritima
		Artemisia maritima
		Juncus maritimus
		Statice reticulata
		Obione portulacoides var. *parvifolia*

(Table from *Scolt Head Island*, 1934, p. 133.)

The relation of the tides to the physical characters of the marshes is important, and special attention has been given to the water-table. In Norfolk this is influenced (*a*) by the daily movements of the tide, and (*b*) by the longer period movement of spring and neap tides. The first, as might be anticipated, is only felt near the creeks because of the resistance offered by the soil to water movement. The second, however, can be traced over the whole marsh because it is acting over a much longer period of time. At this point reference to Fig. 113, which shows the variety of substructure in a single marsh, is helpful. It must be remembered that the shingle of lateral ridges often stretches for a considerable distance under the marshes, and also that sand plays a large part in their composition. The penetration of the water through the mud, silt, sand, and shingle is a complicated process.

During flooding and non-flooding tides, a marked rise and fall of the water-table in the sand beneath the mud is noticeable and the range of this movement is greatest near creeks. In the surface mud, and sometimes in the upper sandy levels, an aerated layer is present which persists during spring tides, although at such periods it may be compressed. Most of this enclosed air appears to be contained in cavities due to decaying roots in the lower parts of the surface mud. If this be the case, the air in question will not be of the same composition as the atmospheric air[1] and the latter in its turn will not penetrate the marsh easily on account of the physical nature of the mud.

A number of factors influence the movement of the water-table from day to day. They include (1) the height of the previous tide which must determine the initial level of the water-table, and which is of most significance at high springs; (2) the varying resistance of the strata of which the marsh is composed; (3) the strength and direction of the wind, which in its turn can influence powerfully the upper limit to which tidal waters flow on a marsh area; (4) the height of the particular tide observed; (5) the elevation of the marsh under observation; (6) the distance from, and the proportions of, the nearest creek (plainly a large creek will influence the water-table more effectively than a small one); (7) the difference in height at any given time between the tidal waters and the water-table. In the marsh itself movement of water may take place in various ways. Soil resistance and distance from a creek will govern the rate of lateral seepage, and this movement will alter in direction according to the movements of flood and ebb tides. Downward drainage will be similar in character, except that it will take place continuously, although it may be masked by lateral seepage during spring tidal cycles. Its rate will again depend upon the composition of the marsh. Surface evaporation on a fine day in summer may be considerable as far as residual water is concerned, but as a primary factor it comes into operation mainly during and after a flooding tide. On a marsh area transpiration, often so important in dry-land vegetation, is nearly always masked by tidal phenomena. Depending upon the elevation of the marsh, the height of the tide, and the strength and direction of the wind, vast quantities of water may be carried on to a marsh very rapidly. To the other factors mentioned by Chapman brief allusion alone is necessary. The salinity of the North Sea near the East Anglian coast is about 30 to 32, but in the marshes the salinity of the soil water may rise to 180 during a period of non-flooding tides in the summer months. The salinity of the marshes in spring is an important factor in germination, and colonization of bare marshes may be largely controlled by this factor. The rainfall is small, and averages between 20 and 25 in. on the coast of Norfolk, Suffolk, and Essex. (On the actual *coastlands* of Cardigan Bay, in contrast, the average is between 40 and 50 in., and in specific places considerably higher.) The Norfolk marsh soils show high percentages of silt and clay which have substantial effects

[1] Analysis shows that the composition is different.

on aeration and drainage. Evaporation in summer is marked, so that salt crystallizes on the surface and on the algae: it is operative mainly in controlling the distribution of the seaweeds. The effects of temperature are limited, and probably explain the localization only of some algae and of *Zostera nana*.

(7) A COMPARISON OF EAST, SOUTH, AND WEST COAST MARSHES

It will be clear from what has been written that the vegetation of a salt marsh runs through a definite cycle of evolution. Tansley suggested that salt-marsh vegetation should be regarded as a formation, or, in other words, an area of vegetation that represents a climax in the conditions in which it exists. Chapman,[1] who has studied marsh ecology so extensively, dissents from this view. He believes that 'Salt-marsh vegetation is essentially dynamic, and whether the succession is apparently terminated by a Juncetum maritimae or Juncetum Gerardii depends on a number of factors. In those areas where fresh water flows out to sea there would seem little doubt that salt marsh progresses to reed swamp, and from this development by way of carr to the woodland climax is theoretically possible.' This view has also been expressed by Godwin in his botanical studies of the Fens. Naturally, the relative movement of land and sea will have an important effect. Should the land rise relative to the sea, the halophytic will sooner or later give place to terrestrial vegetation, and possibly this process is active at the moment in marshlands north of the Humber-Mersey line. But, conversely, should the sea rise relative to the land, one of two things may happen. If, on the one hand, the deposition of silt be equal to, or greater than, the rate of movement, the development of the marshland will be normal, and a transition to a land flora highly improbable. Such conditions may obtain to-day, for example, on the marshlands south of the Humber-Mersey. If, on the other hand, deposition be less than the rate of movement, then the marshes will soon disappear below sea level. As Chapman writes: 'The most satisfactory interpretation of the salt marshes would seem to be that which regards them as a number of different seres,[2] all of which are potentially capable of developing to the climatic or edaphic climax, which in the case of Great Britain would be deciduous woodland or coniferous forest in sandy areas. In many places, however, these seres are inhibited by local factors from proceeding towards the climax.'[3]

[1] V.J. Chapman, *op. cit.* VIII, 29, 1941, 69.

[2] The word 'sere' means a series of plant communities making a succession.

[3] *Op. cit.* p. 70.

On the north Norfolk coast marshes grow outwards from the coast and inwards from the offshore bars. The proportion of silt in the former group is usually higher, while the second group definitely form on a sand flat. There are many examples of such marsh growth on the Norfolk coast, but the best, perhaps, are at Thornham. In all probability the landward marsh will grow more quickly because (1) it is likely, on a flat shore, to begin sooner than the seaward, since some time may elapse before conditions are favourable for the formation of offshore bars; (2) the early stages in the bars are very unstable and embryo marsh is very apt to be washed away; and (3) the sandy floor is not a very good soil for seedlings. These three points are well illustrated on those parts of the shore off Thornham and Stiffkey. Fig. 114 represents Chapman's views on the salt-marsh succession in Norfolk. The long column on the left represents that found on the marshes growing outwards from the mainland, while the column next in length shows the changes in a marsh growing landwards from an offshore bar. Incidentally the status of the Spartinetum strictae is not easy to assess.

Although forming part of the east coast, the Norfolk marshes are not quite typical in all respects. Old reclaimed marshes in Essex were inundated by the sea in 1932–3, and are now covered with *Glyceria maritima*, *Salicornia* spp., and, on the lower areas, with *Aster tripolium*. At the south end of the Wash near the Nene outfall, *Glyceria* is the primary colonist on the sandy bottom, but at Ongar on the west bank of the Ouse, *Spartina Townsendii* appeared in 1925 and has spread rapidly. On the south Lincolnshire coast land vegetation seems to be invading the marsh, while *Obione portulacoides* and the rayless form of *Aster tripolium* (var. *discoideus*) do not occur in any abundance north of the Humber. The distinguishing points of the east coast may thus be summarized as follows: (*a*) the local development of *Spartina Townsendii* and *S. stricta*; (*b*) *Glyceria maritima* is only dominant where there is much sand in the soil; (*c*) the flora as a whole is very varied; (*d*) the marshes are usually very muddy, but nevertheless reasonably firm; (*e*) the marshes, when reclaimed, are excellent for grazing on account of the thick mud; (*f*) the algae are abundant; (*g*) *Obione portulacoides* is typical of the marshes, especially along the creeks.

On the south coast marsh development follows a different course. To-day *Spartina Townsendii* is the primary colonist, but before this hybrid appeared in 1870 the other two species of *Spartina* probably fulfilled its role. *Salicornia* is usually absent as a pioneer plant, perhaps because the substrate is not sufficiently firm. The main features of the sere, however, are (*a*) the enormous spread of *Spartina Townsendii*,

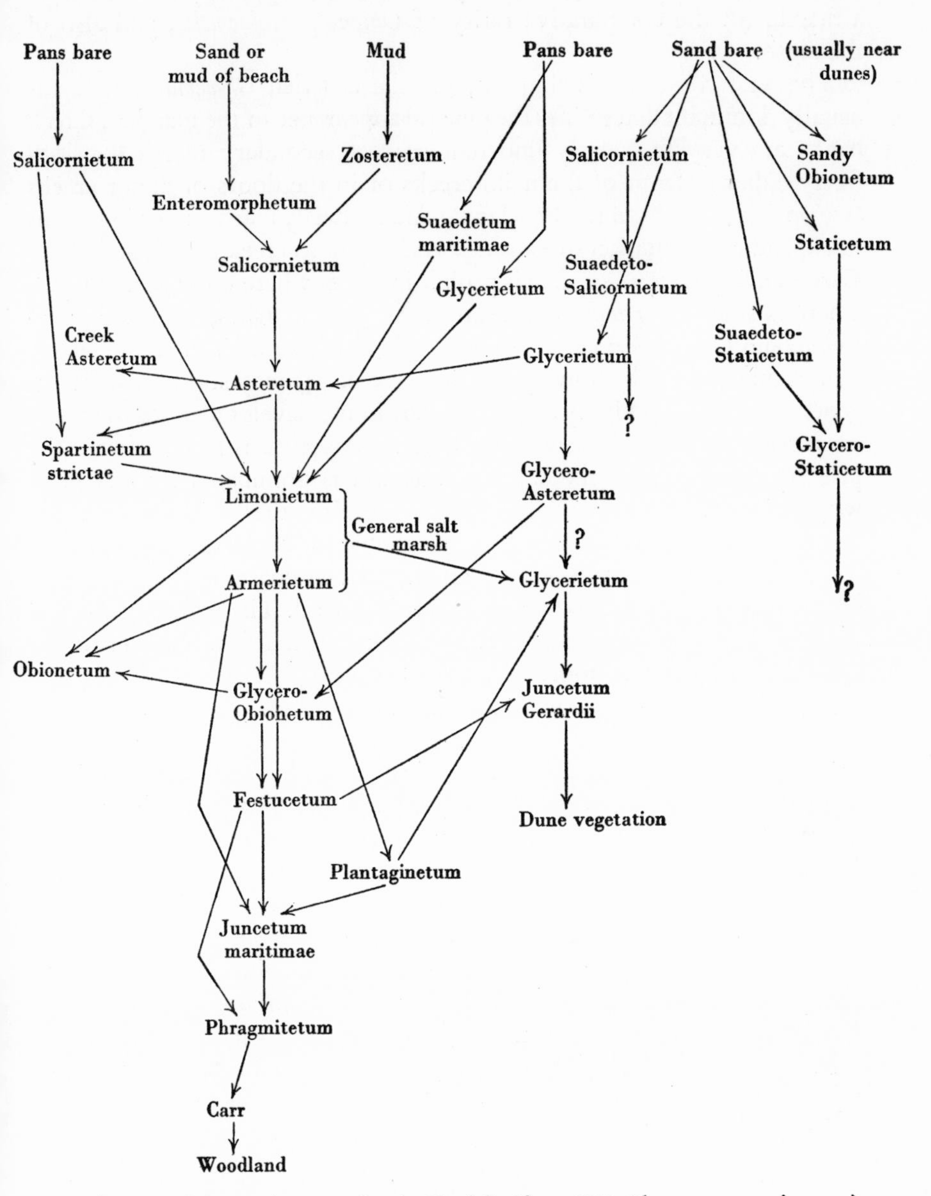

Fig. 114. Salt marsh succession in Norfolk (from V.J. Chapman, *op. cit.* 1941)

(*b*) the common occurrence of very soft mud except near the borders of a marsh, (*c*) the comparative rarity of *Obione portulacoides*, and also of the algae.

The west-coast sere is simplest, and the fact that *Glyceria maritima* is usually dominant determines the general appearance of the marshes. Creek banks are very frequently undercut so that secondary marsh develops freely either in front of the main creeks or in the floors of minor creeks (see pp. 148, 522). On the Penclawdd and Kidwelly marshes grazing seems to explain the limitation of *Obione portulacoides* to creek banks, and along Cardigan Bay *Glaux maritima* is prolific in places where water accumulates. There is some *Spartina Townsendii* in the Dyfi marshes, while *Aster tripolium* and *Obione portulacoides* reach their northern limit of abundance in Morecambe Bay. The Solway marshes (see p. 74) may still be rising. In general, soil plays a highly important part in the development of the west-coast seres: where it is very sandy *Glyceria maritima* is nearly always the primary colonist, and the whole succession is dominated by this plant which gives the marshes their grazing value. It is the sandy nature of the western marshes that renders them unsuitable for enclosure (see p. 91).

It is clear, then, that nowhere on the British coasts is the succession simple, but it is most complex in eastern England. The nature of the soil must, it seems, fundamentally control the vegetation of the seres: the east-coast marshes are more or less of firm clay, the southern ones of a soft silt, and those on the west coast mainly of sand.

The growth and development of salt marshes has thus been outlined. The differences between one locality and another have been shown, and in earlier chapters there has been some discussion of reclaimed areas. Reclamation is practicable and even easy once a marsh has attained sufficient height. The main needs are an enclosing sea wall and the construction of a simple drainage system to get rid of the salt from the soil: the two pieces of engineering provide a new and valuable addition to our countryside. Reclamation schemes, although very important and quite common, seldom cause much comment. To the casual observer they call up unfortunately the mental picture of merely flat and dull areas. To the agriculturist, however, they are valuable projects, and to the geographer they represent the culmination of a long natural process of very great interest.

Note. Only a few references have been given in footnotes. For Salt Marsh Ecology in general see V.J. Chapman, 'Studies in Salt Marsh Ecology I, II, III', *Journ. Ecology*, 26, 1938, 144; IV, V, *ibid.* 27, 1939, 160; VI, VII, *ibid.* 28, 1940, 118; VIII, *ibid.* 29, 1941, 69. These papers form the most modern and comprehensive study of the ecological problems. More specific reference to north Norfolk will be found in the same author's contribution to *Scolt Head Island*, 1934. All these papers contain useful bibliographies. General problems of salt marshes are also discussed in A.E. Carey and F.W. Oliver, *Tidal Lands*, 1919, and in A.G. Tansley, *The British Islands and their Vegetation*, 1939. Local details, together with a general treatment of salt pans, will be found in a paper on 'The Salt Marshes of the Dovey Estuary', R.H. Yapp, D. Johns and O.T. Jones, *Journ. Ecology*, 5, 1917, 65. Physiographical features of marshes are discussed in the references given in the footnotes (Steers) and also in *Scolt Head Island*, 1934. Much of the important pioneer work of F.W. Oliver will be found in Blakeney Point Reports, *Trans. Norfolk and Norwich Nat. Soc.* 1913–29: see also F.W. Oliver and E.J. Salisbury, *ibid.* 9, 1913 ('The Topography and Vegetation of...Blakeney Point...'). *Spartina* problems are discussed by F.W. Oliver, *Ann. Applied Biology*, 7, 1920. References to particular marshes will be found in many parts of this volume.

* (See page 528.) In June 1947, I was able to make a further series of measurements of the accretion. Since the experiment began in September 1935, the total period covered is 141 months. In the table below the accretion is given for this period, and the station numbers correspond to those on Fig. 111 *a*, *b*, *c*.

Lower Hut Marsh		Upper Hut Marsh		Missel Marsh			
1	6·5 cm.	1	2·4 cm.	1	lost	24	11·5 cm.
2	7·5	2	lost	2	8·5 cm.	25	11·0
3	8·0	3	2·0	3	7·5	26	9·8
4	7·0	4	1·4	4	12·5	27	10·0
5	8·7	5	1·2	5	10·8	28	12·0
6	eroded away	6	1·5	6	9·0	29	11·0
7	7·0	7	2·5	7	9·4	30	12·0
8	7·6	8	1·0	8	10·6	31	11·0
9	8·5	9	1·7	9	9·4	32	10·5
10	8·4	10	3·5	10	14·0	33	11·5
11	8·0	11	4·0	11	lost	34	12·5
12	6·4	12	6·0	12	10·0	35	10·5
13	6·5	13	lost	13	lost	36	lost
14	9·0	14	lost	14	10·0	37	lost
15	7·4	15	6·8	15	10·0		
16	7·3	16	lost	16	11·3		
17	6·5	17	lost	17	11·5		
18	8·8	18	3·8	18	10·0		
19	9·0	19	N.R.	19	9·0		
20	5·8	20	1·5	20	11·0		
21	6·0	21	N.R.	21	10·5		
22	8·5			22	10·5		
23	eroded away			23	11·5		
Average	7·5 cm.	Average	2·80 cm.			Average	10·6 cm.

Chapter XV

THE CLASSIFICATION OF COASTS

Although much new work on the matters discussed in the first three chapters of the earlier editions of this volume has appeared in recent years, the basic facts and principles there outlined hold good.

In Chapter 1 brief reference was made to Johnson's classification of shorelines. It has been superseded by others which are more elaborate. It may well be asked what is the value of classifying coasts. Answers would vary and perhaps it may be suggested that the greatest value is that the classifications are not comprehensive in all ways. This is by no means a derogatory point of view, indeed it is the reverse. A careful and detailed classification deserves study because it offers the student of shoreforms a scheme to criticize *constructively*. It is only too easy to say that a classification is inadequate or in some respects even wrong or misleading. It is far more difficult to improve on some of them—and by so trying the student will realize the complexity of the subject and how difficult it can be to add to or improve a given scheme. Coastal classifications usually apply to the world, or to large areas. For that reason alone they seldom do full credit to our own small island, which has had a long and complicated geological history and which is situated in an area subjected to complex isostatic movements.

It was mentioned on page 9 that in Johnson's scheme all the coasts of our island could be included in his submerged class, and that most coasts of the world, including our own, have to be put in his compound group. Newer classifications are usually more comprehensive and take the nature of the adjacent land into full consideration. They are also usually genetic. Four important ones may be briefly outlined.

F. P. Shepard (*Submarine Geology*, New York, 1948) suggests the following scheme:

I. *Primary or Youthful Coasts and Shorelines*, configuration primarily the result of the action of non-marine agencies.

- A. Shaped by terrestrial erosion and drowned by deglaciation or downwarping.
  - (1) Drowned river valley coasts (rias).
  - (2) Drowned glacial erosion coasts.

B. Shaped by terrestrial depositional agencies.
   (1) River deposition coasts: (*a*) deltaic, (*b*) drowned alluvial plains.
   (2) Glacial deposition coasts—partially submerged moraines or drumlins.
   (3) Wind deposition coasts (prograding dunes).
   (4) Mangrove coasts (and ? Saltmarshes, J. A. S.).

C. Shaped by Volcanic activity: lava flows, collapse of craters, etc.

D. Shaped by diastrophism: fault and fold coasts.

II. *Secondary or Mature Coasts and Shorelines*, configuration primarily the result of marine agencies.

A. Shorelines shaped by marine erosion.
   (1) Sea cliffs straightened by erosion (a wave cut bench is held to separate this group from fault and fold coasts).
   (2) Sea cliffs made irregular by erosion.

B. Coasts and shorelines shaped by marine deposition.
   (1) Those straightened by bars across estuaries.
   (2) Those prograded by wave and current deposition.
   (3) Those with offshore bars and longshore spits.

C. A. Cotton (*17th Int. Geog. Congress*, *Abstract of Papers*, 1952, 15), put forward a relatively simple classification:

I. *Coasts of Stable Regions.*

A. Those showing features of recent submergence.

B. Those mainly characterized by features of earlier submergence.

C. A general group including, for example, volcanic and fiord coasts.

II. *Coasts of Unstable Regions.*

A. Those coasts affected chiefly by recent submergence.

B. Coasts of emergence resulting from recent diastrophic action.

C. Fault and monoclinal coasts.

D. A group similar to I C.

H. Valentin (*Die Küsten die Erde*, Gotha, 1952) suggested a classification in which advance or retreat of the shoreline is the dominant factor, and published two maps of the world showing the application of his views to all coasts. (See the following table.)

J. T. McGill's Classification (*2nd Coastal Geography Conference*, Baton Rouge, 1959, 1, and map published separately, Office of Naval

H. Valentin's Classification

Prograding coasts	Emerging coasts			Sea bottom coasts
	Outbuilding coasts (Aufgebaute Küste)	Organically formed	By plant	Mangrove coasts
			By animals	Coral coasts
		Inorganically formed	By sea	Slight tidal effects
				Stronger tidal effects
			By river action	Delta coasts
Retreating coasts	Submerging coasts	Glacial	Erosion	Erosion following valleys, etc.
				Erosion less controlled by valleys and spreading over wider areas
			Accumulation	
		Fluvial	Young folds	
			Old folds (including rias)	
			Flat coasts	
	Coasts suffering erosion (Zerstörte Küste)			Cliff coasts

LAND FORM CLASS			Structure		Ice	Glaciers	Running Water	Wind	Coral	Volcanism
					Principal Agent Shaping Landforms					
LOWLANDS		Constructional Plain	Dominantly flat-layered		Ice Plain	Ice deposition Plain	Alluvial Plain Delta Plain	Dune Plain	Coral Plain	Lava Plain
LOWLANDS		Destructional Plain	Flat-layered	Sedimentary Rocks		Sedimentary Plain				
LOWLANDS		Destructional Plain	Flat-layered	Volcanic Rocks		Volcanic Plain				
LOWLANDS		Destructional Plain	Complex			Complex Plain				
LOWLANDS	LEVEL	Plateaux	Flat-layered	Sedimentary Rocks		Sedimentary Plateau				
LOWLANDS	LEVEL	Plateaux	Flat-layered	Volcanic Rocks		Volcanic Plateau				
LOWLANDS	LEVEL	Plateaux	Complex			Complex Plateau				
UPLANDS	SLOPING	Hills	Flat-layered	Sedimentary Rocks		Sedimentary Plateau: Remnant Hills				
UPLANDS	SLOPING	Hills	Flat-layered	Volcanic Rocks		Volcanic Plateau: Remnant Hills				
UPLANDS	SLOPING	Hills	Complex			Complex Hills				
UPLANDS	SLOPING	Mountains	Flat-layered	Sedimentary Rocks		Sedimentary Plateau: Remnant Mountains				
UPLANDS	SLOPING	Mountains	Flat-layered	Volcanic Rocks		Volcanic Plateau: Remnant Mountains				
UPLANDS	SLOPING	Mountains	Complex			Complex Mountain				
UPLANDS			Flat-layered to locally complex		Ice Plateau			Dune Hills	Coral Plateau	Lava Plateau Volcanoes

Research, Washington, 1958) is elaborate. The scheme on page 551 shows how he categorizes the major coastal landforms, within a zone of 5–10 miles of the sea.

All these classifications have their strong and weak points; there is no need to criticize them in detail. It is not easy to apply them to Britain. The Sussex–Hampshire coast, for instance, is a submerged coast, it is partly an alluvial plain coast, it is partly shaped by folding, it is modified by accumulations of beach in various forms. Other parts of our coastline could be similarly analysed and it would not be difficult to show that they do not easily, if at all, fit into any particular classification. That, in one sense, is immaterial; classifications are interesting in themselves, and should provoke a greater interest in a given coast. Unfortunately, the complex phenomena of our island render it virtually impossible to do full justice to our own coasts on world maps such as those of Valentin and McGill. The latter, for example, shows East Anglia as an ice-deposition plain, with a sedimentary plain on its western side, a short symbol indicates some barrier beaches and some cliffs on the north of Norfolk, and a sign offshore that the tidal range is 10 ft. or more. Since all this is printed in a space not greater than a square half inch it is remarkable that it is so accurate. There is, however, no suggestion of the drowned nature of the coast, or of the differences between Essex and Norfolk. Valentin shows the same coast in his first map as a flat-layered coast with some cliffs, together with a sea-built area where there are strong tidal effects. In his second map it is classified as a presumably sinking coast. Here again, a great deal of information is shown, but since the whole of the British Isles on Valentin's maps is enclosed within, approximately, a square inch adequate representation is out of the question.

To analyse world maps in this way is perhaps unfair, and doubtless if their compilers had unlimited space small areas would be given more detailed treatment. But that in itself would not overcome the difficulty of fully depicting all that a coast shows of its history and evolution. It is an interesting exercise to apply these classifications in detail to England and Wales. By so doing one is forced to consider very carefully the significant factors of a stretch of coastline, and so learn more about it. A coast is far from static. Erosion is cutting back many parts, others are extending by the growth of sand or shingle banks and marshes, much is still slowly sinking partly as a result of isostatic movement, partly because of a slight recent rise in sea level following the melting of ice in polar regions. Parts, despite eustatic movements, are rising more quickly than is the sea level. There are abundant instances of raised beaches and submerged forests, and to-day's movements are but a continuation of those responsible for these features. The position of our islands on the margin of Europe, on the margin of the Quaternary Ice Sheets, means that we are in a place where isostatic movements are not all in the same direction. Moreover, the structure of our islands is complicated: three powerful fold systems have each had a great effect, and each succeeding system has to some extent involved its predecessor. It is relatively easy to classify many hundreds of miles of some continental coasts; the intricate history of our own makes the problem unusually difficult.

Chapter XVI

WAVE ACTION AND BEACH FORMATION

The effects of wave action on beaches have received much attention since the Second World War. This is partly because of the continued interest in the subject for its own sake, partly because of the founding of the Hydraulics Research Laboratory at Wallingford (it began work in 1951) where not only investigations into particular coastal defence and other problems are undertaken but also fundamental research, and much more because of the impact of the great storm of 1953 on the nation. This storm focussed attention on the significance of coastal researches of all kinds. Smaller model tanks for observing wave action, like that in the Department of Geography at Cambridge, have also given opportunity for a good deal of research. At Cambridge Dr C. A. M. King began her work in this field, and in 1959 there appeared her *Beaches and Coasts*. This book has not only significantly added to our knowledge of the whole problem, but has welded the geographical and engineering approaches most successfully. There is no need to follow Dr King at any length since the main theme of this book is the nature and evolution of our own coast. Nevertheless, Dr King's volume forms an essential commentary on many aspects, and not least on beach features and longshore drift.

In Chapter III attention was called to W. V. Lewis's views on wave action, especially his constructive and destructive waves. In general King accepts those views in relation to shingle and sand beaches. Lewis emphasized that wave frequency was important in its effects; to him waves of low frequency were constructive, and those of high frequency destructive. He also considered variation in wave length was a factor. King adds that it is the change in the steepness of waves, height remaining the same, that is more important. Steep storm waves are often entirely destructive on a sand beach, and if the sand is coarse the amount removed is greater. Nevertheless, storm waves may build up a high and long-lasting shingle ridge above the reach of normal waves, although at the same time much shingle from lower levels is taken offshore. Usually, sand beaches have gentle profiles, whereas shingle ones are much steeper. But a steep beach also implies that waves breaking upon it use their energy in a narrow zone, and this means a greater mobility of the beach. King shows that not only does the slope of a beach increase with the coarseness of the material of which it is built, but also both as the wave length decreases, and the wave steepness increases. Much depends on the relation of the force of the swash to the backwash; if the backwash is greater then the slope will be less.

On many of our beaches the shingle stands at the top of the beach, and often there is a clear boundary between shingle and sand. Shingle is churned up into suspension only at the place, the breakpoint, where the waves break. Elsewhere it is moved along the bottom by rolling or sliding, and is thus much influenced by the swash which drives it landwards, and so pushes it to the top of the beach. This same phenomenon can be matched in model tanks. If a beach is wholly formed of sand of a more or less uniform size, or of a mixture of different sizes of sand, it is unusual to find any variation in the arrangement of the sand in different seasons of the year, or under varying wave conditions. Nevertheless, T. Scott (*Inst. Eng. Research*, Wave Research Laboratory, Univ. of California) found in a tank experiment that originally well-sorted sand rearranged itself in such a way that the coarse grains moved towards the beach, and the finer grains in the opposite direction.

Experiments made in model tanks in which material of varying sizes was used showed that the beach profiles made under wave action simulated those found in nature. If the tank waves were steep a beach profile analogous to a storm profile on a natural beach was produced, whereas flat waves gave rise to a 'summer profile', which may be steeper than the storm profile: it is a swash bar (see below). Associated with these profiles are certain types of bar. When steep waves were running it was found that a bar or ridge formed at the line along which they were breaking. At this time the sand was found to be moving landwards on the sea or outer side of the bar, and in the opposite direction on the inside of the bar. On the other hand, the crest of the bar never reached above water level. When the bar reaches its maximum height the waves breaking on it are deformed, and form again inside the bar. When this stage was reached, these new waves were able to move the sand *inside* the main bar in a landward direction. At the same time the effect of the waves on the outer side of the main bar was to cause a deepening of the water there, and so, in turn, to bring about a landward shift of the bar itself. In general, the position of the original bar depends upon the height of the waves since that will determine where they break. If by chance smaller waves follow large ones a second bar may be formed, but if the water level falls the bar is destroyed.

The swash bar is a totally different feature. It forms in front of the breakpoint, and is produced by the action of the swash and backwash of the waves. Flat waves push sand shorewards, and in a tank this can be seen to form a bar near the top of the beach. Unlike the breakpoint bar it can grow upwards to the upper limits of the swash. The waves producing the bar are not steep, and the level to which it may rise is governed by the height of the waves forming it. If it is washed by waves of greater steepness, which break on it and do not swash over it, its front slope is eroded.

In nature the breakpoint bar is represented by the submarine bar. These are best developed in tideless seas, but may occur on oceanic shores after storms. Often there are two or three such bars parallel to one another, the positions of which change from time to time with changing wave conditions. In tank experiments it was found that two bars could form if large steep waves were followed by small ones. 'The same mechanism can account for the presence of several bars on the natural profile; the outermost bar is formed by the largest storm waves, which are large enough to break in this depth of water; the smaller bars, in shallower water inside it, are formed by the intermediate sized storm waves, while the inner bar is formed by the short, steep waves which will affect the beach much more frequently. They may be generated by the sea breezes which blow onshore during summer afternoons' (King, *op. cit.* p. 334). The inner and smaller bar is more mobile than the major one. Measured changes have been made of bar movement, for example by W. W. Williams at Sidi Ferruch, in connexion with landing troops. Bars of this type do not reach above the surface of the sea.

On many sand beaches ridges and runnels are found (see pp. 101, 693). They are particularly characteristic of tidal seas. Gresswell has described these features on the Lancashire coast. They seem to stay in one position for long periods of time, and do not move systematically landward; even if they trend parallel to the shore. Yet they shift from time to time, or one may die out and be replaced by another in a rather different place.[1] They seem to disappear during onshore storm conditions, and reform in calmer weather. A remark made by King concerning bars of this sort 'that nearly all the foreshore surface was rippled with the exception of the seaward faces of the ridges which were composed of very firm smooth sand, the ridge crests on the other hand were often of soft sand, which had

[1] Beach ridges often trend at an angle to the Lincolnshire shore: on any one profile they appear to move steadily landward as a result of their longshore movement.

clearly not been firmly packed by wave action' certainly agrees with the writer's observations at Scolt Head Island.

In tank experiments bars of this type are built by constructive waves in front of their breakpoint, and thus a falling sea level leaves them unscathed. Like ridges exposed on a sandy beach at low water, they have gentle seaward, and steep landward slopes. Thus these ridges have characteristics of a swash bar, and are the result of the constructive action of waves. King maintains that they are best developed in relatively enclosed seas, where there may be a great abundance of sand and that they are absent from beaches influenced by long swells which run on the most exposed beaches and allow a flatter equilibrium gradient. Around our own coasts they belong particularly to the North Sea and the Irish Sea, and not to the south-western peninsula.

A further stage of development is the barrier beach. Constructive waves may under some conditions, especially where the tidal range is not too great, build up a sand ridge above mean sea level. It is more difficult to picture a shingle ridge forming wholly in this way because (see p. 562) of the difficulty of waves moving shingle directly onshore. Hence, in shingle barriers (see, for example, Scolt Head Island) lateral transport of shingle is an essential factor. Moreover, shingle ridges which rise above sea level are usually formed by steep storm waves which are also destructive. Yet these waves often throw shingle well above the parts of the beach reached by normal waves. In this sense barrier beaches are, as it were, a further development of ridges and runnels, but because of their greater significance they are treated more fully on other pages of this book.

Chapter XVII

MOVEMENTS OF BEACH MATERIAL

In recent years considerable advances have been made in the tracing of the movements of sand and shingle both on- and offshore. The knowledge so gained has enabled us to form a far more accurate picture of what happens on a beach, and has already pointed the way to quantitative assessments of the amounts of stones or sand moved. The means by which this information has been obtained are threefold: by radioactive tracers, by fluorescent tracers, by new methods of marking pebbles with paint that can sustain considerable abrasion for a length of time sufficient to allow of adequate observations. To all three may be added another technique—that of aqualung or skin diving. Several countries have experimented in one or more of those methods. Since we are concerned only with England and Wales it will suffice to call attention to the comprehensive paper (Techniques of Marking Beach Materials for Tracing Experiments) by C. Kidson and A. P. Carr (to appear in the *Journ. Amer. Soc. Civ. Engs.*, Harbours and Waterways division). A good bibliography accompanies this paper, which also contains an analysis of the methods used in various experiments at home and abroad. The literature on the subject is increasing rapidly.

One of the first experiments in this country was made in the Thames estuary by the Hydraulics Research Station in conjunction with the Port of London Authority and the Atomic Energy Research Establishment. It may be assumed that any subsequent experiment in this country using radioactive tracers necessarily involves the last named establishment. In this particular experiment the movements of mud and silt were investigated. A preliminary trial was made in 1954. In the full-scale investigation the isotope scandium-46 was selected; it has a half-life of 85 days, and the density of the tracer was similar to that of the mud. It was designed to throw light on the transport of silt in the estuary. The tracer was put down 26 miles below London Bridge, an area in which accretion does not occur, on 5 July 1955 at 45 min. after high water.[1] Immediately after the injection detectors were used, and on this same day increases of activity

[1] The details and precautions necessary in making these experiments, and the way in which the tracers are detected by geiger or other form of counter are omitted. They are described in the papers cited.

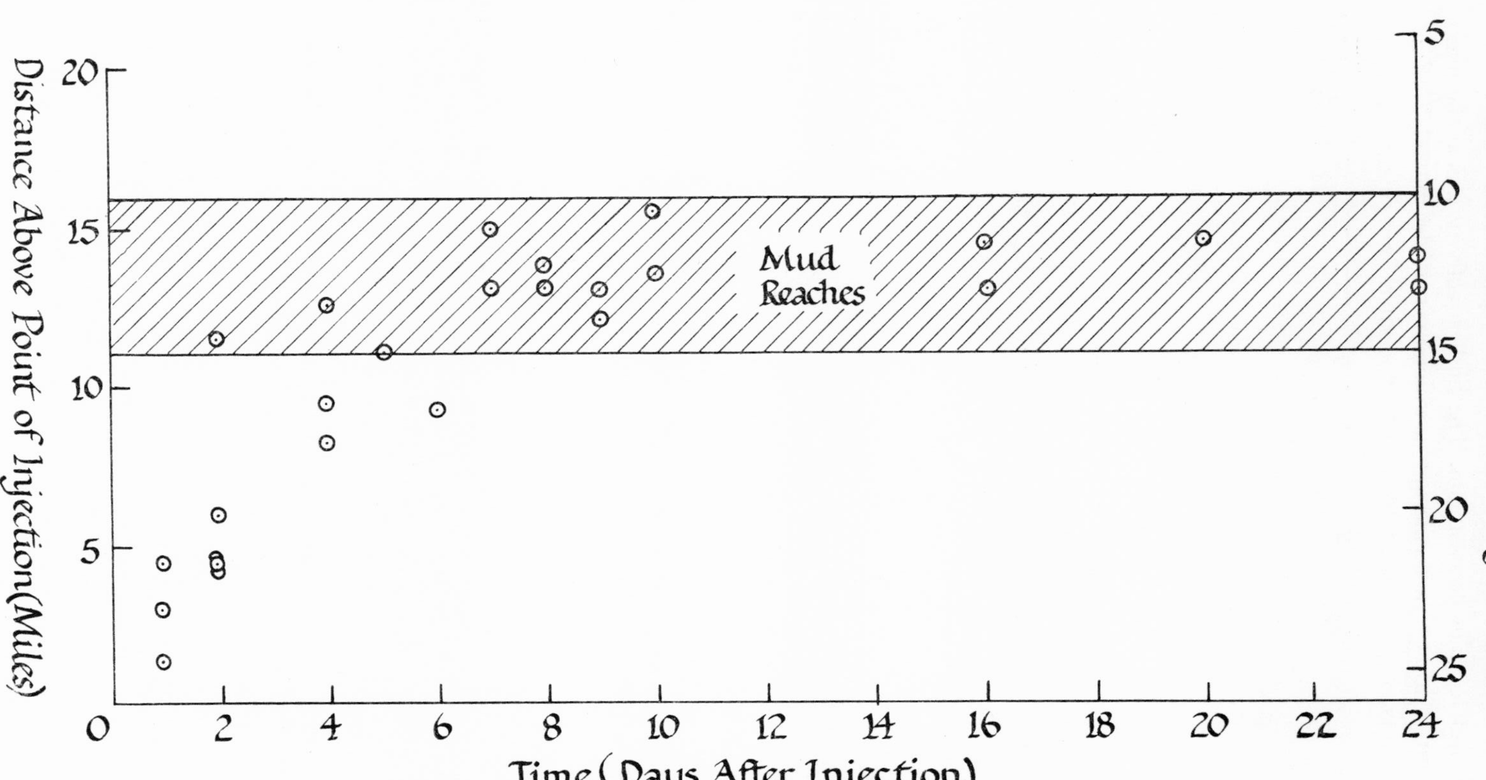

Fig. 115. Positions above injection point where radioactive tracers were detected at various times after injection in the Thames Estuary (D.S.I.R. *Hydraulics Research*, 1955)

were found 10 miles downstream. The highest activity was at the injection point. Fig. 115 shows the way in which the particles moved (*Hydraulics Research*, 1955, D.S.I.R.). In this experiment movement was up or

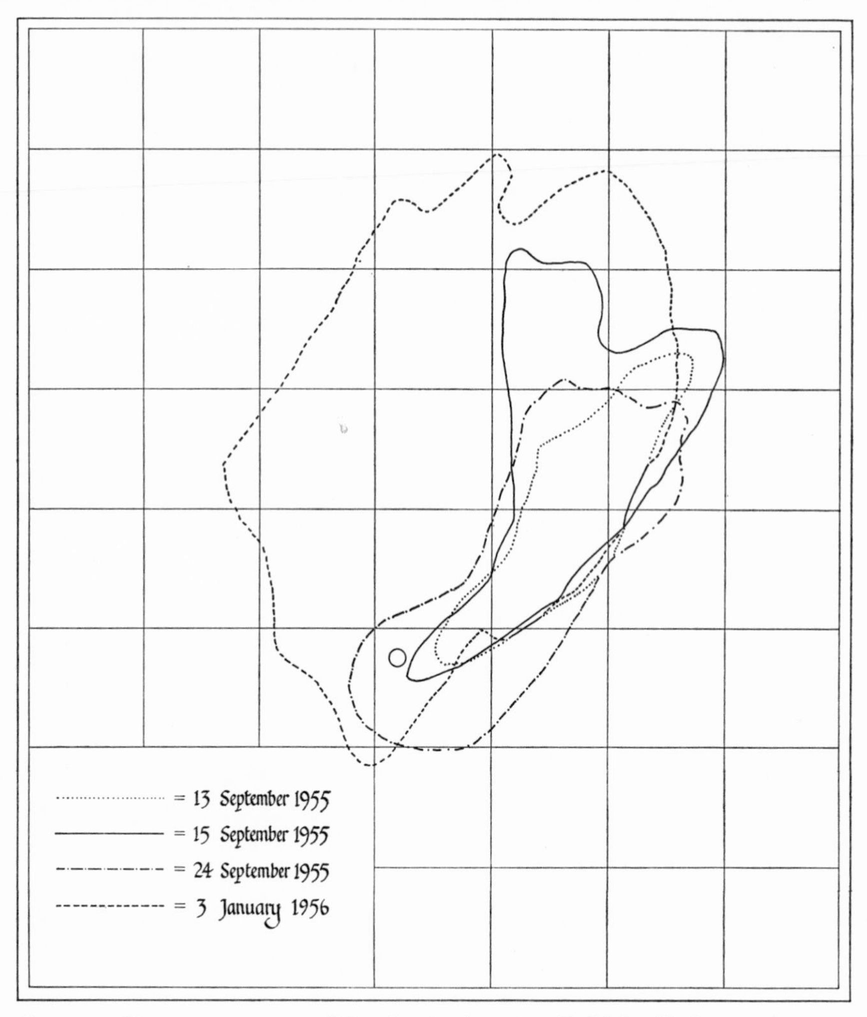

Fig. 116. Tracer experiment off Sandbanks (based on D.S.I.R. *Hydraulics Research*, 1956). The side of a small square represents 200 yards. The small circle represents the original dropping place on 12 September 1955

down channel, and the difficulty of making observations was somewhat less than in the open sea where movement may be in any direction.

Another experiment conducted by Hydraulics Research took place off the Sandbanks peninsula in Dorset. Scandium-46 was used, but on this occasion the movement of sand was under investigation. The sand on the sea bed hereabouts is quartz sand with a mean diameter of approximately 0·22 mm. The tracer material was boron-free soda glass with 1·5% scandium oxide, and its specific gravity was similar to that of the quartz sand. Current and tidal observations were made of the part of the sea to be studied, and the injection took place on 12 September 1955 at a point 2,800 yards offshore from the Sandbanks beach. Surveys of the area were made at intervals for 4 months and the general nature of the move-

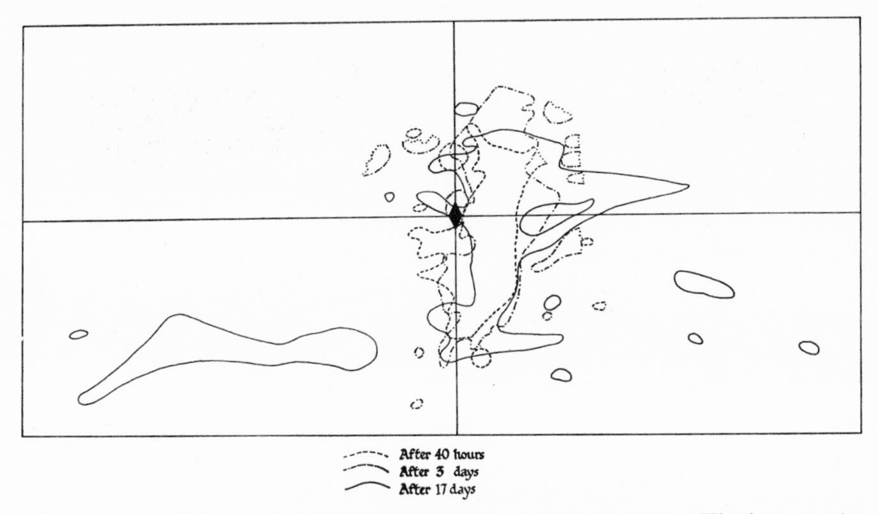

Fig. 117. Tracer movements off Liverpool (after H.R.S. *op. cit.*). (The longer sides of the rectangles represent approximately 8,000 ft.)

ment detected is shown in Fig. 116. It was not until south-easterly gales occurred in early November that appreciable movement took place. 'Briefly, the results indicate that the tidal currents in the area were too weak to induce movement of beach material, and that this movement only took place to a significant degree when wave action occurred under the influence of the wind' (D.S.I.R. *Hydraulics Research*, 1956).

A third experiment carried out by the Hydraulics Research Station took place in Liverpool Bay in 1958 (see also p. 582). This concerned the capacity of the Upper Mersey estuary which, despite heavy dredging, was becoming less. The site of the experiment was outside the mouth, between Newcomb Knoll and Little Banks. Once again scandium oxide was used, and a special glass with a density of 2·55 was prepared as a tracer. The tracer was put down at high water neaps, on 25 August 1958, and searches were made for it over about 2 months. Figure 117 shows the general nature of the movement of the material far better than a long discussion. Slight complications occurred because some naturally radioactive silt was found near the injection point. It was established as a result of the experiment that fine sand put down near the injection point does not long remain there; that the net movement was up the estuary with the flood and that the movement is quick, up to 500 ft. per tide (*H.R.S. Rept. of Experiment carried out in Liverpool Bay in 1958*, D.S.I.R.).

Let us now turn to the detection of pebble movement. The first experiment in this country was made by the author in association with D. B. Smith of Harwell. The object of the investigation was to find out if there was any onshore movement of pebbles to the beach at Scolt Head Island (see pp. 358 et seq.). The supply of pebbles to this beach is obscure. Since, in 1956, the method of placing the tracer, barium-140 with a half-life of 12 days, in a hollow in the pebbles had to be adopted, fragments of Permian sandstone broken down and rounded to approximate to the size of the average beach pebble were used together with a number of hand-made cement pebbles. Their specific gravity was about 0·2 less than that of the natural flint pebbles which were diffiult to bore with holes $\frac{1}{2}$ in. deep and $\frac{1}{8}$ in. in diameter. In all about 1,200 pebbles were used.

They were dumped on the sea floor on 5 April 1956, at a point (see Fig. 119) about 500 yd. seaward of normal high water mark at the Headland. The depth of water varied from about 12–25 ft. according to the state of the tide, and 16–20 ft. may be taken as the average depth during the experiment. The sea floor here is firm sand with some shingle. The dump was marked with a buoy, and checked by triangulation from the land. The weather was squally from time to time, and the night following the dumping the wind reached force 5 to 6 on the Beaufort Scale. Several sweeps were made up to the middle of May and a fairly consistent picture was obtained. Most of the pebbles moved inshore and to the west, the maximum amount of movement being 260 ft. One stone seems to have moved 450 ft. to the north-west. No pebble was traced as far as the beach. The investigation showed that some movement of pebbles, stirred up by waves and (?) helped by tidal currents, can take place under conditions that, even if at times squally, were certainly not rough. But there was no proof that in this way the beach may be adequately fed from offshore supplies (J. A. Steers and D. B. Smith, *Geogr. Journ.* 122, 1956, 343).

This experiment was followed by one at the south end of Orford Ness under the direction of C. Kidson who also took part in that at Scolt (C. Kidson, A. P. Carr and D. B. Smith, *Geogr. Journ.* 124, 1958, 210). Here, too, barium-140 was chosen, but a different method of marking the pebbles was adopted. It was found that if flint pebbles with a surface layer of ferric oxide were used, they could be treated in such a way that the tracer could be made to adhere to their surfaces. Thus the same pebbles as those on the beach were used. In this experiment some stones were dumped at a point 2,200 ft. offshore in about 25 ft. of water, and others were put down on the beach itself. Although several careful searches were made, no movement of those put down offshore was detected. Whilst no storm swept the coast, winds of 18–26 knots were experienced, so that wave action at times was by no means negligible. On the beach the pebbles moved readily, the direction depending upon the winds blowing at a given time. The general direction of growth of Orford Ness is to the south, but during the experiment there were light winds from the south-east for a few weeks. Figure 118 shows where the pebbles, about 2,000 in all, were deposited at low water on 23 January 1957. The south-east winds caused a steady northward movement. After 4 weeks one pebble was 1¼ miles to the north of its origin, and the mean of those found was 600 yd. In the 5th week the wind changed and blew from north and north-east, but never exceeding 15 knots. Nevertheless, the movement of pebbles was reversed, and marked pebbles were found not merely south of the injection point, but also on the knolls in the haven. Others had crossed the haven in which tidal currents reach 6 or 7 knots at springs. How they reached the knolls and Shingle Street was made clear by another investigation (C. Kidson and A. P. Carr, *Geogr. Journ.* 125, 1959, 390) in which both radioactive tracers and special paint were used.

For the offshore work in this second trial barium-140 was used and lanthanum-140, with a half-life of 40 hr., was chosen for the observations on the beaches. Pebbles marked in this way could be traced for about 1 week. All the shingle used was taken at random from the beach, and represented local conditions exactly.

The material, 1,000 pebbles, was dumped offshore on 6 February 1959, at the same place as in 1957. Another 1,000 were dropped at site 2 (see Fig. 118). Other pebbles were put down at sites 3 and 4. Unlike 1957, some movement took place at all stations (first table, p. 565) and in all directions at each station 'with a tendency to slightly greater travel shorewards and northwards'. There is good reason to believe, however, that

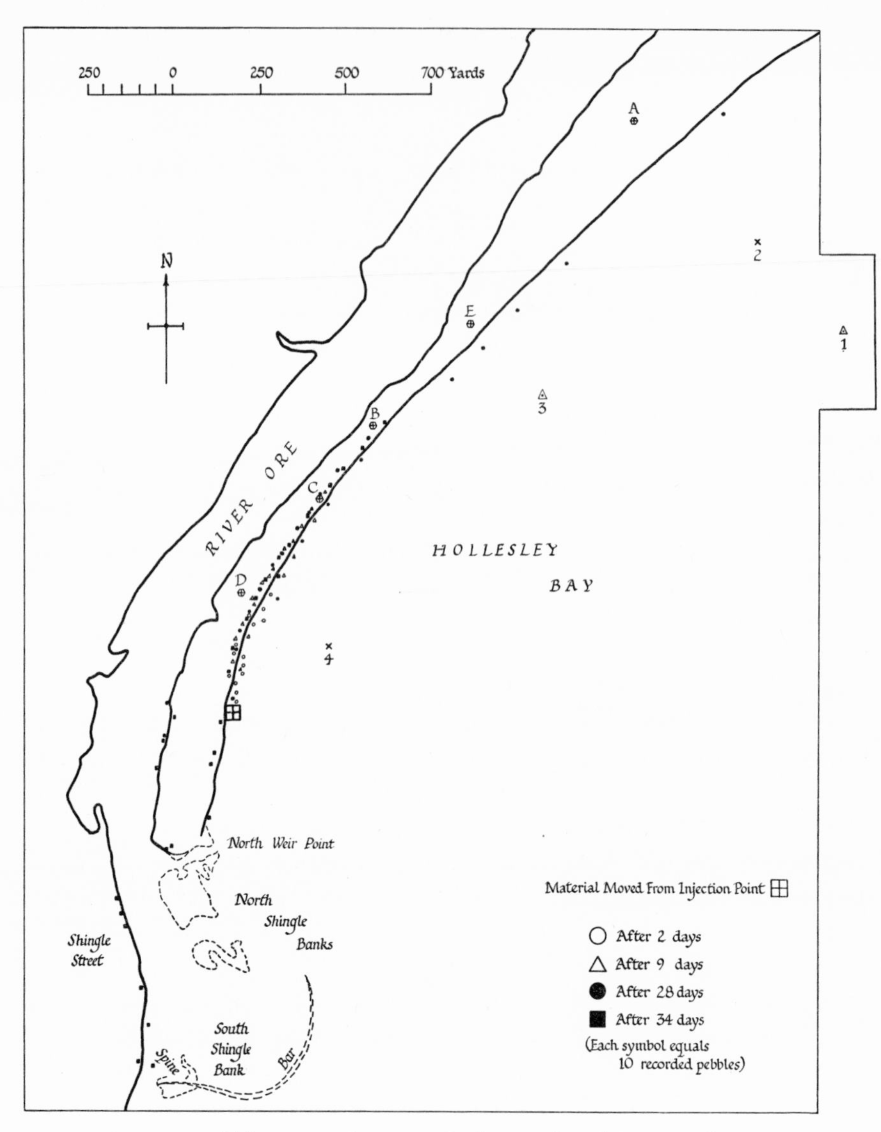

Fig. 118. Pebble movement on Orford Ness (based on C. Kidson, A. P. Carr and D. B. Smith, *op. cit.*)

none of this material reached the beach during the time the observations were made. Weather conditions varied from calm to gale.[1]

Offshore experiment

Site no.	Distance offshore (ft.)	Duration of experiment (1959)	Maximum dispersal (ft.)	No. of radioactive pebbles
1	2,200	6.ii to 9.iv	134	1,000
2	1,060	6.ii to 9.iv	151	1,000
3	700	25.ii to 11.iv	110	1,000
4	665	25.ii to 11.iv	108	1,000

The beach experiments were designed to explore the way in which shingle crossed the estuary. Five stations were chosen (see Fig. 118, and table below) on North Weir Point and the Knolls. At Station 1, North Weir Point, the pebbles moved for the most part up-river, but a few reached the shingle bank to the south. Only small waves occurred throughout this period, and they, working with the flood tide, were sufficient to account for this drift. Pebbles put down on the north shingle knoll worked slowly over the bank to its steep river slope. As far as could be seen there was no movement by current action along the steep river faces of the banks, and it was concluded that pebbles reaching this knoll tended to remain there. 'Shingle which succeeds in crossing the estuary must leave the spit well north of the distal point, and travel round the outer edges of the shingle banks and across the bar. Once on the North Shingle Banks it is effectively removed from circulation for a long period, unless winds from the south return it to the zone where longshore travel is possible.'

Shingle Bank Experiment

Injection site no.	Duration of experiment (1959)	Maximum dispersal (ft.)	No. of radioactive pebbles
1	6.ii to 10.ii	332	600
2	26.ii to 4.iii	0	600
3	9.iii to 12.iii	168	100[1]
4	12.iii to 13.iv	238	85[1]
5	7.iv to 13.iv	222	600

[1] Marine painted pebbles.

[1] Dr Kidson comments that in his view the movements observed were rather in the nature of 'a settling down' of the pebbles.

Radioactive substances present certain disadvantages in tracing coastal drift. They are expensive, and what is more serious, may be dangerous so that great care has to be taken in handling them and, if they should be deleterious to health, in preventing their reaching popular beaches. Moreover, since they take time to prepare they are not well adapted to rapid change of plan should it become necessary.

For these reasons fluorescent tracers are becoming more popular. One of the important requirements of engineering work on beaches is a reasonable quantitative assessment of the volume of material moved. This can be roughly estimated by the pile-up on one side of a groyne or pier and in similar ways, but more exact methods are essential. Experiments made in wave basins at the Hydraulics Research Station showed that 'if the drift was D tons per day, constant in one direction along a beach and T tons of tracer material were added each day at any point on the beach, then at a point down-drift of the injection point where the concentration had become constant, the concentration of tracer C would be $C = T/D$, since T tons of tracer were added to each D tons of drift, and $D = T/C$'. This was true of tracer material like that of the beach, and if the tracer was thoroughly mixed with the beach material, and the experiment continued long enough for the concentration to become constant. In nature, waves approach beaches from varying directions; longshore drift is not always in one direction; and there may be local peculiarities. Nevertheless, experiments on real beaches show that in practice 'the theoretical results based on these assumptions give values for the net annual drift which are tolerably close to the actual values'.

These theoretical views and the practical application to Rye beaches are discussed by W. J. Reid (*The Dock and Harbour Authority*, 41, 1961). Fluorescent tracers can be used for sand or shingle. When sand is used ordinary beach sand is mixed with a fluorescent dye and a plastic glue. Heavy concrete stones 'in which were embedded small fragments of fluorescent plastic' were used for shingle. Many different dyes can be used, but Reid found rhodamine B, primuline, and uvitex best. These show red, green, and blue under an ultra-violet lamp. Sand or stones thus marked can be fed to beaches, and easily picked out at night, and at low water, by means of a portable generator and an ultra-violet lamp.

The experiment at Rye was a quantitative one. Known amounts of tracer were put down each week for a year, and counts were made at certain intervals at known distances from the injection point. A square frame (10 × 10 ft.) was placed at random on the beach at the observation point, and the treated pebbles, which showed up brilliantly, were counted. Since we are here only concerned with the nature of the method there is no need to go into details of calculation which can be found in Reid's paper. The important point is that it offers a method that promises to give useful and reliable quantitative results.

Aqualung or skin diving can be usefully applied to many aspects of coastal work. It has been used successfully in the exploration of now submerged archaeological sites, and is being applied to the investigation of erosion features, beaches, caves, etc., now below water, as well as to an exploration of the geology of the sea bed adjacent to the coast of St Agnes' Head and Cligga Head in Cornwall (see p. 228). There are many practical difficulties in this work quite apart from the training required to become adept in the art. In certain seas, for example the Mediterranean, a high sun and clear water can make submarine observations relatively easy. Around our coasts, especially sandy ones, the stirring up of bottom material by tidal currents even when the surface is but slightly, if at all, disturbed by waves, renders observation difficult. Even a bright sun may not minimize this difficulty much. Moreover, there are the problems of keeping position and direction which are far more difficult under water, especially when a tidal current is running. There is no doubt that aqualung work can help coastal investigations in many ways, but it would be stupid to underrate its difficulties and consequent limitations.

An interesting preliminary experiment was made off Scolt Head Island (C. Kidson, J. A. Steers and N. C. Flemming, *Geogr. Journ.* 128, 1962, 49). Dives were made at the twenty-eight points shown on Fig. 119. In this way the nature of the sea-bed could be assessed, and in a far more accurate manner than would have been the case by taking samples with grabs. The direction, for example, of the sea-bed ripple marks in sand indicated a dominant sand movement to the east. At some of the diving stations shingle was found. Since this, especially the smaller stuff, often carried growths of fragile creatures it was concluded that there was little movement there. The depth of water is 30–40 ft. at high water springs at these places.

At certain stations marked shingle was put down. This was observed directly by divers and so it was possible to watch directly any movement that took place. At the three inshore stations, B, C and D on Fig. 119, no movement occurred in June 1960, although the pebbles were partially buried in sand. In May 1961 when the sites were re-examined no trace of the pebbles was found; it was assumed that they were buried in sand.

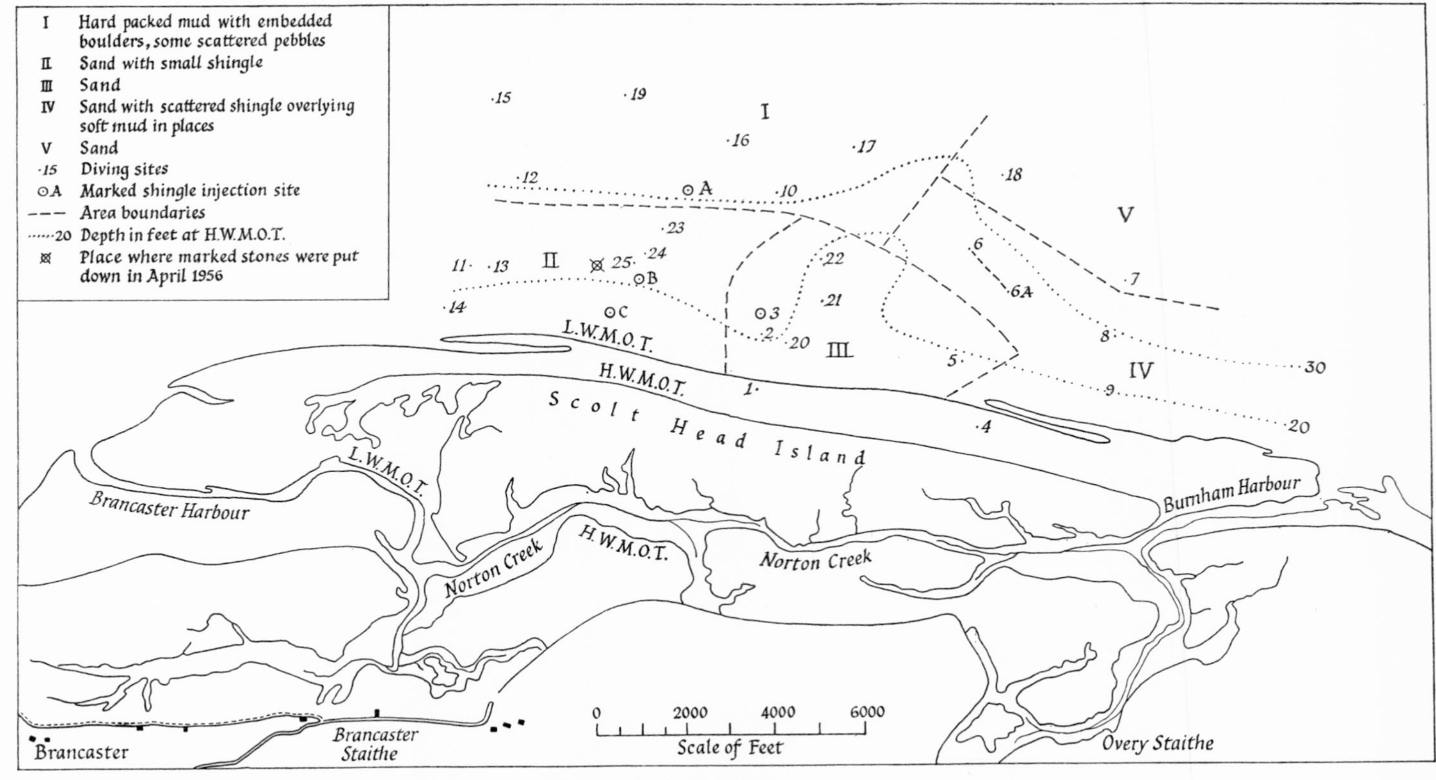

Fig. 119. Scolt Head Island (after C. Kidson *et al.*, *op. cit.*)

At the outer site, A, some movement was found which was consistent with that of the radioactive pebbles put down in 1956, and discussed above, page 562. Most of the pebbles remained close to the station, and although gales had swept the island in the winter, the most distant pebble was only 200 ft. from the station of origin. Moreover, many of the pebbles put down at A in 1960 were incrusted with organisms 11 months later, a fact indicative of but limited movement (see p. 567 for aqualung investigations near St Agnes' Head, north Cornwall).

Work of this type can be extended. It seems more likely to be of use in special cases or localities, but that it is possible to combine it with other methods may well be its greatest use in investigating drift and related problems.

A suggestion made in 1927 (*Geogr. Journ.* 69) has been magnified almost to a hypothesis—the apparent reversal of shingle drift near Sheringham (p. 345). The shingle at Sheringham is often piled up on the eastern side of groynes, thus indicating a westward motion. Although shingle is often characteristic of the top of the beach, it does not follow that because it accumulates on one side of a groyne the drift of the whole beach is in the same direction. Kidson has argued cogently that drift may vary not only at a particular place, but over a considerable length of coast. To some extent his experiments at Orford (p. 565), and the northerly drift that for some time was noticed at Slaughden support this (p. 677).

In the *East Midland Geographer* (No. 16, 1961, 3) Kidson has suggested that, if the direction of drift can change for perhaps some weeks on end over a long stretch of coast, it would be possible for the shingle of Blakeney (which is continuous with that at Sheringham and Cromer) to have been derived from more or less the whole length of the glacial cliffs. This is indeed a possible supply, and we know that the cliffs have been subjected to continuous and serious erosion. But, unfortunately, the argument does not go far enough. That such a supply might apply to Blakeney may be argued, but as far as can be said, it cannot apply to Scolt Head Island. West of Blakeney Point there is extremely little shingle, and west of Wells it is virtually absent from the beaches until Scolt is reached. West of Scolt itself there is also little, and even at Hunstanton it is relatively scarce. Moreover, the harbour entrances at Blakeney, Burnham Overy, and Brancaster Staithe make westward travel of shingle more difficult. The Lincolnshire beaches are generally free from shingle, and there is little if any possibility of any other existing beach or cliff source whence it could come. Much, probably all, of the shingle is derived from glacial deposits. At Blakeney the Eyes (p. 660) may have given a fair supply in the past. See also p. 659.

The problem, therefore, still remains; it is made more rather than less difficult by the experiments so far carried out with tracers, which do not seem to support an onshore movement of shingle. Farther round the coast, on Caister Shoal, A. H. Stride (*Dock and Harbour Authority*, 40, 1959, 145) shows that sand movement on and from the shoal was southwards and roughly parallel with the coast; certainly not to the coast. Careful analysis of the main movements of material on- and offshore, between Happisburgh and Weybourne, would be worthwhile.

Chapter XVIII

GEODETIC LEVELLING AND VERTICAL MOVEMENTS ON THE COAST

There have now been three geodetic levellings of England and Wales[1]—completed in 1860, 1921, and 1956. It was found impossible to base rigid deductions on land movements relative to sea level on the differences between the first two (see p. 496), although there seemed to be evidence of an upward movement in the north, and a downward one in the south. This evidence also appeared to correlate with that derived from a study of recent raised beaches.

A somewhat similar disappointment, from this point of view, derives from a comparison of the second and third levellings. The apparent movements of land relative to the sea are not in agreement with conclusions obtained from tide gauge records at Newlyn, Felixstowe, and Dunbar. Apart from minor short-term variations, which we may neglect, the tide gauge records imply a slight rise of mean sea level at each station.

(1) Observatory		(2) Ht. of Observatory B.M. (Newlyn datum) (ft.)	(3) Ht. of Observatory B.M. above Observatory Zero (ft.)	(4) Ht. of MSL trend above Observatory Zero (ft.)	(5) Ht. of MSL trend above Newlyn Datum (ft.) 4–(3–2)
Newlyn	1918	15·588	25·000	9·434	+0·022
	1950	15·588	25·000	9·637	+0·225
Felixstowe	1918	11·499	25·000	13·482	−0·019 ±0·14
	1950	11·475	25·000	13·656	+0·131 ±0·10
Dunbar	1918	12·840	25·000	13·009	+0·849 ±0·16
	1950	13·415	25·000	13·045	+1·460 ±0·12

[1] R.C.A. Edge, 'Some Considerations Arising from the Results of the Second and Third Geodetic Levellings of England and Wales', *Int. Assoc. Geodesy*, Toronto, 1957; and Major J. Kelsey, 'Matters Arising from the Completion of the Third Geodetic Levelling of England and Wales', *Commonwealth Survey Officers Conference*, Cambridge, 1959.

This rise is a combination of two movements, an absolute rise of sea level and the isostatic movement of the land at each station. The preceding table gives a summary of results (1950 is taken as the final date since after that year the Dunbar gauge was abandoned owing to silting).

There is no significant variation between the figures for Newlyn and Felixstowe. The small rise of sea level indicated is consistent with the presumed rise of sea level produced by the melting of the polar ice. This is estimated at something between 3 and 12 cm. per century. The figures in the table are slightly higher and may be partly the outcome of isostatic movements of the land.

Dunbar, at first sight, presents a different picture, an apparent elevation of mean sea level of about 1·2 ft. above that of the south of the country and, even more anomalous, a rise of nearly $\frac{1}{2}$ ft. relative to Newlyn between 1918 and 1950. Primarily this looks like a rise of the land at Dunbar, relative to Newlyn of +0·575 ft. (= *c.* 5 mm. a year). There is some reason to suppose that isostatically there has been a greater change of level at Dunbar than either at Newlyn or at Felixstowe, but not of this amount. The levelling also suggests an apparent rise of sea level at Dunbar relative to Newlyn of 0·408 ft., a figure which seems too big.

The most probable cause of these considerable differences between north and south is thought to be some systematic error in the precision levelling. One suggestion is 'that the direction of illumination of the staff might affect the readings and that the tendency of the northerly staff to be illuminated by direct sunlight more frequently than the southerly staff might introduce systematic error into lines running generally north and south' (Major J. Kelsey, 1949). Those interested in coastal studies can but regret the difficulties inherent in precision levelling and hope that in the future we may be able to rely on them to establish the exact amount of vertical movement.

Chapter XIX

EARTH EMBANKMENTS

Although it is not intended to discuss sea defences in this book, some attention may be paid to the extensive earth embankments around many of our estuaries and similar places. They account for more than 90% of the defences against the sea in low-lying areas, and between the Humber and Dover there are about 1,200 miles of these banks. Their interest from our point of view is twofold. First, they are for the most part built of material immediately available, mud, silt, and sand in the adjacent marshes, and secondly their preservation depends to some extent on the proper planting of selected maritime plants. Their response to the flood of 1953 showed that not only were many inadequate, but also that no one knew sufficient about their structure in relation to seepage, drying, cracking, and response to vegetation, to advise fully on their reconstruction. They were considered at length by the Waverley Committee (*op. cit.*) and since 1953 a great deal of research has been done on them.

An examination of the many breaches made in the banks by the 1953 storm showed that failure took place in various ways. In more exposed sites direct frontal attack by waves occurred, but relatively little damage resulted. This was partly because exposed banks were usually protected by concrete or other facing, which was sometimes displaced. On an unprotected bank direct erosion took place, but even this seldom, if ever, led to collapse. The waves, even the water level, over-topped many banks and surface scour of the inside of the bank was often severe. There is no doubt that this was a cause of much destruction, but how basic it was is unknown despite careful examination of many breaches. The fundamental cause of failure was, without question, slipping or slumping of the landward face of banks as a result of seepage. It is difficult to distinguish the effects of this from direct erosion of the landward face, but several considerations suggest that it was the more important cause.

The banks are built of marsh mud and other materials immediately available, and the mud or clay shrinks greatly on drying. Hence, deep and sometimes wide cracks develop in the banks. If, then, there is a high level of water on the front of the bank, produced by a storm or surge, enough water may penetrate the cracks, which exist throughout the bank, to induce a downward directed drag on the landward side. Slips, perhaps small, could then easily take place, and even a general lowering of a large part of the bank. Thus failure could take place without any water spilling over the bank. It is true that overspilling water and failure resulting from seepage in fissures can, and almost certainly do, work together. If the bank is built largely of sandy clay, it may be pervious throughout, and so allow water to penetrate during a high sea level. Small slips can easily occur in such circumstances, and a movement of the whole bank is possible. At Dartford Creek one breach occurred of a different type. There the upper part of the bank rested on a layer of much more permeable material which was penetrated by the water which lifted the whole bank and led to violent disruption.

It is not practicable to witness a break formed by natural means, and consequently much of our knowledge must be by inference from the study of ruptures made in a storm. But the Building Research Station has advanced our knowledge a great deal by studying a typical fissured clay bank at Coryton, Essex.

A steel sheet-pile cofferdam was constructed on the river side of a 60-ft length of disused bank, and water was pumped into this cofferdam from a collecting basin dug on the landward side. A series of experiments were made in which typical tides were reproduced with peaks below the crest level of the bank. The quantity of water seeping through the bank, and the pore-water pressures developed in the body of the bank, were measured. Careful watch was kept for movement of the bank. Progressively higher tides were applied until the crest level was reached and finally the water was allowed to trickle over the top of the bank. No movement was observed until this stage was reached and then a shallow slip suddenly occurred in the fissured clay on the backslopes. This left a vertical face about 2½ ft deep near the rear crest, and within a few minutes the thin barrier of soil at the crest was quickly broken through, resulting in an initial breach 2½ ft deep at the centre of the slipped section. This would have developed into a major breach had the water level been maintained. The test has provided valuable confirmation of earlier predictions and also a starting point on which to base further experiments to study improvements works. (*Building Research*, D.S.I.R., 1956, p. 34.)

There are many other matters of interest appertaining to these banks. They include the building of light parapets to prevent splashing over the banks; the width of the bank, especially the summit, since a narrow bank can be destroyed by landward slips more quickly than a wide one; and the weight of the bank and its settling, and therefore loss of height. Coupled with this is the difficulty of making proper allowance for the downward movement of parts of our coast relative to sea level. Apart from these, and perhaps other, general considerations there are always local factors to assess resulting from degree of exposure, material available for making the banks, economic conditions, and perhaps the enterprise of the local or other authority concerned.

The second matter of importance from our point of view is the vegetation, natural or planted, on the bank. In this respect a bank is, in a sense, an upward continuation of the mud or other deposit fronting it. But from the defence point of view, vegetation on a bank has three main functions, protection against weather, protection against minor wave action and, to a small degree, protection against overspill. Plants that can stand up to these conditions should possess many virtues. They must be tolerant of salt, spray, waves, strong winds, extremes of weather and climate, they should be able to resist diseases, and at the same time have a minimum adverse effect on the bank itself. The ideal plant is unknown! Often the faces of banks are sown or turfed, the sowing probably taking place under a screen of thatch of twigs or reeds (cf. The Culbin Sands). Occasionally the grass on the bank is cut or grazed; the first is expensive and the second usually difficult to arrange. Irish Perennial Rye-Grass is often sown; like rye-grasses in general it easily germinates and spreads quickly. Other grasses commonly used are *Agropyron pungens* and *A. repens*. These, however, do not always find favour. At first they may give a good even growth on a bank, but after some years, unless carefully tended, they become tussocky, coarse, and uneven, and a coat of loose vegetable matter forms on the bank. If they are carefully tended, these disadvantages are removed. On the other hand, if these or other plants root deeply they cause drying by transpiration and so lead to difficulties of another kind. On many older banks *Festuca rubra* and *Agrostis stolonifera* have been reasonably successful. Fertilizers may be used, but introduce more expense.

There is, therefore, plenty of scope for research in this field. It is not merely a question of finding the right vegetation to cover the bank to best advantage, but also to find those plants, grasses, which, whilst having the best effects on the outer side of the bank, do least damage within. In other words, a careful study of root systems and their relations to soil structure and moisture content is required.

The development of salt marshes, dunes, and shingle ridges is discussed in many different parts of this book. All may be defences against erosion, and in each vegetation plays a part, of major importance in marshes, but much less so on shingle (cf. p. 348, Blakeney Point). In each environment further study from the point of view of using these natural features as a means of defence against marine erosion is required. (A useful modern summary on the uses and limitations of vegetation in shore stabilization is given by C. Kidson in *Geography*, 44, 1959, 241.)

See Note added in Proof, p. 718.

Chapter XX

THE SOLWAY TO THE DEE

Although research is in progress on certain physiographical aspects of the Solway coast, little has been published in the last 10 years concerning it, apart from some archaeological papers dealing with Roman sites (Volumes 55 and 57 of the *Transactions of the Cumberland and Westmorland Archaeological and Antiquarian Society*). (See note, p. 718 *post.*)

In Volume 30 of the same journal, B. Blake comments on the site of the Roman Fort at Burrow Walls, near Workington. The fort stands on a gentle slope on the north of the Derwent river, and when it was built it is almost certain that it was washed by a meander of the Derwent. The evidence of this is found in the erosion under the fort, and in a deposit of silt, sand, and marsh. The relation of this deposit to Oyster Bank (see Fig. 120) and to the present coast, suggests that at that time the Derwent flowed out to sea at St Helen's Colliery, and that the meander encircled the spur of North Side. It is uncertain if this former mouth took the place of the present mouth, or if it were a secondary one.

The Roman fort at Ravenglass (E. Birley, *op. cit.* 58, 1959, 14) is on a bluff and faces the joint mouth of the Irt, Mite, and Esk. Little has been lost by erosion in nearly 2,000 years, since only the western rampart and the intervallum have disappeared. The fort was originally about 140 yd. square.

Walney Island has been studied in some detail by J. Melville (*Barrow Nats. Field Club and Photo. Soc.* 8, 1956, 25). The suggestion is made that in the Roman period this part of the coast may have been 6–8 ft. lower, relative to sea level, than it is now. If so, Walney probably consisted of three or four separate islands which later became joined by shingle drift. This is in no way contradictory to Spencer's view (p. 86) that the island is an esker ridge.

On several occasions the sea has swept across the island. There are well authenticated records in 1771 and 1796; there were inundations in 1546, 1552, 1553, 1561, and 1564, and others in 1840 and 1852. Erosion remains serious. At Cow Leys Bank, between Lamity Syke and Hillock Whins, there is a silt deposit of the 25-foot sea. The old cultivated land is covered with blown sand, now grass grown. Erosion by the sea reveals

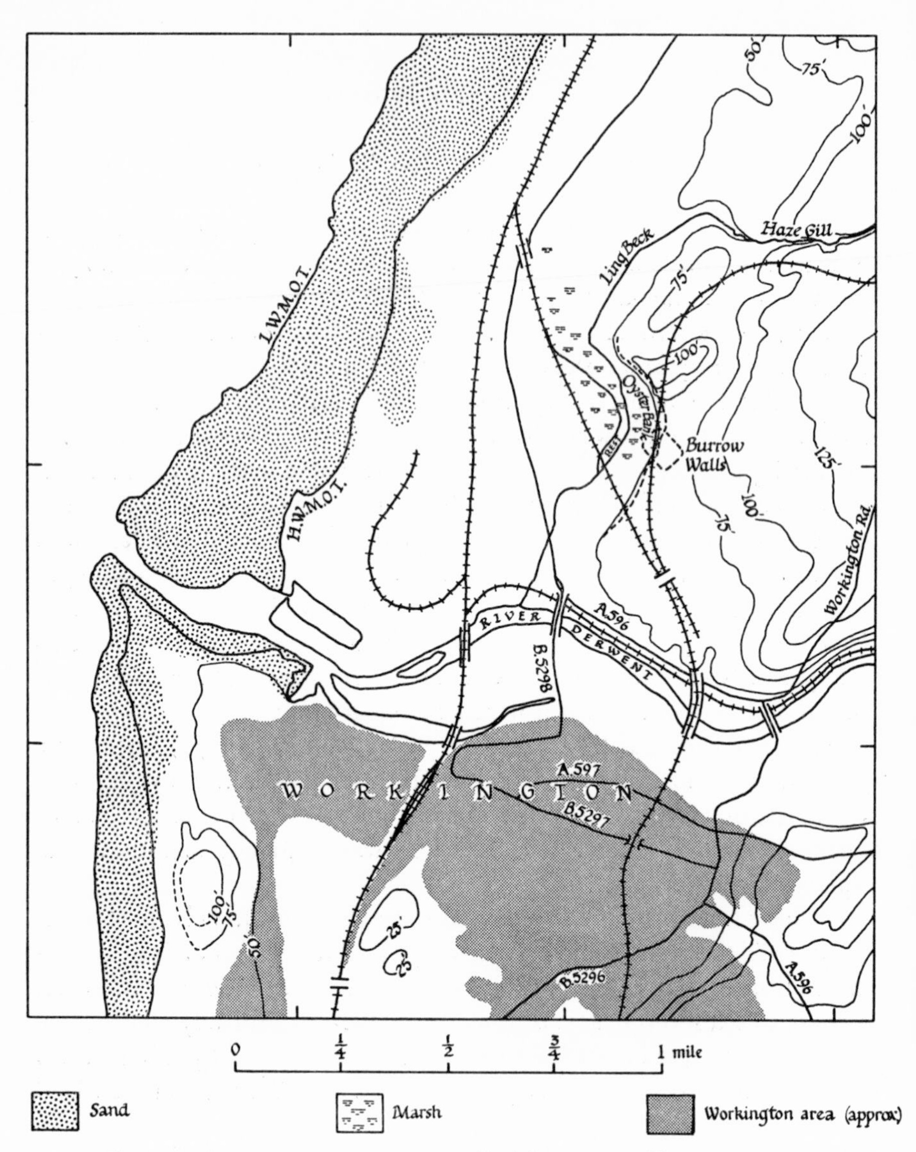

Fig. 120. Roman Fort at Burrow Walls (after R. L. Bellhouse, *op. cit.*)

locally the old plough ridges. The extent of change produced by erosion is made clear by Melville when discussing the changes at the south end of the island 'where Trough Head cottage used to stand...(In) 1833 (it) was about 250 feet from high water, but by 1840, after the sea had swept right across the island, the high water mark was only 140 feet from the cottage. The shoreline remained practically stationary until 1879, when further wastage took place, leaving only 90 feet. By 1902 the cottage was completely ruined, and half of it washed away. Now the shoreline is more than 80 feet inland of where the house stood.

The island lengthens by southerly beach-drifting, and is liable to be washed over in storms—that of 1954 did so at Cross Lane—and Melville suggests that unless heavily protected it may well be two islands within the next hundred years. There are no appreciable changes on the channel shore.

At Sandscale and Roanhead, Pearsall's work on the vegetation has been followed up by G. Wilson (*Proc. Barrow Nats. Field Club and Photo. Soc.* 8, 1956, 36), whose paper contains a useful and checked list of plants. He notes the relative absence of *Agropyron* at Roanhead and the occurrence of *Vicia lutea* (Yellow vetch) and the minute *Radiola linoides*. Between the dunes there are slacks, with *Salix repens*. All stages between bare sand and a dense cover of *Salix* are found. In the wetter slacks the dampness is not the result of incursions of salt water, consequently they resemble freshwater ponds and marshes, and no unusual plants occur.

Morecambe Bay is attracting attention from coastal workers. R. K. Gresswell (*Inst. Brit. Geogr., Trans. and Papers*, 1958) emphasizes that we are still largely ignorant about the origin of the bay itself, and his paper is an account of part of a scheme to investigate the general problem. Gresswell is particularly concerned with the raised beaches and associated features, and he traces the old shoreline. The Duddon, Greenod, and Kent estuaries and shores are flat, and the former surface is below 20 ft. O.D. and often peat-covered. In tracing the boundary, three types of slopes were recognized—precipitous slopes, cut in solid rock and usually along formerly glaciated valleys; steep slopes, nearly always showing some modification by ice; and degraded slopes. Locally the boundary is obscure and has to be inferred rather than plotted from direct evidence.

In the several estuaries, as well as in some of their ramifications no longer covered by salt water, the general level of the infill is $+15\frac{1}{4}$ ft. O.D. This is very constant, despite the fact that some places are widely separated, and suggests that the deposits are estuarine rather than lacustrine. The infill has been bored near the sides or heads of estuaries, and the conclusion is reached that apart from occasional peats and grass-sedges and similar intercalations, the whole depth of clay represents continuous deposition.

During raised beach times the coast must have been more irregular and picturesque than it is now. The Kent estuary is now relatively simple, but then there were several rocky islands on its western side. The same is also true of the Greenod valley. Humphrey Head is still reached by the waves, but in raised beach times the attack must have been more severe, and much of the existing cliffs, as well as the raised beach platform where it is cut in rock, are of that date. The cliffed drumlins south of Flookburgh tell the same tale.

The transgression of the sea during which these features were produced can be dated by pollen analysis as between Zones VI and VII or the Boreal-Atlantic transition. The beach is now a trifle more than 15 ft. O.D. and is to be correlated with the 25-foot beach of Scotland and the rather lower one near Silloth and Maryport. Traces of this level in Morecambe Bay are nearly all in the form of deposits in the various estuaries. These flats would doubtless have been covered at high water, and so the then high tide level must have been about 18 ft. O.D., a figure in rough accordance with a wave cut bench on Winder Moor at 19 ft.

Gresswell's paper was followed a few years later by a discussion on climate, vegetation, and sea level changes in lowland Lonsdale (F. Oldfield, *ibid.* 1960). He agrees in broad outline with Gresswell's conclusions, and shows that the trangression in Silverdale Moss culminated in Zone VII *a*. Oldfield traces a notch at the base of the Carboniferous Limestone cliffs from Arnside to Leighton Moss, except near New Barns Bay. North-east of Arnside Moss, and south of Leighton Moss it is cut in the New Drift. 'Its form is that of a wave-cut notch backed by a steep vertical cliff and fronted by an abrasion platform which slopes gently seaward. For the most part, this cliff line is shielded by a width of saltmarsh stretching seaward from its foot, but at Blackstone Point it is not thus protected.' The Arnside–Silverdale district must have been very beautiful when the sea penetrated into many of the present mosses. It did not reach the small lake, Haweswater, but Arnside Moss was reached through a gap on its north side.

How much has the sea level fallen in the district since the post-glacial maximum? The fact that it floods a much smaller area is not in itself clear proof of fall, but the notch on Blackstone Point and the highest part of the natural marsh at Silverdale are regarded as raised.

Farther south along the Lancashire coast, Gresswell (*Sandy Shores of South Lancashire*, 1953) has continued and elaborated his work referred to on pages 101 et seq., and has shown that the raised beach can be traced in this district also, where it is named the Hillhouse coast. It is usually marked by a low cliff, and characteristic examples may be seen at Hesketh Bank, Holmes Wood, and Martin Bog. The Hillhouse cliffs represent the maximum of what is regarded as a eustatic rise of sea level. The retreat from this maximum, in Gresswell's view, was the result of the isostatic uplift of this part of England, although there was probably a eustatic element also present. However, as the sea retreated a sandy, and later a muddy, beach was built. The sand of this beach, together with that which has been blown inland, forms the Shirdley Hill Sand, so named by de Rance. The Shirdley Hill Sand was followed by the Downholland Silt, but the recession of the sea was not apparently quite regular and at times the silt slightly advanced over the sand, so that the whole succession of deposits is typical of regressive overlap on a receding shoreline. But as the sea fell trees, especially silver birch and oak, gradually extended over the exposed floor, and afforested it. It is these trees which form the well-known submerged forest at Leasowe and other places. The forest was not extinguished by a re-advance of the sea (below), but because the drainage of the exposed flats deteriorated so that peat bogs and fens were formed. Peat now covers large areas, seaward of the Shirdley Hill Sand and Downholland silt, between Hesketh Bank and Crosby, and is in turn often buried under blown sand. Much of this land is ploughed, and affords a good soil.

This sequence of events implies that the 25-foot beach is earlier than the submerged forest (see p. 488). Gresswell thinks that, to produce conditions favourable to the forest, the sea fell about 40 ft. Later it re-advanced, probably as a result of an isostatic downward movement of the land. It is thought that the rate of the rise of sea level may have been of the order of $1-1\frac{1}{3}$ ft. per century for perhaps 3,000 or 4,000 years. But tidal observations at Liverpool to-day in no way suggest that this continues, or was in fact the rate in recent centuries. Nevertheless, since peat beds are being eroded at Hightown and Blundellsands to-day, and were being eroded but recently at Formby and on the north coast of the Wirral, it means that the recent advance of the sea is still going on.

Inland of the Hillhouse coastline the country rises from 25 to about 100 ft., and consists largely of boulder clay, often covered by Shirdley Hill sand. This 25–100-foot surface, the Scarisbrick, Mawdsley, and Simonswood level, is underlain by solid rock, which is continuous with the rock surface westwards. It is taken to represent a pre-glacial beach. Inland the level is followed by a rise, at nearly 100 ft., which is regarded as an old cliff. It is especially clear at Clieves Hill and Parbold; elsewhere it is degraded. Other platforms succeed inland, but have no direct bearing on the present theme.

Some recent investigations have been made on siltation in the Mersey (*Hydraulics Research*, 1960). In the nineteenth century the Mersey seems to have been in regime; there were great fluctuations of as much as 50 million cubic yards in perhaps five years, but they were about a mean value, and there was no progressive change. After 1906 conditions altered and a loss of 90 million cubic yards occurred. This and related problems were investigated in a model of the river and estuary made at the Research Station at Wallingford. Fortunately there existed about 1,000 charts of the channel above Eastham. These had been made each month since 1867, by the Upper Mersey Navigation Commissioners. During nearly all this time the channel remained near the Cheshire side, but moved across to the north bank perhaps once in every two to five years. There it stayed for three to eight months, and later returned to the southern shore again. But after 1891 it stayed on the Lancashire side. This was thought to follow from diversions in the River Weaver, the tipping of slag, and the effect of the piers of the Runcorn transporter bridge.

Experiments showed that the shifts of the low water channel above Eastham decreased when the capacity of the estuary deteriorated. This was ascribed to stabilizing the low water channel near Runcorn. Study was also made of training walls alongside the main sea channel. These increased the velocities of both flood and ebb currents in the channel and so deepened it. Material dredged from the inner parts of the estuary was dumped at sea, although it had, in the first place, to travel inwards from the sea! 'It has been found that among the factors causing the movement are the density currents which assist the upstream movement of sediment in the layers close to the bed. The principal source of the sediment deposited in the upper estuary and in the sea channels is the area of sand banks behind the West Crosby revetment. The North Wales coast is not a major source....' Experiments with radioactive tracers, as well as observations of suspended material in the channels, clearly proved that the material from the upper estuary which is dumped behind the Great Burbo Bank is transported again to the upper estuary! Thus the problem is what to do with the dredged material. Since it amounts to 19 million cubic yards annually it is not easy to deposit it onshore anywhere near the neighbourhood from which it is taken.

Chapter XXI

WALES

On the north coast of Wales there are some shingle distributions which are not readily explicable. E. Neaverson (*Proc. Llandudno, Colwyn Bay and Dist. Field Club.* 28, 1939–47, 17) argues that much of the shingle must come from offshore, even that to east of the boulder clay cliffs near Tan Penmaen Head. The Dulas and Clwyd are both deflected, the latter by a bank called Horton's Nose. Beyond are the sands at Rhyl, and a nearly complete absence of shingle for two miles. East of Rhyl is an interesting stretch of coast. The nearest boulder clay exposures are at Llandulas, which is on the far side of the Clwyd. Hence, Neaverson thinks that littoral drift cannot be regarded as an adequate explanation. He concludes that the only possible source is from offshore, a view which seems to be supported by the observed fact that shingle accumulation increases after gales. But what does this mean? Is it an increase of shingle above low water mark, the result of constructive waves, and liable to be combed down below low water mark when conditions change? Profiles across the beach should be extended below low water before any definite conclusions are formulated.

Erosion increases eastwards, partly because of a sea wall which led to scouring at its lee end, and so to the disappearance of a belt of dunes about 20 yd. wide. Primarily, however, the erosion is the result of beach-lowering caused in part by channels which run more or less parallel with the coast. Neaverson thinks that these channels are forced inland by wind-blown sand, since they act as traps for sand, even at low water. The sea-ward side of the channels fill in this way, and more and more water is forced over to the landward side. Since, too, ordinary lowering of the beach level near the dunes continues, and because no new supplies counteract it, a considerable depth of water, especially at high tide, may reach the dunes and lead to rapid erosion. Rather similar conditions extend as far as Prestatyn, but beyond that place, and at the Point of Air, the dunes are relatively stable. There is, nevertheless, a good deal of local movement.

The dunes at Newborough Warren, Anglesey, have been studied by D. S. Ranwell (*Journ. Ecology*, 46, 1958, 83; 47, 1959, 571; 48, 1960, 117 and 385). The Warren is made up of dune ridges and parabolic dune units, with erosion hollows and slacks lying between them. The dunes reach the 50-foot contour, and where they rest on rising ground may reach 100 ft., but this is not the depth of the sand. Wind records are kept at the lighthouse on Llanddwyn Island, and Ranwell notes that on an average of twenty-one days a year, the wind reaches force 7. At Holyhead the wind resultant is approximately 28° east of the true north; this corresponds closely with the orientation of the parabolic dunes at Newborough.

In his investigations between August 1952 and September 1955, Ranwell found, as a result of measurements, that erosion had occurred on the windward faces of two dune ridges and accretion on their lee slopes, implying an inward movement. This is only to be expected. But there was an interesting difference between the coast ridge and the one farther inland. In the former, the slope of the windward face varied between 3° and 13° from the horizontal, and averaged 7°. The corresponding face of the landward dune was steeper; the variation was from 7° to 13°, and the average about 9°. The contrast was ascribed to the greater holding power of vegetation as the dune moves inland. In other words, crest erosion decreases relatively to the erosion of the windward face, and the dune becomes steeper. The amount of inward movement of the inner dune as a unit is not great; it varies from 5–10 to 22 ft. a year in different transects.

Marram grass apparently can just withstand burying by 3 ft. of sand in a year. The leaves of *Ammophila* are 2–3 ft. long, and if not more than half-buried in a gale, and if sufficient time elapses for the plant to adjust itself to the new conditions before another gale occurs, then it can survive. Excess of sand has the effect of making the *Ammophila* leaves grow longer, and also promotes the growth of vertical rhizomes from axillary buds. This implies that the upward growth of the grass will be most favoured on the landward side of a clump near or at the crest of a dune. Consequently, strong winds, causing some erosion, promote a landward travel of the *Ammophila* itself.

In all there are about five square miles of blown sand at Newborough, much of which is afforested. Near Rock Ridge flints of Tardenois and Mas d'Azil age have been found. These cultures (*c.* 7000 to 5000 B.C.) are usually associated with sandy shorelines of one kind or another, and also with temporary dwellings, and so the inference seems justified that a rather similar coast has existed hereabouts from the Boreal period until to-day. There is, unfortunately, little historical evidence available. Owen quotes from Ministers' accounts 1152/4, 1909:

About one third of the land of the manor was damaged by storm on the feast of St. Nicholas, 6 December 1331, when 186 acres were destroyed so thoroughly by the sea and inflow of sand as to render it useless for agriculture evermore. The same thing happened along the whole south coast of Anglesey through Aberffraw to Rhosneigr; at ebb tide the sand on the shore dried in the strong wind and got blown inland.... Many families were driven out of their crofts into the town of Newborough from land between Llandwyn and Newborough.

It is known that sand was encroaching in the reigns of Elizabeth I and Charles II, and that at various times marram has been planted to hold the sand. On the other hand, a study of maps from the end of the seventeenth century onwards suggests no real change of the coastal margin. Ranwell suggests that the Warren began to form in the fourteenth century and that earlier there was a sandy coast in this district. To test this view, he sought for indications that might exist of pre-fourteenth-century dunes. Air photographs show some parabola-shaped low mounds in agricultural land about ½ mile inland from the present limit of the Warren. Pits were dug in them. There were found some 70 cm. of reddish-brown top soil with a little blown sand, overlying 30 cm. of yellow-brown sand, rounded like that of the Warren to-day. The lower sand rested on boulder clay in which was abundant carbonized material. The implication is that the clay was clad in vegetation before the lower sand accumulated on it. The lower sand layer may be the last vestige of an old dune system which was sufficiently stable for crofting before the accumulation of the overlying layers. These, 70 cm. thick, probably took some time to gather and develop. Since it is a well-drained area, it is thought that it would have required more than the 500 years for the present dunes to have reached their present state. Probably the lower sand was deposited some considerable time after the last glaciation.

The dune slacks at Newborough are interesting. A rough classification of slacks is: (1) those where the free water-table is always more than two metres from the soil surface (= dune associes), (2) dry slacks where the free water-table is between one and two metres below the surface, and (3) wet slacks where the water-table never falls below one metre from the surface. Further study reveals the probability of a cyclic interchange of dune and slack associes. The plant associes at Newborough resemble those of other Welsh dunes and those at Braunton Burrows, Southport, and Keiss and Dunnett in Scotland, but differ from the acid slacks of Winterton (Norfolk).

Myxomatosis had a considerable effect on Newborough. In the three years following the incidence of the disease, there was a marked increase both in growth and flowering of grasses and sedges, and a consequent reduction in low-growing herbs. The rich moss flora continued to be profuse.

At the present time the final stage of succession 'is a dry-slack *Calluna-Salix* heath, which developed under the influence of continued leaching and in spite of heavy rabbit-grazing pressure up to 1954 when myxomatosis practically exterminated the animals'.

Recent work in the Lleyn peninsula by J. B. Whittow (*Proc. Geol. Assoc.* 71, 1960, 31) has thrown light on two of the lower raised beaches.

He discusses the pre-drift raised beach platform which is about 10 ft. above present high water mark. It can be traced between Nevin and Aberdaron, on Bardsey Island, and on Ynys Gwylan. The relationship between this platform and the modern wave cut bench is not simple. It is locally notched by waves which reach it in exceptional storms, and there are stacks and a good deal of vegetation on it.

At Porth Oer the platform cuts contorted pre-Cambrian rocks, stands at about 20 ft. O.D., and is continued seawards across modern sea-stacks. The best example of the platform is at Porth Colmon (see Fig. 121). Whittow advances the interesting suggestion that the Porth Dinllaen peninsula is a series of stacks above the platform, now tied to the mainland by drift.

That the beach is not found elsewhere in Lleyn may be because it has been destroyed by erosion or because it is hidden under drift. The alleged occurrence of a fragment at Criccieth, first commented on by Fearnsides, fails to convince Whittow of its authenticity. In general, the platform resembles that of the *Patella* beach, but there is no means in Lleyn of dating it precisely. It is probable that it is older than the Würm (= Weichsel), and may be as old as the Mindel (= Hoxne)-Riss interglacial, 'and there is no reason to suppose that it is not much older'.

In Porth Neigwl there is a conglomerate at the foot of Creigiau Gwineu which closely resembles a raised beach deposit. There is no accompanying wave cut beach. It is not easy to compare this beach with the earlier one, and the fact that the base of the conglomerate is about 5 ft. below the general level of the pre-drift beach is not a satisfactory means of differentiation. However, the fossil content of the conglomerate points to a post-glacial rather than a pre-glacial age, and some of the drift covering it has slumped from above. There is no precise means of dating the beach; it may be the equivalent to the post-glacial beach at 16–17 ft. described by Gresswell in south Lancashire.

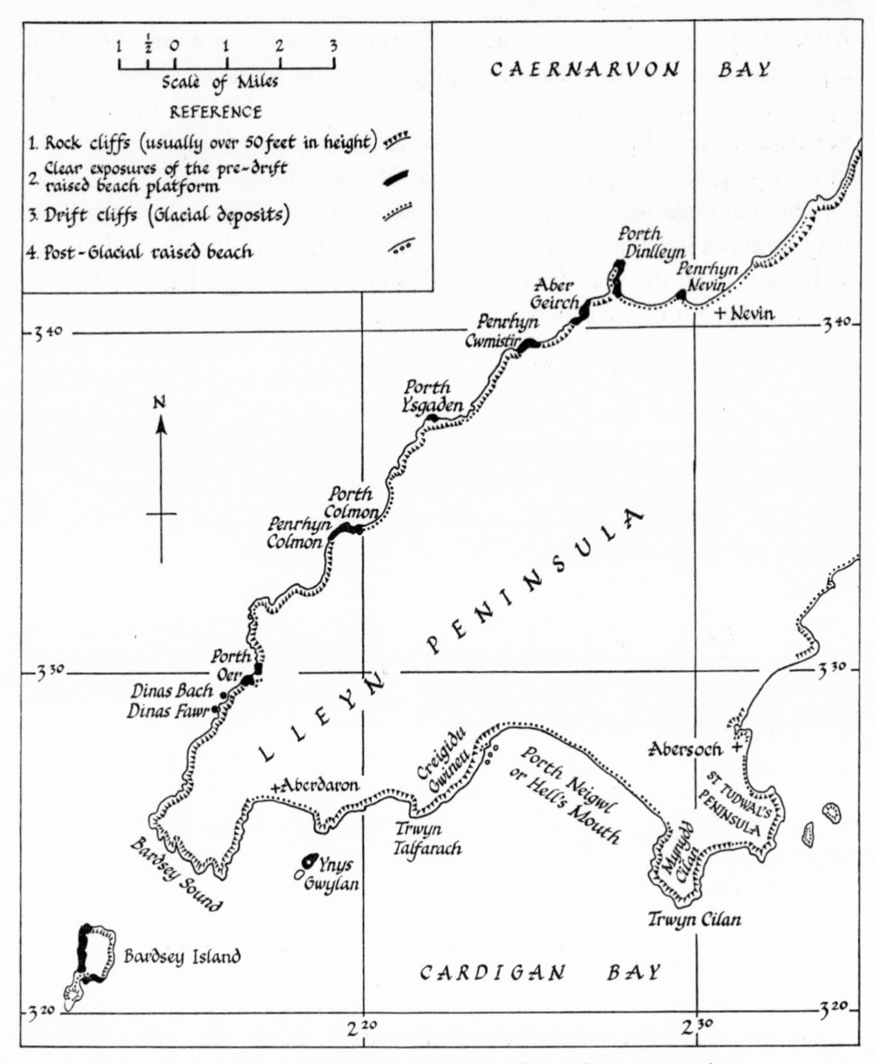

Fig. 121. The raised beach platform of the Lleyn peninsula (after J. B. Whittow, *op. cit.*)

Some interesting work has been done on the forms of cliffs between the Dyfi and New Quay (A. Wood, *Liverpool and Manchester Geol. Journ.* 2, 1959, 271).

For about 30 miles, from Borth to New Quay, the solid rocks which reach the coast consist of rapid alternations of fine grained sandstones and darker poorly cleaved mudstones. Collectively they are known as the

Aberystwyth Grits, and they are fairly uniform and resistant. An inland dip is associated with seaward sloping cliffs; the angle of slope depends upon the jointing. If the beds are horizontal, the cliffs are more or less vertical. If the dip is seaward, there is landslip topography and a cliff face sloping at an angle similar to that of the dip.

The main feature of interest is the coastal bevel. The foot of the bevel may be at sea level, at the cliff top, at any intermediate height, and may be well back from the sea. It is the equivalent of the hog's-back of Devon, and of what Guilcher called a 'fausse falaise'.

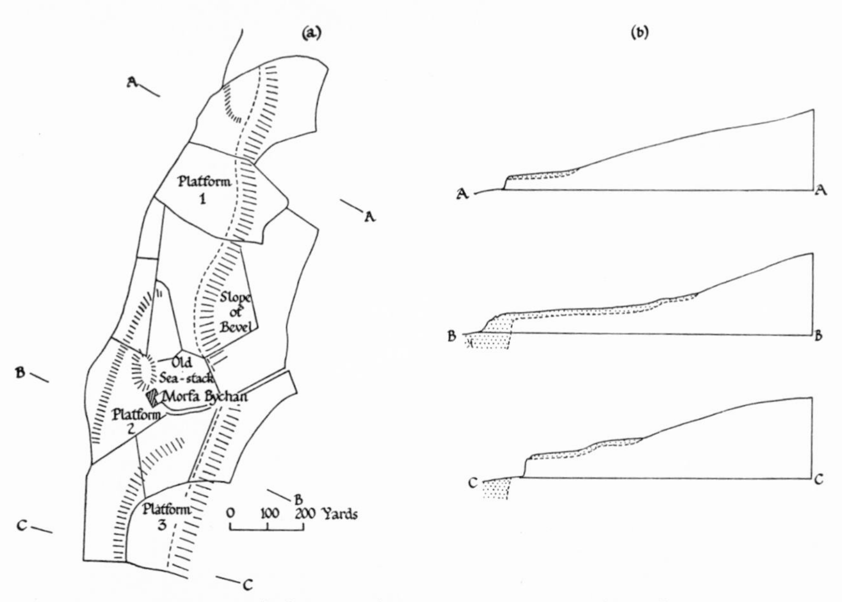

Fig. 122. Platforms and cross-sections at Morfa Bychan (after A. Wood, *op. cit.*)

Figures 122*a*, *b*, refer to the stretch of coast about four miles south of Aberystwyth. The plan and sections show the various platforms and their relations to the coastal cliff and to the spreads of boulder clay. In section A of Fig. 122*b*, the inner edge of platform 1, a raised beach, coincides with the foot of the bevel. A little farther south the boulder clay spreads over the coastal cliff (section B). Two or three hundred yards farther on (section C) the cliff is in solid rock. On the coast itself the boulder clay is plastered over the solid cliff 'as though it had slipped over the cliff edge'. A beach platform occurs only where the sea cliffs are in solid rock, and it is probable that the buried cliff was cut when the sea stood rather lower than it now does. In section B the surface covered with boulder clay rises inland, and there is a second platform (2), and even a third (3) a little farther south. The third platform, of which the solid rock base is nowhere exposed, is at about 180 ft. O.D. Still farther south the cliffs are cut in boulder clay and the fossil cliff runs inland. There are traces of higher platforms at about 315 ft. O.D. in this section. About 600 yd. south-west of Monk's Cave there is 'a particularly significant section above an unnamed headland...where the coastal bevel runs down to 25 ft. O.D. before being cut off by recent erosion. This must mean that at some period in the past any traces of the three platforms recognized to the north were eroded away, prior to the formation of the lower part of the coastal bevel' (**94, 95, 96, 98**).

Behind Llanrhystyd and Llanon the bevel is some way inland, in rear of a gently sloping boulder clay surface. Seawards the boulder clay dips to below high water mark, and a storm beach has been built up between the sea and the fields. Towards Aberarth the coastal bevel is inclined at about 32° and is cut by nearly vertical, high, sea cliffs. North of Aberarth the cliffs are faced with boulder clay, the rock beneath sloping at 45° towards the sea. Locally the old cliff fronts the sea, is cut by recent erosion, and slopes seaward at 77°–88° in marked contrast to that part swathed in boulder clay. Nearer Llanon there is a relatively low fossil cliff hidden beneath boulder clay, and in this neighbourhood two raised beach platforms at approximately 88 and 185 ft. can be seen. Near Craig y Delyn the higher cliff tops are spurs of the Coastal Plateau. Between Clarach and Aberystwyth the cliff form was discussed by Challinor (p. 152). The bevel is here high above steep modern cliffs. At Aberystwyth the bevel swings inland below the National Library which itself rests on a platform at 185–190 ft. The castle is on another platform at 58 ft. O.D. At the southern end of the stretch of coast discussed by Wood, in the north-west corner of Little Quay Bay, the fine cliffs have been produced by interglacial erosion, and only recently exhumed and trimmed by the present sea.

The bevel is regarded as an old sea-cliff, and it is contended that if the bevel is fronted by a concave slope there was once a raised beach there. Platforms, beaches, have been recognized at 315, 185, 88–95, 70, 45–48, and perhaps 20 ft. above sea level. The cutting of a lower one has often partly removed one or more of those above it.

The fossil cliffs are usually not quite as steep as the modern cliffs, and sometimes much less as, for example, near Aberarth. It is probable that there was but a short period of time between the withdrawal of the sea from the steep cliffs and their subsequent covering by boulder clay; but the lesser slope of the others suggests a longer period of subaerial denudation before they were covered by boulder clay. This marked variation in the slopes of fossil cliffs may mean that low level marine erosion occurred at two periods widely separated in time.

If the coastal bevel is an old cliff, it must often be composite and cut at different times. This follows from the fact that bevel slopes may join laterally and become one, uniformly inclined. If the bevel approaches or even reaches sea level it is probably relatively young, and Wood concludes that it is polygenetic, a constant slope to which cliffs become degraded.

Another interesting point concerns the amount of post-glacial marine erosion. It is suggested that if the present boulder clay slopes are continued seawards, and any such reconstruction involves uncertainty, they would reach a position indicating that there has been at the most only a few hundred yards of erosion in the last 5,000–6,000 years, i.e. since the sea attained, more or less, its present level. Locally there has been little or no erosion. In solid rock the width of the platform indicates the recession of the cliffs. Although this may not exist where fossil and recent cliffs coincide, an average width is 250 ft., implying an amount of erosion of about ½ in. a year. (See Symposium on Cliffs, *Geogr. Journ.* 128, 1962, 303.)

To the accounts given of Llys Helig and Cantre'r Gwaelod (p. 148) may be added a brief reference to Caer Aranrhod, a supposed sunken city in Caernarvon Bay. About ½ mile from the shore between Dinas Dinlle and the Afon Llyfni, and about 3 miles north of Clynnog Fawr is a reef of stones, the supposed site of Caer Aranrhod. A critical and interesting analysis of this and the other sunken lands and cities has been compiled by F. J. North (*Sunken Cities*, University of Wales, 1957).

North refers to certain visits to 'Caer Aranrhod' and particularly to those of Ashton in 1909 (*loc. cit.* note p. 116) and the Hon. F. G. Wynn in 1912. Both writers differ considerably in what they 'saw'. 'It would seem that, as in the case of the descriptions of Llys Helig, these (accounts) of the castle or fortress of Caer Aranrhod also vary because they are attempts to read some kind of regularity, regarded as indicative of human design, into an adventitious accumulation of stones.' Like Llys Helig there is no doubt that the stones at Caer Aranrhod were glacially transported and are the remains of a moraine, a drumlin-like feature, or of a spread of boulder clay.

In the Royal Commission on Coast Erosion, evidence given on behalf of the Caernarvon Harbour Trust states: 'When the Romans occupied

Britain they had on the coast of Carnarvon Bay a town now called Caer Aranrhod, which had a paved road, still existing, leading direct to their great Roman station, Segontium, now Carnarvon. The said town of Caer Aranrhod, visible at low water, spring tides, is now one mile from the existing sea beach, showing the rate of erosion since Roman times.' This is pure imagination, but is interesting because it speaks of changes resulting from erosion rather than from change of level. Incidentally, the tales and older views concerning Caer Aranhrod are not in any way concerned with an inundation like that associated with Cantre'r Gwaelod.

The origin and evolution of the landscape in Wales, including the coast, has been discussed intensively in recent years. O. T. Jones (*Quart. Journ. Geol Soc.* 107, 1952, 201) made it the subject of his Presidential Address, and was concerned primarily with the interior of the country and with the higher surfaces, and the development of the drainage. He refers, however, in an important section of his address, to the coastal peneplain along Cardigan Bay, especially near Borth, Aberystwyth, and the Mynydd Bach. He notes that there are considerable variations in level 'and in the absence of lines of separation at any consistent level between areas of different altitude, there is no justification for regarding this coastal tract as a product of marine erosion, though it has some claims to be regarded as one at least of Ramsay's type areas of a plain of marine denudation. Its levels seem to be controlled to some extent by the varying resistance to erosion of the Aberystwyth Grit Group' (see also pp. 181, 182).

On pages 181–3 an account is given of various platform levels around the Welsh coast. In recent years two important contributions have been made to this subject. E. H. Brown in *The Relief and Drainage of Wales* (Cardiff, 1960) has synthesized his various papers in a connected account, and T. N. George (*Science Progress*, 49, 1961, 242) has reviewed earlier writings, and given his own overall picture. Both differ in certain fundamental aspects from Jones, but are in general agreement on their own common ground.

We are not here concerned with the higher platform levels in Wales, except indirectly. O. T. Jones supposed that the summit peneplain was warped down from 1,700 to 1,900 ft. in central Wales to 200 ft. on the coast of south Wales, whereas Brown and George both appear to regard the whole series of plateau levels, from the summit down to the *Patella* beach, as one great staircase. There are, of course, many other important differences of view, not least concerning the sub-aerial or marine origin of all the levels. However, from the coastal point of view we are justified in beginning our account with the 600-foot level of Brown. This is an average height, and there may be two platforms between 500 and 700 ft. At the back of the 600-foot level it is often possible to trace a discontinuous bluff, which is interpreted as a former cliff. Brown contends that levels above the cliff (i.e. about 700 ft.) are sub-aerial in origin, whereas the lower ones are definitely marine. There is no fundamental disagreement amongst all who have investigated the features about the marine origin of the lower levels. Seaward of the 600-foot bluff there are numerous hills and rises which were islands when the bluff was being formed.

The extent of this former coast, and its marked parallelism to the present coast are shown on Fig. 123. The Caledonian grain of both is remarkable. The 600-foot platform and cliffs can be traced almost all round the coast of Wales, and may have been most fully developed in the south-west. North of the Dyfi the platform is much reduced; the summit levels in Anglesey and the Lleyn peninsula suggest a wide platform in that area. On the north coast it is well preserved in Denbigh and Flint, but farther west and in parts of Caernarvonshire it may be obscured by the effects of glaciation. It must also be remembered that during each succeeding lower stand the sea could, and doubtless did, destroy any higher levels in part.

Between Newport and Swansea the platform is cut in the coalfield scarp, and there are no large remnants left. Near Swansea and in the Gower peninsula and around Carmarthen, it is extensive and cuts across folds in such a way as to demonstrate conclusively that it is in no sense a structural feature. Austin Miller (p. 183) regarded the 600-foot level west of Carmarthen as a sub-aerial surface. Brown thinks this is unlikely in view of the now known extent of the level around all Wales, and because in the particular area discussed by Miller there are many breaks of slope on spurs which suggest strand lines. The 600-foot coast can be traced around Mynydd Presely. There was a noteworthy embayment in what is now the Teifi valley. The absence of any appreciable gradient for about 20 or more miles in the valley features corresponding to this level suggests that the then river was engaged in lateral planation, and consequently that the 600-foot stand of the sea was prolonged. Near Aberystwyth remnants of the platform are particularly clear and cut across folded strata of Llandovery age. Farther north traces are scarce; there is no evidence of its recurrence in either the Mawddach or the Dyfi valleys, but a few traces may exist between them. Brown does not explain this, and in view of the resistant rocks in this neighbourhood, and of the effects of glaciation it is not easy to do so.

Comment has already been made on Lleyn and Anglesey. The level

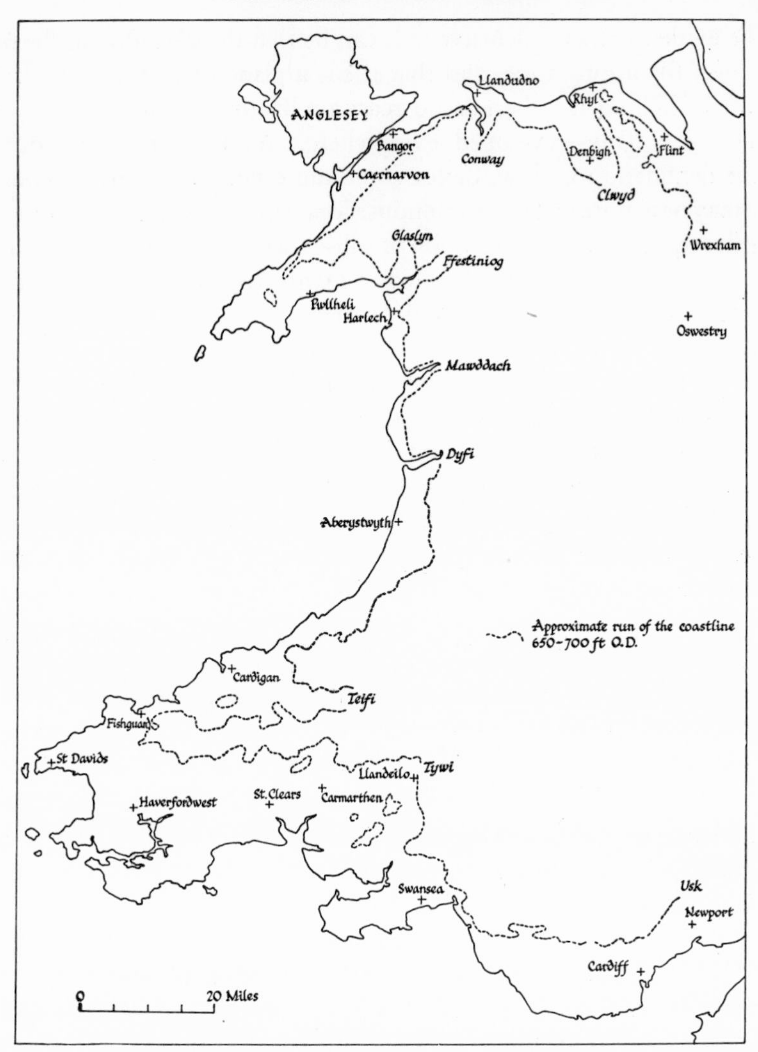

Fig. 123. Rough outline of 600–700-foot coast
(after E. H. Brown, *op. cit.*, simplified)

appears to be preserved on the Great Orme, and it can be traced up the Conway valley to Dolgarrog. The lower courses of the Conway and Dulas, and of the Ystwyth in Cardigan Bay, all suggest they extended themselves down the platform as it emerged, whereas higher up the rivers are adapted to structure. The level is remarkably well-preserved between the Clwyd and the Dee where it is cut in the Carboniferous Limestone.

The evidence is as conclusive as it can be that the sea stood at the 600-foot level for a long time, and that it cut a plane of marine denudation and a cliff behind it. There is no reason why the remains of this plane should be equally developed everywhere. As George points out, if marine denudation is slow, or for some cause not particularly favoured, there may be a more or less continuous slope from perhaps 600 ft. to some lower level, without any indication of subsequent stands of the sea. All the platforms below 600 ft. are accepted as marine by all authorities, and the general absence of beach material on them is the result of the erosive effects of ice.

Many occurrences of these lower platforms have been described (see pp. 181–3) but no complete correlation has been made. Brown suggests that the 400-, 300- and 200-foot levels were cut in the Pleistocene, probably during interglacial periods. But the staircase seems to descend to the *Patella* beach and to that of the present day. George notes the interesting fact that some levels may be directly inherited from Triassic and Liassic seas.

> ...the spectacular unconformity of Trias on Carboniferous Limestone to be seen south of Barry, and the similar stepped overlap of Lias well displayed on the coast in the neighbourhood of Southerndown, and on the flanks of the Mesozoic Cowbridge 'island', are proof of the nature of the sub-Mesozoic floor that, stripped of its Mesozoic cover, simulates closely the kinds of platform, seemingly of relatively recent origin, along the present-day Glamorgan coast.

The surfaces which now seem to be established in Wales find many counterparts in the south-western counties of England. How far correlation is justified is yet to be proved. Not everyone will agree with Brown's views which are in certain fundamental ways different from those of O. T. Jones. On the other hand, George has shown that the assumption of a former Mesozoic cover over all of Wales is one that is difficult to justify. This bigger question, however, need not concern us here. The coastal scene in Wales depends greatly on the plateaux, which affect the form of the cliffs and involve complicated movements of sea level.

Chapter XXII

THE BRISTOL CHANNEL, THE SOMERSET LEVELS, AND THE COAST OF EASTERN SOMERSET

(1) THE BRISTOL CHANNEL

Despite lack of unanimity amongst those who have written in recent years on the origin and evolution of the surface configuration of Wales, and perhaps, but to a lesser degree, of that of the south-western peninsula, there is less diversity of view about the history of the Bristol Channel, at any rate since the early Tertiary period. Some reference must, however, be made to Wales and to parts of England if we are to see the Bristol Channel in its proper setting.

Since the war increasing emphasis has been laid upon planation surfaces on both sides of the channel, and there has been increasing scepticism about the north-westward extent of the Mesozoic rocks, especially the Chalk, and the implied epigenetic origin of much of the river system of central England.

A study of the Palaeozoic rock floor of Britain reveals (see Fig. 124) old highland surfaces in the Pennines, Wales, and Cornubia and an extensive buried surface, the London platform, underlying the south-eastern Midlands. Between these rigid masses are found the Mesozoic basins of Cheshire, Severn, and Wessex. There is probably another beneath Cardigan Bay. Thus, Wales and the south-western peninsula are now surrounded by areas which subsided during the deposition of the Mesozoic rocks. The London platform (i.e. the London–Brabant massif) is now hidden under about 1,000 ft. of newer rocks. It was, however, a prominent feature before it was worn down by erosion which extended from the Trias until the Albian. During this time the other massifs were also being worn down, and eventually the peneplain thus formed was attacked by the waves and finally disappeared under the waters of Mesozoic seas. The western massifs were subsequently raised in the Tertiary period and, judging from the record of the rocks in the south-east, the upward movement took place probably in two stages, at the end of the Cretaceous and in the mid-Tertiary.

In the south-east of England Brown gives reasons for postulating an early Tertiary peneplain. This surface is most surely identified under the

English Channel, where it is covered by Tertiaries. It is also found under the superficial beds in the Bovey Tracey basin (Devonshire) and at Flimston in Pembrokeshire. In the south-east of England the peneplain underlies Eocene beds, but where it is exposed it is, with some consider-

Fig. 124. Generalized map of the post-Hercynian surface (after E. H. Brown)

able degree of certainty, regarded as of sub-aerial origin. However, it is more difficult to identify in the west. Brown thinks the highest summit levels of Wales are part of it, and does not consider that the summit peneplain (1,700–2,000 ft.) is warped: it, too, is regarded as an early Tertiary feature.

In the south-west (see Fig. 125) the early Tertiary peneplain appears in Dartmoor and Exmoor. Perhaps the highest summits of Exmoor accord with an easterly tilted surface. Possible warping in the south-west may be associated with the Miocene folding (see p. 191). But further research may modify this view.

However, it is now generally assumed that the high level planation surfaces of Wales, Cornwall, and Devon are all unwarped and, therefore, all later in age than the postulated early Tertiary peneplain. Granting this, the Bristol Channel is a down-warp in the peneplain, and so presumably related to the Miocene folding.[1]

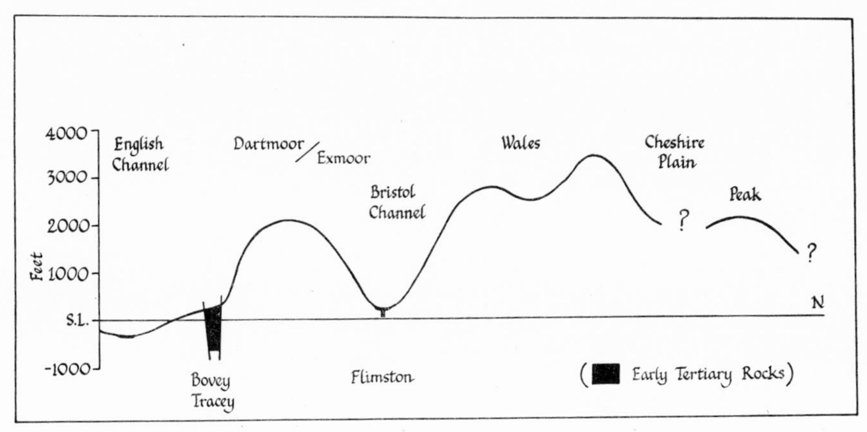

Fig. 125. Deformation of early Tertiary peneplain in southern Britain (after E. H. Brown)

(2) THE SOMERSET LEVELS

In the last decade a great deal of important work has been done on the Somerset Levels (H. Godwin and A. R. Clapham, *Phil. Trans. Roy Soc.* 233, 1948, 233; H. Godwin, *Phil. Trans. Roy. Soc.* 233, 1948, 275, and 239, 1955, 161; H. Godwin, *Proc. Prehist. Soc.* 26, 1960, 1; H. Godwin, *Proc. Roy. Soc.* B, 153, 1960, 187 (The Croonian Lecture)). During the last glacial period the sea level was perhaps 300 ft. lower than it is to-day. As a consequence deep valleys were cut between the hill ranges of the Quantocks, Poldens, Brendons, and Mendips. In Boreal times, when the

[1] Some papers directly or indirectly concerned with the Bristol Channel are: O.T. Jones, *Quart. Journ. Geol. Soc.* 107, 1951, 201; E.H. Brown, *Zeit. für Geomorphologie*, 4, 1960, 264, and *The Relief and Drainage of Wales*, 1960; W.B.R. King, *Quart. Journ. Geol. Soc.* 110, 1954, 77; T.N. George, *Science Progress*, 43, 1955, 291, and 49, 1961, 242.

level of the sea rose rapidly as a result of the melting of the ice, the flat lands of Somerset were inundated.

Although most of the investigations relate to the area near Meare and Glastonbury the general nature of the flooding and of the sediments associated with it are presumably similar over the whole of the Levels. We shall, therefore, deal first with the Meare district.

During the marine transgression an estuarine clay was laid down, the surface of which is now a few feet above mean sea level. Occasional peats occur in this clay, and pollen analysis has established that they belong to Zone VI. The surface layer of the clay marks the Boreal–Atlantic transition, i.e. Zones VI–VII. The clay is almost impermeable, and eventually sedge fens and reed swamps formed on it in the lower part of Zone VII. A change of climate, however, took place during the accumulation of the deposits associated with this zone. Hence Zone VII is subdivided into VII A and VII B.

In Zone VII B there later developed raised bogs which persisted until two or three centuries ago. In both Neolithic and Bronze Age times these bogs grew slowly, and were characterized by *Calluna* and *Sphagnum*. At the end of the Bronze Age the surface became much wetter, especially at the beginning of Zone VIII. This, in Godwin's view, is related to the damper climate of the Sub-Atlantic period, and he dates the change at about 500 B.C. The more humid conditions are clearly shown in the vegetation; *Cladium-Hypnum* fen with *Myrica* spread rapidly, implying river floods of lime-bearing waters from the Mendips and Poldens. Somewhat later peat followed the *Cladium-Hypnum* association. However, at about A.D. 50 there was a second flooding which led to a recrudescence of *Cladium* and *Hypnum*, although to a lesser degree than on the earlier occasion.

Once again *Sphagnum* succeeded *Cladium* and *Hypnum*, and continued to flourish until late Romano-British times. At Shapwick Heath bog growth then ceased, but no reason can be given for this. One or two centuries earlier, about A.D. 250, there began a marine transgression on the seaward side of the levels. This led to the penetration of marine and brackish waters into the valleys and, on the seaward edge, caused some erosion of the bogs. These were later covered, for a distance of two or three miles from the sea, by clays which reached +20 ft O.D., and were settled, in part, in Romano-British times.

It is interesting to follow this transgression up the valleys, especially in the Polden–Wedmore district. This broad valley must then have

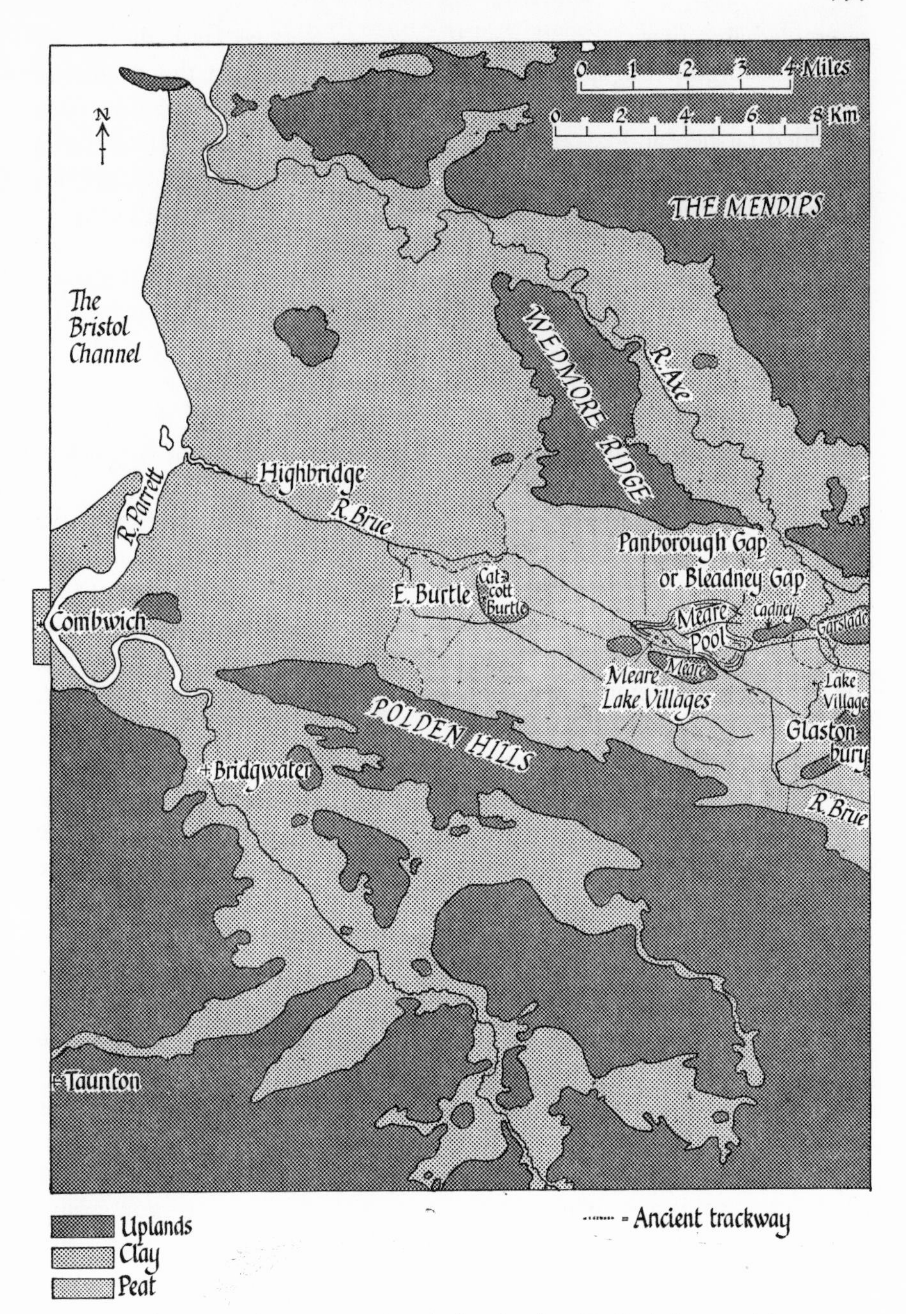

Fig. 126. The Somerset Levels (partly after H. Godwin, *op. cit.*)
(For Cadney, read Codney)

resembled an estuary, and allowed the clays to reach as far as the former islet of Edington Burtle (Fig. 126). However, between the Wedmore ridge and the Mendips, along the valley of the Axe, the transgression extended farther, and covered much of the early peat lands. The clay is now found all along the valley, and extends through the gap at Panborough into the valley between the Wedmore ridge and the Poldens as far as Godney. It is almost certain that the River Bure, instead of taking its present direct course from near Glastonbury to the sea near Highbridge, then flowed in a winding course west of Glastonbury to a gap in the little ridge at Godney. From that place it continued to the Panborough gap and the River Sheppey, and eventually to the sea along the valley of the Axe. This view is supported both by general evidence and that derived from a study of aerial photographs. This state of affairs held good in mediaeval times.

The clay just reached the site of Glastonbury Lake village, and the north-eastern part of Meare pool. This pool seems to have originated as a result of raised bogs effectively adding to the height of a low ridge of Lias near Westhay. There is no doubt that the pool antedates the clay. 'The fresh-water clay which forms the present surface is apparently related to the period after digging of the artificial channel of the present River Bure. Information is lacking as to the eastward extent of the lake in prehistoric time, but in Early Iron Age time its seems likely that raised bogs intervened between it and the open water at the Glastonbury Lake Village site' (Godwin, 1955). Thus, in Meare pool and neighbourhood there is the upper fresh-water clay, of medieval date, at the surface, an estuarine clay of Romano-British age, between +8 and +20 ft O.D., and a basal clay of estuarine origin. This clay is mainly Boreal, but extends into the Atlantic period, and fills all the levels. The fact that clay of Romano-British date does not reach the Meare district from the north-west suggests that at that time there was no natural estuary along what is now the River Brue.

There is less information about conditions in the Parret valley. At Combwich, about 4 miles below Bridgwater, there is an interesting section. The surface of the coastal clay, i.e. that laid down in a transgression of about A.D. 250, is approximately at the same level as that of mean spring high water, or 18–20 ft. O.D. The clay belt is some miles wide, and Bridgwater stands on it. The Combwich section shows 14 ft. of this surface clay, then a peat bed 8 in. thick, and below that a bed of blue clay 6 ft. thick. Below is a thin peat bed (less than 1 in.) and then more blue clay. The surface clay, although it offers no direct evidence of marine origin, is nevertheless regarded as brackish in view of its position. The lower clay is certainly marine, and at Combwich its surface is about +6·5 ft. O.D. It is part of the same transgression we met at Shapwick where its surface is approximately +4 ft. O.D. The clay is rather more marine in character on its seaward side (Fig. 127).

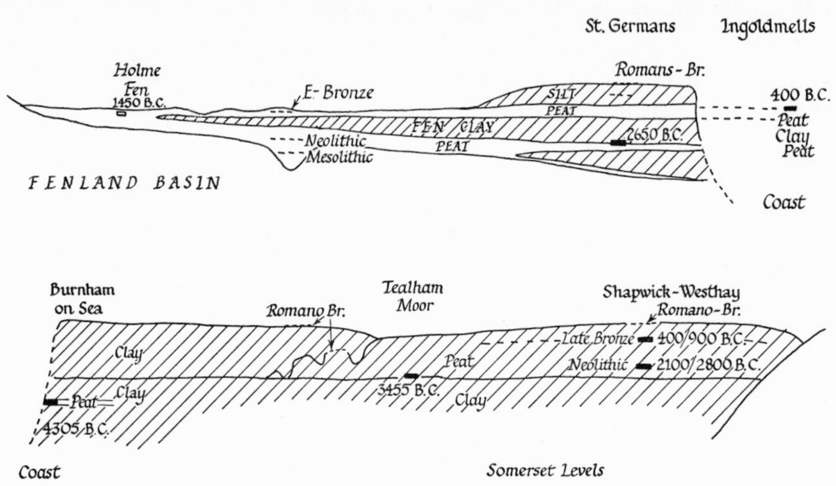

Fig. 127. Stratigraphy of the Fenland and Somerset Levels (after H. Godwin, *op. cit.*)

Remains of ancient trackways have been found at two distinct horizons in the levels. They were built at times when the rise of water level made it difficult to go dry-shod from one island to another. The earlier group has been ascribed, on radio-carbon evidence, to the period 2800 to 2000 B.C.; i.e. the Neolithic. The later group is Late Bronze Age, 900 to 450 B.C. Pollen analysis of key sites supports these conclusions.

In the Somerset Levels there was (1) an eustatic rise of sea level which culminated about 3500 B.C. On the surface of the clay then deposited (2) peat grew and raised bogs formed during the Neolithic, Bronze, and Iron Ages. Later, in Romano-British times there was (3) another invasion of the sea which eroded the seaward parts of the peat, and eventually covered the coastal areas and part of the peat with a second layer of clay. No change of level of an eustatic nature occurred, unless of smaller amplitude than that ending about 3500 B.C., until Romano-British times. The sequence of events and deposits in the Somerset Levels is not the same as that in the Fenland of eastern England.

(3) THE COAST OF EASTERN SOMERSET

On the coast itself some interesting work has been carried out by Kidson (*Inst. British Geogrs.* 1960) under the aegis of the Nature Conservancy. Near Lilstock there are cliffs which reach about 80 ft., but at Hinkley Point they are only 5 or 6 ft. high. Between these two places an extensive platform of Liassic limestone occurs on the foreshore. Farther east there are four main masses of shingle on the mainland, and a fifth on Stert Island. From Wall Common to Stert Point there is a prolific growth of *Spartina*, first introduced here from Poole Harbour in 1928.

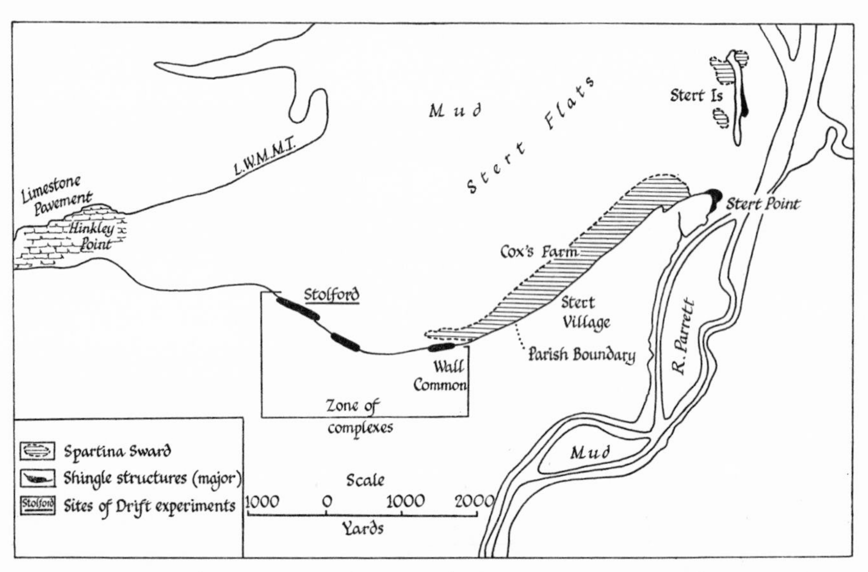

Fig. 128. Bridgwater Bay (after C. Kidson, *op. cit.* 1961. Coastline east of Hinkley Point from Nature Conservancy Surveys, 1956–60)

Outside the shingle and *Spartina* are the extensive mud flats on either side of the Parret estuary. Factors of the utmost significance in the study of this coast are the great tidal range and the short time when the waves are in contact with the shingle masses (Fig. 128) (**70, 71**).

The wide abrasion platform implies past as well as present erosion. A survey by Chilcott in 1819 affords some rough measurements, and shows that erosion has occurred fairly recently all the way from Hinkley Point to Stert Point. Nevertheless, deposition of shingle at Stolford and Wall Common has more than compensated the effects of erosion.

The shingle is derived from the cliffs and from the platform; the shaley beds help to maintain the great mud flats. In general, the size of the shingle decreases eastwards. From Lilstock to Hinkley Point the individual boulders range from those a foot in long diameter to those of the order of 6–8 in. But from Hinkley Point to Stert Point the average size is only about $1\frac{1}{2}$–2 in. in long diameter, although some larger stones occur throughout. The fall in average size is a measure of the relative ease with which the Lias yields to erosion. As far as Wall Common about 80% of the shingle is limestone; farther east the amount of sandstone material increases, so that near Stert Point only 10–20% is formed of limestone. The sandstone pebbles vary in size and are resistant as well as insoluble. It seems that the relative proportions of limestone and sandstone pebbles have also varied in the past. Near Wall Common, for example, some of the older ridges show a much higher proportion of sandstone material than occurs in the present beach ridge.

The general movement of beach material is to the east, but it is extremely slow, partly because of the factors already noted, and partly because of the sheltered nature of this part of the coast. It is only in storms and at exceptionally big tides that noteworthy movement of beach material takes place. Kidson and Carr made some careful observations and experiments, and showed that movement was limited to a narrow zone near high water mark. From Wall Common to Stert Point there is, to all intents and purposes, no movement. The pebbles at Stert Point cannot now be replenished from the west, and must have arrived in this locality some time in the past.

The positions of the shingle masses (see Fig. 129) are governed by the inability of the waves at the present time to move them farther eastwards. Moreover, the *Spartina* has now spread so widely that it is a most effective barrier to movement of beach material. The period during which the waves can work on the shingle at Wall Common is about 27 hr. a lunar month. Farther east, the waves can only *reach* the shingle at high water in storms or during equinoctial spring tides. Thus the shingle ridges are regarded 'as depositional forms marking a stage of maturity in the coast line' (Fig. 129).

Kidson and Carr (*Proc. Bristol Nats. Soc.* 30, 1961, 163) have analysed the movement of shingle in some detail. 'Even six years after the marked material was laid down on the beach the farthest travelled marker from Lilstock Harbour was found only 7,500 feet from its point of origin. At Hinkley Point, where movement was more rapid than at any other site, the farthest travelled marker averaged only 80 feet per month.' The larger shingle moves both at a different speed and locally even in a different direction from smaller shingle. 'The larger material travels relatively quickly towards the end of the beach ridge or shingle complex of which it is part. There it is held up and the more slowly travelling smaller material begins to overhaul it.'

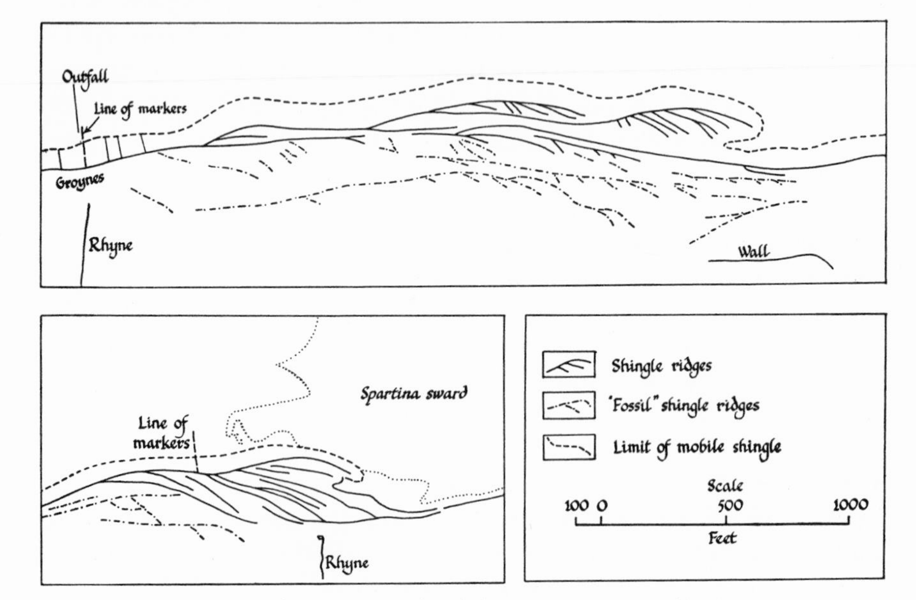

Fig. 129. Bridgwater Bay, detailed site maps. (*a*) Stolford Farm, (*b*) Wall Common (after C. Kidson, *op. cit.* 1961)

[The Draining and Reclamation of the Somerset Levels between 1770–1833 are described by M. Williams in *Trans. Inst. Brit. Geogrs*, No. 33, 1963, 163.]

Chapter XXIII

THE SOUTH-WESTERN PENINSULA

(1) EXMOOR

It was in relation to the coast of Exmoor that Newell Arber first introduced the term hog's-back cliffs. His daughter, M. A. Arber, has studied these cliffs not only in Exmoor, but elsewhere. Newell Arber did not discuss the evolution of this cliff form. Balchin (see p. 614) noted that they occur commonly where the coast and strike of the rocks are approximately parallel and put forward the view that they originated when the climate was far more severe, and sea level stood lower than it does to-day relative to the land. Miss Arber thinks this form of cliff is directly related to geological structure. 'Where the true hog's-back is developed on the grandest scale, near Combe Martin, the regional dip is southerly and thus inland, so that the seaward slope from the crest is simply an escarpment cutting across the edges of the beds. Inland from the crest is a gentler dip slope into a valley running more or less parallel with the coast, the opposite (southern) side of the valley being another escarpment' (M. A. Arber, *Geogr. Journ.* 114, 1949, 191). This implies that the steepness of hog's-back cliff depends upon the pre-existing topography, apart from that part of the cliff directly eroded by the present sea, and explains why the Hangman Grits weather into higher hills than do the Ilfracombe beds. Since, however, the profile must depend mainly on the angle of dip, lithology will have little direct influence. A clear distinction between flat-topped, bevelled (see p. 588) and hog's-back cliffs is difficult to draw, and many transitional forms can be found.

Scott Simpson (*Proc. Geol. Assoc.* 64, 1953, 14) suggests that the Valley of the Rocks is a continuation of the valley of the East Lyn, which once extended farther west, where remnants of it can be traced at Lee Abbey and on the 'flat' of Crock Point. Both Newell Arber and Scott Simpson (who does not refer to Arber's work) believe that the river which formerly flowed in the Valley of the Rocks was dismembered by sea-erosion. Scott Simpson, because he extends East Lyn so much more to the west, suggests that there were stages in this process. The first, at or beyond Heddon's Mouth, must have been early since the steep gradients have had time to migrate upstream. The Woody Bay capture led to the cutting of the steep valley, now dry, at Lee Abbey. A third capture is postulated at Wringcliff. 'The steep gradients produced by this capture are similarly preserved in the Valley of the Rocks to-day as a result of the last capture at Lynmouth (**67**). The Lynmouth capture is so recent that the West Lyn still preserves the phenomenal fall of 400 feet in its last mile' (Scott Simpson, *op. cit.*).

The relatively sheltered position of much of the coast of Exmoor makes it difficult to accept certain other views put forward by Scott Simpson. He recognizes that the rocks are all resistant, but postulates a band of much less resistant rocks in front of the present coast, which were quickly eroded. Their presence was ascribed to a fault which carried down soft Mesozoic rocks against the harder Devonians (**68**).

It is not unreasonable to suggest that such a basin as that of Porlock Vale once extended westwards past the Foreland. There may well have been a time when the coast east of Combe Martin Bay swung many miles to the north and may even have joined the South Wales coast on the east side of Swansea Bay. If then the coast, retreating eastwards, encountered the Lias Clays within the Trias basin postulated, it might have quickly extended towards Bridgwater and brought virtually the present coastline into existence as a marine fault-line escarpment.

This is speculative and prompts the question—Why is it necessary to invoke the view?

The great floods at Lynmouth in August 1952 are of particular interest in that they showed how great is the power of what in normal times is an insignificant stream. When the climate was more severe, the tree covering less, and when there was still much snow on high ground such as Exmoor, we can appreciate far better how the great gorges were cut, that now form the picturesque drowned mouths and similar features of the south-western peninsula, if we examine briefly what happened at Lynmouth a few years ago.

On 15 August 1952, after a wet period during which the ground became saturated, there was rain nearly all day, but most fell between 8 p.m. and 1 a.m. on the 16th. In all 9·1 in. were recorded at Longstone Barrow and 7·58 in. at Challacombe. Kidson (*Geography*, 38, 1953, 1) calculated that about 300 million tons fell on the 38 square miles drained by the two Lyns. For a short period the discharge at Lynmouth was 23,000 cusec., or almost as much as the record figure for the Thames. Nearly 40,000 tons of boulders invaded Lynmouth, many weighing 10 tons or more. A deltaic deposit on the right bank of the East Lyn was estimated to contain 200,000 cubic yards of boulders. Higher up the river there were landslides, and fallen trees temporarily blocked the river. When these dams gave way the greatest damage was done lower down.

This was an exceptional storm, but there have been a number of severe storms in other parts of the country in recent years. The great spread of boulders on the foreshore at Lynton may well be the result of such storms in the past, although marine erosion and local cliff falls have doubtless helped. The storm emphatically showed how, even to-day, deep gorges can be deepened and widened (see also C. H. Dobbie and P. O. Wolf, *Proc. Inst. Civ. Eng.* 1953, 2, pt. 3, 522).

(2) BRAUNTON BURROWS

The Sandhills at Braunton Burrows have received some publicity in recent years since they have excited the interest, from very different points of view, of the services, conservationists, and scientists. A full account of their botany and ecology was published in 1959 (A. J. Willis, B. F. Folkes, J. F. Hope-Simpson and E. W. Yemm *Journ. Ecology*, 47, 1 and 249.)

Some dunes reach the 100-foot contour, and the authors state that the highest crests are usually a little inward from the beach, and farther landward the dunes, apart from a few prominent hills, are lower. This is refuted by Kidson (*R.I.C.S.*, December 1960, p. 3) who has mapped the area in detail. He shows that the landward ridge is higher. There are many mobile dunes, and an abundant sand supply on the beach. The parallel arrangement described on page 216 still holds. The best development of dunes is in the central section; in the south two ranks can be distinguished, and in the north the dunes are lower.

The 'dynamic character' noticeable in the dunes in recent years is a more or less constant trait, and not wholly the result of military activities, although doubtless accentuated by them. Many of the dunes are parabolic in form; these are advancing and those to the west are said to be catching up and aggregating with those already to landward—another point contradicted by detailed surveys by Kidson covering 1950–60. Comparison of Admiralty Charts of 1839 and 1953 show that the seaward margin, apart from minor irregularities, has not appreciably altered in more than a century.

The unstable nature of the burrows is a permanent feature resulting from their situation between the open ocean and the Taw estuary. This fact is made apparent by the difficulty found by *Ammophila* in stabilizing the surface of high dunes. It is thought that bracken will spread and gradually replace dune pasture, although bracken is unable to survive in exposed and mobile places on the dunes.

The upper surface of the water-table in the dunes is shaped like a flat dome or shield. It fluctuates readily with rainfall, and the low-lying areas are often flooded. The response of the vegetation suggests that factors other than the rise and fall of the water-table affect plant life in the slacks and hollows.

The general relationship between topography and vegetation is apparent in the following table:

TOPOGRAPHICAL FEATURES

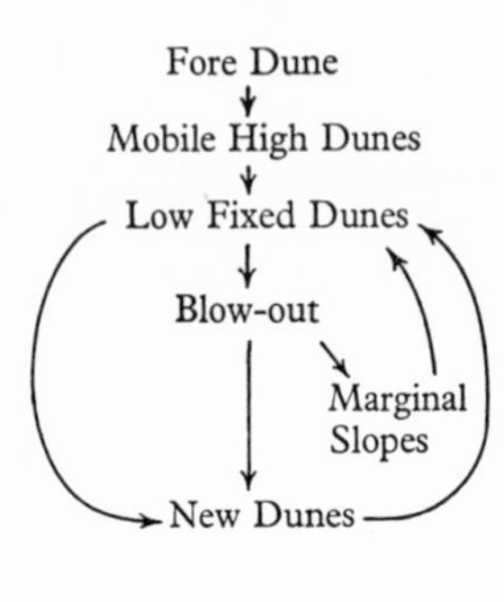

VEGETATIONAL DEVELOPMENT

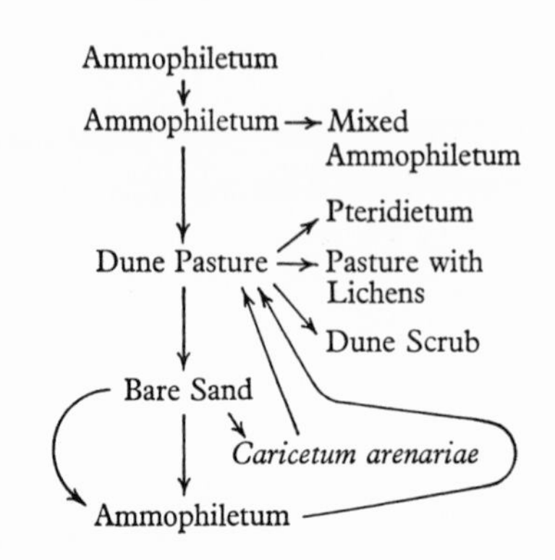

In the slacks and hollows the ecology is more intricate:

(*a*) Open communities in the slacks near the sea.
(*b*) *Plantago–Leontodon* communities.
(*c*) Damp pasture: *Carex–Hydrocotile* communities.
(*d*) Rush communities:
Juncetum maritimi.
J. acuti.
Holoschoenatum vulgaris.
(*e*) *Salicetum repentes.*
(*f*) *S. atrocinariae.*

Some of the slacks orginated as blow-outs and may be as much as 30 ft. O.D. This does not imply any change of level.

(3) BARNSTAPLE BAY

Near Westward Ho! there are some interesting features discussed by E. H. Rogers (*Proc. Devon Arch. Exploration Soc.* 3, 1937–41, 109). On the cliffed coast to the west of the town there are remnants of raised beach, presumably the *Patella* beach, and platform. West of the pebble ridge there is a series of deposits, blue clays, peat, a midden, and an earlier ridge of large pebbles, all of which imply a complex history in terms of levels of land and sea. Section 1 in Fig. 130 shows (1) soil creep overlying, (2) the Head (see p. 479), and below that deposit the pebbles of the raised beach on the old rock platform (4). This is now some 30 ft. above (5) the present pebble beach which rests on a 'modern' rock

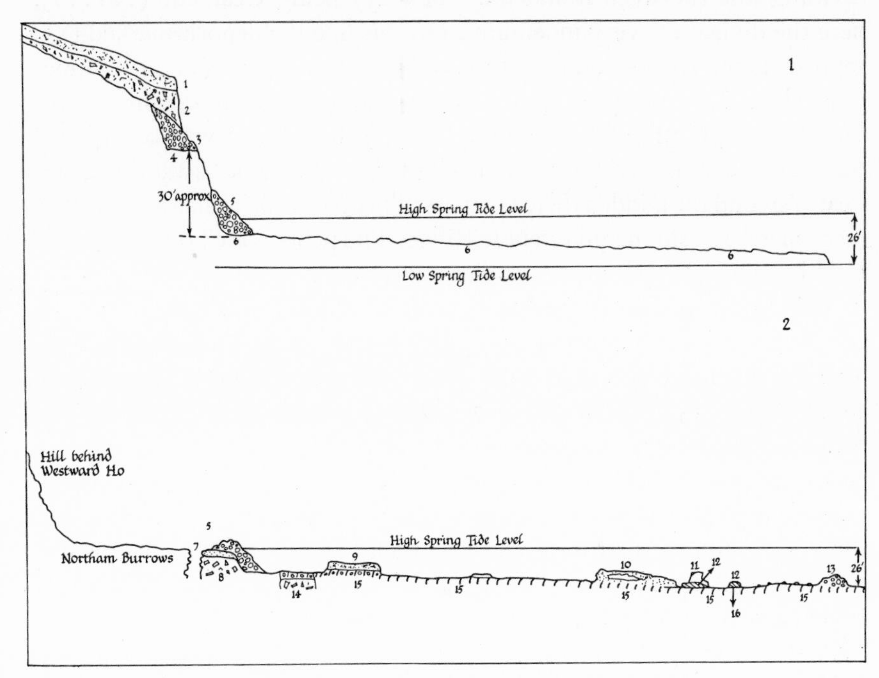

Fig. 130. Sections at Westward Ho! (not to scale). No. 1 section, at west end of raised beach; no. 2 section, at south end of present pebble ridge. The figures on each section are explained in the text. (Based on E. H. Rogers, *op. cit.*)

platform (6). The origin of this platform is still uncertain (see p. 480). The second section is not a direct continuation of the first, since it is taken at the south end of the present pebble ridge and runs roughly east and west, whereas the former runs more nearly north and south. The two together make a composite picture of the deposits on the rocky and sandy foreshores at Westward Ho! In the second section (5) is the present pebble ridge resting partly on (7) the top soil of Northam Burrows and (8) a brown clay with many angular rock fragments, not unlike the Head.

Farther west is an outcrop of peat (9) the Inner Peat, 12–15 in. thick and resting on (15) the blue clay which, at this place is studded with many rounded pebbles. The deposit numbered (14) is a brown clay, only very occasionally seen at this particular locality, but is presumably the same as (8). Farther seaward is another outcrop of peat (10), the Outer Peat. It is probably part of the same formation as (9), the Inner Peat, but in the outer peat the 'stone contents are mixed and rounded, and instead of the dividing line (between it and the blue clay) being clear cut (......), here the division is very indefinite'. On this line the deposit marked (11) represents the last tree stump of the old forest rooted on (12), the upper blue clay. This seals the (16) kitchen midden, which contains shells, bones, teeth, flintflakes, and cores. At (12) the midden was sandwiched between the upper and lower blue clays; it is also underneath the Outer Peat. Beyond the midden there is a mass of pebbles which Inkerman Rogers recognized as a submerged Pebble Ridge. The pebbles are bound together by a brown clay with many angular and unworn stones.

The raised beach (3) implies a change of level of about 30 ft.; the submerged pebble ridge (13), if it is correctly interpreted one of 26 ft. The lower blue clay (15) points to subsidence, and (16), the midden, suggests that it was built at a time when the area was dry ground. The forest (11) means an upward movement of the land of 15 ft. or more above sea level to allow the growth of oak. The upper clay (12) and the peats (9 and 10) indicate subsidence; that associated with the peats led to the destruction of the forest, and eventually to the levels of the present day. The full significance of these deposits still needs to be worked out. Pollen analysis was applied, but unfortunately gave disappointing results, the pollen apparently having been almost wholly destroyed.

Not far away, at Lower Yelland, on the south side of the Taw, about two miles above Instow, there is a stone row. The ground on which it stands is level, and 10 ft. below high spring tide mark. The general ground section for the whole row is: 1 in. of tidal sand and silt, resting on 3 in. of a tough blue clay, which covers a well-defined old land surface, a mixture of red earth and shillet. The row was built after the deposition of the blue clay, since there is a packing of clay round all the exposed stones. The dating of the row depends upon flint artifacts only; no pottery has been found. Mesolithic people occupied the old land surface below the clay, which implies a submergence. The later Neolithic folk lived on the site when it re-emerged. Bronze Age people probably succeeded the Neolithic, and occupied the site until the early part of the Middle Bronze Age when submergence began again, eventually to produce the conditions we have to-day. The row itself is almost certainly a Bronze Age construction.

Changes of sea level in Barnstaple Bay are confirmed on very different evidence. Seismic surveys have been made of parts of the Taw and Torridge valleys, and also of the Erme in south Devon (P. B. McFarlane, *Geol. Mag.* 2, 1955, 419). The Taw and Torridge are apparently graded to −85 ft. O.D. The buried channel of the Taw closely follows the present channel, and did not cut direct through Braunton Burrows. Seismic soundings in the sands prove an almost uniform thickness of 30 ft. All three rivers have flat-bottomed rock channels which suggest by their

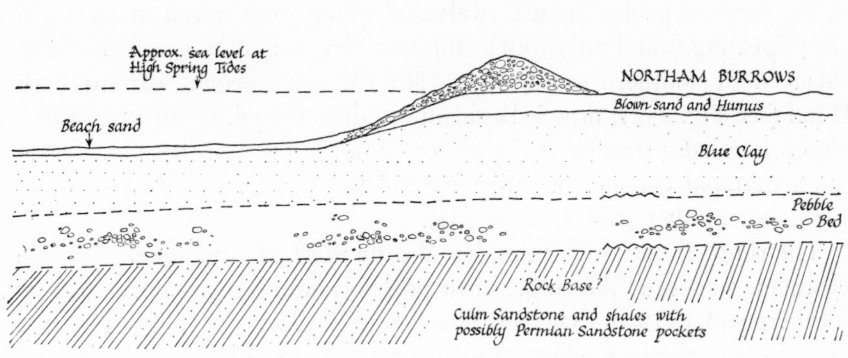

Fig. 131. Westward Ho! Beach and underlying beds (after A. Stuart and R. J. S. Hookway, *op. cit.*)

nature and position a sea level change of the order of 150 ft., and that the shoreline was about eight miles farther seawards in Barnstaple Bay than it is now. This finding is not inconsistent with what is known of the Tawe valley in South Wales.

In a Report to the Coast Protection Committee (Special) of the Devon County Council (September 1954) A. Stuart and R. J. S. Hookway discuss both erosion at Westward Ho! and the changes in the pebble ridge. Figure 131 shows the beach and underlying beds. The lower pebble bed is probably the seaward part of the old raised beach. The patches of submerged forest referred to on page 610 rest on the gentle slope of the beach sand and blue clay. Stuart and Hookway believe that the present pebble ridge came into existence as a result of the sea advancing not only on to a new cliff and eroding it, but still more because of its encroachment on the pebbles of the raised beach. The pebbles of the old and present beaches are both derived from the Culm sandstones.

The origin of the ridge is unknown. The earliest reference seems to be in Risdon's *Survey of Devon* (seventeenth century), but no definite historical evidence is forthcoming until the early part of the nineteenth century. The first good survey is that of 1861. Then the crest of the ridge stood 500 ft. farther seaward than it did in 1950; in that time 70 acres of land have been lost. The two authors contend that, despite former conflict of opinions, the part of the ridge near the sand hills is a product of the last century, and the area is still one of accretion. The growth seems to have started about 1860, but for what reason is unknown. This northward growth has been accompanied by a pushing back of the ridge farther south, and also by its loss of height. The northern extension shows itself in the form of pebble ridges, twelve of which were traced in 1952 (in 1884 Spearing found only four). But northern accretion is unfortunately balanced by southern erosion. In 1861 the ridge began near the Bath Hotel (see Fig. 132); now it begins more than 300 yd. north-east of that place. It recedes steadily at the present time; in the past there have been times of rapid change. In 1877 it retreated 35 ft; more than 50 ft. in 1878, and another 30 ft. in 1896 (**64**).

The ridge, in recent decades, has also diminished in volume. Reliable observers measured the ridge, most probably, but not certainly, near its southern end, in 1869, 1879, and 1884: these measurements all suggest that up to 1884 it stood 7 ft. above the Burrows and 21 ft. above the beach, and was about 160 ft. wide. In 1954 it was 80–85 ft. wide, 11–15 ft. above the sands, but still 7 ft. above the Burrows. An incidental result of these changes is that neap tides now no longer reach it. The shape of the ridge in plan remained consistent in the past; it is straight between the cliffs and Sandy Mere, which has been a point about which it has pivotted. Defence works have caused changes; it still pivots in the north, but in the south it retreats *en echelon* with the sea wall.

Experiments with marked pebbles have shown that they usually move east and north at a rate depending on the force and direction of approach of the waves. They move most rapidly over other pebbles, and very slowly on the sands. Occasional, but exceptional, movements of 100 ft. a day have been recorded; average rates are small. The rate of supply cannot be measured but seems to have varied little, if at all, during the last 100 years. The exhaustion of the raised beach supply has been balanced by pebbles derived from the cliff which increases in height as it is cut back.[1]

[1] There is some reason to think, but none to prove, that the average size of the pebbles is somewhat less to-day than in the past.

The section of the beach (Fig. 131) shows that an apron of pebbles runs down the front of the ridge on to the blue clay. This apron probably has a great effect in the preservation of the ridge. If it were absent the waves would have far more erosive effect on the soft clay. Stuart and Hookway maintain that the best way of strengthening the ridge is to

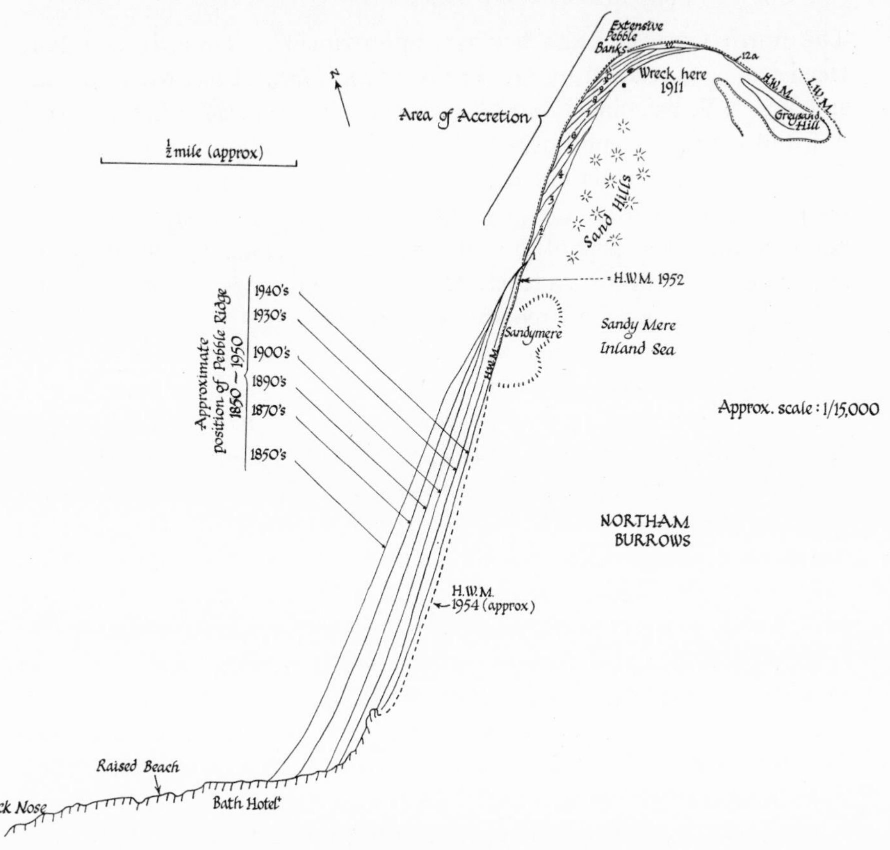

Fig. 132. The pebble ridge and Northam Burrows (after A. Stuart and R. J. S. Hookway, *op. cit.*)

hold the apron with groynes and so increase its thickness. In 1935 a breach was made near the sandhills, but was 'self-healed' in 3 days. Today 'there is totally insufficient material anywhere on the apron for 500 yards north of the groynes to enable this (i.e. "self healing") to be done without depleting the apron completely over a considerable length, with imminent danger of other breaks following'. If a serious breach occurred, and immediate repair were impossible, probably all the southern end of the Burrows would soon disappear, and there would be unpredictable effects in the estuary.

(4) CLIFFS OF NORTH CORNWALL

The north Cornish coast between approximately Boscastle and Park Head has been studied by G. Wilson (*Proc. Geol. Assoc.* 63, 1952, 20), and W. G. V. Balchin (*Trans. Roy. Geol. Soc., Cornwall*, 17, 1946, 317).

From Pentargon and Boscastle in the north to Treligga about six miles farther south the cliffs are cut in the Trevena platform (p. 257), and often drop sheer from the platform to the sea. Many of the cliffs are bevelled, the upper part of the cliff being a steep slope, convex upwards, over a vertical drop to sea level. If the surface is more or less uniformly inclined, hog's-back cliffs replace bevelled ones.

The structure of the area is more complicated than Dewey supposed (pp. 224 et seq.). Between Smith's Cliff and Trebarwith there are several overthrust slices, produced by movements coming from the south-east. After the thrusting had taken place, further disruption occurred in the form of approximately parallel normal faults, with downthrows to the west and north-west. Additionally, there was some gentle warping produced by the Davidstow anticline. In the northern section, roughly Smith's Cliff to Boscastle, there is no thrusting. The rocks showing in the cliffs in this section range from the Culm (in the north) to the Woolgarden phyllites, but in Bossiney Haven the dip changes from northerly to westerly on account of the Davidstow anticline passing through Bossiney Haven. All the beds are unmoved in the north, but in the south they are either overlain by, or faulted against, the thrust masses.

As a result of thrusting and normal faulting a great number of planes of weakness have been produced. Usually, the thrusts are nearly horizontal and often inconspicuous, whereas the waves have eroded freely along the shear-zones formed by faulting. The same earth-movements have also led to the development of numerous joints, many of which guide erosion and help to produce the intricacies of the coast. Near Tintagel there are joints related to the thrusting; these strike N. 30°–35° W.; at right angles to this direction is a second set, the longitudinal joints, which have local importance. Thirdly, there are many oblique joints, but especially numerous are those striking north and south, i.e. at 30°–35° east of the direction of thrusting.

Between Pentargon and Penally Point the dip is to the north, the cliffs are of the hog's-back type, and their height is controlled by the Trevena platform. The effect of jointing is profound at and near Pentargon. Balchin shows that Pentargon Seal Hole, Little Pentargon, and Eastern Blackapit are all closely associated with jointing. Wilson points out that the cliff at the head of Pentargon is a joint plane, and the points of the headlands enclosing the inlet on the north are truncated by joint planes. On either side of the mouth of the Valency, the jointing affects the form of the cliffs, and both Penally Point and Eastern Blackapit are cut off by fractures, and the blow hole is cut along joint planes. South of Boscastle there are many irregularities in the coast. Grower Gut owes its rectangular shape to jointing. In the hog's-back cliffs joints and faults, one set of which dips north-west at 45°, and the other set, dipping more or less at right angles to the first, have helped to produce the rectangular pattern of this coast. Nearer Short Island the interaction of wave erosion and steeply dipping joints have given rise to stacks, reefs, and gullies. This island and Long Island are separated from the mainland by erosion along joints. Normal faults are responsible for Saddle Rocks and Trambley Cove. North-north-west joints, along which erosion has been effective, have led to the inlets known as Trewethet Gut, Rocky Valley, and another unnamed one about 200 yd. farther south. The walls of Bossiney Haven and and Elephant Rock are joint-controlled. But between Bossiney Haven and Lye Rock, Wilson calls specific attention to an exception—a rounded headland in altered volcanic rocks. Lye Rock, the Sisters, and Willapark all owe their shapes largely to jointing, and the same is true of the passages between the two islands, and between the islands and the mainland. Balchin shows the tendency of the Tredorn phyllites to weather easily along joint planes (see map Fig. 133).

Smith's cliff, Barras Nose, and the island present greater complications. Some of the prominent features, and their relationship to faults, thrusts, and joints are shown on Fig. 133. It is in this part of the coast that thrusting becomes important, and the caves fault zone plays a big role in the formation and evolution of Tintagel Haven, which is, in one sense, the estuary of the Trevena brook, which now reaches the beach by a fall rather more than 40 ft. high. This stream at one time continued westward across Blackarock, and the slope above the iron gate, and the tiny stream between Barras Nose and Blackarock was a tributary. These are also lines of weakness, and erosion, fluvial and marine, working on them helped to form the haven. The joint pattern around the island reveals much of its

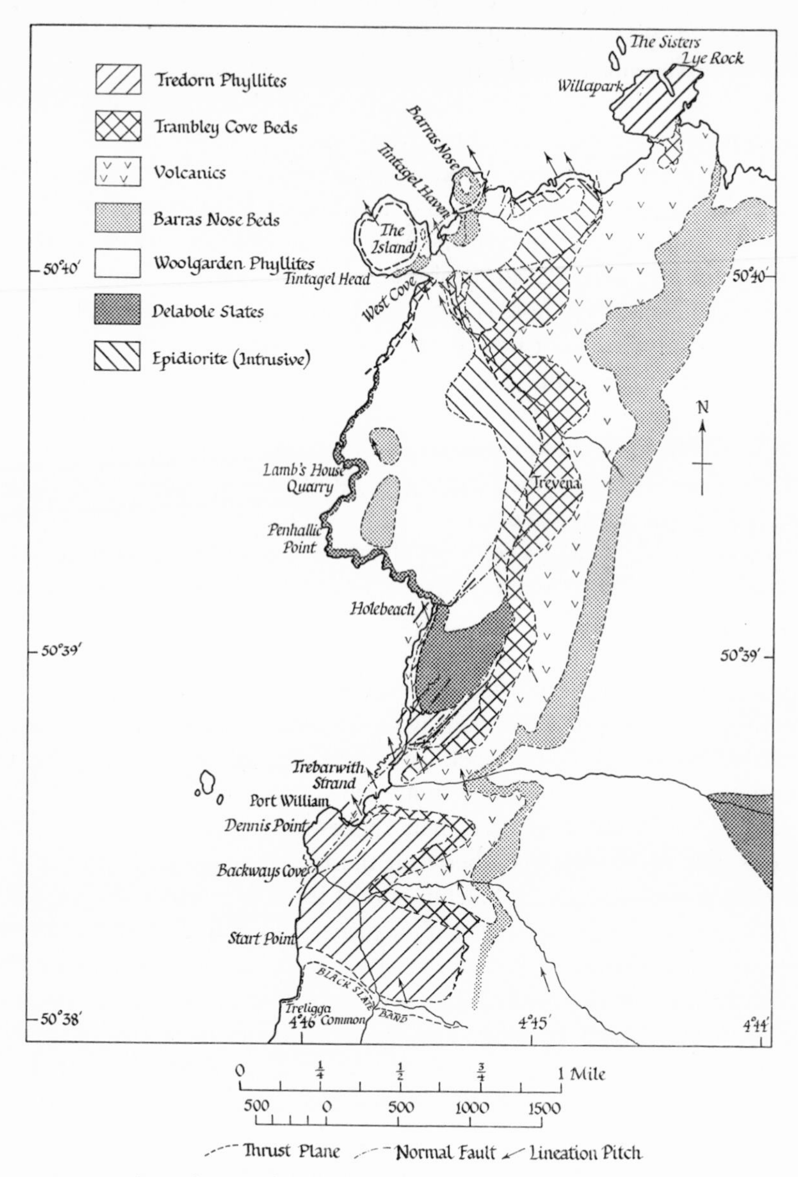

Fig. 133. Geological map of north Cornish coast, showing prominent features and their relationship to faults, thrusts and joints. (Based on G. Wilson, *op. cit.*)

structure. The outer cliffs at sea level, like those on Barras Nose, are in sheared volcanic rocks. The resistant Woolgarden Phyllites coincide with the narrower part of the haven. The low ridge between island and mainland is eroded along fault planes. It is probable that when the castle was built, 1145?, parts of it stood on the isthmus. Additions to the castle were built on the mainland in 1235–40, and a bridge joined the two parts. It is believed that in the sixteenth century the gap between island and mainland was much as it is now. Under the island there is a natural tunnel between Merlin's Cave and West Cove; this follows a normal fault. There is a similar cave near Tintagel Head. 'There is little doubt that the Island, a salient in the coastline, owes its existence to the fact that the base of its cliffs to the south, west, and north, is composed of resistant flat-lying volcanic rocks. On these rest the phyllites which, as a klippe, form the mainmass of the headland' (Wilson, *op. cit.*) (**58**).

To the south the general outline of the coast is simple, and many headlands between the island and Jacka Point approximate to a straight line joining those two places. Hole Beach and Trebarwith Strand, however, form a marked recession from this line; yet along the cliff at the back of this inlet the base is in the resistant volcanic series. This dips gently to the west, and the jointing is widely spaced. The southern end of the inlet, Port William, shows the relation of marine erosion and normal faulting very well. Although each individual fault is accompanied by a cave, the whole re-entrant from Penhallic Point to Dennis Point is probably to be ascribed to erosion acting in part along oblique faults (see Fig. 133) which to-day still affect its outline. Now, however, the less resistant beds have been removed, and the waves are attacking the harder volcanic series.

This change of direction of the fault system relative to the coast, at right angles at Barras Gut, oblique at Trebarwith, becomes approximately parallel with the coast south of Dennis Point. Thus, the cliffs at Treligga Common are fault-line cliffs and have probably receded but little. Wilson thinks that similar conditions prevail as far as Port Isaac Bay.

In Port Isaac Bay the rocks are mainly slates, purple, green, and grey in colour and suffer active erosion, as can be seen in the steep rock-strewn cliffs and in the coarse sand and abundant debris on the foreshore. The rocks are much folded, and the waves are able to exploit minor areas or patches of weakness. Balchin thinks that the higher cliffs north of Tresungers Point imply that all the Rosken level has been destroyed, and that the cliffs are now in the Trevena platform. At Bounds Cliff erosion has reached the Treswallock surface, and cliffs often exceed the 400-foot contour and are amongst the highest in Cornwall. There are several streams reaching the sea in Port Isaac Bay, and these break up the cliff into segments. Near Port Isaac, valleys showing traces of more than one cycle of erosion cut into the Rosken level and because their lower parts are now drowned, give rise to minor harbours at Port Quin, Port Gaverne, and Port Isaac.

Varley Head, a salient, is made of purple and green slates which have been let down by a major fault. Reedy Cliff is in pillow lavas and Kellan Head shows grey slate under lava, which often shows pillow structure. The same lava occurs in Trevan Point and the headlands enclosing Port Quin. More to the west, near and at Com Head, the grey slates reappear, and the coast is cut back; the cliff's are hog's-backs. At Lundy Bay there is a good example of a sea-breached valley (see also p. 221), and another breach may later occur at Markham's Quay. Cliff Castle and Rumps Point bring in another feature, the greenstone intrusions which are conspicuous elsewhere on the Cornish coast. In between the outcrops of greenstone, erosion has cut small re-entrants in the slates. Pentire Point is in pillow lavas, and stands in marked contrast to the coast of slate cliffs. The lavas produce craggy outlines 'in which masses of lava protrude through intermittently grassy and rock strewn slopes' (Balchin, *op. cit.*).

Padstow Bay, enclosed by the greenstone headlands of Pentire Point and Stepper Point, is carved in Upper Devonian slates, but the more conspicuous features on its shore, St Saviour's Point, Ball Hill, Dennis Point on the west, and Burniere, Carlyon, Brea Hill, and others on the east, are greenstone. The Pleistocene deposits at Trebetherick are discussed on page 479 (**57**).

Between Stepper Point and Trevose Head the topography, coastal and inland, is relatively subdued. The slates of which much of the area is built weather readily and in Balchin's view the local hill tops are all cut by the Rosken surface at about 280 ft. Greenstone forms the headlands on the north side of Trevone Bay (where there is a fine blowhole), Cataclews Point, Trevone Head and the Quies. On the other hand, Harlyn, Newtrain, and Trevone Bays are in the slates. Southward from Dinas Head as far as Carnewas Island, the coastline shows differential weathering to perfection. All the headlands are in igneous rock, and the softer slates give rise to the bays. The Staddon grits outcrop from Carnewas Island to Pendarves Island, Park Head is greenstone, Porth Mear is in slates, and the Trescore Islands imply rapid erosion which increases northwards to and beyond Minnows Island where the cliffs are cut in Upper Devonian slates. In Constantine Bay there are features of archaeological interest, but here we need only note traces of a 10-foot raised beach and extensive sand hills behind a wide modern beach.

The discussion of the cliffs of Exmoor and those of north Cornwall make this a proper place to say something on the more general question of cliff evolution.

M. A. Arber (*Geogr. Journ.* 114, 1949, 191) uses three terms, flat-topped, bevelled, and hog's-back cliffs. These are generally accepted as useful descriptive terms. The first, flat-topped cliffs, may be simply the result of the sea cutting into a land which has been planed down by marine, or possibly sub-aerial, action at some earlier time. Balchin's work on Exmoor and north Cornwall has shown that in the south-western peninsula there are several surfaces of planation, and in its simplest form the height of a flat-topped cliff will depend into which surface the sea has cut. The other factors that may modify the cliff form can be neglected for the moment.

Hog's-back cliffs (see p. 605) are regarded by Arber as directly dependent upon geological structure, and since only their bases are being cut by the sea, the upper slopes were probably produced by sub-aerial weathering. Balchin notes that in north Cornwall these cliffs occur when the coast is parallel with the strike of the rocks; the dip being either towards or away from the sea. He also suggests that they were developed under a far more severe climate, when there was abundant solifluction and the formation of head. Hence, they were probably formed at a period of low sea level (**68**).

Bevelled cliffs are those which show a curve from the coastal plateau to the vertical cliff below. In relevant *Geological Survey Memoirs* (Land's End, 1907; Padstow and Camelford, 1910) the suggestion is made that the bevel is Pliocene in age, although the precise date is uncertain. Gullick notes the independence of the bevel and folding; Flett associates it with dykes leading to slips and the breaking down of the cliff edge. In west Wales the form is ascribed to sub-aerial weathering. M. A. Arber concludes 'where marine and subaerial erosion are more nearly balanced and where there are no dominant structural planes in the cliffs, a bevelled cliff results...'. Cotton (*Geol. Mag.* 88, 1951, 113) has suggested that these cliffs are of two-cycle origin (see below).

Wilson's (*op. cit.*) detailed analysis of the Tintagel cliffs has shown that there are many youthful features in them, but nevertheless the marine erosion to which they are subjected has been of long standing. The way in which the waves are working on planes of weakness of different origins has already been discussed. If fault planes are parallel, or but slightly oblique, to the cliffs, erosion will soon lead to collapse of large rock slices, and if similar faults occur a little farther inland, the process will be

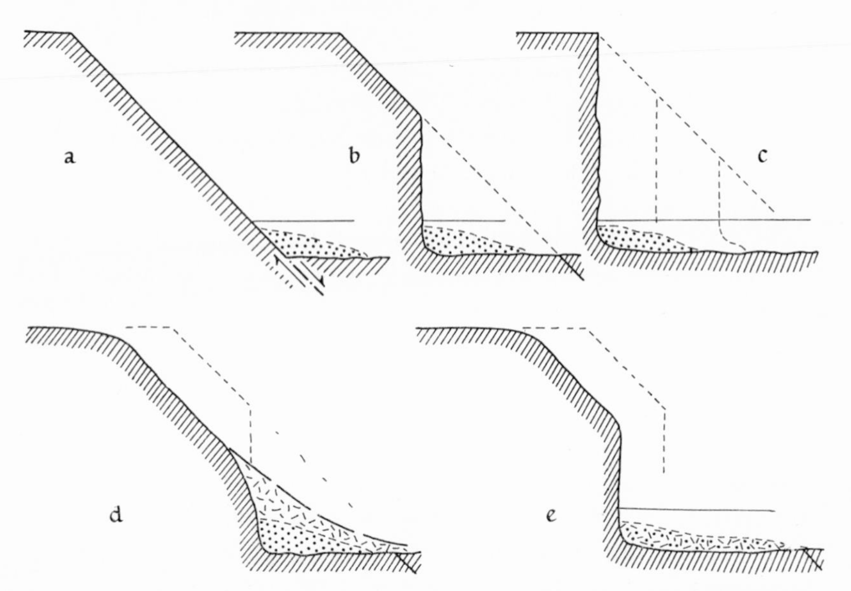

Fig. 134. The development and retreat of a fault-line cliff: (*a*)–(*c*) by marine erosion; (*d*) and (*e*) by two-cycle process, as suggested by Cotton (1951). (*a*) The stripping of an inclined fault plane; (*b*) wave erosion develops a slope-over-wall cliff; (*c*) continued wave action destroys the slope, and a flat-topped cliff is formed; (*d*) sub-aerial weathering during a period of lowered sea level results in the retreat of the cliff and the accentuation of head on top of marine deposits; (*e*) return of the sea rejuvenates or 'freshens' the lower face of the cliff. (Based on G. Wilson, *op. cit.*)

repeated. The cliffs between Tintagel Island and Treligga are of this nature; the fault plane is inclined seawards, and the effects of direct erosion may be shown in the three figures (Fig. 134).

The second figure, slope over wall, Wilson describes as a hog's-back cliff. It may perhaps be regarded as a particular instance of that type of cliff, since here the sea is working on a fault plane and not a dip slope. If Cotton's two-cycle scheme is adopted, then the process may be represented as in *d* and *e*, based on Wilson's interpretation: '. . . when the sea-level fell during a period of glaciation, the cliff would be subjected to sub-aerial periglacial erosion. This would result in flat-topped cliffs becoming bevelled, and in original hog's-back cliffs being eroded back or retreating parallel to themselves' (Wilson, *op. cit.*). In this scheme the initial slope is retained, although in the end the slope might become continuous. But if there was a rise of sea level the buried part of the old cliff would be once again subjected to wave attack.

Cliffs of all three types occur in several parts of Devon and Cornwall and elsewhere along the English and Welsh coastline. Bevelled profiles have not been emphasized, but they are common, and good examples occur at Nanjizal, near Land's End, Port Isaac and the Lizard (see also pp. 588–90).

The sea bed close inshore between St Agnes's Head and Cligga Head has been investigated by aqualung divers (R. H. T. Garnett, *Roy. School of Mines Journ.* 1960, and *Journ. R.I.C.S.* January 1961). The area may be roughly defined as that contained within three points, Bowden Rocks, St Agnes's Head, and Cligga Head. Of this about 40% consists of bed rock; over the remainder there is a cover of sand, about 6 in. thick at its outer limit, increasing landwards to perhaps 6 ft. as a maximum. But the sand layer varies both in thickness and in place with the seasons. At all diving places free from sand it was found possible to measure dip and strike, and care was taken to make sure that the rock was *in situ*. It was found that where the contours followed the coastline, the bed rock was hidden beneath sand; where there was no relation of contours to coast, the bare rock was exposed. Correlation of results showed that the metamorphic aureole of the Cligga granite is greater than could be inferred only from its land outcrop. The sea bed explored appears to be part of a platform of erosion, and variations in its height are produced by folding in the killas, differences in the degree of metamorphosism, and the intensity of quartz veining.

(5) DUNES AND ARCHAEOLOGY

The sea beaches of north Cornwall are presumably fed by the wear and tear of rock material derived from subaerial and marine action on that part of the county north of the main watershed. An analysis of eighteen localities by A. Stuart (*Trans. Roy. Soc. Cornwall*, 17, 1937–48, 13) revealed more than sixty different minerals. The petrological details need not concern us, but some of his general conclusions are of interest. Quantitatively, the beach sands correlate with local sources, but the absence or rarity of staurolite and kyanite in the south implies that Pliocene material plays a small part in the modern sands. In the north there is more, yet somewhat rare, non-local debris which suggests derivation from glacial deposits. On account of the broken nature of the whole coast, longshore drift is negligible in many places. Bude and Westward Ho! are to some extent exceptions.

There are dunes in several localities, including Braunton, the Camel estuary, Constantine Bay, Perranporth, and Gwithian. They are fed from the adjacent beaches. Those in Constantine Bay are largely formed of shell sand, which drifts readily. In such places sand has often overwhelmed buildings. Perranzabuloe (= Sanctus Piranus in Sabulo) in Penhale Sands, north of Perranporth, illustrates the relation between sand inundation, archaeology, and legend. The tale is that long ago a great sandstorm destroyed the parish and buried the shrine, or oratory, of St Piran. Not much more than a century ago it was found again by accident. After the first church was buried a second was built, and was threatened in the same way. However, the parishioners took it down stone by stone, and rebuilt it as the third church at Lamborne. The foundations of the second church were kept, and a cross was set up to mark its site.

It is historical fact that the oratory was built at the head of a small and fertile valley in either the sixth or the seventh century. It was, however, an oratory and not a parish church. Blown sand sooner or later caused a nearby stream to alter course and threaten the oratory. A protective dam was therefore built in the eleventh century. The relics of St Piran were placed in the oratory and became the object of a mediaeval pilgrimage. Various architectural changes were made in the building. It was kept clear of sand up to the Reformation, but in the time of Elizabeth I the roof was removed and the place was neglected and was soon surrounded by sand. Camden (1586) refers to it as 'on the sands' and Wilson, in *The English Martyrologe*, 1608, says it was still visible. Soon after, it was lost until 1787 when natural movements of the sands exposed a gable. It was excavated in 1825 and 1843, neglected until 1892, and finally restored in part in 1910.

The original parish church was erected between the ninth and eleventh centuries at a place about $\frac{1}{3}$ mile east of the oratory, and where it was thought to be safe from sand. It was added to in the twelfth and fourteenth centuries, and kept free from sand until early in the seventeenth century by a brook which prevented the sand from reaching the building. Local mining operations eventually diverted the stream, and it was only a matter of time before the sand encroached. Although efforts were made to hold it, it nevertheless prevailed. Parts of the church were removed, and the remainder was visible as late as 1848. The present church dates from 1805.

There has also been some recent archaeological work at Penhale Sands (J. R. Harding, *Antiq. Journ.* 30, 1950, 156), an area associated with the legend of the lost city of Langarrow or Langona, 'a large city which stretched from Crantock to Perranporth'. The sea, the flat land and the mines afforded wealth to the people, and the place flourished. Then 'Sin became the familiar habit of the people and divine retribution followed in the form of a violent sandstorm'. This is said to have buried the city and its people. There is no doubt that encroaching sand gradually made the district uninhabitable. Harding suggests that the tale refers to the tin trade which flourished especially at the end of the Bronze Age and beginning of the Iron Age.

A little farther north at Kelsey Head the sands bury middens. There are remains of embankments on the south-west of the head, and Harding suggests that together with the earthworks marked on the Ordnance map, they would have surrounded the whole headland. These too are of Iron Age date.

Excavations in the sands at Gwithian, the east side of St Ives Bay, are affording interesting results (C. Thomas, *Proc. West Cornwall Field Club*, Appendix to Vol. 1 (N.S.) 19, 53–6 and *ibid.* 1953–54, p. 59). Part of the land covered by these dunes was the site of a settlement which was deserted 'within about a century of 1000 A.D.'. There is little doubt that sand choked the tidal creeks and made the area uninhabitable. There is little material, either ornaments, pottery, or domestic objects found, and the suggestion is that the desertion of the village was planned and the inhabitants took their possessions with them, and doubtless went to the adjacent high ground. The main site was a mass of small houses of Dark Age date at the base of Godrevy Towans. St Ives Bay is almost ½ mile to the west, 'but at the period in question a tidal estuary existed, flooding a small creek on the north side of the bluff, and on the south side continuing up the Red river valley at least as far as Reskajeage. The site would thus have been a long low sand peninsula....'

On the other side of the Hayle River dunes became a menace to Lelant church in the seventeenth century.

(6) THE LIZARD

In the second edition of the Lizard memoir (*Mems. Geol. Surv.* 1946) Flett and Hill discuss the coast and the form of the cliffs. The height of nearly all the cliffs is governed by the planation at 200 ft., but locally, to the south of Mullion, they are higher. The finest and boldest occur where the serpentine faces west or south and receives the full attack of the Atlantic waves. Jointing sometimes plays a prominent part in their detailed sculpture. The Rill, the Horse, and Pengersik are noteworthy. On the east side, beyond Kennack and up to Coverack, there are fine cliffs, but usually they show a sloping top, and the vertical part of the cliff seldom exceeds 100 ft. This form is ascribed to the occurrence of many dykes of gneiss, epidiorite, and gabbro, which weather unequally, and lead to landslips and the wearing down of the cliff edge. On the west coast dykes are less common, but where they occur they add to the beauty of the cliff scenery as can be seen at Kynance and Pentreath (**47, 48**).

The hornblende-schists usually produce cliffs even more varied in feature. Since grey lichen grows abundantly on them, they have not the sheen and colours of the serpentine cliffs. The foliation in the schists from Polurrian to Mullion and also at Predannock Head is almost vertical, thus allowing the formation of sharp ridges and crests. On the other hand, at the Lizard, Pen Olver, and Bass Point the foliation is nearly horizontal, or but slightly inclined, so that the cliffs take on a different aspect, and can often be described as battlemented. Near Cadgwith and Polbarrow, the same schists, with a fine parallel foliation, dip landward and so produce vertical cliffs.

The gabbro gives rise to another type of cliff. The rock decomposes readily; it is coarsely crystalline and felspar is abundant in it. Between Coverack and Porthoustock high cliffs are rare, and instead there are many sloping and grassy banks, covered with bracken and brambles, and running down almost to water level. On these slopes there are often large grey blocks of rock and a few knobs *in situ*. The situation of these cliffs, in a relatively sheltered place, is also a factor in their evolution, since the less effective wave action does not favour the building of vertical cliffs.

The coast from Porthleven to Polurrian is in the killas, which is locally much veined with quartz, and also sheared. Nevertheless, the bedding planes are generally preserved. The cliffs are varied, cut up by faults and fractures so that caves and re-entrants are common. At Jangye-

Ryn the killas is much folded, and the cliffs accordingly more intricate and interesting. There is, on the Killas coast, usually a beach of small pebbles, most of which are flint.

Elsewhere, in the Lizard peninsula, beaches are for the most part only found in coves, and the best are on the east side. They are often partly encumbered by big blocks fallen from the cliffs. In many of the coves, there is a sheet or wash of head on parts of the surrounding cliffs. The fact that the head covers much of the cliff top from Black Head to Porthoustock partly accounts for its subdued outline. The deposits of head on Lowland Point, at Polnare Cove, and on Nare Point are well known. On the west coast, about 1 mile west of Porthleven harbour, there is a large block of microcline-gneiss. It weighs about 50 tons, and is not moved even in severe storms. It is an erratic, its source is unknown, and it was stranded on the shore by floating ice. There are other erratics near to it, most of which have probably been washed out of the raised beach.

(7) ST AUSTELL

C. E. Everard (*Trans. Royal Geol. Soc. Cornwall* (for 1959–60), 19 (iii), 199–219) has written an interesting paper on the evolution of the coast near St Austell. He has been able to trace the changes in relation to mining and china clay working and the accumulation of waste. Up to 1800 tin streaming contributed much material; between 1780 and 1890 metalliferous shaft-mining was important, and since 1840 china clay quarrying. The waste products of the first and second are easily distinguishable from the third.

Investigations were made at Pentewan, Carlyon Bay, and Par. At Par, which was examined in detail, Everard concludes that 'the original spit at Par Green was completed between 1700 and 1750 and must have been very largely the product of the tin-streaming period. The silting of the estuary above the spit, artificially accelerated though it was, depended on detritus from both streaming and mining and was complete by 1805. The third phase depended upon the growth of the china clay industry and saw the accumulation of nearly 1,400 feet of sand seawards of the original spit in 150 years.'

The direction of local beach drifting is also analysed, and the changes in the coast at Par illustrated in maps, mainly of the nineteenth century.

(8) HALLSANDS

The interesting piece of coast at Hallsands, a few miles north of Start Point, has been reconsidered by Robinson (*Geogr. Journ.* 127, 1961, 63) who has been able to assess the offshore conditions in greater detail than the earlier writers were able to do. This part of Start Bay shows a gently sloping shelf of rock at about 6–8 fathoms. The outer edge of the shelf flattens out at about 22–24 fathoms into the bed of the Channel. The Skerries Bank rests on this shelf. In the immediate neighbourhood of Hallsands the dominant deposit on the sea floor is sand; shingle is rare except on beaches. Apart from the shingle all materials can be moved by tidal currents, and in Start Bay the ebb current runs both longer and stronger than the flood. Hence the ebb stream is most effective in the patterns of banks and channels on the sea floor (**42**).

Hansford Worth, in his studies of Hallsands between 1903 and 1923, concluded that the beach there had never recovered from the removal of material used in the dockyards. A comparison of the cliff edge at the Coast Guard cottages marked on the 6 in. map of 1907, and its position in 1957 shows that it had receded up to 20 ft. in that time. A little to the south there is a raised-beach platform in front of the cliff. This stands at 20–24 ft. above O.D., and is overspread by a thick layer of beach, and at two points by beach deposits. The original village of Hallsands was built on this shelf.

The present beach at Hallsands is made of a variety of materials which Robinson groups as follows: local materials, mica schist 2% and quartzite 16%; non-local materials, chalk flints 73%, Dartmoor felsite 5%, Dartmoor slate 2%, and granite 2%. The nearest chalk is in the Haldon Hills 25 miles away, and on the bed of the Channel, 20 miles farther east.

> When Hansford Worth first levelled his profiles in 1903 it was across a beach which had already been considerably depleted of shingle. Although no direct measurement had been taken previously, the extent of the covering could be estimated from photographs.... Wilson's rock, previously almost covered by the shingle of the beach, now stood as a stack fourteen feet high, indicative of a lowering of about twelve feet in this part of the beach. Although the amount of lowering varied along the whole length of the beach, it everywhere exceeded ten feet. (Robinson, *op. cit.*)

Robinson also made several measurements and analyses of profiles. These show not only that seasonal fluctuations are small, but also that there is no available supply of material to re-build the beach to its pre-1887 magnitude. Robinson believes that there is no dominant direction of longshore drift at this place, although changes in the morphology of the offshore bottom support the view that the ebb stream moves material. Nevertheless, there is no evidence that there is much, if any, transport of materials to the shore by wave action. The bank shows no consistent movement shorewards, and Robinson and Worth both agree that there is no replenishment from this source. The formation of the original and remaining beach is therefore unknown. 'Whatever the origin of the unusual constituents of the beach shingle, it is clear that the supply was not unlimited and probably for sometime past the amount of shingle has not increased on any of the Start bay beaches....The beach at Hallsands, therefore, can best be regarded as a feature of some antiquity which, once denuded, either naturally or artificially, cannot be replaced.' Hallsands demonstrates most effectively how dangerous it is to tamper with a beach, how wrong it is to make assumptions about the drift of beach material without a full investigation, and how important it is to study each part of the coast intensively and not apply general ideas too readily.

(9) DAWLISH WARREN

In 1950 the inner warren at Dawlish was a low sand ridge, with its highest point at the distal end and the lowest near to the place where it springs from the mainland. It was relatively stable, covered largely with grass and gorse, and used as a golf course. The outer warren was narrow and consisted of a broken line of sandhills. Kidson (*Inst. Brit. Geogrs.* 1950) analysed both features and noted the great changes in the outer warren, the retreat of which blocked the outlet of Greenland Lake. The Warren extends to Pole Sand, which may be regarded as its continuation. The changes in the other sand banks in the estuary, Bull Hill, Shelley, and Starcross sands, may have affected the evolution of the Warren.

In 1938 the outer warren had advanced eastwards so as to make the channel between it and Exmouth only 253 yd. wide. But during this growth, it was suffering erosion on its seaward side, a fact brought home to all by the gradual destruction of bungalows which finally disappeared in the storms of 1944 when, however, the channel was 625 yd. wide. Kidson established that this kind of change was characteristic, erosion on the seaward side and alternate advance and retreat of the far end (Fig. 135).

Martin (pp. 252, 253) found that the whole Warren, including Greenland Lake, had decreased in size from 275 acres in 1787 to 151 acres in 1888. The loss was almost entirely in the outer warren, and he suggested

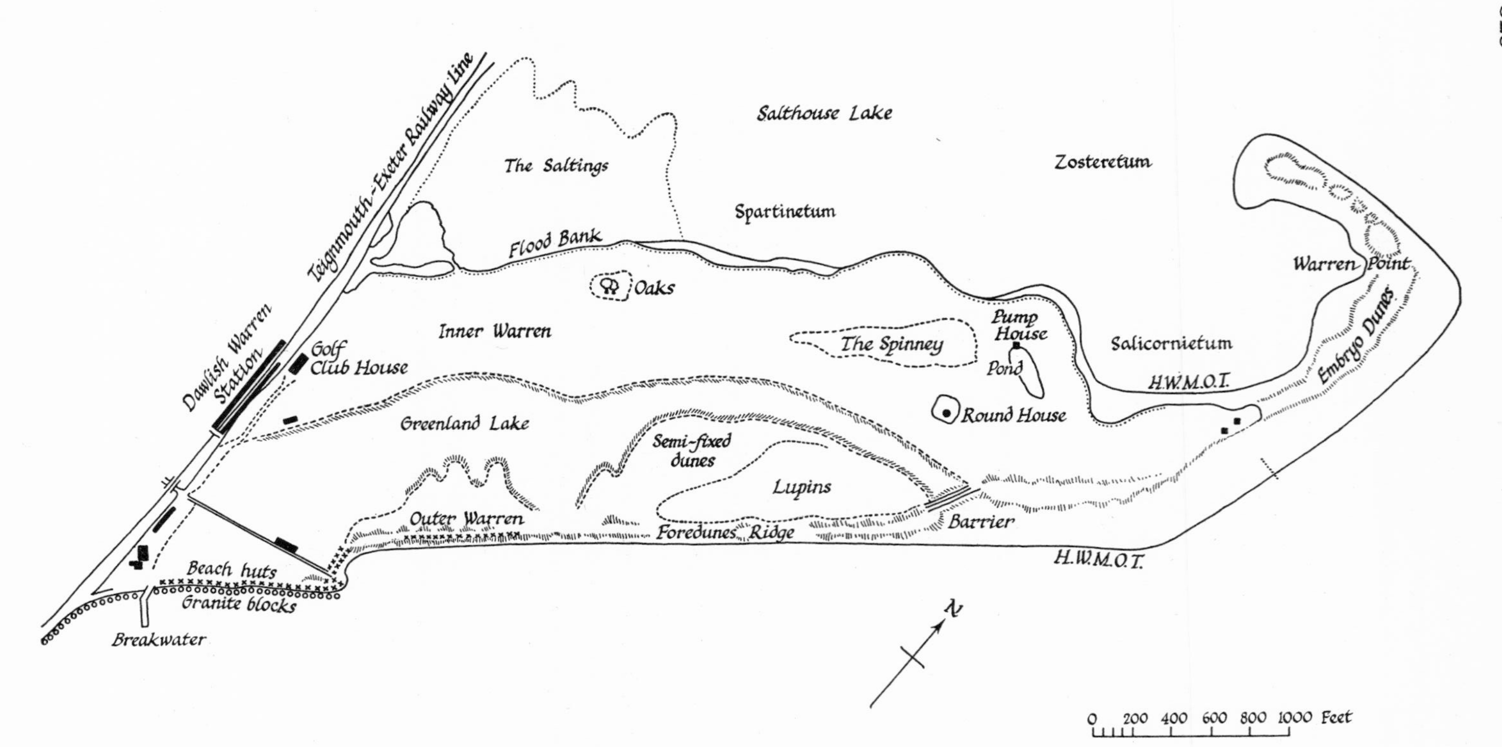

Fig. 135. Sketch map of Dawlish Warren (after C. Kidson, *op. cit.*)

that it would disappear about 1953. The whole feature has extended from west to east, as a result of beach drifting, and its general orientation owes much to the approach of waves over the long fetch to south-east, something to waves in the estuary, and also to the river current. Additionally, sand blowing from the exposed beaches helps to feed the Warren.

Kidson seeks the origin of the double spit in wind action, and maintains that there is no evidence that the inner warren was formed before the outer.

> Sand blown from the outer warren seems to have piled up on the mud flats behind the spit. The small streams which before the building of the railway, drained across the mud-flats immediately behind the outer warren into the river Exe, were able, with the help of tidal scour to keep their course clear. The blown sand, however, would pile up and simulate a second spit in the position of the inner warren. This...is supported by the fact that the inner warren is lowest in the south-west and highest in the north-east. The prevailing winds are the south-westerlies.

The point made on page 254 that the railway has had relatively little effect on the Warren is supported by Kidson who thinks that at a much earlier stage when sea level was lower relative to the land, Langstone Point would have been a much larger and more prominent headland. With rising sea level, the erosion of this headland may have been accentuated and rapid. There would also have been deposition of sand in the river. The combination of both processes would have favoured the the building of a spit. But in course of time it is argued that erosion on the headland would decrease and so the sand supply to the Warren would lessen until to-day it is insufficient to maintain it. Indeed, in storms, the inner warren can be directly attacked by the sea. Various steps have been taken to accumulate sand, but it is doubtful if they can be of lasting value.

A study of the ecology of Dawlish Warren following the surveys of A. J. Lee (1947) and C. Kidson (1950) was made by T. J. Wallace (*Rept. and Trans. Devon Assoc.* 85, 1953, 87). The outer warren was formerly more stable and there were at the time Wallace wrote a few semi-fixed dunes on which there was a dense patch of *Lupinus arboreus*, a native of California. The inner warren is unstable, and Lee showed that it was built on a clay bed rising above high water, but this clay has not been found underneath the outer warren. In Greenland Lake the floor was a mixture of plants associated with dunes, dune slacks, salt marshes, as well as mesophytes. In winter, when the lake is usually at its maximum, it catches more blown sand than at other times. There are also some saltings between the railway embankment and the north side of the inner warren. In 1935 the local golf club introduced *Spartina townsendii*.

Chapter XXIV

THE SOUTH COAST

(1) CHARMOUTH

Some notes by W. D. Lang (*Proc. Dorset Nat. Hist. & Arch. Soc.* 75, 1953, 151, and also *ibid.* 74, 110; 76, 110; 77, 40), concern the cliffs and mud-flows which often stretch over the beach near Charmouth. New material may be deposited on an old flow; it may or may not extend so far as the old one. Rain may set it in movement, and may also cause the older flow to advance. Observations suggest that the deeper parts of the flows are relatively static, and that the seaward face may slip rather than flow. A flow is therefore not eroding the gulley in which it rests, but successive flows are filling up the gulley. Cracks develop in the flows and allow water to penetrate and so renew movement. When the base of a flow remains more or less stationary, sliding and other movements may take place in the upper part. In this way, if local circumstances are favourable, 'a series of steps tilted down hill more steeply than the general slope' may be produced. As in other slips, back-tilting is often observed. Occasionally the foreshore is almost blanketed by falls and slips.

(2) CHESIL

A great number of observations were made on Chesil Beach by W. V. Lewis and groups of students. It is hoped that these will in due course be arranged and discussed. Lewis's tragic death in America in June 1961 is a loss felt in many ways, not least by all who are interested in our coast.

In 1954 W. J. Arkell published (*Proc. Dorset Nat. Hist. & Arch. Soc.* 76, 141) some interesting observations on the detailed forms on the Fleet side of Chesil Beach. On 26 November 1954 there was a severe storm. Reliable witnesses confirmed that at the Abbotsbury end the waves spilled over the top, but most of the water came *through* the shingle. At the back of the beach, there are a number of small gullies in the shingle. They are separated by 'spurs' of stones, and water from the seaward side bursts through the shingle and passes through these ravines, locally named caverns. One of them was about 10 ft. deep and 180 ft. wide after the storm.

There followed another storm on 4 December, and a careful inspection of its effects was made by Lt.-Col. Drew and Dr Dalton. Dr Dalton noted that

The highest part of each 'ravine' was a sub-conical pit, up to ten feet in diameter at the top, and up to four feet deep. Initially slightly narrower than the diameter of its pit, each ravine gradually broadened till it formed a delta at the creek (The Fleet). The dimensions of the ravines in general were about two feet across and two feet deep in the centre. Their bottoms were of washed pebbles embedded in mud. In one case a considerable amount of pebbles had been carried out beyond the delta, forming an island of about fifteen feet diameter, and projecting about one foot above the water. From our observations I consider that the sea has forced its way through the Chesil Bank in many places, chiefly at the base but also at higher levels (**30**).

Arkell visted the beach in June 1955; the ravines and associated details were still clear, although pebbles had gathered in some of the hollows seen by Dalton. At the Portland end sand had locally been brought up from below. Hereabouts the main beach is advancing on to dunes, but water had burst up from below bringing sand and pebbles with it.

Why are there caverns or cans? Why is not seepage approximately the same everywhere along the beach? Arkell makes a comparison with inland seepages at, for example, the junction of a pervious sand and a lower bed of clay. In such a case there are usually springs at particular points. These may correspond with local details of structure or of lithology, or perhaps to kinks in the contours. However, once a spring has started, it is likely not only to stay, but to grow larger, and to make a definite channel. Perhaps, then, when storms are pounding the Chesil, and when the beach is full of water, the water will run out where the beach is narrowest—into bay heads on the back-slope. To form a bay head in the first place, it must be supposed that storms, which from time to time throw stones over the crest, form spurs which will enclose a 'bay'. Once drainage begins into a bay, it will enlarge and form a typical can.

Why the water should rise from below in such places is even less certain. Arkell suggests that if downward cutting in the can or ravine reduces the thickness of pebbles below a critical limit, water may force its way up more easily than elsewhere. This is the more likely since along the foot of the backslope of the Chesil there is a certain amount of consolidation as well as growing vegetation. In some such way as this the forcing up of sand may be explained.

(3) DOUBLE SPITS ON THE SOUTH COAST

On pages 286 et seq. the two spits partly enclosing Poole Harbour are described. A. W. Robinson (*Geogr. Journ.* 121, 1955, 33) makes a comparison between this pair of spits, and those that enclose Christchurch Harbour, and those at Pagham Harbour. It is convenient to consider all three together. No satisfactory discussion of these features existed before Robinson published his paper. He has added to our knowledge of the three places, but it is improbable that he has produced a complete answer.

At Poole the spits enclose a narrow and deep channel in which currents reach 5 knots at springs. There are good surveys of this area (see p. 288) especially those of Mackenzie 1785, Sherington 1849, and Parsons 1878. An ordnance map shows at a glance that the southern spit, South Haven Peninsula, is wide and that the northern one, Sandbanks, is narrow and set well back relative to South Haven Peninsula, although it continues the general curve of Poole Bay. It has been subject to erosion, a wall was built and undermined, and in 1896–98 groynes were put in and succeeded in gathering a beach. They did so by holding up drift from the south-west, and stone groynes built only a decade or so ago to replace the older ones show the same result—that material accumulates on the side nearer the entrance to Poole Harbour. Robinson suggests that this north and eastward drift holds all the way from the harbour as far as Hengistbury Head which is the true break between Poole Bay and Christchurch Bay.

At Christchurch the southern spit is again the greater, and its lengthening (see p. 293) suggests a north and easterly direction of drift, but Robinson also thinks that large waves approaching directly on shore have a considerable effect. There is no evidence of a counter westward drift, and a groyne built east of the beach cafe at Friar's Cliff in 1954(?) testifies to an easterly drift. This does not entirely disprove any westward movement. Haven House (i.e. the northern) spit rests in part on a shingle flat thought to be an original fragment of the southern spit. Robinson argues that a former, and still deep, western branch of the river channel *within* the harbour, 'may have been occupied by the river whose exit (at that time) lay along the depression in the grounds of Sandhills House'. If this were the case, there would have been a single and long spit beginning at Hengistbury Head, and the shingle on which the northern spit now stands was part of it. If this spit were breached in a storm, a 'double' spit would have been formed, and the old mouth abandoned. Robinson admits this is based on conjecture; it may well be true, but it is hardly a parallel to the Poole Harbour spits which appear never to have been one single formation.

Pagham Harbour is the last vestige of an open strait between Selsey and the mainland. Silting and reclamation have produced the present conditions. It also differs from the two previous instances in the absence of a river draining into the harbour. The 'Armada' map of 1587 suggests that there were two spits, but in 1672 there was a single embankment. McKenzie's map of 1786 is similar. Between this date and 1843 the southern bank extended 1,800 ft. to the north-east. This tendency, as well as an inward movement of the whole bank, continued until about 1876 when it reached right across the entrance. It was breached in 1910. Robinson contends that the erosion of the low cliffs near Selsey Bill, the inward rolling of the bank, and the preservation of the smooth curve of the coast from Selsey Bill to Bognor are closely related phenomena. The changes at Pagham have been rapid and since the bar can be breached, the harbour entrance can be, and has been, at various times in the north, centre, or south of the embankment.

Robinson believes that in all three places local conditions favour the formation of a single bar across each harbour. This is partly the result of longshore drift, partly of direct wave action, especially storm waves. Later breaching may give two spits. Only at Pagham is there a complete cartographic record of events. 'There is no reason why it should not have occurred at an early date at Poole and Christchurch, although direct eivdence in support is lacking. Subject to these limitations, the hypothesis can be put forward as an alternative to the counter-drift theory.' That an alternative is wanted is certain, the one suggested is helpful but not sufficiently comprehensive. South Haven Peninsula is a different structure from any of the others. It is sheltered by the Foreland, but why is there so much more progradation there than elsewhere? The observations of the general easterly drift are valuable, but why is the South Haven Peninsula so much farther eastward than the Sandbanks spit? Moreover, proof of any supply of pebbles and other beach material from offshore is needed. Recent work on radioactive and other tracers (see pp. 562 et seq.) appears not to favour the assumption of direct supply from offshore.

In a paper shown to me in typescript (*Zeitschrift für Geomorphologie*, 7, 1963, 1–22), Kidson takes up the origin of double spits. He considers four examples: Braunton Burrows and the Popple at Westward Ho!, Dawlish Warren and Exmouth Point, Orford Ness (North Weir Point) and Shingle Street, and Stert Point (Fenning Island) and Stert Island.

All of these pairs represent spits which are growing towards one another, and in no case can breaching be claimed as an explanation, unless the cutting off of the southern end of Orford Ness and the accumulation of pebbles at Shingle Street is so regarded. Exmouth Point is certainly not 'cut off' from the Warren, and the growing together of the spits in Bridgwater Bay is historical fact. Drift and counter-drift are probably more common than is often assumed, and much must be allowed for local conditions (see, for example, p. 362, Burnham Harbour and Scolt Head Island).

(4) SOLENT, SPITHEAD, AND ISLE OF WIGHT

Recent investigations by C. E. Everard (*Inst. Brit. Geogrs.* 1954, 41, and *Papers and Proc. Hants. Field Club*, 19, 1957, 240) have thrown more light on the origin of the Solent and of the separation of the Isle of Wight from the mainland.

Everard traced drift gravels in the Hampshire Basin from 420 to 60 ft. below Southampton Water. They imply a polycyclic landscape, and it is probable that in addition to recording a general fall in sea level, interrupted by still-stands, they also point to estuarine conditions with some fluviatile phases. Figure 136 shows that the older and higher gravels are in the north and west, and that the shoreline, at any rate from the 185-foot stage, has moved east and south. The drainage of the New Forest area is local; the Beaulieu and Lymington rivers are in no sense the trunks of beheaded rivers. The 70-foot stage suggests the birth of many small streams. The sea continued to fall, the rivers to incise their valleys, so that eventually the Solent itself became a river valley which was cut well below present sea level. At this stage the general pattern of the mainland and Isle of Wight rivers was broadly that pictured many years ago by Clement Reid (see p. 297 and Fig. 64).

It is assumed by Everard that the low sea level at the time of the Solent River coincided with the last major glaciation, presumably the late Pleistocene. The post-glacial rise of sea level produced the present conformation. Apart, however, from the flooding produced in this way, there was also the breaching and eventual destruction of the Isle of Wight–Purbeck barrier. Reid at first thought this took place in the late Pliocene, but later considered a mid-Pleistocene date more probable. This was also the view of Bury and White. Green assumed a lower Taplow date. 'He believed that the lower Taplow terrace of the Avon divided down stream, the lower and more steeply sloping sub-stage

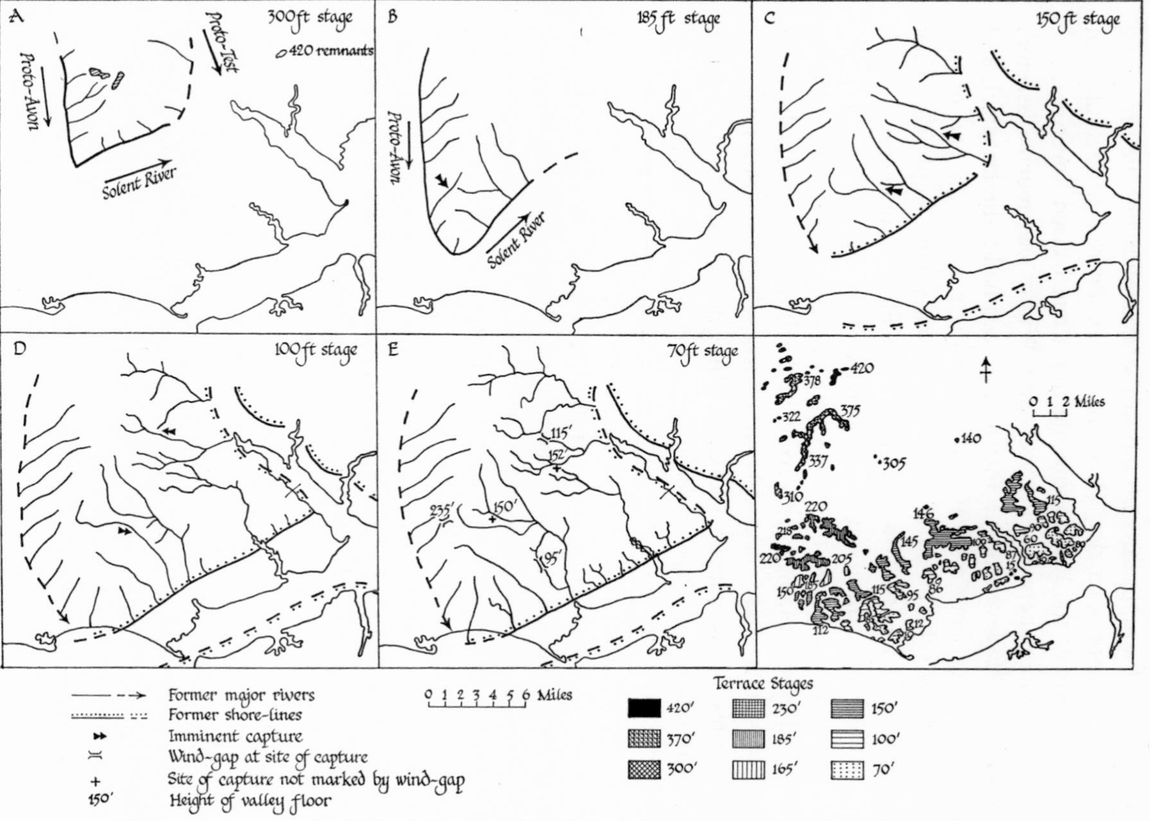

Fig. 136. The evolution of the drainage of the New Forest (after C. E. Everard, *op. cit.* 1957)

representing an increase in the gradient of the river. This he suggested was the result of a more direct route to the sea, following the breaching of the Wight–Purbeck barrier' (Everard, 1954). There is little doubt that in the later parts of the Pleistocene the rivers had incised lower courses, and that the ridge between Purbeck and Wight was much dissected. The oscillating sea levels of the Pleistocene each had a similar effect, and probably there was not a great deal left of the ridge at the end of that period. But the submerged terraces in various valleys imply that they were laid down when the lower parts of the valleys were deeply incised, and that there was still a southern barrier at that time. When, in post-glacial times, the sea rose, the rivers became marine channels, the ridge was attacked by waves, the sea probably broke through or merely drowned lower passes in it, and it was destroyed in a relatively short time.

Two notes on Keyhaven (A. C. G. Huggate, 1958, and B. Martin, 1955, *Milford-on-Sea Record Society*) record some historical changes and offer a brief account of the salt marsh and shingle vegetation in that locality. A little farther eastwards, at the mouth of the Beaulieu River, M. Human (*Wessex Geogr. Year*, 1960–61, **22**: a publication of the Department of Geography at Southampton University) has discussed for the first time the Warren Farm spit which is marked on the 1909 edition of the 6 in. map. The seventh edition of the 1 in. map shows it but does not name it. It is formed of ungraded shingle, and there are several laterals running, in general, south-west to north-east. The map (Fig. 137) also shows an older spit, 100 yd. or more inland. It is thought that the modern spit is quite independent of the older one, which is now starved of shingle (**16**).

The modern spit is narrow, and beyond the Creek (see map) has grown rapidly, in contrast to the slower growth of the older part. Moreover, there are no laterals beyond the Creek. At the present time (1961) the distal end almost passes Needs Oar Point. Throughout this tapering end shingle is thrown on to the marsh, and in front there is a small cliff. At the far end there is much fine shell material (**15**).

To-day the spit extends north-eastwards from an area of gravelly marsh. It is suggested that, after the earlier spit had formed, *Spartina* colonized the foreshore, causing accretion, and a new spit (the present one) formed on the outside of the marsh. This view is thought to be supported by the fact that in recent months there has been some erosion in that part, and a clay bench, similar to that at the western end, has been revealed. The origin of the shingle is uncertain.

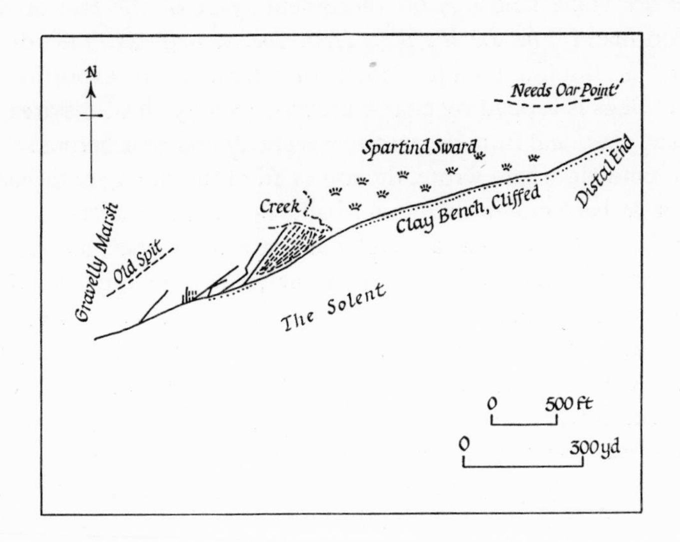

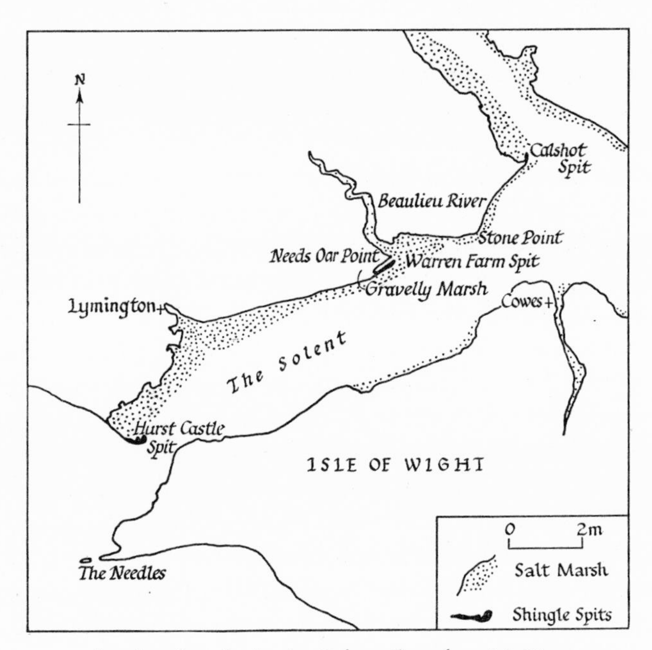

Fig. 137. The shingle spits in the Solent (based on M. Human, *op. cit.*)

There are some landslips on the Solent coast of the Isle of Wight. R. F. Moorman (*Proc. I.o.W. Nat. Hist. Soc.* 3, 1938–45, 148) describes conditions at Bouldnor Cliff. North of Cranmore an exposure of the Hamstead Beds is capped by coarse gravels. Nearly all of the area slopes gently seawards, and in moist weather is sticky and treacherous.

At the time Moorman wrote, the top 35 ft. of the cliff were the steepest, sloping at perhaps one in two-and-a-half. Below was a ledge-like feature about 20 ft. wide. Thence a small mud stream reached another ledge 70 ft. below. This second ledge was about 40 ft. wide and 300 yd. long. It was the starting point of mud streams, especially near its western end. These streams, 20–30 ft. wide at their sources, cut channels in their downward courses and, partly as a consequence, their widths lessen.

The main flow, however, started about 150 ft. below the main ledge, and at first was some 90 ft. wide, but lower down it had cut a channel. Locally it crossed harder outcrops which led to mud-falls. But above and below these slopes, the flows of half-dried mud gave rise to features comparable to seracs and crevasses on glaciers. In wet weather when the rate of flow is high, its weight causes bulging along its banks. When it contracts in dry periods these bulges may be likened to lateral moraines. It crossed the beach at its lower end, and pushed its snout more than 100 ft. into the sea. A note referring to J. F. Jackson is added: '...the great mud-flow as a whole forms a comparatively permanent feature of Bouldnor Cliff, as he remembers that it appeared to be much the same at present as on the occasion of his first visit to the Isle of Wight in 1913'.

In the same volume (p. 50), G. W. Colenutt contributed some notes on coast erosion in the island. The effect of direct wind action on sandy cliffs near Chale, and the enlargement in this way of Ladder and Chale Chines is noted. At Whitecliff Bay little had changed up to 1904, but the building of the fort on Bembridge Down led to the removal of much shingle from the beach, and so to serious erosion of the cliffs. At Priory Bay, St Helens, Oligocene clays slide over the wall built below Nodes fault. This is the result of land water. Since 1882 there has been active erosion of the cliffs west of Wootton Creek, where the coast is formed of clays, marls, and limestones of the Osborne Beds. There are now sea-walls. There is also erosion and land-slipping at Gurnard and Thorness, part of the same general problem, discussed above, at Bouldnor.

(5) SUSSEX

C. Perraton (*Journ. Ecology*, 41, 1953, 240) discusses the salt marshes on the Hampshire–Sussex border. The streams draining into the several harbours[1] are unimportant, and carry but little load. Yet the harbours are silting up, and marshes are developing. There is no doubt that *Spartina townsendii* plays an important role. Nevertheless, there is some erosion in the harbours (p. 297), which are slowly extending in area, although decreasing in depth. Marsh is growing outwards from their shores, and the following zones can be recognized:

(1) Bare mud with Algae.
(2) *Spartinetum townsendii.*
(3) Obionetum–Spartinetum.
(4) General Salt Marsh.
(5) *Beta maritima* community.
(6) *Agropyretum pungenti.*

Of these, the Spartinetum is the most conspicuous. The drainage channels in it are much branched, and since water from the areas covered with vegetation continues to run down these channels even at low water they are continually scoured, and the major ones remain free from plant growth. The outer limit of the Spartinetum is often marked by a steep cliff, some 60 cm. high, 'along a line which probably represents the ecological limit of the species at the time of colonization. Accumulation of silt...has raised the Spartineta to their present levels. The existence of this cliff suggests that seaward extension ceased some years ago.'

The general succession of marsh development is:

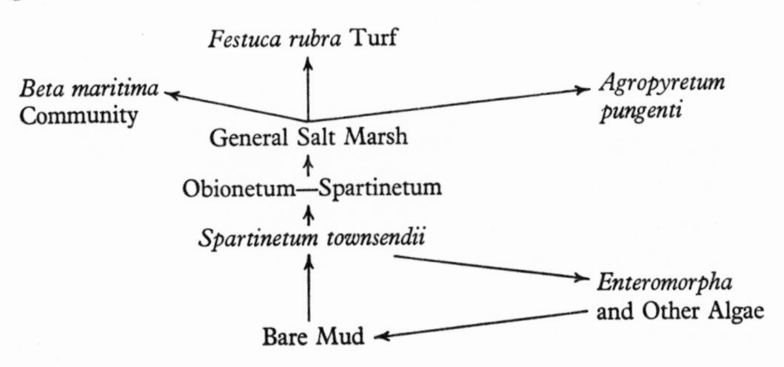

[1] J. H. Andrews, (*Sussex Arch. Colls.* 92, 1954, 93) refers to a manuscript map of Chichester Harbour in the possession of the Society. Its history is unknown, but seems to have been compiled in connexion with a coast defence scheme. It is dated about 1750.

R. G. West and B. W. Sparks (*Phil. Trans.* B, 243, 1960, 95) investigated certain interglacial deposits at Selsey, Stone (on the Beaulieu river), and at Arromanches in France.

The fossil deposits at Selsey, which indicate widespread freshwater and estuarine conditions, are earlier than the raised beach. They are exposed in several places in the low cliffs near Selsey where, to the west of the Bill, the shingle reaches 25 ft. O.D. This is part of the shingle, regarded by Palmer and Cooke (p. 309) as belonging to the 15 ft. raised beach, and occurs locally between Littlehampton and Brighton. Smith, at Brighton, noted that the platform on which the shingle rested at Black Rock, sloped upwards in an inland direction. The shingle seen by West and Sparks is regarded by them as of the same age as that at Black Rock, but is considerably lower down on the platform. Taking all the facts into consideration, West and Sparks suggest that mean sea level at the time of the formation of the beach was perhaps 25 ft. above the present. Apart from the inclination of the platform and its effect on the height of the shingle, it is probable that variations also occur as a result of some shingle being laid down during the transgressive phase, and some during the subsequent regression.

The evidence at Stone is similar, but the shingle is less rounded. It has been suggested that since the locality is in enclosed water, the shingle is reworked terrestrial material, and so little rounded by wave action.

R. S. Waters (*Būiletyn Peryglacjalny*, No. 9, Lódz, 1960, 1963) has commented on the higher raised beach at Slindon. There marine sands and pebbles rest on a rock-cut platform at 98–100 ft. O.D., and are overlain by unstratified gravel. Waters interprets the sequence as follows (cf. p. 309).

(1)	Deposition of marine sands and pebbles	Mindel-Riss interglacial
(2)	Solifluction of fossil soil material; deposition of loess in an adjacent area	Riss I
(3)	Weathering of head (= orange brown layer) and loess	Riss I/II
(4)	Solifluction of fossil soil and weathered loess	Riss II
(5)	Formation of frost-soil structures	Würm
(6)	Weathering of frost-disturbed head	Post-glacial

Shoreham Harbour continues to interest historians and physiographers. (H. C. Brookfield, *Sussex Arch. Colls.* 88, 1949, 42, and 90, 1951–52, 153). To-day it is a flourishing pool for vessels up to about 600 tons. In mediaeval times and until 1760 a narrow entrance restricted the tidal flow, but in 1800 restoration took place, and at springs the tide then rose and fell 8 ft.

at Shoreham bridge, and 14 ft. in the entrance. Beach drifting, however, still forced the entrance farther east, and in 1815 a new entrance was made near the 1760 mouth; this was opened in 1818 (see 1816 on Fig. 65, p. 307).

H. Cheal (*The Story of Shoreham*, 1921) and F. G. Morris (*M.A. Thesis*, London, 1931) argue that the prosperity of the Adur estuary in early mediaeval times, its later decline, and the erosion in the fourteenth and fifteenth centuries would not have been possible had there then been a protecting shingle bar. But an open harbour in this part of the Sussex coast is most improbable and, as far as known, the river has always hugged its left bank as a result of deflexion by shingle. Cheal suggests that the harbour of Old Shoreham was in a creek behind the town, and that the High Street is part of an old road from Brighton 'through the lost village of Aldrington, to a ferry across the Adur at New Shoreham'. Brookfield thinks that it is unlikely that any road from Brighton which, at that time, was unimportant, would connect with a ferry at New Shoreham, when there was already a suitable road to a shorter ford at Old Shoreham. The Armada survey of 1587 shows a tidal creek on the east side of Old Shoreham at the Ham. Presumably it was on this creek that Cheal assumed the port to be situated. Moreover, the creek also implies that Shoreham stood on a little peninsula of firm land.

On Yeakell and Gardner's map of 1778–83 there is marked a belt of marsh behind the shingle from Goring to the harbour and then eastwards nearly to Hove. Remnants of this belt now only occur at Lancing and west of Worthing, where it is being obliterated by the inward movement of the shingle bar. Records of storms, for example, *Nonae Rolls*, 1292 and 1340, and the great storm of 1666, show that much damage occurred on this coast, possibly as the result of flooding of reclaimed land. Brookfield's thesis is that there must have been some kind of offshore bar, or barrier beach, on both sides of the Adur, and reaching from at least Goring to Brighton. Only at the eastern and western limits of this beach did solid land face the sea. On the other hand, if the bar existed why has there been erosion at Shoreham? Before answering this, one or two other matters must be considered. At Bramber, about four miles inland, a medieval bridge was discovered in 1839. This had four big stone arches, cutwaters, and subsidiary bridges to east and west. Such a structure was not built to span the present river, but a strongly running stream which drained a large tidal area above the bridge. The ebb and flow of such a river would have had a much greater effect on the mouth, and might well have been able to maintain a direct outlet to the sea. The entrance may also have been wide, even if diverted somewhat to the east by beach-drifting. But in course of time reclamations were made, the tidal compartment of the river was constantly reduced, and the shingle bar was driven landwards. This all points to less scour, and the possibility of increasing eastward diversion of the mouth. In time, such a state of affairs would direct the erosive force of the river against Shoreham and the reclaimed land and marshes thereabouts. If we suppose that this took place partly at times of storminess, or even during a slight rise of sea level, we have conditions favourable to serious erosion and submergence such as Cheal and Morris suggested.

At Newhaven there has been some slipping of the cliffs on the seaward side of Castle Hill (W. H. Ward, *Proc. 2nd Internat. Conf. Soil Mechanics and Foundation Engineering*). Recent movements have been limited to the upper part above the chalk cliff where there is a small outlier of Woolwich and Reading Beds. The hill seems to have been occupied from the late Bronze Age until about A.D. 250, because there was a hill fort there. There is no record of occupation between A.D. 250 and *temp.* Elizabeth I. Since then there have been several forts, including the present one, built in 1864.

At the end of January 1943 there was much rain, and slope failures occurred in three places. Three weeks later the surface soil was still wet and soft, and was pouring over the edge of the main cliff in gulleys scoured in slightly hardened Thanet sand. All this had gathered on the beach. Since so much superficial sludging had occurred, the original detachment of a block of chalk from the upper cliff and its arrival in the form of mud on the beach was hard to trace. Blocks in general 'flounder forwards and tilt backwards'. The general nature of the movements in relation to the structure of the cliff can be seen in Fig. 138.

Between Newhaven and Seaford Head is a slight embayment, now much altered by man (F. G. Morris, *Geography*, 16, 1931, 28). Two spurs, Hawth Hill and Blatchington Hill are (? were) being cut by the sea, but in mediaeval times there was much shingle and it lay up to 100 yd. farther south, enclosing the river. In the sixteenth century the mouth of the Ouse was as far east as Seaford Head. Part of this ancient channel is the Mill Creek near Newhaven, and much more of it was traceable until 1795. The various works constructed at Newhaven deprived the coast to the east of new supplies, and gradually erosion set in. More than once in the nineteenth century the shingle spit was breached. The shingle that still exists east of the harbour is probably carried there by south-easterly winds and waves.

Seaford is first noted as a port at the beginning of the thirteenth century, but because of silting, French raids, and the increasing size of

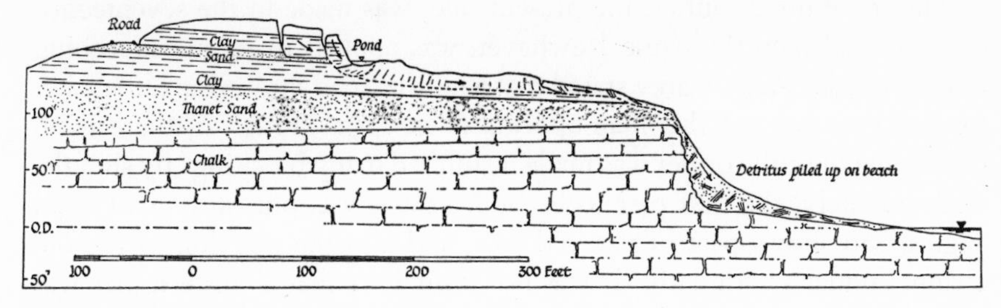

Fig. 138. Generalized section at Castle Hill, Newhaven (after W. H. Ward, *op. cit.*)

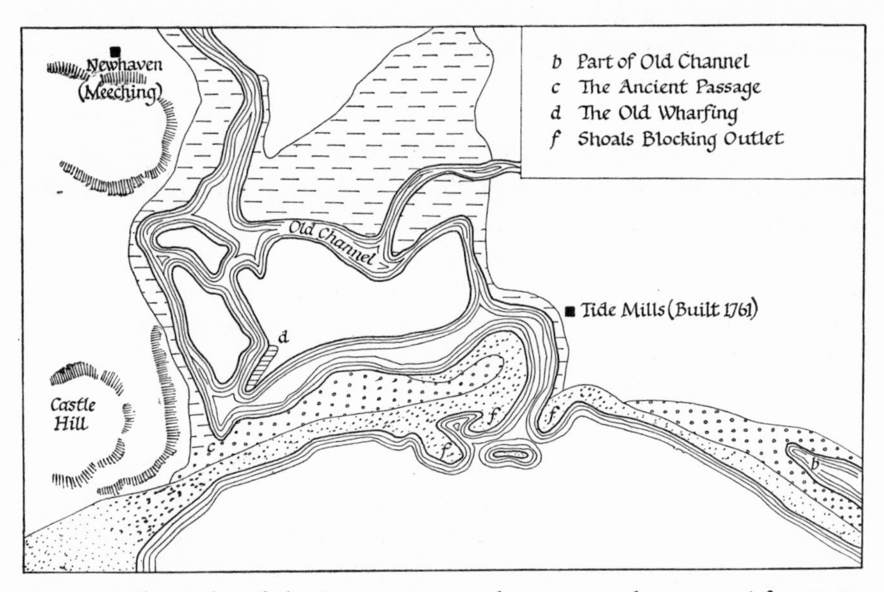

Fig. 139. The outlet of the Sussex Ouse in the seventeenth century (after F. G. Morris, *op. cit.*). (*b*) part of the old channel; (*c*) 'the ancient passage'; (*d*) the old wharfing; (*f*) shoals blocking outlet.

ships, by the end of the sixteenth century it was moribund. Probably, too, the shingle bar made matters increasingly difficult. Newhaven is first mentioned in 1565, and in 1566 the name of Meeching harbour occurs. Newhaven and Meeching are not synonymous. 'Evidently a small settlement grew up near the new outlet to which the name Newhaven was given, and the fact that the names refer to different places strengthens the supposition that this outlet was not the existing one but was farther to the east....' It may well have been near the tide mills.

The more direct outlet, the present one, was made in the seventeenth century, and then the name Newhaven was applied to it, and Meeching fell out of use. A. E. Carey stated that a map of 1620 showed an outlet near the present one, but this map appears to be lost. However, there is no doubt that once the outlet was made it suffered many blockings by shingle. Unfortunately, the first pier to be built (1664) was put on its east side! Figure 139 shows part of this old pier, and also the position of the old channel under Castle Hill. It was not until 1731 that a western pier was built; this fixed the position of the harbour, and despite difficulties caused by beach drift, has remained until now. But it was not until 1845 that the groyne to the west was extended beyond low water mark. This really created the modern harbour. Bishopstone was finally blocked when the tide mills were built.

(6) KENT

There is little to add on the evolution of Romney Marsh. A short paper by J. H. Andrews (*Sussex Arch. Colls.* 94, 1956, 35) discusses Rye harbour in the time of Charles II, and a manuscript map of 1665 is reproduced and shows two entrances. There was only one channel in 1677. It is suggested that the Rother was once navigable to Bodiam castle in the walls of which are mooring rings.

The discussion on page 330 is still open! G. Ward (*Arch. Cant.* 65, 1952, 12) in a paper on the Saxon history of the town and port of Romney recalls the various mouths of the Rother, states that Limen and Rother are the same, and that Rother was not popular until the sixteenth century. The most interesting point he makes is: 'There is one feature absent from the Romney of 740 which some people have been willing to fit into the picture. This is the old channel which existed before it silted up with mud and became the Rhee wall on which a main road runs to-day. This channel was a canal cut in the thirteenth century when the great estuary was filling up. It was not dreamt of in Saxon days.' There is great scope for a comprehensive physiographical, archaeological, and ecological survey of the whole of Romney Marsh and Dungeness.

In Sandwich Bay A. H. W. Robinson and R. L. Cloet (*Proc. Geol. Assoc.* 64, 1953, 69) carry the story of the Stonar shingle bank a stage farther. It is regrettable that no fulls exist on this bank to give any clue of its evolution. It is almost entirely formed of rounded flints varying from 1 to 12 in. in diameter. The larger shingle is mainly on the seaward side of the bank. The contribution made by Robinson and Cloet is particularly interesting because it brings into consideration the Brake Sands and possibly the Goodwins. The Brake Sand is about $1\frac{1}{2}$ miles offshore, and in some ways resembles the Stonar Bank in shape and composition. The parallel, however, should not be pushed too far. It is assumed that at the close of the Ice Age the sea floor hereabouts had an uneven surface, and that the deposits left on it were shaped by wave and current action into north–south trending banks. One of these *may* have been a proto-Brake Sand which was pushed westwards to become the Stonar shingle bank. The fact that the bank is thinner in the north is ascribed to the deeper water there, the deep entrance to the Wantsum. An alternative suggestion is made that the Stonar Bank originated in the Goodwins area. 'An anti-clockwise rotary movement of these sands would induce a corresponding landward movement of the proto-Stonar Bank, if the channel between the two was to remain at a constant width.' In either case, it is assumed that the bar moving toward the former estuary would simulate a bay bar. It is also claimed that recent movements studied on the Brake Sand support a possible onshore movement of this type.

Unfortunately, the Stonar Bank cannot be dated; finds of coins merely serve to show that it existed in Roman times. The argument adumbrated by the two authors would be greatly strengthened if we knew more about the movement of shingle under water.

R. L. Cloet (*Geogr. Journ.* 120, 1954, 203) has analysed the Goodwin Sands from the hydrographical point of view. He notes that Lyell (see p. 338) unintentionally misled many people because he took no notice of a boring made in 1849 which reached the chalk at 78 ft., but did *not* pass through any clay. Cloet, by making use of all available surveys, defined that part of the sands which has remained more or less stable over a period of little more than 100 years. If any former island (Loomea, p. 338) is postulated, it is likely that it was in this stable part. With present sea level it is as certain as can be that no island ever existed since little, even of the most probable island area, is covered by less than 2 fathoms. Before the Dunkirkian transgression there may have been an island, probably like the banks off Great Yarmouth. Even so it may have been liable to flooding at high tides. '...the existence of a bank uncovering at low water, as at present, may have applied also in Roman days, which bears out Holmes' theory (p. 338) that "an obstacle existed here in Caesar's time"'.

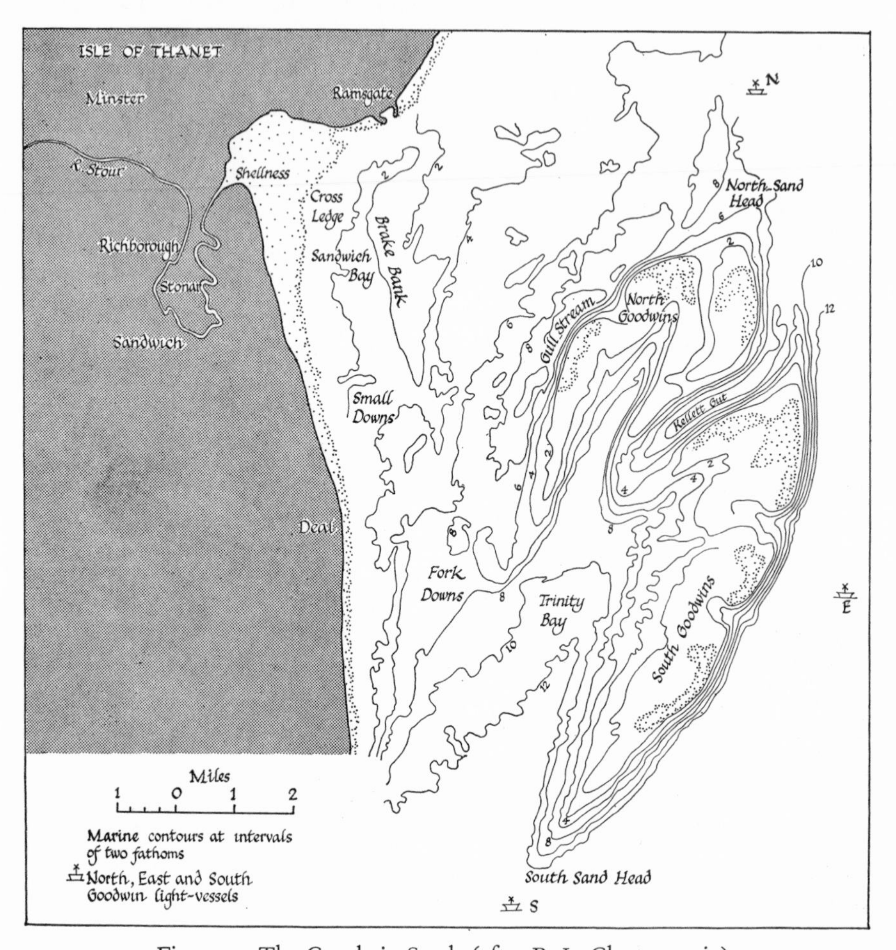

Fig. 140. The Goodwin Sands (after R. L. Cloet, *op. cit.*)

The 1849 boring passed through no clay. Clay has, however, been reported from other borings. Cloet's evaluation of all the available evidence suggests that only nodules of clay or clayey sand were found, but never a true bed of clay. Soundings and bottom samplings also suggest that chalk is not far below the surface of the bottom. Soundings, and geological structures on the mainland, imply that there is a ridge of chalk continuing the Minster anticline of the southern slope of Thanet. The ridge seems to curve south-eastwards and disappears just north of the Gull stream. Contours also point to its continuation under the Goodwins.

Reference to Fig. 140 shows that the Kellett Gut almost divides the Goodwins into two parts. Occasionally this channel has ceased to exist (e.g. 1865). Older charts often show the sands in two parts, but Waghenaer in 1584 shows them as one, yet on charts of 1666, 1736, and 1779 the gut, or swatch, is shown. In 1795 it reached the chalk, since chalk *in situ* is marked on the chart. The gut existed until about 1850; there was no trace in 1865, and it reappeared about 1910. Closely associated with the Goodwins is the Brake Sand which is still moving coastwards, in such a way as to conform, in general outline, with the curve of Sandwich Bay.

The main flood and ebb channels are shown on Fig. 140, and they carry a great amount of sediment.

> The resulting movement of the whole of the sand bank due to the unequal outward movement of the Fork and South Sand Head is...a counter-clockwise rotation. This in turn is responsible for the Brake being pushed westward by the Gull Stream channel trying to maintain the debit of water through it. Although the Small Downs flood channel west of the Brake is able to move more sediment northwards, it is weaker than the Trinity Bay flood channel.

The interplay of these streams, their varying strengths and effects relative to one another, nonetheless seem to balance one another. It is for this reason that despite the considerable amount of local movement, there is no tendency for the sands as a whole to move either towards the Channel or the North Sea.

In the north Kent marshes J. H. Evans (*Arch. Cant.* 66, 1953, 103) investigated the relation of archaeology to sea-level changes. Near Chatham dockyard the general sequence of deposits is:

	Thickness (ft.)	Depth (ft.)	O.D. Level (base of bed)
Alluvial Clay	10·0	10	1·0 ft. above
Peat	1·0	11	O.D.
Alluvial Clay	18·0	29	18·0 ft. below
Peat	2·0	31	20·0 ft. below
Alluvial Clay	14·5	45·5	34·5 ft. below
Gravel	0·5	46·0	35·0 ft. below

In this and similar sections it is assumed that the clays imply subsidence and the peats elevation. Throughout the lower Thames the nature of the evidence is similar, but the level of the lower peat bed will vary with the gradients of the old river beds. At Burntwick Island it is found at −68 ft. O.D., and at Chatham at −34 ft. O.D., a fall of 34 ft. in six miles. There is no need to assume that the peat beds are continuous, or that their thickness is constant.

The Roman occupation level in the Thames-Medway is between O.D. and 1 ft. O.D., and corresponds with the upper peat bed. The little evidence that is available suggests that in the Saxon period (seventh to

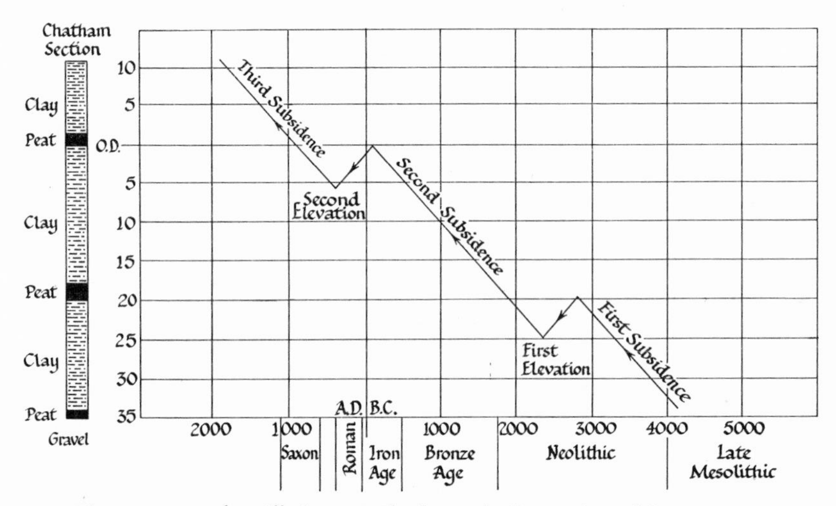

Fig. 141. Level oscillations at Chatham; the line is that of deposition at Chatham, and not at sea level (after J. H. Evans, *op. cit.*)

tenth centuries) the marshes, although higher than now, were probably rather below the level in Roman times. Certainly lands now flooded were then good grazing land. In the mediaeval period, at any rate from the thirteenth century, the low-lying lands caused anxiety, and many banks were built. These had to be raised at frequent intervals. In the late mediaeval period, and to the present day, the land has continued to sink and bank raising has been essential.

Since the peat bed which equates with the Roman period contains birch, hazel, elm, yew, and some oaks it is clear that when these trees flourished it must have been well above high tide level. This may be generally dated at 100 B.C. This surface probably coincided with high water mark of ordinary spring tides between A.D. 900 and 1200, say A.D. 1000 approximately. Since A.D. 1000, using that date only as a rough guide, a minimum of 8 ft. and a maximum of 10 ft. of silt have been deposited on the upper peat bed. Assuming a constant rate of deposition, this means 10·1–12·4 in. a century, This is also interpreted as the rate of movement. The Roman period in Britain extended from A.D. 50 to A.D. 400. If the figures given above may be used in a rough arithmetical calculation, it follows that at about A.D. 400 the land surface in north Kent stood 15–16 ft. above the present. This is in general agreement with other evidence from the lower Thames.

Evans summarizes his views in Fig. 141. The peats are shown as implying elevations. In the year A.D. 1900 the level of Chatham marsh was +11 ft. O.D.; A.D. 1000 is taken as the horizon of the top of the upper peat bed (+1 ft. O.D.), and A.D. 400 is assumed as the date of the peak of the last elevation. 100 B.C. is made to coincide with the base of the upper peat bed. If these same rates are assumed, and the graph is continued backwards in time, the result is shown in Fig. 141. How far the peats mean actual elevation as distinct from cessation of subsidence is uncertain. The general inferences deducible from the graph are:

(1) Silt deposition began at Chatham on the buried channel gravels about 4000 B.C.

(2) Since that time subsidence has been of the order of a foot a century.

(3) The peats are assumed to mean elevations. The top peat bed is Romano-British in age; at that time the land surface was 5–6 ft. above high water springs.

(4) Since then there has been a more or less continuous subsidence, and in the twelfth and thirteenth centuries walls became necessary and have had to be raised from time to time.

At Canterbury F. Jenkins (*Arch. News Letter*, 5, 1954, 34) made some observations in a storm drain near the river and found at about 4 ft. from the surface a brownish-yellow loam, covered by black silty mud. Native pre-Claudian type potsherds were found in the loam. The deposit was above water level in about the first century A.D. and there has been a rise of at least 5 ft. relative to sea level since. 'Hence, it can be stated with some confidence that the eastern branch of the River Stour within the walls of the Roman city was non-existent in Roman times, and was the outcome of natural change in more recent times....' (See also 'The Saxon History of the Wantsum', G. Ward, *Arch. Cant.* 51, 1944, 23.)

The same, or a similar, sequence of events is found near Whitstable where M. W. Thompson describes a group of mounds on the Seasalter Level (*Arch. Cant.* 70, 1956, 44). Some reach the unusual height of

7–15 ft. or more. They are mediaeval and represent old salt works. The area was embanked in 1325. If the sea wall was now moved the mounds would be flooded, as indeed happened in 1953. Their unusual height is perhaps explained by severe floods which are known to have occurred just before and after the wall was built. The pottery associated with the mounds is that characteristic of the thirteenth century. There are also large mounds on both sides of the Wantsum valley near Reculver.

Chapter XXV

THE WASH TO THE THAMES

There has been a considerable amount of research on the coast from Hunstanton to the Thames since the war of 1939–45.

At Thornham J. F. Peake (*Trans. Norfolk and Norwich Nats. Soc.* 19, 1960, 56) has added to our knowledge of the salt marshes. Perhaps the most significant plant at Thornham is *Phragmites communis*. Its presence suggests conditions resulting either from a higher surface or a supply of fresh water, or a combination of both factors. Levelling suggested that height changes alone were insufficient to explain its distribution. However, there is a freshwater spring in the south-western part of the marsh, and a small stream in the south-east, and the water from these flows over a layer of ferruginous and impervious clay close to the ground surface. Near the sea wall, in an artificial dyke, the *Phragmites* is luxuriant.

A survey of *Phragmites* showed that where it is most luxuriant it is growing on a considerable thickness (120 cm.) of *Phragmites* peat; it is far less dense on the ferruginous clay. This may be because there is less available water in the clay, and also because the clay is less easily penetrated by the *Phragmites* rhizomes. Between the reed bed and the dyke, where there is thick clay, the vegetation reverts to that of a normal salt marsh with but occasional *Phragmites*. If a layer of mud rests directly on the clay, the reeds survive and flower, but less than on the peat. The appearance of the *Phragmites* on the surface, and the evidence of a certain number of shallow bores made in the marsh, suggest that the peat band was formed in an old creek system which was blocked so that it ceased to be invaded by sea water. Air photographs show that the most lush growth of *Phragmites* follows winding belts indicative of former creeks.

[In the following pages on the marshland coast the more important changes and features are discussed. There have also been slow continuous alterations, such as the gradual inward migration of Scolt Head Island and Blakeney Point, and the building of new ridges at Holkham. These are described in *Trans. Norfolk and Norwich Nats. Soc.* 17, 1951, 206, where also a map showing the main alterations in the coastline up to 1948 is given.]

At Scolt Head Island investigations have been made to try and account

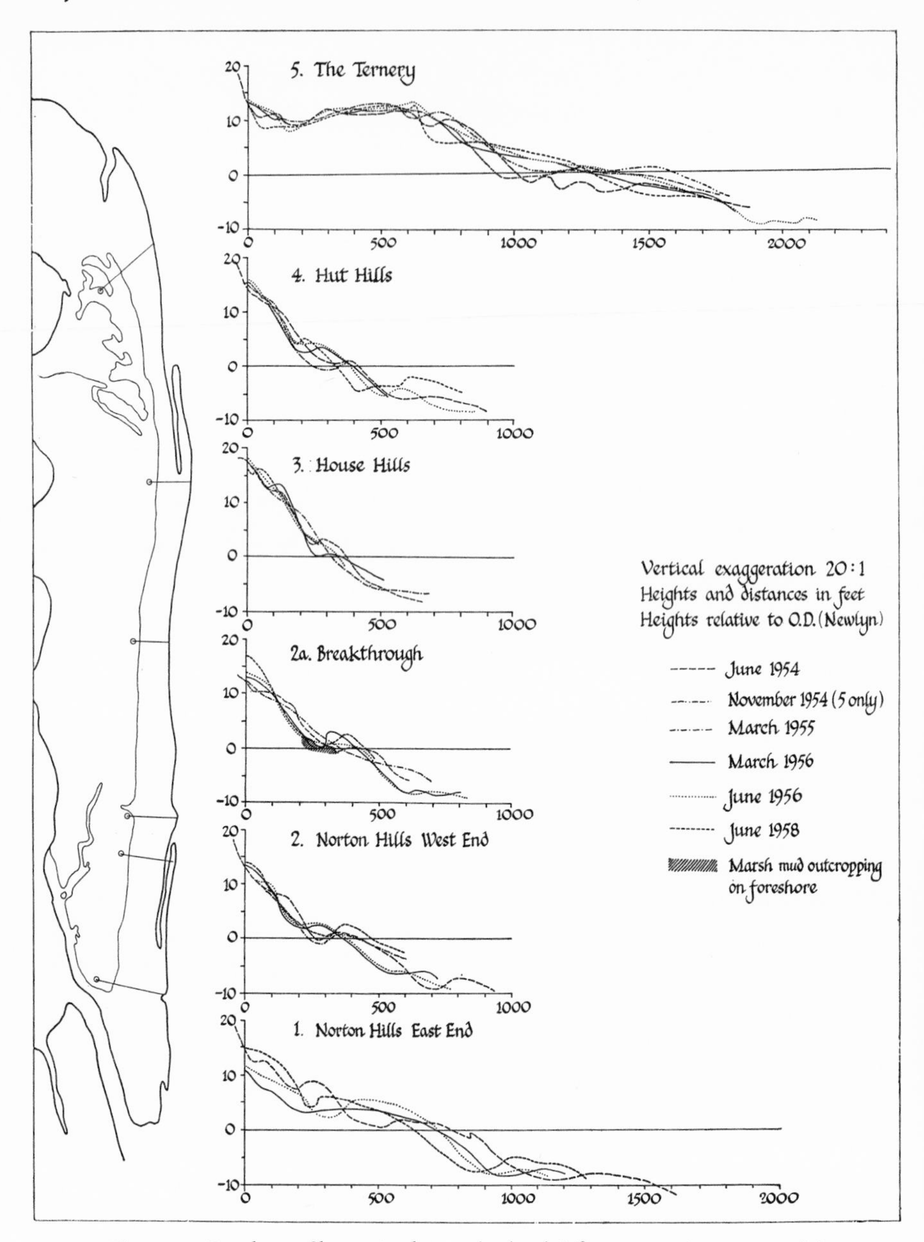

Fig. 142. Beach profiles at Scolt Head Island (after A. T. Grove, *op. cit.*)

for the supply of beach material; to prevent repetition these are discussed on pages 562 et seq. A. T. Grove (*Scolt Head Island*, Heffer, 1960), made a series of profile measurements of the beach. Figure 142 shows the positions of the profiles and also the changes measured in each. The beach in the middle part of the island is steeper than near the two ends. In many ways the nature and variations of the profiles recall those on the Lincolnshire coast (see pp. 692–6). Grove points out that the undulations in the profiles represent alternating fulls and lows which shift in position from time to time. A sweep zone may be defined here as readily as on the Lincolnshire coast (p. 694). Up to 1958 the profiles showed (*a*) that the beach opposite the break-through of 1953 has been built up about 5 ft., and (*b*) that the outcrop of marsh-mud (profile 2*a*) has been cut back about 200 ft. in 4 years, 1954–58. Further work on this interesting problem and on the volume of sand moved along the beach is being carried out (**151**).

The ecology of the salt marshes was revised by V. J. Chapman (*Journ. Ecology*, 47, 1959, 619) in 1957. His new maps should be compared with those made in 1932 and 1933 if a full appreciation of the changes is to be made. Missel Marsh which, before the war, was a typical *Salicornietum strictae*, with many bare areas of mud, had by 1957 been transformed into an *Asteretum tripolii*, and plants of the next stage, the General Salt Marsh, were invading it. This process continues and now (1962) almost the whole marsh is covered, including its upper parts.

In 1933, the youngest marsh, a Zosteretum, was low and covered by almost all tides. It is now higher, and less frequently covered, and many patches of *Spartina townsendii* are spreading. The most spectacular change is seen in the rather older marsh at the north-eastern end of the Cockle Bight. What was a Zosteretum, fringed by *Salicornia* and annual *Suaeda* on its sandy northern margin, is now thickly covered with *Spartina*, and the parts formerly characterized by *Salicornia* are now areas of General Salt Marsh, or covered by *Puccinellia* and *Aster*. The adjacent closed marsh, Salicornia marsh, has evolved from a *Salicornietum strictae* to an Asteretum, and *Spartina* is invading it, and the algae, *Pelvetia libera*, and *Bostrychia scorpioides* replace the older Chlorophyceae and Fucus communities. Anchor Marsh is now in the General Salt Marsh stage.

Those who have known the island for the last 30 or 40 years will have noticed that not only in the places described, but nearly everywhere else, and particularly in and near parts of Norton Creek, the marshes are more lush and have grown upwards and outwards. *Spartina* is invading many places, notably mud banks in Norton Creek.

Station	Missel Marsh 1937	Missel Marsh 1947	Missel Marsh 1957	Lower Hut Marsh 1937	Lower Hut Marsh 1947	Lower Hut Marsh 1957
1	0·20	NR	—	0·70	6·5	11·0
2	1·25	8·5	20·0	1·18	7·5	—
3	1·25	7·5	17·0	1·42	8·0	—
4	1·75	12·5	22·0	1·05	7·0	—
5	1·78	10·8	—	1·40	8·7	—
6	1·75	9·0	—	1·50	NR	—
7	1·48	9·4	—	1·15	7·0	12·0
8	1·88	10·6	—	1·37	7·6	13·0
9	1·70	9·4	14·0	1·68	8·5	—
10	2·50	14·0	23·0	1·72	8·4	14·0
11	NR	NR	—	1·45	8·0	12·0
12	1·67	10·0	—	1·80	6·4	10·0
13	1·00	NR	14·0	1·50	6·5	—
14	1·72	10·0	20·0	1·32	9·0	15·0
15	1·48	10·0	—	1·50	7·4	14·0
16	1·55	11·3	15·0	1·40	7·3	12·0
17	1·45	11·5	—	1·30	6·5	11·0
18	1·60	10·0	—	1·52	8·8	14·0
19	1·68	9·0	—	1·65	9·0	17·0
20	1·72	11·0	19·0	1·12	5·8	—
21	1·80	10·5	23·0	1·35	6·0	—
22	1·63	10·5	20·0	1·52	8·5	—
23	1·60	11·5	17·0	1·12	NR	
24	2·05	11·5	—			
25	2·47	11·0	—			
26	1·85	9·8	18·0			
27	1·68	10·0	—			
28	1·70	12·0	—			
29	1·70	11·0	—			
30	1·95	12·0	19·0			
31	1·65	11·0	—			
32	1·60	10·5	17·0			
33	1·50	11·5	—			
34	1·60	12·5	20·0			
35	0·75	10·5	22·0			
36	NR	NR	—			
37	NR	NR	—			

Numerous small runnels and salt pans (Missel Marsh, stations 23–37)

All measurements in centimetres from 1935, the year in which the experiment began. A line represents a creek, e.g. between 8 and 9 on Missel Marsh.

This change in appearance is supported by fact. In 1957 it was still possible to make some measurements of the rate of accretion at most of the stations investigated in 1937 and 1947. The table on page 654 shows the changes on Missel and Hut marshes far more effectively than any description (see also Figs. 111, 112, 113).

Spartina townsendii is now a menace in parts of the Norfolk marshland coast. It was introduced experimentally by Oliver at Blakeney in 1925 (see p. 657), and there was some near Burnham Overy before the last war, and doubtless there were other patches. At Blakeney (see map p. 658) it covers extensive areas and at Scolt and other places it is spreading rapidly. Since it has arrived at Scolt some experiments have been made in the Cockle Bight to see if it is possible to kill the plant. The following paragraphs have been written by Dr D. S. Ranwell and are incorporated with his permission.[1]

Mature *Spartina* marshes on the south coast show evidence of 'die-back' in some areas (e.g. Lymington) and Goodman *et al.* (1959, 1960, 1961) have demonstrated that 'die-back' soils are highly reducing and of high sulphide content. Although the evidence strongly suggests that 'die-back' is caused by a toxic reduced inorganic ion in the soil there is no direct evidence that the sulphide ion is itself responsible for 'die-back' and the cause remains obscure. 'Die-back' is at present of relatively limited occurrence and most of the younger *Spartina* marshes like those at Scolt Head Island are still expanding vigorously.

The vigorous powers of colonization and growth of *Spartina* have enabled it to spread along the foreshores of west coast holiday resorts in some areas where it is an unwelcome invader. Elsewhere, as at Scolt Head Island, it is invading and replacing saltmarsh communities which form valuable food resources for grazing birds like the Brent Goose (Ranwell,

[1] Relevant papers are: P.J. Goodman, E.M. Braybrooks and J.M. Lambert, (1959): 'Investigations into "die-back" in *Spartina townsendii* Agg. I. The present status of *Spartina townsendii* in Britain', *J. Ecol.* 47, 651–77; P.J. Goodman (1960): 'Investigations into "die-back" in *Spartina townsendii* Agg. II. The morphological structure and composition of the Lymington sward', *J. Ecol.* 48, 711–24; P.J. Goodman and W.T. Williams (1961). 'Investigations into "die-back" in *Spartina townsendii* Agg. III. Physiological correlates of "die-back"', *J. Ecol.* 49, 391–8; D.S. Ranwell and B.M. Downing (1959): 'Brent goose winter feeding pattern and *Zostera* resources at Scolt Head Island, Norfolk', *Animal Behaviour*, 7, 42–56; D.S. Ranwell and B.M. Downing (1960): 'The use of Dalapon and substituted urea herbicides for control of seed-bearing *Spartina* (Cord Grass) in inter-tidal zones of estuarine marsh', *Weeds*, 8, 78–88.

1959). For these reasons, the need for methods of controlling the spread of *Spartina* has arisen. Experiments by the Nature Conservancy in the Bridgwater Bay National Nature Reserve on the Somerset coast showed that certain herbicides like Dalapon and substituted urea compounds were effective in killing *Spartina* (Ranwell and Downing, 1960). These studies were extended at Scolt Head Island by Mr B. Ducker of the Nature Conservancy who confirmed that Dalapon[1] at 50–70 lb./acre gave a high kill of *Spartina* and that Ureabor[2] at 2,000 lb./acre and Fenuron[3] at 50 lb./acre gave 100% kill of *Spartina* in trial plots. Other more laborious methods of eradication have been used with success at Scolt Head Island. These include digging *Spartina* out by hand and smothering clumps with linoleum after the top growth has been cut.

The great surge of 1953 did relatively little damage on the marshland coast of Norfolk. At Scolt Head Island the dunes near the Hut were cut back about 60 ft, and all the lower dunes between Hut Hills and Norton Hills were covered. The sea broke through at the western end of Norton Hills at a place where a minor break occurred in 1938 and where during the war artillery fire from a mainland range had weakened the dunes. This break, and a much smaller one a little farther west, were repaired and sand has blown up and more or less covered the wire fences. Every now and again wind or sea, or both, re-expose parts of these defences, a movement likely to be accentuated in the future. This need imply no danger, since the dunes as a whole slowly shift landwards. The Norton dunes were also cut back in the surge.

At various parts of the sea wall between Burnham Overy and Gun Hill, and especially along the wall facing Wells Harbour, breaches were made and the marshes flooded. The height reached by the flood is marked on a pillar on the site of the former station at Holkham. Beyond Holkham the surge eroded dunes, but for the most part passed harmlessly over the marshland. At Blakeney dunes were cut away and opposite Salthouse the main shingle ridge was breached. This, together with the overtopping of the shingle, led to serious loss of life and damage at Salthouse. The whole bank of shingle was pushed inland, and its encroachment on the marshes at Blakeney is still noticeable (**153**, **154**).

Blakeney Point was remapped by E. C. F. Bird and J. F. Wain in 1961; it was also mapped in 1953, after the surge, by D. Brearley and M. D.

[1] Dalapon = sodium salt formulation of 2,2-dichloropropionic acid.

[2] Ureabor = sodium borate +3-(*p*-chlorophenyl)-1,1-dimethylurea.

[3] Fenuron = 3-phenyl-1,1-dimethylurea.

Chisholm. The 1953 map showed the encroachment of the shingle on to the marshes, and indicated the main areas of developing marsh on North Side. The 1961 map (Bird and Wain) indicates many changes between 1953 and 1961. That the Far Point has altered in shape may be taken for granted, but the growth of marsh is more significant. The map (Fig. 143) shows in broad outline the present distribution of vegetation, and the growth of *Spartina* is emphatic. Bird and Wain also added a considerable amount of detail on South Side, and their map gives a useful picture of the distribution of sand, shingle, and mud there, as well as showing the trend of the narrow sand-shingle ridge between Morston and Blakeney. On this side of Blakeney Channel the extension of *Spartina* is noteworthy.

The flowering plants of Blakeney Point have been listed by D. J. B. White (*Trans. Norfolk and Norwich Nats. Soc.* 19, 1960, 179). The account is fuller than those previously published, and contains much of interest for the general reader as well as for the botanist. White confirms the tolerance of *Agropyron junceiforme* to salt in the soil and to direct and prolonged inundation by the tide. Although it is generally the pioneer in dune formation at Blakeney, Marram (*Ammophila arenaria*) may also act in this way 'if numerous individuals are present to ensure a good deposition of sand'.

Oliver first put in a few plants of *Spartina townsendii* in the spring of 1911, but these were eaten down by rabbits. In January 1925 he sowed a few seeds near the Watch House. The marsh then was one of wet and sloppy mud. By October 1927 the seeds had become well-developed plants and Oliver tried to eradicate them. Nevertheless, they survived and are incorporated in the *Spartina* meadow to-day. There is no doubt that some of the *Spartina* has spread to this area from elsewhere. On the other hand, *Spartina maritima* has not been reported in recent years although it used to flourish in and near the Channel.

The shingle ridge and sand flats at Blakeney have been the subject of research by J. R. Hardy (*Trans. Inst. Brit. Geogrs.*, Summer 1964, and Section E, British Association Meeting, Norwich, 1961). He challenges the usual views about the evolution of the spit and makes some interesting suggestions. In 1927 the possibility of a divergence in the travel of beach material near Sheringham was suggested; to the west of that place shingle moved mainly westward, and to the east it moved eastwards. It was not envisaged that there was a 'point' of divergence, but rather a short stretch of coast along which movement might vary, but at either extremity of which the drift was predominantly to west or east. Hardy

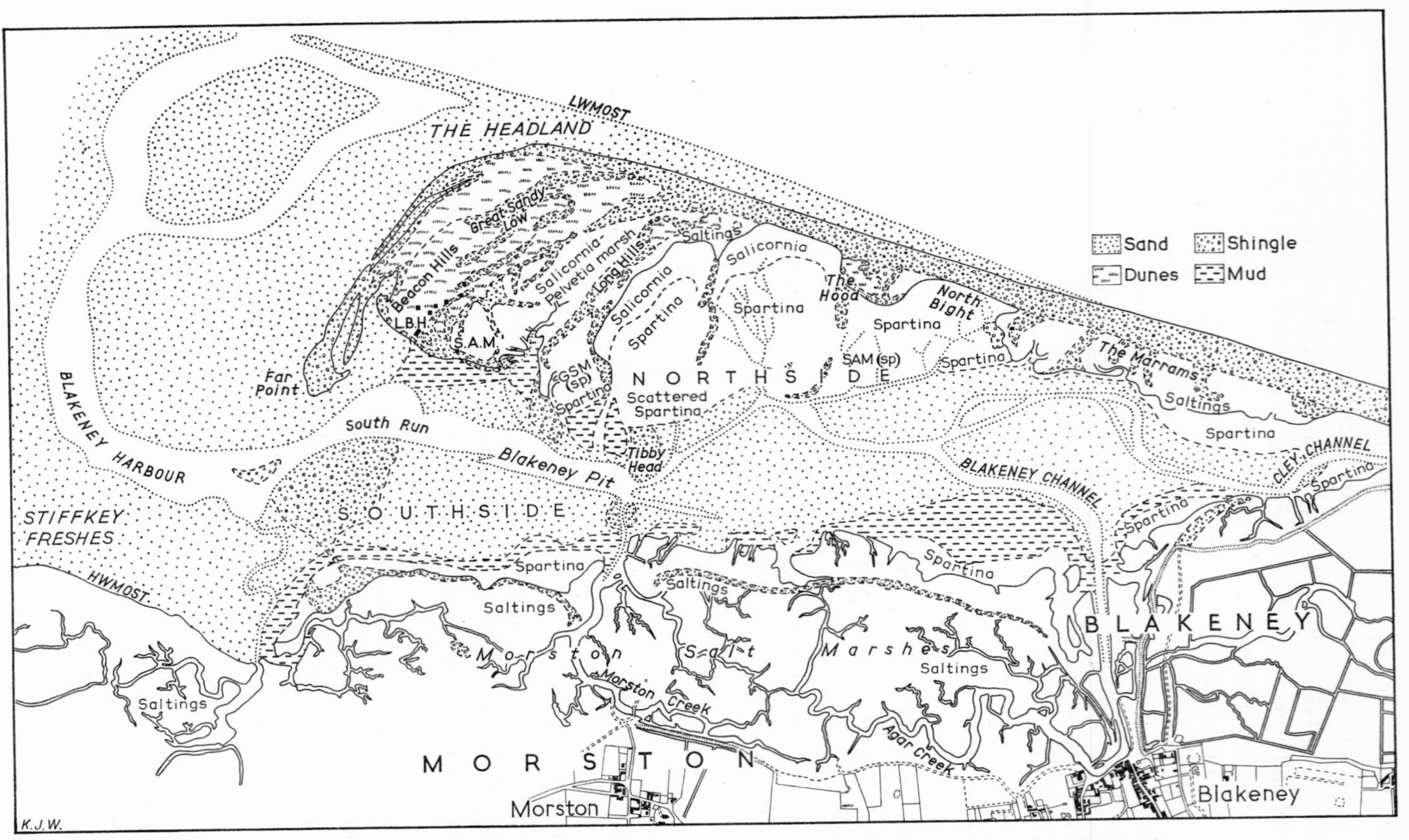

Fig. 143. Blakeney Point, Norfolk (revised October 1961 by E. C. F. Bird and J. F. Wain)

criticizes this view, and has carried out a careful analysis of the effects of wave incidence in the Cromer–Sheringham district. He used shingle distinctively marked so that its movements could be traced. He concluded from his experiments, made at Cley, Weybourne, and Sheringham, that shingle moved eastwards except when winds blew from between north-east and south-east. 'Thus a divergent movement was never found. With the present (1961) prevalent westerly winds of higher frequency at all strengths, the drift of shingle therefore seems to be eastwards along the whole spit. This is consistent with the known present direction of movement at Sheringham' (typescript of British Association address).

Hardy maintains that an examination of the distribution of shingle supports his view. The map (Fig. 143) shows the wide sand flats off the Headland and there is almost certainly no supply of shingle from offshore. He maintains that the Weybourne–Sheringham cliffs and the Eyes (see pp. 352 and 660) are the only possible sources. Following certain calculations he has made on the volume of drift he concluded that it is 'preferable to suggest that the (Blakeney) beach may be losing shingle gradually over its whole length by a slow net movement eastward towards Sheringham' (*op. cit.*). Moreover, he maintains that the grading of the shingle suggests the same conclusion. All three lines of evidence point, in his view, to an eastward movement. The shingle on Blakeney main beach is, in the author's view, markedly ungraded as a whole: that on the seaward slope may be less noticeably so.

The sand presents a different problem. It can be put into suspension by wave action, and can move with tidal currents. There is no reason to disagree with Hardy's contention that the main movement of the sand is probably eastward; in fact the suggestion made on page 350 anticipates this view. His assertion that the shingle, piled up at the top of the beach and on the recurves, is 'little more than a superficial cover', is open to argument. It is a matter that can easily be investigated, and even if the thickness is only a few feet, it nevertheless is a most important element of the beach, however much sand there may be below it or lower down toward the sea. There is abundant sand near and to the west of the Headland; dunes are readily built, and sand bars form offshore. If Hardy's contention is correct the point must be wasting away, and if so it is surprising that it does not retreat measurably in relatively short periods.

It is difficult to agree with Hardy's views in so far as the shingle is concerned. There may be periods, often prolonged, when the general movement on a beach may be reversed (see p. 563, Orford). But the structure and historical records of Blakeney Point clearly indicate a westward growth. It may perhaps have reached its maximum and, if conclusive volumetric measurements were possible, it might be possible to show that it is now wasting away or at least, that it is static. Hardy himself really gives the answer when he writes 'It seems possible, however, that as the spit grows by the addition of sand, shingle may move westwards in a spell of easterly winds....A few more pebbles might remain after each spell of easterly winds.' This seems to me to be what has happened; I differ from Hardy only in thinking that over the time the spit has been forming westerly movement of shingle, as a result of winds and waves from an easterly quarter, has prevailed. It may well be that at the present time, as his experiments appear to show, westerly winds and an easterly movement of beach material are more characteristic. But to assert that this has been always the case presupposes a supply of shingle from a totally unknown and improbable source. That there may be opposite movements of sand and shingle on the same beach is not a new suggestion.

In the library of the War Office there is an interesting document dealing with defences against the Armada, 1588. B. Cozens Hardy (*Norfolk Arch.* 27, 1941, 250), following a brief historical account by O'Neil, comments on the topographical significance of the document. It is accompanied by a map, not to scale, which shows Weybourne and Cley churches, and a fire beacon on what is probably Muckleburgh Hill. Baconsthorpe Castle, now in ruins, is marked. At this date, there were no enclosed marshes at Blakeney, Cley, and Salthouse. Fortifications, entrenchments, are indicated at Weybourne, and between them and the sea is marked a wide area of salt marsh: nearer to Cley this passes into 'The greater Salt Watter'. There are two forts shown, one at Weybourne, presumably at the cliff end, and the other at 'Black Key to garde the entry at Clay Haven'. This may mean Blakeney, and as Cozens Hardy says 'the only places where it could have been placed were the two eyes or islands at the beach end of Cley channel, known as Thornham Eye on the Blakeney side, and the Eye (as it is still called) on the Cley side'. He suggests the Eye may then have been called the Black Eye. The fort is shown surrounded by water. The map marks Wiveton Bridge, and at that time the sea flowed at high tides as far as Glandford Mill. It is interesting that in 1588 just as in the two world wars Weybourne, with its deep water close inshore, was regarded as a danger spot.

The glacial cliffs between Weybourne and Happisburgh present problems of interpretation in terms of ice advances and interglacial stages which have not yet been completely answered. As a result of more recent work, particularly by R. G. West and colleagues, not only the nomenclature but also the significance of the various deposits occurring in the cliffs, have been altered. Briefly, the succession of events now generally recognized in East Anglia is:

Hunstanton Glaciation (= Hessle of Yorkshire and Lincolnshire)
Ipswichian Interglacial
Gipping Glaciation
Hoxnian Interglacial
Lowestoft Glaciation { Lowestoft Till; Corton Beds (Mid-Glacial Sands); Cromer Till }

[The North Sea Drift contains the Cromer Till proper, the Norwich Brickearth, the Contorted Drift and the Corton Beds: correlations between the Norfolk coast and the tripartite succession of southern Norfolk and Suffolk are uncertain.]

The time sequence of events, however, is of less moment in the present context than that of the relation of the cliffs to erosion both by marine and atmospheric agencies. In this respect what is said on pages 374 and 375 holds goods.

The great chalk erratics which occur in the cliffs are interesting. Some are very large. For about 1,000 yd. between East and West Runton there is a great tabular mass of chalk, mostly horizontal, but locally folded and fractured. It carries a capping of crag and even a fragment of the Cromer Forest Bed series. Near East Runton this mass occupies nearly all the cliff, and is much cut by thrusts, so that the beds are repeated. The chalk is similar to that in the foreshore beds, and was carried to its present position by ice moving from the north or north-west.

A large mass was noted at Overstrand in 1878, and in 1906 ten others were recorded, although in 1925 artificial sloping of the cliff largely obscured them. Since 1939 the easternmost and largest mass has been re-exposed beneath the site of the former Overstrand Hotel. 'It is almost 300 feet long, and its western end appears to exhibit a 40-foot vertical section of chalk dipping at 15 deg. south, but like the Runton erratic, turns out to consist of a series of overthrust repetitions of the same set of beds...' (N. B. Peake and J. M. Hancock, *The Geology of Norfolk*, Norwich, 1961, p. 325). At Trimingham there are some masses on the foreshore which gave rise to considerable controversy. These, too, are

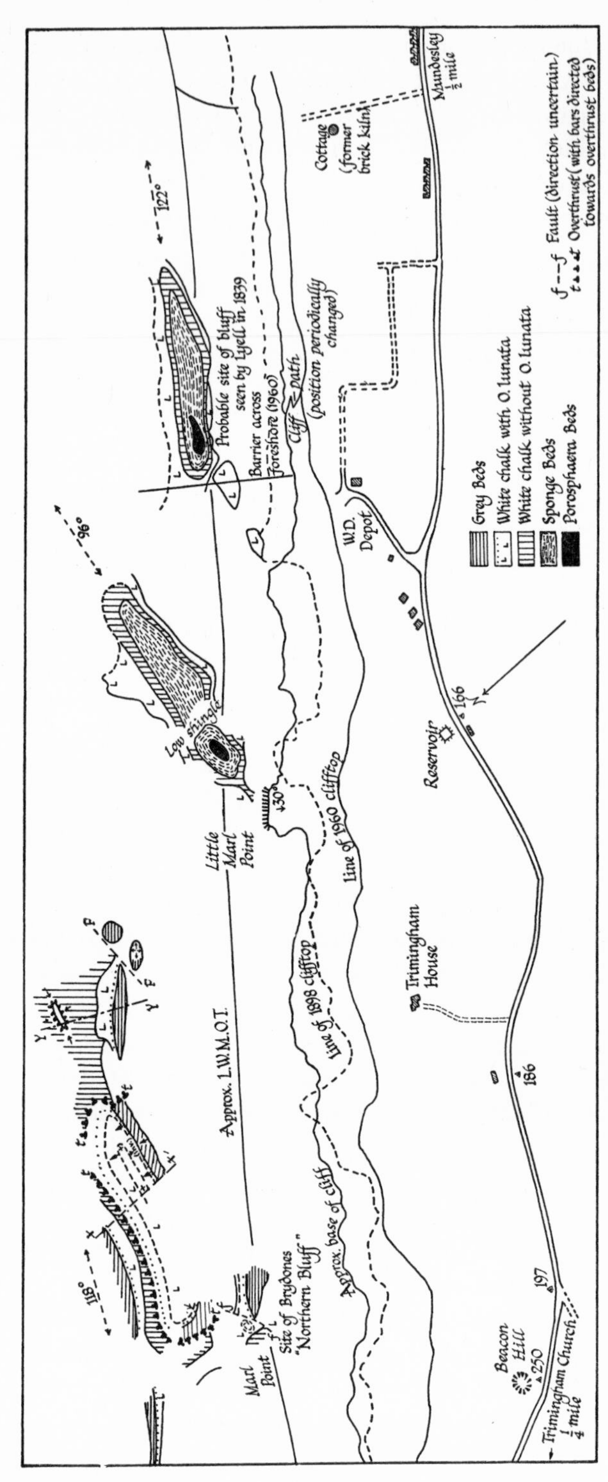

Fig. 144. The Trimingham foreshore (after N. B. Peake and J. M. Hancock, *op. cit.*)

folded and cut by thrusts. In Lyell's day two at least of them formed bluffs, one of which (the northern one) was 318 ft. long. By 1906 it has shrunk to three separate parts, and finally disappeared in February 1907. Now the exposures are below normal low water mark, and only occasionally seen (Fig. 144).

At Sidestrand a mass was seen by Brydone in 1937, and in 1948 had the form shown in Fig. 145. The 1953 storm cut deeply into it and began to expose another one. 'By 1959, the central mass had been eroded back.... During the night of December 8th, following heavy rains and gales, the greater part of this mass slid bodily seawards for a distance of 240 feet—pushed forward by the soft glacial material which slid down in its wake. This lends support to the idea that it does indeed rest upon till which acted as a lubricant.'

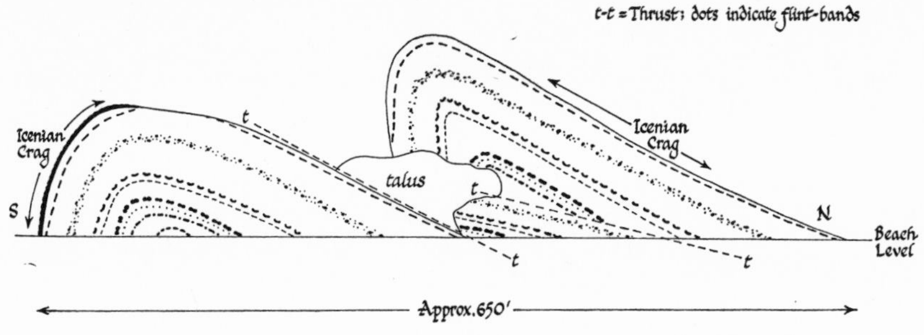

Fig. 145. Sidestrand western mass, viewed from ENE, i.e. looking along core of monochine from shore. (Diagrammatic only—not to scale; Pleistocene beds omitted. After N. B. Peake and J. M. Hancock, *op. cit.*)

There is now general agreement that all of them are erratics, but it is not known how they were taken from their original outcrops.

Near Sheringham there are occasionally exposed on the beach tubular chalk stacks. They are often covered by the beach pebbles, and are made of a limestone much harder than the normal chalk in this district. Stratigraphically they occur at the unconformable junction of the chalk with the Weybourne crag. This boundary is 15 ft. above ordinary high water mark at Weybourne and sinks gradually to the east, so that at Sheringham it corresponds with high water mark, at West Runton it approximates to low water mark, and at Cromer the junction is occasionally exposed at low water of spring tides (**155** and **156**).

The stacks have been discussed by T. P. Burnaby (*Proc. Geol. Assoc.* 61, 1950, 226). Figure 146 is based on two in his paper, and shows in

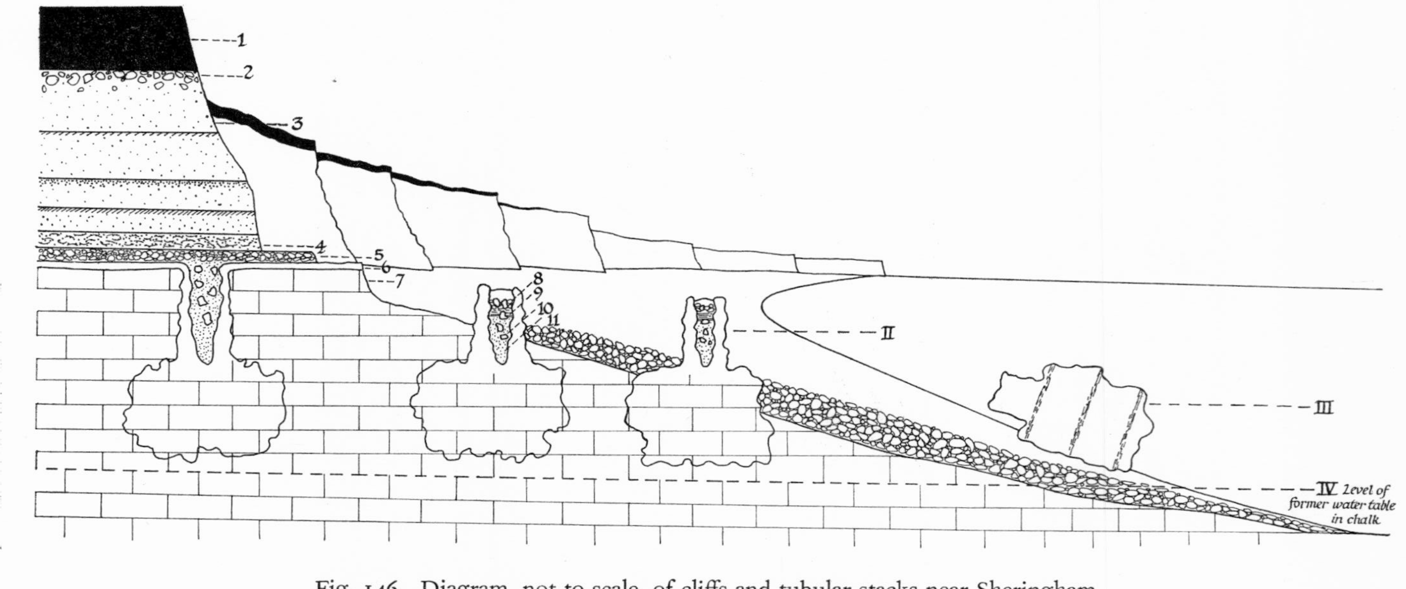

Fig. 146. Diagram, not to scale, of cliffs and tubular stacks near Sheringham (after T. P. Burnaby, *op. cit.*)

(1) Glacial drift
(2) Ferruginous gravel
(3) Laminated sands
(4) Shelly band
(5) Ferruginous basal conglomerate
(2)–(5): Weybourne Crag
(6) Lithified crust of chalk
(7) Chalk
(8) Recent beach pebbles
(9) Recent grey clay
(10) Sand and gravel
(11) Brown sand
(10)–(11): of Weybourne Crag

diagrammatic form the cliff section and the relation of the crag to the chalk. The stacks originated as solution pipes when the chalk surface formed dry land. The pipes seem to have penetrated about 10 ft. into the chalk. Their walls are very hard as a result of the cementation of the chalk with crypto-crystalline calcite. They were formed just as are the solution pipes so frequently seen in chalk quarries under a cover of gravel or other superficial deposit. Their lower limit of 10 ft. below the chalk surface was probably controlled by the then level of the water-table.

Marine erosion of the chalk exposes these pipes from time to time as suggested in Fig. 146. The likely position of some not yet reached by the sea is indicated. The contents of the hollow are shown. Burnaby first noticed the stacks in August 1947 when there was no means of observing their underground structure. In the following April 'a large boulder of chalk was found lying on the beach below the tubular stacks, with the remains of two of them still attached to one side (formerly uppermost). It became evident that each tubular stack (or close group of two or more) is attached to a large irregularly shaped basal block of hardened chalk' (*op. cit.* p. 228).

The surge of 1953 was severely felt on the glacial cliffs, and at Bacton the sea overflowed the low cliff into part of the village. Sea walls, locally sand-covered have been built along this part of the coast. Farther east, beyond Happisburgh, where the cliffs give way to a narrow belt of dunes, the surge removed about half the width of the dunes, and at Sea Palling broke through and caused serious loss of life. The sea wall at Horsey (p. 378) has been extended, and a pull-over has been made at Sea Palling so that a low path through the dune belt cannot be made and so allow another inundation. Along this part of the coast the beach is relatively narrow, and a low is usually formed parallel with the sea.

The sea wall reaches within a short distance of Winterton Ness (pp. 378, 380). Figure 147 shows its general nature. The contrast between the coast north and south of the ness point is striking. To the north dunes fringe a wide sandy area reaching almost to East Somerton. This flat is made up of sandy pasture and woods, low-lying tracts of marsh and fen carr, and a belt of slacks or damp hollows. The dune belt, now protected by a wall, narrows northwards, and the slacks increase in size in that direction. Since the dunes have been trained and replanted, the purely natural blow-outs and other features that existed before the 1953 flood are either modified or eliminated. There are still some blow-outs and hollows at a rather higher level which *may* have originally been slacks. There are

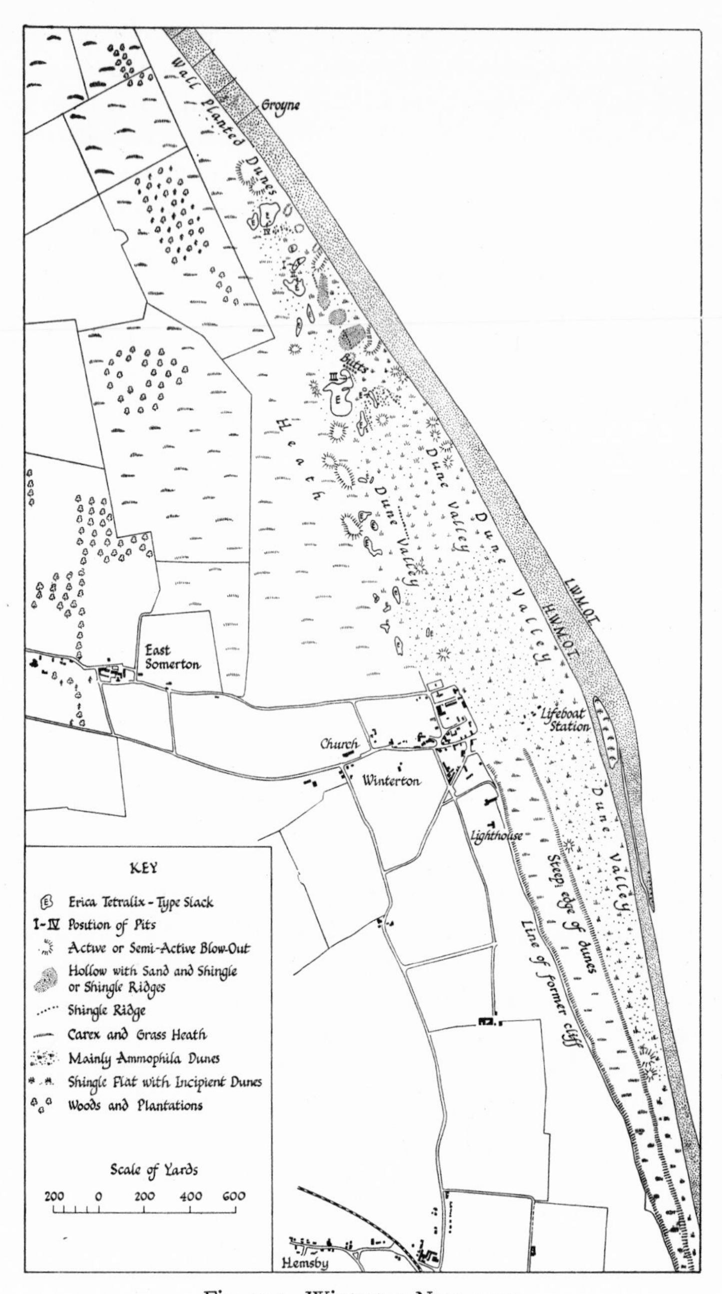

Fig. 147. Winterton Ness, 1952

also occasional ridges of shingle, and the foreland probably consists of a series of shingle and sand ridges covered by dunes and enclosing low areas. This pattern is most noticeable near the lifeboat station (see map).

South of the ness is a great valley between the old cliff and the dunes. Unfortunately historical information about the evolution of this part of the coast is meagre. It is probable that the cliff behind the dunes was reached by the sea up to the beginning of the eighteenth century. In 1616 the Commissioners for sea breaches refer to Winterton as if the cliffs were then washed by the sea. Blomefield, in Volume 5 of his *History of Norfolk*, writes 'About January 15, 1665, the hightides washing down the cliffs here (i.e. Winterton)....See London Gazette, November 20th, in 1665'. In a copy of the *Terrier* of all glebes and houses, etc.,...belonging to the rectory and parish church of Winterton, 28 June 1770, we find

Item 26. Is a piece of land lying in Hemsby field, let to John Parker of Hemsby. N.B. In the old *Terrier* it is thus described:

Which abutts on the lands of the Lord of the Manor of Hemsby to the south, and on the sea to the north-east, and lies between the lands of Mrs. Tilney, widow, in the east, and the Lord of the Manor of Winterton to the west; and contains two roods good; although the sea washes it away yearly, by under-mining the bottom of that and other lands (Extract from the *Terrier* of 1735).

The evidence derived from a study of charts and maps implies that material travelling along the coast had been gathering at Winterton for some considerable time, especially to the north of the village. It is unlikely that there were any dunes in front of the cliffs south of the village before the end of the seventeenth century or even later. In fact, there seems to have been some recession of the coast after Greenville Collin's survey of 1689. There is no convincing explanation why the dunes should form in such a way as to leave a deep low between them and the old cliff, but there may have been a sandy spit or bank, or even a shingle bank, on which they could form (**158**).

The vegetation of the dunes north of the ness is interesting and differs in some ways from that at, for example, Scolt Head Island and Blakeney Point. There are four main types of habitat—dunes, slacks, old shingle ridges, sandy heath. These were described by Jensen (J. A. Steers and J. A. P. Jensen, *Trans. Norfolk and Norwich Nats. Soc.* 17, 1953, 259). The northern foredunes have been altered by sea walls and planting. Near the ness there is accretion, and a wide belt of low, parallel, ridges 3–4 ft. high. Higher ridges characterize most of the dune belt, and in general the vegetation is homogeneous. The rare *Corynephorous canescens* is found, and near Hemsby there is abundant *Hippophae rhamnoides*.

The slacks, with two exceptions in 1953, were all characterized by *Erica tetralix*. 'Its distribution, and that of its most frequent associates, is obviously related to (*a*) differences in the ground water level, (*b*) the age of the slack, (*c*) minor variations in surface level, (*d*) the amount of blown sand entering the slack, and (*e*) the nature of the substratum and its degree of leaching.'

The following table shows the main species found in the slacks:

Aulacomnium palustre	o	*J. conglomeratus*	f-a
Betula pubescens (seedlings)	f	*J. effusus*	f-a
Calamagrostis canescens	o	*J. squarrosus*	f
Calluna vulgaris	o	*Lotus uliginosus*	o
Carex arenaria	f-a	*Osmunda regalis*	o
C. nigra	f	*Polytrichum commune*	f
Erica tetralix	a	*Potentilla erecta*	f
Hydrocotyle vulgaris	o	*Salix atrocinerea*	o
Juncus acutiflorus	f	*S. repens*	o
J. articulatus	f	*Sphagnum* sp.	o
J. bulbosus	f		

All imply acid and wet conditions. Pits dug in certain slacks revealed slight variations. The two given below will indicate the general nature of the sections:

Pit I (see map)

0–2 in. Peat with occasional sand grains
2–3 in. Brown, humus-stained sand
3–28 in. Mainly greyish sand
$>$28 in. Layer of gravel

Water occurred at a depth of 6 in.

Pit III, in a somewhat drier area

0–1.5 in. Humus and dark-stained sand
1·5–4 in. Yellowish sand
4–5 in. Humus and dark sand
5–32 in. Pure sand

Water was found at 32 in.

Slack vegetation is better developed at Winterton than elsewhere on the Norfolk coast. Usually the slacks represent phases of damp heath with *Erica tetralix* dominant. If sand is blown into them they soon pass into a type of dune grassland with *Carex arenaria* and sometimes *Calluna vulgaris*. Continuous leaching causes the heath to pass into an acid bog with *Sphagnum*.

Shingle ridges occur in some hollows; those nearest the sea are the youngest. This age differentiation is also reflected in the vegetation. The inner ridge carries a closed community, whereas there is an open community on the outer one. Jensen suggests that they may be 200–250 years old and possibly comparable to the Hood at Blakeney; both carry *Corynephorous cancescens*.

On the Heath *Carex arenaria* is abundant. Other species include *Agrostis canina*, *Festuca ovina*, *Corynephorous*, and also *Galium hercynicum*. There are also small patches of *Calluna vulgaris*. Grass heath also characterizes the great valley south of the village.[1]

The cliffs of Flegg (p. 380) have been investigated since the 1953 surge (C. Green, G. P. Larwood and A. J. Martin, *Trans. Norfolk and Norwich Nats. Soc.* 17, 1953, 327) which, by the erosion it caused, left a clean face. Fourteen sections were analysed between Hemsby and just south of California. All of them show Norwich brickearth at the base, followed by Corton sands. The 3 southern sections are capped by clay and sands; number 4 was obscured in its top part; and 5–11 inclusive are capped by chalky boulder clay. The remainder were obscured. The brickearth and Corton sands are much contorted and disarranged in sections 5–9. The disturbance begins just north of number 4, at California, and is ascribed to two tongues of ice merging at this place. The whole cliff offers little resistance to erosion where it is unprotected.

The general evolution of the spit on which Yarmouth stands and its relation to the former estuary of the Bure–Yare–Waveney and to the Broads have been discussed in some detail. On page 383 a quotation is given from a private note from J. N. Jennings. Since that was written two memoirs of great significance have appeared. In the first of these (*R.G.S. Research Series*, No. 2, 1952, The Origin of the Broads) Jennings gives a full account, supported by pollen analysis, of the deposits in the Bure and Ant valleys. The correlation table on page 670 sums up his conclusions in terms of relative movements of sea level and of vegetation changes in association with climatic fluctuations. He clearly had in mind the possibility that, Breydon Water excepted, the individual broads might have had an artificial origin and stated that part of Barton Broad had indeed been ascribed to peat cutting. This and three other factors all pointed

[1] Those interested in Winterton may wish to consult, by permission, the Management Plan, prepared by B. F. T. Ducker, for the Nature Conservancy.

towards such an origin: (1) the frequent sharp margins of many broads; (2) in some broads (Salhouse and Barton) the muds reach a lower level than the *Phragmites* clay and *Phragmites* peat of the ronds; and (3) mediaeval documents seem to ignore the broads just at the time when the meres of Fenland were commonly mentioned because of their fisheries.

ALLUVIAL STRATIGRAPHY – MIDDLE BURE & ANT VALLEYS			PREVALENT STATE	RELATIVE MOVEMENT OF LAND AND SEA	FOREST ZONES	BLYTT-SERNANDER CLIMATIC PERIODS
Seawards and near rivers	Inland and near valleysides where effects of transgression:- large	small				
Brushwood peat PHRAGMITES PEAT	Where Broads have shrunk: Brushwood peat, Cladium, Phragmites, Typha and Scirpus peats.	Brushwood peat	Locally dry	? SLOW SUBMERGENCE	VIII - MODIFIED	RECENT
				STANDSTILL		
UPPER CLAY (markedly estuarine to brackish)	FORMATION OF BROADS NEKRON MUDS (freshwater)	CAREX AND PHRAGMITES PEATS	WET	SUBMERGENCE	VIII ---?--- TRANSITION	SUB-ATLANTIC (cold and wet)
Erosion channels cut BRUSHWOOD PEAT			DRY	STANDSTILL OR SLOW SUBMERGENCE	VII - VIII	SUB-BOREAL (warm and dry)
Phragmites peat and detritus muds LOWER CLAY (slightly brackish) Phragmites peat & detritus muds	COARSE DETRITUS NEKRON MUDS (freshwater)	Only in deep channel incised into buried valley	WET	SUBMERGENCE	VII	ATLANTIC (warm and wet)
					?	?

Correlation table for the alluvium of the Ant and middle Bure valleys (the thickness of the zones bears no relation to duration of time) (J. N. Jennings)

In a subsequent memoir (*R.G.S. Research Series*, No. 3, 1960) in which Jennings was a joint author with Lambert and others, the artificial origin of the broads as mediaeval peat cuttings is demonstrated. In this sense, then, we are not concerned with them since they are not coastal forms. We are, however, very much concerned with the valleys in which they lie, and with Breydon Water, the only completely natural broad.

Green and Hutchinson, part authors of the 1960 memoir, call attention to an important find made when the power station on Yarmouth spit was being built. Excavations for the inlet and outlet culverts necessitated deep trenching in the gravels and sands of the spit. The authors named the beds beneath the spit the Red Beds, fine, compact, reddish sands, current-bedded and sterile. They suggested that these beds were lacustrine in origin and possibly of Pleistocene age. Their surface was eroded and uneven. A layer of beach ballast rested on the Red Beds.

From the body of this beach came a quantity of pottery fragments of various periods. The most significant potsherds were, of course, the latest in date. These were not very water-worn and were all datable to the thirteenth century A.D. It is archaeologically certain, in view of the quantity of material preserved, that this layer of beach-ballast must have been deposited...during this century. Furthermore, after the early days of the excavation, it was observed that, wherever the shingle seemed suitable, the surface of the beach carried a colony of acorn barnacles (*Balanus balanoides*). That this was a true colony of the acorn barnacle was evidenced by its density and mode of occurrence at the surface. The longer pebbles carrying the shells were not scattered throughout the layer, as would have been the case if they had been derived from a colony elsewhere. The occurrence of a colony of this species is accepted by zoologists as evidence of an intertidal zone (Yonge, C. M. 1949, *The Sea Shore*, London, 112). It seems, therefore, that here, at a greatest observed depth of −17·5 feet (5·3 m.) O.D., but mainly above this level, a surface lay above the level of low water spring tides in the later part of the thirteenth century.

Above the beach ballast there was a thick band of dark-brown silty sand containing potsherds of the thirteenth century. Certain fragments were of a distinctive pattern and allowed a precise dating. Colonies of mussels (*Mytilus edulis*) were found on the silt. These mussels provoked some difference of opinion about possible changes of level. One authority held the view that this mussel *to-day* does not colonize below the extreme low water of spring tides on the Norfolk coast; others, however, think that elsewhere it may be found some feet below that level. On the mussel beds beach material continued to accumulate up to a level of +11·0 ft. O.D. Later the level was raised by blown sand to about +17 ft.

Green and Hutchinson regard the barnacles as undoubted evidence that sea level at that time was lower than it is to-day. They gave careful consideration to possible local explanations—e.g. slumping—but concluded that the deposits could not be interpreted except by involving a change of sea level relative to the land. They think the change of level began a little before the great flood of 1287. If so, the extreme low-water level of the twelfth–early thirteenth century now lies rather more than 17·5 ft. (5·3 m.) below O.D. This level must reflect the maximum of the post-Roman emergence, and this emergence of the land may therefore be known as the 'Saxo-Norman Marine Regression', a term used for the complex of relative measurements in Anglo-Saxon times, culminating at about the time of the Norman Conquest or soon after.

The section is unfortunately no longer available for inspection, and we may have to wait some time before any corroborative evidence is found in similar places. Every opportunity is taken to examine likely excavations, but it does not follow that even under very similar conditions a comparable series of deposits will occur. The present writer has had the advantage of several discussions with the finders of the section and is in general agreement with their interpretation. It is difficult to suppose that the alterations of level of land and sea were confined to the Yarmouth area; on the other hand, they need not necessarily have extended all along the east coast. Green and Hutchinson have made an important discovery, and we must hope that they, or others, will soon find corroborative sections.

W. W. Williams (*Geogr. Journ.* 122, 1956, 317, and *R.I.C.S. Journ.* April 1953) has examined the interesting shingle headland of Benacre Ness. With it must also be considered Benacre and Covehithe broads and the cliffs at Covehithe (**160**).

Along most of this piece of coast erosion is serious except at the ness itself, which consists of a crescentic-shaped area of sand and shingle about $1\frac{1}{4}$ miles long and rather less than $\frac{1}{4}$ mile wide. To its east and north-east are two shoals, Barnard and Newcome, which are constantly moving. Williams maintains that hereabouts the general direction of material eroded from the Covehithe cliffs is northward and there is no doubt that the ness itself is moving northwards. Figure 148 based in part on Williams's paper (1956) shows the movement. Williams thinks that the present ness is a direct successor of Easton Ness shown on Saxton's map of Suffolk. There is the possibility of confusion here. Suckling (*History of Suffolk*, 1848) speaks of Easton not as a ness but as a lofty promontory washed by the sea and forming the most easterly point of England.[1] A ness in the position shown on Fig. 148 might have been an accumulation feature, but both at Southwold and between Buss Creek and Easton Broad there are cliffs which have long suffered erosion. If a ness had existed in front of them erosion would have been largely modified. A cliff headland may well have existed hereabouts and it may or may not have been the easternmost point of England. That a shingle ness may have existed farther north, near the present Easton Broad, is certainly possible, and if so it may have travelled northwards as far as to the present Benacre Ness.[2] Hodgkinson's 'ness' of 1783 was perhaps a local gathering of shingle and sand. The positions marked (Fig. 148) by the dates 1783, 1865, and 1826 are inconsistent, as Williams himself notes, amongst themselves. In the absence of proof it seems more reasonable to regard the present ness as a unit since approximately 1826, and positions indicated to the south as local accumulations of beach material. This is a matter of opinion, but a major or even

[1] '...and is supposed to have terminated eastwards in a lofty promontory, which has...been wasted by the sea...its extreme cape must have extended more than three miles from the present cliffs...' (Suckling, *loc. cit.* 1848).

[2] A measurement made in July 1962 shows that the point is still moving northwards.

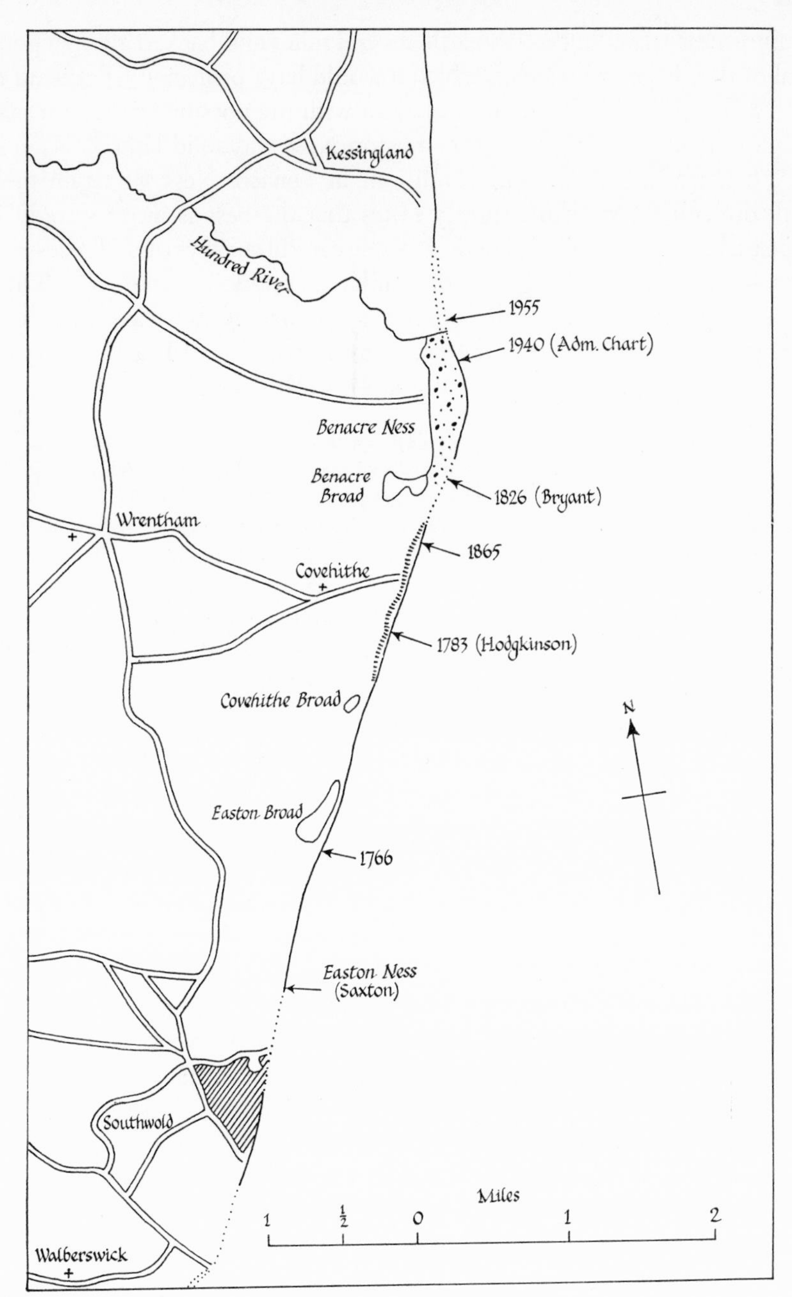

Fig. 148. Sketch map of east Suffolk coast, showing Easton Ness and Benacre Ness (partly based on W. W. Williams, *op. cit.*)

a minor ness travelling northwards several miles and backed along a good deal of this distance by a cliff, which it would have protected, is difficult to envisage, especially if taken into account with the possibility—and it may be more than that—that Saxton's Easton Ness was solid land.

When we turn to present conditions at Benacre Ness we again meet with difficulties. Williams rightly notes that the ness is more shingly in appearance than in reality. He made some careful surveys of cliff recession, of ness growth and changes, and of adjacent sea-floor conditions. These were supported by careful observations on current and wave incidence. Along this part of the coast the prevalent movement of beach material is to the south, e.g. Yarmouth Haven and Orford Ness.

There is one theory which appears to satisfy all the conditions. It is reasonable to suppose that a northward transport of material occurs during the recession of surges (see below). During the advance of the surge, wave action may well be increasing in intensity; the greatest damage to the cliffs is likely to be at the maximum height of the surge which may reasonably be supposed to coincide with the height of the storm. If we suppose that at this time a large mass of material has fallen, it will come under the action of an abnormal north-going pull as the level of the sea falls, and the water retreats whence it came; it does not go directly to the ness, but to a point a little seaward of it, and from this point it is at a later stage pushed up the beach.

This may apply to-day; it is difficult to see how it could have applied in earlier times when it is assumed that the ness existed in front of the cliffs at Covehithe and so protected them.

In the discussion of Williams's paper, R. C. H. Russell advanced a relatively simple mechanism for this undoubted northward shift of the ness point to-day. Figure 149 shows a simplified ness on a straight coast along which material travels north to south. The direction of dominant wave approach is shown. 'Along the northern face of the promontory, because it faces directly into the waves, the drift will be zero and along the southern face of the promontory, because it makes a very oblique angle with the waves, the drift will be greater than normal...along the northern shore there will be accretion...and along the southern shore there will be erosion....Accordingly, the promontory moves northwards.'

The cliffs at Covehithe suffer severe erosion. Williams saw them just before and immediately after the 1953 surge. He points out that the high level of the sea plus the much greater wave action at the time meant that here (and elsewhere) the waves were no longer expending their energy on the beach, but on the cliff face which, because of its nature, offered no effective resistance. The cliff, 40 ft. high, was cut back 40 ft., and where it was only about 10 ft. high, the sea encroached 90 ft. Williams estimated that in 24 hr. some 300,000 tons of material disappeared from Covehithe cliffs. The two broads were also flooded, and some hours after the maximum of the surge waves were running over the shingle banks which

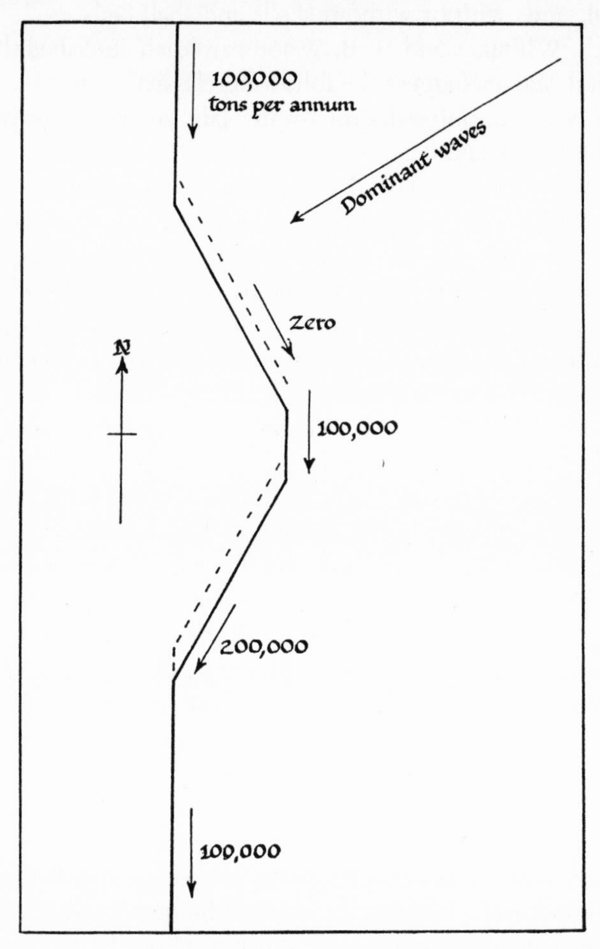

Fig. 149. Diagram of simplified ness on a straight coast (material travels north to south) (after R. C. H. Russell, *op. cit.*)

are about 10 ft. above mean sea level. It is undoubtedly in conditions such as these that the maximum erosion takes place on the east coast. The 1953 surge was particularly severe, but lesser ones are frequent.

The cliffs between Easton Broad and Southwold have been examined by G. P. Larwood and A. J. Martin (*Trans. Suffolk Nats. Soc.* 8, 1952, 54,

159). We may assume that the same beds continue northwards to Covehithe. Three main units were recognized:

(3) Uppermost sands with gravels.

(2) Variable blue-grey clay unit.

(1) Basal sands with or without shells and shell beds.

Prestwich, Whitaker and H. B. Woodward each described this section. Larwood and Martin suggest the following classification:

(3) Pebbly series, unfossiliferous water-lain sands and gravels.

(2) Chillesford Clay.

(1) Norwich Crag, sands and shell beds.

The thickness of the beds varies. None is resistant to sea erosion, and throughout recorded history this has always been a receding coast. An interesting comment on erosion is made by A. Beaufoy (*Suffolk Nat. Soc.* 9, 1954, 56, 372). On 1 January 1956 a round brick pillar, 30 ft. high, and rising from the beach several yards from the cliff face was visible. It was an old well dug for a naval camp in the First World War, when there were two or three fields between the well and the cliff edge. It was destroyed by waves in March 1956.

Between Covehithe and Southwold erosion has long been active. Measurements were made in the summer of 1950 and again in July 1962. The great storm of 1953 had profound effects especially at and north of Covehithe. At 5145 E. 7820 N. the rate has been about 6 ft. a year since 1925 (6 in. edition O.S.). The houses that were built just north of this point in 1935 have been pulled down since the cliff edge had reached the one nearest the sea. At the end of the road passing Covehithe church the average rate has been at least 13 ft. a year. On the other hand the most easterly point of Benacre Ness in 1925 was nearly in line with the east–west road at Beach Farm. In July 1962 it was approximately 2,000 ft. farther north, and almost in the same meridian as in 1925.

The surge of 1953 temporarily made Southwold an island again, and wrought a good deal of havoc amongst the bungalows south of the cliff. On the other hand, although flooding occurred on the marshes between Southwold and Dunwich, little change took place at Dunwich itself. The cliff was scarcely affected; at its southern end, overlooking Minsmere Level perhaps a foot or two was eroded away. Yet this cliff, as far as the material of which it is composed is concerned, is as vulnerable as that at Covehithe and Easton Bavents. It appears to be the case that along this coast the sea attacks in a series of bites. North of Kessingland there is still a small and shallow re-entrant in the cliff which is protected by a fairly wide beach, but at Kessingland and at Pakefield erosion has been serious, and is now restrained by elaborate defence works.

At Slaughden, south of Aldeburgh, the storm flattened the beach, and for a time waves ran over it into the river. But the beach was not severed. It was soon rebuilt artificially, and its appearance even to-day reflects the bull-dozing to which it was subjected. If it had been left to nature, it would probably have remained intact and would have been built up gradually. But for some time it would have remained relatively low, and a weak defence in time of storm. Along this part of the coast, which is defended by groynes, temporary northward movement of beach material has been noticed, sometimes for several weeks. That the whole beach is gradually being inrolled is demonstrated by the way in which it had encroached on to the Martello Tower. This was built *c.* 1804 and was then well above high water mark.

The southern part of Orford Ness is now a National Nature Reserve under the aegis of the Nature Conservancy. Some interesting experiments of the movement of radioactive pebbles have been made here, and are discussed on page 563. At Shingle Street R. T. Cobb (*Ann. Rept. Field Studies Council*, 1956/7, 31) has examined the nature of the beach and lagoons, and has also made observations on the movement of pebbles (cf. p. 565). Shingle Street cannot, however, be dissociated from North Weir Point and the southern part of Orford Ness (**163**).

A study of old maps and charts suggests that the entrance to the river shows a cyclic development. A cycle appears to begin when conditions are more or less as at present (1961), i.e. one or more shingle banks partly obstructing the haven, and continuing the general line of the spit to the south. In course of time they became incorporated in the growing spit which may extend roughly as far as the Martello Tower AA. This has happened on two occasions. A violent storm eventually destroyed the southern part of the bank and left some shingle islands. The cycle then recommenced.

The lagoons at Shingle Street fall into two groups. Those to the south of the houses are tidal and floored with alluvium, now often obscured by shingle. They were once part of the river bed. Cobb states that they were traceable in embryo form in 1902; in 1924 they were semi-permanent and shingle was gathering on the beach; in 1945 they were being filled, and in 1948–52 shingle was being rolled inward. North of the Beacons (see Fig. 150) the lagoons vary in origin. Three may have originated as borrow-pits either when the sea wall was built or when shingle was required for

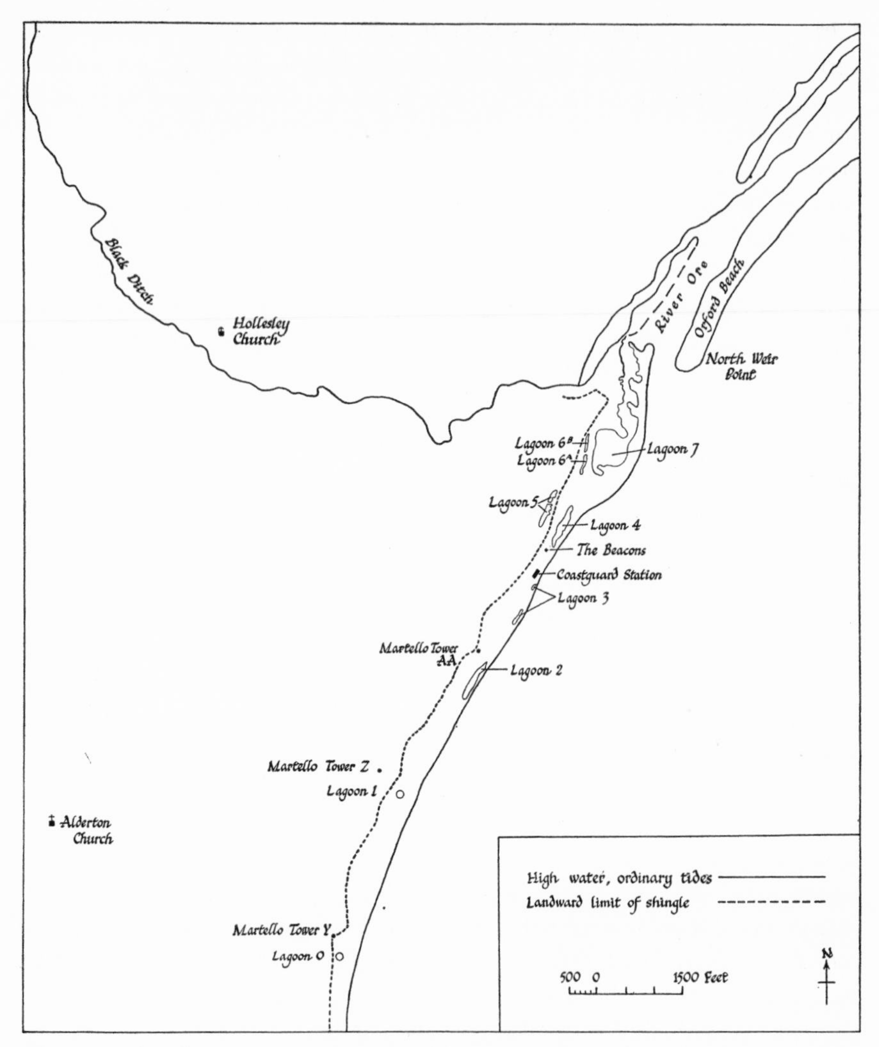

Fig. 150. Shingle Street. Lagoons 0 and 1 indicated by general position only, the other lagoons roughly as in 1955. The sketch is composite, and based on several maps by R. T. Cobb, *op. cit.*

various purposes in the Second World War. No. 4 has changed little. No. 7 opens through a winding channel to the river. The shingle ridges on its western side were built from the north, whereas on the other side of the lagoon the ridges have grown from the south.

Cobb agrees with the view expressed on page 389 that North Weir

Point is the most probable source of supply of shingle The shingle has to traverse the river mouth, in the centre of which there is a fairly steep-sided trench some 3 fathoms deep. This trench shallows seaward where there is an 'arcuate mass of shingle joining North Weir Point with Shingle Street'. It is along this that the shingle is assumed to travel, and when it reaches Shingle Street some is taken northward by combined wave and current action and some continues to the south. Cobb maintains that on this hypothesis it is possible to explain recent changes at Shingle Street, i.e. accretion north of the Beacons from 1920 until about 1953, when the rate increased. Since 1924 high water mark at the coastguard station has moved inwards more than 500 ft.

Little new work has been published on the Essex coastline. A short paper by A. H. W. Robinson (*Essex Naturalist*, 29, 1953, 1) describes the coast and speculates on the origin of some features noted on pages 389–92. Robinson calls attention to the importance of landslips between Harwich and Clacton, to the erosion on that part of the coast, and to the development of the spit running north-north-west from Colne Point. There has been some erosion near this point and also a lowering of the foreshore.

The 1953 surge caused great damage and serious loss of life in Essex. It topped sea walls and inundated enclosed marshes at Jaywick and elsewhere. The most serious flooding was at Canvey Island. This was caused by the failure of the walls in Benfleet Creek. Canvey Island was enclosed in 1623 by Dutch engineers although an attempt was made to enclose it in Roman times. Two-Tree Island was partially inned at the same time. The adjacent mainland marshes were embanked by 1653. On the existing natural marshes in this district, salinity and rapidity of drainage appear to be the main factors affecting the vegetation (J. E. Myers, *Essex Naturalist*, 29, 1952–56, 155). The well-drained marginal zones are almost entirely covered by *Halimione*; the inner parts are dominated by *Puccinellia*, *Suaeda maritima*, and *Aster tripolium*, together with various *Salicornias*. *Spartina townsendii* was noted in isolated patches in the central zone, but farther east it spreads over wide areas. The vegetation is, in general, that characteristic of east coast marshes.

The Red Hills of Canvey are all inside the sea wall. Records show that prior to the building of the wall the island was wholly salt-marsh, used for grazing, and interspersed with low mounds barely visible at spring tides. E. Linder (*Essex Naturalist*, 27, 1940–46, 48) suggests that the hillocks mentioned by Camden are the Red Hills of to-day. These mounds (see

p. 397) were occupied by Romano-British people in the first and second centuries A.D. The fact that hearths are now several feet below the surface implies a gradual sinking over the last 2,000 years. Some of the sites show occupation at different times from the first century A.D. to *temp.* James I, and many hearths are now well below spring tide level. Some land at Northwicke was used for arable in 1557, about 70 years before the embanking; this now stands at 11 ft. O.D. Linder concludes that these and other facts 'are consistent with the occupation at successive periods of a burnt earth surface subsiding at an average rate of about eight inches per century...'.

An interesting sequence of deposits was found at Brentford, where the Brent joins the Thames (R. E. M. Wheeler, *Antiquity*, 3, 1929, 20). On the surface were the usual modern relics, although material belonging to the eighteenth and nineteenth centuries was absent, having been scoured away. There was a well-marked layer appertaining to A.D. 1600–1700 below which a medieval layer was found. Below, there was clear evidence of Romano-British occupation, including the remains of a pile-dwelling. Lower down and around this hut, and still more in gravel below present low tide level, fragments of pottery were found which were confidently assigned to the Hallstadt period (*c.* 1000 to 500 B.C.). Farther out in the stream Bronze Age implements were dredged up. This sequence is comparable to that found near Southend.

It is relevant here to say something of the regime of the Thames estuary. Two factors have had considerable effects on the flow of water since early times. The gradual rise of mean sea level, produced partly by a downward movement of the land and partly by a very slight eustatic rise in sea level, has caused the acceleration of the tidal wave up the estuary. This, in turn, implies a greater tidal discharge, and so a deeper channel. The other factor, which operated from the reign of King John until 1832, was old London bridge, which consisted of nineteen arches and a drawbridge, the largest span being 30 ft. It was 20 ft. wide at first, but was later broadened, and houses were built on the bridge. They partly overhung the water and were supported by struts. The bridge had a serious effect on the tidal regime. After its removal the range at and above London was increased up to 25%.

The increase in discharge caused by removal of the bridge altered the regime of the whole estuary. Upstream of the old bridge very rapid deepening of the channel took place, to such an extent that two or three other bridges were endangered. Downstream the effects will have tapered off but must have been perceptible as far as Woolwich Reach and the Mud Reaches. The first channel changes of this kind took place very rapidly, but it is considered that final regime in the upper reaches has not even yet been attained; indeed, there are signs that slow scouring of the bed is still in progress, although other causes must be held to account for this. (Sir C. C. Inglis and F. W. Allen, *Mins. Proc. Inst. Civ. Engs.* 7, 1957, 827.)

The same authors suggest that a direct cut through Broadness would be an improvement, because it would mean a greater range at springs at London Bridge (about 8 in. at first) and a general increase in river depths. It would lead to an upstream movement of the Mud Reaches. Much material dredged to improve the channel was dumped at the Black Deep. It is now known that instead, as was hoped, of this material going out to sea, most of it was swept upstream again, much of it falling in the Mud Reaches. Dredged material should not be dumped in the estuary, but ashore. The regime in the estuary is a delicate balance between accretion and erosion; tampering with conditions at one place upsets the balance and leads to difficulties elsewhere.

Chapter XXVI

HOLDERNESS AND LINCOLNSHIRE

(1) HOLDERNESS

The erosion of the Holderness coast has been studied by H. Valentin (*Die Erde*, 1954, 296). Earlier measurements were made at isolated places, but Valentin examined the whole stretch of boulder clay cliff and made 307 measurements at intervals of approximately 200 m., during the summer and autumn of 1952. Since he used fixed objects, houses, walls, fields or other boundaries which existed in 1852, a direct estimate of erosion along the 61½ km. of the coast and covering 100 years was possible. The following table shows the amount lost by parishes:

Number	Parish	Yearly cliff retreat (m.)	Length of cliff in parish (m.)	Yearly surface loss (m.²)	Average cliff height (m.)	Yearly loss (m.³)
1	Bridlington	0·36	4,875	1,755	13·5	23,693
2	Carnaby	0·49	1,350	662	9·9	6,549
3	Barmston	0·52	5,700	2,964	7·0	20,748
4	Ulrome	1·57	1,175	1,845	7·1	13,098
5	Skipsea	1·44	3,600	5,184	10·8	55,987
6	Atwick	1·13	3,075	3,475	16·6	57,681
7	Hornsea	0·84	4,175	3,507	12·6	44,188
8	Mappleton	1·45	6,600	9,570	17·4	166,518
9	Aldbrough	1·24	4,600	5,704	18·0	102,672
10	East Garton	1·14	1,925	2,195	22·6	49,596
11	Roos	0·85	5,275	4,484	17·0	76,224
12	Rimswell	0·85	1,950	1,658	13·4	22,211
13	Withernsea	1·08	3,600	3,888	10·6	41,213
14	Hollym	1·36	2,250	3,060	11·9	36,414
15	Holmpton	1·50	2,000	3,000	15·3	45,900
16	Easington	1·89	9,375	17,719	13·6	240,975

These figures can be analysed so as to demonstrate the effects of local conditions. In the table on page 683 the coast is divided into four sections. The relatively small loss in section 1 is partly explained by the promenade at Bridlington. Before this was built, erosion near what is now the present southern end of the promenade averaged about 1·38 m. a year.

After the promenade was built small dunes formed in front of the cliff, but about 3 km. farther south there was serious erosion. In section 2 there was a general loss, except at Hornsea where there is a promenade. In the third section erosion was serious. The worst stretch is between Easington and Kilnsea, where in places erosion is as much as 2·75 m. a year.

Number	Section	Yearly cliff retreat (m.)	Length of section (m.)	Yearly surface loss (m.^{2})	Average cliff height (m.)	Yearly loss (m.^{3})
1	Sewerby–Earl's Dike	0·29	8,100	2,357	11·0	25,927
2	Earl's Dike–Hornsea	1·10	13,650	15,015	11·8	177,177
3	Hornsea–Withernsea	1·12	24,250	27,160	16·2	439,992
4	Withernsea–Kilnsea Warren	1·75	15,525	27,200	13·2	359,040
Whole coast (averaged)		1·20	61,500	72,000	14·0	1,000,000

There is thus a general increase of erosion in a south-easterly direction. Valentin estimates that the average recession of the whole cliff was 120 m., or an areal loss of 720 hectares (1778 acres) of land in the 100 years ending in 1952 (**144**).

The drainage of land water and normal wave erosion are important causes of this loss. Storms and surges such as that of 1953 do serious damage in a short period of time. The greater exposure to wave action in the southern part of Holderness is also significant. In the north, Flamborough Head and the Smithic sandbank protect the coast. In the south there is no such protection and the depths offshore are greater, the 10 m. line being only 600 m. seaward from Dimlington.

Figure 151, based on Valentin's work, illustrates certain changes on this stretch of coast. Any such attempt implies the making of big assumptions of average loss, and should also allow for changes of sea level that have taken place. A downward movement of the land relatively to sea level increases erosion, and probably this has been the dominant trend. On Fig. 151 Valentin has inserted for each of his 307 stations both a cliff profile and an estimate of the erosion. The scale of these two features is exaggerated (the profiles $\times 300$ and the erosion $\times 30$). Therefore, as far as the erosion is concerned, the figure, in addition to indicating the outer edge of the coast about 100 years ago, also attempts a very rough

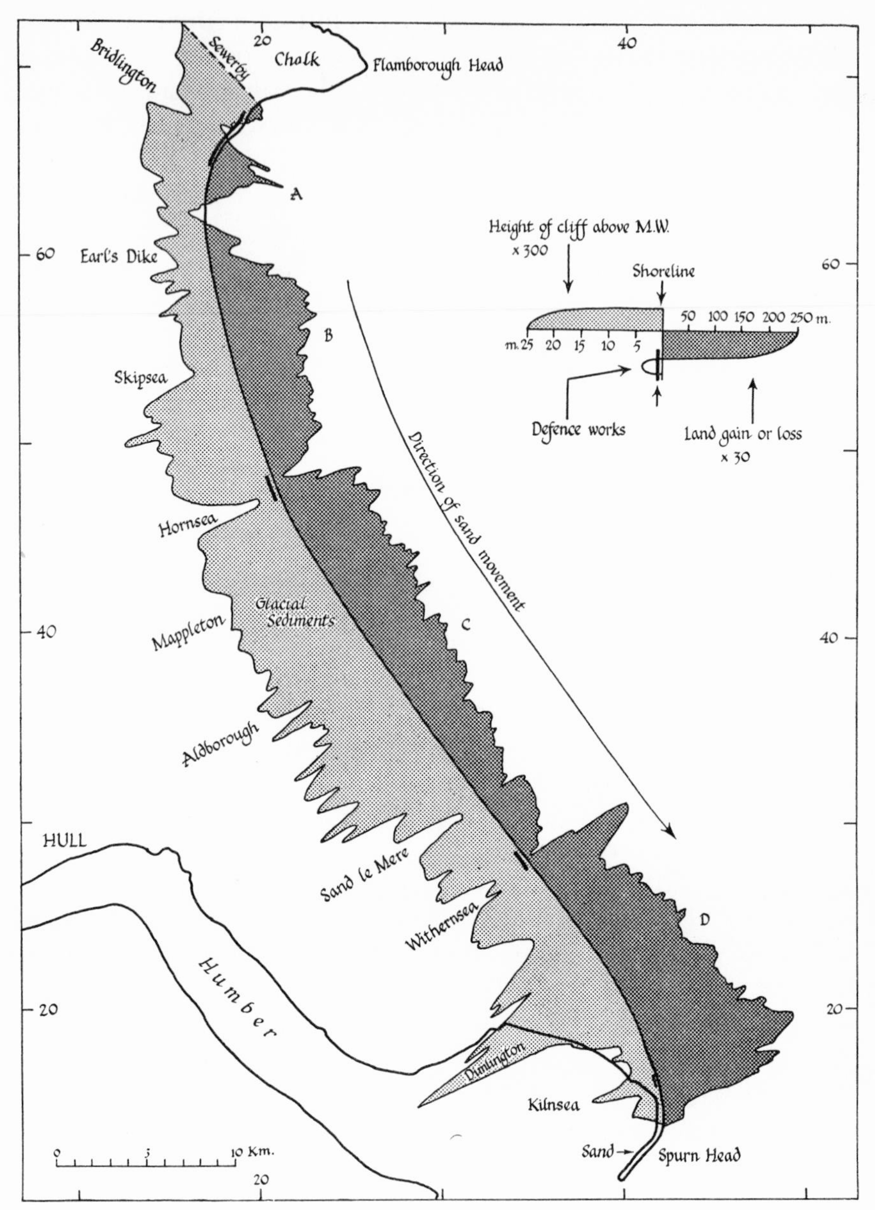

Fig. 151. Cliff profile and loss and gain in Holderness (after H. Valentin, *op. cit.* 1954)

indication of the position of the coast some 3,000 years ago. This is speculative and not in agreement with the figures given by Sheppard (pp. 411 et seq.) for loss since Roman times.

Valentin also suggests that minor changes of sea level have influenced erosion in the last few decades. He investigated a large number of records from tidal stations around the North Sea, and concluded that between 1852 and 1887 the sea level, relative to eastern England, was a little low, hence erosion was less, averaging in Holderness 1·53 m. a year. From 1889–1908 was a time of rising sea level and increasing erosion, reaching 1·77 m./year. From 1908 to 1926 there seems to have been little change and between 1926 and 1952 the level fell slightly and so erosion was again less. Theoretically, slight changes of this type are interesting, but it is doubtful if they have any real significance in this context. A severe storm, a surge, or even strong onshore winds coinciding with a spring tide, are each likely to have far greater effects.

(2) SPURN HEAD

The military occupation of Spurn Head ended in 1959. Now it is a Nature Reserve under the care of the Yorkshire Naturalists' Trust. A programme of field work is planned and an indication of its scope can be found in *The Naturalist*, 1950–57 (p. 75). G. H. Ainsworth in a brief account of the ecology of Spurn distinguishes six regions: (1) the marshy meadowland on the right of the road from Kilnsea to the Warren; (2) the Warren itself; (3) the main ridge, subdivided into north, central, and south parts; (4) the salt marsh on the Humber side; (5) the Humber mudflats; and (6) the sea shore. In (1) the plants are characteristic of moist places with a stiff clayey soil. The main ridge protects the Warren (2) from sea winds, and so that locality is richer in plants, but was much disturbed during the war; parts are covered by small dunes. The main ridge (3) is dune-covered and contains plenty of water. The sea buckthorn occurs. Rabbits were abundant and caused the disappearance of much marram. The dunes are fed by the sand from the shore (6). The movement of Spurn westwards is illustrated in (4). In a north-north-westerly gale in 1849 a break, 320 yd. wide, was made about $\frac{1}{2}$ mile north of the lighthouse. At high water perhaps 12 ft. of water stood in it. It was later decided to close the breach by a chalk wall. To-day this wall is about 100 yd. from high tide mark at its southern end, and 40 yd. at its northern end. 'Sand dunes have formed near to the Humber, and the Humber water at high tides flows through an opening in this dune barrier at its northern end, and covers the land between the dunes and the chalk wall. This is now a saltmarsh with a channel about 2 ft. deep and a yard wide running down the middle.' The plants are those common to the east coast. The mud flats (5) are rich in animal life. *Spartina townsendii* has been planted; some sewage reaches the flats.

The history and evolution of Spurn Head are being studied by G. de Boer, who allows me to make use of unpublished work. The first known reference to Spurn is in Alcuin's life of St Willbrord, born in A.D. 567 or 568. The saint lived on the headland. After the battle of Stamford Bridge in 1066, the Scandinavian survivors left from Ravenser (= Hrafn's sandbank). From 1235 there is a reasonably continuous record of events, and some ground for thinking that Spurn passes through a cycle of events in two to two-and-a-half centuries. This may be related to the severe erosion of Holderness (**145**).

The south-eastern tip of Holderness is wedge-shaped. The rapid erosion by the sea implies that this tip retreats north-westwards; at the same time the length of Spurn increases (see Fig. 152), and sooner or later the neck is broken through. Should that happen, the headland of Spurn is destroyed, and a new cycle may begin. The present cycle started in the early years of the seventeenth century. A lighthouse (Angell's) was built in 1674; in 1760 another light (Smeaton's) was required because of the much greater length of the headland at that time. However, this decade coincided with the second, erosion, phase of the cycle, and six small lights were built between 1778 and 1851, and the neck was finally breached in 1849 (see Fig. 152). This breach was sealed and groynes preserved the headland, and prolonged its life to the present day.

de Boer thinks that what is known of this last cycle also represents fairly well what happened on earlier occasions, when no artificial means of securing the headland were available. The lighthouse (Angell's) of 1674 was not the first to be built on Spurn. R. Reedbarrow erected one in 1428, which suggests that by that date Spurn had grown considerably since the destruction of Ranvenser Odd, the headland near Hrafn's sandbank, in 1235–1370. de Boer's views of the cycles are shown in Fig. 152. The scheme is not to be taken to mean an exact repetition of events as, for example, in the form of the distal end of the point, but as an indication of their probable trend.

Old Den is a shingle bank inside Spurn. It is first shown on G. Collins's chart of 1684, and was much larger in 1849, thus it suggests a connexion

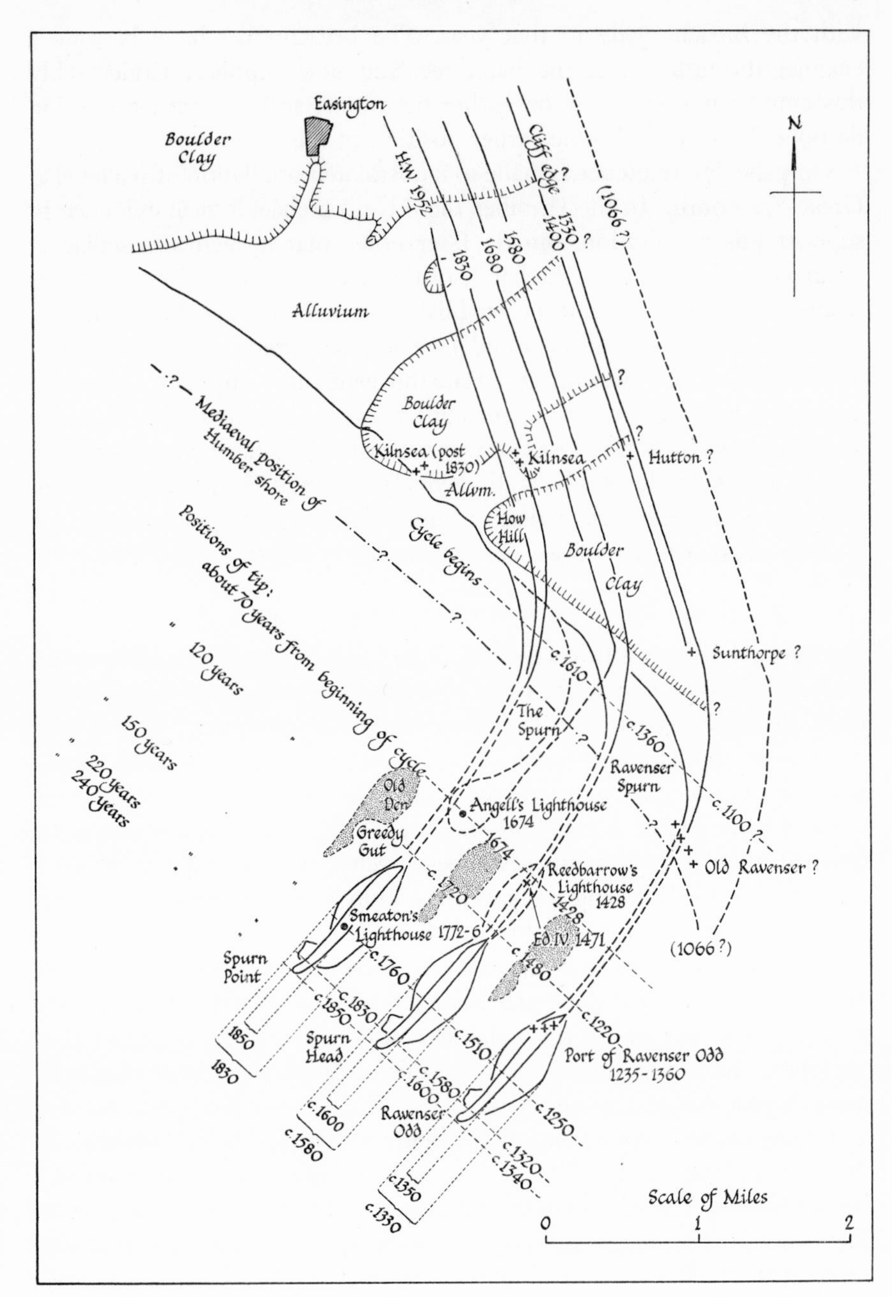

Fig. 152. The historical evolution of Spurn Head. Based with permission on G. de Boer's unpublished work

with the breach made in that year. The breach may have become a channel through which the bank received new supplies. Conceivably this could have happened on earlier occasions, and on that assumption de Boer shows the possible earlier position of the bank.

On page 671 reference is made to a postulated oscillation of sea level at Great Yarmouth. In the Humber there is no archaeological evidence to support this assumption, but de Boer notes that if such an oscillation occurred in the Humber, the banks of the river would have stood higher relative to water level, and the establishment of Frismersk, East Somerte, Penisthorp, and Orwithfleet would have been favoured. 'The chronicles of Meaux Abbey tell how, during the late thirteenth and fourteenth centuries, a losing battle was fought against increasing tidal inundation, but the settlements were finally lost. It was only after 1650, when the reduced rate of rise of sea-level which allowed silting in the Humber to catch up once more, that the reclamation of Sunk Island became practicable.'

(Spurn Head and its Predecessors. *The Naturalist*, Oct.–Dec. 1963, p. 113.)

(3) THE HUMBER FENLANDS

A great deal of light is thrown on this area by an important paper by A. G. Smith (*New Phytologist*, 57, 1958, 19). He has examined the deposits by methods similar to those used by Godwin in the Fenland.

Between Hatfield Moors and Thorne Waste a branch of the Don originally ran east to the Trent, and was joined by the Idle which made the western boundary of Hatfield Moors (p. 418). The Idle was diverted to the Trent in the first part of the seventeenth century, and at about the same time the eastward flowing branch of the Don was dammed near Thorne. Then the drainage of the region first became effective. To the south of Hatfield Moors is a raised bog, through which Lindholm Hillock (Heathy Island) projects. This island was under cultivation in 1727 when the bog itself was impassable. Between Hatfield Moors and the Trent is the Isle of Axholme. Before draining took place this probably resembled Lindholm and, in some ways, the Isle of Ely.

The sequence of deposits in Hatfield Moors and other localities in North Lincolnshire is given in Fig. 153. In the Moors the boundary between zones VII and VIII is at 150 cm. Zone VII is subdivided, and at this division *Plantago lanceolata* is important. This plant, together with fragments of charcoal, may imply the clearing of forest by fire, and since weed pollen begins at about the same time, the obvious conclusion is that this horizon marks the beginning of agriculture in the district.

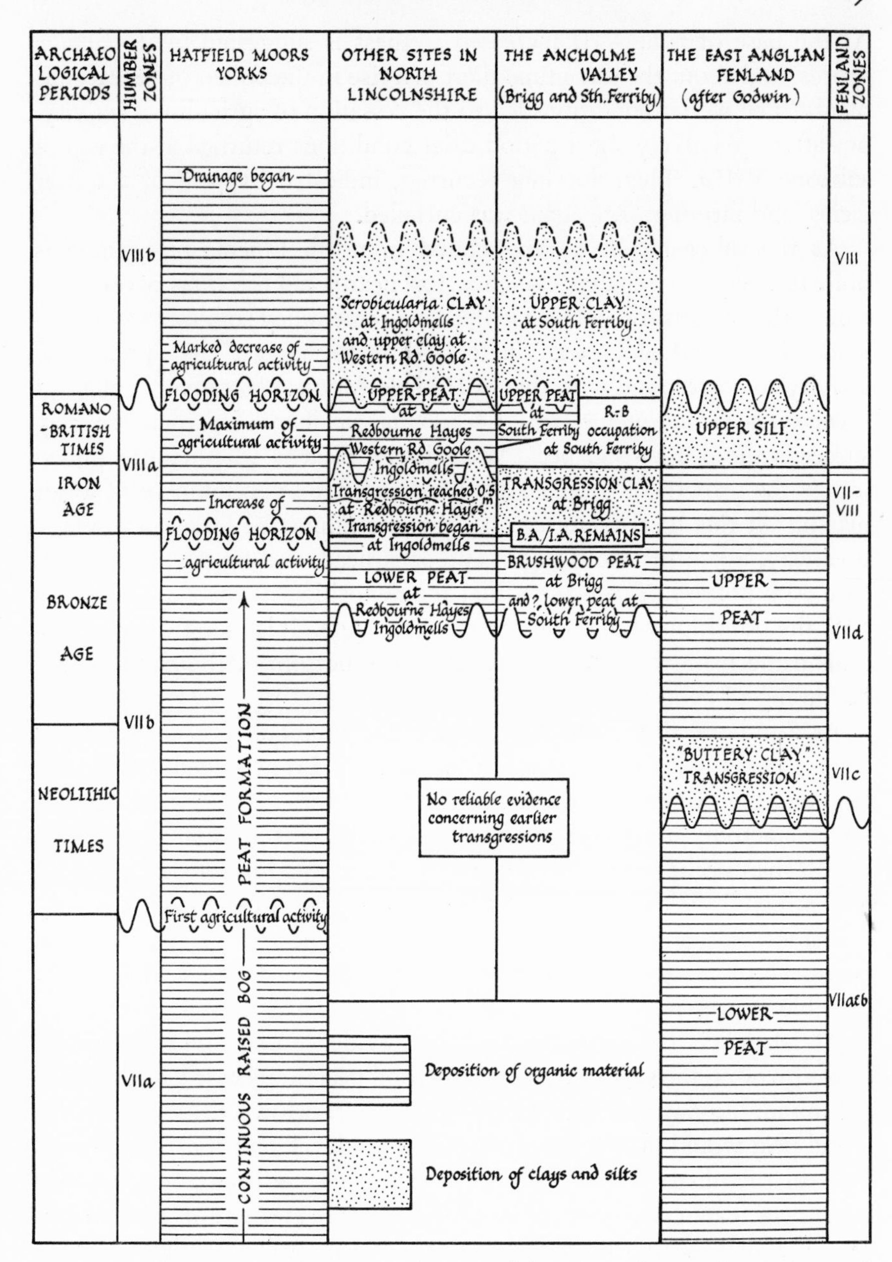

Fig. 153. 'The organic and inorganic deposits of the sites in the Humber region, and the stratigraphical sequence from the East Anglian Fenland are set out in relation to the pollen zone schemes and the archaeological periods. The diagram illustrates the difference of timing of the marine transgressions in the two areas. Wavy lines are used where horizons cannot be dated precisely' (A. G. Smith)

At the base of zone VIII increased wetness is suggested by *Sphagnum* spores, and about the same time there is a rise in the pollen of herbaceous plants. The wetter conditions led to the cessation of agricultural activity, but after a relatively short period drier conditions returned at the top of subzone VIII*a*. Then flooding recurred, indicated by peat of a wetter facies, and farming once again was curtailed.

At the other localities investigated in north Lincolnshire, there is unfortunately no reliable evidence for nearly all the time covered by zone VII. The brushwood peat in the Ancholme valley suggests fen woodland with abundant *Alnus*. There is little to suggest any agricultural activity to correspond with that at Hatfield Moors. Moreover, pollen of *Compositae*, *Umbelliferae*, and *Chenopodiaceae* increases when the clay succeeds the brushwood peat. What is not clear is whether this pollen means the growth of ordinary field weeds or the development of a salt marsh. At this level there have been some important finds—a canoe, a trackway, a late Bronze Age pin and spearhead, and fragments of Early Iron Age pottery. These show that the deposition of the clay coincided with the transition of the Bronze to the Iron Age. It is thought that all the finds came from the top of the peat or the bottom of the clay, i.e. the beginning of zone VIII.

At South Ferriby, which is close to the Humber, it seems that the transgression which laid down the clays at Brigg ended before the beginning of Romano-British times. Pollen traces suggest some agricultural activity in the top part of the clay, and corresponding in time to the Roman occupation. Somewhat later there was further deposition of silt, which is taken to mean either a new transgression, or possibly occasional flooding from the Humber estuary. The upper clay at Western Road, Goole, is probably of the same age, but no rigid correlation is possible at present.

At Island Carr (Brigg) there is good evidence of a rise in sea level. Trees grow on the surface of the brushwood peat; at its base land molluscs are found. To-day it is between -3 and -9 ft. below O.D. The brushwood peat would not have formed below sea level, and consequently sea level in this district in the Bronze Age must have been approximately 9 ft. lower than it is to-day. This figure implies (*a*) no serious compaction of the sands beneath the peat, and (*b*) a minimum value for the suggested change in sea level, since if the tides in the Humber were then as big as they are now, a greater change of level is possible. Smith suggests that the later manifestation of the transgression at Redbourne Hayes 'indicates that some relative rise of sea level took place during the deposition of the clay, and it is not unreasonable to suppose that the transgression was caused, if not entirely, by this kind of change rather than by a change of tidal regime'.

Figure 153 relates the north Lincolnshire sequence of deposits to those of the Fenland. Whilst there is a broad similarity, there are also differences. Near the Humber, submergence was taking place from the beginning of the Iron Age. There may well have been some differential movement between the Fenland and the Humber during the late Bronze and early Iron Ages. Smith points out the possibility that although submergence may have begun in the Fenland at the beginning of the Iron Age, it could not be regarded as a transgression.

At South Ferriby the Romano-British site could have been in use at its present level of +5 ft. O.D. if it were in some way or another protected from the highest tides.

The archaeological finds at the base of the clay come from the Bronze Age–Iron Age transition and make it more than likely that this level represents the opening of the Sub-Atlantic period. The most recent dating of this transition is 400 B.C. in this country. Hence the second flooding of Hatfield Moors was of this date, and a marine transgression in the Ancholme Valley began at about the same time. Furthermore, there was some flooding at South Ferriby at the time of the uppermost flooding horizon, even if this, at South Ferriby, did not reach the proportions of a transgression.

(4) THE LINCOLNSHIRE COAST

The storm surge of 1953 had profound effects upon the coast of Lincolnshire, much of which is man-made; extensive sea walls are virtually continuous from the north part of Skegness to the northern end of Mablethorpe. Erosion has long been serious in Lincolnshire, and A. G. B. Owen (*Lincs. Historian*, No. 9, 1952, 330) summarizes much of what is known on this topic in the past. The stretch between Mablethorpe and Theddlethorpe to the southern parts of Skegness has always been the most vulnerable. At Mablethorpe itself there were originally two parishes, and St Peter's Church stood about 1 mile north-east of St Mary's which is still intact. It is recorded that St Peter's was seriously damaged in 1287. Although details are lacking, it is known that severe flooding took place in, for example, 1335, 1430, 1443, and 1529. An account of 1602 implies that St Peter's Church disappeared 50–60 years earlier. At Trustthorpe it is generally assumed that a church has been lost. In 1834 it was thought that it stood about $\frac{1}{4}$ mile from the coast. Much the same is true of Sutton. It is significant that the present Ordnance map shows that parts of Sutton, Trusthorpe, and Mablethorpe are less than 7 ft. above mean sea level, or more expressively, 3 ft. *below* ordinary high water mark (**146**).

Huttoft and Anderby have been relatively unaffected by erosion. A severe storm in 1571 did much damage at Chapel St Leonards. Here, and elsewhere on the Lincolnshire coast, Armstrong's map of 1776–78 is helpful. Around Chapel St Leonards he marks considerable areas of 'salt marsh', but no trace of them is shown in the Ordnance map of 1824. There is reason to think that these marshes may have been $\frac{1}{2}$ mile wide. At Ingoldmells there were sea walls as early as 1272, and there are records of land lost in the first half of the fifteenth century. Commissions of Sewers in 1430 and 1626 refer specifically to loss hereabouts, and a storm in February 1735 caused great damage and flooding. Owen is of the opinion that the original Ingoldmells Point was lost by erosion, and that the name was transferred some 200 yd. farther south to the drain outfall called in 1824 Ingoldmells Out End. He also suggests that there may have been a ness at Skegness. There has certainly been much erosion where the parishes of Ingoldmells, Winthorpe, and Skegness meet. Despite the fact that Leland refers to a 'great haven' at Skegness, Owen thinks that local pride exaggerated its importance. But a haven of sorts did exist and prospered in the mid-fifteenth century when it had both a coastal and a Baltic trade. There are several records of storms and damage at Skegness; in 1526 the church and much of the parish were submerged, and erosion. seems to have continued for another 100 years or more. In 1639 dunes at Hobthirst Hill in the King's Low Marsh were destroyed. However, after approximately this date, erosion on the south side of Skegness ceased, and presumably conditions similar to those existing to-day came into being.

The 1953 storm demonstrated not only what can happen in a few hours, but also what has happened, with varying degrees of intensity, many times before on our east coast. Before the storm there were extensive sandy beaches in Lincolnshire. The sand (see p. 421) was not always thick, but it hid from view the erosion in the clay[1] below. The storm swept the sand away and exposed the substratum over wide areas. Some of the beach material was swept inland, and some houses alongside the coast were half-buried in it; some was carried seawards. The sea walls were breached and great damage was done to houses by direct wave attack;

[1] The clay is often the marsh clay; peat is often exposed.

many more were affected by flooding. King and Barnes made an invaluable series of measurements of the beaches soon after the storm. They surveyed 23 profiles most of which were resurveyed eight or nine times, and some even 27 times, so that we have an accurate picture of the changes. Figure 155 shows that the beaches are narrower and steeper in the mid-parts of the coast. At and beyond the northern half of Mablethorpe and south of Skegness the profiles are higher and flatter, and the erosion was less severe, although dunes were cut away in the north. Figure 154 shows the amounts gained or lost on these profiles. The height of the top of the beach varies, and the extensive engineering works, walls, and groynes, that were erected soon after the storm have had no small effect on the replenishment of the beaches. At Mablethorpe and Trusthorpe the gains have been considerable since there is an abundant supply of sand to the north. Since, however, measured profiles did not exist before the storm, it is impossible to say if all beaches have fully recovered.

> On eight lines of profile in June 1955, there was more sand on the upper beach than had been recorded by any of the earlier surveys, but on ten lines a maximum had been reached at an earlier date. A maximum was found on the whole beach on ten lines of profile in June, 1955, but a maximum had been reached earlier on another seven lines.... The continued accretion may be partly due to the effect of the new groyne system... but these beaches are presumed to be still in process of recovering.[1]

An analysis of the profiles suggests that four fairly distinct types were present: (*a*) Sandy beach with ridges and runnels, commonly backed by dunes. This type usually wedges out both landward and seaward (see no. 2 in Fig. 155). (*b*) Beach with ridges and runnels, exposing the clay base in places. There are often sea walls at the back of this type. The surface of the clay base may be convex in the upper part, and flattens lower down. At Trusthorpe the break in slope occurred at 0 ft. O.D., and at −2 ft. O.D. at Chapel Point North. King and Barnes think that the profile is partly explained by wave erosion of the clay base, which is deeply grooved perpendicular to the coast, thus implying erosion at those times when the beach was stripped in storms. (*c*) Beaches with a smooth profile on which the clay base has been exposed. The profile at Mablethorpe is regarded as anomalous, since there has been no indication of ridge development. This may be because there is a powerful outfall on the lower beach. (*d*) Beaches with a fair amount of shingle. Shingle is relatively uncommon in Lincolnshire, and there is more near Ingoldmells than elsewhere. This leads to the greater steepness of the middle and upper beaches there-

[1] The central and northern profiles appeared to have reached equilibrium by 1957.

abouts. At this point it is relevant to describe the sweep zone, a concept introduced by King and Barnes. It may be defined as 'that portion of the vertical plane perpendicular to the coastline, within which movement

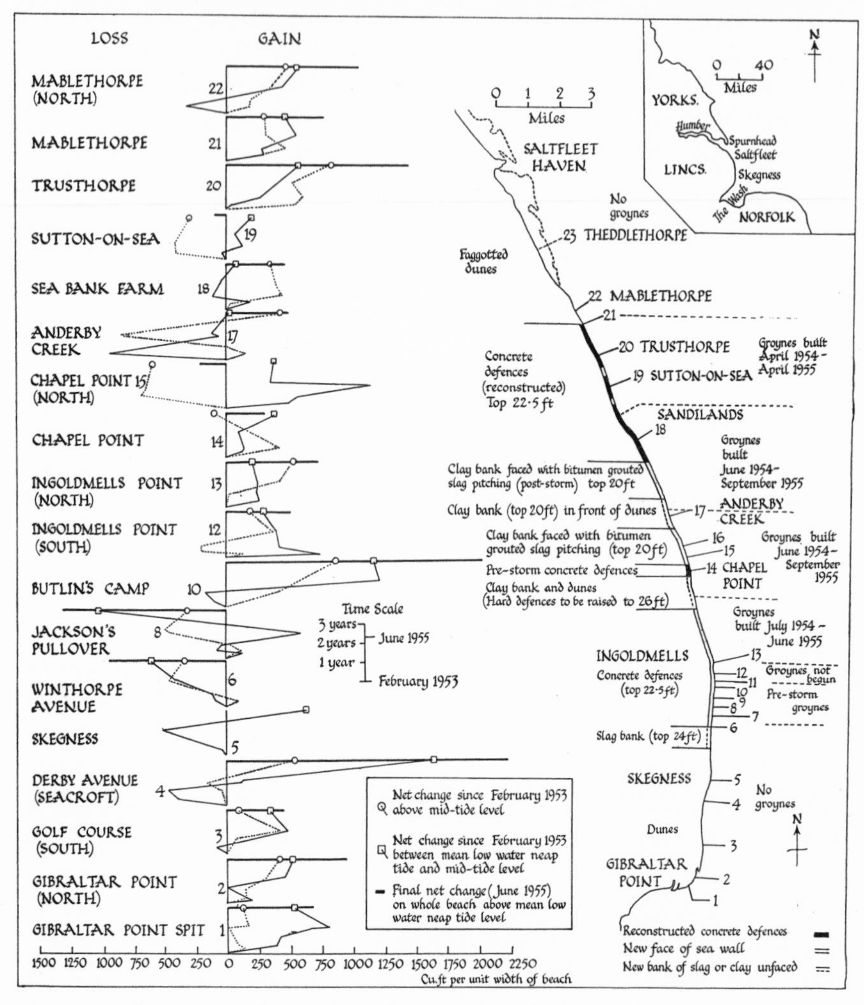

Fig. 154. Changes in volume of beach material on the lines of profile since February 1953, and coastal defences in 1955 (from C. A. M. King and F. Barnes, *op. cit.*)

of material may take place under wave action'. In practice the profile is limited seawards by low water neaps, or possibly a little lower.[1] Move-

[1] The sweep zone concept would be strengthened if, by soundings or other means, the profiles could be extended farther seawards.

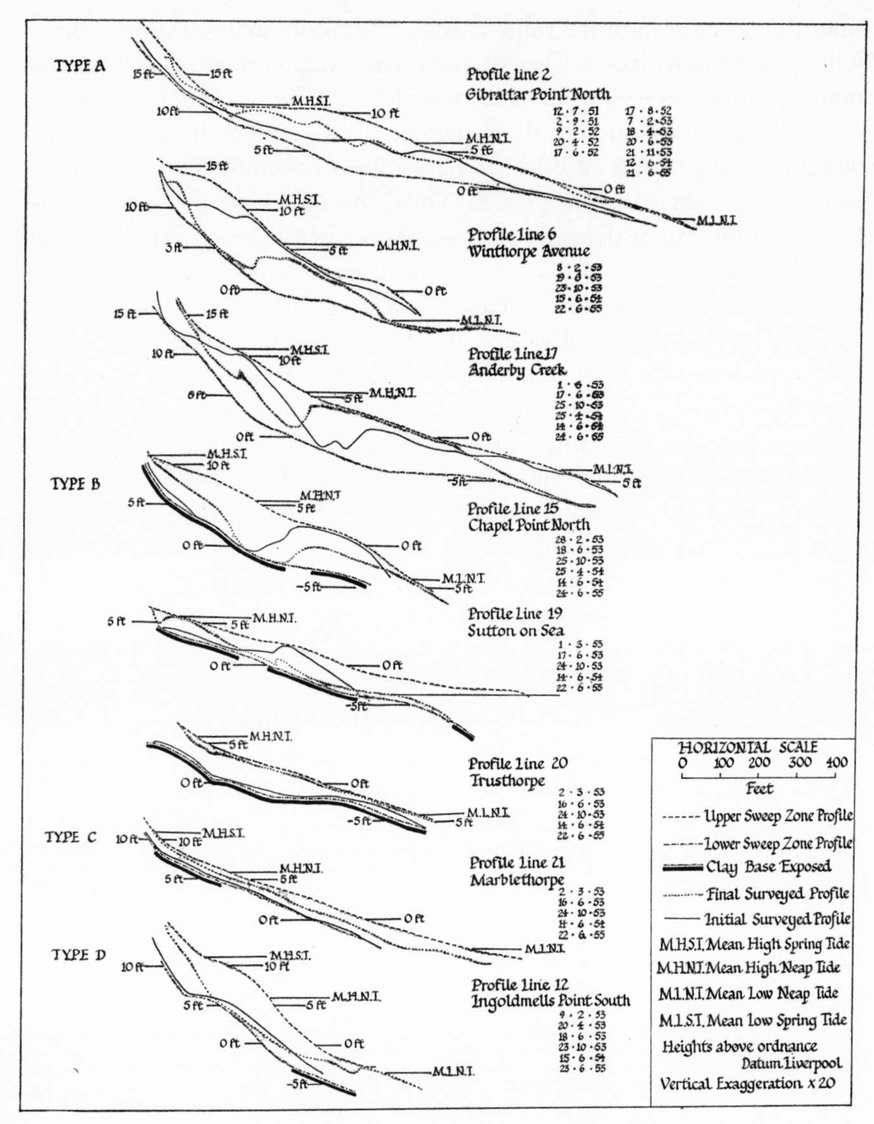

Fig. 155. Sweep zone profiles, Lincolnshire
(from C. A. M. King, *op. cit.* 1959, p. 201)

ment of material will go on farther seawards, but it is difficult but not insuperable to make any accurate measurments of it. The base of the profile is composite, as is also the upper profile. Both of these are found by superimposing successive surveys along a given line. The top line is a smooth curve through the ridge crests; the bottom line is a similar curve joining the runnel bases. Clearly, the greater the number of profiles, the more accurately the sweep zone can be shown. The sweep zone as shown can only refer to relatively short periods. Over a much longer time the beachward migrations of offshore banks must have effect on the waves, and so on changes in the profile. Thus, the nature of the sweep zone must to some extent depend on the character of the beach profile, and if the solid base is exposed that, in the short term, is its base.

The sweep zone is not so much a new concept but rather a more logical and quantitative means of assessing beach changes. That a beach often builds up in normal weather, and is combed down in storms is a commonplace and, in a sense, is the kernel of the sweep zone concept. The sweep zone is a help in considering beach changes, and although it exists on any beach, it is of particular significance to those of Lincolnshire. If the profile is low on a sandy beach, the export of sand from that beach is limited. On the other hand, a high profile means that a considerable quantity of sand can be removed from it by natural means.

Beach profiles were analysed again in 1958 and 1959 (*E. Midland Geogr.* 12, 1959, 46). At Seacroft, between Skegness and Gilbraltar Point, the profile showed a marked growth both upwards and outwards. It illustrates how the beach grows by the development of beach ridges. Farther north the groynes have helped, although they naturally have a selective effect. Profiles 15, 19, and 21 show increases, but 17, at Anderby Creek, and 18 show only small changes since groynes to the north affect the supply of beach material. Similarly, the concrete points of Ingoldmells and Chapel Point seem to have had adverse effects. The conclusion is reached that the movements in the Lincolnshire beaches are closely related to defence works in the north, and to the displacements of offshore banks in the south.

(See also *Zeit für Geomorphologie*, Neue Folge, 8, 1664, 105.)

(5) GIBRALTAR POINT

Since the war Gibraltar Spit and Point have been the scene of much physiographical and ecological work under the aegis of the University of Nottingham and particularly by King and Barnes (*E. Midland Geogr*, 15, 1961, 20).

The spit is, in one sense, the southern end of the beaches dealt with in the previous section. In the 4 years ending 1959, it was found that the relatively static area of the spit was in accord with the behaviour of the beach in front of it. The sweep zone remained more or less stationary from 1952 to 1959. But just to the north of the spit a steady seaward growth has been traced. This affects conditions farther south because it decreases the amount of material at the top of the beach which can be moved southwards by wave action.

Figure 156 shows the general nature of Gibraltar Spit and Point; together they make a link between the Lincolnshire coasts of the North Sea and the Wash. The point is the distal part of the spit. If the growth of the whole feature is studied from maps and other records over the last two centuries, it is apparent that near Gibraltar the coast has extended eastwards rather than southwards. In fact, a number of separate spits have developed and disappeared. The present feature (1957) is a young one and its crest is in much the same position as the low water mark of a century ago. Armstrong's map of 1779 shows a long spit running southwards from near the old inn at Gibraltar, and in line with the western dune ridge of 1957. In 1790 it was washed by the waves, and deflected the Steeping river for about a mile (**148**).

In the earlier part of the last century there were movements in the offshore banks which enabled south-easterly waves to attack the point. The Inner and Outer Knock banks still served as a protection farther north. Salt marshes were present in 1870 and it is thought that they were in existence in the 1820's; the drainage line north of the point indicates this. The Ordnance maps of 1904 show that the eastern ridge of dunes had formed by that time, and that there had also been a southward shift of the Outer Knock which had encouraged some erosion south of Skegness.

> The present frontal dunes near the Point have been built upon an independent shingle bank separated from the inner lines of the eastern dune system by a former saltmarsh now covered with sand. These dunes were not present when the six-inch O.S. map of 1904 was surveyed and their vegetation suggests that they may not be more than about 25 years old. Their establishment as permanent features indicates that the beach in front of them has been raised by marine deposition, and this process must have been associated with the initiation of the present spit to the south, which can hardly be older than the dune line. (*E. Midland Geogr.* 8, 1957, 24.)

The spit, whilst resembling places like Scolt Head Island, Wells, and similar localities is distinct in many respects. It consists of alternating layers of shingly sand and sandy shingle, and the amount of shingle in the form of layers increases on the seaward side. Sand is easily washed inwards over the seaward ridge, and in this way the upper beach is added to. Additions also occur by the welding on to the outer beach of other ridges travelling in from seawards and from the north. These additions are not

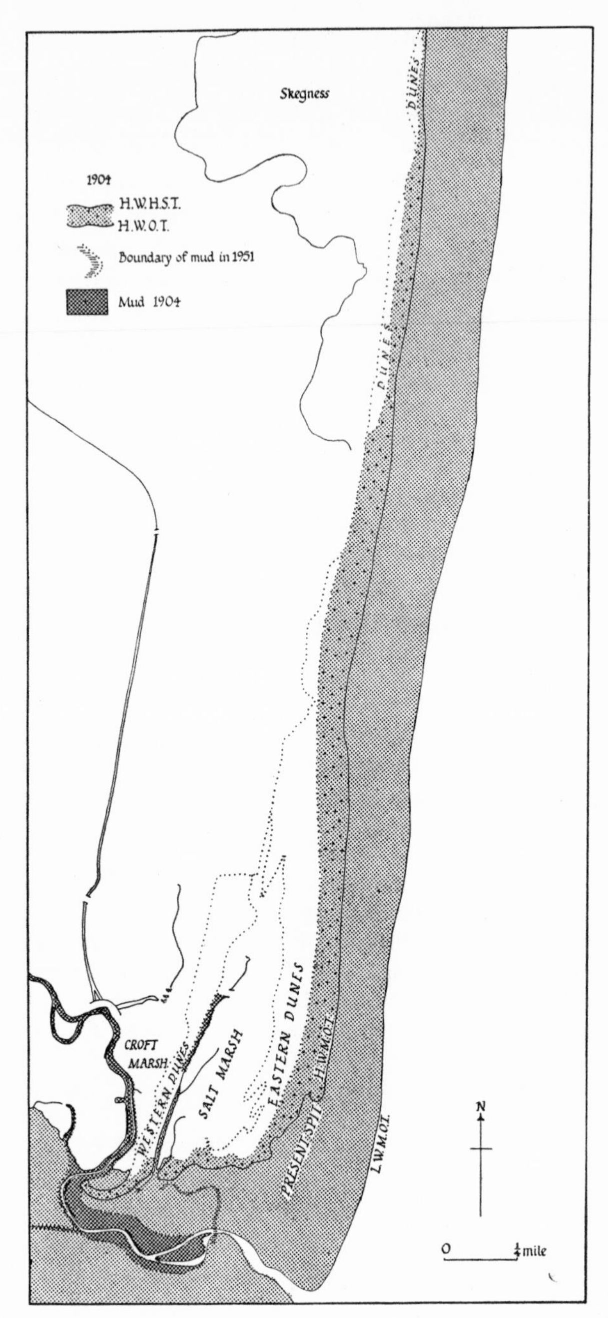

Fig. 156. Gibraltar Point, position of spit indicated (outline based on Ordnance Survey 1949)

necessarily permanent. Although temporary widening occurs in this way the spit chiefly broadens as a result of material thrown over on to the protected landward side. This results almost wholly from wave action and to some extent from wind blown sand. Just as on any other comparable foundation, beach-drifting (in this case from the north) is the main agent by means of which the beach is replenished, although at times there may also be a bodily southward movement of an already formed ridge. If the material coming from the north exceeds that carried away to the south, there is a temporary widening. These changes depend upon the direction of approach of the dominant waves, which in their turn are influenced by the height, shape, and position of the offshore banks, which are thought to show a slow southward migration. In the last quarter of a century the coast has grown outwards between Skegness and Gibraltar Point, and this has had some effect on the alignment of beaches.[1]

'Former spits...appear to have undergone a cycle of development in which accumulation, widening and lengthening at an irregular rate were followed by the development of a hook form and shortening, and this in turn by a final phase of gradual decay as shelter on the eastern side diverted incoming material from the spit. This cycle is related to the way in which the coast has built out eastwards.' There is also an east to west shore at the southern end of the area of accretion. 'There is no reason to suppose that the present spit is not undergoing a similar cycle of development' (*E. Midland Geogr.* 8, 1957, 31).

The growth and evolution of the spits and beaches have created conditions favourable to the development of salt marshes. The enclosure of Croft marsh dates back to the sixteenth century, and some reclamations are said to date from 1608–10. Since the travel of beach material is southwards ridges bend round to the west at their distal ends, and practically enclose the marshes, which are floored with laminated sand and mud which soon becomes colonized by plants, especially *Salicornia maritima*. The mature salting (Fig. 156) is really a wide strip marsh of earlier date, and protected by the storm beach, south of which a new marsh is growing.

Some interesting measurements of accretion recorded between 1951 and 1959, have been obtained on the New Marsh. Accretion reaches 3 ft. near the inner edge of the distal end of the spit, elsewhere more than a foot has accumulated in 8½ years. In the mid-part of the marsh there was an

[1] On these sandy beaches the dominant waves are long swells from the north, and *not* storm waves.

accretion of almost 1 ft. in the lowest vegetation zone; this amount decreased outwards, and on the landward fringe was replaced by erosion. But the figure of approximately 1 ft. in the vegetation is more than three times the amount measured on Scolt Head Island (p. 654) and in the Dyfi estuary (p. 533). Near the main creek silt has thickened at the rate of 5 cm. a year. The careful investigations made on these marshes show, *inter alia*, the great effect of *Spartina townsendii* as a trapper of silt.

(6) THE FENS: RECLAMATIONS IN PRE-DRAINAGE TIMES

East Fen, West Fen, and Wildmore Fen lie behind the coast between Wainfleet and Boston. Kimmeridge clay underlies Wildmore and West Fens, which consist of silts; peat occurs in East Fen. In all the drainage is poor. The villages of Stickford, Stickney, and Sibsey stand on a low ridge[1] which is the boundary between West and East Fens. The former, with Wildmore Fen, stands about 5 ft. above mean sea level, whereas East Fen is but 1 ft. above that level. There is another line of villages along the coast and situated on silt. In Armstrong's day the differences between West and East Fens were more marked than they are now. In East Fen there were a number of lakes called deeps; the surrounding fen was inundated only in winter. There were no lakes in West Fen. This part of Lincolnshire was one of the last to be drained (G. J. Fuller, *E. Mid. Geogr.* No. 7, 1957, 3).

Some work by H. G. Hallam (*The New Lands of Elloe*, No. 6, *Papers*, University College, Leicester, 1954) has increased our knowledge of the Wash area considerably. He thinks that the Roman Bank was the main one of all Elloe, at least by the late thirteenth century (see Fig. 96). Near Holbeach it may have been built piecemeal, but by the time of the Conquest it seems to have been one bank. The earliest reference to it is in 1182–88. It is argued that the position of the salterns relative to the bank throws light on its age. In medieval times the method used for salt-making in this district necessitated the tide flowing 'over the greva or sandacre from which the salters took the salt-saturated sand. The cotes or hogars still stand along the sea-banks of Holland....' The dating of the salterns throws light on the age of the bank. In Fleet, the salterns are outside the main bank, and in Domesday Fleet is recorded as having two

[1] According to A. Straw (*E. Mid. Geogr.* No. 9, 1958) the moraine at the limit of the Newer Drift.

salterns. Thus it appears that the sea bank is pre-Conquest. The Fleet and Holbeach parts of the bank enclosed an estuary and protected a great deal of new land from tidal flooding.

Outside the bank there are many traces of pre-Conquest activities. There is no doubt that salt-making flourished in the twelfth century and the salterns near Fleet (in the park of Hovenden House) and alongside the bank towards Gedney are regarded as undoubtedly those recorded in the Domesday Survey. On the other hand, the earliest recorded enclosure on the seaward side of the bank appears to date from the end of the eleventh or from the beginning of the twelfth century. This was in Sutton and Lutton. Others were made in Holbeach in 1142–75 and many more a few years later.

Now, and in the past, accretion has usually been rapid near the outfall of rivers, and it appears to have been markedly so from the Conquest to the end of the thirteenth century. The accumulating silts were frequently enclosed, the size of the enclosure varying perhaps with individual or communal activity. Elloe itself was largely reclaimed in this period, and an important bank was built which was the major line of defence against the sea up to the time of Charles II. Even more important during these centuries were the extensive enclosures that were made southwards into the fens: Hallam estimates that about 50 square miles were enclosed between 1170 and 1240. The same process was taking place in other parts of Holland and even at Wainfleet. It is thought that the tidal area enclosed amounts to about 100 square miles. This was additional to that reclaimed directly from the sea, and emphasizes that by no means all the drainage in the Fenland took place in the seventeenth century.

The general relations between the settlement and enclosure of the silts and the movement of the coastline is shown in Fig. 151. Bordering on the peat fens is a belt of Romano-British settlement in which there are abundant remains of a salt industry. The sites have been revealed by air photographs. In this belt there are low clay areas which grade southwards into the peats, and higher parts and more silty ones which gradually merge into the homogeneous silts nearer the Wash. There is often quite a considerable difference in height between these two; the former may be as low as 5 ft. O.D., the latter 8–10 ft. and locally even 12 ft. O.D. The main centres of occupation were closely associated with numerous small water courses, which eased the problem of transport for both fuel and salt.

As time went on the settlement crept seaward. By the second and third centuries the silts were more and more occupied. Salt production apparently

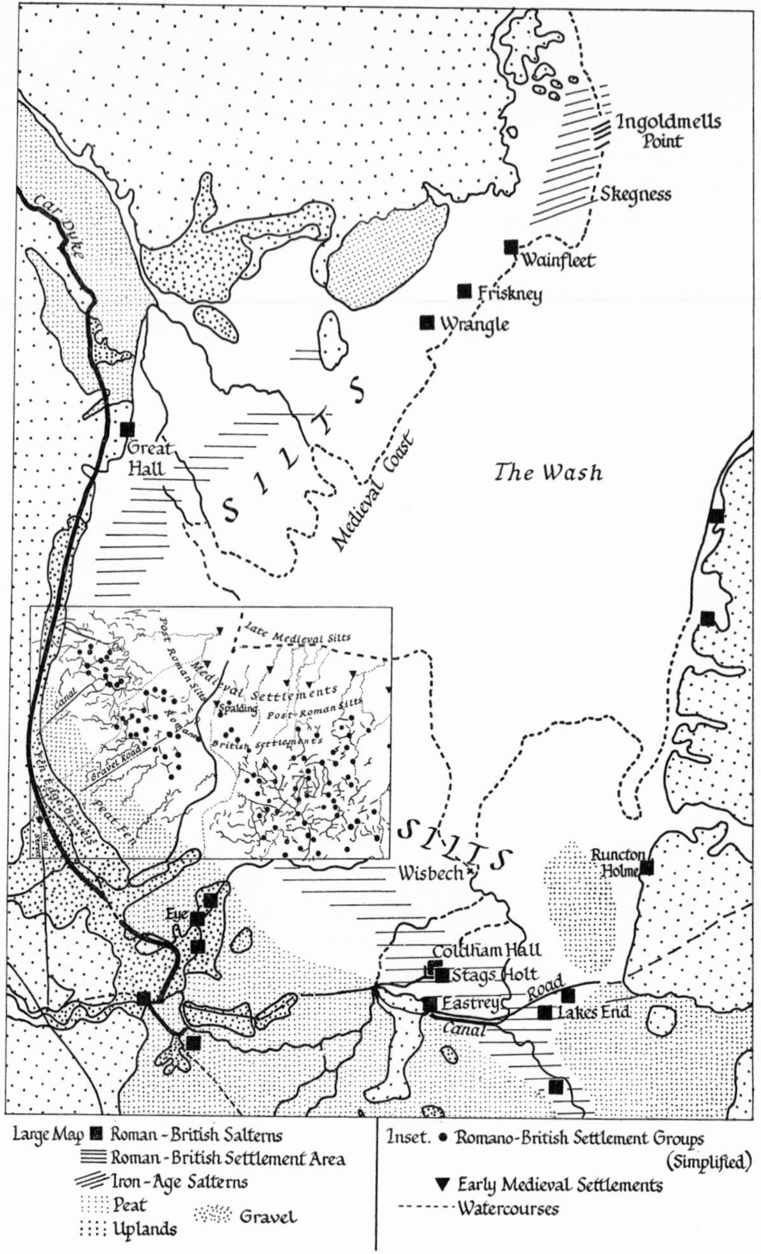

Fig. 157. General relations between Fenland settlement and enclosure of the silts and movement of the coastline (based on S. H. Hallam, *op. cit.*)

lessened, and gave place to pastoral activities. Since accretion is more rapid east of the Welland, and because this also seems to have been so in past ages, the successive belts of settlement, Romano-British, Post-Roman, early, and late Mediaeval are each wider in the east than they are in the west, especially north of the Welland (see S. H. Hallam, *Lincs. Archit. & Arch. Soc. Reps. & Papers*, 8, 1959–60, 35; and *Antiquity*, 35, 1961, 152).

(7) MODERN RECLAMATIONS: THE WASH

Modern measurements along the south bank of the Wash afford striking figures of the rates and nature of the accretion. Inglis and Kestner (*Mins. Proc. Inst. Civ. Engs.* 11, 1958, 435) investigated this problem especially between the outfalls of the rivers. The rise in mud level was recorded at a number of sampling sites, cross-sections were taken, and the seaward extension of the marsh measured from O.S. maps and Admiralty Charts. Near the Witham outfall 'the rate of 45 feet per ten years between 1828 and 1871 (increased) to 136 feet per ten years after completion of the Witham Outfall Cut, in the period 1887–1903. It reached its peak rate of 350 feet for ten years between 1903 and 1917/18, and dropped to 170 feet in ten years between 1917/18 and 1952.' The Witham outfall has had a great effect. More than 6 miles away from it the advance of the shore has been small. It appears, then, that engineering works, even if not in any way intended to cause accretion, nevertheless have that effect near the outfalls. At Butterwick there was an advance of 260 ft. in ten years; locally advances up to six times as great have been recorded. Near the Nene there have been noteworthy changes. Its old channel was abandoned in 1830, a new cut was taken through reclaimed land, and the outfall was made to the west of the estuary. The area east of the new channel changed rapidly. It silted up, became grass-covered, and was reclaimed up to Sharpe's Bank in 1867. Not until after 1900 were additional smaller enclosures made, in 1917, 1924, 1952, and 1954 (**149**).

At Wingland there has been rapid accretion. This place lies between the outfalls of the Nene and Ouse. In Fig. 158 curves (*b*) and (*c*) at first resemble curve (*a*). Then a sudden change occurs. In 1852 a cut (Marsh cut) was made along the left bank of the Ouse, and it had the effect of concentrating flow near the Wingland foreshore. This led to extension of the marshes there, whereas little if any change occurred near the Ouse. Curves (*c*) and (*b*) show not only a slackening, but also retreat of the saltings, the result of erosion related to the new low-water bed of the

Ouse in its effect on the more seaward and north-eastward parts. The effect is first noticed on curve (*c*) and later on (*b*). From about 1904 new trends began. Silting and reclamation on the east of the estuary reduced the flow in the estuary itself and eventually had the same effect on the New Channel. 'As in the case of the erosion phase, which slowly spread from

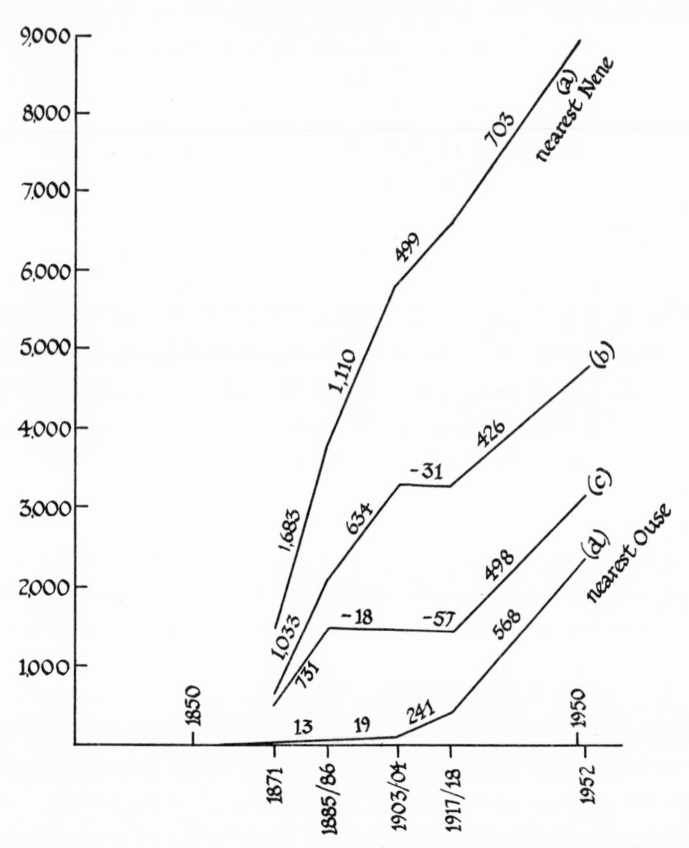

Fig. 158. Diagram showing growth of saltings at Wingland. The figures on the lines indicate the rate of advance in feet each decade (after Sir C. C. Inglis and F. J. T. Kestner, *op. cit.*)

curve (*d*) to curve (*c*) and then to curve (*b*), accretion appeared first in curve (*d*) right in the Ouse, and gradually spread, some years later, to the next seaward curve (*c*) and finally affected even that toward the Nene, curve (*b*).'

Measurements were also made of the vertical rates of accretion. At Wingland a mean rate of 3·07 in. per year was obtained, or about

$2\frac{1}{2}$ times as much as that found at Butterwick. Accretion was continuous in the upper parts of the foreshore, but lower down periods of accretion and erosion alternated. Engineering works, either because they reduce the volume of flow, or because they dissipate energy as a result of altering the natural flow, cause greater deposition, less scour and consequently considerable accretion. Under natural conditions there is usually maintenance of the regime, scour and deposition more or less balancing each other. Even a small engineering work may cause a big change.

Inglis and Kestner also investigated changes in the Wash. The area of the Wash is about 20 square miles, and since the mid-years of the seventeenth century some 75,000 acres have been reclaimed. In the last *complete* survey of the Wash (1917–18) the total area covered at high water was 183,000 acres and 95,000 were exposed at low water. Saltings occur all round the Wash; seawards they pass into mud flats, and beyond there is a belt of sand. In the deepest areas some mud patches are found. Experiments were carried out to trace the movement of sand. To do this plugs were put down in selected places. Small cores, 2 in. long, were taken from the sand flats and replaced by 'a creamed suspension of silica flour'. There was no doubt that surface movement of sand was taking place, but there were no appreciable changes in configuration over periods of a few years.

Possible longer term variations from surveys made in 1828, 1871, 1917–18, and a partial survey of 1943 were also investigated. But little change was noted, and that only after a careful study of contours in some particular locality. Boston South Channel, in 1828, joined the outfalls of the Welland and Witham and before it reached Lynn Deeps there was a bar across it. Rather similar detailed changes were noted in the channels between Pandora and Ferrier sands. Even if no permanent change is measurable, it does not mean that there is no movement, but rather that oscillatory movements are balanced, so that no change is apparent. Inglis and Kestner express the matter thus: 'An observer flying over the Wash at low water today, equipped only with the 1828 map, would have little difficulty in identifying most of the features shown on that old chart. He would recognize the characteristic shapes of Gatt, Roger, and Long Sands...Thief, Seal, Pandora, and Ferrier Sands. The main features can still be distinguished on the...chart of 1693, although the map is somewhat distorted.' (See also F. J. T. Kestner, The Old Coastline of the Wash, *Geogr. Journ.* 128, 1962, 457, and especially the discussion following the paper.)

(8) THE LAST GLACIATION AND ITS RELATION TO THE EAST COAST

In 1951 Farrington and Mitchell (*Proc. Geol. Assoc.* 62, 1951, 100) described the drift country north of Flamborough Head. They concluded that, on the basis of topographical details, small steep-sided valleys and hollows and the general freshness of the features, the area was characterized by the Newer Drift. In contrast, they believed that south of Flamborough Head and in North Lincolnshire, the topography was older in appearance. In Norfolk, the Cromer Ridge was regarded as older than the drift north of Flamborough. The irregularity of the Cromer Ridge on its northern side was ascribed to erosion and not original deposition. At Holkham (Norfolk) the Hunstanton brown boulder clay was also considered old. Farringdon and Mitchell take as their standard for freshness the appearance of the Midland general glaciation of Ireland.

They contend that the moraine north of Flamborough Head is a recent end moraine marking the southern limit of the last major ice advance in this country, and that it is to be correlated with the end moraine in south Ireland and that in Jutland.

Valentin studied Holderness (*Abhand. Geogr. Inst., Freien Univ. Berlin*, 1937). On the continent the end moraine of the Weichsel glaciation is conspicuous, and Valentin thinks that the key to unravelling the English glaciation is to try and trace the corresponding moraine on this side of the North Sea. To do this he considers the features on the North Sea floor (Fig. 159) and since some of these are akin to small valleys in Holderness, he draws a line marking the limits of the east English glacier at what he regards as the maximum of the last glaciation. This line excludes north Norfolk, Lincolnshire, and most of inland Holderness and Flamborough Head. In short, coastal Holderness and a strip on the north side of Flamborough Head are regarded as fresh glacial topography, whereas the rest is subdued and older, and probably Warthe in age.

A more recent interpretation is that by Suggate and West (*Proc. Roy. Soc.* B, 150, 263, 1959). These authors are familiar with eastern England and have studied the effects of glaciation on it in detail. The full arguments for their conclusions need not be given here, except to say that they depend on careful stratigraphical sequences and radio-carbon dating of key deposits. Farrington and Mitchell and also Valentin based their arguments largely upon the fresh or old appearance of the glacial topography. This is an argument of equivocal value since freshness is a relative term and what is, in fact, still fresh glacial topography can be easily obscured by the filling of hollows by peat mosses and other forms of post-glacial sedimentation. This, Suggate and West maintain, applies equally well to Holderness and Lincolnshire. Valentin emphasized subglacial channels, now gravel-filled.

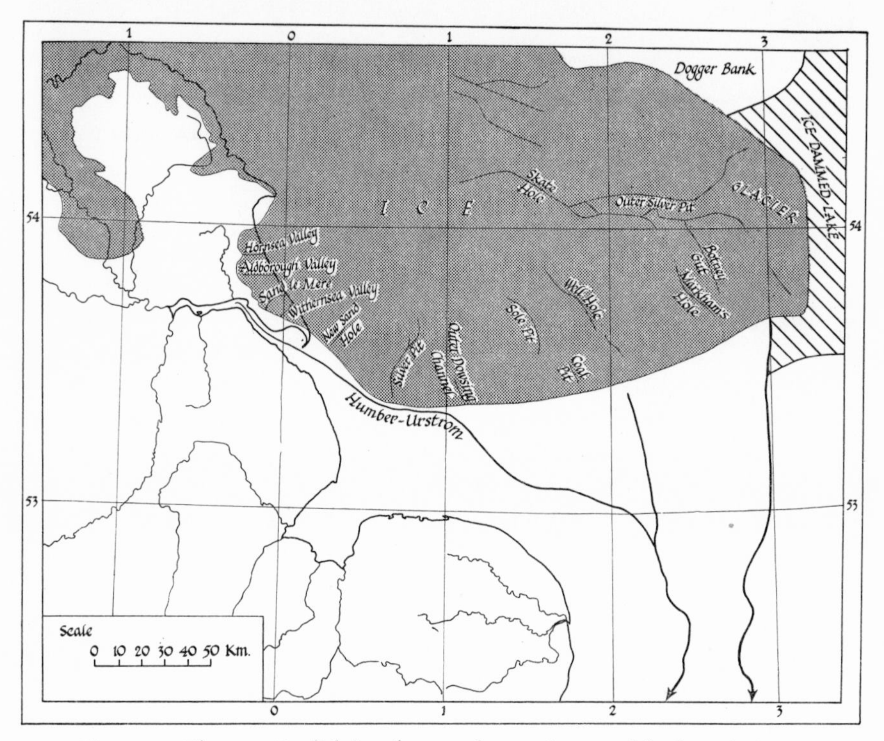

Fig. 159. The east English ice sheet at the maximum of the last glaciation (after H. Valentin, *op. cit.*)

...they were probably developed in eastern Holderness because the slope of the pre-Hessle topography was in the direction of ice movement, whereas in eastern Lincolnshire and, indeed, north of Flamborough Head the slope was not so alined with the ice movement. In north Norfolk the maturity of the Hunstanton boulder clay has been overstressed by Farrington and Mitchell. Hummocks and minor closed hollows near Holkham, and an excellent example of an esker in Hunstanton Park are features that could hardly have been preserved through an interglacial and a glaciation. (Suggate and West.)

Thus, taking into account both topography and stratigraphy, as well as certain late glacial deposits in north-eastern Lincolnshire, Suggate and

West are convinced that (see Fig. 160) the last glaciation in England included Holderness, eastern Lincolnshire, and northern Norfolk. In the Fens, boundaries of deposits are obscured as a result of aggradations since glacial times.

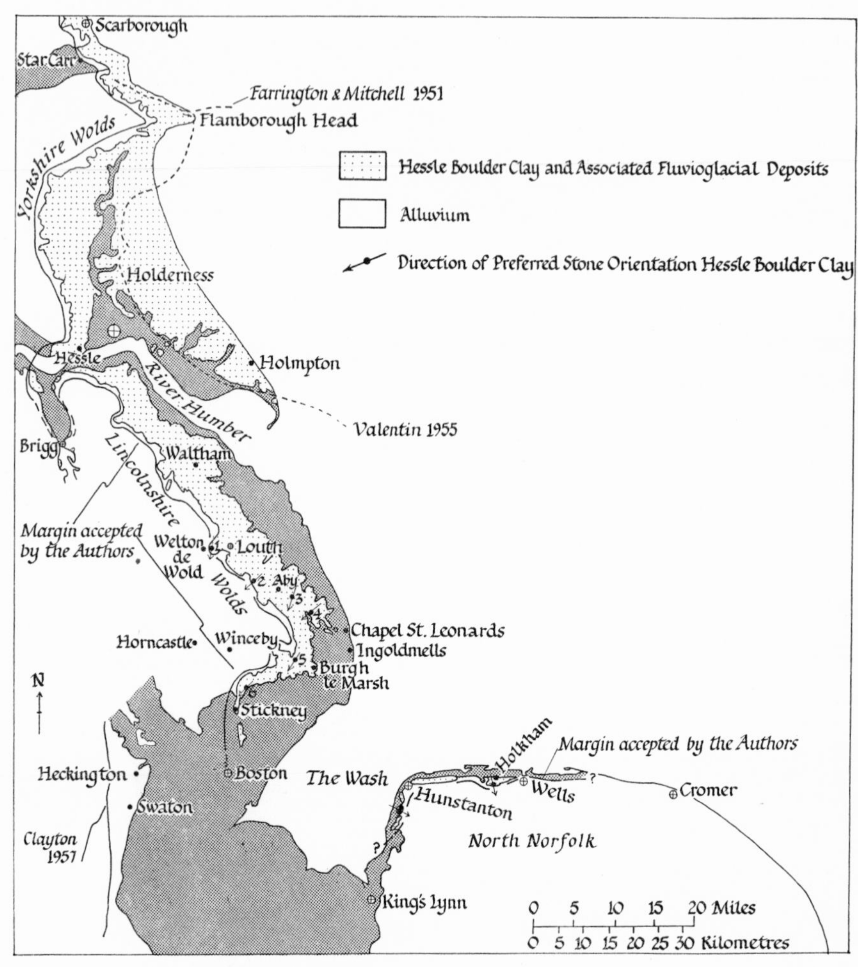

Fig. 160. Sketch map of coast of Holderness, east Lincolnshire and north Norfolk (based on Suggate and West, *op. cit.*)

In Norfolk the Hunstanton brown boulder clay is not known to rest on the Gipping, and so its assumed 'last glacial' age is based on purely lithological and morphological grounds. The only chalky boulder clay in Lincolnshire is regarded as Gipping, and in one place (Welton-le-Wold) the Hessle clay rests on it. In Holderness the Hessle rests on pre-Quaternary rocks near the Wolds and on purple boulder clay farther east. The age of this purple clay is uncertain, and all that can be said of the Hessle is that it 'was deposited by the last ice to reach the area'. Since the Hessle occurs in both Holderness and Lincolnshire it is logical to regard the occurrences as of the same date and there is no good reason for not regarding the Hessle and the Hunstanton brown boulder clay as equivalent. These are suggestions and not proved facts; the matter is still an open one. In east Lincolnshire A. Straw (*E. Mid. Geogr.* No. 9, 1958) shows two stages of advance of the Newer Drift; the earlier reaching the Stickney moraine, the later the Hogsthorpe moraine. They are identified on morphological and other grounds, and support the views of West and Suggate.

[Nomenclature of the British glaciations has changed since the earlier editions of this book. The table given on page 408 is misleading. The Norfolk sequence is renamed as follows:

Hunstanton Brown Boulder Clay
Gipping (= Upper Chalky Drift)
Lowestoft (= Great Chalky Boulder Clay = ?Norwich Brickearth)

In Yorkshire the Hessle overlies purple boulder clay of unknown age, and the Hessle also occurs in Lincolnshire. Any further correlation is, at present, completely unjustified.]

Chapter XXVII

THE NORTH-EAST COAST

Beach conditions in Marsden Bay, which is small and sandy, and hemmed in by cliffs of Magnesian Limestone have been analysed by C. A. M. King (*Inst. Brit. Geogrs.* 1953). The cliffs are reached by waves at high water and their retreat is constant, but slow. Since the bay is enclosed by headlands, it is thought unlikely that sand can reach it by beach-drifting. If this is so, sand movements in the bay are either entirely local or possibly there is some transport from the zone offshore. The bay faces north-east, and the maximum fetch relative to it is 1,200 miles. Three factors are important on this beach: (1) the rather coarse sand of which it is composed allows more rapid percolation of the swash and, therefore, a reduced backwash; (2) to counterbalance this the gradient is fairly steep, a feature which implies that the backwash is intensified; and (3) the period or length of the waves playing on the beach.

King expressed her conclusions in terms of wave types associated with on- and offshore winds.

If there is a swell with long waves and the wind is onshore, the waves are also high and fairly steep. These are destructive waves, and cause a combing down of the beach, and the seaward transport of a great amount of sand. On the other hand, a swell acting in conjunction with offshore winds produces low waves, which are constructive and carry sand up the beach. If conditions are the product of local onshore winds, the waves may be high and short. These steep waves are destructive, especially on the higher parts of the beach, and cause a movement of sand to the lower parts. These and the two preceding sets of circumstances tend to make low gradients. If, however, local conditions have caused onshore waves to meet an offshore wind, the waves are low and flat with little energy. Thus they are mildly constructive in their effects, but work over a relatively narrow zone which they steepen. Thus, the general conclusion, as far as Marsden Bay is concerned, is that 'the prevailing offshore wind is associated with constructive waves while the dominant onshore wind has a destructive effect'. Although no specific work has been carried out on other and roughly comparable beaches of the north-east coast, perhaps one may assume that, given similar conditions, there would be comparable changes in their beach profiles.

In and around the Tees estuary R. Agar (*Proc. Yorks Geol. Soc.* 29, 1954, 237, and 32, 1960, 409) has collated the evidence of a number of borings, and by an analysis of the sequence of deposits revealed by them has thrown much light on recent changes of sea level. The estuary is floored by Triassic marls and sandstones, and the lowest part runs west of Seaton Carew, i.e. north of the existing estuary. Between Middlesbrough and Billingham the deepest point reached was −80 ft. O.D., part of a narrow cut produced by post-glacial erosion (Fig. 161).

All the area is drift-covered and there may be a possible correlation of the boulder clays with those farther south. Since, however, there is at present little else than colour to connect them, it is better not to build on this. Resting on the drift there is a deposit, formerly referred to as blue clay, usually thin, and seldom more than 30 ft. thick. It seems to have been laid down as a continuous sheet immediately after the deposition of the boulder clay. It is laminated and, locally, small stones and clean sand occur. At one place a boulder of Shap granite, about a ton in weight, was found. This implies floating ice at the time of deposition. The sandy and gravelly layers suggest floods. Probably an ice-dammed lake existed over the area in which Middlesbrough is now situated. The sand patches near the margin of the clay are found usually between 55 and 87 ft. O.D. Certain sections in the clay have established it as a late glacial deposit, and small post-glacial valleys cut through it. The marginal sand deposits associated with the clay imply shore conditions. At this time the water level must have been about 75 ft. O.D.

A fall in sea level ensued, in this district effective to about −150 ft. O.D. This produced deep valleys in the deposit and also in the underlying solid rocks. A subsequent rise of sea level up to approximately −60 ft. O.D. is indicated by the sand and gravel deposits in the erosion channels. The succeeding grey alluvium, which at one point is 84 ft. deep (where it is sandy gravel), represents a continuing rise in sea level in the expanding estuary. The upper parts of this deposit contain much organic matter, and a hollowed-out canoe was found in it. The rise of sea level indicated by this bed continued until the water level was at −16 ft. O.D. relative to present sea level. A peat of Atlantic age succeeds the alluvium and indicates that it was formed when sea level was fairly steady between −16 and −10 ft. O.D.

The marine sands associated with the raised beaches are only a foot thick, but imply a continued rise of sea level up to a maximum of 30 ft. O.D. *Mytilus* and *Cardium* are found just as on a modern beach. This 30 ft. level corresponds with the raised beach at Saltburn, a feature now

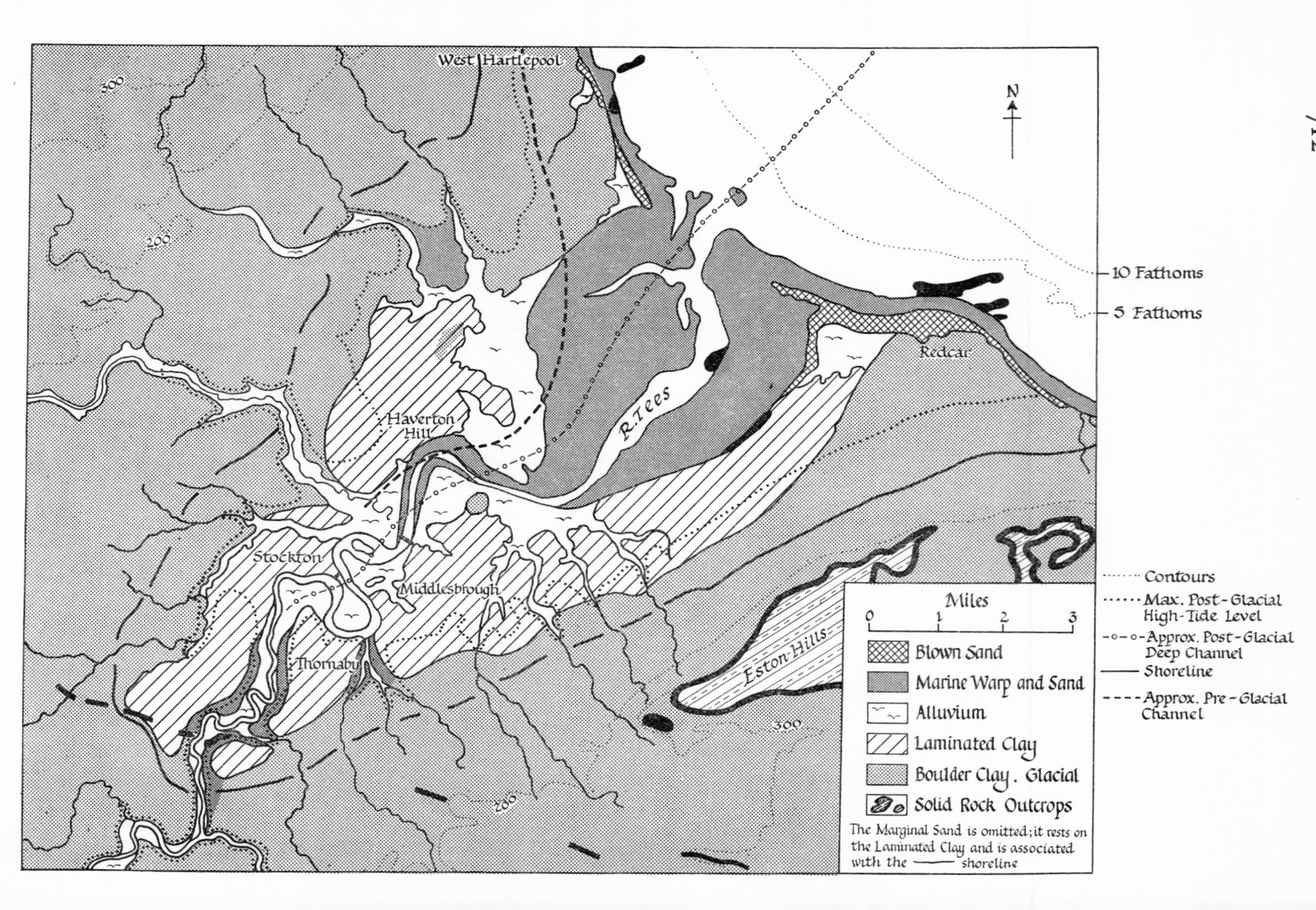
West Hartlepool
N
10 Fathoms
5 Fathoms
Redcar
R. Tees
Haverton Hill
Stockton
Middlesbrough
Thornaby
Eston Hills
300
200
300
200
Contours
Max. Post-Glacial High-Tide Level
Approx. Post-Glacial Deep Channel
Shoreline
Approx. Pre-Glacial Channel
Miles
0 1 2 3
Blown Sand
Marine Warp and Sand
Alluvium
Laminated Clay
Boulder Clay. Glacial
Solid Rock Outcrops
The Marginal Sand is omitted; it rests on the Laminated Clay and is associated with the shoreline

regarded as substantiated by the deposits found in Middle Beck. It is part of the 25-foot beach of Scotland and the Solway. A drop in sea level followed in Sub-Boreal times, and at the same time the Brown alluvium began to collect in beck valleys. It is distinctive in colour, and Agar suggests that it has been produced as a result of clearance and cultivation in historic times. This and other minor fluctuations in sea level are regarded as similar to those outlined by Godwin (see p. 491). To-day the estuary beaches and margins are formed of dark silt and sand with patches of yellow sands and true sandy beaches.

Agar extended his work along the coast as far as Ravenscar. He made some valuable observations on cliff topography in relation to rock resistance and marine erosion and concluded, as far as this coast is concerned, that in order of decreasing resistance the rocks could be arranged as follows: sandstones, sandy shales, shales, boulder clays, sands and gravels. The coast is generally high, and ridges and valleys run approximately at right angles to it, allowing erosion to produce prominent headlands alternating with bays in drift. The sea-bed immediately offshore is a wave-cut platform, the history of which is long and complex. 'Between the 20 and 30 fm. contours the solid rock is probably Lower Lias, but covered by sand and gravel of unknown thickness sloping up at a very small angle of about 1 in 800. From 20 fm. to the foreshore, wherever the sea-bed is Lower Lias, the cover is mainly of sand and the slope steepens gradually to about 1 in 100. Where the harder Middle Lias crops out opposite headlands, the gradient is about 1 in 30, and there is little sand cover.' The foreshore, *sensu stricto*, is often formed of shale scars, which vary in size from that of a boulder to a great mass of debris. The width of the foreshore varies; it may be 500–1,000 ft. in bare shale, but only perhaps 100 ft. if it is cut in the hard Middle Lias. The foreshore can be subdivided into two parts. The lower, below L.W.M.M.T. and up to the mid-tide limit slopes at about 1 in 80, the upper extends to the cliff near to which it may have a gradient of 1 in 8. The boundary between these two zones is occasionally picked out by patches of conglomerate, boulders, and pebbles which are assumed to indicate all that remains of a former beach. The blocks below the conglomerate are often perched (see p. 716).

The cliff-foreshore junction is usually about 7 ft. O.D., but varies about 4 ft. plus or minus about this level. Immediately above the notch the cliff face is vertical or slightly concave. This is the part that bears the major attack of the waves, and Agar argues that the rate at which this notch recedes governs the speed of erosion of the whole cliff. It seems that the cliff face above the limit of usual wave attack retreats *pari passu* with the notch, but slopes at an angle varying with the nature of the rock. The slope is about 70° in Lower and Upper Liassic shales, but is vertical (or nearly so) in the harder beds of the Middle Lias and Deltaic series.

Cliff erosion over the full height of the cliff occurs as a general rule only where there is *no* capping of the Deltaic series. Elsewhere the cliff profile shows steeper (*c.* 70°) slopes below, and a gentler slope above. The upper slope is often grassed over, inclined at an angle between 30° and 40°, seems to be nearly static, and is undoubtedly an old feature. Its angle of inclination and its position may have remained much as they are now for the whole period of post-glacial time. Landslips are not uncommon, and where erosion has revealed that which lies beneath them, the notch is usually prominent. But there are places where the notch is buried by deposits, the nature of which suggests that the landslip did not occur whilst the notch was being actively eroded by the sea. In fact, the nature and appearance of these particular deposits recalls those in the old cliff at Sewerby (p. 477). There are, however, no raised beach features in the cliff face, nor is there anything until Whitby is reached to compare with the 30-foot beach at Saltburn. On either side of the harbour at Whitby, the old town is built on a bench at approximately 28 ft. O.D., and the Esk Valley flattens out at Sleights at about 25 ft. O.D.

Estimates of erosion have been made at sixteen places along this stretch of coast. Measurements have been taken since 1892 to the cliff foot and in five localities also to the cliff top (i.e. O.S. 1/2,500 map). Taking the average of all, Agar estimates the recession of the cliff foot, where it is formed of shales of medium resistance, at 30 ft. per century, and where it is in glacial drift at 92 ft. per century. Since the cliff foot is by no means always easily defined on account of falls, beach deposits, and other factors, the 30 ft. per century 'represents a standard rate which is only attained under...favourable conditions'. Usually, the rate is less since boulders and growing sea-weed cover much of the shale foreshore, and many of the glacial cliffs are in relatively protected places. For these reasons the estimated rates of erosion are:

	Erosion (ft. per century)	
	Cliff top	Cliff foot
Headlands	4	12
Bays only	13	22
Whole coast	7	16

Thus, the bays are still being cut more rapidly, and so for the time being the indentation of the coast is increasing. There are two places of particular interest where erosion can be measured. On Hunt Cliff there is a Roman signal station. About half of this has disappeared over the cliff, and even if the cliff edge itself originally formed one side of the station, the total recession has been about 100 ft. Near the Abbey, on the East Cliff, erosion has been very high for solid rock, about 62 ft. per century. This results from the marked jointing in the rocks at that locality.

Erosion varies with changing sea level. In dealing with this aspect, Agar ventures on to less certain ground. He assumes that, apart from minor fluctuations, present sea level was fixed about 3500 B.C. But since then there have been considerable isostatic movements of the land. Taking both factors into consideration, Agar thinks that north-eastern Yorkshire has been rising, relative to the sea, at the rate of about 6 in. in 100 years. This figure is related to the height of the 30-foot beach at Saltburn. The isostatic rate is assumed constant for all post-glacial time. The absence of beaches in south Yorkshire is thought to be because this land movement dies out in that direction. The nature of the coast in any case precludes the maintenance of benches or beaches.

There is a good wave-cut platform along the north-eastern part of the Yorskhire coast. If sea level were to fall a few feet cliff erosion would soon cease and the waves would spend their energy on this platform. Agar also contends that the assumed rapidly rising sea level from about 8000 to 5000 B.C. would have had little effect. Higher sea levels than at present between 5000 B.C. and A.D. 1000 have not cut the present platform and cliffs, and so it is taken for granted that these features are comparatively recent.

Figure 162 sums up Agar's history of this coast. In the Eemian interglacial there was cliff recession, and the larger sandstone boulders falling from the Lower Deltaic Series remained on the foreshore. With the onset of the Würm, sea level fell and the shore platform was subjected to subaerial denudation, and landslips occurred. In the Würm he assumes that the blocks fallen from above were not removed. In the post-glacial period the sea began to rise rapidly to −16 ft. O.D. and much drift, with its vegetation cover, was washed away. Then a relative pause occurred between −16 and −10 ft. O.D. in the movements of sea level during which peat formed in the Tees Estuary and, it is suggested, there was erosion of drift and also of the lower platform so that the sandstone boulders became perched. Then followed a more rapid rise of water level, eventually reaching a maximum of about +30 ft. O.D., which was succeeded by fluctuations, usually in a downward direction until about A.D. 1000. Then, until now, the sea has remained more or less at its present level.

These views are provocative, and they do not prove the suggested origin of the perched blocks, nor need it be assumed that the landslips

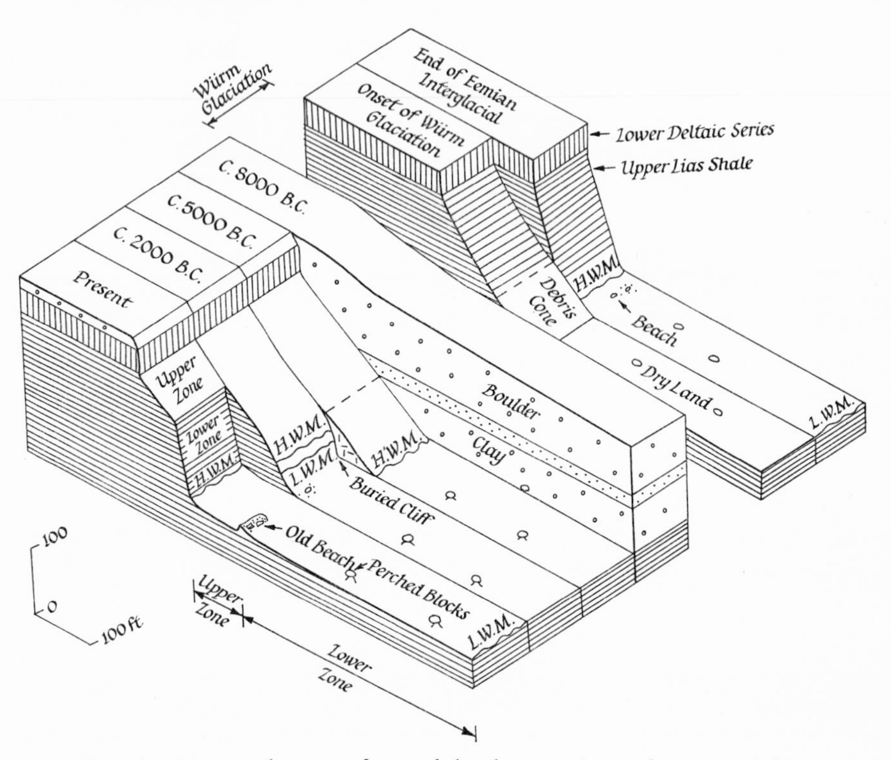

Fig. 162. Suggested stages of coastal development in north-east Yorkshire (after R. Agar, *Proc. Yorks. Geol Soc.* 32, 1960, 409)

were solely interglacial. But implicit in Agar's thesis is the view that the present cliffs are not entirely formed under existing conditions. The value of the more speculative part of his work is considerable, since cliffs are all too frequently taken for granted. Far more work is needed before we can be at all certain of their true origin.

Another author has also written on the cliffs and coast around Whitby (J. E. Hemingway, *A Survey of the Whitby Area*, 1958), and he also found the Lias cliffs resistant to erosion. He ascribed this not to hardness but to the nature and distribution of joints. The three sets of joints in the lower cliff just east of Whitby pier control the form of the cliff and also account for its erosion. At higher levels, above the zone of splash, cliff weathering depends upon lithology as well as jointing. At East Cliff the Lias is vertical and succeeded by a slope of 80° in the Lower Deltaic Series, above which the sandstones of the Deltaic Series and the Eller Beck Bed form cornices.

There may be a broad, bare, wave-cut platform, or one largely covered by great blocks of sandstone, the result of old cliff falls. The nature of the platform must vary with the rocks; near Staithes it is beneath Lias shales and ironstone, whereas at the Scaur and Jet Wyke it is in front of Middle Jurassic cliffs with but little sandstone. Hemingway comments on cliff falls, and notes that some may be sufficiently close to one another, so that they unite. Vegetation grows on them, and they become hardened to some extent by iron. Some may last for centuries, but sooner or later surges, such as that in 1953, demolish them. This contrasts with the view held by Agar (p. 715) who apparently ascribed them mainly to the beginning of the Würm glaciation.

In cliffs formed of glacial deposits, five main factors in their stability or instability are of great but unequal importance. In Runswick Bay the heavy load of material causes a good deal of squeezing out of the moist lower layers. On all cliffs, especially of incoherent materials, rain-wash is important, and marine erosion, whilst constant, is particularly effective in storms, surges, and at all times when wind blows onshore during a high tide. Land water is another factor of significance, especially in glacial beds which vary so much in lithology in short distances. Surface cracks or joints help the percolation of land water and lead to greater erosion.

Over a period of four to five years the complex process produces at the cliff foot an irregular fan of slipped debris ranging in size from unaltered blocks several yards in length to squeezed remoulded clay and mud. As a heterogeneous mass it may rise to 50 feet or more above the cliff foot. It produces, however, a phase of relative stability when the weight of the foundered clay temporarily retains the upper cliff. The continuous marine erosion at the toe of the debris, as well as some flowage and sliding within the mass, removes the fan in a period usually less than ten years, and the cycle recommences. (J. E. Hemingway, *op. cit.*)

NOTE ADDED IN PROOF

The morphology of the marshes in the upper Solway is discussed by D. R. Marshall (*Scot. Geogr. Mag*, 78, 1962, 81). They are of the sandy or sandy-mud type characteristic of western England and Wales, and under natural conditions they present a lawnlike sward. Their composition is very constant—about 90 % fine sand in nearly all parts. Most of it is shore sand, but some is directly derived from red sandy boulder clay. The main direction of drift is eastward, and wave action is undercutting the marshes in several places. This means that more material is carried eastwards by the drift. The marshes are associated with a raised beach at 18–25 ft. O.D.; this too is suffering some erosion. The composition of the beach closely resembles the modern marshes, and Marshall suggests that it was formed some 5,000–6,000 years ago, in much the same way as the present marsh. The raised beach is in two distinct terraces, at *c.* 18 and *c.* 23 ft. O.D., but farther east the higher step reaches *c.* 27 ft. The modern marshes also show a slight slope up-stream, the result of tidal flow and prevalent westerly winds. The modern marshes are also terraced, but it is more than probable that the terraces are erosion features and are not the result of changes of level.

	Acreage (1856)	Acreage (1946)
Dumfriesshire		
Kirkconnell	50	480
Caerlaverock	480	1,020
Priestside	320	340
Kenziels	100	80
Browhouses	50	40
Redkirk	70	50
Cumberland (1846)		
Rockcliffe	1,640	1,750
Burgh	1,700	1,320
Newton	530	1,390
Skinburness	1,100	1,250

Rates of accretion have been measured over a short period—1·25 in. a year at 15 ft. O.D. in the Puccinelletum, but much less at higher levels 'there being an average of 0·5 in. per annum at 16 ft. O.D., whilst above that level, the dust (coal dust was spread to form a datum for measurement) was barely covered in two years'.

The horizontal extent of the marshes is considerable in places. In some localities there is erosion. The table shown opposite is of interest.

The plants are similar to those of west country marshes, and the nature and development of creeks and pans not unlike those at, for example, Talsarnau (p. 130). Aggradation was locally helped by the building of a rubble wall by the Nith Navigation Commission in the estuary of the Nith, but the wall is now broken down and relatively ineffective.

INDEX

This index should be used in conjunction with the contents list at the beginning of this volume. Headings of chapters and sections are generally not given in the index